Chollada Yesuwan
Seattle, WA
July 1988

FOUNDATION ENGINEERING

FOR DIFFICULT SUBSOIL CONDITIONS

FOUNDATION ENGINEERING
FOR DIFFICULT SUBSOIL CONDITIONS

Leonardo Zeevaert

Second Edition

VNR
VAN NOSTRAND REINHOLD COMPANY
New York Cincinnati Toronto London Melbourne

Library of Congress Catalog Card Number: 82-1877
ISBN: 0-442-20169-9

Manufactured in the United States of America

Published by Van Nostrand Reinhold Company Inc.
135 West 50th Street, New York, N.Y. 10020

Van Nostrand Reinhold Publishing
1410 Birchmount Road
Scarborough, Ontario MIP 2E7, Canada

Van Nostrand Reinhold Australia Pty. Ltd.
17 Queen Street
Mitcham, Victoria 3132, Australia

Van Nostrand Reinhold Company Limited
Molly Millars Lane
Wokingham, Berkshire, England

15 14 13 12 11 10 9 8 7 6 5 4 3 2 1

Library of Congress Cataolging in Publication Data

Zeevaert, Leonardo, 1914–
Foundation engineering for difficult subsoil conditions.

Includes bibliographies and index.
1 Foundations. 2. Soil mechanics. I. Title.
TA775.Z45 1982 624.1′5 82-1877
ISBN 0-442-20169-9 AACR2

PREFACE TO FIRST EDITION

Throughout thirty years of professional practice in such difficult subsoil conditions as those encountered in the seismic area of Mexico City, the author has had the benefit of observing and designing many large foundations. The new concepts and working hypotheses given in this book are based on this experience, in order to achieve better designs on a rational basis, reducing practical rules that in the past have resulted in poor performance of building foundations. In the engineering profession it is necessary to investigate continuously the physical laws of soil behavior and soil masses, to be able to eliminate the guesswork supported by empirical generalizations. Statistics, however, is a valuable research tool in investigating the general trend of the phenomena and an aid to establish theories and working hypotheses when deviations from the statistical laws established are understood and carefully observed.

Several good books on soil mechanics, foundations and engineering geology have been written, in which the foundation engineer can study the general aspects of design and construction in foundation engineering. The scope of this book is to supplement this literature with basic technical fundamentals, pointing out the problems that may be encountered in practice when the foundation is involved with difficult subsoil conditions. Therefore, the writer assumes the reader is acquainted with the current literature on this subject.

Foundation engineering is not an exact science. Nevertheless, sufficient precision is required to assure a successful foundation design and construction. This goal is achieved when the behavior in the field complies within the predictions and factors of safety used, thus obtaining a satisfactory performance without sacrificing economy. Difficult subsoil conditions may be defined as those encountered in soil sediments of medium to very high compressibility and medium to very low shear strength extending to great depth, and in those where the hydraulic conditions play an important role, as well as when the soil deposits are found in areas subjected to strong ground motions induced by earthquakes. Under these environmental conditions, the foundation engineer is compelled to use all the knowledge and experience he has gained in soil and foundation engineering, sampling and testing of materials. The aspects of engineering geology in recognizing the engineering characteristics of the subsoil used for foundations are of primary importance, since it is recognized that the behavior of a small soil sample is not representative of that of the entire deposit or strata encountered. It should be kept in mind that the foundation engineer has to work with soil deposits that are far from being isotropic and homogeneous. Therefore, his understanding of the behavior of the subsoil can only be complete after considering the real conditions that may be expected from a geological point of view.

Allowance should be given in all engineering designs, using a factor of safety to cover the deviations of the theories and working hypotheses, the mechanical properties of the material, and construction procedures that may also deviate to a certain degree from design considerations. The selection of a factor of safety should be based on the knowledge the foundation engineer has obtained from the environmental conditions and forces involved, namely, the geological and physiographical conditions, hydraulic and mechanical properties of the sediments, as well as the functional requirements of the project for which the foundations should be designed. All these elements should be made compatible with the economy of the design; therefore, the precision required in the calculations is summarized by the ability of the foundation engineer to manipulate the laws, theories and working hypotheses that may be available in soils and foundation engineering to a degree to which he has gained confidence from experience. This book specially emphasizes this approach as strictly necessary to be able to perform a rational and successful design.

In order to avoid mentioning "approximate method" throughout this book, the author wishes to point out that actually in civil engineering and mostly in foundation engineering, there is not such a thing as an "exact method or theory." All the methods proposed in this book have a degree of accuracy, or shall we say, an uncertainty acceptable from the practical engineering point of view. Nevertheless, it is true that some methods are more reliable than others for the problems encountered in practice. The uncertainty of a particular method is covered by the corresponding factor of safety, which as mentioned before, should also cover not only the so-called theory, but also the deviations of any other environmental forces found under field conditions. Therefore, foundation engineering requires experience of field behavior and of the deviations obtained from the theoretical design calculations. Moreover, one should not forget that theories and methods of design in civil engineering are subjected to further investigations, as more experience is gained with time. Therefore, theories have to be established under simplified assumptions covering, in the best possible manner, the mechanics expected under real conditions. Often, because of the nonuniform characteristics encountered, it would be a waste of time—or rather an illusion—to try to approximate the solution of a problem to an unreal accuracy. The decision depends on the ability of the foundation engineer to visualize the problem and make a good estimate that will enable him to obtain sufficient precision and economy in the design. Nevertheless, it should be kept in mind that during construction the design expectation may be somewhat altered. Construction methods should go together with theoretical design, and the factor of safety selected accordingly.

Chapter II has been prepared as a review of the mechanical properties of difficult soils, advancing some concepts of approach, mainly in the field of fine sediments exhibiting intergranular viscosity. The methods exposed have been used by the author satisfactorily for several years. They have suffered theoretical adjustments since first published to obtain better correlations with behavior observed in the field.

In deformation problems, the soil should be considered a two-phase material. The solid phase represented by the skeleton structure and the liquid phase represented by the water should be studied separately. This implies knowledge of the stress-strain-time properties of the materials and of the stress dissipation in the soil mass due to load application, as well as of the state of hydraulic pressures and their changes imposed during construction or other environmental conditions. Chapters II and III have been prepared to review these concepts, providing the practicing foundation engineer, in Chapter III, with stress nets to facilitate estimates of stress changes. The theoretical background to trace flow nets in different foundation problems is also reviewed. The use of well systems to dewater excavations is treated. At the end of Chapter III, the problem on stability and bearing capacity is discussed. Bearing capacity factors for deep foundations are given based on current theoretical considerations; the result given, however, is not more than another theoretical essay on bearing capacity complying with the experience of the author.

In Chapters IV, VI and VII an attempt is made to introduce the foundation engineer to the complex field of subgrade reactions. This may be considered where the foundation and structural engineers meet. Furthermore, the author believes, from his experience, that soil mechanics and foundations cannot be divorced from design of the foundation structure, since there must exist compatibility between these two branches of civil engineering. The unit foundation modulus, also called the "coefficient of subgrade reaction," is a variable function of the geometry of the loaded area, the subgrade reaction distribution, and the mechanical properties of the subsoil for the stress level applied. The foundation structural problem becomes very complicated when the foundation structure is in itself a statically indeterminate structure. The only means to solve these complicated problems in a practical manner is by means of simplified working assumptions, reducing the unknowns to a number that may be handled by current methods. The methods given in the book may be used by the experienced foundation engineer. Nevertheless, since all of them give only particular solutions, they will only serve as a guide to establish a school of thought. The final assumptions and methods of calculation, however, call for the skill and experience of the foundation and structural engineers involved in the solution of the particular problem, to establish the best and most practical procedures. Computer programs may be written to facilitate and speed up the calculations.

The ground surface subsidence occurring in difficult subsoil conditions and the implications of this phenomenon in civil engineering works cannot be overlooked, since in most occasions, difficult and complex problems may be encountered. The illustration and deduction of working hypotheses to evaluate these problems and their effects in foundation engineering may be explained more simply by means of a case history, as used by the author in Chapter V.

The behavior of friction piles is an important item in foundation engineering, mainly in those problems related with negative skin friction in piles and piers. Chapter VIII has been devoted to explain the mechanics and use of friction piles, based on an ultimate skin friction theory. The methods of calculation are also given; their applications are studied in Chapter IX for the friction pile compensated foundation, and in Chapter X for negative friction on point bearing piles and piers. These methods of calculation have been used extensively by the author with satisfactory results, and are published for the first time to their full extent in this book.

The process of performing excavations is an important factor in the future behavior of foundations requiring deep excavations. The water flow induced by deep pumping produces changes in the effective stresses in the soil mass, affecting the stability and deformation during excavation. The approach to these problems is treated in Chapter XI; however, the reader should be acquainted first with Chapters III and VII.

There are many places in the world with difficult subsoil conditions subjected to destructive earthquakes, where it is necessary to investigate the behavior of foundations to be able to perform a rational and safe design. For this purpose, the foundation engineer should investigate the probable behavior of the subsoil mass under strong ground motions. Chapter XII was prepared with the aim of introducing the foundation engineer to seismic foundation engineering. With this in mind, the author has taken the case history of Mexico City, where field information on strong earthquakes is available. The contents of sections 3, 4, 5 and 6 of Chapter XII are given for the first time in this book. They may be taken as an advance and guidance from investigations carried on in this subject.

Although the foundation engineer is compelled to generalize the subsoil conditions to be able to produce workable and practical methods of computation, this generalization should be made on a sound and rational basis using all the power of soil mechanics he has at his disposal, considering, moreover, that in nature there is no such thing as an isotropic subsoil condition. The mechanical properties of soils are more complex than any other engineering material. Therefore, the only means is to use the closest representative theories and working hypotheses that may be compatible with the behavior observed in the field, and from there establish the most simple correlation satisfying the statics of the problem. The development of theories is necessary to establish the basis of comparison with real behavior in the field, and accordingly, screen out inconsistencies with the aim of obtaining more reliable and technical methods of approach.

The bibliography in soil mechanics is very extensive at present, and has grown considerably in each country where basic research is carried on. The obtention of published material and the thorough study and selection of its contents, with the barrier of languages, is becoming a gigantic task beyond the possibilities of an individual. Therefore, the author wishes to apologize if some important publications on the subject treated in this book have escaped his attention. The selected bibliography given to each chapter is intended only to contribute in the understanding of the corresponding chapter.

The main content of this book is the compilation of the work of the author during his professional practice, which has been gradually added to by experienced colleagues in the field to whom the author is greatly indebted, mainly on the intergranular viscosity of soils, the critical stress in preconsolidated soils and hardening, the plastic theory to estimate friction in piles, the dewatering of excavations to reduce heave, the injection of water outside excavations to reduce settlements, and the drifting forces on underground elements, motivated by strong ground motions due to earthquakes. The author is highly indebted to his nephew, Mr. Adolfo E. Zeevaert, C. E., M.Sc., for his great help and interest during the preparation of the manuscript, in the calculation of graphs and tables, checking formulas and practical illustrative examples used in the text, and in the Appendices. The author wishes also to extend his appreciation to his secretary, Mrs. Diana A. de Balseca, for the arduous task she has taken in typing the manuscript, and finally, to the editor, whose interest in this book contributed in a presentation beyond the aim of the author.

Mexico, D. F.

Leonardo Zeevaert, Ph.D., C.E.
Professor of Soil Mechanics and Foundations at the Faculty of Engineering, U.N.A.M.
Consulting Civil Engineer.

PREFACE TO SECOND EDITION

In the eight years since the appearance of first edition, and through its use in the courses given by the author at the Graduate School of Engineering of the U.N.A.M., the author has improved the content of several chapters. These improvements have been included in the second edition to make it more explicit and practical for graduate courses and foundation engineering practice.

All the chapters, however, have been revised. In Chapter II, new and more precise formulas are given to estimate vertical displacement due to the intergranular viscosity phenomenon. The basic principles, however, have been retained until future investigations may show a more accurate and practical method to be used.

Chapter III has been extended to include, in the solid phase, formulas to calculate ground stresses for surface rectangular loaded areas and for different values of Fröhlich's concentration factor. Also, theoretical methods of calculating the reduction of piezometric water levels in stratified subsoils and of estimating the depressed water table in well groups for excavation purposes have been added.

A completely new Chapter VI has been written to include the most recent practical methods developed by the author regarding soil-structure foundation interaction considering the importance of knowing the approximate value of the subgrade reactions in foundation structural design. (See L. Zeevaert, 1980, ISE.)

Chapters IV, V and VII to XI have been revised, and more on soil-structure interaction has been added to Chapter X.

Chapter XII has been enlarged to include a practical and rational method of estimating the loss of bearing capacity in loose cohesionless soils during strong ground motions induced by earthquakes. A method is included for computing the seismic rocking phenomenon of box type foundations for tall buildings supported on stratified subsoil conditions. At the end of the chapter, a general method is given for estimating the seismic soil-pile interaction behavior, including illustrative numerical examples.

Finally, in Appendix E, new numerical examples for Chapters VI and VII are presented with the purpose of illustrating the methods of computation for soil-structure interaction given in Chapter VI.

The author has considered that nowadays the practicing foundation engineer is getting more and more involved in matrix algebra calculations he can perform with his desk computer, therefore more matrix algebra has been used in the book. With this in mind the author has given ready to use algorithms and methods of computation that will permit the practicing foundation engineer to write his own programs to expedite his calculations with an approximation compatible with the practical problems involved. Especially interesting along this line, are the calculations to estimate the ultimate skin friction in piles, subsoil seismic behavior, the soil-structure interaction of compensated mat foundations, the seismic rocking phenomenon and the behavior of piles, piers or vertical shafts subjected to strong ground motions.

In the soil-structure interaction problems, the foundation engineer should carefully select the secant stress-strain parameters for the increment of stress and stress levels involved, as described in Chapters II and VII.

The author is indebted to Miss Eloisa E. Rey, C. E., M.I., for her great help and interest in assisting the author to revise the new additions, formulas and examples for the second edition, and to the editor for his interest that this book should continue to be up-to-date, and serve the advanced student and professional practicing foundation engineer for consultation in his every day work.

Mexico, D. F.

Leonardo Zeevaert
Professor of Soil Mechanics and
Foundation Engineering
Faculty of Engineering, U.N.A.M.

CONTENTS

FOUNDATION ENGINEERING

FOR DIFFICULT SUBSOIL CONDITIONS

▸I◂ INTRODUCTION

I.1 TYPICAL FOUNDATIONS

The art of designing the best and most economical foundations for a project greatly depends on a careful investigation by the foundation engineer. A study should be made of the environmental factors and the compatibility of the subsoil engineering conditions with the type of foundation structure on which the loadings are to be supported. Hence, as a first approximation, the foundation engineer should consider the qualitative index and mechanical characteristics of the subsoil at the site at which the project will be constructed. This preliminary knowledge will permit him to judge the behavior of the subsoil under applied load, and after analyzing the probable behavior of different types of foundation structural systems in conjunction with the project requirements, he will be in the position to select the proper foundation.

The purpose of this chapter is to visualize the selection of the type of foundation, reviewing the typical foundation structures that may be used in conjunction with the subsoil conditions to be encountered, to fulfill the requirements of total and differential settlements. It must be borne in mind, however, that in the design of a foundation there are two important mechanical items to be considered: first, the bearing capacity of the soil for the applied load; and second, whether the total and differential settlements are compatible with the foundation structure selected, type of superstructure and architectural demands of the project. As an example of total and differential settlements, the case of widely spaced footings used for light flexible roofs may be mentioned, where one may allow large differential settlements, in contrast with other problems like installation of machinery or equipment, where the differential settlements are often restricted to very small values. Therefore, the foundation engineer should investigate the differential settlements that may be permitted for different problems of building design, and also the magnitude of the total settlement not damaging adjacent construction.

The specification of total and differential settlements is studied carefully for each problem in question, as the allowances can vary a great deal, depending on the mechanical limitations of the project in question, as well as on adjacent buildings and public utilities. In other words, one could say that for a certain specific building, a total settlement of 30 cm may be allowed, provided that there is no damage and differential settlements for certain predetermined spans between columns do not exceed ½ cm. This specification appears to be bold, since one could say also that a total settlement of 30 cm is large, even if no damage takes place. If the total settlement, however, could be forecast and the building is isolated in an area away from other buildings and no damage of any property is expected, then there is no reason to allow large settlements in the design, provided also that the connections of public utilities going into the building are taken care properly, and the foundation structure is designed in such a way that differential settlements in the building will not produce damage to the construction. If such is the case, the functional requirements of the project are fulfilled and the foundation may be considered to work under satisfactory conditions.

The foundation engineer experienced in soil mechanics and engineering geology, as well as with the behavior of foundation structures and building design, is able to visualize, as a first step, which foundation to select for the problem in question. Once he has selected the optimum type of foundation to be used, then he may investigate quantitatively its behavior. The selection should always be the most economical type of foundation that can be used, fulfilling the requirements of allowable total and differential settlements in conjunction with the subsoil condition encountered.

In order to give the foundation engineer the first approach in the philosophy of selecting a foundation, the principal types of foundations will be discussed, and the relation they have with different subsoil deposits from which the probable behavior may be forecast. In this approach, the foundation engineer is assumed to be acquainted with the index and general mechanical properties of soils and with the general behavior of different types of foundation structures.

1.1 Isolated Footings

Footings are understood formed by a rigid rectangular base of stone or concrete of dimensions: width B and length L, in which the ratio of L/B will not exceed 1.5. The foundation structure will support the column load. The bearing capacity of the footing may be estimated, and its dimensions selected; thereafter, a forecast of the settlement is made.

To illustrate the case of footing foundations, consider a building with nine columns (Fig. 1-I.1) supported on isolated footings. In this case, the footings will work independently of each other. Therefore, it is required that the differential settlements between footings will not exceed the allowable total and differential settlement requirements. The differential settlements may be reduced selecting

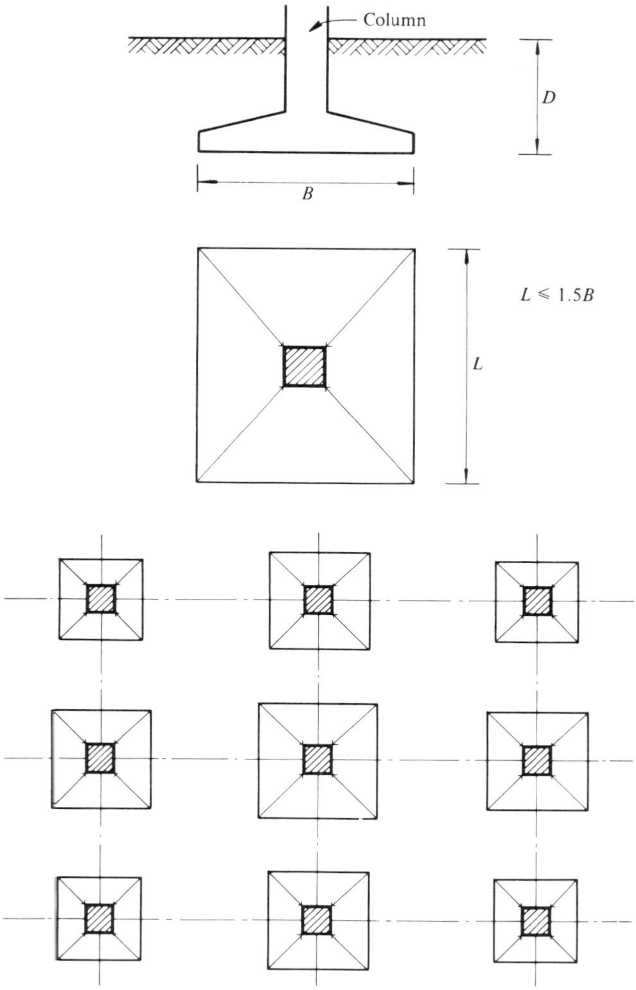

Fig. 1-I.1 Single footings.

properly the area of the footings, and at times, using the stiffness of the superstructure. From the structural point of view, however, the superstructure should not be allowed to take high secondary stresses induced by the differential settlements of the footings, except in very special cases. Single footing foundations, in general, will be used only in soils of low compressibility and in structures where the differential settlements between columns may be controlled by the superstructure flexibility, or including in the design of the building joints or hinges that will take the differential settlements and/or rotations, respectively, without damaging the construction.

1.2. Continuous Footings

When it is necessary to control within certain limits the magnitude of differential settlements between columns supported on footings, and when soil deposits of medium or low compressibility are encountered, it is recommended to use continuous footings. They may be defined as resisting elements joining columns together by foundation beams.

Continuous footings are arranged by joining two or more columns together with beams. The vertical differential displacements may be controlled via beam stiffness (Fig. 2-I.1). The selection of the foundation beams, either running in one direction or the other along column rows, depends largely on the layout of the column loads,

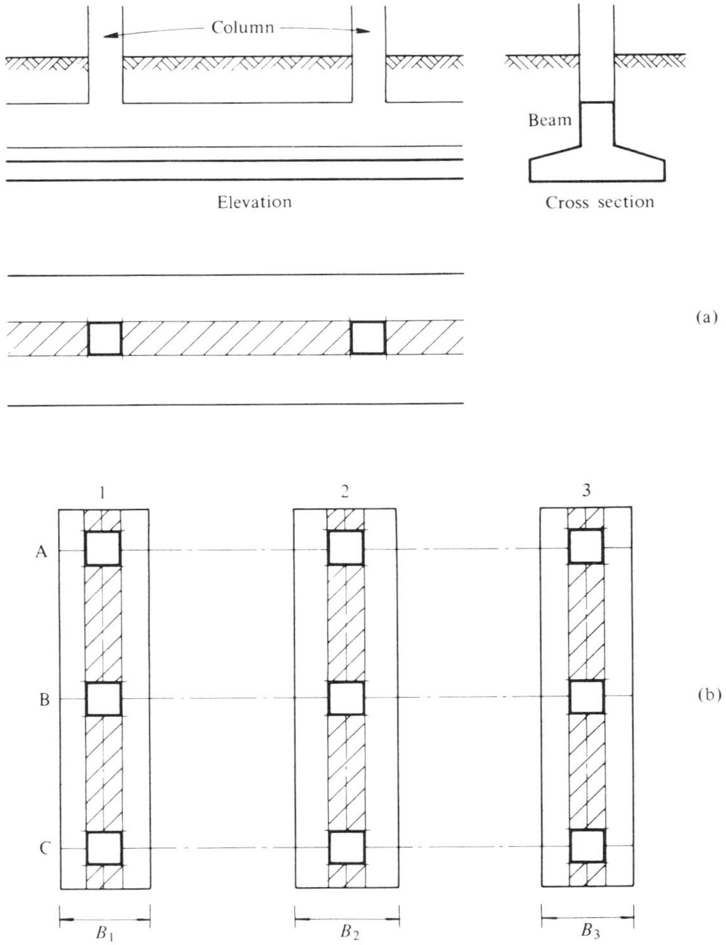

Fig. 2-I.1 Continuous footings.

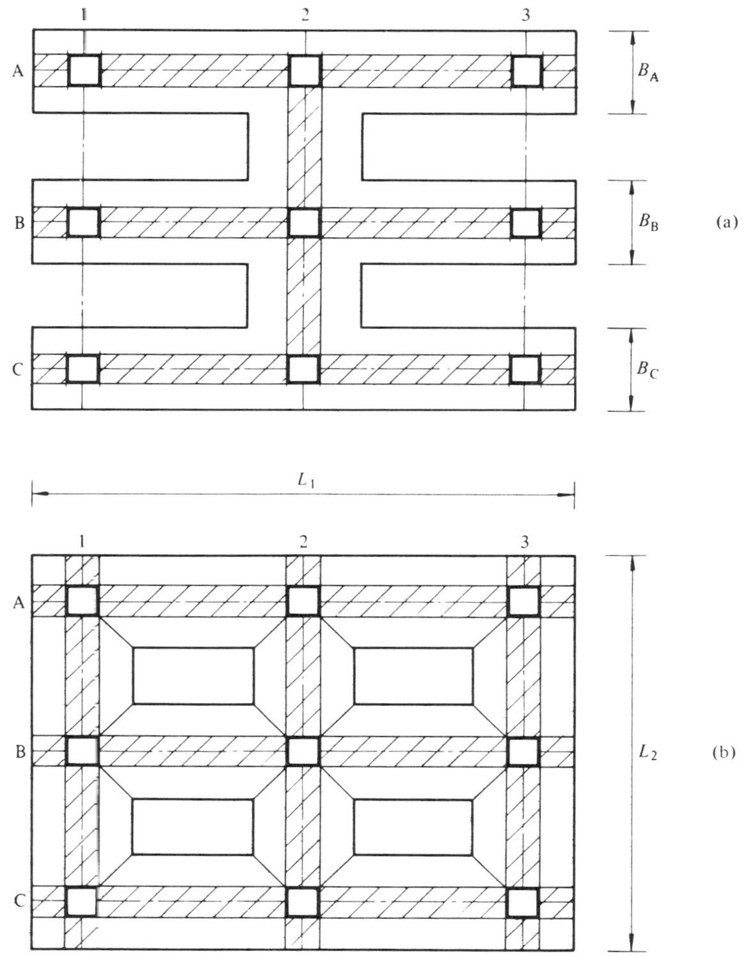

Fig. 3-I.1 Continuous footings.

and other functional requirements concerning the structural and architectural design of the project.

For heavier loads, and when the project calls for stiffness in both directions (namely, along column rows A, B and C and also along rows 1, 2 and 3), the foundation is given stiffness with beams in both directions (Fig. 3-I.1). In this case, it may be observed that the footing slabs will cover practically all the foundation. This type of foundation using continuous footings is advantageous in soils of medium compressibility, where it is necessary to control differential movements between columns. The foundation beams are designed with the necessary stiffness to fulfill the differential settlements requirements.

1.3 Raft Foundation

When the loads are so large that continuous footings will occupy close to 50% of the projected area of the building, it is more economical to use a continuous mat covering the entire area, as shown in Fig. 4-I.1. The total load in this case may be assumed uniformly distributed in the area covered by the building. The soil reaction is determined on the basis of a safe bearing capacity. The total and differential settlements may be investigated considering the stiffness of the raft or foundation slab is a matter of economy, compatible with the allowable differential settlements.

Flexibility is important to obtain economy; however, restrictions in differential vertical displacements between columns may call for certain slab stiffness, either by making it thicker or by placing foundation beams joining column rows. The beams can be designed with the required stiffness to reduce differential displacements. This type of foundation may be used generally in soil deposits of medium compressibility; however, in certain instances, the surface raft foundation may be used in soils of high and very high compressibility, where large total settlements may be allowed. This type of foundation may be used efficiently in reducing differential settlement.

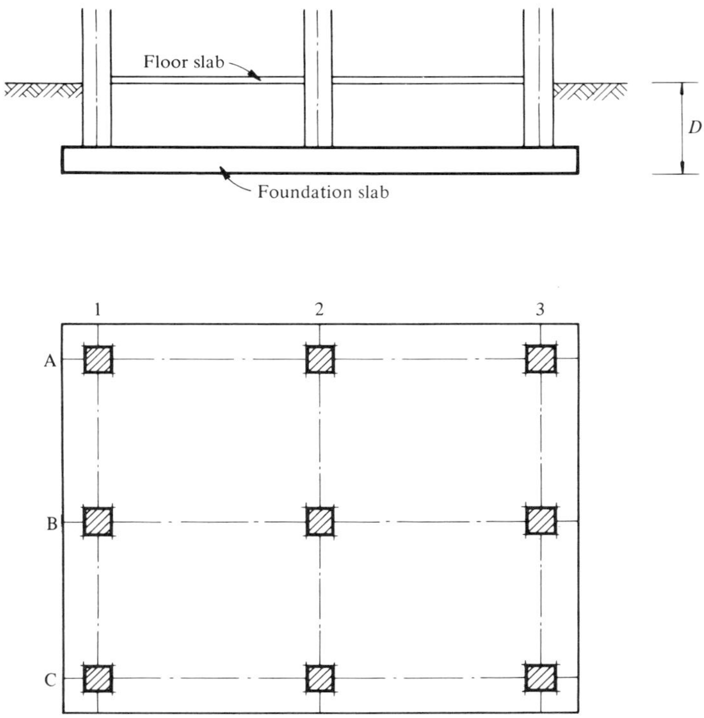

Fig. 4-I.1 Mat foundation.

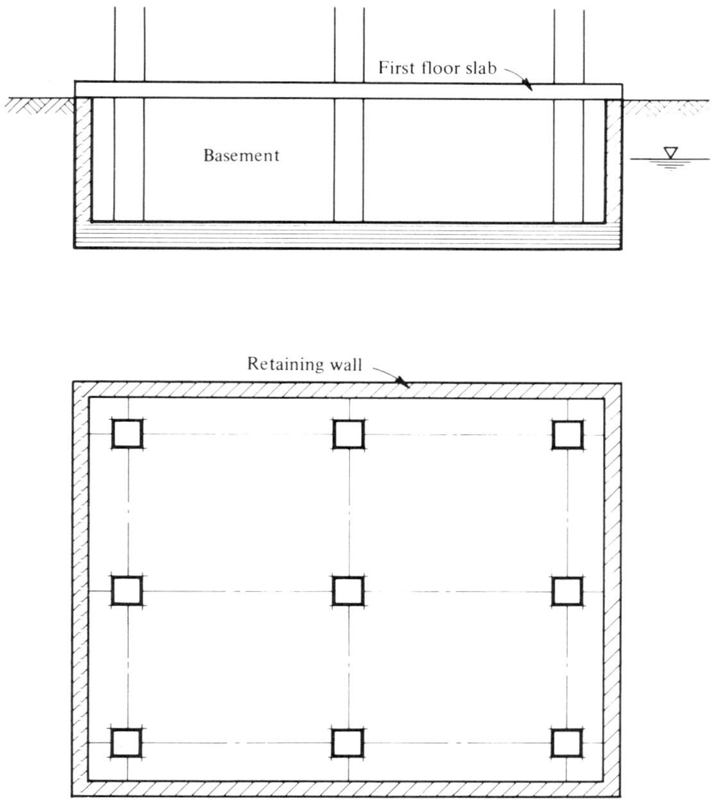

Fig. 5-I.1 Compensated foundation.

1.4 Compensated Foundations

In soil deposits of medium, high and very high compressibility and low bearing capacity, compensated foundations are indicated. This type of foundation requires a monolithic box foundation, as shown in Fig. 5-I.1. When the water table is close to the ground surface, water proofing is necessary to use the buoyancy effect in designing the foundation. In the design of compensated foundations, it should be borne in mind that the soil should be considered as a material of two phases, namely: a solid and a liquid phase. Therefore, in a compensated foundation, the compensation is made by adding two effects: (1) substitution of the submerged weight of solids, and (2) the buoyancy effect by the weight of liquid displaced. Both effects are used to equalize the total weight of the building. The volume of the concrete box forming the foundation structure and basements will displace a weight of liquid that, according to Archimedes' principle, will contribute in floating the foundation up to this value, reducing the load applied to the solid phase. The load taken by the solid phase will, however, deform the soil because of the change in effective stresses induced in the soil structure. It should be investigated from the point of view of bearing capacity of the soil and total and differential settlements, as previously discussed for other foundations.

A compensated foundation, as shown in Fig. 5-I.1, is designed usually with a stiff foundation structure; therefore, differential settlements are controlled rather easily. The foundation structure is designed either as a flat slab system or a slab-beam foundation system, joining the columns with beams in both directions. This type of foundation, owing to the characteristics of the soils where it is constructed, implies the necessity of knowing more accurately the stress–strain–time characteristics of the material, in order to evaluate settlements because of changes in effective stresses in the solid phase. The load of the building will be compensated by means of an excavation sufficiently deep to permit the obtention of the necessary load capacity, and the reduction of the vertical displacements to magnitudes that will be satisfactory.

The differential settlements will be controlled giving the necessary stiffness to the foundation structure. The mechanical behavior of the foundation is controlled by the solid phase because of changes in effective stresses in the ground. The magnitude of settlements in this type of foundation depends mainly on the ability of the foundation engineer in keeping the prestressed condition of the soil mass when the excavations are performed, and thereafter as the soil is reloaded. For design, it is important to know the basic concepts related with the hydrodynamic flow of water in the subsoil. The dewatering of the excavation should be designed in such a way as to preserve the original effective stresses. Therefore, the future behavior of the foundation will be a function of the process of making the excavation and of the way the hydraulic conditions are controlled in the subsoil.

Theoretically, if one could make a substitution of load without changing the effective stresses and hydraulic pressures, no vertical displacements would take place at the ground surface. Therefore, the fundamental concept of this type of foundation is to achieve the minimum change in effective stresses during excavation and construction of the foundation structure.

1.5 Compensated Foundations with Friction Piles

When a compensated foundation as described is not sufficient to support the load with the allowable total settlement, in spite of designing the foundation with sufficient stiffness to avoid detrimental differential settlements within the foundation itself, friction piles may be used in addition to the concept of compensation. This case may be present in deposits of high or very high compressibility extending to great depth. The piles will reinforce the upper part of the soil where a higher compressibility is encountered. The applicability of this foundation calls for a soil that varies from very high compressibility at the upper part of the deposit, to medium or low compressibility at the bottom (Fig. 6-I.1).

The total settlement of this type of foundations depends greatly on the way the friction piles are driven, their spacing and length, the procedure used to perform

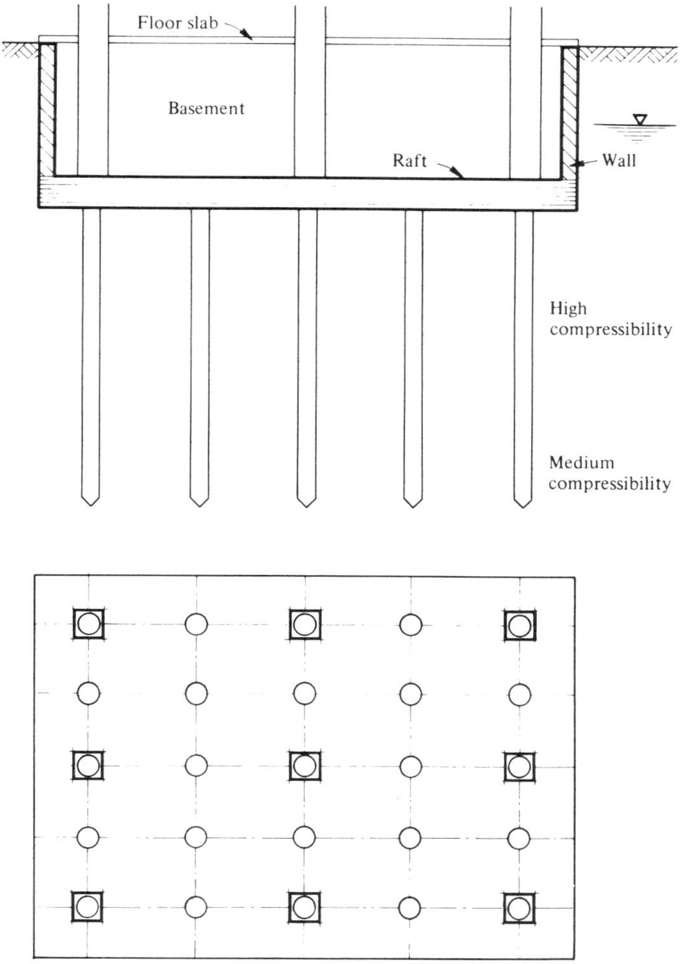

Fig. 6-I.1 Compensated friction pile foundation.

the excavations, and the control of the hydraulic conditions of the subsoil. To reduce total and differential settlements one should observe always the fundamental concept of minimum change in effective stresses. The benefit of the piles is mostly achieved when they are driven before the excavation is made, making them work under tension forces during the excavation, thus preserving the confined original state of stress condition in the subsoil.

1.6 Point Bearing Pile Foundations

When the loads to be supported are higher than those a compensated friction pile foundation can take, then it will be required to find a deep-seated hard stratum of low to very low compressibility and high shear strength, where piles can be driven to point bearing. One can distinguish two main cases of point bearing pile foundations (Figs. 7 and 8-I.1).

The first case is recognized when the hard stratum of convenient thickness is found underlain by materials of medium compressibility. In these cases the piles should be evenly distributed as shown in Fig. 7-I.1. After solving the problem of point bearing of the piles in the hard stratum, there still exists the problem of finding if the lower compressible soil stratum will have a safe bearing value, and also if the total and differential settlements will be within the allowable values specified for the foundation in question. This type of foundation should be designed with sufficient stiffness to control differential settlements.

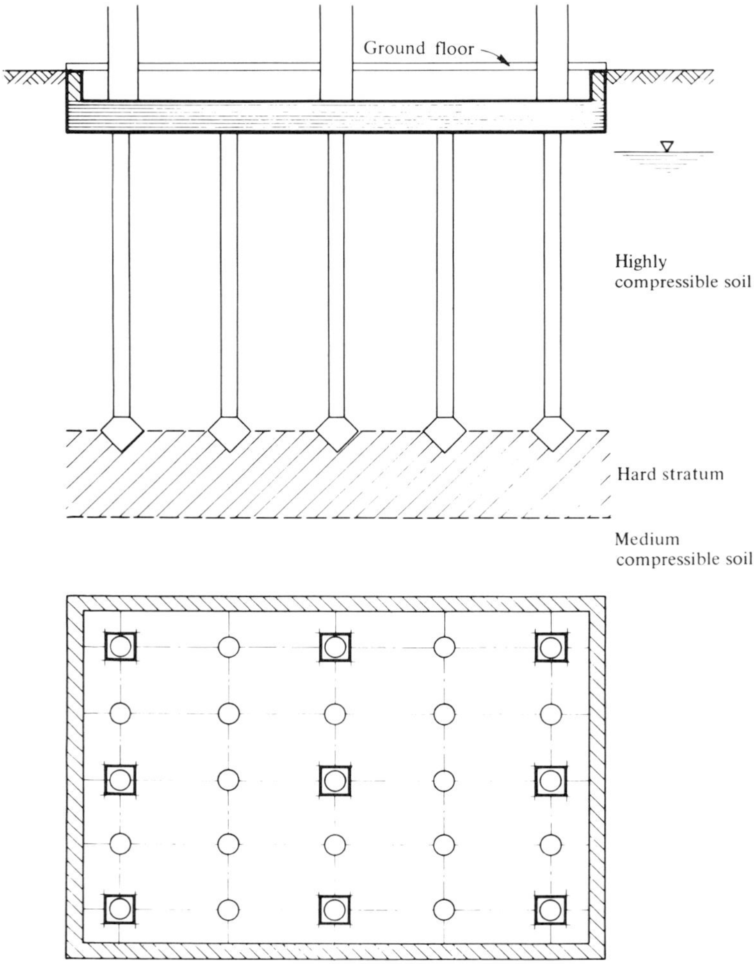

Fig. 7-I.1 Point bearing piles.

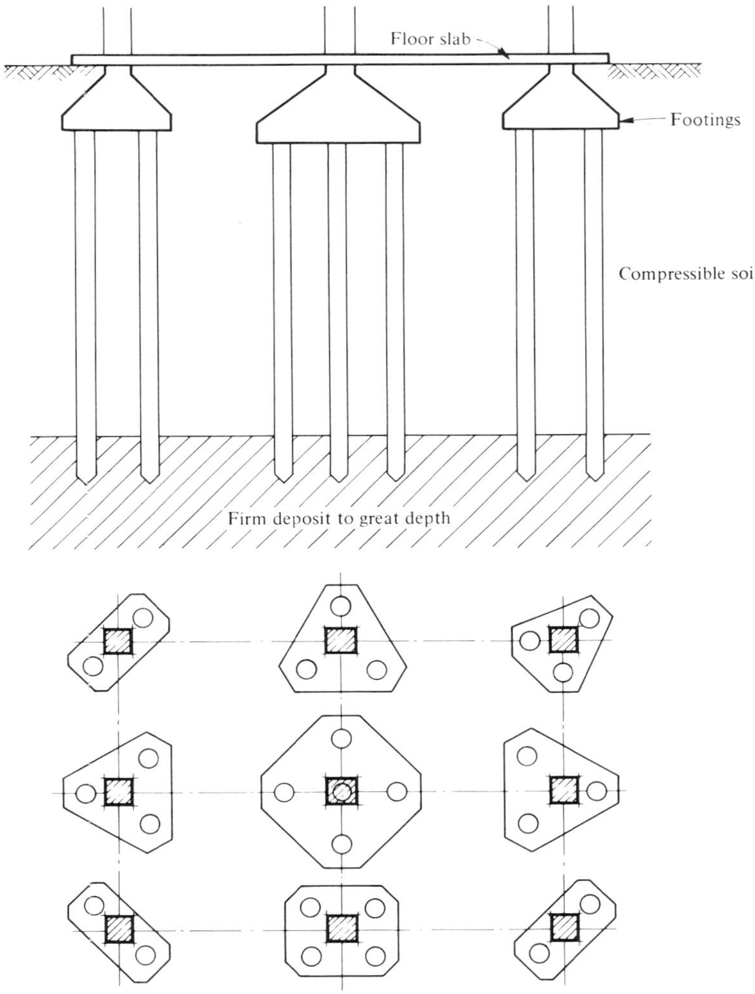

Fig. 8-I.1 Point bearing piles in groups.

The second type of pile foundation is recognized when the point bearing piles rest in a firm deposit of low compressibility extending to great depth (Fig. 8-I.1). In this case, it is economical to use groups of piles to solve the foundation problem. The columns will rest on single footings supported on the piles. The piles driven in the firm stratum develop lateral friction contributing to the total bearing capacity. The bearing capacity of the piles will depend mainly on the mechanical properties on shear strength of the deposits in which they are driven, on the spacing of the piles, on the length of penetration into the bearing stratum, and on the state of density and confinement of such stratum. The point bearing piles may be driven in groups or using a uniform distribution, depending on the compressibility of the deposits underlying their points. The point bearing capacity of the piles may be increased if special points are designed, improving the mechanical characteristics of the deposits into which they are driven.

In the design of point bearing piles, the phenomenon of negative skin friction and the effect this phenomenon produces in the confinement of the deposit where these elements are bearing should be taken into consideration. The phenomenon of negative friction is extremely important and may be recognized when there is a downward relative movement of the compressible soil mass with respect to the firm stratum where the point bearing piles are driven, originating on them drifting forces. This phenomenon implies a load transfer of part of the weight of the soil mass to the piles, and consequently, a reduction of the vertical confining stresses on the stratum where the piles are bearing. Sometimes, the downward dragging forces may be large, forcing the point bearing piles to penetrate into the supporting stratum. Therefore, it is important to consider properly the phenomenon known as negative friction. The total and differential settlements of these foundations may be estimated computing the compression of the strata underlying the point of the piles.

1.7 Pier Foundations

Pier foundations are used to support very heavy loads in buried soil deposits of very low compressibility (Fig. 9-I.1). Their load capacity is a function of the mechanical properties of the soil under the base of the pier, and of the confining stress of the bearing stratum. Actually, the bearing capacity of such an element is determined as a deep-seated isolated footing.

The piers, column-like elements cast in place, in most cases carry high loads of 500 ton or more; therefore, the compressibility of the deposit on which they are resting should be very low, in order that they may be recommended. Pier shafts may be used from diameters of about 1 m to larger diameters. The bearing capacity and the base dimensions are also a function of the procedure used to perform the excavation, and of the way the hydraulic conditions are handled. The density of the material where these elements are bearing may be altered during excavations if an upward water flow is produced. Specially important is the case when the material is a cohesionless fine sediment or when the cohesion is small, in which case it is necessary to perform the excavation using a pneumatic system, introducing air under sufficient high pressure to balance the flow of water toward the bottom of the excavation, preserving the natural confining and density conditions of the bearing stratum. Usually, if precautions are taken in the installation of these elements, the settlements will be very small. The settlement, however, may be estimated knowing the stress-strain characteristics of the strata encountered under the base of the piers. The negative friction on these elements may take large proportions: hence, it should be estimated.

When these rigid elements are used in seismic regions to support loads through deposits of high and very high compressibility, it is necessary to investigate the effect

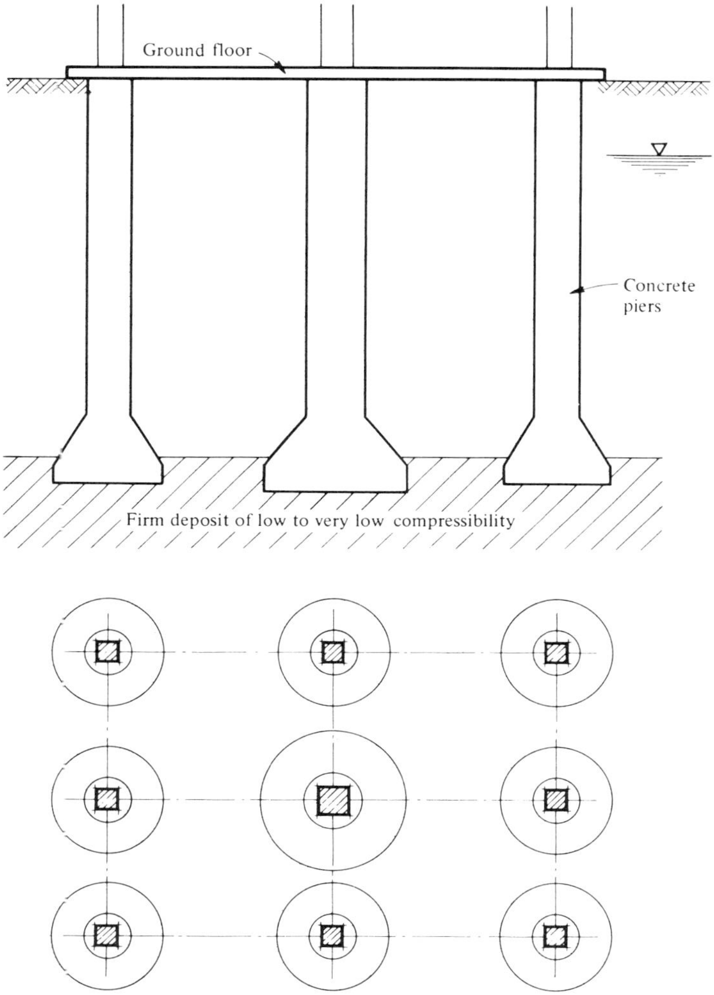

Fig. 9-I.1 Pier foundation.

of the horizontal motion of the soil mass during earthquakes. The horizontal drift forces against the piers because of soil displacement should not be overlooked. In occasions, rigid elements have been damaged because of the strong horizontal motions produced by the earthquakes.

1.8 Sand Pier Foundations

The solution of foundations using sand piers or sand piles is shown in Fig. 10-I.1. This type of foundation is used to increase the load capacity of the soil by reducing

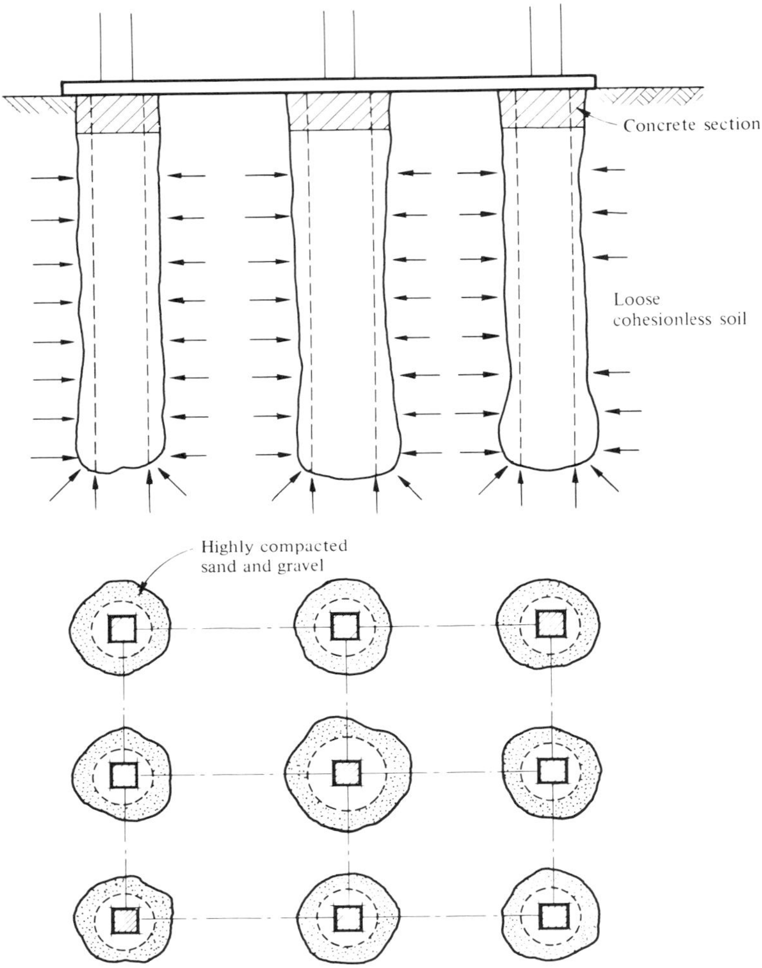

Fig. 10-I.1 Sand piers.

its compressibility and increasing its shear strength capacity properties. This type of pile may be used in loose or medium dense sand deposits. The improvement of the subsoil is a function of the volume of sand introduced at the time these elements are installed. Usually first a hole is driven in the ground, then sand is introduced and highly compacted in layers, using a heavy ram. The sand element will take the load because of the lateral confinement given by the subsoil. The deformation of these elements may be estimated by means of the stress-strain properties of the sand used, considering the pier as a long sand cylinder laterally confined by the soil. This type of foundation is only recommended in places where the cost of cement is very high, and good aggregates to fabricate concrete are difficult to obtain.

Summary. The nominal types of foundations previously described are usually recommended for specific typical subsoil conditions. Combination of these types, however, may be used in occasions, when the subsoil engineering properties permit their use, and the allowable total and differential settlements are fulfilled. The selection of the type of foundation and foundation structure depends on the ability of the foundation engineer to recognize the mechanical behavior of subsoil materials encountered in nature.

I.2 SUBSOIL SEDIMENTS

The selection of the type of foundation, as described early in this chapter, should be made by the foundation engineer after studying, first, the index engineering properties of the subsoil materials at the specific site in question; and second, the mechanical behavior of the type of foundation structure to be used, the loads to be supported, and the allowance on total and differential settlements. It is also important for the foundation engineer to consider the layout of the building, and behavior requirements from the architectural and structural point of view. In seismic areas, it will be necessary to study the earthquake forces and their effects in the subsoil mass.

Unconsolidated sediments where building foundations are supported may be classified from a practical point of view in six large groups, namely: residual, eolian, alluvial, lacustrine, marine and piemont deposits. The volcanic and glacial origin deposits may be classified within the above-mentioned six groups, the difference being only because of the pyroclastic or clastic characteristics, respectively. Erosion and transportation agents—water, water-vapor, wind and gravity—are the same.

2.1 Residual Soils

Residual soils are the product, in situ, of the disintegration and chemical alteration of the lithological components of parent rock because of weathering. The granulometry of residual materials may be very variable, from large fragments to gravel, sand, silt, clay and colloids. Therefore, density and cementation may be very variable. Organic matter may also be present. Weathering may reach deep into the parent rock as in the case of tropical and subtropical zones. Low densities may be found in the upper part of the subsoil due to eluviation. The properties of compressibility may be high, and in some cases, very high with low shear strength. Generally, residual soil profiles are encountered in thicknesses of a few centimeters to several meters, depending on the climate and physiographical environment of the region. In humid regions, deep soil profiles are encountered with medium to high compressibility and low shear strength. The hydration of the aluminum silicates produce clay minerals. In case of volcanic areas, the minerals may be of the expansive type, if they contain the clay mineral montmorillonite. In semiarid regions, the material is more stable and the thickness of the soil profile is smaller. In dry climates, caliche is formed due to illuviation giving a better foundation material.

The expansive properties of these soils may be important to consider in the selection of the type of foundation. Expansion is expected when the clay is exposed to water, and adsorption of water molecules takes place in the clay mineral structure. The expansion may be as high as 5%. It is extremely important, in those cases, to investigate the mechanical properties for different conditions of humidity of the soil. The type of foundation to be used in residual soils is difficult to predict. Residual soils are surface materials used generally to support single or continuous footings. The residual soils may be found with low to medium compressibility. The topsoil in horizon A containing organic matter should not be used to support foundations, since the organic material is sensitive to small changes of humidity and oxidation, and thus changes in volume of this soil cannot be predicted in a rational manner.

Total and differential settlements should be carefully considered. When the soil is of the expansive type, a raft foundation or a short piles foundation may be contemplated. The bearing capacity of residual soils may vary from 0.5 kg/cm^2 to over 4 kg/cm^2.

2.2 Eolian Deposits

Materials transported by the wind build up deposits of sediments to which the foundation engineer should give special consideration. Some of these sediments form dunes, loess, loessial type deposits, eolic beaches and large volcanic dust deposits. These sediments in nature may be found with medium to high compressibility. They may be encountered in nature with low relative density, cohesionless or slightly cohesive. The eolian deposits are characteristic of arid regions, and the water table is encountered at great depth from the ground surface. Eolian deposits, however, show the peculiarity of changing mechanical properties upon saturation due to changes in the water table or seepage conditions. Therefore, when their natural humidity conditions are changed, they suffer sudden compaction; they are also known as collapsible soils. After the mechanical change has taken place, they behave with medium to low compressibility and take the name of modified eolian deposits. If the relative density is found medium to high, satisfactory bearing capacity may be obtained under well confined conditions. Where eolian deposits are used in their natural state, it is necessary to determine properly their mechanical properties of shear strength and compressibility, and the possibility that under certain conditions, they might be subjected to an increase in their natural humidity. If such is the case, damage of structures supported on them may be expected. When the material retains indefinitely its original humidity, then single or continuous footings may be used, and for heavy loads, a raft foundation may be indicated. It is very seldom necessary to use deep foundations. Modified eolian deposits may be considered of better quality; hence any one of the foundation types previously discussed may be used depending on the magnitude of the loads and bearing capacity encountered. The allowable bearing capacity ranges from 1 kg/cm^2 to over 4 kg/cm^2. The settlement, however, should be estimated in accordance with the stress–strain characteristics of the soil encountered, and environmental hydraulic conditions at the proposed site. Usually, no special problem of differential settlements is encountered except in loose sediments, and when important changes of humidity take place under the foundation area. When this is the case the use of piles may be recommended.

2.3 Alluvial Deposits

Alluvial sediments are those deposited by water in movement; their grain size varies from large rock fragments, like those encountered in river beds, to gravel, sand, silt and some clay. They are in general well graded and may be found with medium to a very compact state. The finer sediments assume medium compressibility, and low to a very low in coarser sediments. When alluvial deposits are well confined, the foundation problems are minimum, except for very large loads, or when special water subsoil conditions have to be overcome. In general, single footings may be used.

In sandy clayey silts, continuous footings or raft foundations are indicated. On river planes where the finest alluvial sediments are encountered, compensated foundations may be used, and in occasions, the use of piles or piers may prove to be necessary.

2.4 Lacustrine and Marine Sediments

Fine and very fine sediments like silts and clays are deposited when running water comes to rest, like in lakes, marginal lagoons, estuaries and deltas. These deposits may be encountered with medium to high, and very high compressibility. They may be encountered with contents of colloidal organic matter, or they may be totally composed of organic material like peat. The stress-strain behavior is complicated if compared with other sediments. They exhibit intergranular viscosity in their mechanical behavior. The stress-strain-time relationships should be investigated to be able to estimate settlements. Because of their very low permeability the process of consolidation is important, since retardation of the deformation because of hydrodynamic processes cannot be overlooked. Compensated foundations with or without friction piles may be used in compressible deposits extending to great depth.

2.5 Piemont Deposits

Piemont deposits are sediments that accumulate at the foot of mountain slopes because of avalanches, slides or instability of the slope surface material. These deposits contain materials of all kinds and grain size, including vegetation in large fragments and fine organic matter. The compressibility and shear strength are very variable. The support of columns has to be investigated one by one; usually the safest foundation is to use piers excavated to a depth where firm support is encountered.

2.6 Recent Volcanic Deposits

The nonconsolidated volcanic sediments belong to a special group because of their great variety. The pyroclastic materials may be encountered in detritus, avalanches, and from large rock fragments to very fine volcanic dust. From the sedimentation point of view, however, the compressibility and shear strength may be closely classified within the eolian, alluvian and lacustrine sediments depending on the agent of transport present and the physiographical environment where they are sedimented, respectively. When they suffer weathering, they may be classified in the bracket of residual soils. They may be characterized by their variable shear strength and properties of the grains, from soft pumice to fragments of basalt and volcanic fine glass, and from dust and silt size up to coarse sand grain size. The grains are usually of angular shapes that in well confined conditions give high shear strength.

In case of volcanic detritus, a great variety of materials may be found, from molten lava, bombs, large rock fragments, sand and fine dust. The approach, in these cases, is similar to that described for the piemont deposits. Foundations are variable in depth and size, from shallow to deep footings or piers; the support of each column load or bearing wall should be investigated.

2.7 Glacial Deposits

In the glaciated areas the study of sediments and their deposits require special treatment, however, similar to the pyroclastics in the volcanic areas. They may be classified as eolian, alluvian, lacustrine and residual soils as already described.

I.3 TOTAL AND DIFFERENTIAL ALLOWABLE SETTLEMENTS

The allowable magnitude of the vertical displacements is vital information for the foundation engineer. With this knowledge and information on subsoil conditions, it is possible for him to select, from the economical point of view, the proper type of foundation. The foundation engineer, however, is concerned with the magnitude of settlement to be considered, and who is going to be responsible to specify its magnitude, since it is expensive to reduce total and differential settlements. The decision will depend on the different parties involved in the project, mainly, the project architect or engineer, the structural and mechanical engineers, the tenant, the owner, the building authority and the insurance company. On this respect, a brief analysis will be made on the contribution each one of the above-mentioned parties performs in deciding the magnitude of total and differential settlements to be used to design foundations for the project under consideration.

The owner is not concerned about the amount of total and differential settlements, provided his investment is safe and not demerited by failure to work under certain predicted conditions, and if he will not incur expenses because of damaged adjacent private and public property.

The occupant will start claims when the total and differential settlements affect his interests because of poor performance of the building, with respect to total and differential settlements that may require excessive maintenance, in which case the owner is also involved.

The structural engineer, usually, is not very much concerned about total and differential settlements provided their magnitude does not imperil the stability of the structure he has designed, and fulfills the architectural demands.

The architect is interested in preserving the aesthetics of the building, the architectural finishings, and functional requirements as well as those established by the mechanical installations.

The building authority is concerned to avoid claims due to misbehavior of foundations, and is required to prevent damage to other parties and public utilities on account of poor performance of the building. The building authority, being responsible for the public welfare, has to observe the expected behavior of the approved project. Therefore, the building authority may codify that total and differential settlements should be such that no damage be brought to adjacent buildings, public utilities and to the project itself. The building authority will hold responsible the designer and owner if the specifications are not fulfilled.

The insurance company will not insure any building against damage to itself or neighbors, or public property because of total and differential settlements if he is not convinced that the foundation engineer has performed a safe design. The insurance company looks for the probability of a good bargain.

The opinion of the public as a critical layman and professional is important, and has a considerable weight in classifying a building with large total settlements, or with problems of differential settlements; his criticism will result in a public devaluation of the building, in which case the owner will be affected.

The magnitude of the allowed total and differential settlements of a building should be obviously the maximum values that may be permitted to satisfy all parties mentioned above, and still be compatible with the economy of the foundation of the project in question. It is expensive, however, to achieve small total and differential settlements. Moreover, it must be recognized from foundation behavior point of view, that total settlements have a completely different meaning than differential settlements within the different loaded areas. The above statement may be easily visualized since a building with a rigid foundation exhibiting a very large total settlement will not be necessarily damaged in its structural or architectural design, provided the total settlement is uniform or reasonably uniform, and if the structure keeps its verticality. On the other hand, small total and differential settlements may damage the structure or installations very strongly, in some cases without chance of repair. Following this reasoning, it may be also mentioned that the type of foundation structure selected for certain subsoil conditions will contribute in establishing the magnitude of the total and differential settlements for a specified project. The discussion on settlements will be divided into two parts: (1), total settlements, and (2) differential settlements.

Thus the magnitude of allowable settlements depends greatly on the functional requirements of the project, the subsoil conditions and the environment. Under functional requirements the following may be stated:

(a) Architectural requirements, to obtain proper behavior of mechanical installations in and out, and within the project itself. Architectural facing materials, window frames, curtain and partition walls, and floor distortion.

(b) Structural requirements from the point of view of stability, and function of the superstructure, economy and limitation in the induction of unforeseen high secondary stresses in the structural frame.
(c) From the point of view of public utilities, no damage should be allowed to sidewalks, sanitary conduits, hydraulic and electric intakes.
(d) Aesthetic appearance should be preserved, in order to assure the investment undergoes normal depreciation.

The above-mentioned requirements should be fulfilled for every type and location of a building or project to satisfy the several parties mentioned before. From the above discussion, it will be recognized that it is not possible to set forth a general specification concerning magnitudes for total and differential settlements, if economy is the main question. Examples will illustrate this fact.

A long warehouse with no adjacent construction is designed with a simple roof supported on two column rows. The magnitude of total and differential settlements is required to design and select the foundation type on certain known subsoil conditions. Therefore, the following is required:

(a) Total allowable settlement of footings in each column row.
(b) Differential allowable settlement between column rows. This phenomenon may cause certain rotation of the roof and tilting of the walls; however, no damage should take place.
(c) Differential allowable settlement between columns in each row of columns. In this case, if the wall is made of rigid material, allowance in differential settlements is narrower than if the wall is, for instance, of corrugated sheet metal.
(d) Settlement induced in the footings because of storing load on the floor area of the warehouse. This information is very important and in some cases may govern the design of the foundation and layout of the load, as is the case of heavy loads on highly compressible soils. The foundation engineer is compelled to select a foundation that will not be affected by such settlement; for instance, the use of piles or piers might be necessary.

From this simple example, it may be seen that it is practically impossible to select the magnitude of the total and differential settlements if the foundation engineer is not well acquainted with subsoil conditions, and the behavior he may expect of the different foundations and structural systems, as per the structural design requirements of the superstructure and the preservation of the aesthetics of the building set as a norm by the project architect or engineer. Therefore, it is clearly seen that each case will have its own characteristics and specifications for the magnitude of total and differential settlements, to assure a proper foundation design. The agreement on allowable settlements is a function of a joint decision taken by several different parties, namely: owner, tenant, structural engineer, mechanical engineer, project architect or engineer, building authority, public, soils and foundation engineers on the economy and available foundation types.

In order to illustrate the solution of the problem, let us assume again the warehouse. The building will be assumed isolated from other buildings, making the problem simpler. The walls will be 5 m high and the roof span 30 m. Trucks will not enter the warehouse, but will use one of the long sides of the building as an unloading dock. The subsoil is of medium compressibility; therefore, from the very beginning it is recognized that in order to gain economy, one should design a flexible structure and allow the largest possible settlements. The walls will be floating brick panels between column rows; they will permit a small differential settlement. Columns will be separated at 5-m centers, and the structural engineer has set allowances of differential settlements between columns, in order that secondary stresses in the superstructure will remain within economical limits. He claims he can allow a maximum differential settlement on the order of 20 cm between column rows, and the design of the 30-m span roof will not be affected. On the other hand, he specifies 1-cm differential settlement between adjacent columns in a row.

The architect will accept differential settlements of 1 cm between columns in a row, but he will not accept the differential settlement of 20 cm; he claims that the walls should not rotate more than 3/10 of 1% of their height, in order that the building preserve the aesthetics. More rotation might be detected easily by the naked eye. Therefore, the allowable differential settlements between column rows should not exceed 10 cm as a maximum. This magnitude does not affect the tenant provided the unloading dock will not settle more than 5 cm and differential settlements in the floor area will not produce slopes higher than 1/3 of 1% between different loaded areas in the warehouse. Under these specifications, the tenant can operate easily the lift trucks and pile up his products.

From this argument, it may be seen that the tenant is allowing a total settlement on the order of 5 cm for the unloading dock on one side of the building, and he will not object to 10-cm differential settlement provided he does get slopes in the floor less than 1/3 of 1%. On this agreement, it may be found that the opposite column row could settle as much as 10 cm and nobody will object.

It may be noticed in this simple example that the owner, the building authority and the insurance company have not contributed so far in the discussion. They are satisfied if the tenant is satisfied, and if the foundation engineer can assure the required behavior of the building and comply with the demands of the architect, structural and mechanical engineers. Under these conditions, the public will not criticize the behavior of the building, since the naked eye will never notice that any settlement has taken place. The solution of the problem calls for the experience of the foundation engineer in selecting the most economical and functional type of foundation and floor of the warehouse, for the soil conditions encountered.

Take as a second example a high raised building with heavy column loads. The structural engineer has specified a differential settlement between columns not larger than 1.5 cm, otherwise costly high secondary stresses will be induced in the superstructure. On the other hand, the architect will not object to this magnitude set by the structural engineer, since spans are 800 cm on column centers, and the slopes in the floor because of this maximum differential settlement will not be sensitive. It was agreed, however, to use a factor of safety in order that differential settlements within the building do not exceed 65% of the magnitude just mentioned.

The proposed building is adjacent to another building, which increases the importance of the problem of total settlement. The building authority advises the following in order to fulfill the city code: the total settlement agreement should be of any magnitude provided no damage is produced to adjacent buildings and public utilities, like sidewalks and service intakes. The adjacent buildings will be permitted to tilt only 1/4 of 1%. Permanent tilt should be also investigated because of horizontal forces induced by earthquakes and wind. Furthermore, space should be left on each boundary line so that leaning will not surpass adjacent property. Otherwise, the owner will pay for the damage originated. The insurance company will not insure the building if it is not sure that the building foundation was properly designed to fulfill the above-mentioned requirements. The foundation engineer, in conjunction with the structural and mechanical engineers as well as the project architect, will investigate the rigidity and foundations of the adjacent buildings to find out the magnitude of the differential settlement that as a maximum may be permitted without affecting the stability of the structure and the aesthetics of the buildings, and consequently demeriting their value. Otherwise, a serious claim from the owners of the adjacent buildings can be expected. The maximum specification of total settlement should be agreed with the building authority. Assume it was agreed the adjacent buildings should not rotate more than 1/4 of 1%, because it was concluded by the structural engineer and the architect that the buildings may rotate as a rigid body without damage. If the adjacent building has a front of 20 m, then the settlement should not exceed 5 cm at the edge of the new construction. Upon agreement of this magnitude, the foundation engineer and architect will provide special precautions in the design of the building to satisfy the building authority and adjacent owners.

Settlement and damage to adjacent property may take place also during excavations to construct the foundation of the building, especially when it is necessary to perform deep excavations for one or more basements. In this case, allowable total settlements of adjacent buildings should be again investigated to be able to design economically shoring protections, and the methods for the excavation and construction of the foundation structure. The allowable settlements in this case should be determined by a careful study. The soils, foundation and structural engineers and the project architect, in joint meeting with the professional representative of the affected parties, should reach an agreement based on no damage or very minor damage to the building, but not affecting the structural stability and aesthetics of the adjacent property. The magnitude of the allowable settlements is investigated from the point of view of the type of construction and foundation in question, but in any event, protection of the building will be limited to an agreed maximum rotation, to avoid major damage in walls and architectural items.

As a third example, take the problem of a factory where heavy machinery is involved and very critical and important problems may arise concerning total and differential settlements. Assume the construction of a paper mill. The machine usually required for these factories is formed by a long train of independent parts joined together mechanically. The pulp enters at one end of the machine and the ready made paper comes out at the far end. Total and differential settlements are very critical. The designer of the machine, in this case, will specify the amount of the maximum allowable settlements. A differential settlement in the longitudinal direction is fixed and should not exceed, in the majority of the cases, 2/10 of 1%. In the transverse direction, however, the limitations are extremely critical, as any excess of transverse rotation of the foundation from the maximum allowable, may produce large wearing of bushings, large frictional forces, and an uneven paper production. Rotation in the transverse direction should not exceed 1/10 of 5/100 of 1%, depending on the speed of the machine.

Total settlements may be larger provided the differential settlements remain below those specified. They will depend, however, on the way the power is transmitted laterally to the machine, intakes of pipe lines and other installations. In this case, and for each particular design, the mechanical engineer of the manufacturer has to specify, primarily, the displacements that may be allowed.

From the examples and discussion presented above, we conclude that the magnitude of allowable total and differential settlements cannot be fixed or generalized in building codes, making a permanent numerical specification to cover all construction cases. It is concluded that the selection of the total and differential settlements is highly dependent on the functional requirements, environmental conditions, and economy of the specific project, and on the judgement of the different responsible parties involved in the design.

The city authority should be informed of the requirements to be fulfilled by the foundation engineer to comply with the city code and general philosophy on the behavior expectation, and to guarantee that damage will not be produced. The owner and tenant are consulted to agree on the decisions. The allowable settlements under consideration may be further reduced, if possible, in benefit of better behavior of the project; however, this practice may increase unnecessarily the cost of the foundation. Under this condition, the insurance may be obtained at the lowest rates.

Once the allowable total and differential settlements are agreed, the foundation engineer should design the foundation required to comply with the agreement. He will consult with the soils engineer on the magnitude and precision of the soil studies to be performed, and must obtain all the necessary information on the stratigraphy, hydraulic conditions at the site, and soil mechanical properties, to assure a rational foundation design and guarantee the behavior of such foundation to the other parties involved in the design and construction.

1.4 SUMMARY

For the proper design of foundations, the foundation engineer will require all necessary elements to be able to decide, in a technical and economical way, and from the point of view of construction, the type of foundation that will be used for a specific building, hence:

1. Project information and agreements
   (a) Static and dynamic column loads of the superstructure.

   (b) Type of superstructure with respect to flexibility and possibility of permitting vertical displacements.
   (c) Allowable differential settlements between columns or walls and loaded areas, and permissible maximum permanent tilt, pertaining to the proposed project.
   (d) Restriction of total settlements with respect to the project itself, and in relation with adjacent construction and mechanical installations going into the building.
   (e) Study of foundations conditions of adjacent buildings. Allowable total and differential settlements that these buildings may take without damage during construction of the proposed building and thereafter.
2. Engineering properties of the site and environments
   (a) Stratigraphy.
   (b) Hydraulic conditions encountered in the subsoil at the moment of the design, and possibility of future changes. Ground surface subsidence.
   (c) Mechanical properties of shear strength, compressibility and permeability of the subsoil materials to a depth where it may be estimated that the influence of excavation and loads of the foundation are negligible.
   (d) Knowledge of the dynamic properties of the soil in seismic regions, to estimate soil displacements and accelerations during earthquakes in conjunction with the foundations and superstructure of the building.

With the information pertaining to the first and second requirements, the foundation engineer will select the type of foundation more convenient for the specific building and justify its economy, and proceed to the following studies:

(a) Stability analysis of the foundation to determine the load capacity for the type of foundation selected, and of the excavation from the mechanical and hydraulic point of view.
(b) Analysis of vertical and horizontal displacements produced because of excavations, that is to say, heave of the bottom of excavation, lateral displacements and subsequent total and differential settlements taking place during and after the building load is applied on the ground.
(c) Analysis of the behavior of the foundation because of forces induced in seismic and high velocity wind regions.
(d) The analysis of the calculated displacements will indicate the procedure to be used to design the construction, that is to say, the performance of the excavation and load application to obtain the minimum possible displacements and comply with the allowable total and differential settlements.
(e) Design of the structural foundation system and process of construction made compatible with the excavations, the control of settlements and hydraulic pressures during construction of the foundation.
(f) Program of observations during construction and thereafter, of vertical and horizontal displacements and piezometric water elevations to assure the behavior forecast of the foundation design, and to be able to perform any corrections that might be necessary during construction.

SELECTED GENERAL BIBLIOGRAPHY

Baver, L. D. (1940). *Soil Physics*, John Wiley & Sons, Inc. New York.

Feld, Jacob (1965). Tolerance of Structures to Settlement, *Journal of Soil Mechanics and Foundations, ASCE. Design of Foundations for Control of Settlement*, Evanston, Ill.

Feld, Jacob (1966). The Factor of Safety in Soil and Rock Mechanics, *Proc. VI ICOSOMEF*, Vol. III, p. 185, Montreal.

Golder, H. Q. (1971). The Allowable Settlement of Structures, *IV Panamerican Conference on Soil Mechanics and Foundation Engineering*, Vol. I, p. 171, San Juan, Puerto Rico.

Jenny, H. (1941). *Factors of Soil Formation*, McGraw-Hill Book Company, Inc., New York.

Jumikis, A. R. (1964). *Mechanics of Soils*, D. Van Nostrand Co., New York.

Jumikis, A. R. (1967). *Introduction to Soil Mechanics*, D. Van Nostrand Co., New York.

Kézdi, A. (1964). Bodenmechanik, 1964, *VEB Verlag für Bauwesen*, Berlin.

Krynine, D. P. and Judd, W. R. (1957). *Principles of Engineering Geology and Geotechnics*, McGraw-Hill Book Company, Inc., New York.

Leonards, G. A. (editor) (1962). *Foundation Engineering*, McGraw-Hill Book Company, Inc., New York.

Little, A. L. (1961). *Foundations*, Edward Arnold, Ltd., London.

Longwell, C. R. and Flint, R. F. (1955). *Introduction to Physical Geology*, John Wiley and Sons, Inc. New York.

Meyer-Abich, H. (1953). *Geología General*, Instituto Tropical de Investigaciones Científicas, San Salvador, C. A.

Mikhejev, V. V., Polshin, D. E., Tokar, R. A. and Ushkalov, V. P. (1961). Foundation Design in the U.S.S.R., *Proc. V ICOSOMEF*, Vol. I, Paris.

Peck, R., Hanson, E. W. and Thornburn, T. H. (1953). *Foundation Engineering*, John Wiley and Sons, Inc., New York.

Polshin, D. E. and Tokar, R. A. (1957). Maximum Allowable Nonuniform Settlement of Structures, *Proc. IV ICOSOMEF*, Vol. I, p. 402, London.

Schultze, E. and Muhs, H. (1967). *Bodenuntersuchungen für Ingenieurbauten*, Springer Verlag, Berlin.

Skempton, A. W. and Mac Donald, D. H. (1956). *The Allowable Settlements of Buildings*, Institution of Civil Engineers, pp. 727–768.

Taylor, D. W. (1948). *Fundamentals of Soil Mechanics*, John Wiley and Sons, Inc., New York.

Terzaghi, K. (1925). *Erdbaumechanik*, Vienna, F. Deuticke.

Terzaghi, K. (1936). Settlement of Structures, *Proc. I ICOSOMEF*, Vol. III, p. 79, Cambridge.

Terzaghi, K. (1943). *Theoretical Soil Mechanics*, John Wiley and Sons, Inc., New York.

Terzaghi, K. and Peck, R. (1967). *Soil Mechanics in Engineering Practice*, 2nd Edition, John Wiley and Sons, Inc., New York.

Todd, D. K. (1967). *Ground Water Hydrology*, John Wiley and Sons, Inc., New York.

Trask, P. D. (editor) (1950). *Applied Sedimentation*, John Wiley and Sons, Inc., New York.

Tschebotarioff, G. P. (1951). *Soil Mechanics, Foundations and Earth Structures*, McGraw-Hill Book Company, Inc., New York.

Verdeyen, J., Roisin, V., Nuyens, J. (1971). *Applications Dela Mecanique De Sols*, Ed. Dunod, Paris.

Winterkorn, H. F. and Fang, Hsai-Yang (1975). *Foundation Engineering Handbook*, Van Nostrand Reinhold Co., NY.

Zeevaert, L. (1964). The Engineering of Large Structures, Chapter 7 of the *World of Engineering*, J. R. Whinnery, editor, McGraw-Hill Book Company, Inc., New York.

Zeevaert, L. (1980). *ISE-Interacción Suelo-Estructura de Cimentaciones Superficiales y Profundas Sujetas a Cargas Estaticas y Sismicas*, Publ. LIMUSA, Mexico, D. F.

▸II◂ MECHANICAL PROPERTIES OF SOIL

II.1 INTRODUCTION

The foundation engineer, based on information gathered from subsurface explorations including the hydraulic conditions of the site, selects the type of foundation satisfying the specifications of behavior. Once this has been done, the final foundation design may be divided into the following main steps:

1. Bearing capacity problem.
2. Vertical and horizontal displacements.
3. Design and stability of excavations.
4. Soil-structure interaction.
5. Analysis and design of the foundation structure.

Subsurface explorations will be required to determine the stratigraphy and hydraulic conditions at the site and the index and mechanical properties of the soils. For bearing capacity, it will be necessary to know these engineering properties of the subsoil to select the proper parameters of shear strength. To estimate the total and differential settlements concerning the different foundations involved in the project, it is necessary to investigate the stress-strain-time characteristics of the soil materials. In cases of low and very low compressibility of soils, the mechanical properties may be assumed from their index properties; likewise when the precision required for the determination of the settlements is not very important. In the case of cohesive materials of medium to very high compressibility, it is required to obtain undisturbed samples to determine in the laboratory the mechanical properties of shear strength and compressibility closely representative of the different soil strata in the field.

Any type of foundation requires the design of excavations. For this purpose, the foundation engineer should investigate the permeability characteristics of the subsoil strata and the hydrodynamic conditions of the site to determine the state of hydraulic pressures necessary to evaluate the stability of excavations. The stress-strain-time characteristics are necessary when performing excavations to calculate horizontal and vertical displacements. The foundation should be designed in such a way as to keep the project within the allowable horizontal and vertical displacements and avoid damage to public and adjacent property, and also to preserve the original undisturbed subsoil conditions while the foundation structure is being constructed for the project in consideration.

Moreover, the state of hydraulic pressures in the liquid phase will be required to estimate the effective stresses and to design pumping wells, well points or any other means to dewater the excavations, estimating, from the permeability properties of the soil strata, the water discharge in the excavation during construction of the foundation. The change in effective stresses in the solid phase because of excavation and loading is determined with reasonable accuracy by the theory of elasticity. Furthermore, up to date, this is the only procedure to calculate the change in effective stresses. The design and calculation of the foundation structure calls for knowledge of the contact stress distribution. For this purpose, it will be necessary for the foundation engineer to consider the unit foundation modulus, which is a function of the stress-strain-time characteristics of the soil, relaxation properties and load distribution applied at the foundation grade elevation. In some cases the design may be simplified assuming a uniform reaction distribution. Knowing the loads on the columns and the subsoil reaction, the structural engineer is able to calculate the foundation structure.

The mechanical properties of the materials—permeability, shear strength and stress-strain-time characteristics—are determined by means of careful laboratory tests. The precision of the results obtained from laboratory tests performed on undisturbed samples should be taken carefully into consideration, since in spite of all precautions when taking the samples from the subsoil, it is not possible to obtain samples completely representative of the subsoil conditions. Therefore, the factors of safety to be used in the design of foundations have to be compatible with the precision in obtaining the stratigraphical, hydraulic and mechanical properties of the soil. Factors of safety to be used are important and depend on the confidence the foundation engineer acquires concerning the accuracy of the information given to him by the soil mechanics engineer or the laboratory technician.

The factor of safety is very important since the economy of the foundation depends on its value; it will be discussed in further chapters in relation to the subject under consideration. In order to use the lowest allowable factor of safety in the design of foundations, the foundation engineer should investigate in the most accurate possible way all the engineering characteristics of the subsoil at the site and the environmental conditions that may affect the foundation design he has in mind. Only well determined mechanical properties of the subsoil materials, with proper knowledge of stratigraphy and hydraulic conditions of the site in question, will permit the foundation engineer to use, in difficult subsoil conditions, the minimum possible factors of safety compatible with the type of foundation selected. The result will be a successful and economical foundation.

II.2 PERMEABILITY

The permeability is a mechanical property of soil that permits the foundation engineer to estimate the discharge of water he has to foresee when pumping from excavations during foundation construction. The water table may be found close to the ground surface or even covering it, like in foundation problems along shore lines. In case of fine materials like silt and clay, the permeability is important to estimate the consolidation of soil layers.

The water discharge per unit area may be expressed as

$$\frac{Q}{A} = v \qquad \text{1-II.2}$$

in which Q is the volume per unit time and A is the area perpendicular to the water flow. Therefore, the discharge per unit area is equal to the velocity. The velocity, however, as expressed by 1-II.2 is a virtual velocity that does not represent the real velocity of water in the capillary conduits of soil. If n is defined as the porosity of soil, the pore space area will be $A_v = nA$, and we can establish the pore velocity v_p as follows:

$$v_p A_v = vA \qquad \text{hence: } v_p = \frac{v}{n}$$

In granular noncohesive soils, the value of n is from 0.35 to 1; therefore $v_p > v$. The value of v_p is not the effective velocity in the soil conduits, but represents only an average pore velocity. Since the pores assume variable sizes within the soil skeleton, the water pore velocity assumes unpredictable values. In some instances the water velocity may be very much larger from the average value v_p, producing internal scour and transport of fine material from one pore space to another. Therefore, the magnitude of the internal velocities may affect the permeability of the soil.

For laminar flow, Darcy found that

$$v = k \cdot i \qquad \text{2-II.2}$$

in which k is defined as the coefficient of permeability, and i is the hydraulic gradient defined as the change in head per unit length along the water flow path, hence:

$$i = \frac{\partial h}{\partial l} \qquad \text{3-II.2}$$

If i is multiplied by γ_w, the unit weight of water, we obtain the pressure gradient defined by:

$$i_p = \frac{\partial u}{\partial l} \qquad \text{4-II.2}$$

The coefficient of permeability may be written as follows:

$$k = \frac{K}{\eta} \tag{5-II.2}$$

in which K is the permeability which is a function of grain shape, size and porosity. The quantity η is the viscosity of water, which is a function of temperature. From the foundation engineering point of view, the value of η may be considered practically constant as this value changes very little for the range of temperature that may be encountered in the subsoil. The value of k is determined in the laboratory by means of permeameters, and reduced at 20°C for which the value of $\eta = 10.3 \times 10^{-6}$ g · sec/cm².

Investigations show that the coefficient of permeability changes with the void ratio as follows:

$$\frac{k_1}{k_2} = \frac{e_1^2}{e_2^2} \tag{6-II.2}$$

If the void ratio of a soil in a loose state is reduced to the medium dense state, that is, reducing the void ratio to one half, then the permeability is reduced four times. Hence, compaction to reduce flow of water is an important factor in a given soil. Hazen investigated the coefficient of permeability for clean sand filters finding the following empirical expression:

$$k = C \cdot D_{10}^2 \tag{7-II.2}$$

in which k is the coefficient of permeability in cm/sec, C is a coefficient that varies from about 50 to 150; on the average a value of 100 may be taken. The effective diameter D_{10} in centimeters is determined by sieve analysis tests. It represents the diameter at which 10% of the material passes this diameter. From formula 7-II.2, it may be recognized that a change in the effective diameter is very sensitive to the coefficient of permeability. Figure 1-II.2 shows the range of values of k that may be expected as a function of the effective diameter D_{10}. Formula 7-II.2 may give an accuracy of ±50%. The value of the parameter C depends greatly on the shape characteristics of the grains. For soil deposits with the same grain shape characteristics, the following relation may be used approximately:

$$\frac{k_1}{k_2} = \frac{[D_{10}^2]_1}{[D_{10}^2]_2} \tag{8-II.2}$$

Relation 8-II.2 may be useful to study the permeability of a deposit if only a few direct permeability tests are available, from which the value of C may be determined, and then by simple sieve grain size determinations, the values of k may be estimated for the same type of sediments.

The coefficient of permeability is one of the mechanical properties of soil most difficult to determine with accuracy. In sands, from fine to coarse, where sieve analysis may be performed, the coefficient of permeability may be determined by means of direct permeability tests in conjunction with sieve grain size determina-

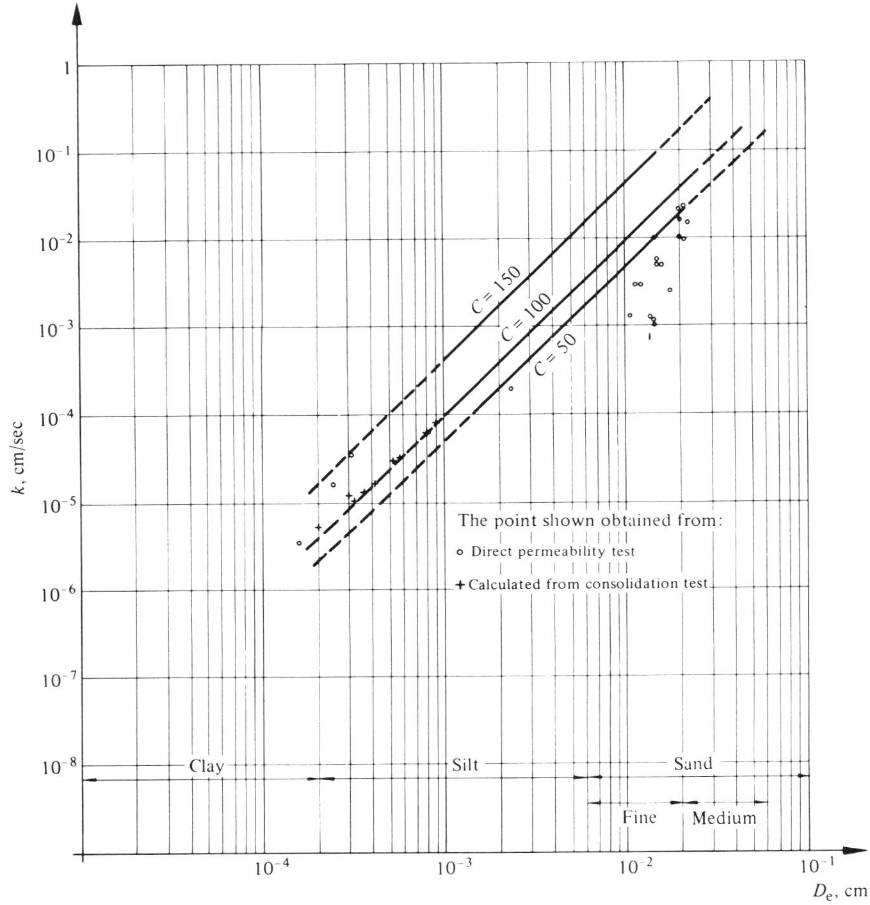

Fig. 1-11.2 Coefficient of permeability $k = CD_{10}^2$, according to Hazen.

tions as stated before. For finer materials like silt and clay, the permeability is very difficult to determine. The effective diameter in these soils loses its physical meaning. For practical purposes, in case of water discharge problems, silt and clay may be considered impervious. By means of direct permeability tests in fine silty sand in conjunction with sieve analysis, the smallest value of the permeability that is possible to determine is about $k = 4$ to 5×10^{-3} cm/sec. The only laboratory means to determine these small permeabilities is by means of the variable head permeameters on undisturbed samples representing the subsoil deposit. When permeability is lower than $k = 10^{-6}$ cm/sec as in clayey silt and clays, the materials may be considered impervious. Fat and colloidal clays may reach values up to $k = 10^{-9}$ or 10^{-10} cm/sec. In these cases, the knowledge of the permeability is valuable for problems of consolidation in clay layers, silty clay and clayey silts. For these problems the concept of the coefficient of consolidation is introduced.

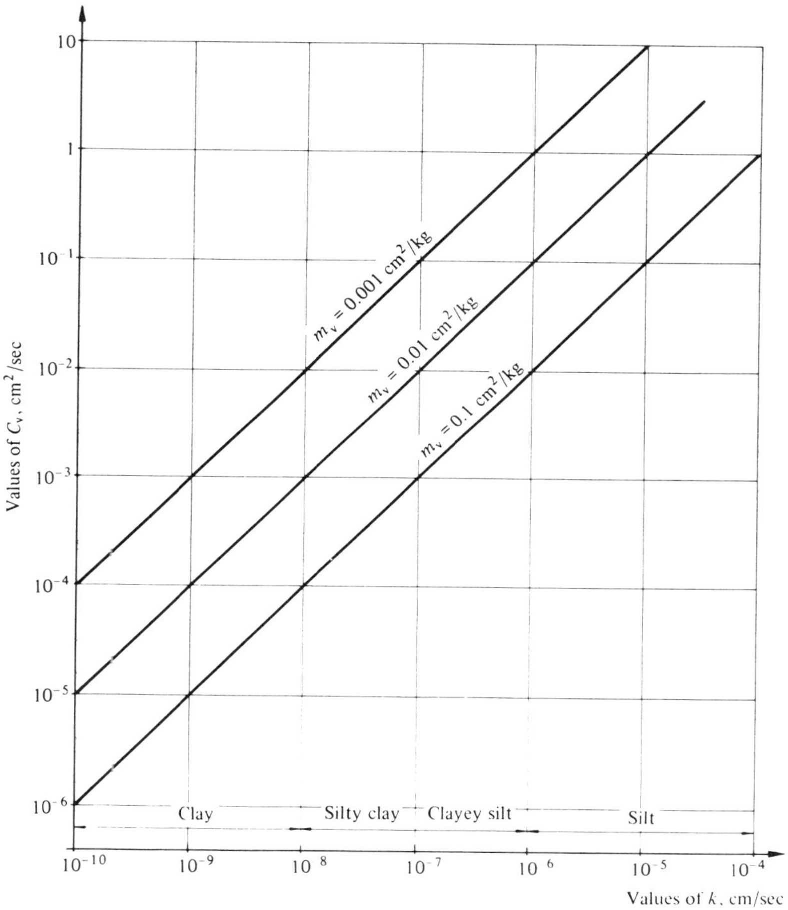

Fig. 2-II.2 Coefficient of consolidation vs. coefficient of permeability.

This value is directly proportional to the coefficient of permeability (Fig. 2-II.2). From Terzaghi's theory of consolidation (see Section II.3.9) we have

$$c_v = \frac{k}{m_v \gamma_w} \tag{9-II.2}$$

in which c_v is the coefficient of consolidation in cm^2/sec, m_v the coefficient of unit volume compressibility in cm^2/kg, and γ_w is the unit weight of water in kg/cm^3. Therefore:

$$k = c_v \cdot m_v \cdot \gamma_w \tag{10-II.2}$$

The value of c_v is also very sensitive. Take, for instance, a clayey silt with $c_v = 10^{-3}$ cm^2/sec. If in the undisturbed state $m_v = 0.01$ cm^2/kg, then:

$$k = 10^{-3} \cdot 10^{-2} \cdot 10^{-3} = 10^{-8} \text{ cm/sec}$$

Now assume the material is of high sensitivity when completely disturbed at constant water content, the skeleton structure of the soil is disturbed, and by the same token, the coefficient of volume compressibility is increased to m_v = 0.10 cm^2/kg. Since the void ratio has not changed during remolding we may assume that k remains approximately the same; hence c_v will take the change and will be reduced to $c_v = 10^{-4}$ cm^2/sec. Upon remolding, however, the value of k is also reduced to some extent since the new arrangement of the ultrafine grain fraction may clog the soil conduits. The change in c_v will affect the rate of consolidation very substantially, since from the theory of consolidation

$$\frac{c_{v1}}{c_{v2}} = \frac{t_2}{t_1} \qquad 11\text{-II}.2$$

where t_2 and t_1 are times taken at a specific average degree of consolidation. Therefore, in case of sensitive clayey silts and clays, the damage in the skeleton structure has large bearing on the coefficient of consolidation (see Section II-3.9). Very highly sensitive clays may show coefficients of consolidation in the remolded state on the order of 0.1 to 0.01 of that obtained in the undisturbed state.

II.3 STRESS-STRAIN-TIME RELATIONSHIPS

3.1 General Concepts

The mechanical properties of stress-strain and the time effect in unconsolidated fine sediments is one of the most difficult and complex problems for determination and interpretation in soil mechanics. The study may be made from the point of view of rheological concepts in conjunction with the phenomenological behavior observed. The modulus of unit deformation, coefficient of unit volume compressibility and the rigidity of the soil under dynamic conditions are necessary to calculate approximately the behavior of foundations. The foundation engineer must forecast total and differential settlements after he has succeeded in assigning a safe bearing load on the soil. It is not only the problem of settlements for a building, but also the importance of soil displacements in general. When performing excavations, the heave at the bottom and horizontal displacements should be estimated, as well as the induced displacements in adjacent soil and construction. Accordingly, these displacements should be calculated, since they may be critical in buildings and machinery installations.

Soil material is formed of a skeleton structure of rigid grains, which may or may not have large cohesion, depending on the content of the cementing materials. The voids or cavities left by the structure of the soil may be found saturated or partially saturated with water. Therefore, it is concluded that the properties of stress-strain-time of soil is a function of several factors, namely:

1. State of stress in which the material was compacted in the past.

2. Type of skeleton structure.
3. Intergranular cohesive forces or cementation.
4. Form, dimensions and strength of the grains.
5. State of density at which the soil may be encountered.
6. Degree of saturation.
7. Permeability.

The compressibility of a soil may be correlated with the modulus of unit deformation or strain modulus, defined by the following relation:

$$\lim_{\Delta\sigma \to 0} \frac{\Delta\epsilon}{\Delta\sigma} = M \qquad \text{1-II.3}$$

The strain modulus is represented by the tangent to the stress-strain curve (Fig. 1-II.3). The general characteristics of this curve are typical of any soil material. It may be observed that, as the stress level increases, the strain modulus increases, until close to failure its value becomes very large.

The total strain for stress level σ_i may be calculated as follows:

$$\epsilon_1 = \sum_{i=1}^{n} M_i \cdot \Delta\sigma_i \qquad \text{2-II.3}$$

In the particular case in which the value of M is constant, and equal for loading or reloading, then:

$$\epsilon_1 = M \cdot \sigma_1 \qquad \text{3-II.3}$$

The value of M in this case represents a constant of proportionality, corresponding to an ideal elastic solid. In reality, upon unloading the material after reaching stress level σ_1, one observes that the strain ϵ_1 is not completely recovered

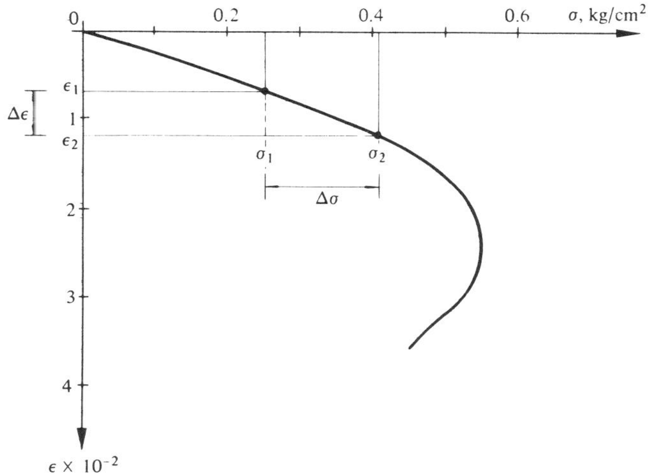

Fig. 1-II.3 Stress-strain curve.

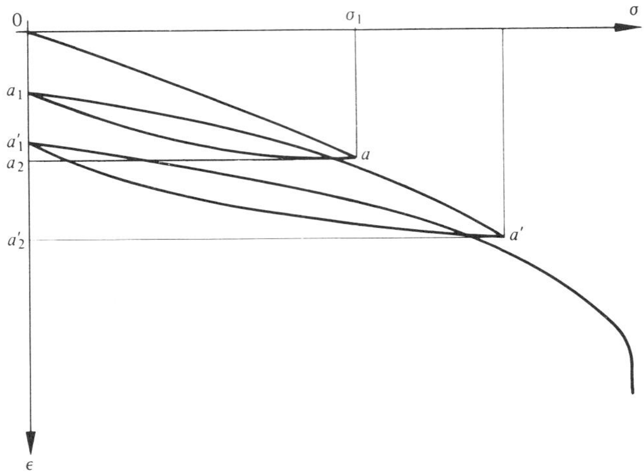

Fig. 2-II.3 Stress-strain behavior.

(Fig. 2-II.3); therefore, the material remains deformed to certain amount oa_1 that represents structural rearrangement. The recovered part a_1a_2 represents the reversible phenomenon or response of elastic characteristics. The magnitude depends on the type of material, and on the influence the above-mentioned factors may have on the structural soil system. In general, it may be observed that the hysteresis loops remain parallel to each other upon unloading and reloading at increasing stress levels and up to a stress of about 2/3 of the failure stress level. This phenomenological observation is important, since it demonstrates that the elastic response strain modulus is practically independent of the stress level at which total unloading takes place, except close to failure.

The time effect in the deformation is of extreme importance for certain materials, mainly those that contain the fine and very fine fraction, like silts and clays. Therefore, after applying a rapid increment of stress $\Delta\sigma$ on the material, an increment of strain $\Delta\epsilon$ takes place immediately (Fig. 3-II.3). As time goes on, however, it is observed that the deformation continues. The rate of strain $d(\Delta\epsilon)/dt$, representing the slope of the curve $\Delta\epsilon$ vs. t, reduces with time depending on the type of soil material, and approaches a small value after a long time. When the material is permeable, after the elapse of time t_1, with rapid unloading it is observed that certain deformation is recovered also immediately, because of the response of the elastic elements contained in the material. Upon reloading at t_2, it is observed that an immediate deformation takes place followed by a time aftereffect as mentioned before. The general configuration of the deformation-time curve is shown in Fig. 4-II.3. From the above-mentioned phenomenological behavior, one should recognize the importance of studying the laws of stress-strain-time behavior of test specimens in the laboratory and correlation with soil deposits in the field.

It has been stated that the soil contains a skeleton structure with special properties to consider and understand. The structure may be arranged in such a way that

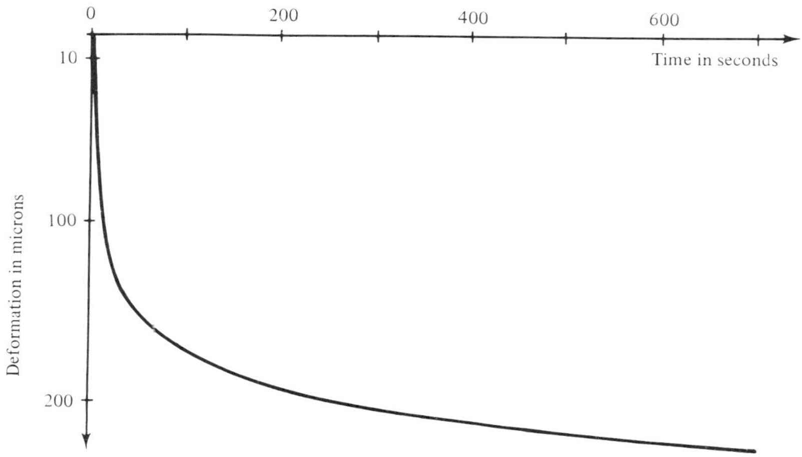

Fig. 3-II.3 Time effect in deformation.

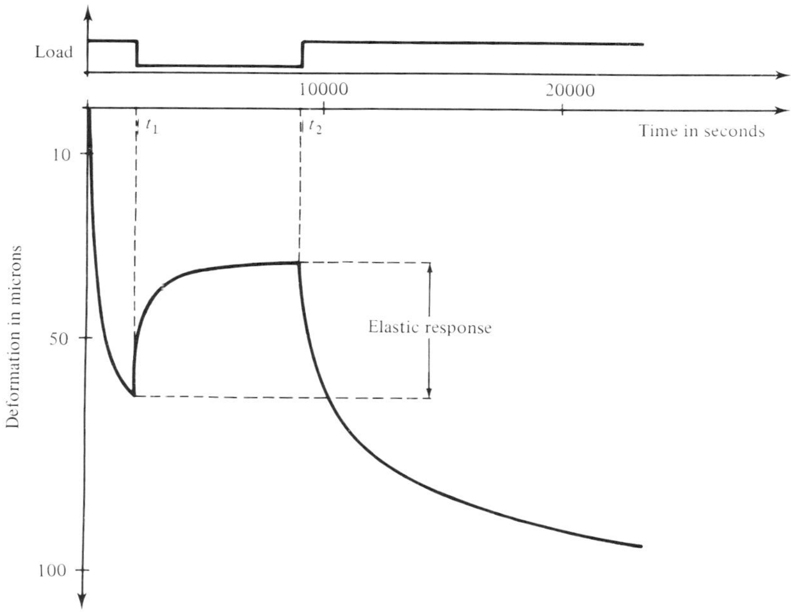

Fig. 4-II.3 Deformation as a function of time.

on some occasions large elastic behavior may be observed, that is to say, the material may be considered to contain a large number of elastic elements per unit volume of soil, giving good static stability to the soil under stress. The elastic elements building up the skeleton structure give stability that depends, however, on the stress level to which the soil is subjected and on the extent these elements have been used. For a small stress level, the grains do not suffer important relative displacement; hence the behavior may be considered highly elastic. When the skeleton structure of the soil is subjected to higher stress levels, the unstable grains suffer relative displacements that are not recoverable as the load is removed. The relative displacement of the grains is purely a plastic phenomenon shown by the deformation remaining after unloading. A large part of this behavior takes place instantaneously, and at the same time, the elastic deformation takes place. The plastic deformation is followed immediately by a viscous intergranular behavior that implies strain vs. time. When the material is saturated and confined under a certain state of stresses, the deformation suffers a hydrodynamic retardation because, for a volume change to take place, the gravitational water must be expelled from the pores of the soil within the skeleton structure.

The plastic behavior may be better noticed in soils with coarse grain characteristics where the viscous intergranular phenomenon has no important proportions. For example, when performing a test in loose sand, it is observed that upon loading the material suffers a rapid elasto-plastic deformation. When the material is unloaded after reaching certain deformation, it may be observed that a small part is recovered. The large permanent deformation observed in these materials is caused by the relative displacement and rearrangement of cohesionless grains that have the tendency to occupy more stable positions upon loading. During this movement, it is also possible that rupture may take place in the intergranular contacts of sharp and soft grains. Therefore, it may be concluded that the immediate elasto-plastic strain of a soil material when loaded is the following:

$$\epsilon_{ep} = \epsilon_e + \epsilon_p \qquad \text{4-II.3}$$

That is to say, the total strain is the sum of the elastic and plastic strains, respectively.

In materials of fine grain size in which the grains are bonded together by, for example, clay minerals, observations show that the strain is a very strong function of time. Upon stress application σ and after obtaining the elasto-plastic strain ϵ_{ep} in not fully saturated materials, the deformation continues with time. This implies that the intergranular viscous properties of the fine material enclosing the larger grains retards the relative displacement of the grains. It may be observed in practice that under a sustained load, the rate of strain decreases with time and depends highly on the amount of viscous intergranular elements per unit volume that the soil contains. When soil is compressed in a confined state, it changes in volume and becomes more compact and therefore less compressible. If the material is saturated, there exists a hydrodynamic resistance opposing volumetric change.

From the above reasoning one may conclude that when an increment of stress is applied on a representative specimen of soil, the strain is a function of time; however, deformation must come to an end if the material has sufficient elastic elements linked with plastic elements, and if the elastic elements have not been completely damaged. The magnitude and the rate of strain under sustained load depends on the degree of confinement at which the soil is subjected. Therefore, rheological models well selected in conjunction with the observation of phenomenological behavior may offer a useful tool, helping to visualize and interpret the stress-strain-time relationships of soil materials.

Therefore, it may be concluded that for a given soil, the strain modulus is a function of the confined stress level σ_c to which the sediment may be subjected and of the time the state of stress acts on it. Therefore

$$M_{\sigma,t} = M(\sigma, t) \quad \text{5-II.3}$$

which may be written also in the form

$$M_{\sigma,t} = M_\sigma(\sigma) \cdot M_t(t) \quad \text{6-II.3}$$

Therefore, we can write

$$\Delta\epsilon = M_\sigma(\sigma) \cdot M_t(t)\,\Delta\sigma \quad \text{7-II.3}$$

To perform displacement calculations in soil it is necessary to know the phenomenological laws governing the above-mentioned functions.

3.2 The Elastic Unit

The elasticity of materials was investigated for the first time by Hooke in the seventeenth century, (1670). He defined elasticity as the property of an ideal solid to recover its form completely after it is unloaded. From here, the following theorem well known to engineers may be stated, known as Hooke's law: "The distortion of an elastic body is proportional to the stress applied to such body, and the phenomenon is reversible." That is to say:

$$\gamma = \mu\tau \quad \text{8-II.3}$$

Here μ is a constant representing the coefficient of proportionality and measures

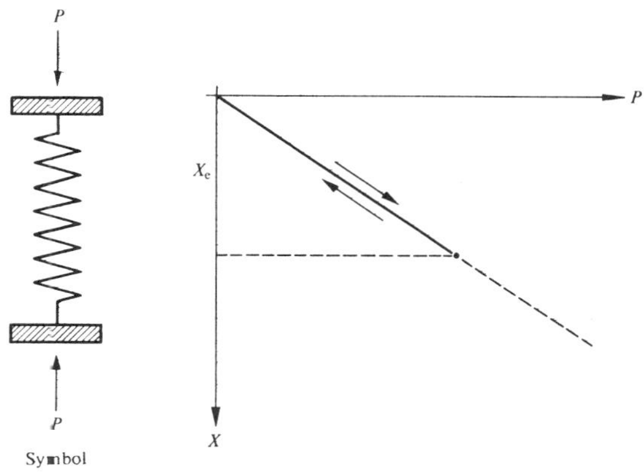

Fig. 5-II.3 The Hooke unit.

the distortion of the material to applied shear stress. This law may be applied also to linear deformation: the linear strain is directly proportional to applied stress, i.e., $\epsilon = M\sigma$. The behavior may be represented by an ideal spring not suffering permanent deformation upon unloading, independently of the frequency at which the load is applied (Fig. 5-II.3).

3.3 The Plastic Unit

A body is ideally plastic when upon applying a determined load it deforms immediately, but after unloading does not recover its original form, that is to say, the body preserves the total deformation caused by the original load.

To illustrate this behavior let us imagine a rigid body (A) of weight W placed on a surface (Fig. 6-II.3). The body A will be subjected to a horizontal force P_h which will produce no movement with respect to the base until the friction resistance P_s between the two bodies is overcome. The limiting resistance according to the friction law of Coulomb (1783) is

$$P_s = \mu W \tag{9-II.3}$$

When this resistance is reached the body starts moving, and upon unloading the displacement x_p remains constant.

If we link an elastic element with a plastic element in series, a new mechanical unit is obtained known as St. Venant body (1800) (Fig. 7-II.3). When load P is applied to this body, it will behave elastically until the value P_s is reached, corresponding to the limit resistance; thereafter, the body enters into a plastic state, that is to say, when the resistance between the two bodies is overcome, the body continues deformation in an increasing and continuous way. Upon load removal, the material offers elastic response because of the spring. This behavior is very close to the one observed in materials like steel that follows very closely St. Venant's

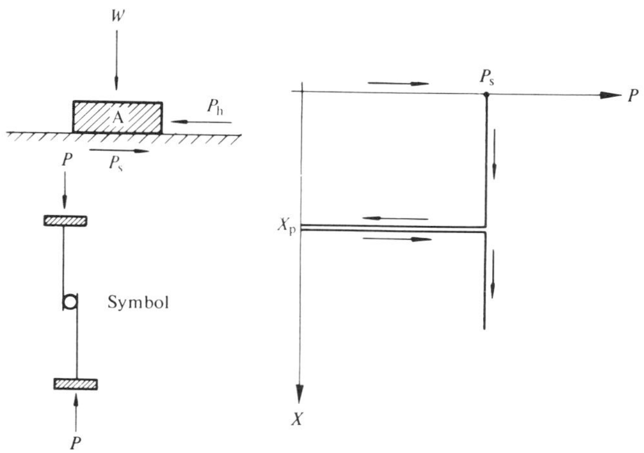

Fig. 6-II.3 The Coulomb unit.

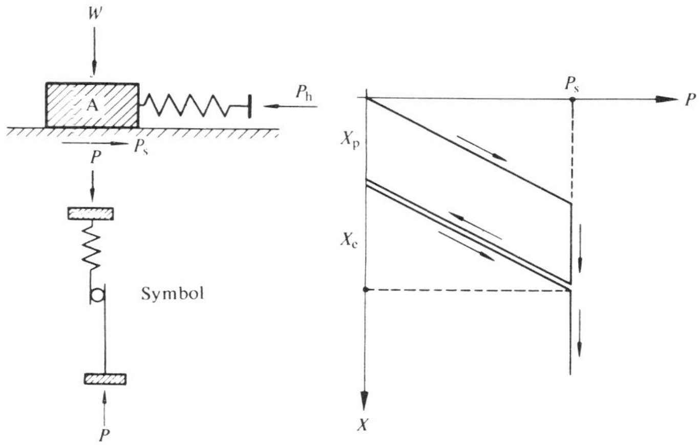

Fig. 7-II.3 The St.-Venant unit.

body, up to stresses close to the hardening zone where the phenomenological law of stress-strain is different.

3.4 The Elasto-Plastic Unit

One St. Venant's mechanical unit, however, is not satisfactory to explain elasto-plastic behavior of soils. Furthermore, it is observed that upon reaching load P_s the body enters in movement, developing dynamic friction that is less in value than the static friction. Therefore, the body accelerates itself coming to a plastic uniform movement with time, as shown in Fig. 8-II.3. In order to obtain elasto-plastic equilibrium, it is necessary to consider a large number of St. Venant's units in parallel, forming as a whole an elasto-plastic body where static equilibrium may be obtained before reaching the limiting plastic loads of all their elements. Figure 9-II.3 shows such a model that explains in general terms elasto-plastic behavior. This model formed by a series of St. Venant's bodies in parallel may be assumed

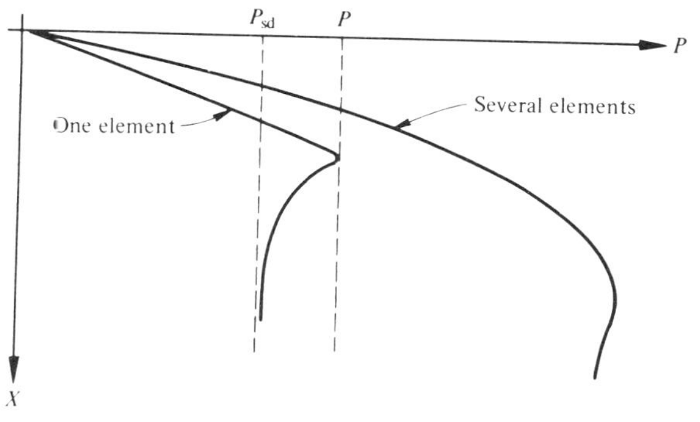

Fig. 8-II.3 Plastic behavior, St.-Venant unit.

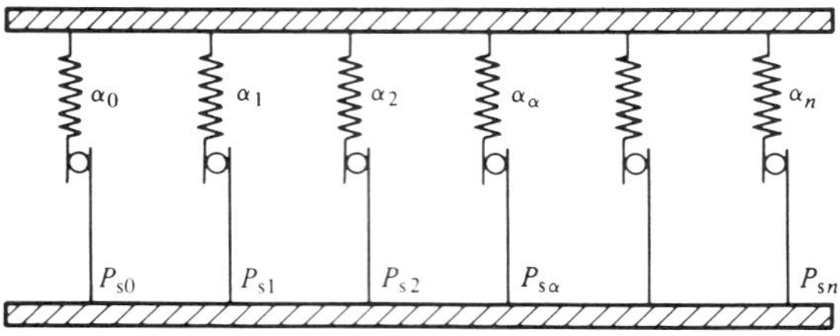

Fig. 9-II.3 St.-Venant elasto-plastic elements in parallel.

with springs having different moduli of deformation, and the plastic Coulomb's elements having different plasticity strength limits, varying from zero to some large value. In this way, upon load application, the St. Venant's elements will be utilized gradually overcoming one by one their plastic strength. For an intermediate stress level, one may obtain elasto-plastic equilibrium, and after load relief elastic response is obtained as a fraction of the total deformation.

When tests are made in soil materials, one observes that when increasing the load level the strain increases gradually, until close to the rupture load it increases indefinitely (Fig. 10-II.3). Also one may observe that for small levels of stress the material behaves more like an elastic solid. The model in Fig. 9-II.3 represents approximately the observed phenomenological behavior from which it may be concluded that

$$\frac{d}{d\sigma}(\epsilon_{ep}) = f(\sigma) \qquad \text{10-II.3}$$

in which $f(\sigma)$ should satisfy the following conditions:

1) For small stress level, $d\epsilon/d\sigma$ is well defined, as well as for $\sigma = 0$, $\epsilon = 0$;
2) When the ultimate load σ_u of the soil is reached the rate $d\epsilon/d\sigma$ and the strain ϵ are very large.

Several mathematical functions of σ may be investigated representing these limiting conditions. The simplest function that may be studied to obtain a particular solution representing such behavior is

$$\frac{d\epsilon}{d\sigma} = b + \frac{\sigma}{\sigma_u - \sigma} a$$

By integration

$$\epsilon_{ep} = b\sigma + a\,\{(\sigma_u - \sigma) - \sigma_u \text{ Log } (\sigma_u - \sigma)\} + C$$

From this expression it may be observed that for small values of σ the value b represents the initial elastic modulus: $b = \alpha_e$. Also setting $\alpha_p = a$ one may write:

$$\epsilon_{ep} = \alpha_e\sigma + \alpha_p\sigma\left(\frac{\sigma_u}{\sigma} \text{ Log } \frac{\sigma_u}{\sigma_u - \sigma} - 1\right)$$

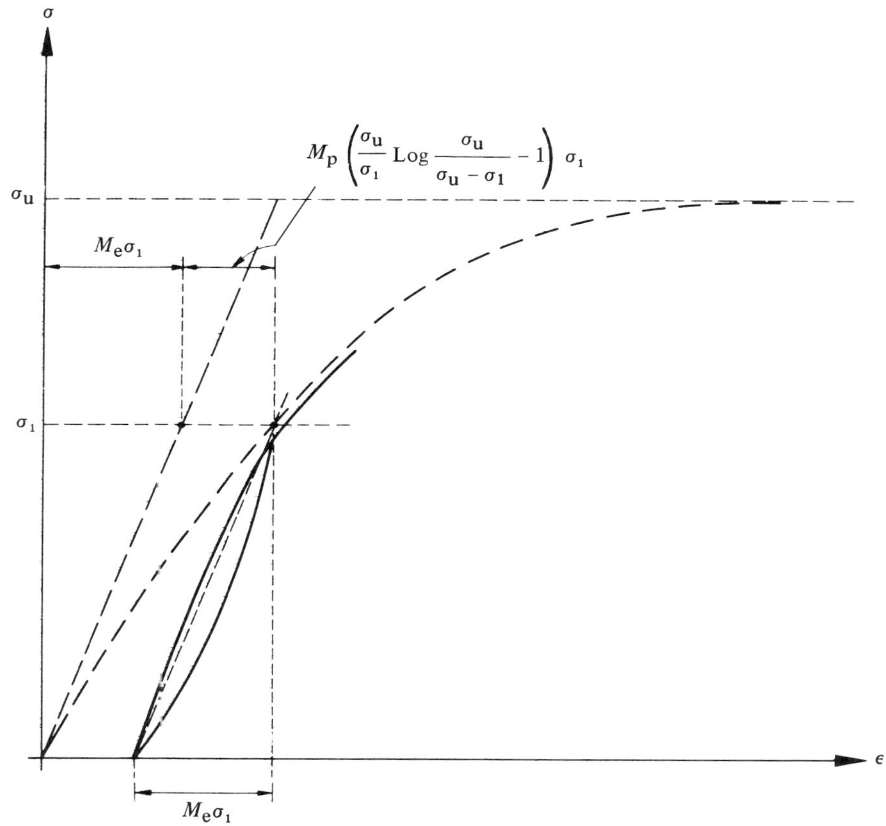

Fig. 10-II.3 Stress-strain curve.

The value of α_p measures the plastic phenomenon, since when $\alpha_p = 0$ only elastic behavior takes place. This law illustrates that the plastic behavior is not important until a certain level of stresses has been reached, upon which important relative displacements between grains take place; finally, the strain has the tendency to become very large when the applied load reaches the rupture stress σ_u (Fig. 10-II.3).

The following simple expression may be expressed in terms of the strain moduli for elastic and plastic strain, respectively:

$$\epsilon_{ep} = \left\{M_e + M_p \left(\frac{\sigma_u}{\sigma} \text{Log} \frac{\sigma_u}{\sigma_u - \sigma} - 1\right)\right\} \sigma \qquad 11.\text{II}.3$$

This formula expresses the probable phenomenological elasto-plastic behavior of stress-strain for certain state of confinement stress held constant during the process of the test. This discussion, however, may be taken only as a qualitative indication of the rheological elasto-plastic behavior based on phenomenological observations.

3.5 The Strain Modulus

The determination of this mechanical property is achieved in the laboratory in test specimens representative of subsoil materials. The test specimens should be of undisturbed materials when the soil has cohesion. When the material is cohesionless, the determination of this mechanical property is more complicated, since it will be necessary to study various states of compaction, and estimate from these results the strain modulus that will correspond to the state of compaction and state of stress at which the soil is encountered in the field. The test specimens of undisturbed materials are difficult to obtain by means of sampling, in spite of using the best sampling tools. The samples suffer a disturbance which must be taken into consideration in the interpretation of the test results. The undisturbed samples giving better results for cohesive materials are those obtained from block samples. This procedure is, however, sometimes impossible to follow when the material to investigate is deep seated and below the surface water table. Several procedures have been developed to obtain undisturbed samples; however, discussion of these methods is beyond the scope of this book, but may be found in the literature on the subject.

The investigation of the stress-strain properties of soil materials may be generalized accepting that the material has different mechanical properties only in two directions, that is, in the direction normal to the planes of stratification and parallel to them. Under these conditions, let us call: M_z = linear strain modulus in the vertical direction, and M_h = linear strain modulus in the horizontal direction. Let Fig. 11-II.3 represent an element of the soil to which an increment of stress $\Delta\sigma_z$ is applied, hence the increment of strain will be $M_z\Delta\sigma_z$, and in perpendicular planes XZ and YZ the induced strain is

$$-\nu\Delta\sigma_z M_z$$

in which ν is Poisson's ratio, and will be considered as a fixed value valid for the horizontal and vertical directions, respectively. The stress $\Delta\sigma_x$ will produce an increment of strain $M_h \cdot \Delta\sigma_x$ in the x-direction and $-\nu M_h \cdot \Delta\sigma_x$ in the perpendicular directions. In the same way we find expressions when stresses are increased in direction of σ_y.

From the above reasoning, it may be concluded that when incrementing stresses in the three directions, the increment of strain corresponding to one single direction may be expressed as follows:

$$\Delta\epsilon_z = M_z\Delta\sigma_z - \nu M_h\Delta\sigma_y - \nu M_h\Delta\sigma_x \qquad \text{12-II.3}$$

and for directions x and y;

$$\Delta\epsilon_y = M_h\Delta\sigma_y - \nu M_h\Delta\sigma_x - \nu M_z\Delta\sigma_z \qquad \text{13-II.3}$$

$$\Delta\epsilon_x = M_h\Delta\sigma_x - \nu M_h\Delta\sigma_y - \nu M_z\Delta\sigma_z \qquad \text{14-II.3}$$

After simplifying and arranging terms conveniently, we may write for the three perpendicular directions the following:

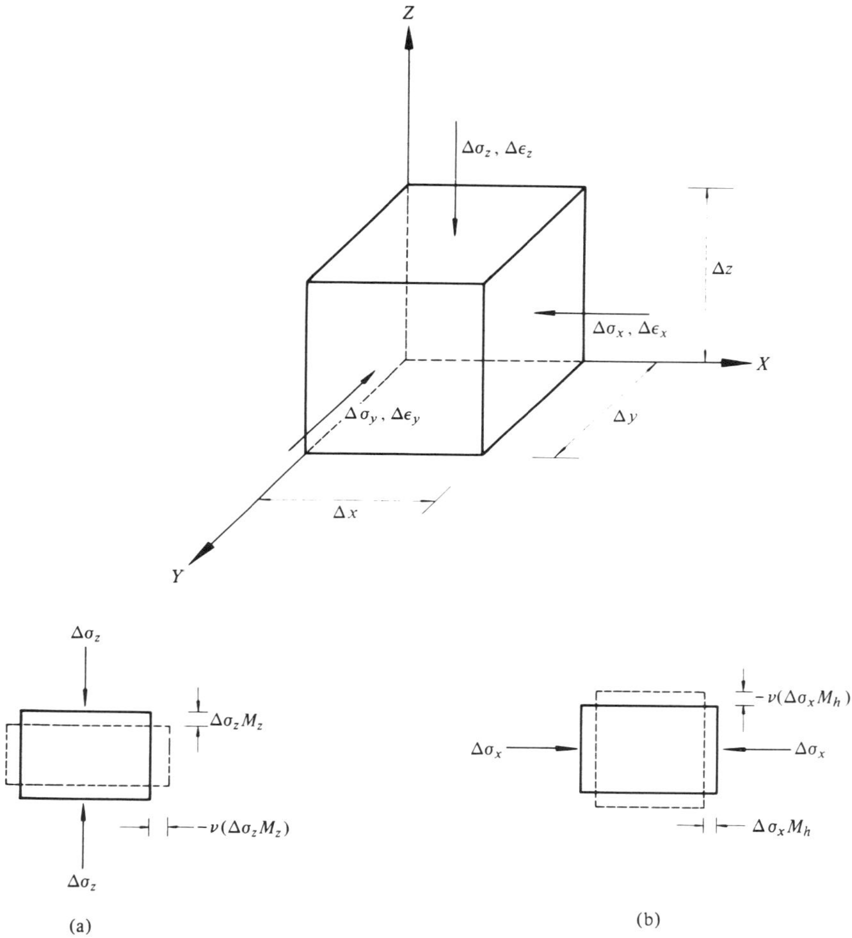

Fig. 11-II.3 State of stresses and strains at a point.

$$\Delta\epsilon_z = \left[1 - \nu \frac{M_h}{M_z} \cdot \frac{\Delta\sigma_x + \Delta\sigma_y}{\Delta\sigma_z}\right] M_z \cdot \Delta\sigma_z \qquad \text{15-II.3}$$

$$\Delta\epsilon_y = \left[1 - \nu\left(\frac{\Delta\sigma_x}{\Delta\sigma_y} + \frac{M_z}{M_h} \cdot \frac{\Delta\sigma_z}{\Delta\sigma_y}\right)\right] M_h \cdot \Delta\sigma_y \qquad \text{16-II.3}$$

$$\Delta\epsilon_x = \left[1 - \nu\left(\frac{\Delta\sigma_y}{\Delta\sigma_x} + \frac{M_z}{M_h} \cdot \frac{\Delta\sigma_z}{\Delta\sigma_x}\right)\right] M_h \cdot \Delta\sigma_x \qquad \text{17-II.3}$$

From the above expressions, several cases of deformation at a point may be analyzed.

First Case. When the material is subjected to increment of stresses $\Delta\sigma_z$, $\Delta\sigma_y$, $\Delta\sigma_x$ and the conditions of strain are not restricted, that is to say, the strain takes place freely. Then equations 15, 16, and 17-II.3 represent the increment of strain and they are exclusively a function of the applied increment of stresses and the mechanical properties of the material.

Second Case. When unit deformation is zero in one horizontal direction, that is, $\Delta\epsilon_y = 0$ and $\Delta\epsilon_x \neq 0$, then the following condition may be obtained from 17-II.3:

$$1 = \nu\left(\frac{\Delta\sigma_x}{\Delta\sigma_y} + \frac{M_z}{M_h}\cdot\frac{\Delta\sigma_z}{\Delta\sigma_y}\right) \qquad \text{18-II.3}$$

from which the ratio of increment of stresses will read as follows:

$$\frac{\Delta\sigma_y}{\Delta\sigma_z} = \nu\left(\frac{\Delta\sigma_x}{\Delta\sigma_z} + \frac{M_z}{M_h}\right) \qquad \text{19-II.3}$$

and

$$\frac{\Delta\sigma_y}{\Delta\sigma_x} = \nu\left(1 + \frac{M_z}{M_h}\cdot\frac{\Delta\sigma_z}{\Delta\sigma_x}\right) \qquad \text{20-II.3}$$

By substitution in formulas 15 and 16-II.3 and arranging terms, expressions for plane strain are finally obtained:

$$\Delta\epsilon_z = (1+\nu)\left[1 - \nu\left(1 + \frac{M_h}{M_z}\cdot\frac{\Delta\sigma_x}{\Delta\sigma_z}\right)\right] M_z\cdot\Delta\sigma_z \qquad \text{21-II.3}$$

$$\Delta\epsilon_x = (1+\nu)\left[1 - \nu\left(1 + \frac{M_z}{M_h}\frac{\Delta\sigma_z}{\Delta\sigma_x}\right)\right] M_h\cdot\Delta\sigma_x \qquad \text{22-II.3}$$

Third Case. The material is confined to zero lateral deformation in both horizontal directions, that is to say, $\Delta\epsilon_x = \Delta\epsilon_y = 0$. Using formulas 16 and 17-II.3, the ratio of increment of stresses will be:

$$\frac{\Delta\sigma_y}{\Delta\sigma_z} = \frac{\nu}{1-\nu}\frac{M_z}{M_h}, \quad \frac{\Delta\sigma_x}{\Delta\sigma_z} = \frac{\nu}{1-\nu}\frac{M_z}{M_h} \qquad \text{23-II.3}$$

and substituting their value into expressions 15-II.3 the following formula is found:

$$\Delta\epsilon_z = \frac{(1+\nu)(1-2\nu)}{(1-\nu)}\cdot M_z\Delta\sigma_z \qquad \text{24-II.3}$$

From 24-II.3 it may be seen that the unit vertical deformation does not depend on the ratio M_h/M_z; however, it is a function of Poisson's ratio. These conditions may represent in nature a soil deposit or sediment extensively loaded at its surface, that is to say, when the thickness of the stratum is small in magnitude if compared with the loaded area. This condition also represents the case in nature of a soil deposit very compressible in the vertical direction but highly stratified in the horizontal direction, and when the horizontal strata constitute very rigid materials that do not permit horizontal displacements. From expressions 13 and 14-II.3, it may be shown that for $\Delta\epsilon_x = \Delta\epsilon_y = 0$; $M_h = 0$ and $\nu = 0$. This condition applies when the reinforcing horizontal sheets are very close to each other; then 24-II.3 takes the value: $\Delta\epsilon_z = M_z \Delta\sigma_z$.

To illustrate numerically these cases of anisotropy, let us consider a value of $\nu = 0.25$ and a ratio of the strain modulus $M_h/M_z = 1/3$. Also suppose that the ratio of the increment of stresses in the horizontal and vertical directions is $\Delta\sigma_x/\Delta\sigma_z = \Delta\sigma_y/\Delta\sigma_z = 1/2$. Under these circumstances, it is found for the first case:

$$\Delta\epsilon_z \doteq 0.92\, M_z \cdot \Delta\sigma_z$$

In case the material is confined to zero horizontal deformation, using the same values

$$\Delta\epsilon_z \doteq 0.835\, M_z \cdot \Delta\sigma_z$$

therefore it may be concluded that depending on the type of confinement, the strain may be different for the same value of the linear strain modulus M_z. Figure 12-II.3 shows the value of:

$$\frac{(1+\nu)(1-2\nu)}{1-\nu} = \nu_c \qquad \text{25-II.3}$$

from which it may be noticed that in case of total confinement and $\nu = 0.50$ the strain is zero, that is to say, the material cannot deform in the vertical direction when lateral displacements are totally restricted.

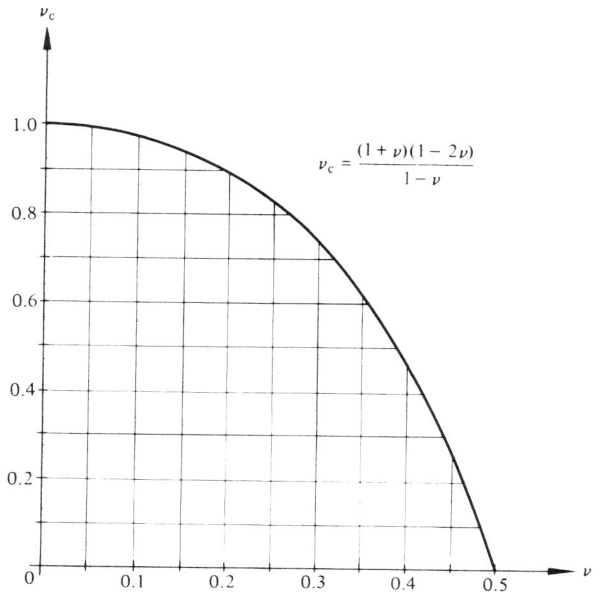

Fig. 12-II.3 Theoretical ratio of coefficient of unit volume compressibility to linear strain modulus.

In order to clarify this situation, let us study the volumetric compression of a sediment:

$$\Delta\epsilon_v = \frac{\Delta V_v}{V} \qquad \text{26-II.3}$$

Here, ΔV_v is the change in volume in the voids of the material subjected to certain increment of stresses, and V is the total volume. When the change in stresses $\Delta\sigma_z$, $\Delta\sigma_y$, $\Delta\sigma_x$ takes place, the material suffers a volumetric strain:

$$\Delta\epsilon_v = \Delta\epsilon_z + \Delta\epsilon_y + \Delta\epsilon_x \qquad \text{27-II.3}$$

Therefore, the volumetric strain is equal to the sum of the linear strains in the three perpendicular directions. If the values of the increments of strain given by formulae 15, 16 and 17-II.3 expressing the increments of strain in the three directions are substituted in expression 27-II.3, the following expression is found for the volumetric strain:

$$\Delta\epsilon_v = (1 - 2\nu)\left[1 + \frac{M_h}{M_z}\,\frac{(\Delta\sigma_y + \Delta\sigma_x)}{\Delta\sigma_z}\right] M_z \Delta\sigma_z \qquad \text{28-II.3}$$

from which it may be seen that for an incompressible material when $\Delta\epsilon_v = 0$, Poisson's ratio should take the value $\nu = 0.50$.

The unit deformations, volumetric and linear, may be compared for the case of a totally confined material. The stresses necessary for confinement are given in formulas 23-II.3, therefore substituting their values in the volumetric strain:

$$\Delta\epsilon_v = \left[\frac{(1 + \nu)(1 - 2\nu)}{1 - \nu}\right] M_z \Delta\sigma_z \qquad \text{29-II.3}$$

This expression is identical to formula 24-II.3 from which it may be concluded that for conditions of confinement to zero lateral deformation $\Delta\epsilon_v = \Delta\epsilon_z$. Furthermore, the coefficient of unit volume compressibility defined by Terzaghi in the theory of consolidation corresponding to a confined compression test to zero lateral deformation is defined as follows:

$$m_v = \frac{\Delta\epsilon_v}{\Delta\sigma_z}$$

Using the above expression, the following ratio between the coefficient of unit volume compressibility and the linear strain modulus may be found:

$$\frac{m_v}{M_z} = \nu_c \qquad \text{30-II.3}$$

From Fig. 12-II.3, it may be seen that for values of Poisson's ratio of 0.42 corresponding approximately to a confined clay deposit, the ratio $m_v/M_z = 0.39$.

Table 1-II.3 gives an idea of values that may be encountered for the coefficient of unit volume compressibility and Poisson's ratio for different sediments.

TABLE 1-II.3

Compressibility	m_v cm^2/kg	ν	K_0	Sediment
Very high	greater than 0.1	0.43 to 0.35	0.75–0.60	Lacustrine clays and silts
High	0.1–0.02	0.35 to 0.30	0.60–0.45	Clays and silts, lacustrine sandy silts. Residual soils. Loose volcanic dust
Medium	0.02–0.005	0.30 to 0.25	0.45–0.33	Compact clays and silts, fine eolian sediments. Residual soils and volcanic semi-compact sediments. Fine alluvium
Low	0.005–0.002	0.25	0.33	Sand, compact silts, alluvial soils. Compact and well graded sediments
Very low	less than 0.002	0.25	0.33	Sands, gravelly soils. Compact alluvial sediments, cemented and well graded

In foundation engineering problems, however, the ground surface is loaded in limited areas and the stresses in the ground are calculated by the theory of elasticity (see Section III.2). At a certain depth below ground surface the vertical strain due to the change in stresses may be calculated with the formulas given for Cases I, II and III, respectively, depending on the assumptions that could be made concerning the lateral confining conditions of the subsoil strata. Nevertheless, Case I represents general conditions to calculate vertical displacements more precisely than the ones used by current methods; however, the value in the brackets of formula 15-II.3 should be investigated for the change in the state of effective stresses taking place in the subsoil because of the load applied at the ground surface, namely the change in the horizontal stresses induced by the vertical stress change (see Section III.2). The probable anisotropy of the soil should be also included in calculating the equivalent strain modulus. For large loaded areas and stratified soil deposits, it may be found that the use of the value of m_v is sufficiently accurate for most practical purposes.

3.6 The Compressibility of Fine Sediments

It has been mentioned that the soil may be considered as a skeleton structure of solid grains bound together by cohesive materials. This skeleton has cavities or voids that may be filled with air, water or the combination of both elements. If a representative sample of the soil is subjected to an increment of stress, its volume will decrease. The reduction of volume is due to the void change in the skeleton structure.

The reduction in volume due to compression of the solid grains or water is so small that for practical purposes it may be omitted. The large change in void volume is due mainly to the relative movement of the grains. They are forced to occupy new and more stable positions, making the material more compact. Therefore, the compressibility of the material will be exclusively a function of the facility with which the grains may change in position, and of the type and rigidity of the skeleton structure of the soil. A material with a honeycomb structure will be more compressible than that with a simple type structure, and less compressible than a material possessing a skeleton structure of the flocculent type with a large content of two-dimensional type grains.

That portion of the volume change due to grain displacement produces inelastic behavior. The structure of the soil when subjected to small distortions behaves elastically to a certain stress level, as observed in cohesive soils where elastic expansion takes place upon stress removal. In soil mechanics it is current to correlate the stress with the volumetric deformation of the material to express the compressibility in terms of unit stress and void change. The change in void ratio expresses the change in volume that the soil undergoes when subjected to an increment of stress. To perform compressibility tests representing strata in the field, a well known device designed by Terzaghi is used, known as the oedometer or consolidometer. The test specimen is introduced carefully in a metal ring that confines the material to zero lateral deformation during the test (Fig. 13-II.3). Porous stones are placed at the upper and lower part of the test specimen, in order to permit the expulsion of the gravitational water contained in the sample, thus permitting volume changes. To perform the test, increments of vertical stresses are applied and the vertical displacements are recorded for each one of these increments, obtaining the change in volume for each increment with time. The volu-

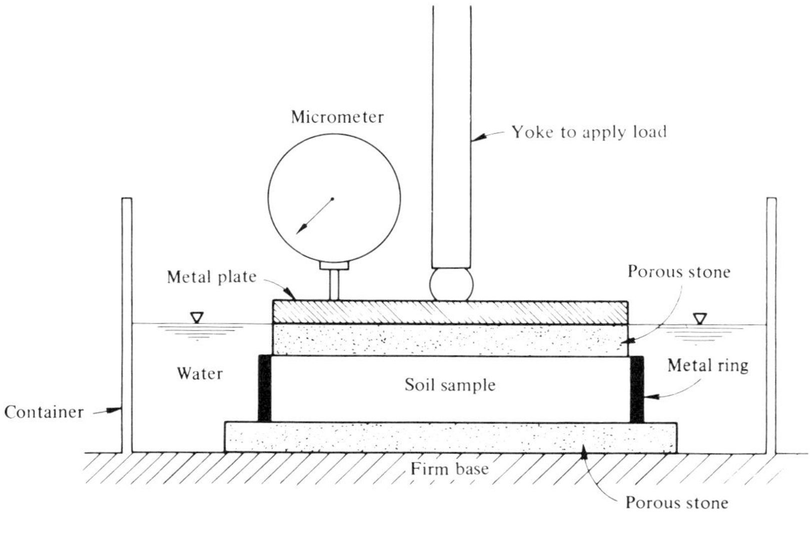

Fig. 13-II.3 Oedometer.

metric strain vs. time behavior is known as the process of consolidation of the soil for a specified sustained increment of stress at which the material is subjected.

The consolidation process is the hydrodynamic phenomenon of water expulsion from the pores of the material when this is saturated with water. Each increment of stress will produce a reduction of the voids; therefore, the material becomes less compressible. The result is a concave compressibility curve as shown in Fig. 14-II.3. The slope of the curve is defined as:

$$\lim_{\Delta\sigma \to 0} \frac{\Delta e}{\Delta \sigma} = a_v \qquad \text{31-II.3}$$

The value of a_v is defined as the coefficient of compressibility. The change in

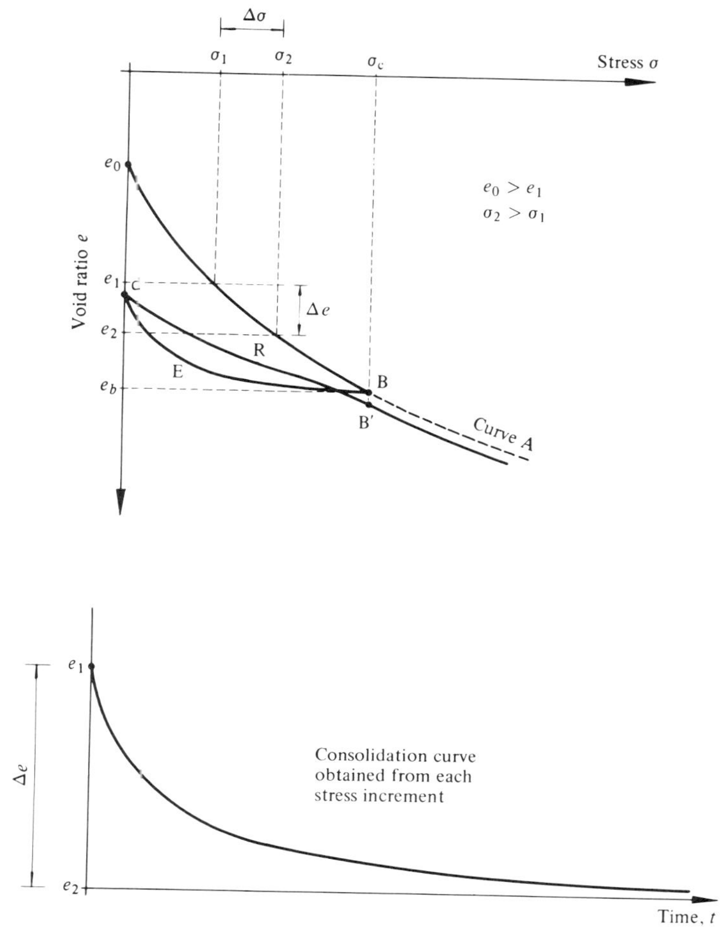

Fig. 14-II.3 Compressibility and consolidation curves.

volumetric strain in terms of void ratio is:

$$\Delta\epsilon_v = -\frac{e_2 - e_1}{1 + e_1} \qquad 32\text{-II.3}$$

When the stress corresponding to a certain void ratio represented by point B is removed by decrements, then curve BEd is obtained; it has the characteristic to be almost horizontal close to point B, and upon total stress removal it becomes sharply inclined close to point d. If later the stress is reapplied in increments, curve dRB′ above the expansion curve is obtained. In the neighborhood of point B corresponding to stress σ_c the curve changes its curvature passing under point B, and subsequently, it will follow close to curve A, but underneath. It is understood that curve A represents the virgin curve that would be obtained if the cycle of hysteresis of unloading and reloading were not performed. By definition:

$$m_v = \frac{\Delta\epsilon_v}{\Delta\sigma} \quad \text{or} \quad m_v = \frac{\Delta e}{(1 + e)\,\Delta\sigma} \qquad 33\text{-II.3}$$

The value m_v is the coefficient of unit volumetric compressibility defined by Terzaghi.

The strength of the grains forming the skeleton structure is important in the compressibility, since at the contact points high stresses are generated that may produce rupture, mainly when the grains have sharp edges, are flat, or come from soft rocks. The phenomenon produced by the high concentration of stress in the sharp edges of the grains increases the plasticity of the material. This may be easily observed in macroscopic grains and large rock fragments. The rupture of the grains in soft and altered rocks is common in fills where large rock fragments with sharp edges are used, as is the case of rock embankments used for several purposes in engineering structures. The evaluation of the compressibility of masses built of large rock fragments is difficult to determine, since it is impossible to reproduce tests in the laboratory on a sufficiently large scale to obtain satisfactory results representative of the real field behavior. Therefore, it is concluded that the only way to study the compressibility of macroscopic materials is by means of observations in the field. For this purpose, it is required to install instruments from the start of construction to measure the displacements in the vertical and horizontal directions. The results will permit the engineer to estimate the compressibility of masses formed by large rock fragments.

The very fine sediments containing clay minerals show generally high and very high compressibility. The observed phenomenon of stress-strain behavior shows high intergranular viscosity that should be taken carefully into consideration. Because of the thixotropic properties of clays, the phenomenon of hardening or solidification is important to consider in the mechanical properties of volume compressibility. These materials are generally encountered saturated in nature, or with a high degree of saturation; usually they come from lacustrine or marine environments. They may contain organic gases diluted in the pore water; therefore, during the obtention of undisturbed samples to be studied in the laboratory, the gases are liberated because of the reduction in hydrostatic pressure in the pores of the material at the depth the sample is taken. When this happens, the sample looses its saturation. This phenomenon may take place also in practice; when performing excavations the effective stresses and water pressures are reduced in the subsoil, and elastic response takes place rapidly.

The other category of clayey materials encountered in the ground is the product of weathering of rocks, that is to say, residual soils that occupy horizon B of the soil profile; they are, in general, not saturated. The compressibility of these materials may be studied with the oedometer. The problem is reduced to the study of stress-strain relationships normal and parallel to the planes of stratification. The lacustrine and marine very loose sediments start to acquire consistency when the water content is reduced to the liquid limit because of the overburden load of other sediments. An oedometer test in clay starting with a water content close to the liquid limit shows in most of the cases a compressibility curve close to a straight line in semilog scales (Fig. 15-II.3). Upon load increment, the material becomes more compact and the void ratio decreases; therefore, the relative consistency increases and the material becomes less compressible. The slope of the curve defines, as

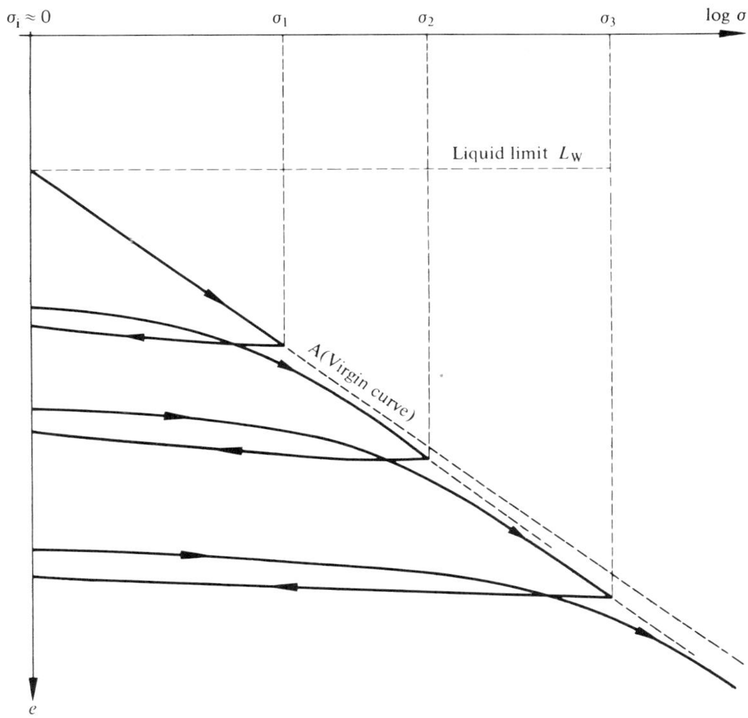

Fig. 15-II.3 Hysteresis loops in compressibility curve.

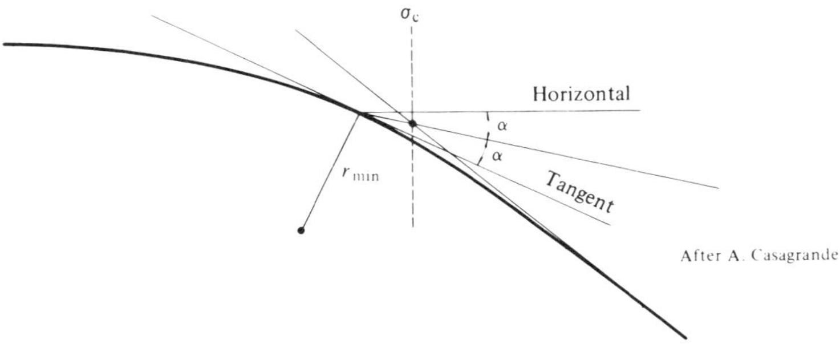

Fig. 16-II.3 Stress at which the material was preconsolidated.

mentioned before, the volume compressibility coefficient that decreases as the stress level increases. When during the process of loading and upon reaching an effective stress σ_1 the material is unloaded and loaded again, parallel hysteresis cycles are obtained. The reloading curves do not again reach the virgin curve A that represents the normally consolidated material. During the process of load re-application and close to the point at which the stress was removed, the compressibility reloading curve suffers a deflection and the curve continues under and approximately parallel to the so-called virgin curve. From the reloading curve, at the break, using an approximate construction proposed by A. Casagrande as shown in Fig. 16-II.3, the stress at which the material was mechanically preconsolidated may be approximately determined. In sediments containing active clay minerals the phenomenon of solidification and its effects on compressibility may be investigated by means of compressibility tests made in the oedometer (Fig. 17-II.3).

3.6a Normally Loaded and Preconsolidated Type Sediments. Assume a compressibility curve in a clayey test specimen at the natural water content is initiated up to stress level σ_0. If at this point the process of loading is suspended for a long time, the void ratio continues decreasing with time until the value of e_1 is obtained for which subsequent changes may be neglected. Notice that the change in volume $(e_0 - e_1)$ corresponds to deformations that are originated in the material by the phenomenon of integranular viscosity, which is a function of time. After a long time, solidification takes place and equilibrium is reached. Upon reloading in increments of stress applied every 24 hours, it may be observed that the material up to reaching stress level σ_b appears less compressible than if compression had taken place along virgin curve A and the process of loading had not been interrupted (Fig. 17-II.3). Furthermore, the value of σ_b corresponds to a critical compressive stress that in this case appears to the right of the virgin curve. The effect is that of an apparent mechanical precompression. The inclined branch that extends beyond σ_b has the tendency to become asymptotic to the virgin curve A of the normally consolidated material. From these experiments, it may be concluded that the structure of the soil after a long time of rest develops firm contacts, making the

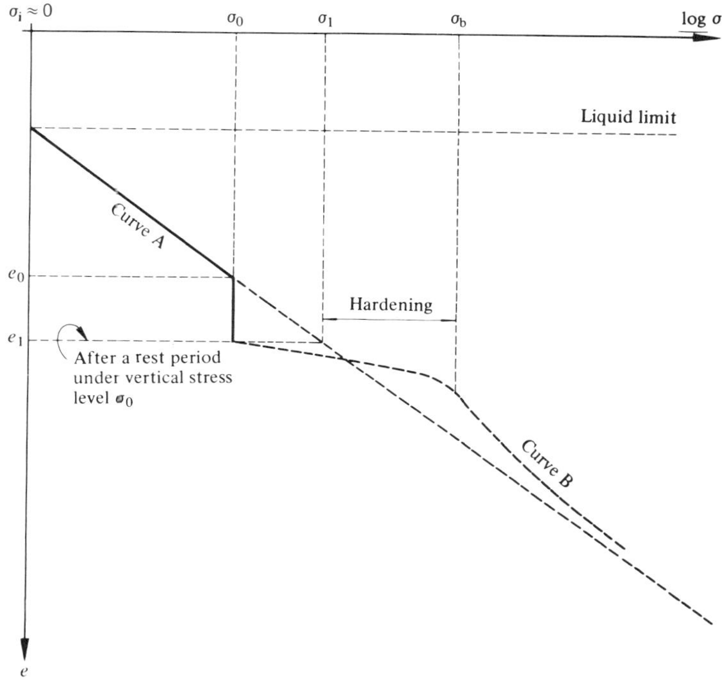

Fig. 17-II.3 Hardening of clay.

material less compressible and stronger under sustained effective stresses. This phenomenon takes place, mainly, in clays and clayey silts containing active minerals, and appears less strong when the clay minerals are of low activity. From the above reasoning and the results of this type of test, the existence of the solidification may be clearly understood. The result of this type of test may be applied to investigate the time dependence of solidification under intergranular stresses applied to samples obtained in their natural state. Figure 18-II.3 shows the results of this type of test performed in a montmorillonitic silty clay typical of the subsoil of Mexico City, and an illite silty clay sediment of the Bay of New Haven, Connecticut, U.S.A.

Thus we see the importance of the break in the compressibility curve in clayey materials. The effective stress corresponding to the break of the curve (critical stress σ_b) defines, in these sediments, a structural skeleton breakdown that considerably changes the compressibility properties of the material, as well as the value of the coefficient of consolidation defined by Terzaghi; this value may be expected to decrease from 10 to 100 times in highly sensitive soils.

The application of these basic concepts in natural deposits allows us to make important decisions in foundation engineering; the engineer can select with certainty the coefficient of volume compressibility to be used for analysis of displacements

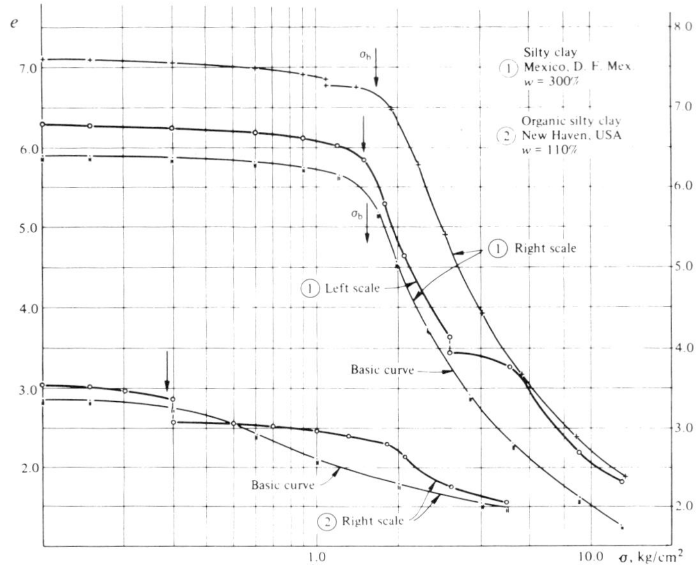

Fig. 18-II.3 Hardening tests of clay.

and settlement calculations. In order to illustrate the problem, imagine (Fig. 19-II.3) that the average state of vertical effective stress in a clay stratum is σ_0 and its void ratio e_0, represented by point O in coordinates σ, e. Under these conditions, if an increment of effective stress is applied for the first time in history on this clay stratum, the compression may take place in two different ways.

1) If the material is of the normally consolidated type, the compression takes place along curve A, which is a continuation of the virgin curve of the natural process of mechanical compression that the deposit suffered during its formation in nature; this may be observed mainly in loose silt sediments.
2) The compressibility curve follows curve B up to a critical stress σ_b, where a break in the curve takes place, after which the compressibility follows the inclined branch of the curve to the right. In this case the material may be recognized to be of the preconsolidated type or cemented in such a way that it has gained strength by any of the following processes:

A. By physicochemical hardening because of clay minerals or any other cementing agent.
B. By a mechanical process due to higher effective pressures applied in the past:
   a) larger weight of sediments on the deposits that have been removed by erosion;
   b) high effective stresses motivated during the history of the deposit, because of drawdown of the piezometric water levels, which at present are reestablished;

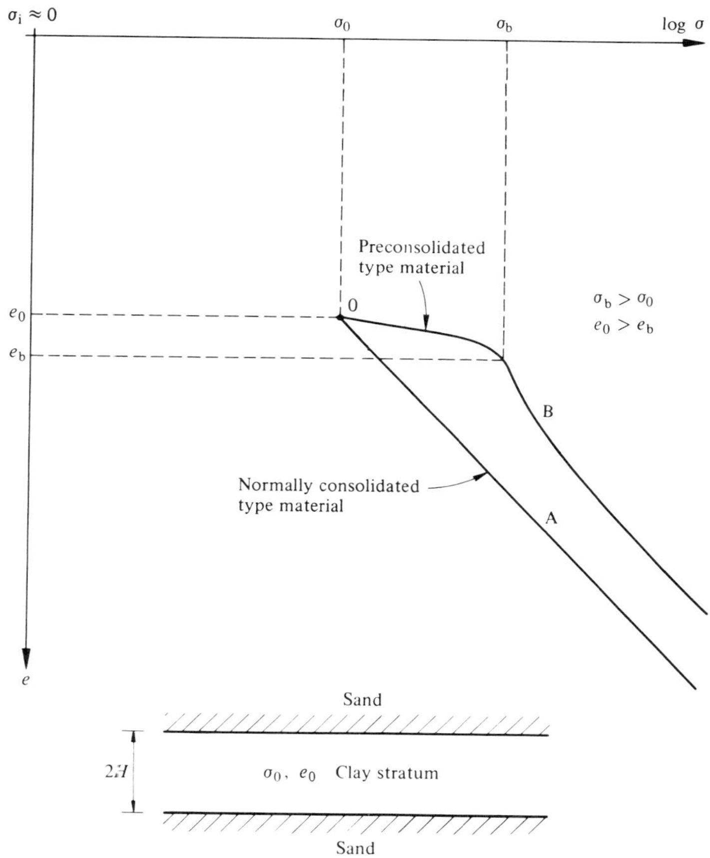

Fig. 19-II.3 Compressibility under field conditions.

c) larger effective stresses originated by the phenomenon of capillary tension, as the material dried during certain period of its formation.

Nevertheless, whatever the motive by which the soil has hardened up to the critical compressive stress σ_b, the material will be less compressible in the stress region $(\sigma_b - \sigma_0)$, if compared with a normally consolidated soil.

From conventional oedometer tests on undisturbed soils representative of the natural deposit, one can determine whether the sediment is normally consolidated or of the preconsolidated type. Two recompression tests are shown in Fig. 20(a)-II.3; one test refers to a normally consolidated soil specimen and the other to the preconsolidated type. The overburden effective or intergranular stress σ_{0z} is shown, corresponding to the depth at which the representative sample was taken. This stress may be calculated knowing the stratigraphy, the volumetric weight of the material and the hydraulic conditions in the subsoil. Using this concept of

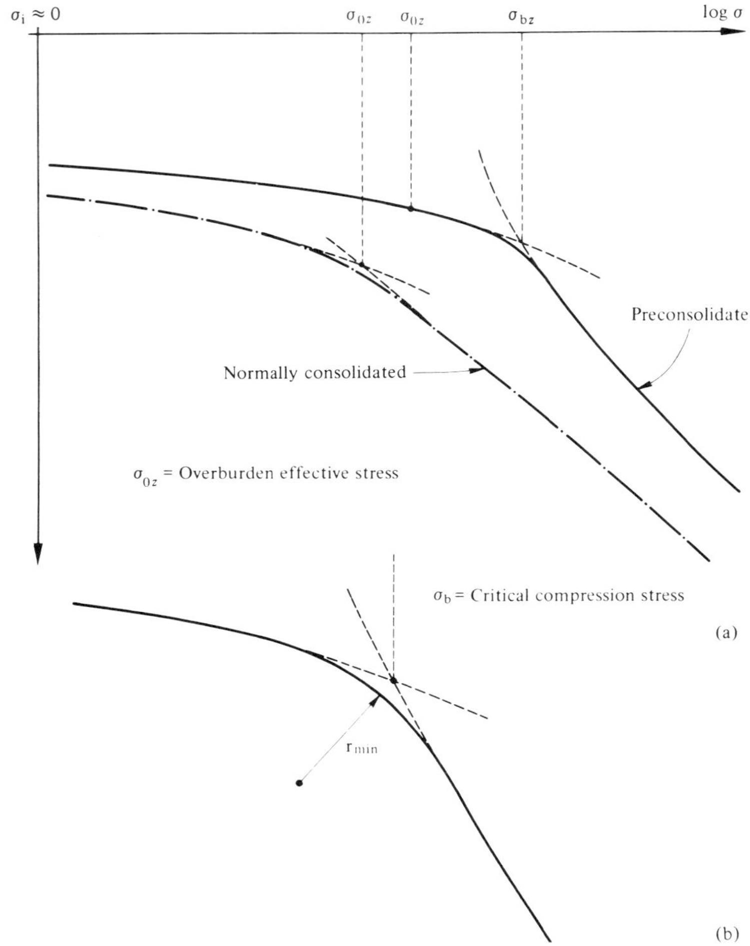

Fig. 20-II.3 Compressibility curves.

critical stress, it may be observed that for a normally consolidated material $\sigma_{bz} \leqslant \sigma_{0z}$. Theoretically $\sigma_{bz} \doteq \sigma_{0z}$; however, in practice this does not happen because of hysteresis and also certain disturbance taking place in the soil when the material is extracted from the ground. For materials of the preconsolidated type, it is observed that always $\sigma_{bz} > \sigma_{0z}$. The determination of the critical compressive effective stress may be found by extending the curvatures of the two branches of the compressibility curves, that is to say, extending the flat branch to the right and the steep branch to the left (Fig. 20(b)-II.3).

3.6b Expansive or Swelling Soils. The compressibility properties of expansive clays are of practical importance in soil mechanics and to the foundation engineer. The materials that fall into this category may be found, mainly, in nonsaturated residual soil deposits of fine material that contain clay minerals of the montmorillonite or illite type, which are sensitive to changes in humidity. Water adsorption occurs in the active clay minerals when decompressed and in contact with water, and as a result they expand; when they dry, they shrink strongly. If a foundation is placed on this type of expansive material and thereafter the environmental changes increase the humidity in the subsoil, the stratum containing this type of material suffers strong vertical displacements damaging the construction they are supporting. To estimate the magnitude of the expansive properties of these materials in a confining state, as they may be found in nature, compressibility tests in the oedometer may be performed; the results of these tests are shown in Fig. 21-II.3. Curve A represents a compressibility test performed on an undisturbed material test specimen with its natural humidity, and curve B shows the compressibility of the same material when saturated by capillary action before performing the test. By means of these two curves, the expansion of the material under a certain state of stress in nature may be determined. The maximum expansion takes place when the material is saturated under unloaded conditions or with a very small initial stress. If a test specimen with its natural humidity is loaded to stress σ_2 on curve A, and afterwards saturation of the material is permitted by capillarity, the material suffers an expansion passing from void ratio e_1 to e_2. The change in strain because of saturation for stress σ_2 will be:

$$\Delta\epsilon_s = \frac{e_2 - e_1}{1 + e_1} \qquad \text{34-II.3}$$

Notice that there will be a neutral stress σ_n for which the material will not suffer expansion or contraction upon saturation.

The values of $\Delta\epsilon_s$ may be plotted against the vertical effective stress at which the soil was saturated, as shown in Fig. 21(b)-II.3. It must be mentioned, however, that nonsaturated materials of residual origin become more compressible upon saturation. Therefore, the maximum compressibility of these soils may be estimated by means of compressibility tests in the oedometer on previously saturated samples. For certain gain in the natural humidity, the expansion characteristics may assume intermediate values between limiting curves A and B as shown in Fig. 21-II.3.

3.6c Collapsible Soils. The mechanical properties of compressibility of eolian fine sediments are of interest to the foundation engineer. Their mechanical behavior requires special treatment and interpretation. The fine sediments deposited by the wind, like nonvolcanic deposits and those of volcanic origin transported by wind and steam, contain silt and fine sand. The eolian sediments contain also little clay and calcium carbonates that bind together grains in the skeleton structure. They are very susceptible to changes in humidity, and are known as collapsible soils. When a sediment of this type is saturated, as the water table approaches the ground surface or when a water current passes through them, the clayey cementant

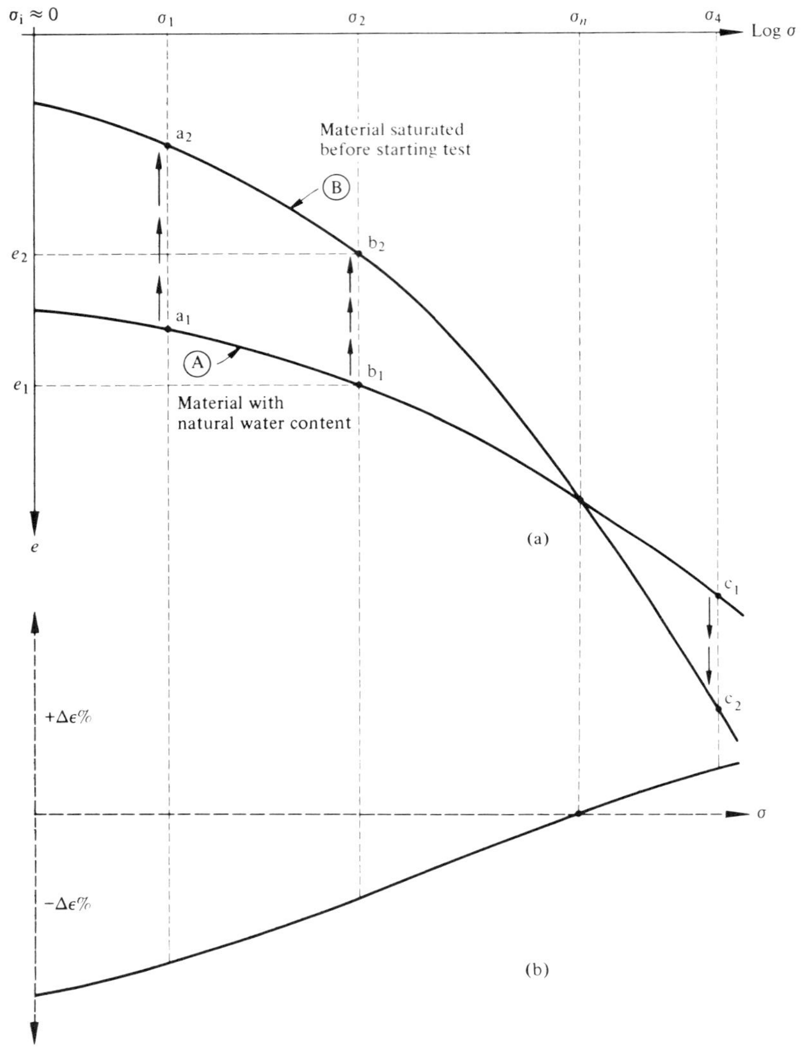

Fig. 21-II.3 Swelling clay.

reduces its strength and the calcium carbonates diluted in water are eluviated. Under these conditions, the material suffers structural collapse and changes its mechanical properties of compressibility. For this reason, it is necessary to investigate their properties for both natural humidity and saturation. Because of water seepage through them, these sediments modify, and thereafter take the name of modified eolian sediments. The fine eolian sediments may be found protected from weathering in nature by a surface crust of impermeable materials like clay products. This crust protects the entrance of large quantities of water,

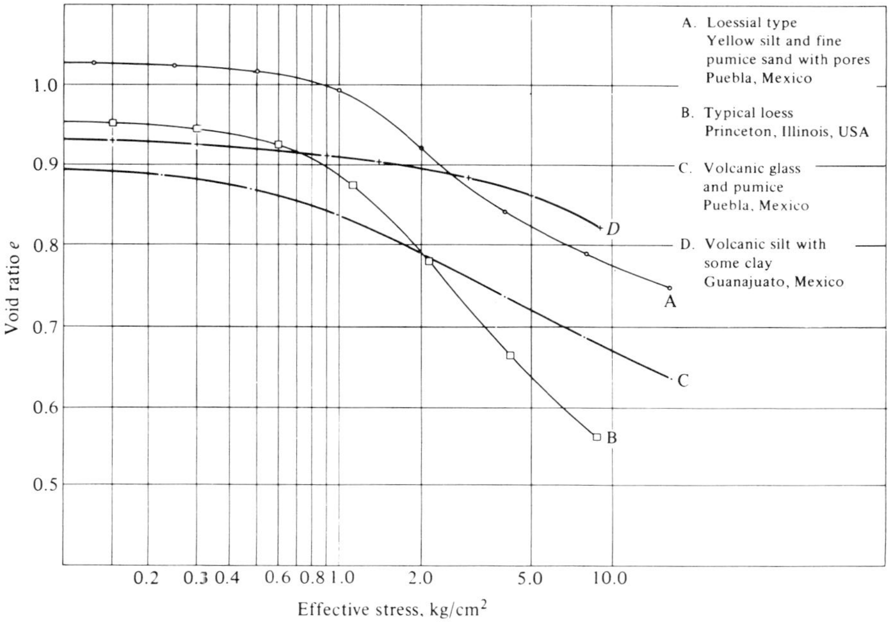

Fig. 22-II.3 Compressibility curves of fine eolian materials.

thus preserving their natural humidity. For this reason, even in climates of medium pluvial precipitation, fine eolian deposits are modified either not at all or only partially. Typical compressibility curves performed on these materials are shown in Fig. 22-II.3. Curve A is a material of the State of Puebla, Mexico, with a conventional void ratio of 1, classified as of the loessial type, containing silt and pumice fine sand slightly cemented; its color is light brown with voids left by roots. This is a material of medium compressibility in its natural state and may suffer a structural collapse resulting in strong subsidence when saturated under applied load. Curve B shows a typical loess of limonitic color of the State of Illinois, U.S.A. Curves C and D show eolian sediments of volcanic origin, mainly formed by volcanic fine glasses, slightly cemented with clay; their color is limonitic. The strain modulus of fine eolian soils and, consequently, the coefficient of volume compressibility change with the confining pressure. For a zero confining pressure, the material has certain consistency and the initial strain modulus may be determined from simple compression tests.

To illustrate the behavior of these materials, Fig. 23-II.3 shows a study made on three test specimens of exactly the same material. Curve A shows the stress-void ratio behavior and structure collapse when the material is saturated by capillary action under stress σ_1. The same may be observed with curves A2 and A3 corresponding to test specimens saturated under higher stresses σ_2 and σ_3, respectively. Notice that when the test is continued after saturation and collapse, the compressibility curves obtained have the tendency to join to a single curve, that is to say, to be asymptotic to curve B. Curve B was obtained with a test of exactly the

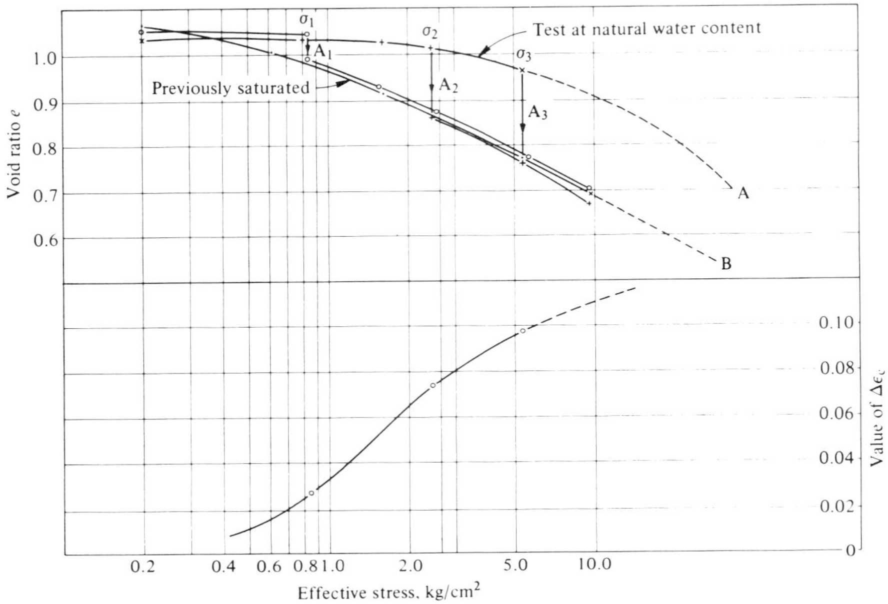

Fig. 23-II.3 Compressibility curve for loessial type material.

same material but previously saturated. Conclusions from these experiments show that the phenomenon expressing the change in compressibility by saturation is confined between curve A of the material in its natural state and curve B of the material modified by capillary saturation immediately before the test. Therefore, from limiting curves A and B in Fig. 23-II.3 the unit strain may be determined for the stress at which the collapse takes place:

$$\Delta\epsilon_c = -\frac{\Delta e_c}{1 + e_c} \qquad \text{35-II.3}$$

where e_c is the initial void ratio corresponding to the moment at which the material is saturated under stress σ_c. The values of the change in strain may be plotted against the corresponding effective pressure (Fig. 23-II.3).

As an example, assume a foundation creates a load of 2.0 kg/cm^2 on a fine eolian silt stratum 2 m thick. The stratum in question has an average effective overburden stress of 1.0 kg/cm^2. Hence, the total average stress will be 3.0 kg/cm^2. The eolian material has compressibility properties as shown in Fig. 23-II.3; therefore, for 3.0 kg/cm^2 a value of $\Delta\epsilon_c = 0.080$ is obtained. Consequently, the compression of the stratum upon saturation will be:

$$\delta_c = 0.080 \times 200 = 16.0 \text{ cm}$$

Thus the importance of this phenomenon in eolian fine sediments cannot be underrated.

3.6d Compaction and Desiccation. When materials of borrow pits are used to construct compacted fills, they should satisfy grain size conditions and cementation adequate to permit a good compaction. The result is the obtention of a material with low to very low compressibility and high shear strength. It is required that the material be well graded and permeable. The clay content should be very low, exclusively to permit sufficient cementation of the coarse material and permit a good compaction. The quantity of silty clay that may be allowed under this condition of the nonexpansive type should not be larger than 5%. The methods employed to compact borrow pit materials in the field may be found elsewhere. The compressibility tests to determine the stress-strain characteristics of these materials are the same as already described in previous chapters. It is important to investigate the compressibility and the shear strength when they have to support heavy foundations. The hydraulic conditions at which they should work have to be carefully considered, in order to select the type of compressibility tests to determine their stress-strain characteristics, to facilitate the investigation of the possible vertical displacements because of load application. When compaction is performed in well graded materials, the stress-strain behavior as a rule shows high elasticity within the range of stresses at which the compaction was performed. The phenomenon of intergranular viscosity will be, in general, of small importance. The compaction is performed under minimum water contents that should permit the fine fraction to become practically saturated, and acts as a lubricant making the grains of the macroscopic fraction accommodate themselves in the most dense possible conditions. The water content used for compaction is approximately that corresponding to the fine and very fine fraction passing sieve No. 40, and corresponds approximately to the plastic limit.

When a material containing active clay minerals or organic colloids suffers desiccation, a contraction takes place because of the effect of the capillary forces and serious problems may arise under these circumstances. As water is lost, changes in soil volume take place. Therefore, cracks, settlements and displacements at the surface of the soil occur that damage the construction supported on them. To investigate quantitatively the phenomenon due to desiccation, tests of linear contraction may be performed. A test specimen is prepared with the corresponding material with a length of approximately 15 cm (Fig. 24-II.3). The material is left to air-dry and the change Δl in length with time is recorded. If the original length of the test specimen is l_1, then the unit strain taking place will be:

$$\Delta\epsilon_l = \frac{\Delta l}{l_1} \qquad \text{36-II.3}$$

The value of $\Delta\epsilon_l$ is plotted against the water content for the same time of observation. The water content is taken from other test specimens in parallel to the one under observation to determine the loss of humidity with time. The tests are performed perpendicular and parallel to the planes of stratification or compaction.

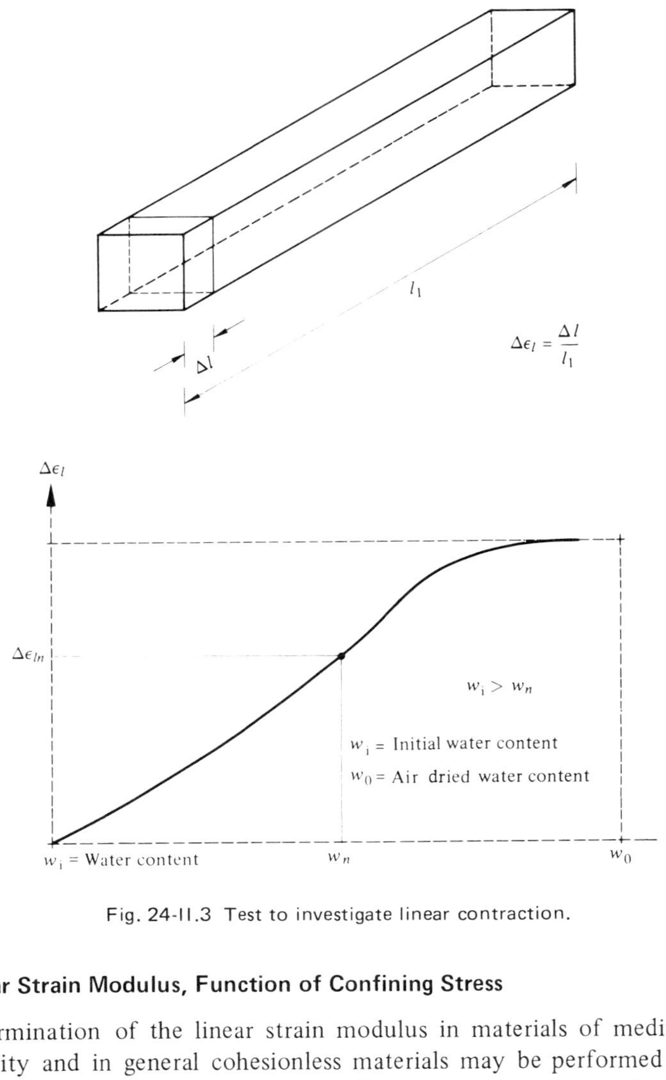

Fig. 24-II.3 Test to investigate linear contraction.

3.7 Linear Strain Modulus, Function of Confining Stress

The determination of the linear strain modulus in materials of medium to high permeability and in general cohesionless materials may be performed with an instrument confining the soil to a state of stresses without restricting lateral deformations. For this purpose, a cylindrical test specimen is used with a height of 2 to 2.5 times its diameter. The test specimen is placed in a sealed chamber known as the triaxial compression chamber. It is shown schematically in Fig. 25-II.3. The test specimen is covered with a rubber membrane firmly fixed to the heads of the instrument, where the soil sample is resting. The chamber is basically built of a cylinder with a translucent wall fixed by means of two metal plates in such a way as to form a watertight chamber in which a liquid may be subjected to pressure. Through valve A a liquid is introduced into the chamber and kept under pressure, therefore subjecting the test specimen to a volumetric stress. The description and

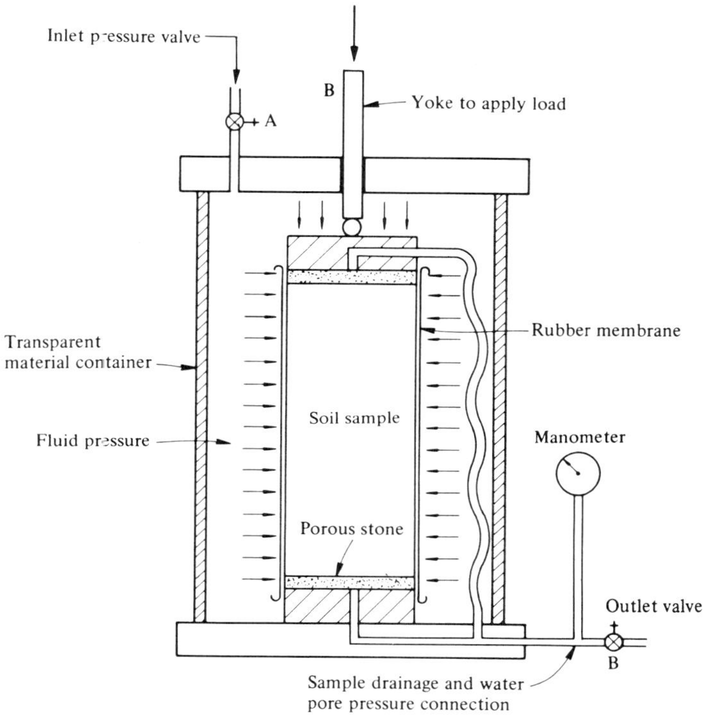

Fig. 25-II.3 Triaxial compression chamber.

technique of the triaxial chamber may be found explained in detail elsewhere. The test specimens may be subjected to any desired fluid volumetric pressure or overall confining stress. After having performed this, valve A is closed; however, valve B is left open in order to permit the release upon load application of any hydrostatic excess pressure in the sample. The determination of the strain modulus has to be performed under zero hydrostatic excess pressures. The volumetric confining stress σ_c is kept constant during application, by means of yoke B, of an effective increment vertical stress $\Delta\sigma_z$, that is known as the deviator stress because it introduces shear stresses in the test specimen. Under these conditions, the total vertical effective principal stress will be: $\sigma_1 = \sigma_c + \Delta\sigma_1$, and the horizontal stress will be equal to the confining principal stress: $\sigma_3 = \sigma_c$.

Let us assume that the test specimen is stabilized to a volumetric confining stress σ_c. Under these conditions, we apply an increment of volumetric stress $\Delta\sigma_c$. Therefore, in accordance with equation 28-II.3, the volumetric strain will be:

$$\Delta\epsilon_v = (1 - 2\nu)\left(1 + 2\frac{M_h}{M_z}\right) M_z \Delta\sigma_c$$

If the material is not saturated, that is to say, if Poisson's ratio ν is different than 0.5, immediately a value for $\Delta\epsilon_v$ is obtained. Otherwise, it is necessary to wait a certain time to obtain the total volumetric change, permitting the expulsion of the water from the pores of the soil leaving valve B open (Fig. 24-II.3) in such a way as to have always zero hydrostatic excess pressures. Since the fundamental equation in saturated soil establishes the condition that

$$\Delta p = \Delta\sigma + \Delta u$$

The applied pressure is always equal to the sum of the intergranular stress and the hydrostatic pressure. Therefore, in order that formula 28-II.3 may be applied, $\Delta u = 0$. Considering the material isotropic, that is to say $M_h/M_z \doteq 1$, then:

$$\Delta\epsilon_v = 3(1 - 2\nu)\, M_z \Delta\sigma_c$$

If the change in volumetric strain could be measured with precision after the application of $\Delta\sigma_c$, it would be possible to determine the value of the linear strain modulus M_z, knowing previously the value of Poisson's ratio. The value of ν for clean sand may be on the order of 0.25 and for clays may reach up to 0.43. Close to the state of stresses producing plastic flow, the value of ν approaches 0.5.

The value

$$3(1 - 2\nu)\, M_z = K_v \qquad \text{37-II.3}$$

represents the unit bulk compressibility. From the test this may be determined from:

$$\frac{\Delta\epsilon_v}{\Delta\sigma_c} = K_v, \quad \text{hence} \quad M_z = \frac{K_v}{3(1 - 2\nu)}$$

Nevertheless, because it is practically impossible to measure accurate changes in volume in the test specimen, it is necessary to determine directly the value of M_z applying only vertical increments of stress $\Delta\sigma_z$ and observing the linear strain $\Delta\epsilon_z$, according to expression 15-II.3:

$$M_z = \frac{\Delta\epsilon_z}{\Delta\sigma_z}$$

The test is performed in the following way (Fig. 26-II.3). The test specimen is placed in the triaxial apparatus with initial void ratio e_0, an initial confining pressure σ_{c1} is applied and the test specimen is allowed to stabilize under this volumetric stress and new void ratio e_1; thereafter, a small vertical increment $\Delta\sigma_{z1}$ of effective stress is applied, from which we obtain the relation between $\Delta\epsilon_{z1}$ and $\Delta\sigma_{z1}$, giving us the secant strain modulus M_{z1} corresponding to confining stress σ_{c1} and initial void ratio e_1. Then the test specimen is unloaded from the increment $\Delta\sigma_{z1}$ and the volumetric confinement stress increased to $\sigma_{c1} + \Delta\sigma_{c1} = \sigma_{c2}$. After the specimen is stabilized under this new volumetric pressure, a new small vertical increment of stress $\Delta\sigma_{z2}$ is applied obtaining the relation between $\Delta\epsilon_{z2}$ and $\Delta\sigma_{z2}$, corresponding to the confining stress σ_{c2}, kept constant during the

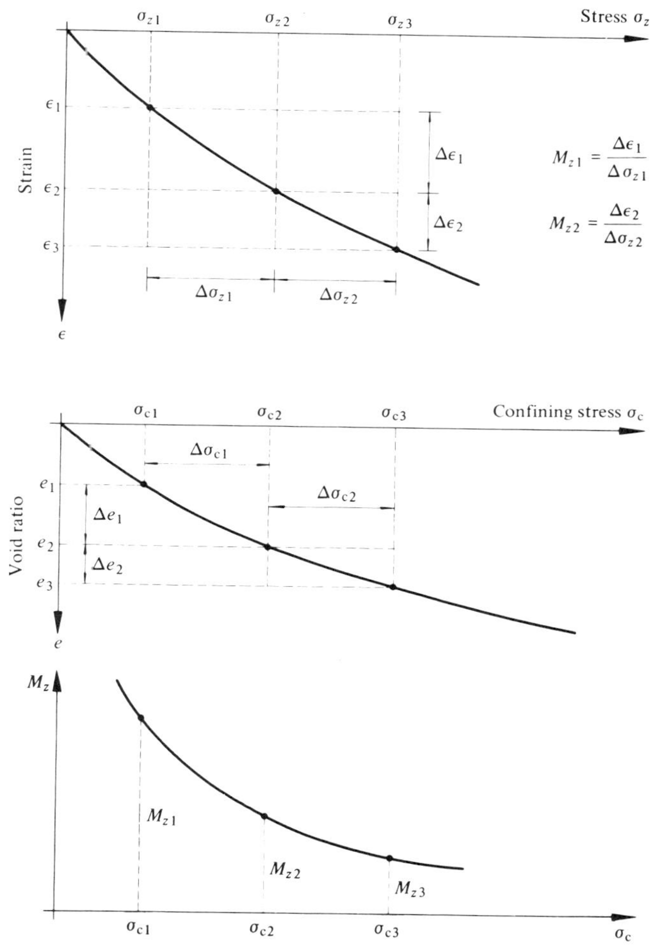

Fig. 26-II.3 Laboratory determination of the strain modulus.

increment. The next strain modulus M_{z2} corresponding to the confining stress σ_{c2} and void ratio e_2 is calculated. In this way, the test is continued to determine for each step the value of the strain modulus M_z for the range of stresses desired.

The increment of effective vertical stresses applied in each step should be planned so that the shear stresses induced in the test specimen never reach more than 1/2 of the shear strength of the material. The purpose is to minimize the shear stresses induced during the test, in order not to influence significantly the properties of plastic deformation in the test specimen. The values so obtained are plotted as shown in Fig. 26-II.3 from which M_z vs. σ_c and M_z vs. e may be studied. The change in void ratio Δe may be calculated, using formula 28-II.3 for $\Delta\sigma_x = \Delta\sigma_y = 0$, from the following expressions:

$$\Delta\epsilon_v = (1 - 2\nu) M_z \Delta\sigma_z$$

But

$$\Delta\epsilon_v = \frac{\Delta e}{1 + e}$$

Hence:

$$\Delta e = (1 - 2\nu)(1 + e) M_z \Delta\sigma_z \qquad \text{38-II.3}$$

From these results, it may be found that the strain modulus is an important function of the confining stress and initial void ratio of the test specimen, that is, of the state of compaction. Therefore:

$$M_z = f(\sigma_c, e) \qquad \text{39-II.3}$$

The variation of the values of M_z with the confining stress σ_c may be investigated plotting them either in log-log scales or semilog scales, where they usually will appear as straight lines. Figure 27-II.3 shows these values plotted on a double logarithmic scale from which the phenomenological law of M_z is determined:

$$M_z = C_0 \cdot \sigma_c^{-n} \qquad \text{40-II.3}$$

This law indicates that as the confining volumetric stress increases, the value of the strain modulus decreases, and when the value of σ_c approaches zero, the value of M_z approaches infinity. This implies that in the case of slightly confined cohesionless granular materials like sand and silt, the strain modulus is very large. To give an idea of average values of the constants C_0 and n, Table 2-II.3 has been prepared for well graded sands.

From the table it is concluded that the average for a well graded sand in a semicompact state will be $C_0 = 0.0045$ and $n = 0.5$. Therefore, the following formula may be used:

$$M_z \doteq \frac{0.0045}{\sqrt{\sigma_c}}, \quad \frac{\text{cm}^2}{\text{kg}} \qquad \text{41-II.3}$$

The sediments in nature may be found confined to certain state of stresses. Therefore, their compressibility is a function of the volumetric state of stress of confinement and compaction. The state of compaction of cohesionless sediments may be encountered in nature from a loose to a very dense state. Table 2-II.3 shows the variation of the constants C_0 and n with the state of compaction of the soil. At certain depth, the sediment is subjected, in nature, to a vertical stress σ_z and a horizontal stress σ_h; hence the equivalent volumetric stress is

$$\sigma_c = \tfrac{1}{3}(\sigma_z + 2\sigma_h)$$

or

$$\sigma_c = \frac{1 + 2K_0}{3}\sigma_z$$

Here $K_0 = \sigma_h/\sigma_z$ is called the ratio of effective stresses at rest.

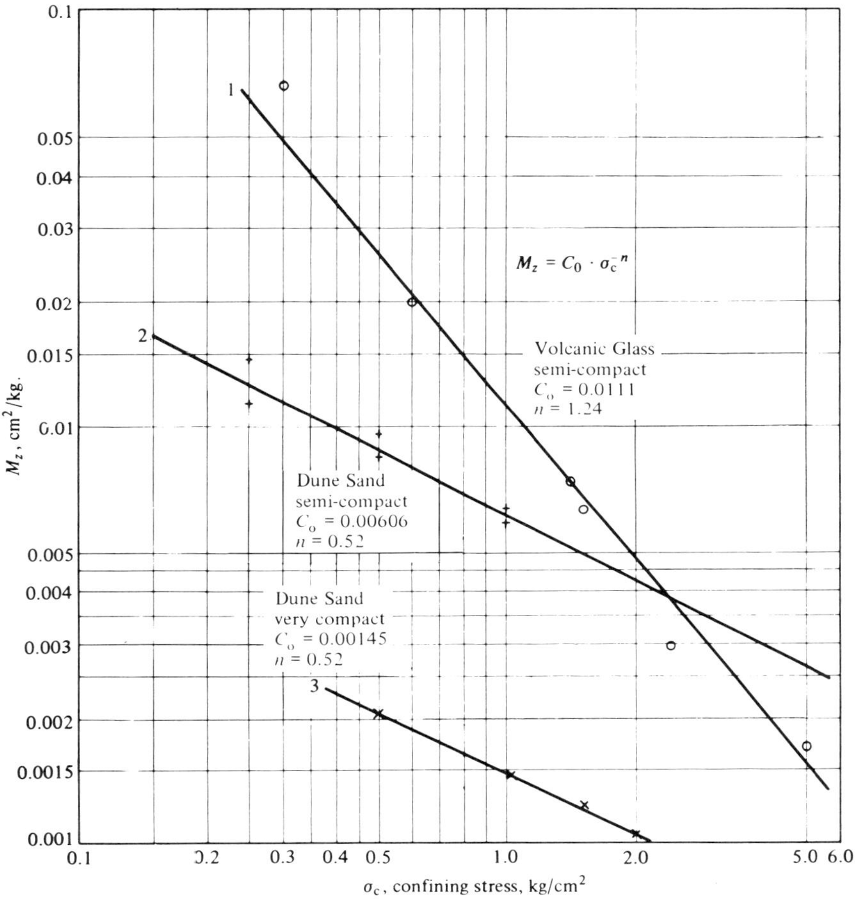

Fig. 27-II.3 Linear strain modulus for cohesionless soils.

TABLE 2-II.3

State of Compaction	D_r	$C_0 \times 10^{-3}$	n
Very loose	0.2	10	0.65
Loose	0.2–0.4	10–6.0	0.65–0.60
Semicompact	0.4–0.6	6.0–3.0	0.60–0.50
Compact	0.6–0.8	3.0–2.0	0.50–0.45
Very compact	>0.8	< 2.0	0.45

The values of M_z vs. σ_c may be plotted also on a semilog scale if the behavior is observed to be a straight line (Fig. 28-II.3); then the slope of the line is

$$n = \frac{\log \dfrac{M_{zo}}{M_z}}{\sigma_c - \sigma_{co}} \qquad \text{42-II.3}$$

Hence, the phenomenological law expressing the variation of the strain modulus with the confining stress is

$$M_z = M_{zo}\, e_N^{-n_c \sigma_c} \qquad \text{43-II.3}$$

where $n_c = 2.3n$. When $\sigma_c \to 0$, the strain modulus approaches a finite value M_{zo}. This implies that for zero confining stress in case of cohesive soils, there is an initial strain modulus. Here e_N is the base of the Napierian logarithms. For most practical purposes the value of n_c in 43-II.3 may be taken as unity. The value of M_{zo} corresponding to the initial modulus under zero confinement stress may be approximately determined by means of simple unconfined compression tests at the time the natural consistency of the material is determined.

In order to correlate the state of stress in the field with the triaxial and oedometer tests, it is necessary to study the theoretical relationships existing between them. For this purpose, assume the vertical effective stress in the field at certain depth z is increased in $\Delta\sigma_z$, and accordingly, the stresses in the horizontal direction are increased to $\Delta\sigma_x$ and $\Delta\sigma_y$, because of the application of $\Delta\sigma_z$. The vertical strain at that point is (see Section 3.5.)

$$\Delta\epsilon_z = \left[1 - \nu \frac{M_h}{M_z} \frac{\Delta\sigma_x + \Delta\sigma_y}{\Delta\sigma_z}\right] M_z \Delta\sigma_z \qquad \text{15-II.3}$$

Therefore, the equivalent strain modulus under imposed increment of stresses is:

$$\overline{M}_z = \left[1 - \nu \frac{M_h}{M_z} \frac{\Delta\sigma_x + \Delta\sigma_y}{\Delta\sigma_z}\right] M_z \qquad \text{44-II.3}$$

The value of M_z is that of the linear strain modulus for the initial volumetric stress σ_c in the field. Expression 44-II.3 may be used to correct the M_z-values to field conditions, where $\nu, M_h/M_z$ and $(\Delta\sigma_x + \Delta\sigma_y)/\Delta\sigma_z$ are known values. Furthermore, as in the oedometer test, under confined conditions to zero lateral displacement

$$\Delta\epsilon_z = \nu_c \cdot M_z \Delta\sigma_z = m_v \Delta\sigma_z \qquad \text{24-II.3}$$

In the laboratory M_z and M_h are determined in triaxial tests as previously described, and m_v in the oedometer: $m_v/M_z = \nu_c$.

As an example assume $M_z = 0.02\ \text{cm}^2/\text{kg}$, $\nu = 0.3$, and $M_h/M_z = 1/2$. From load applied in the field and by the theory of elasticity assume it is found that: $(\Delta\sigma_x + \Delta\sigma_y)/\Delta\sigma_z = 1$. Therefore, $\overline{M}_z = (1 - 0.3/2)M_z$ or $\overline{M}_z = 0.85 \times 0.02 = 0.017\ \text{cm}^2/\text{kg}$. On the other hand, under totally confined conditions to zero lateral

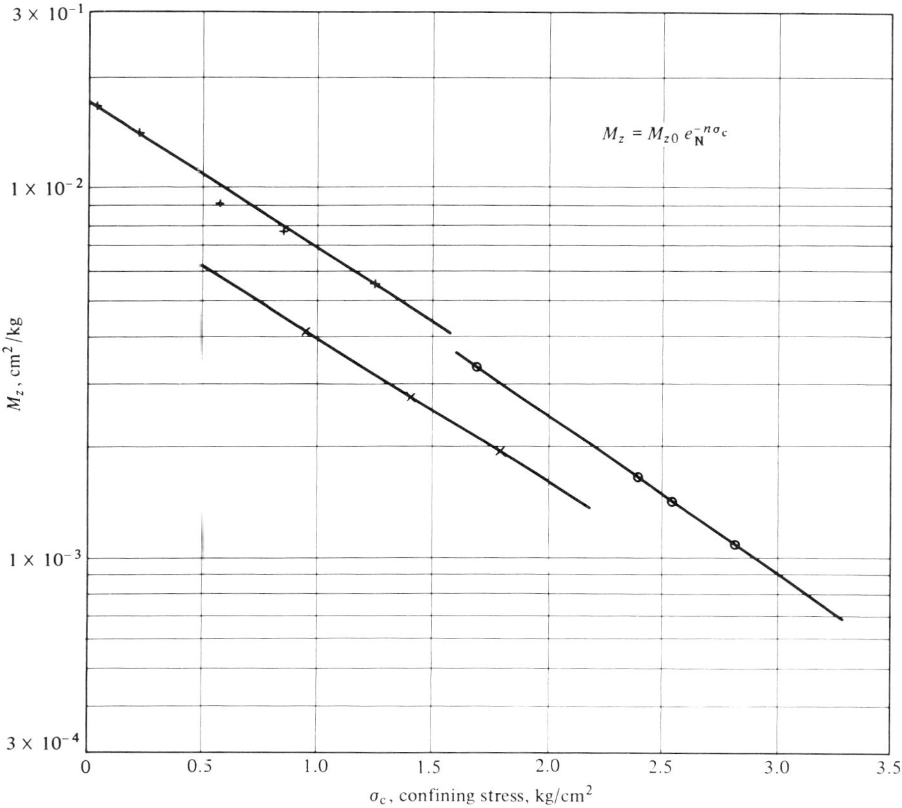

Fig. 28-II.3 Linear strain modulus for cohesive soils.

displacement $\nu_c = 0.73$. Hence, $m_v = 0.73 \times 0.02 = 0.0146$ cm²/kg. This value may be used only in the field for strata confined to practically zero lateral deformation, in case of strata loaded in a large area as compared with their thickness. When the value of m_v is determined in the oedometer test and assuming ν_c, the value of $\overline{M}_z$ may be obtained; hence:

$$\overline{M}_z = \left[1 - \nu \frac{m_h}{m_v} \cdot \frac{\Delta\sigma_x + \Delta\sigma_y}{\Delta\sigma_z}\right] \frac{m_v}{\nu_c} \qquad \text{45-II.3}$$

Again, assume that oedometer tests gave $m_v = m_h = 0.05$ cm²/kg for a soft uniform clay, for which a value of $\nu = 0.42$ is selected. The increment of stress of a footing loading the soil in a limited area gives at depth z an average ratio of $(\Delta\sigma_x + \Delta\sigma_y)/\Delta\sigma_z = 1.2$. From Fig. 12-II.3, $\nu_c = 0.38$. Then using 45-II.3:

$$\overline{M}_z = [1 - 0.42(1)1.2]\,\frac{0.05}{0.38} = 0.0652 \text{ cm}^2/\text{kg}$$

Hence it is concluded that under a footing on partially confined conditions, the strain modulus is larger than it would be under perfect lateral confined conditions, when $\Delta\epsilon_x = \Delta\epsilon_y = 0$.

3.8 Linear Strain Modulus, Function of Time

In practice, it is important for the foundation engineer to study the effects taking place with time after load application. For this purpose, it is necessary to investigate the linear strain modulus at a certain confinement stress σ_c. In pervious or nonsaturated test specimens, after stabilization with σ_c, the test is performed applying a vertical increment of stress $\Delta\sigma_z$, obtaining an immediate elasto-plastic deformation, measured by the strain $\Delta\epsilon_{ep}$ (Fig. 29-II.3). Therefore, under this condition the elasto-plastic strain modulus will be:

$$M_{ep} = \frac{\Delta\epsilon_{ep}}{\Delta\sigma_z} \qquad 46\text{-II.3}$$

The final value of the strain under sustained stress, however, will be a function of time. The values given by point a represent the strain immediately after the load is applied (Fig. 30-II.3). Nevertheless, as time elapses under sustained increment of stress, the material will show further deformation because of visco-plastic behavior. The strain modulus will be greater and may be expressed by:

$$M_z = \frac{\Delta\epsilon_{ep} + \Delta\epsilon_{vp}}{\Delta\sigma_z}$$

Fig. 29-II.3 Immediate elasto-plastic deformation and visco-plastic behavior.

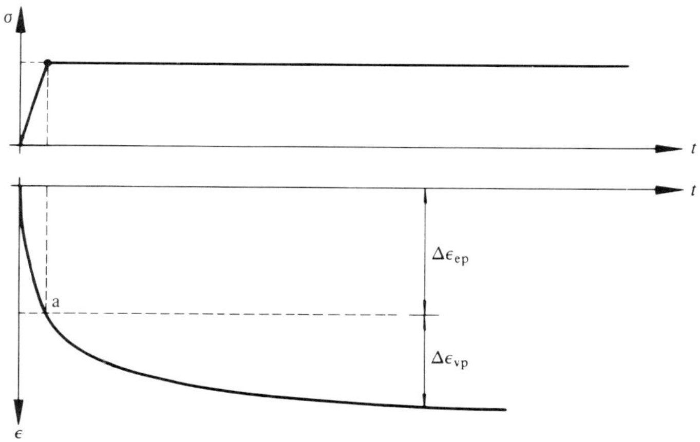

Fig. 30-II.3 Strain as a function of time.

or arranging terms:

$$M_z = M_{ep}\left[1 + \frac{\Delta\epsilon_{vp}}{\Delta\epsilon_{ep}}\right] \qquad \text{47-II.3}$$

Here M_{ep} represents the strain modulus for the instantaneous elasto-plastic deformation, and $\Delta\epsilon_{ep}$ represents the strain under this behavior. The value $\Delta\epsilon_{vp}$ represents the additional visco-plastic strain after time has elapsed, that is, the additional strain because of intergranular viscosity. To measure this additional strain, a long time must be allowed for this to reach an ultimate value of sufficient accuracy to be considered ultimate from the point of view of foundation engineering. Therefore, calling

$$\kappa_v = \frac{\Delta\epsilon_{vp}}{\Delta\epsilon_{ep}} \qquad \text{48-II.3}$$

the vertical strain modulus will read:

$$M_z = M_{ep}(1 + \kappa_v) \qquad \text{49-II.3}$$

The values of κ_v may be investigated for different types of nonsaturated soil materials under special triaxial or oedometer tests (see Table 3-II.3).

In the preceding article it was found that the ratio of the linear strain modulus determined under free lateral confinement and the coefficient of unit volume compressibility as defined by Terzaghi is:

$$\frac{m_v}{M_z} = \nu_c \qquad \text{30-II.3}$$

By means of the above expression a formula may be obtained to calculate the vertical displacement of a stratum of thickness $2H$ subjected to an increment of

TABLE 3-II.3 Range of approximate values of κ_v

Material	Values of κ_v
Clays Clayey silts	greater than 1
Sandy silts Silty sands	1–0.5
Fine sand	0.5–0.2
Well graded sand	0.2–0.1
Rock angular fragments	0.5–0.25

stress $\Delta\sigma_z$ in terms of the strain modulus for the immediate elasto-plastic behavior M_{ep}, Poisson's ratio ν and κ_v; hence:

$$\delta = \nu_c \cdot M_{ep}(1 + \kappa_v)(2H)\Delta\sigma_z \qquad \text{50-II.3}$$

The value of M_{ep} may be calculated by either one of the following two phenomenological laws:

a) for cohesive materials: $M_{ep} = M_0 e_N^{-n_c \sigma_c}$

b) for cohesionless materials: $M_{ep} = C_0 \cdot \sigma_c^{-n}$ 51-II.3

In formula 50-II.3, the coefficient of unit volumetric compressibility is expressed by:

$$m_v = \nu_c M_{ep}(1 + \kappa_v)$$

or

$$m_v = m_{ep}(1 + \kappa_v) \qquad \text{52-II.3}$$

The linear strain modulus, as explained in preceding paragraphs, is determined in the triaxial chamber for nonsaturated, semipervious and pervious materials. In fine impervious saturated materials like clays and silty clays the value of m_{ep} is determined with the oedometer, by means of consolidation tests. From the consolidation curves obtained in this test, the visco-plastic and hydrodynamic phenomena may be better observed and determined (see Section II.3.13), specially when the change in void ratio because of applied stress is large, as in the case of materials of high and very high compressibility.

3.9 The Theory of Consolidation

In previous articles we saw that subsoil materials in addition to elasto-plastic properties possess high intergranular viscosity. This phenomenon is specially important in fine and very fine soils of low and very low permeability, as clays, silty clays and clayey silts. On the other hand, the hydrodynamic lag during volumetric deformation is produced when the pores of the skeleton structure are saturated,

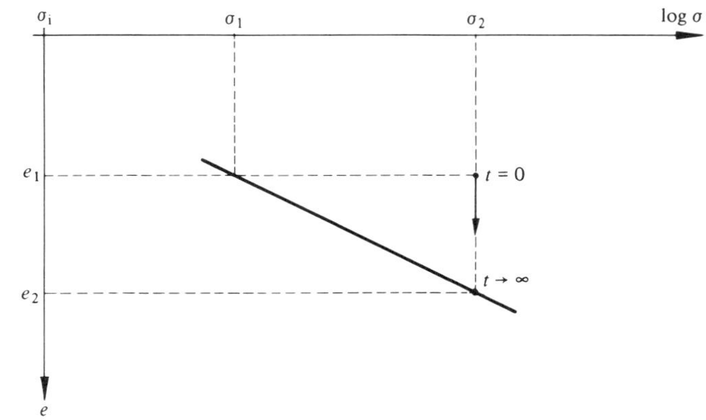

Fig. 31-II.3 Slow reduction of void ratio upon load application.

and it is not possible to obtain a rapid change in volume upon load application, until the gravitational water occupying the pores is expelled from them. Therefore, when performing an oedometer or consolidometer test, the void ratio e_1 cannot be reduced to the void ratio e_2 rapidly upon application of the increment of stress (Figs. 31 and 32-II.3).

The compressibility curve is obtained for stress increments sustained for equal time periods; however, to obtain a limiting compressibility curve, the increments of stress should remain acting a long time, until observed changes in void ratio are unimportant. It is common practice in routine oedometer tests in clay to permit 24

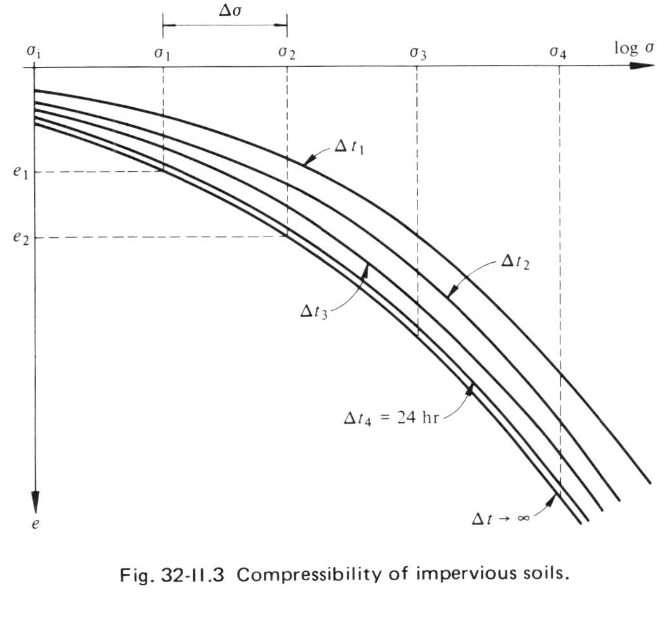

Fig. 32-II.3 Compressibility of impervious soils.

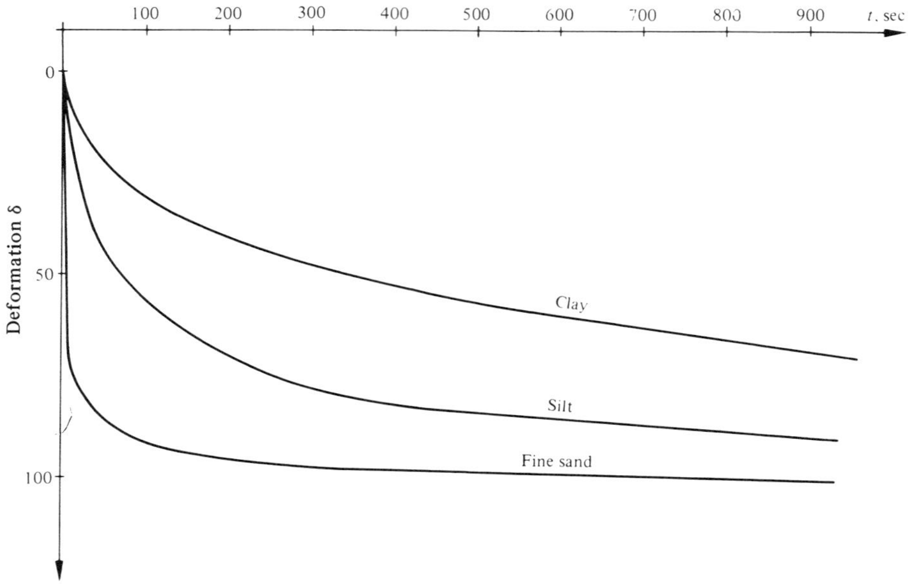

Fig. 33-II.3 Consolidation curves.

hours for each increment of stress. Qualitative consolidation curves for three types of materials subjected to a sustained increment of stress are shown (Fig. 33-II.3). Very fine sand with a coefficient of permeability on the order of 10^6 times larger than clays consolidates rapidly, and silts with a permeability 10^3 times larger than clays, consolidate with an intermediate rate. Observation shows that once the gravitational water has been expelled from the pores of the skeleton structure, an intergranular viscous deformation continues, produced by the relative displacement of the grains in the skeleton structure. This intergranular viscous phenomenon is commonly known in soil mechanics as secondary consolidation.

The theory of consolidation discovered by Terzaghi (1925) explains exclusively the hydrodynamic phenomenon that takes place when gravitational water is expelled from the pores of the resisting elasto-plastic skeleton structure, in which the effect of intergranular viscosity has not been considered. Via the theory of consolidation we seek an approximate relationship for the change in void ratio with time for a saturated soil, and for an increment of stress sustained during the process of deformation.

In the field, a soil stratum may be considered confined to zero lateral deformation when the thickness of the stratum is small compared with the loaded area. In this case, the oedometer test represents approximately the field conditions. The theoretical understanding of the phenomena involved will allow us to adjust the results of the phenomenological properties of the material and to deduce necessary parameters obtained in the laboratory for the resolution of consolidation problems in the field. The working hypotheses in the theory of consolidation are the following:

1. The soil mass is saturated, homogeneous and isotropic.

2. The water and the grains are incompressible.
3. The expulsion of water is only that corresponding to the gravitational water expelled from the pores of the skeleton structure. Darcy's law is considered valid.
4. The change in void ratio or volumetric strain approaches a finite value, as the process of consolidation ends.
5. The stress-strain mechanical properties of the material are only related with those of immediate elasticity and plasticity. Therefore, after the process of consolidation has taken place, elasto-plastic equilibrium is obtained.

Under the above working hypotheses, Terzaghi's theory may be applied. The average percent of consolidation is expressed as follows:

$$U\% = 100\% \left\{1 - \sum_{m=0}^{m=\infty} \frac{8}{(2m+1)^2\pi^2}\, e_N^{-[(2m+1)^2\pi^2/4]\, T_v}\right\} \qquad \text{53-II.3}$$

The above formula is useful to calculate the average percentage of consolidation of a soil stratum confined to zero lateral deformation. The quantity T_v is called the time factor. We write

$$U = F(T_v) \qquad \text{54-II.3}$$

The function $F(T_v)$ may be called Terzaghi's function for theoretical consolidation. In practice, it is usually found that layers of impervious soils are bounded by pervious soils serving as drainage surfaces. Under these conditions, the time factor has the following value:

$$T_v = \frac{c_v}{H^2}\, t \qquad \text{55-II.3}$$

Terzaghi's function has been plotted in semilog scales, from where it may be observed that the curve has the tendency to become asymptotic rapidly for values of T_v equal or larger than unity (Fig. 34-II.3). For small values of T_v (for example,

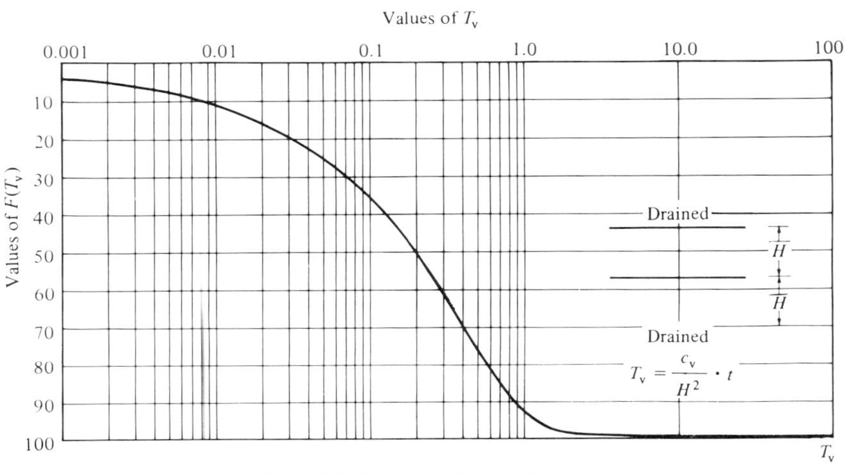

Fig. 34-II.3 Terzaghi's function.

$T_v = 0.01$) the average degree of consolidation rapidly reaches 12%. Using Terzaghi's function, the compression may be calculated as a function of time for a stratum with thickness $2H$. Therefore, the volumetric strain is

$$\Delta\epsilon_v = \Delta\epsilon_{ep} \cdot F(T_v) \tag{56-II.3}$$

in which

$$\Delta\epsilon_v = \frac{\delta_t}{2H}, \quad \Delta\epsilon_{ep} = \frac{\delta_{ep}}{2H}$$

Hence:

$$\delta_t = \delta_{ep} F(T_v) \tag{57-II.3}$$

The value δ_t represents the compression with time of the stratum. After the process of consolidation has taken place, then $F(T_v) \rightarrow 1$, the ultimate compression will be:

$$\delta_{ep} = m_{ep}(2H)\Delta\sigma \tag{58-II.3}$$

Comparisons may be studied, considering two equally drained strata with thicknesses $2H_1$ and $2H_2$, respectively. For the same average degree of consolidation, the times of consolidation are in the following proportion:

$$\frac{c_{v1}}{c_{v2}} \cdot \frac{t_1}{t_2} = \frac{H_1^2}{H_2^2} \tag{59-II.3}$$

When the coefficient of consolidation is equal for both strata, we may write:

$$\frac{t_1}{t_2} = \frac{H_1^2}{H_2^2} \tag{59a-II.3}$$

By means of this simple relation, it is possible to correlate the consolidation in the test specimen with a stratum in the field with the same drainage characteristics.

The next important step in the theory of consolidation is the determination of the coefficient of consolidation:

$$c_v = \frac{k}{m_{ep}\gamma_w} \tag{60-II.3}$$

This is a function of the coefficient of permeability k, the coefficient of unit volume compressibility m_{ep} and the unit weight of water γ_w. The c_v units are given in cm^2/sec. However, since k is impossible to determine in impermeable materials like silts and clays by means of permeability tests, the value of c_v is found from the consolidation curves obtained from the oedometer when performing a compressibility test. The technique of performing oedometer tests is described in detail in elementary soil mechanics literature.

When applying an increment of stress to the test specimen in the oedometer, a consolidation curve is obtained similar to that shown in Fig. 35-II.3. It may be observed that volumetric compression does not end as the theory predicts but when

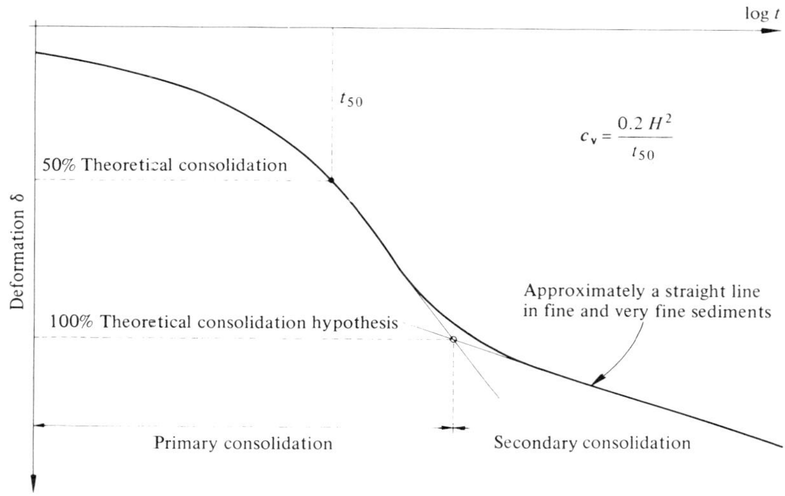

Fig. 35-II.3 Casagrande's method for determining consolidation coefficient.

the theoretical consolidation is about to end—in the majority of the cases in the neighborhood of $T_v \doteq 1.0$—the volumetric compression continues. The inclined branch after the break is known in soil mechanics as secondary consolidation, and in some soils plays a very important part. The hydrostatic excess pore pressure during this process is small, since most of the applied stress has been transmitted to the soil skeleton structure, which, however, continues changing in volume because of intergranular viscosity. The theoretical consolidation representing the change in volume in the skeleton structure plays an important part during the first part of the process. It should be observed in further studies, that the theory of consolidation explains only the phenomenon of hydrodynamic lag which is a function of the permeability and compressibility of the material, and of the stratum thickness.

To calculate the volumetric strain as a function of time, it is necessary to determine the mean value of the coefficient of consolidation c_v for each subsoil stratum. When the material is of low intergranular viscosity, it is possible to use a procedure proposed by Arthur Casagrande. This empirical procedure assumes that, at the intersection of the tangents to the theoretical consolidation curve and the secondary compression, the hydrodynamic consolidation is complete; hence it is approximately 100%, according to Terzaghi's theory. Therefore, the time elapsed to reach 50% theoretical consolidation may be found corresponding to a time factor $T_v = 0.2$. With this information the value of the coefficient of consolidation c_v may be calculated by means of the formula

$$c_v = \frac{0.2H^2}{t_{50}} \qquad \text{61-II.3}$$

here $2H$ represents the thickness of the test specimen at the instant of applying the pressure increment $\Delta\sigma$. The average value of c_v obtained is plotted against the mean stress level, that is to say, $\sigma + 1/2 \cdot \Delta\sigma$.

To give an idea of the importance of the coefficient of consolidation for clayey materials of high and very high compressibility, an example will be given with values in the range of impermeable materials.

Compressibility	cm^2/kg m_{ep}	cm/sec k	cm^2/sec c_v
Very high (silty clay)	0.1	10^{-8}	10^{-4}
High (clayey sandy silt)	0.05	10^{-7}	2×10^{-3}

Consider a stratum 20 m thick drained on both sides. The time to reach 50% consolidation is

$$t = \frac{0.2H^2}{c_v}$$

and using the numerical values we find in the first case 63.5 years, and in the second case 3.2 years.

Thus, the importance of studying the deformation as a function of time is shown. If the stratum in the field is subjected to a stress increment of 0.5 kg/cm^2, then the compression will be:

$$\delta_{50\%} = (0.5)m_{ep}(2H)\Delta\sigma$$

In the first case, the deformation will be 50 cm in 64 years; in the second, 25 cm rapidly. Therefore, the question arises as to which deposit will be more desirable to support the construction from the point of view of settlements in foundation engineering. Evidently, the first case would be preferable, since in five years the value of T_v would be 0.016, with consolidation of only 14%; hence, a settlement of 14 cm would take place, against 25 cm immediate settlement after load application in the second case. A large and quick settlement is likely to produce damage. Furthermore, it is important to correlate Terzaghi's theory of consolidation with rheological considerations, in order to be able to deal with more complicated problems.

3.10 Viscous Unit of Linear Fluidity

The phenomenological laws of linear elasto-plastic viscosity may be studied also by means of simple rheological models to find the significance of the parameters involved in the mathematical expressions.

When the body behaves according to Newton's law (1685):

$$\frac{\partial}{\partial t}\Delta\epsilon_N = \phi_a \Delta\sigma_N \qquad \text{62-II.3}$$

which states that the strain rate is directly proportional to the applied stress, the material represents the perfect Newtonian liquid. Here, the proportionality coefficient measures the fluidity, and is considered constant during the period of deformation under sustained load. Therefore, from Newton's law, the following is obtained:

$$\Delta\epsilon_N = \phi_a \int_0^t \Delta\sigma_N dt \qquad \text{63-II.3}$$

If the load remains constant, then

$$\Delta\epsilon_N = \phi_a \cdot \Delta\sigma_N \cdot t \qquad \text{64-II.3}$$

The graphical representation of this behavior is shown in Fig. 36-II.3, where it is seen that for each value of $\Delta\sigma$ a straight line $\Delta\epsilon$ vs. t is obtained, representing a viscous linear phenomenon. Notice that the reciprocal value of ϕ_a represents the viscosity of the material. Nevertheless, in the application of the viscosity and fluidity concepts to natural sediments, it should be noted from observed behavior that real sediments do not behave with linear fluidity. For example, clays are minerals that under confined conditions show a decreasing rate of deformation with time owing to nonlinear intergranular viscosity. The viscosity of the material increases because the grains get closer together establishing more contacts. Let us assume, however, that the elastic deformation is retarded by the linear viscous phenomenon. This type of behavior may be investigated considering a Hooke unit and a Newton unit in parallel, obtaining what is known as a Kelvin unit (Fig. 37-II.3). The loading condition is the following:

$$\Delta\sigma = \Delta\sigma_H + \Delta\sigma_N$$

and the strain is given by 65-II.3

$$\Delta\epsilon = \Delta\epsilon_H = \Delta\epsilon_N$$

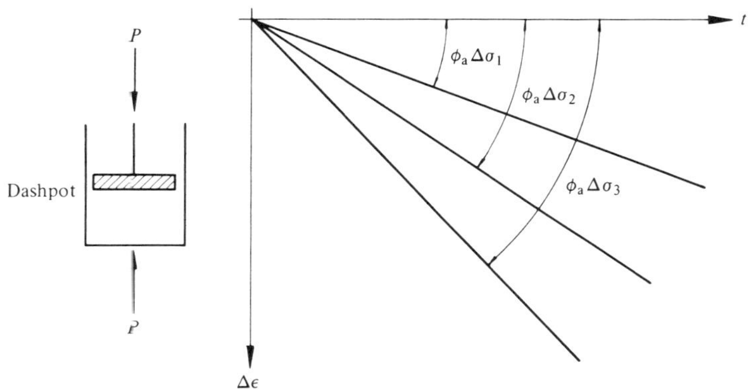

Fig. 36-II.3 Newtonian fluid.

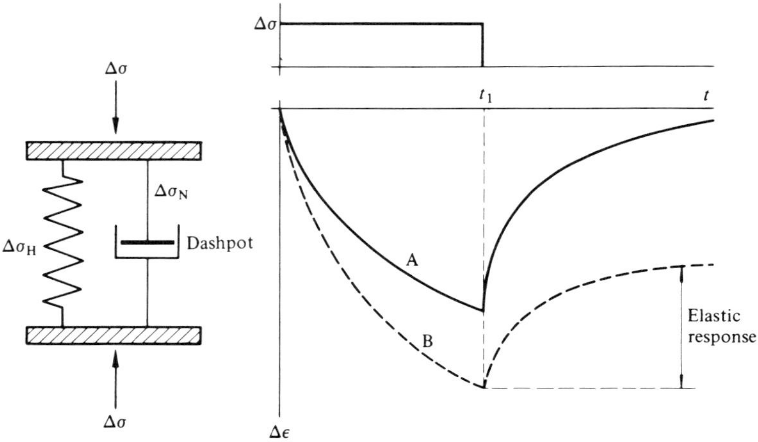

Fig. 37-II.3 The Kelvin unit.

However, since:

$$\Delta\epsilon_H = \alpha_e \Delta\sigma_H$$

and: 66-II.3

$$\frac{\partial}{\partial t}\Delta\epsilon_N = \phi_a \Delta\sigma_N$$

we have, substituting in 65-II.3,

$$\Delta\sigma = \frac{\Delta\epsilon}{\alpha_e} + \frac{1}{\phi_a}\frac{\partial}{\partial t}\Delta\epsilon \qquad \text{67-II.3}$$

The solution of this differential equation is:

$$\Delta\epsilon = e_N^{-(\phi_a/\alpha_e)t}\,\phi_a \int e_N^{(\phi_a/\alpha_e)t}\,\Delta\sigma\,dt + C \qquad \text{68-II.3}$$

From expression 68-II.3 after integration and considering a sustained stress $\Delta\sigma$, the phenomenon of elastic lag either on compression or expansion may be obtained when considering that for $t = 0$, $\Delta\epsilon = 0$:

$$\Delta\epsilon = \alpha_e\,[1 - e_N^{-(\phi_a/\alpha_e)t}]\,\Delta\sigma \qquad \text{69-II.3}$$

This implies that the elastic unit will not deform immediately, but gradually as the linear viscosity permits, such that after a long time the total elastic deformation under stress $\Delta\sigma$ will be reached. Upon removal of the stress, an expansion will take place also as a function of time (Fig. 37-II.3, curve A). This phenomenon is observed in soils as the effect of the hydrodynamic process of expelling the gravitational water from the pores of the material according to the theory of consolidation. Furthermore, the Hooke element may be substituted by a series of St. Venant units representing elasto-plastic behavior, then Kelvin unit will contain elasto-plastic elements capable of taking an increment of stress. The behavior of the Kelvin unit is the same as Terzaghi's rheological model, except that the value of α_e representing the elastic behavior will be changed into α_{ep} as the parameter representing the elasto-plastic behavior under the applied increment of stress. Therefore

$$\Delta\epsilon_{ep} = \alpha_{ep}\ [1 - e_N^{-(\phi_a/\alpha_{ep})t}]\ \Delta\sigma \qquad \text{70-II.3}$$

Upon removal of stress, only the elastic response is retarded by the linear fluidity, as shown by curve B in Fig. 37-II.3.

Nevertheless, the soil may be considered built of a great number of Kelvin elasto-plastic units, hence the strain

$$\Delta\epsilon_{ep} = \Delta\sigma \sum \alpha_n (1 - e_N^{-(\phi_a/\alpha_n)t}) \qquad \text{71-II.3}$$

Calling $\Sigma\alpha_n = \alpha_{ep}$ the elasto-plastic average strain modulus, the above expression may be written in the form of a series of elements per unit volume:

$$\Delta\epsilon_{ep} = \alpha_{ep} \cdot \Delta\sigma\ [1 - \sum \frac{\alpha_n}{\alpha_{ep}} e_N^{-(\phi_a/\alpha_n)t}\ ] \qquad \text{72-II.3}$$

3.11 The Kelvin-Terzaghi Relationships

Since $\alpha_{ep}\Delta\sigma$ represents the total strain when the process is completed ($t \to \infty$), the deformation time factor represented by the brackets may be compared with Terzaghi's theory. From formula 53-II.3 we find the Kelvin-Terzaghi relationships

$$\frac{\phi_a}{\alpha_n} t = (2m + 1)^2 \frac{\pi^2}{4} T_v \qquad \text{73-II.3}$$

and

$$\frac{\alpha_n}{\alpha_{ep}} = \frac{8}{(2m + 1)^2 \pi^2}$$

from which:

$$\frac{\phi_a}{\alpha_{ep}} = 2\frac{T_v}{t}, \quad \text{or} \quad \frac{\phi_a}{\alpha_{ep}} = \frac{2c_v}{H^2}$$

on the other hand: $\alpha_{ep}\Delta\sigma = m_{ep}\Delta\sigma$. Hence $\alpha_{ep} = m_{ep}$, and the average fluidity is

$$\phi_a = 2m_{ep}\frac{c_v}{H^2}, \quad \frac{\text{cm}^2}{\text{kg} \cdot \text{sec}} \qquad \text{74-II.3}$$

Therefore, Terzaghi's function may be also represented by a series of elasto-plastic Kelvin units; the parameters have the relation expressed before.

From the theory of consolidation we have

$$c_v = \frac{k}{m_{ep}\gamma_w}$$

and the value of the linear fluidity is expressed also as a function of the coefficient of permeability and of the thickness of the strata $2H$ drained on both sides:

$$\phi_a = \frac{2k}{\gamma_w H^2} \qquad \text{75-II.3}$$

3.12 Theory of Consolidation When Load Increases Linearly with Time

In practice, during construction of a building, the increment of load is applied gradually on a compressible soil stratum. Therefore, it is of interest for the foundation engineer to investigate the theoretical behavior under these conditions. From the practical point of view we may assume that the application of the increment of stress is linear with time; therefore

$$\Delta\sigma_t = \frac{\Delta\sigma_c}{t_c} \cdot t \qquad \text{76-II.3}$$

in which $\Delta\sigma_c$ is the total applied increment of load per unit surface in the construction period t_c. Hence

$$\Delta\sigma_t = \Delta\dot{\sigma}_c \cdot t \qquad \text{77-II.3}$$

From rheological considerations just found, and with the use of Kelvin elasto-plastic units with linear viscosity, for linear increase of load increment with time we find the consolidation function

$$\Delta\epsilon_t = e_N^{-(\phi_a/\alpha)t}\ \phi_a\ \Delta\dot{\sigma}_c \int e_N^{(\phi_a/\alpha)t} \cdot t dt + C \qquad \text{78-II.3}$$

Considering $\Delta\epsilon_t = 0$ for $t = 0$, we have after integration

$$\Delta\epsilon_t = \alpha\Delta\dot{\sigma}_c t - \phi_a \left(\frac{\alpha}{\phi_a}\right)^2 \Delta\dot{\sigma}_c \left(1 - e_N^{-(\phi_a/\alpha)t}\right) \qquad \text{79-II.3}$$

From the above expression we get the rate of strain

$$\frac{d}{dt}\Delta\epsilon_t = \alpha\Delta\dot{\sigma}_c \left(1 - e_N^{-(\phi_a/\alpha)t}\right) \qquad \text{80-II.3}$$

For a multiple number of units, with the aid of Kelvin-Terzaghi's relationships we write

$$\frac{d\Delta\epsilon}{dt} = m_{ep}\Delta\dot{\sigma}_c \cdot F(T_v) \qquad \text{81-II.3}$$

Hence

$$\Delta\epsilon = m_{ep}\Delta\dot{\sigma}_c \int_0^t F(T_v)\, dt + c$$

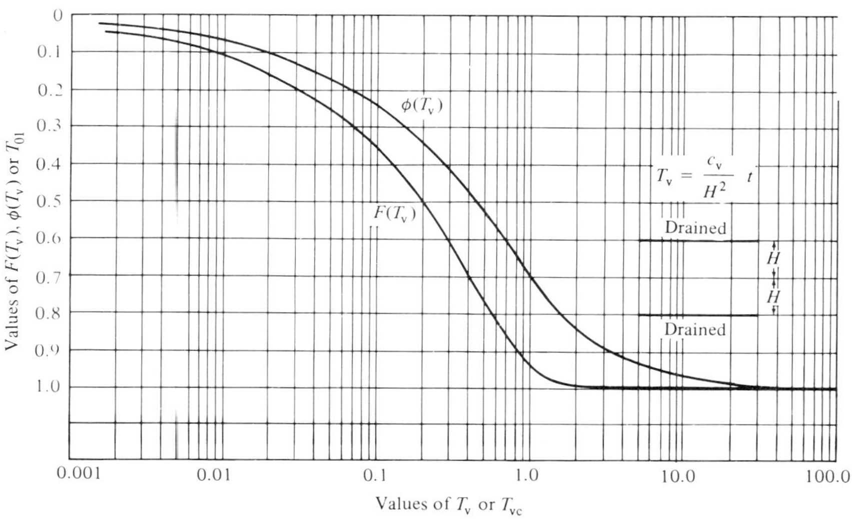

Fig. 38-II.3 Consolidation functions.

The constant of integration may be found considering $\Delta\epsilon = 0$ for $t = 0$; therefore:

$$\Delta\epsilon = m_{ep}\Delta\dot{\sigma}_c \cdot t \left\{1 - \frac{1}{T_v}\sum_{m=0}^{m=\infty}\frac{32}{(2m+1)^4\pi^4}\left(1 - e_N^{-[(2m+1)^2\pi^2/4]\,T_v}\right)\right\} \qquad \text{82-II.3}$$

Calling the value in the brackets $\phi(T_v)$, we may write:

$$\Delta\epsilon = m_{ep}\Delta\dot{\sigma}_c \cdot t \cdot \phi(T_v) \qquad \text{83-II.3}$$

The function $\phi(T_v)$ is plotted against values of T_v in Fig. 38-II.3.

Formula 83-II.3 represents the compression of a stratum when the increment of stress is applied linearly with time. In practice, however, it is required to apply the increment of stress only up to time t_c corresponding to the construction period, thereafter, remaining constant. Therefore, between times: $0 < t < t_c$, expression 83-II.3 is valid. Hence

$$\Delta\epsilon_1 = m_{ep}\Delta\sigma_c \frac{T_v}{T_{vc}}\phi(T_v) \qquad \text{84-II.3}$$

For values of $T_v \geqslant T_{vc}$, after $\Delta\sigma_t$ becomes constant, $F(T_v)$ should be satisfied (Fig. 39-II.3). Hence for the second part of the deformation

$$\Delta\epsilon_2 - (\Delta\epsilon_2)_{tc} = m_{ep}\Delta\sigma_c\,[F(T_v - T_{01}) - F(T_{vc} - T_{01})]$$

where T_{01} may be considered as a retarding time factor. The function is retarded because of the linear application of the increment of stress, that, however, suffers a sharp discontinuity at time t_c. At this time $\Delta\epsilon_1 = \Delta\epsilon_2$. Hence, the value of T_{01} can be determined for $t = t_c$ by means of the following condition (Fig. 39-II.3):

$$\phi(T_{vc}) = F(T_{vc} - T_{01}) \qquad \text{85-II.3}$$

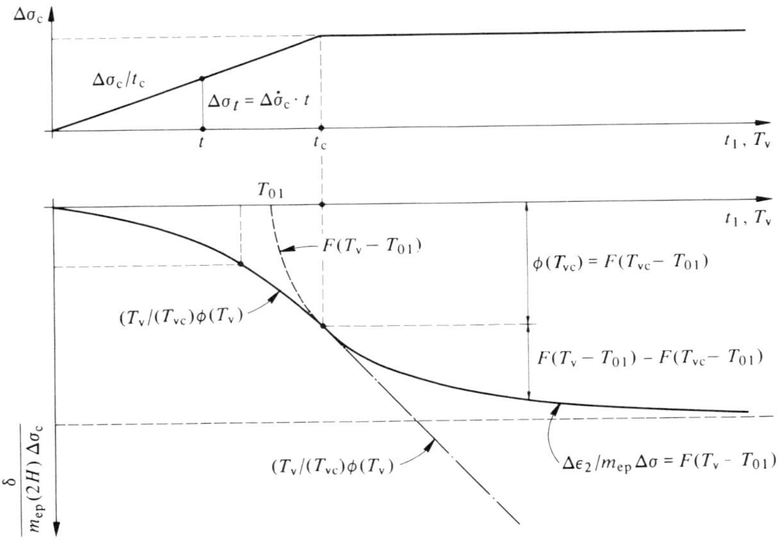

Fig. 39-II.3 Consolidation considering a construction period t_c.

To calculate the consolidation of a stratum of thickness $2H$, when the increment of load is applied linearly with time up to the construction period t_c and thereafter remains constant, we use the formulas:

for $0 < T_v < T_{vc}$

$$\frac{\Delta\epsilon_1}{m_{ep}\Delta\sigma} = \frac{T_v}{T_{vc}}\phi(T_v) \qquad \text{86-II.3}$$

and for $T_v > T_{vc}$

$$\frac{\Delta\epsilon_2}{m_{ep}\Delta\sigma} = F(T_v - T_{01}) \qquad \text{87-II.3}$$

The value of T_{01} may be determined using Fig. 38-II.3. Knowing the value of T_{vc} for a specific case, the value of $\phi(T_{vc})$ is determined, and since $\phi(T_{vc}) = F(T_{vc} - T_{01})$ using function $F(T_v)$ the value of $(T_{vc} - T_{01})$ is found and thus the value of T_{01}. Nevertheless, a formula for the value of T_{01} may be found using the first term of the functions $F(T_v)$ and $\phi(T_v)$, respectively. For values of $T_v > 0.1$

$$1 - \frac{8}{\pi^2} e^{-(\pi^2/4)(T_{vc} - T_{01})} = 1 - \frac{1}{T_{vc}} \frac{32}{\pi^4} (1 - e^{-(\pi/4)T_{vc}}),$$

and, solving for T_{01}, we obtain

$$T_{01} = \frac{4}{\pi^2} \operatorname{Log} \frac{4(e^{(\pi^2/4)T_{vc}} - 1)}{\pi^2 \cdot T_{vc}} \qquad \text{88-II.3}$$

The values of T_{01} vs T_{vc} may be found plotted in Fig. 40-II.3.

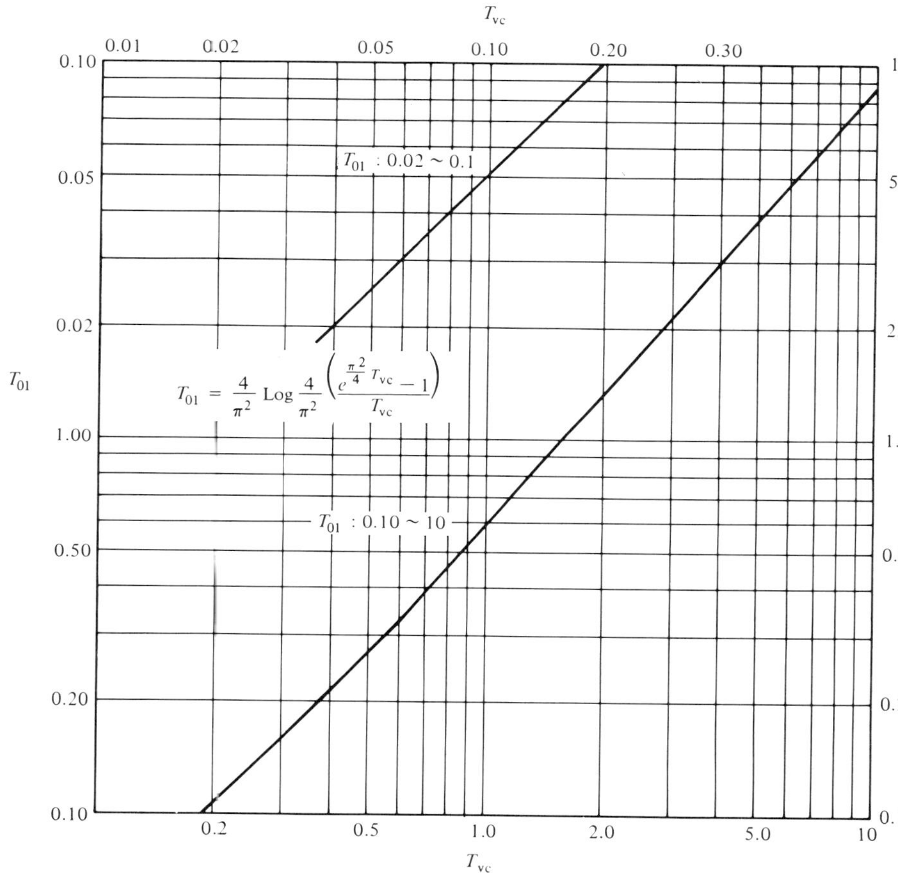

Fig. 40-I .3 Retarding time factor in primary consolidation theory.

3.13 The Intergranular Viscosity in Saturated Soil Sediments

The secondary consolidation observed in materials exhibiting intergranular viscosity, such as lacustrine and marine silty clays and clayey silts, and those containing organic colloids, is of great importance in soil mechanics and to the foundation engineer, to calculate vertical and horizontal displacements in this type of soil deposits. The fact is that the real behavior of the soil appears explicitly, after the first part of the hydrodynamic process has taken place. Thereafter, a viscous intergranular phenomenon is clearly observed. An important fraction of the elastic and plastic deformation takes place implicitly during the hydrodynamic process of consolidation, according to Terzaghi's theory of consolidation, and as the skeleton structure gradually takes effective stresses. Nevertheless, after compression has reached an advanced degree of consolidation, the volume change continues. This phenomenon is known in soil mechanics as secondary consolidation, and may be interpreted as an intergranular viscous phenomenon where the rate of volumetric strain decreases with time, and at the end, motivates the solidification of the material under the applied increment of stress; it takes place with very low hydraulic gradients in saturated soils.

The consolidation of a clay deposit may be divided into two fundamental parts. First, the compression because of the hydrodynamic lag taken by the gravitational water, known as Terzaghi's theory of consolidation, that implies only the retardation of the elasto-plastic strain, and secondly, the compression receiving the name of secondary consolidation represented by a phenomenological law of compression because of viscous effects. In the majority of the cases and in a certain range of time, it is observed that the intergranular viscosity is a linear function of the logarithm of time. The linear logarithmic law was reported for the first time by K. A. S. Buisman (1936) in Holland, from laboratory and field observations (Fig. 41-II.3), and may be written as follows:

$$\delta_t = \delta_B + C_t \log \frac{t}{t_B} \qquad \text{89-II.3}$$

Here C_t represents the slope of the line, and t_B is the time at which it is observed that the linear logarithmic law starts to have effect. The laboratory and field observations show with great approximation the above-mentioned phenomenological law in lacustrine and marine sediments such as clays, silty clays, clayey silts and silts.

The effect of the intergranular viscosity may be clearly detected in consolidation curves of saturated materials when plotted in semilogarithmic scales (Fig. 42-II.3). Curve A shows the behavior observed when the effect of the intergranular viscosity is small; curve B represents an intermediate value, and curve C represents a material with very high intergranular viscosity behavior. From the above-mentioned ob-

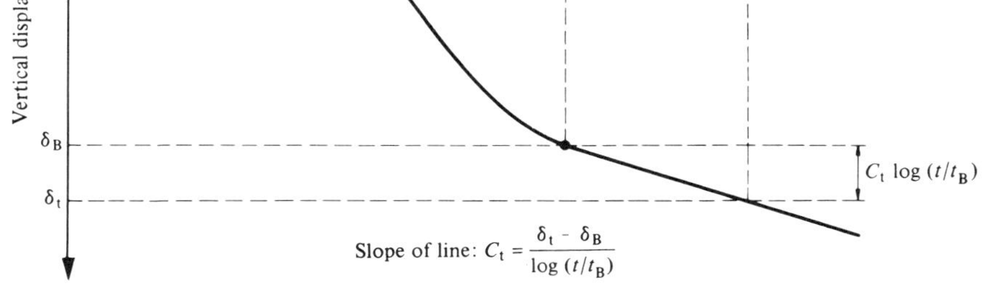

Fig. 41-II.3 Consolidation test.

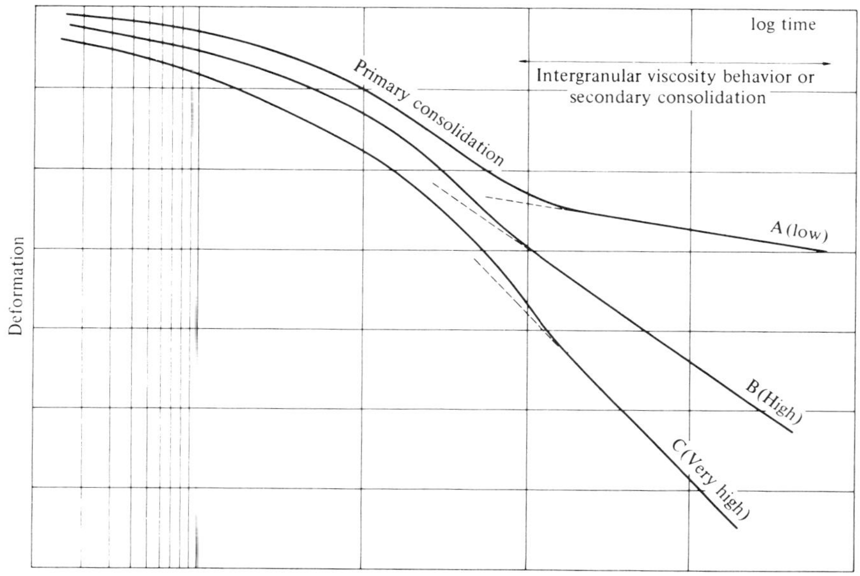

Fig. 42-II.3 Intergranular viscosity as shown by consolidation curves.

served phenomenon, it may be concluded that Terzaghi's theory of consolidation alone cannot express satisfactorily the volumetric change with time of impermeable sediments. Moreover, the simple observational law, 89-II.3, does not permit such an interpretation of the phenomena that would enable us to correlate results from small test specimens in the laboratory with large deposits in the field. Consequently, it is necessary to make a step forward and establish an approximate theory that permits, for practical use, more satisfactory correlations with the phenomenological observations.

A soil sediment is deposited in calm waters and contains from the granulometrical and mineralogical point of view a variety of grains, from very fine sand, silt and clay minerals to organic colloids. The sediment contains also microscopic shells and diatoms. During sedimentation, flocculation takes place because of the well-known electrolytic action present in practically all fine sediments deposited in calm water (Fig. 43-II.3). The clay minerals and colloids adhere to the silt grains, diatoms, microscopic shells and they fur them totally; also they agglomerate in clusters forming grains on the order of the size of fine silt and fine sand. The structure formed by all these different types of grains constitutes the porous skeleton structure of the soil, which is capable of transmitting effective stresses. The pores in the skeleton structure will be called the SS pores; they may be considered larger than the pores found in the agglomerations of clay minerals and colloids, called VP pores (Fig. 44-II.3). During confined compression it is evident that in order that a volume change take place, the total system of pores has to change in volume expelling the gravitational water from them to the drainage surface. The

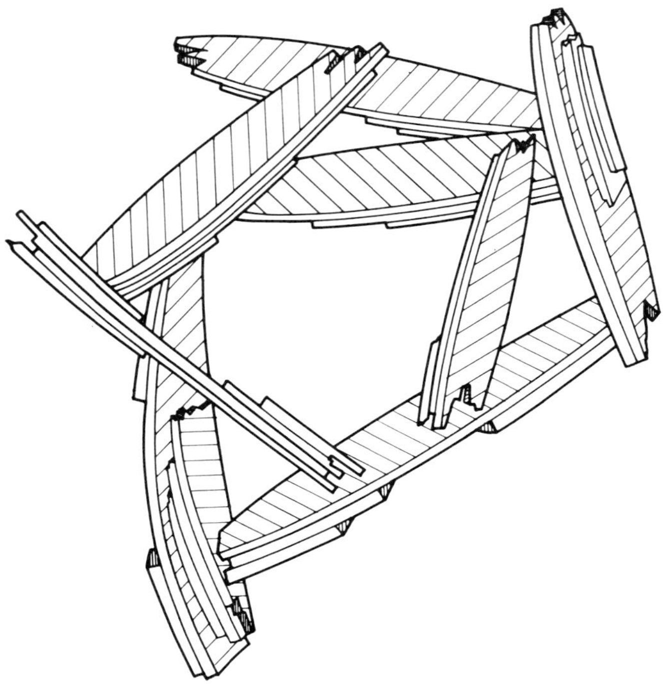

Fig. 43-II.3 Artist's conception of clay mineral grains agglomeration.

assumption is made that the water pressure in the submicroscopic VP pores is consistently equal to or larger than that in the skeleton structure pores SS; hence water flows from the VP pores into the SS pores and from them to the drainage surfaces. From theoretical considerations and from the practical point of view, we conclude that the soil may be considered formed by two basic structures representing a different mechanical behavior, mainly: the part formed by the skeleton structure of the soil with grains of size on the order of silts and larger, representing mainly the elasto-plastic behavior, and the part formed by the agglomeration of clay minerals and colloids exhibiting nonlinear intergranular viscosity. Furthermore, it may be considered for practical purposes that the change in pore volume of the submicroscopic pores begins at the instant the increment of stress is applied.

Applying the above working hypotheses, we conclude that the volumetric strain of the material $\Delta\epsilon_v$ is equal to the sum of the volumetric strain $\Delta\epsilon_{ss}$ in the SS pores of the skeleton structure added to the volumetric strain $\Delta\epsilon_{vp}$ in the submicroscopic pores VP. Thus, we obtain the condition

$$\Delta\epsilon_v = \Delta\epsilon_{ss} + \Delta\epsilon_{vp} \qquad \text{90-II.3}$$

The volumetric strain-time relationships because of elasto-plastic and viscous intergranular behavior in saturated sediments may be studied from the rheological point of view. The object is to establish theoretical relationships in approximate agree-

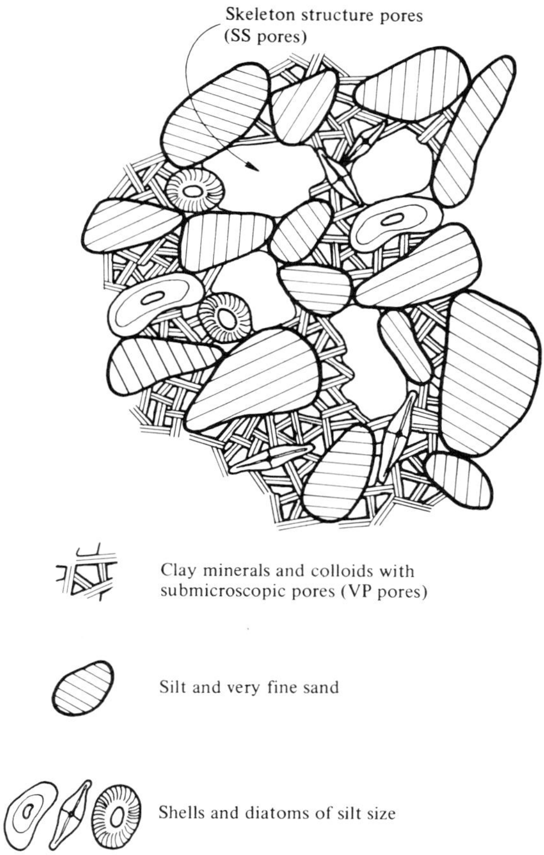

Fig. 44-II.3 Mixed silty clay structure.

ment with the phenomenological observations, thus achieving a more satisfactory correlation of observations in the laboratory with those in the field.

Considering the structural arrangement of the sediment, the rheological concepts representing the soil may be reduced to the behavior of two models: one unit representing the soil skeleton structure, and the other representing the behavior of the clusters formed by the fine and ultrafine clay grains with nonlinear viscous properties. The volume change of these two systems because of a stress increment may be visualized as retarded by linear viscosity in the pores containing gravitational water. The visco-plastic intergranular behavior of nonlinear fluidity in the case of impervious materials containing clay minerals may be interpreted using Newton's law:

$$\frac{d}{dt}\,\Delta\epsilon_{\mathrm{vp}} = \phi \cdot \Delta\sigma \qquad \text{91-II.3}$$

From phenomenological observations, the value of the apparent fluidity is not a constant during the process of time deformation. The quantity ϕ has a definite value at the start of the process and diminishes with time until reaching a final value ϕ_f, which may be very small or even zero. Under these conditions, the simpler mathematical law expressing approximately the nonlinear viscous phenomenon is the following:

$$\frac{d}{dt}\Delta\epsilon_{vp} = \left(\phi_f + \frac{a}{b+t}\right)\Delta\sigma \qquad \text{92-II.3}$$

such that when $t \to \infty$, $d\Delta\epsilon_{vp}/dt \to \phi_f\Delta\sigma$; hence the material becomes an ideal Newtonian liquid, and when $t = 0$, then $d\Delta\epsilon_{vp}/dt$ approaches a finite value. Therefore, expression 92-II.3 satisfies the limiting conditions of the phenomenon. Upon a rapid load application, the viscous intergranular phenomenon may be assumed to start. If the load is held constant during the whole process, then by integration:

$$\Delta\epsilon_{vp} = \left[\phi_f t + a \operatorname{Log}\frac{b+t}{b}\right] \cdot \Delta\sigma \qquad \text{93-II.3}$$

The law stated above representing the nonlinear viscous phenomenon indicates that if $a = 0$, a perfect Newtonian liquid of fluidity ϕ_f is obtained. Hence, the term $\phi_f \cdot t$ represents a steady creep. On the other hand if $\phi_f = 0$, a logarithmic law will represent the phenomenon; its rate of deformation approaches zero as time approaches very large values, and represents under laterally confined conditions the phenomenon of nonlinear viscosity and solidification or hardening, as gradually more firm contacts are developed because the grains get closer together when compaction takes place. At this stage the adsorbed water between clay mineral grains plays a very important role.

3.13a The Z-Unit. The plasto-viscosity rheological law will be further investigated, as the water in the pores retards the phenomenon because of the effect of expelling the gravitational water from them. A new unit is formed having the property of nonlinear viscosity and also linear viscosity representing the hydrodynamic lag because of gravitational water squeezed from the submicroscopic pores. Hence, the gravitational water will retard the volumetric strain during the viscous intergranular behavior. This new rheological model will be designated the Z-unit (Fig. 45-II.3). The equilibrium condition requires at any time that

$$\begin{aligned} \Delta\sigma &= \Delta\sigma_Z + \Delta\sigma_N \\ \Delta\epsilon_{vp} &= \Delta\epsilon_Z = \Delta\epsilon_N \end{aligned} \qquad \text{94-II.3}$$

Furthermore, the following expressions may be written for a nonlinear viscous element:

$$\frac{d}{dt}\Delta\epsilon_{vp} = \left(\phi_f + \frac{a}{b+t}\right)\Delta\sigma_Z \qquad \text{95-II.3}$$

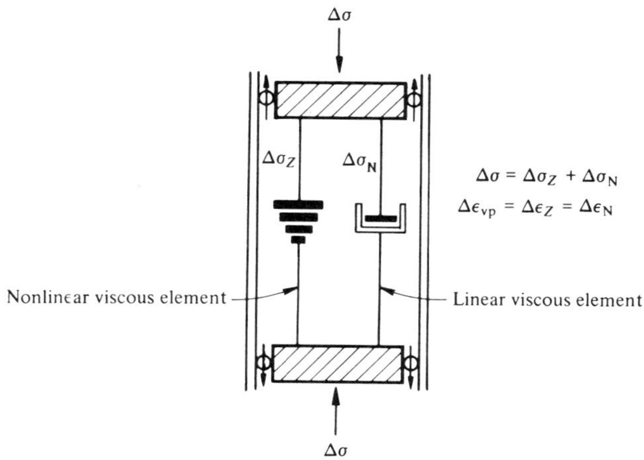

Fig. 45-II.3 The *Z*-unit.

and for the linear viscous element:

$$\frac{d}{dt}\Delta\epsilon_{vp} = \phi_a \Delta\sigma_N \tag{96-II.3}$$

in which ϕ_a is the apparent fluidity of the gravitational water in the pores. After substitution into conditions 94-II.3, we get

$$\Delta\sigma = \left[\frac{1}{\phi_f + \dfrac{a}{b+t}} + \frac{1}{\phi_a}\right]\frac{d}{dt}\Delta\epsilon_{vp} \tag{97-II.3}$$

Performing algebraic operations the following expressions may be written:

$$A = \left(b + \frac{a}{\phi_f}\right)\frac{\phi_f \cdot \phi_a}{\phi_f + \phi_a}, \quad B = b + \frac{a}{\phi_f + \phi_a}$$
$$\phi_{af} = \frac{\phi_f \cdot \phi_a}{\phi_f + \phi_a} \tag{98-II.3}$$

from which the differential equation governing the phenomenon is found:

$$\frac{d}{dt}\Delta\epsilon_{vp} = \left[\frac{A}{B+t} + \phi_{af}\frac{t}{B+t}\right]\Delta\sigma \tag{99-II.3}$$

After integration:

$$\Delta\epsilon_{vp} = [(A - \phi_{af}B)\,\mathrm{Log}\,(B+t) + \phi_{af} \cdot t]\,\Delta\sigma + c \tag{100-II.3}$$

Upon application of an instantaneous load $\Delta\sigma$, $t \doteq 0$ and $\Delta\epsilon_{vp} = 0$; hence calling $(A - \phi_{af}B) = A_1$, the following law may be written:

$$\Delta\epsilon_{vp} = \left[A_1 \text{ Log} \frac{B + t}{B} + \phi_{af} \cdot t\right] \Delta\sigma \qquad \text{101-II.3}$$

Notice that $\phi_{af} \cdot t$ represents a linear steady creep, however, for a laterally confined soil, the value ϕ_{af} may be considered small, hence:

$$A = A_1 = a, \; B = b + \frac{a}{\phi_a}, \; \phi_{af} \doteq 0$$

Therefore

$$\Delta\epsilon_{vp} = a \text{ Log} \frac{b + \dfrac{a}{\phi_a} + t}{b + \dfrac{a}{\phi_a}} \cdot \Delta\sigma \qquad \text{102-II.3}$$

The gravitational water in the pores introduces a time lag measured by a/ϕ_a (Fig. 46-II.3).

The fraction of load taken by the gravitational water at any time is

$$\Delta\sigma_N = \frac{1}{\phi_a} \cdot \frac{d}{dt} \Delta\epsilon_{vp} \qquad \text{103-II.3}$$

Therefore, by differentiation of expression 102.II.3 we write

$$\Delta\sigma_N = \frac{\dfrac{a}{\phi_a}}{b + \dfrac{a}{\phi_a} + t} \cdot \Delta\sigma \qquad \text{104-II.3}$$

For small values of t, the gravitational water takes the load $\Delta\sigma$; theretore $b = 0$.

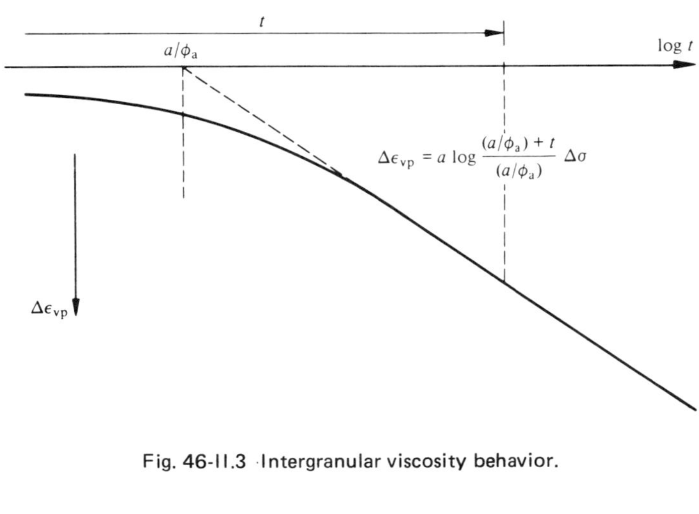

Fig. 46-II.3 Intergranular viscosity behavior.

From the preceding discussion it may be concluded that the two rheological units studied producing retardation of the volumetric strain by viscous linear and nonlinear effects, and for rapid and sustained applied stress, are the following:

a) retardation of the elasto-plastic behavior, with linear viscosity from Terzaghi's theory:

$$\Delta\epsilon_{ep} = m_{ep}\Delta\sigma F(T_v) \qquad 105\text{-II.3}$$

b) retardation of the plasto-viscous intergranular behavior with linear fluidity from the Z-unit:

$$\Delta\epsilon_{vp} = \bar{a}\,\mathrm{Log}\,\frac{\dfrac{\bar{a}}{\phi_a} + t}{\dfrac{\bar{a}}{\phi_a}} \cdot \Delta\sigma \qquad 106\text{-II.3}$$

The above formulas are represented graphically in Fig. 47-II.3. The value of $\bar{a}$ stands for an average parameter in the sum of all the Z-units. From correlations obtained for the Kelvin-Terzaghi models, the parameters may be defined as follows:

m_{ep} = coefficient of unit volume compressibility representing the elasto-plastic phenomenon behavior.

m_{vB} = coefficient of unit volume compressibility including the elasto-plastic and intergranular viscosity behavior at time t_B after which the intergranular viscosity phenomenon becomes explicit, it corresponds to the break B in the consolidation curve represented in semilog scales of strain or displacement vs. time.

$m_t = 2.31\bar{a}$ = coefficient of unit intergranular viscous volume compressibility obtained from the slope of the logarithmic law of base 10 representing the secondary consolidation behavior: $m_t = C_t/(2H)\Delta\sigma$.

$\Delta\epsilon_v$ = total volumetric strain.

$\Delta\epsilon_{ss} = \Delta\epsilon_{ep}$ = volumetric strain because of elasto-plastic behavior in the SS pores.

$\Delta\epsilon_{vp}$ = volumetric strain because of the intergranular viscosity or plasto-viscosity behavior of the VP pores.

$\Delta\sigma$ = increment of stress.

$$\beta = \frac{m_t}{m_{ep}} \quad \text{and} \quad \alpha = \frac{m_t}{m_{vB}}$$

From 74-II.3 the apparent linear fluidity is found as $\phi_a = 2m_{ep}\dfrac{c_v}{H^2}$, and the ratio

$\dfrac{\bar{a}}{\phi_a} = \dfrac{\beta}{4.62} \cdot \dfrac{H^2}{c_v}$; we let $\tau = \bar{a}/\phi_a$.

From the above we derive the important relationship

$$\frac{\tau_1}{\tau_2} = \frac{H_1^2}{H_2^2} \cdot \frac{c_{v2}}{c_{v1}} \qquad 107\text{-II.3}$$

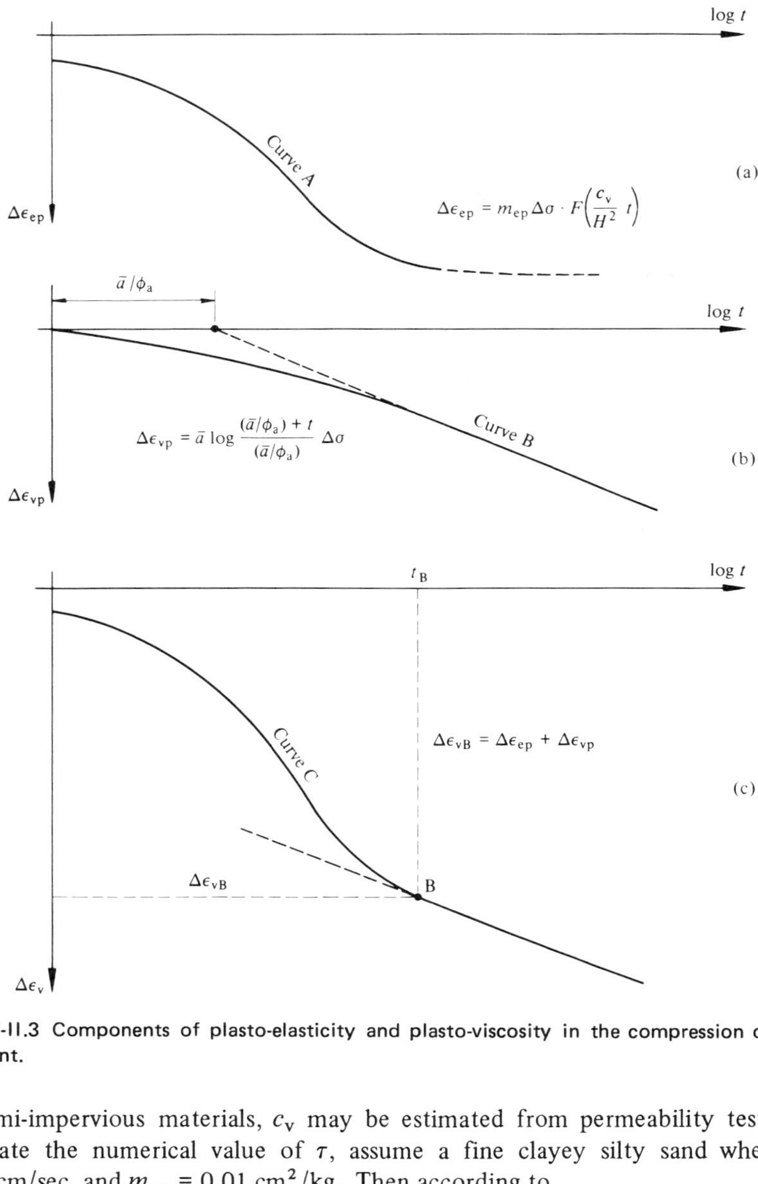

Fig. 47-II.3 Components of plasto-elasticity and plasto-viscosity in the compression of a soil sediment.

In semi-impervious materials, c_v may be estimated from permeability tests. To illustrate the numerical value of τ, assume a fine clayey silty sand where $k = 10^{-6}$ cm/sec, and $m_{ep} = 0.01$ cm^2/kg. Then according to

$$c_v = \frac{k}{m_{ep}\gamma_w},$$

we obtain

$$c_v = 0.1 \text{ cm}^2/\text{sec}$$

Furthermore, if for this material β = 0.3, and the stratum is saturated and has a thickness $2H$ of 2000 cm, then

$$\tau = \frac{0.3}{4.62} \cdot \frac{10^6}{10^{-1}} = 6.5 \times 10^5 \text{ sec} = 0.25 \text{ months}$$

Relationship 107-II.3 may be used to correlate the laboratory tests with the field geometry.

3.13b The Strain-Time Behavior for Rapid Load Application. From our working hypotheses, it was established that the total volumetric strain

$$\Delta\epsilon_v = \Delta\epsilon_{ep} + \Delta\epsilon_{vp}$$

Therefore, after performing proper substitutions in expression 106-II.3, the final general expression may be obtained for the volumetric strain under conditions of rapid application of a stress increment:

$$\Delta\epsilon_v = \left\{ m_{ep}\, F\left(\frac{c_v}{H^2}\, t\right) + m_t \log\left[1 + \frac{4.62}{\beta}\left(\frac{c_v}{H^2}\, t\right)\right]\right\} \Delta\sigma \qquad \text{108-II.3}$$

which may be written in a nondimensional form as follows:

$$\frac{\Delta\epsilon_v}{m_{ep}\Delta\sigma} = F(T_v) + Z(T_v, \beta) \qquad \text{109-II.3}$$

in which $T_v = (c_v/H^2) \cdot t$ is the time factor for a stratum of thickness $2H$, drained on both sides. The first term represents Terzaghi's function and is plotted in Fig. 38-II.3. The second term represents the intergranular viscosity phenomenon

$$Z(T_v, \beta) = \beta \log\left(1 + \frac{4.62}{\beta}\, T_v\right) \qquad \text{110-II.3}$$

This function is plotted for different values of β in Fig. 48-II.3, and expression 109-II.3 in Fig. 49-II.3.

Table 4-II.3 has been prepared to give an idea of the values of β, for different saturated sediments.

TABLE 4-II.3

Intergranular Viscosity	Values of β	Material
Very high High	greater than 0.8 0.8–0.5	Lacustrine and marine clays and silts of active minerals of high to very high compressibility
Medium Low	0.5–0.3 0.3–0.1	Clays and lacustrine silts of medium compressibility
Very low	less than 0.1	In general, materials of low and very low compressibility

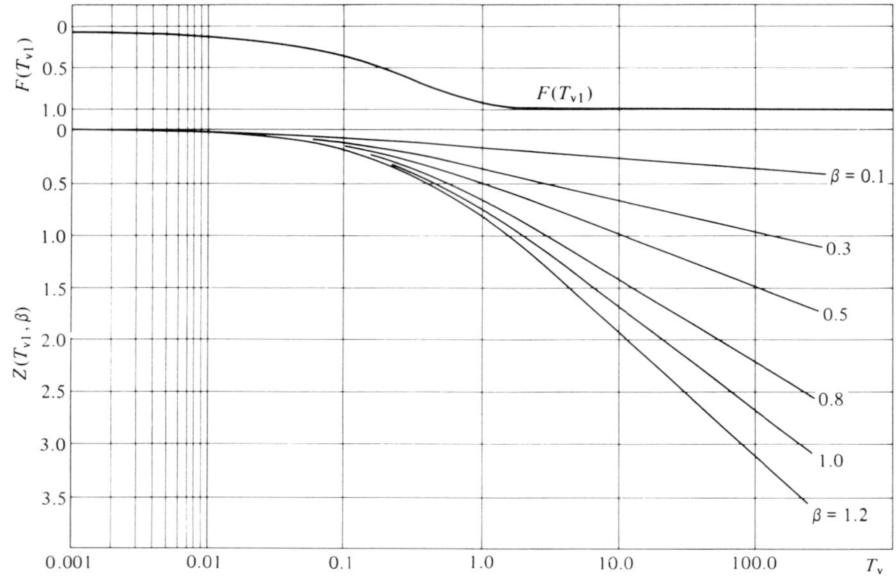

Fig. 48-II.3 Values of $\beta \log [1 + (4.62/\beta)T_v]$.

Relating the values of β with the compressibility curves for soils of medium and high sensitivity it may be concluded that for consolidation stresses larger than the critical stress σ_b, $\beta > 1$, and for smaller values than σ_b, $\beta < 1$. In practice for foundation engineering design purposes the steep branch of the compressibility curve in sensitive soils should not be used, since for stresses larger than σ_b, the compressibility increases several times as compared with the one obtained in the recompression flat branch of the compressibility curve.

3.13c Strain-Time Behavior for Linear Load Application. In practical problems the load is not applied instantly; therefore, the formulas given above serve only to obtain the parameters for the applied stress increments from consolidation tests when the load is applied instantaneously (see fitting methods).

For all practical purposes, one may assume that the stress increment is applied linearly with time up to a certain value $\Delta\sigma_c$, in time t_c; therefore:

$$\Delta\sigma_t = \left(\frac{\Delta\sigma_c}{t_c}\right) t$$

where $\Delta\sigma_t$ is the increment of stress at time t, for $0 < t < t_c$. The value of t_c represents the time at which the total load is reached after the construction of a building. We have

$$\Delta\dot{\sigma}_c = \frac{\Delta\sigma_c}{t_c}, \quad \text{and} \quad \Delta\sigma_t = \Delta\dot{\sigma}_c \cdot t$$

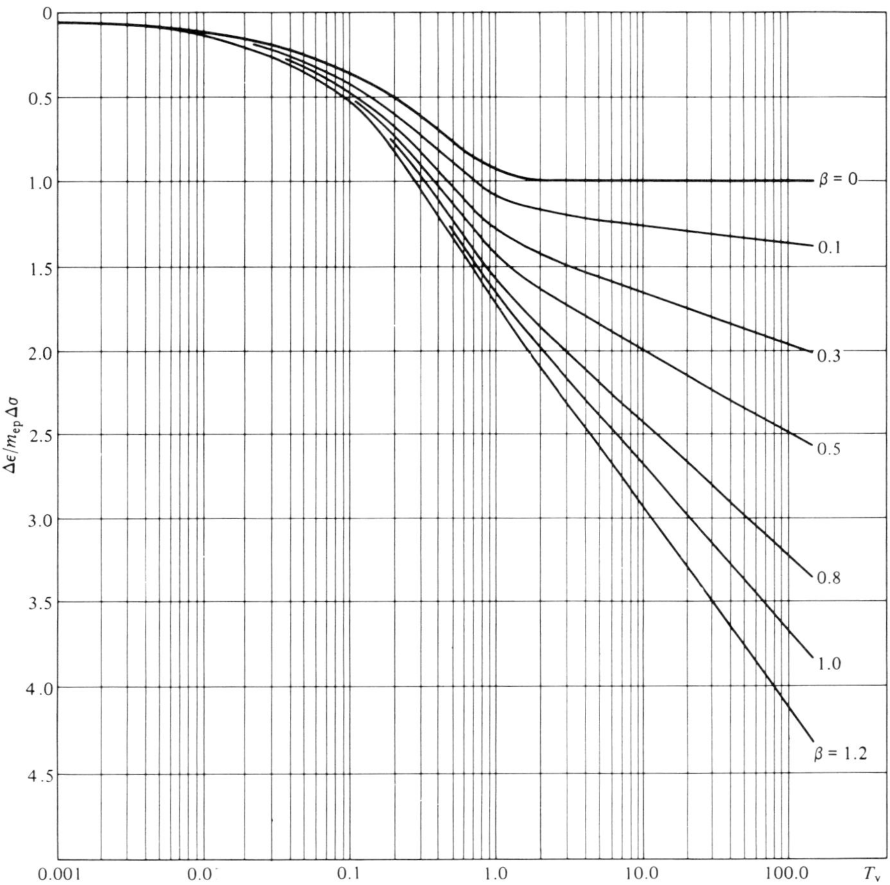

Fig. 49-II.3 Values of $F(T_v) + Z(T_v, \beta)$.

From the theory of consolidation (Section II.3.12) it was found that the elasto-plastic behavior may be calculated as follows:

for $0 < t < t_c$ or $0 < T_v < T_{vc}$:

$$\Delta\epsilon_{ss} = m_{ep}\Delta\sigma_c \frac{T_v}{T_{vc}} \phi(T_v) \quad \text{87-II.3}$$

and for values of $t \geqslant t_c$, that is to say after $\Delta\sigma_c$ remains constant:

$$\Delta\epsilon_{ss} = m_{ep}\,\Delta\sigma_c F(T_v - T_{01}) \quad \text{88-II.3}$$

Here T_{01} may be determined from the condition

$$\phi(T_{vc}) = F(T_{vc} - T_{01}) \qquad 85\text{-II.3}$$

For the Z-unit it is necessary to obtain a formula for the application of the stress increment linearly with time; therefore, from 110-II.3, the rate of strain is

$$\Delta\dot{\epsilon}_{vp} = m_{ep}\beta\Delta\sigma \left[\frac{0.43}{1 + \dfrac{4.62}{\beta} T_v}\right] \frac{4.62}{\beta} \frac{d}{dt} T_v$$

Substituting $T_v = (c_v/H^2) \cdot t$ and $\Delta\sigma_t = \Delta\dot{\sigma}_c \cdot t$ and integrating, we obtain

$$\Delta\epsilon_{vp} = 0.43\beta m_{ep} \left(\frac{4.62}{\beta} \frac{c_v}{H^2}\right) \Delta\dot{\sigma}_c \int_0^t \frac{t\,dt}{1 + \dfrac{4.62}{\beta} \dfrac{c_v}{H^2} t} + C \qquad 111\text{-II.3}$$

After integration, the following formula is obtained for the volumetric strain:

$$\Delta\epsilon_{vp} = \frac{m_{ep}\beta}{2.3} \frac{\Delta\dot{\sigma}_c}{\left(\dfrac{4.62}{\beta} \dfrac{c_v}{H^2}\right)} \left[\left(1 + \frac{4.62}{\beta} \frac{c_v}{H^2} t\right) - \text{Log}\left(1 + \frac{4.62}{\beta} \frac{c_v}{H^2} t\right)\right] + C$$

For $t = 0$, the volumetric strain $\Delta\epsilon_{vp} = 0$, therefore, for $0 < t < t_c$,

$$[\Delta\epsilon_{vp}]_1 = \frac{\beta}{2.3} m_{ep}\Delta\dot{\sigma}_c \left[t - \frac{1}{\dfrac{4.62}{\beta} \dfrac{c_v}{H^2}} \text{Log}\left(1 + \frac{4.62}{\beta} \frac{c_v t}{H^2}\right)\right] \qquad 112\text{-II.3}$$

Since $\Delta\dot{\sigma}_c = \Delta\sigma_c/t_c$, we obtain

$$[\Delta\epsilon_{vp}]_1 = \frac{\beta}{2.3} m_{ep}\Delta\sigma_c \frac{T_v}{T_{vc}} \left[1 - \frac{1}{\dfrac{4.62\, T_v}{\beta}} \text{Log}\left(1 + \frac{4.62}{\beta} T_v\right)\right], \qquad 113\text{-II.3}$$

calling

$$A_1 = 1 - \frac{1}{4.62 \dfrac{T_v}{\beta}} \text{Log}\left(1 + 4.62 \frac{T_v}{\beta}\right) \qquad 114\text{-II.3}$$

The visco-plastic strain for values of $t \leqslant t_c$ may be written in the form:

$$[\Delta\epsilon_{vp}]_1 = \frac{\beta}{2.3} m_{ep}\Delta\sigma_c \frac{T_v}{T_{vc}} \cdot A_1 \left(\frac{T_v}{\beta}\right) \qquad 115\text{-II.3}$$

The value of A_1 vs. (T_v/β) is plotted in Fig. 50-II.3.

The behavior of the volumetric strain for $0 < t < t_c$, when the stress application is linear, is represented in Fig. 51-II.3. At point t_c, the stress increment remains

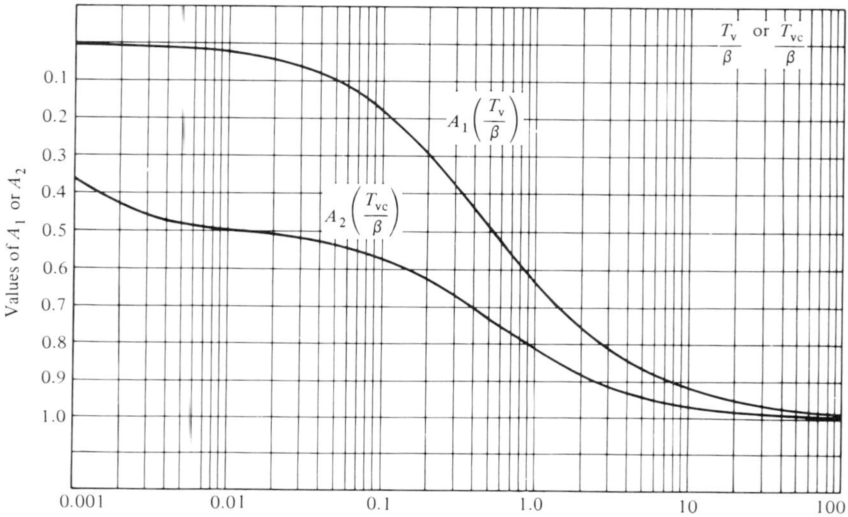

Fig. 50-II.3 Auxiliary functions A_1 and A_2.

constant with time; therefore, the following law applies for values of $t > t_c$

$$[\Delta\epsilon_{vp}]_2 = \beta m_{ep} \Delta\sigma_c \log\left[1 + \frac{4.62}{\beta}(T_v - T_{02})\right] \qquad \text{116-II.3}$$

The value of T_{02} represents the virtual starting time of the logarithmic law and its value may be found from the condition that for $t = t_c$, $[\Delta\epsilon_{vp}]_1 = [\Delta\epsilon_{vp}]_2$, therefore, according to Fig. 51-II.3

$$\log e + \log\left(1 + 4.62\frac{T_{vc}}{\beta}\right)^{-1/4.62(T_{vc}/\beta)} = \log\left[1 + \frac{4.62}{\beta}(T_{vc} - T_{02})\right]$$

calling

$$a_2 = \left(1 + 4.62\frac{T_{vc}}{\beta}\right)^{-1/4.62(T_{vc}/\beta)}$$

we find

$$\log a_2 = \log\frac{1 + \dfrac{4.62}{\beta}(T_{vc} - T_{02})}{2.72}$$

after algebraic operations we obtain

$$T_{02} = T_{vc}\left(1 - \frac{2.72\,a_2 - 1}{4.62\dfrac{T_{vc}}{\beta}}\right). \qquad \text{117-II.3}$$

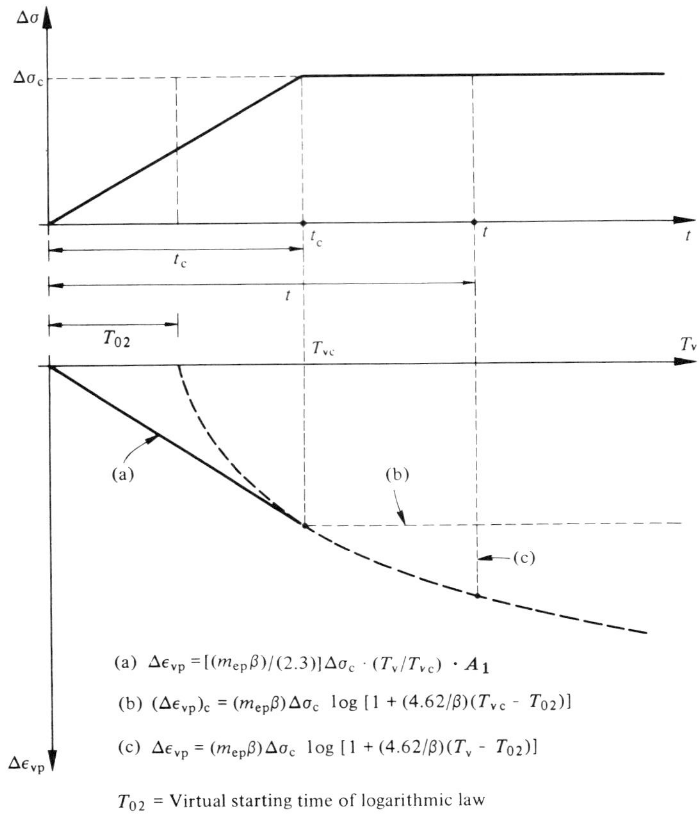

Fig. 51-II.3 Intergranular viscosity deformation for linear load application.

We call $T_{02}/T_{vc} = A_2$, hence:

$$A_2 = 1 - \frac{2.72\,a_2 - 1}{4.62\,\dfrac{T_{vc}}{\beta}}, \qquad \text{118-II.3}$$

from which we write the value of T_{02}

$$T_{02} = T_{vc} \cdot A_2 \left(\frac{T_{vc}}{\beta}\right) \qquad \text{119-II.3}$$

The auxiliary function A_2 vs. T_{vc}/β may be found plotted in Fig. 50-II.3.

Finally, we obtain formulas in dimensionless form to calculate the volumetric strain for the compression of the SS and VP pores:

For $0 < t < t_c$

$$\frac{[\Delta\epsilon_{vp}]_1}{m_{ep}\Delta\sigma_c} = \frac{T_v}{T_{vc}}\left[\phi(T_v) + \frac{\beta}{2.3} \cdot A_1\right] \qquad \text{120-II.3}$$

we call

$$f_1 = \left[\phi(T_v) + \frac{\beta A_1}{2.3}\right] \frac{T_v}{T_{vc}}$$

and for values $t > t_c$

$$\frac{[\Delta \epsilon_{vp}]_2}{m_{ep} \cdot \Delta c_c} = F(T_v - T_{01}) + \beta \log \left[1 + \frac{4.62}{\beta}(T_v - T_{vc} \cdot A_2)\right] \qquad \text{121-II.3}$$

here we call

$$f_2 = F(T_v - T_{01}) + \beta \log \left[1 + \frac{4.62}{\beta}(T_v - T_{vc} \cdot A_2)\right]$$

The formulas thus obtained may be used to calculate the volumetric strain for elasto-plastic-viscous compression in homogeneous stratified fine soil sediments under lateral confinement and when the load is applied linearly with time up to the construction period t_c, thereafter remaining constant for an indefinite period. The parameters m_{ep}, m_t, and c_v are determined in conventional consolidation tests as explained in Section 3.17.

To illustrate the use of the above formulas, assume a value of $c_v = 2 \times 10^{-4}$ cm²/sec. The thickness of the impervious stratum is 600 cm, and the construction period is $t_c = 3$ yr $= 3.15 \cdot 10^7 \cdot (3)$ sec. Then we calculate:

$$T_{vc} = 2 \times 3.15 \times 10^{-1}\,(3)/(3)^2 = 0.21$$

From Fig. 40-II.3 we find $T_{01} = 0.115$. Assuming $\beta = 0.6$; $T_{vc}/\beta = 0.35$ and from Fig. 50-II.3 the value of $A_2 = 0.695$ is obtained, and substituting the values of T_{01} and A_2 in expression 121-II.3 we obtain for $t > t_c$

$$\frac{[\Delta \epsilon_{vp}]_2}{m_{ep} \cdot \Delta \sigma_c} = F(0.07t - 0.115) + 0.6 \log\,[1 + 7.7(0.07t - 0.146)]$$

Now assume the deformation is required for a period of 10 years, for an average value of $m_{ep} = 0.03$ cm²/kg, and stress increment at the end of the construction period of $\Delta\sigma_c = 0.6$ kg/cm²; therefore,

$$\Delta\delta_2 = 0.03(600)0.6\,\{F(0.7 - 0.115) + 0.6 \log\,[1 + 7.7(0.7 - 0.146)]$$

$$\Delta\delta_2 = 10.8\,\{0.81 + 0.43\} = 13.43 \text{ cm}$$

The total settlement at the surface will be the addition at equal times of all the compressible strata affected by the average change in stress $\Delta\sigma_i$, hence:

for $0 < t < t_c$

$$\delta_1 = \sum_1^n [\Delta\delta_1]_i,$$

for $t = t_c$

$$\delta_2 = \sum_1^n [\Delta\delta_2]_i$$

The boundary conditions, however, should be carefully observed in selecting the drainage surfaces. The formulas discussed so far will serve when either one or two boundaries are drainage surfaces. Nevertheless, the case may be encountered of a drained soil strata formed by a series of impervious sediments with variable values of m_{ep}, m_t and c_v. In such cases average values may be used:

$$(m_{ep})_a = \frac{\sum_1^n (m_{ep}d)_i}{\sum_1^n d_i}, \qquad (m_t)_a = \frac{\sum_1^n (m_t d)_i}{\sum_1^n d_i}, \qquad k_a = \frac{\sum_1^n d_i}{\sum_1^n \frac{d_i}{k_i}} \qquad \text{122-II.3}$$

where

$$k_i = (c_v m_{ep} \gamma_w)_i, \qquad \text{hence } c_{va} = \frac{k_a}{(m_{ep})_a \gamma_w}$$

The average values thus obtained are used in settlement formulas 120 and 121-II.3 to calculate the compression of each stratum with certain definite boundary conditions. From formulas 120 and 121-II.3 we may notice that the coefficient of volume compressibility for a stratum of a fine sediment may be expressed as a function of time as follows:

for $0 < t < t_c$

$$m_{v1} = m_{ep} \cdot f_1(\beta, T_v, T_{vc})$$

and for $t > t_c$

$$m_{v2} = m_{ep} \cdot f_2(\beta, T_v, T_{vc})$$

The unit compression of the stratum N of thickness d_i in either case may be expressed in general terms as follows:

$$\alpha_c^N = (m_v d)_i \qquad \text{123-II.3}$$

where α_c^N is again a function of time. The total compression of the stratum due to an average increment of stress $\Delta\sigma_i$ is obtained as follows:

$$\Delta\delta_i^N = \alpha_i^N \cdot \Delta\sigma_i^N \qquad \text{124-II.3}$$

3.14 Intergranular Viscosity in Saturated Soils with Cavities

The saturated materials with intermediate permeability, like silt, fine clayey silty sand and organic clay materials having cavities like root-holes making the material more permeable, require special consideration. They are commonly encountered in lacustrine plains, river banks and marginal lagoons.

In this case, the behavior observed in the laboratory in oedometer tests is as shown in Fig. 52-II.3. The primary consolidation takes place rapidly because of the explusion of water from the semipermeable soil skeleton structure. The end of this phenomenon is registered at point B by a prompt soft break in the consolidation curve, after which the plasto-viscous intergranular phenomenon is clearly present, and may be interpreted by the phenomenological law studied before for intergranular viscosity:

$$C_t \log \frac{\tau_s + t}{\tau_s} \qquad 125\text{-II.3}$$

The determination of C_t and τ_s may be obtained from consolidation tests as shown in Fig. 52-II.3. Let us call t_B the time at which the break in the consolidation curve takes place, for which one can assume $F(T_v) \doteq 1.0$. Hence

$$\delta_B = \delta'_{ep} + \delta''_{ep} + C_t \log \frac{\tau_s + t_B}{\tau_s}$$

From the time vertical displacement behavior shown in Fig. 52-II.3, we conclude that the value $(\delta'_{ep} + \delta''_{ep})$ is obtained in a very short time. Hence, for values of $t > t_B$ one can write without appreciable error

$$\delta = \delta_B + C_t \log \frac{\tau_s + t}{\tau_s} \qquad 126\text{-II.3}$$

Therefore, from consolidation curves in the oedometer, the values of the parameters δ_B, C_t and τ_s may be easily determined.

In order to correlate the test specimen with field geometry, consider the thicknesses of the test specimen ($2H_s$) and the stratum in the field ($2H_f$). From

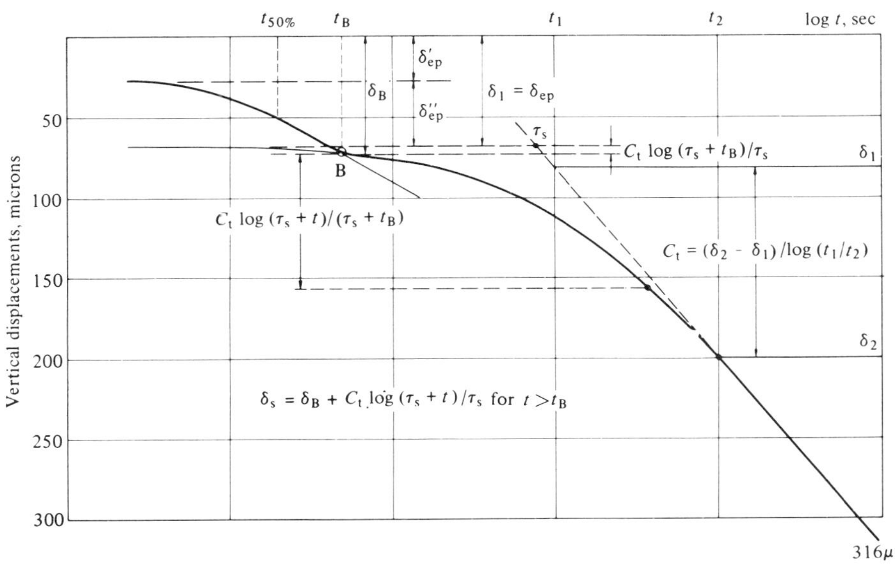

Fig. 52-II.3 Saturated soils with cavities.

section 3.13a, calling c_{vp} an apparent coefficient of consolidation for the intergranular viscosity phenomenon, we obtain

$$c_{vp} = \frac{\beta}{4.62} \cdot \frac{H_s^2}{\tau_s} = \frac{\beta}{4.62} \cdot \frac{H_f^2}{\tau_f}$$

also:

$$\frac{\tau_s}{\tau_f} = \frac{H_s^2}{H_f^2}, \quad \text{and} \quad \tau_f = \left(\frac{H_f}{H_s}\right)^2 \cdot \tau_s \qquad \text{127-II.3}$$

Using the same definitions as before, we have for the change in stratum thickness

$$\delta = \left[m_{ep} + m_t \log \left(1 + \frac{t}{\tau_f}\right)\right] (2H)\, \Delta\sigma$$

Setting $T_{vp} = (c_{vp}/H^2) \cdot t$, we can write in dimensionless form the plasto-viscous phenomenon for rapid load application:

$$\frac{\Delta\epsilon_v}{m_{ep}\,\Delta\sigma} = 1 + \beta \log \left(1 + \frac{4.62}{\beta}\, T_{vp}\right) \qquad \text{128-II.3}$$

Since the main interest to the foundation engineer is to apply these formulas in the approximate case of a linearly applied increment of stress, the results may be used in expressions 120 and 121-II.3, for $F(T_v) = 1$ and $\phi(T_v) = 1$; therefore:

for $0 < t < t_c$

$$\frac{[\Delta\epsilon_v]_1}{m_{ep}\Delta\sigma} = \left[1 + \frac{\beta \cdot A_1}{2.3}\right] \frac{T_{vp}}{T_{vc}} \qquad \text{129-II.3}$$

and for $t > t_c$

$$\frac{[\Delta\epsilon_v]_2}{m_{ep}\Delta\sigma} = 1 + \beta \log \left[1 + \frac{4.62}{\beta} (T_{vp} - T_{vc} \cdot A_2)\right] \qquad \text{130-II.3}$$

The auxiliary functions A_1 and A_2 given in Fig. 50-II.3 may be used to expedite the calculations.

3.15 Intergranular Viscosity in Nonsaturated Soils

When the material is semisaturated, the water from the saturated pores will flow rapidly to the empty voids. Therefore, the value of c_v/H^2 has no significance. In this particular case τ is independent of H, and the plasto-viscous phenomenon may be assumed to start for all practical purposes as soon as the load is applied at time t_B; therefore, approximately:

$$\delta = \delta_{ep} + C_t \log \frac{t}{t_B} \qquad \text{131-II.3}$$

Under field conditions, the value of δ_{ep} has effect as the load is applied in construction time t_c. Thereafter, the intergranular viscosity deformation is observed, and for field conditions

$$\delta = \delta_{ep} + C_t \log \frac{t}{t_c} \qquad 132\text{-II.3}$$

or in dimensionless form:

$$\frac{\Delta\epsilon_v}{m_{ep}\,\Delta\sigma} = 1 + \beta \log \frac{t}{t_c} \qquad 133\text{-II.3}$$

As usual $\beta = m_t/m_{ep}$. The values of m_t and m_{ep} are determined from consolidation curves obtained in the oedometer, as previously described.

3.16 The Use of the κ_v-Value in Soils Exhibiting Intergranular Viscosity

In section 3.8, it was found that the change in thickness of a confined stratum subjected to an average increment of stress $\Delta\sigma$ is:

$$\delta = (\nu_c M_{ep})\,(1 + \kappa_v)\,(2H)\,\Delta\sigma$$

in which $\nu_c M_{ep} = m_{ep}$, hence:

$$\frac{\Delta\epsilon_v}{m_{ep}\,\Delta\sigma} = 1 + \kappa_v \qquad 134\text{-II.3}$$

Assume now a nonsaturated soil exhibiting intergranular viscosity; β was found from tests equal to 0.6, and formula 133-II.3 applies in this case:

$$\kappa_v = \beta \log \frac{t}{t_c}$$

In the field t_c is the time of load application. If t_c is assumed to be one year, for a 10-year period:

$$\kappa_v = 0.6 \log \frac{10}{1} = 0.60$$

From this example, it is shown that κ_v is a function of β, the stress level σ and time:

$$\kappa_v = Z(\beta, \sigma, t) \qquad 135\text{-II.3}$$

Therefore, for each consolidation curve obtained from oedometer tests, for certain stress level σ_i and sustained stress increment $\Delta\sigma_i$, the value of κ_v may be found from the formula

$$\kappa_v = \frac{\delta_{vp}}{\delta_{ep}} \qquad 136\text{-II.3}$$

and the approximate change in stratum thickness with account for intergranular viscosity may be calculated.

The use of the κ_v-value concept is particularly useful in those cases of semipermeable, nonhomogeneous and nonsaturated soils, where the phenomenological law δ vs. t cannot be represented simply by mathematical functions as discussed in this chapter.

3.17 Parameter Determination: Fitting Methods

From the study of the stress-strain-time behavior of fine sediments given in Sections 3.14, 3.15 and 3.16, three typical cases may be recognized for which it is necessary to determine the parameters, namely:

Case I. When the theory of consolidation may be applied in conjunction with the intergranular plasto-viscosity phenomenological law. This condition requires an homogeneous material, as assumed by the theory of consolidation. Sections II.3.9-3.13.

Case II. When consolidation takes place rapidly because of the permeability characteristics. The elasto-plastic deformation takes place rapidly as the increment of stress is applied. The saturated conditions, however, retard the intergranular plasto-viscous phenomenon. Section II.3.13a.

Case III. When, for nonsaturated soils, the elasto-plastic compression takes place at the same rate at which the load is applied. Thereafter, a further vertical displacement is observed with time under sustained load.

From consolidation curves obtained in oedometer tests, these cases may be classified and the parameters determined accordingly.

Case I. The parameters m_t, β and c_v for calculating the volumetric strain in elasto-plastic materials exhibiting intergranular viscosity may be determined from oedometer tests with the following formula:

$$\Delta\epsilon_v = \left\{ m_{ep} F\left(\frac{c_v}{H^2} t\right) + m_t \log\left[1 + \frac{4.62}{\beta}\left(\frac{c_v}{H^2}\right)t\right]\right\} \Delta\sigma \qquad \text{108-II.3}$$

in which $m_t = \beta m_{ep}$.

The parameters are determined for each stress increment at which the material is progressively consolidated, and therefore, the values obtained represent increasing stress levels. A consolidation curve plotted in semilog scales shows a break close to values of $F(T_v) = 1$ (Fig. 53-II.3). Let us represent the coordinates at the break by δ_B, t_B. Shortly after time t_B the deformation law is logarithmic. Therefore, from the consolidation curve the value of C_t can be determined:

$$C_t = \frac{\delta_2 - \delta_1}{\log \dfrac{t_2}{t_1}}, \quad \text{and} \quad m_t = \frac{C_t}{(2H)\,\Delta\sigma} \qquad \text{137-II.3}$$

Furthermore, let us call:

$$\alpha = \frac{C_t}{\delta_B}, \quad \text{or} \quad \alpha = \frac{m_t}{m_{vB}}, \quad \text{and} \quad m_{vB} = \frac{(\Delta\epsilon_v)_B}{\Delta\sigma} \qquad \text{138-II.3}$$

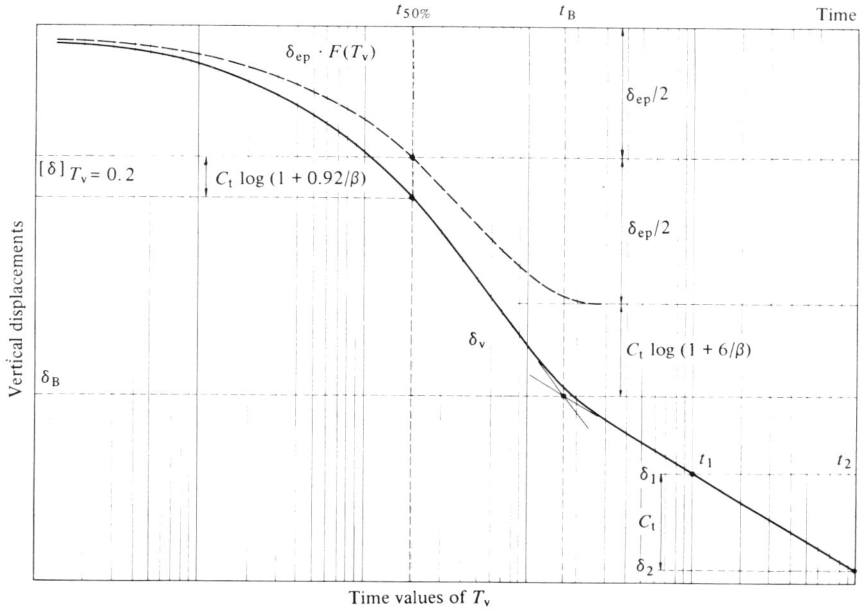

Fig. 53-II.3 Determination of consolidation parameters for Case I.

Therefore, from formula 108-II.3 and considering that:

$$F\left(\frac{c_v}{H^2}\, t\right) \doteq 1 \quad \text{when} \quad T_v = 1.3$$

then at the break, we may write the following:

$$\frac{(\Delta\epsilon)_B}{\Delta\sigma} = m_{ep} + m_t \log\left[1 + \frac{6}{\beta}\right] \qquad \text{139-II.3}$$

Dividing by m_t, we have

$$\frac{1}{\alpha} = \frac{1}{\beta} + \log\left(1 + \frac{6}{\beta}\right) \qquad \text{140-II.3}$$

The above expression may be used as an auxiliary equation to determine the unknown value of β. The value of $\alpha = C_t/\delta_B$ is readily determined from the consolidation curve. The auxiliary equation 140-II.3 may be found plotted in Fig. 54-II.3. Therefore, with the value of α we obtain β and thus $m_{ep} = m_t/\beta$ and $\delta_{ep} = C_t/\beta$.

The consolidation coefficient may be determined for $F(T_v) = 0.5$, hence for $T_v = 0.2$, equation 108-II.3 takes the following value:

$$[\delta]_{T_v=0.2} = \frac{1}{2}\,\delta_{ep} + C_t \log\left(1 + \frac{0.924}{\beta}\right) \qquad \text{141-II.3}$$

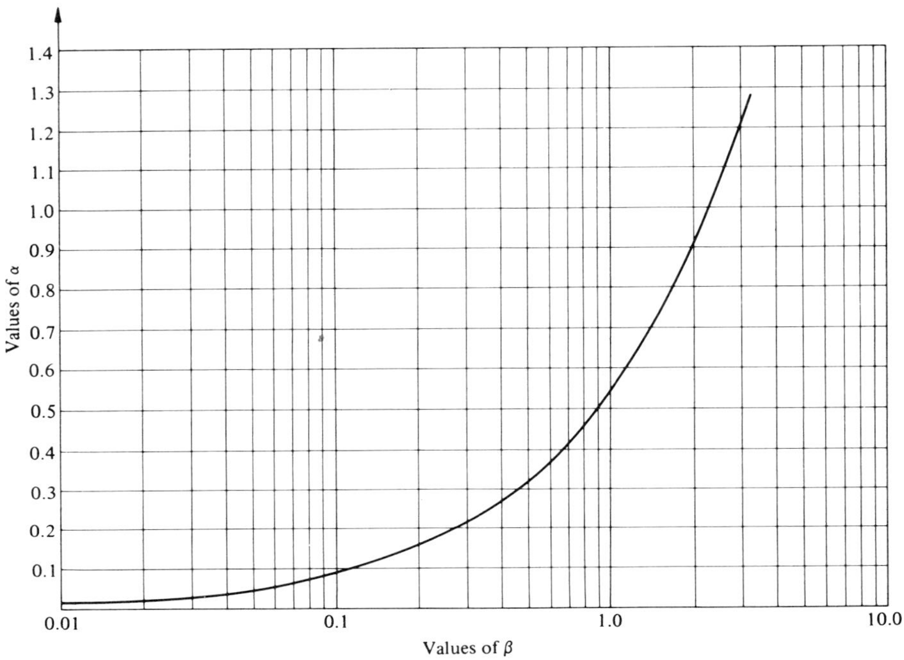

Fig. 54-II.3 Auxiliary function $1/\alpha = 1/\beta + \log(1 + 6/\beta)$.

The time t_{50} is found from the consolidation curve corresponding to δ_{50}, and thus we have the consolidation coefficient

$$c_v = \frac{0.2H^2}{t_{50}}$$

in which $2H$ is the thickness of the test specimen under stress level σ and applied stress increment $\Delta\sigma$.

Since the values of parameters m_t, β and c_v are functions of the stress level, these values are plotted against stress: $\sigma + \Delta\sigma/2$. An example will show the procedure, using consolidation curve C shown in Fig. 55-II.3, obtained in the laboratory for a silty clay of high compressibility, for stress level 0.8 kg/cm^2, under an applied stress increment $\Delta\sigma = 0.30$ kg/cm^2. At a stress level of 0.8 kg/cm^2, $2H = 1.675$ cm. From the laboratory curve $\delta_B = 0.0185$ cm, and $C_t = 0.005$ cm. Therefore:

$$\alpha = \frac{50}{185} = 0.27$$

Using the auxiliary equation (Fig. 54-II.3) a value of $\beta = 0.41$ is found; hence:

$$\delta_{ep} = \frac{C_t}{\beta} = \frac{0.005}{0.41} = 0.0122 \text{ cm}$$

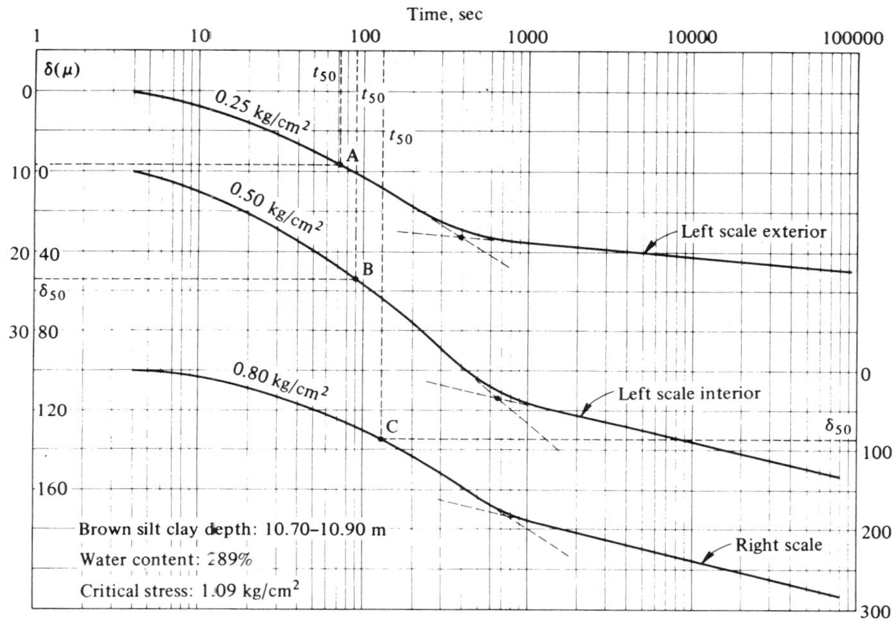

Fig. 55-II.3 Consolidation curves.

Using expression 141-II.3, the value δ_{50} is calculated:

$$\delta_{50} = \frac{1}{2}(0.0122) + 0.005 \log\left(1 + \frac{0.924}{0.41}\right) = 0.00866 \text{ cm}$$

From consolidation curve in Fig. 55-II.3, a value of $t_{50} = 130$ sec is obtained, from which we obtain

$$c_v = \frac{0.2(0.837)^2}{130} = 1.08 \times 10^{-3} \text{ cm}^2/\text{sec}$$

Therefore, the parameters for stress increment $\Delta\sigma = 0.3$ kg/cm^2 and the corresponding average stress level $\sigma + \Delta\sigma/2 = 0.95$ kg/cm^2 are

$$m_t = 0.00995 \text{ cm}^2/\text{kg}, \; m_{ep} = 0.0242 \text{ cm}^2/\text{kg}, \; \beta = 0.41, \; c_v = 1.08 \times 10^{-3} \text{ cm}^2/\text{sec}$$

they are plotted against the average stress level. The consolidation parameters obtained for a compressibility test and for each increment of stress up to reaching σ_b are shown in Fig. 56-II.3.

To illustrate the settlement calculation a simple example is given in Table 5-II.3. Assume the settlements at 1, 2, 3, 5, 10 and 20 years are required for an impervious soil stratum that falls in Case I. The stratum is 6 m thick and is drained at top and bottom. The average stress increment is $\Delta\sigma = 0.3$ kg/cm^2 and will be applied linearly over a period $t_c = 3$ yr. The average overburden stress in the stratum after

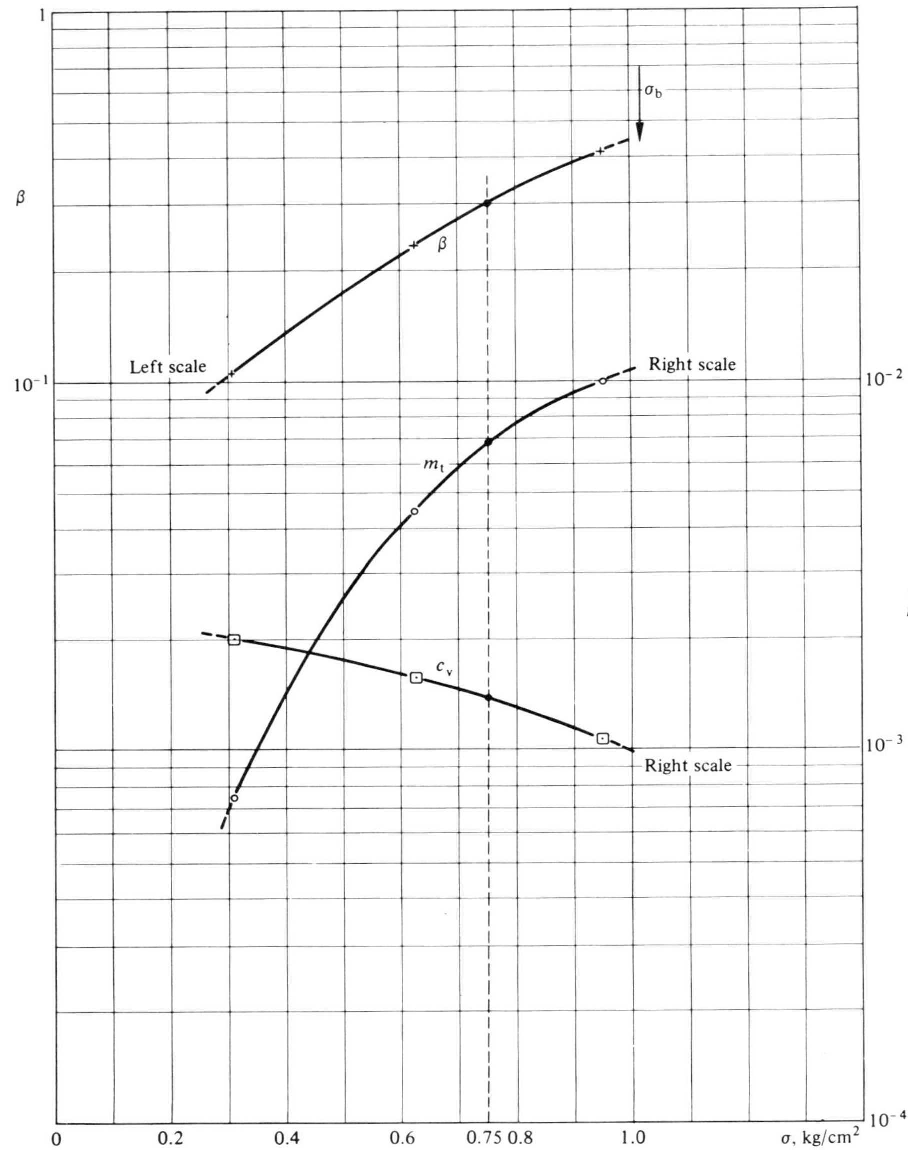

Fig. 56-II.3 Parameters of consolidation.

excavation is $\sigma = 0.6$ kg/cm^2, and hence the average stress level to obtain the parameter is $\sigma = 0.75$ kg/cm^2, all the stresses falling below the critical stress σ_b. Using parameters reported in Fig. 56-II.3, the following values are obtained: $m_t = 6.8 \times 10^{-3}$ cm^2/kg, $\beta = 0.3$, and $c_v = 1.4 \times 10^{-3}$ cm^2/sec (Table 5-II.3).

Case II. The parameters for these conditions are m_t, β, and c_{vp}. The consolida-

TABLE 5-II.3 Settlement example

INFORMATION

$$2H = 600 \text{ cm}, \quad t_c = 3 \text{ years}, \quad \sigma = 0.6 \text{ kg/cm}^2, \quad \Delta\sigma = 0.3 \text{ kg/cm}^2$$

$$\sigma + \tfrac{1}{2}\Delta\sigma = 0.75 \text{ kg/cm}^2$$

$$m_t = 0.0068 \text{ cm}^2/\text{kg}, \quad \beta = 0.3, \quad c_v = 0.0014 \text{ cm}^2/\text{sec}, \quad 1 \text{ year } 3.15 \times 10^7 \text{ sec}$$

Required: Settlement at 1, 2, 3, 5, 10, 20 years.
Using formulas 120 and 121-II.3

for $0 < t < t_c$

$$\delta_1 = (m_{ep}\Delta\sigma \cdot 2H) \cdot f_1, \quad f_1 = \left\{\phi(T_v) + \frac{\beta A_1}{2.3}\right\} \frac{T_v}{T_{vc}}$$

for $t > t_c$

$$\delta_2 = (m_{ep}\Delta\sigma \cdot 2H) \cdot f_2, \quad f_2 = F(T_v - T_{01}) + \beta \log \left\{1 + \frac{4.62}{\beta}(T_v - T_{vc} \cdot A_2)\right\}$$

SOLUTION

$$\delta_{ep} = m_{ep} \cdot \Delta\sigma \cdot 2H = 0.0227 \times 0.3 \times 600 = 4.08 \text{ cm}$$

$$\frac{c_v}{H^2} = 0.0014/9 \times 10^4 = 1.56 \times 10^{-8}$$

$$T_v = 1.56 \times 10^{-8} \times 3.15 \times 10^{-7} \times t \text{ years} = 0.495 \cdot t \text{ years}$$

$$m_{ep} = m_t/\beta = 0.0068/0.3 = 0.0227 \text{ cm}^2/\text{kg}$$

$T_{vc} = 3 \times 0.49 = 1.47, \quad T_{vc}/\beta = 4.90$

From Fig. 40-II.3, $T_{01} = 0.93$

Fig. 50-II.3, $A_2 = 0.85, \quad T_{vc} \cdot A_2 = 1.25$

years t	T_v	$\phi(T_v)$	$\frac{T_v}{\beta}$	A_1	$\frac{\beta A_1}{2.3}$	$\frac{T_v}{T_{vc}}$	f_1	$T_v - T_{01}$	$F(T_v - T_{01})$	$T_v - T_{vc}A_2$	β log	f_2	δ_{cm}
1	0.49	0.52	3.33	0.83	0.11	0.33	0.21						0.86
2	0.98	0.69	6.66	0.88	0.11	0.67	0.54						2.20
3	1.47	0.78	10.0	0.915	0.12	1.00	0.90	0.54	0.78	0.22	0.11	0.89	3.6
5	2.45							1.52	0.98	1.20	0.37	1.35	5.5
10	4.91							3.97	0.99	3.66	0.48	1.47	6.0
20	9.81							8.88	1.00	8.56	0.64	1.64	6.7

tion curves obtained in oedometer tests are as shown in Fig. 57-II.3. To determine the value of m_t and τ_s, the last part of the consolidation curve is approximated to a linear logarithmic law and a straight line is traced, intercepting a horizontal line through the break shown in the consolidation curve, obtaining the value of τ_s. The straight line has the slope

$$C_t = \frac{\delta_2 - \delta_1}{\log \frac{t_2}{t_1}}$$

from which the value of m_t is calculated. The thickness of the specimen at the start of the stress increment $\Delta\sigma$ is $2H$. Moreover, the deformation δ_{ep} at the break is

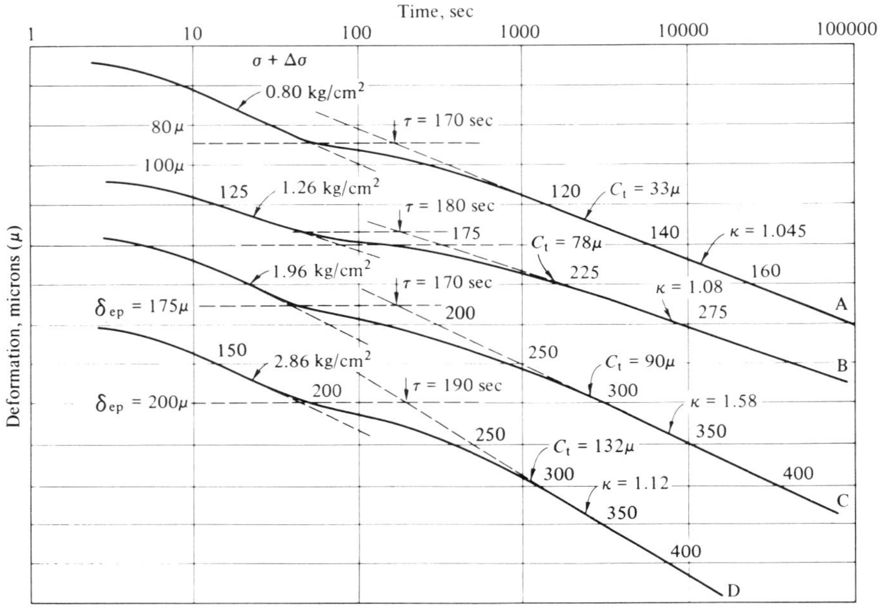

Fig. 57-II.3 Consolidation curves for very fine sediment with root holes.

assumed to represent the elasto-plastic deformation taking place rapidly after load application, because of the permeable characteristics of the fine sediment. Hence $m_{ep} = \delta_{ep}/2H\Delta\sigma$, from which $\beta = m_t/m_{ep}$ is determined.

The parameter τ_f under field conditions is calculated for the specific problem considering the drainage boundary conditions of the stratum in nature. For the same drainage boundary conditions as in the tests

$$\frac{\tau_f}{\tau_s} = \frac{(2H_f)^2}{(2H_s)^2}$$

When in the field the stratum has only one drainage boundary condition, then:

$$\frac{\tau_f}{\tau_s} = \frac{H_f^2}{(2H_s)^2}$$

The value of the apparent coefficient of consolidation for the intergranular viscosity phenomenon is

$$c_{vp} = \frac{\beta}{4.62} \cdot \frac{H_s^2}{\tau_s}$$

This is calculated from τ_s obtained in the laboratory curves for the thickness $2H_s$ of the test specimen at the stress level to which the consolidation curve is obtained.

With the value of m_t, β and c_{vp}, the compression of the strata may be calculated with formulas 129 and 130-II.3. The parameters obtained from the tests are plotted against $\sigma + \Delta\sigma/2$, that is, for the average stress level for which they were determined. A group of consolidation curves for a clayey organic silt with vertical

TABLE 6-II.3

SOIL: Organic silt with cavities
DEPTH: 12 m

Curve	σ, kg/cm^2	$\Delta\sigma$, kg/cm^2	$2H$, cm	$(2H)^2$, cm^2	$\delta_{ep} \times 10^{-4}$, cm	$C_t \times 10^{-4}$, cm	τ_s, sec	m_{ep}, cm^2/kg	m_t, cm^2/kg	β	c_{vp}, cm^2/sec
A	0.42	0.38	2.00	4.00	88	33	170	0.0116	0.00435	0.375	0.000480
B	0.80	0.46	1.99	3.98	150	78	180	0.0164	0.00853	0.52	0.000622
C	1.26	0.70	1.97	3.86	175	90	170	0.0111	0.00572	0.515	0.000633
D	1.96	0.90	1.90	3.76	200	132	190	0.0116	0.00770	0.663	0.000676

$$c_{vp} = \frac{\beta}{4.62} \cdot \frac{H_s^2}{\tau_s}, \quad m_{ep} = \frac{\delta_{ep}}{2H\Delta\sigma}, \quad m_t = \frac{C_t}{2H\Delta\sigma}$$

root holes taken at a depth of 12 m are shown in Fig. 57-II.3. The calculation of the parameters is shown in Table 6-II.3.

Case III. The parameter for this condition is rather simple to determine, as it is only necessary to obtain m_t and β from representative consolidation curves (Fig. 58-II.3). The value of m_t is determined as in Cases I and II. The value of m_{ep} corresponds to the immediate compression δ_{ep} in the test. The values of m_t and β are plotted against the average stress level for which they were determined in the oedometer. The calculations are similar to those shown in Table 6-II.3.

Moreover, it is important to state that for sensitive soils the parameters after the break in the compressibility curves, defined by the critical compressive stress

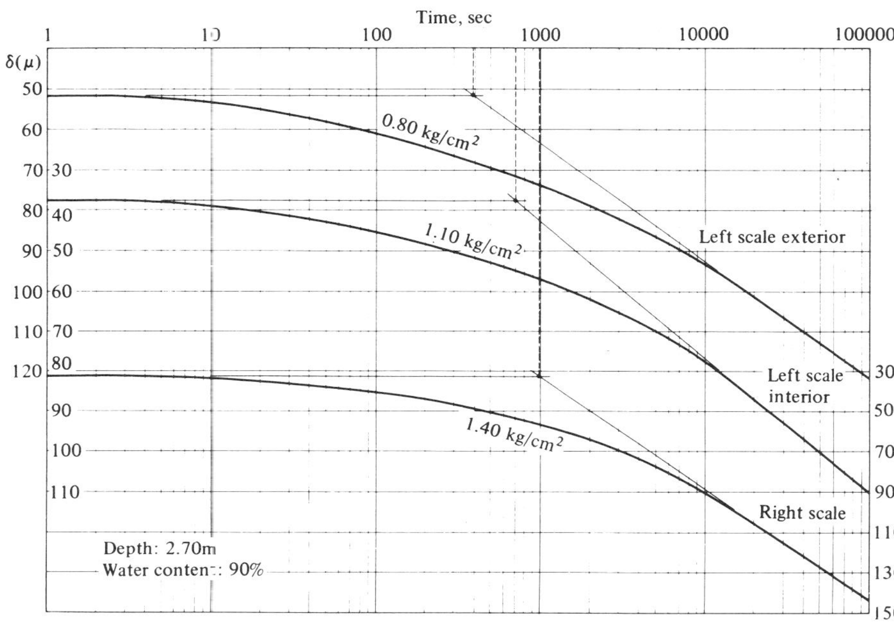

Fig. 58-II.3 Consolidation curves for nonsaturated silt with some clay, with cavities, olive gray.

σ_b, have no practical application in foundation design, since for increments of stress larger than $(\sigma_b - \sigma_0)$ the material shows large compressibility. Therefore the application of the parameters is valuable only for design in the recompression range, corresponding to the flat branch of the compressibility curve (Chapters VI to VII). Nevertheless, in special cases in which the increment of stress added to the overburden effective stress is greater than σ_b, the parameters may be investigated with a method similar to those described.

II.4 SHEAR STRENGTH

4.1 Basic Concepts

The shear strength phenomenological law was discovered by C. A. Coulomb (1773), and has been studied in the last decades by several investigators in different countries, much more than any other soil mechanics property. The following paragraphs contain a brief discussion of the different forms the shear strength takes when calculating the stability of foundations, and some remarks on the diagnosis of specific problems to facilitate the selection of the shear strength parameters.

Coulomb's law reads as follows:

$$s = c + \sigma_n \tan \phi \qquad \text{1-II.4}$$

in which

s = unit shear strength;
c and ϕ = parameters that represent the cohesion and the angle of internal friction of the material, respectively;
σ_n = normal stress on the potential surface of sliding.

The necessary unit force to mobilize the frictional shear strength is:

$$s_\phi = \sigma_n \tan \phi \qquad \text{2-II.4}$$

The value of $\tan \phi$ is the friction coefficient as given by Coulomb between two sliding surfaces at the instant that relative displacement occurs. At this instant the static friction between the grains is mobilized, after which the material enters into motion with a smaller frictional shear resistance. This frictional resistance is called dynamical or residual friction, and is a function of the differential velocity between the surfaces. In case of granular materials, the angle of internal friction is an important function of the dimensions and form of the grains, the skeleton structure arrangement, and the state of compaction of the material. The shear strength because of cohesion in the grains may be defined as that shear strength for zero normal stress, hence:

$$s_c = c \qquad \text{3-II.4}$$

The value of c is a function of the type of cementing materials or clay minerals, and of the intergranular stress at which cementation took place, that is to say, of the loading history of the sediment.

The bond between grains in the soil skeleton structure is given by cementing substances like carbonates or silicates, and by clay minerals. The shear strength in impervious soils is a complicated physical property that, however, may be easily applied when its limitations and mechanical properties are well understood.

4.2 Coulomb-Terzaghi's Equation

Coulomb's phenomenological law may be investigated for nonsaturated granular materials by means of an instrument called the direct shear apparatus, well known in elementary soil mechanics. A detailed description of the technique required to perform this test can be found in manuals and books dealing with laboratory analysis. In practice, one may find different methods of operation, depending on certain modalities and patents of manufacturer's instruments or those of the home-made variety.

Independently of the refinements offered by the different designers and builders of this instrument, the basic principles are the same, and may be represented schematically in Fig. 1(a)-II.4. The instrument is formed of a box cut in two into which the test sample is placed. The lower box is fixed to a firm base in which a porous stone is located to permit the drainage of the soil specimen placed in that box. To the upper part of the box and frame is attached a cable which runs over a pulley. To the end of the cable a load is applied to exert a horizontal force F on the soil specimen. The apparatus is provided with two extensometers, one horizontal to measure the shear distortion of the test specimen, and the other in the vertical position to measure the contractions or expansions taking place during application of the normal stresses and during shear stress application, respectively. With such an instrument one may prove the shear strength phenomenological law for dry and semisaturated soils. The experiment is performed as follows: the soil is placed carefully in the box with the desired density in case of cohesionless materials like sand and remoulded silty sands. In case of undisturbed cohesive materials, it is required to obtain a perfect fit in the lower and upper boxes, respectively. The index properties of the test specimen are determined, such as the void ratio, density and degree of saturation. Thereafter the normal stress σ_n is applied and transmitted to the test specimen by means of the rigid plate. The deformation produced by the vertical stress is measured with the extensometer until the material consolidates and reaches a sufficiently stable condition. Thereafter the horizontal load is applied slowly by adding increments of weight to the end of the cable until failure of the test specimen along plane $(a - a)$ is obtained. A special mechanism, however, may be attached to apply a continuous load at the desired rate. The shear strength s_1 for ultimate shear force F_1 and area A_1 under normal stress σ_{n1} is $s_1 = F_1/A_1$. The values of s_1 and σ_{n1} are represented by a point in the coordinates system s, σ_n (Fig. 1-II.4). Another identical test specimen is installed, and the operation is repeated with a larger normal stress σ_{n2}, which will give upon failure a shear strength s_2, also larger. This new set of values gives another point in the plot s vs. σ_n. The procedure is repeated with test specimens of

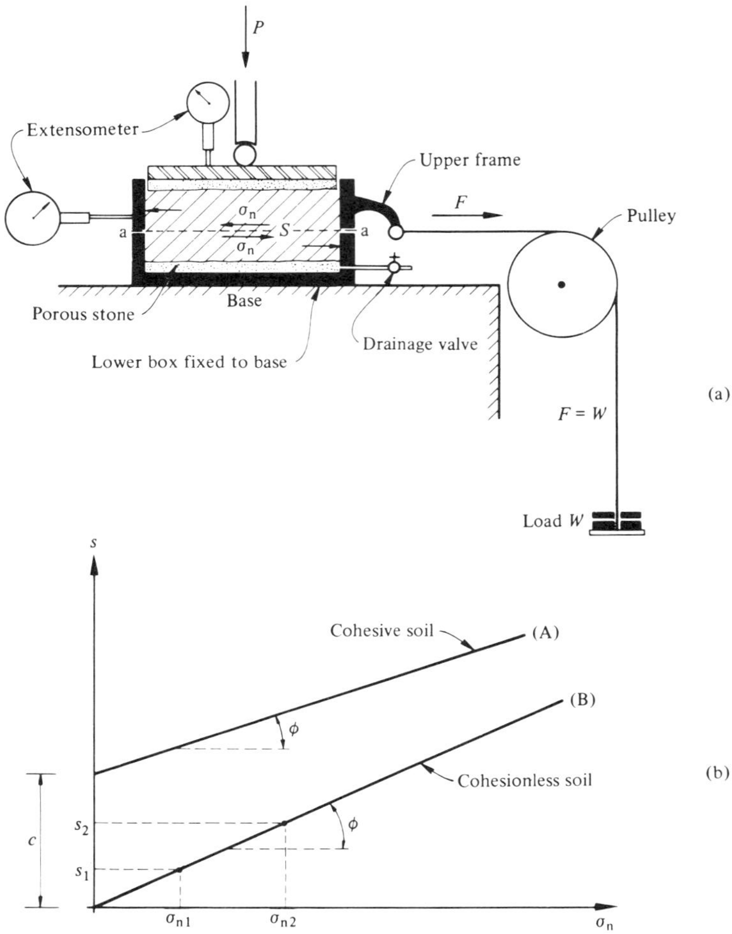

Fig. 1-II.4 Direct shear testing instrument.

equal index properties, obtaining a set of values s_i and σ_{ni} for each equivalent specimen tested. The plotted points are found located approximately in a straight line (Fig. 1-II.4). In the case of materials without cohesion, the equation of the straight line is:

$$s_\phi = \sigma_n \tan \phi \qquad \text{4-II.4}$$

and in the case of cohesive materials:

$$s_{c\phi} = c + \sigma_n \tan \phi \qquad \text{5-II.4}$$

Both expressions represent Coulomb's phenomenological law of shear strength, in which ϕ is the angle of internal friction and c is the cohesion or shear strength for zero normal stress.

Evidently, there are other very important factors to consider during the test which will be discussed in further articles. The most important factor is the change in volume suffered by the test specimen during the test, this is measured with the vertical extensometer. The value ϕ_d represents the real angle of internal friction between the grains, when σ_n is the intergranular stress. It may be noticed, however, that for every set of values s and σ_n a different value of the void ratio is obtained, decreasing as σ_n increases.

When the material is saturated in impervious soils, the applied normal stress is not transmitted immediately to the soil skeleton structure, since the water in the pores takes part of the applied load according to the fundamental consolidation theory; water pore pressures develop when the shear strength is reached in a short time. Therefore, it will be necessary to perform the test at a sufficiently low rate of load application to keep within very small values of hydraulic excess water pressures, obtaining, at any instant during the test, effective intergranular stresses. In further discussions in this article, the intergranular stress will be represented by σ, the water pressure by u, and total pressures by p. The phenomenon of water pore pressures may be studied by means of Terzaghi's theory of consolidation that considers a material as formed of two phases, the solid and the liquid phase. The theorem established for saturated materials reads as follows: the total applied pressure in a saturated soil is at any time equal to the intergranular effective stress added to the water pore pressure:

$$p = \sigma + u \qquad \text{6-II.4}$$

Therefore, if the total pressure remains constant during the time of the test

$$\frac{\partial \sigma}{\partial t} = -\frac{\partial u}{\partial t} \qquad \text{7-II.4}$$

The above expression implies that during the test any change in the effective stress produces an equal and contrary change in the hydraulic pressure. During a test in an impervious saturated material, the effective stresses change continuously from the instant the load is applied. To understand Coulomb's law for saturated materials in which failure can take place during certain hydrostatic excess pressures in the material, one has to correct Coulomb's expression, introducing Terzaghi's fundamental theorem for effective stresses. Therefore, we obtain the Coulomb-Terzaghi equation:

$$s_d = c_d + (p - u) \tan \phi_d \qquad \text{8-II.4}$$

The shear parameters c_d and ϕ_d for a soil may be determined if one knows the value of u at the precise instant of the shear failure, or perform the test with such rate of loading that during the test and at failure $u = 0$. Furthermore, from the index properties of the soils it is concluded that the parameters c_d and ϕ_d defining the shear strength are specific mechanical properties for each type of material, depending on the following factors:

Cohesion:

1. Type of cementing agent: carbonates, silicates, clay minerals.
2. Load history of the sediment, that is, water content, effective stresses at which the material was previously consolidated.
3. Skeleton structure.

Angle of internal friction:

1. Grain size distribution.
2. Shape of grains.
3. Soil density or compaction.
4. Mineralogical composition, mainly: in the case of high stresses, the strength of the grains.

The determination of the shear strength of a sediment is difficult, and the direct shear apparatus cannot be used to best advantage in determining all types of shear strengths representing the soil. The disadvantages of this instrument against the advantages of its simplicity are the following:

1. During distortion of the test specimen confined in the boxes, the state of stresses generated is not uniform along the plane of shear. Moreover, the nominal shear area is reduced. This latter effect, however, may be corrected.
2. The results of slow tests performed in dry or semisaturated soils are approximately valid.
3. In impervious materials, to determine only the value of c, the test has to be performed rapidly in order to avoid consolidation; therefore, the results are affected by the viscous properties of the clay materials.
4. The test cannot be performed slowly and at constant volume, since the water under pressure will escape at the interface of the two boxes, that is to say, at the potential surface of sliding, plane $(a - a)$ in Fig. 1 (a)-II.4.

These difficulties and others may be avoided using the well known compression triaxial chamber (Fig. 25-II.3), by means of which more precise results are obtained, though at the expense of more testing time. This instrument was already described in Section II.3.

4.3 Coulomb-Mohr's Failure Concept

The state of principal stresses at a point ($\sigma_1 > \sigma_2 > \sigma_3$) may be represented by Mohr's circle (Fig. 2-II.4), where σ_1 represents the major principal stress and σ_3 the minor principal stress. The corresponding normal effective stresses in the three coordinate directions are $\sigma_z > \sigma_y > \sigma_x$. For plane stress we may write:

$$\sigma_x = \frac{\sigma_1 + \sigma_3}{2} - \frac{\sigma_1 - \sigma_3}{2} \cos(180^\circ - 2\alpha) \qquad \text{9-II.4}$$

$$\sigma_z = \frac{\sigma_1 + \sigma_3}{2} + \frac{\sigma_1 - \sigma_3}{2} \cos 2\alpha \qquad \text{10-II.4}$$

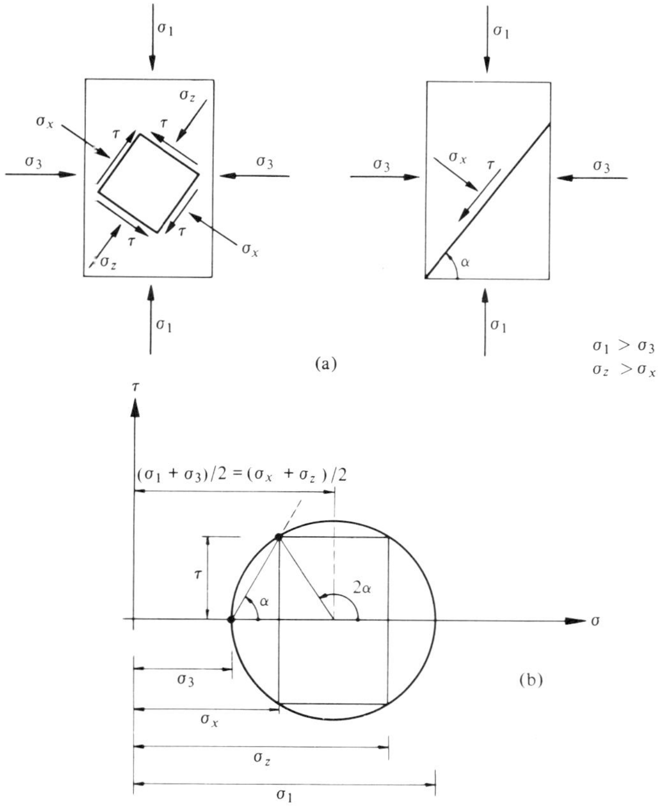

Fig. 2-II.4 Mohr's stress circle.

$$\tau = \frac{\sigma_1 - \sigma_3}{2} \sin 2\alpha \qquad \text{11-II.4}$$

The value $(\sigma_1 + \sigma_3)/2 = (\sigma_z + \sigma_x)/2$ represents the position of the center of the stress circle on the σ-axis. The radius of the circle $(\sigma_1 - \sigma_3)/2$, is the measure of the shear stress. The angle α represents the inclination of the plane where the stresses τ and σ_n are applied. Furthermore, the principal stresses may be also expressed as a function of σ_z, σ_x and τ:

$$\sigma_1 \text{ and } \sigma_3 = \frac{1}{2}(\sigma_z + \sigma_x) \pm \sqrt{\frac{(\sigma_z - \sigma_x)^2}{4} + \tau^2} \qquad \text{12-II.4}$$

It is sometimes desirable to find the maximum and minimum principal stresses σ_1 and σ_3 from known values of σ_x, σ_y and τ, respectively.

The shear strength phenomenological law is presented approximately for all practical purposes, and for a certain range of stress levels, by a straight line in coordinates s, σ_n (Figs. 3, 4-II.4). Now assume that the test specimen is subjected to

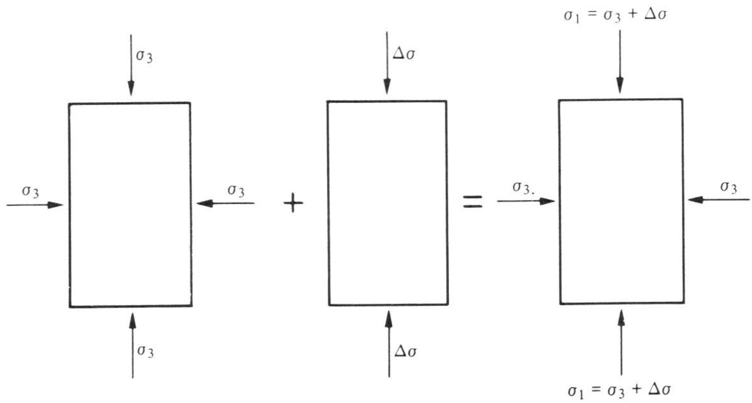

Fig. 3-II.4 State of applied principal stresses.

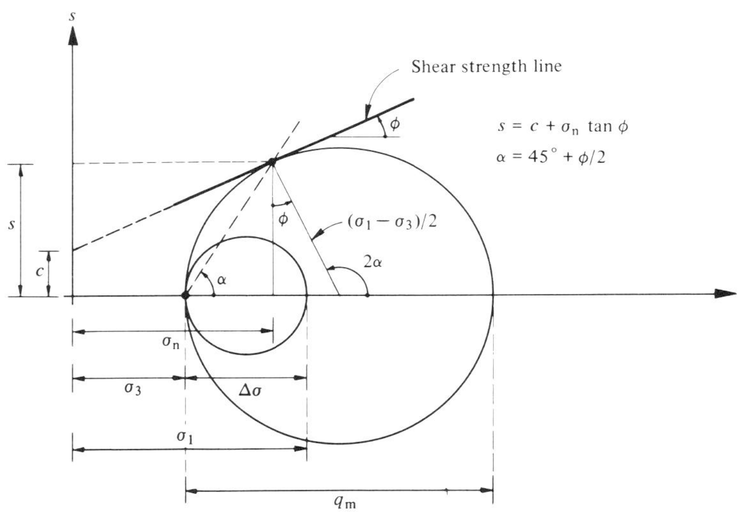

Fig. 4-II.4 Coulomb-Mohr failure concept.

a state of principal stresses equal to σ_3; the graphical representation will be a point with abscissa σ_3 in the coordinate system. Now assume that the vertical stress is increased in such a way as to obtain the principal stress $\sigma_3 + \Delta\sigma = \sigma_1$. The stresses τ and σ_n at an inclined plane α may be obtained from the stress circle.

The vertical stress increment $\Delta\sigma$ inducing the shear stress in the soil is known as the deviatory stress. This stress is increased to its maximum until the soil is brought to failure, when $\Delta\sigma = q_m$. At this instant the stress circle will be tangent to the shear strength law and failure will take place theoretically on a plane with inclination

$$\alpha = 45^\circ + \tfrac{1}{2}\phi \qquad \text{13-II.4}$$

Therefore, the stresses take the following values on the potential surface of sliding:

$$\sigma_n = \frac{\sigma_1 + \sigma_3}{2} - \frac{\sigma_1 - \sigma_3}{2} \sin\phi \qquad \text{14-II.4}$$

$$\tau = \frac{\sigma_1 - \sigma_3}{2} \cos\phi \qquad \text{15-II.4}$$

and calling $\sigma_3 = \sigma_c$ the confining effective stress, the major principal stress $\sigma_1 = \sigma_c + q_m$ and

$$\sigma_n = \sigma_c + \tfrac{1}{2} q_m (1 - \sin\phi) \qquad \text{16-II.4}$$

$$s = \tfrac{1}{2} q_m \cos\phi \qquad \text{17-II.4}$$

The shear strength at different states of stresses and volumetric conditions may be investigated using the compression triaxial chamber, shown in Fig. 25-II.3. Therefore, from Coulomb-Terzaghi's equation 8-II.4, the shear parameters may be determined for the following conditions:

4.3a The drained shear strength. The test specimen is placed in the triaxial compression chamber eliminating the effects of capillary forces. With the drainage system open, valve B, a confinement effective stress σ_{ci} is applied, allowing sufficient time to permit consolidation under this stress so that $u = 0$. Thereafter, a deviatory stress q_{mi} is applied, slowly in order to maintain $u = 0$, until failure. Under these conditions, one obtains a maximum stress circle where the principal stresses defining the failure are

$$\left.\begin{aligned} \sigma_3 &= \sigma_c \\ \sigma_1 &= \sigma_c + q_m \end{aligned}\right\} \qquad \text{18-II.4}$$

The values are plotted as Mohr's circles with center $\sigma_c + q_m/2$ and radius $q_m/2$ (Fig. 5-II.4). The process is repeated with other test specimens with equal index properties and an ultimate stress circle is obtained for each one of them. The envelope to all the stress circles represents the shear strength equation. Therefore, the shear parameters c_d and ϕ_d may be defined as those obtained in a slow and drained test that demands complete consolidation under the applied stresses. Under this condition, the shear strength equation is:

$$s_d = c_d + \sigma_n \tan\phi_d \qquad \text{19-II.4}$$

4.3b The consolidated-undrained shear strength. To obtain the shear strength parameters, a test is performed applying first a confining stress σ_c to the soil specimen and permitting the material to consolidate until $u \doteq 0$. Thereafter, the drained system is closed and a deviatory stress q_m is applied until failure. During the ap-

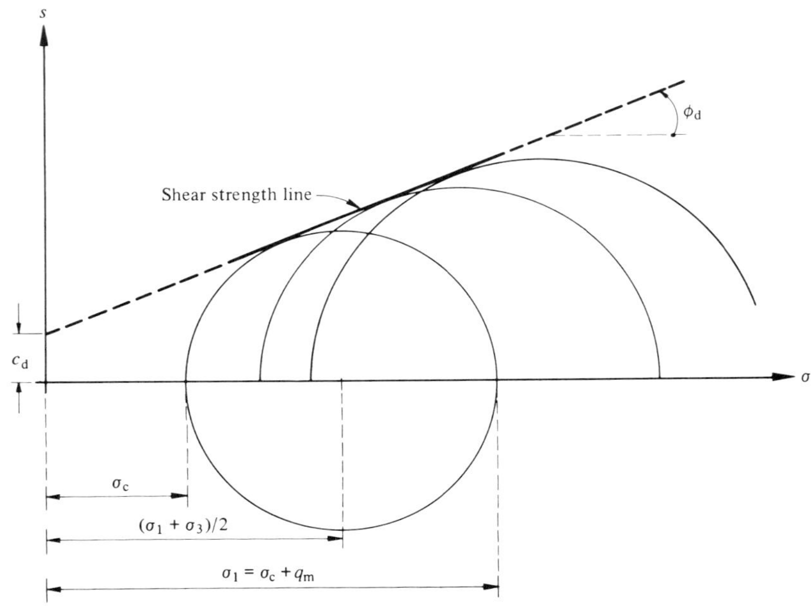

Fig. 5-II.4 Drained shear strength in cohesive soils, $u = 0$.

plication of the value of q_m, the test is made under constant volume. A hydraulic pressure u is generated in the test specimen, that will be proportional to the applied deviatory stress. Therefore, we can write: $u = a \cdot q_m$. In this case, the total applied stresses for a specific state of stress are

$$\sigma_3 = \sigma_c$$

$$\sigma_1 = \sigma_c + q_m$$

The effective stresses with which the test specimen failed are (Fig. 6-II.4)

$$\sigma_3 = \sigma_c - u$$

$$\sigma_1 = \sigma_c + q_m - u$$

From the test results, two stress circles for maximum stresses are obtained, one representing effective stresses when the u-values are known, and the other one representing the total applied stresses. Each one of the maximum stress circles so obtained will have a different equation for the shear strength. If the tests are performed at constant volume and slow application of q_m, in such a way as to eliminate totally the viscous intergranular phenomenon, then the shear strength ordinate at the origin of the applied stresses will have approximately the same value. Therefore, the shear strength law for total applied stresses in a consolidated soil failing at constant volume is expressed by:

$$s_{cu} = c_{cu} + p_n \tan \phi_{cu} \qquad \text{20-II.4}$$

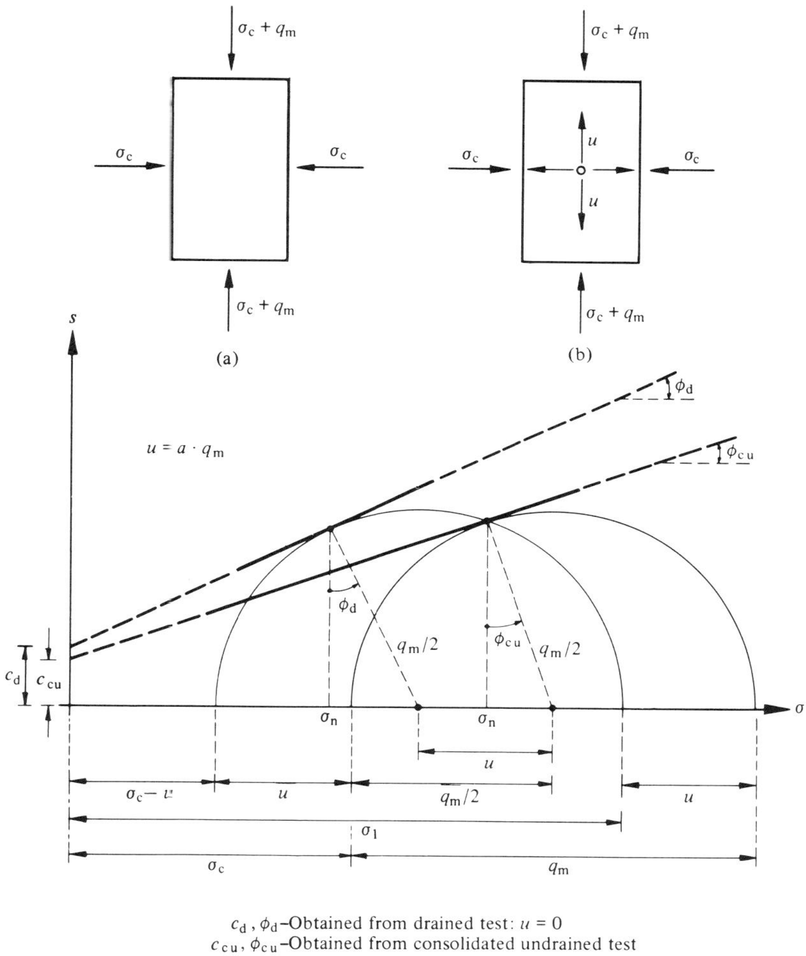

Fig. 6-II.4 Consolidated-undrained shear strength in cohesive soils.

The shear parameters c_{cu} and ϕ_{cu} representing the cohesion and the angle of internal friction are apparent shear strength parameters satisfying this type of test, and correspondingly are valid for shear failure under the same conditions in the field.

4.3c The undrained shear strength. By means of the triaxial chamber one may perform a test under constant volume from the very beginning of the applied stresses to be investigated. The saturated test specimen is introduced into the chamber and the surface capillary forces are eliminated, thus reducing to zero any initial confining stress produced by water surface tension. Thereafter, the drain system is closed and a confining pressure is applied. Under this condition, the applied pressure is taken completely by the water since there is no change in volume. Therefore, no consolidation takes place in the test specimen. The applied stress σ_c is not transmitted as effective pressure. The effect will be that of a hydrostatic pore pressure $u = \sigma_c$. Upon application of the deviatory stress q_m, the vertical principal stress at failure of the test specimen is $\sigma_1 = u + q_m$. The application of q_m, however, generates an additional pore pressure that may be written as $\Delta u = a \cdot q_m$ (Fig. 7-II.4). The test specimen fails with principal effective stresses independent of the initial stress; $\sigma_1 = (1 - a) q_m$ and $\sigma_3 = -aq_m$. Therefore, the shear strength of the material is in this case independent of the applied stresses and may be written as follows:

$$s_u = \tfrac{1}{2} q_m \quad \text{or} \quad c_u = \tfrac{1}{2} q_m \tag{21-II.4}$$

If in nature the material is found consolidated under the stresses σ_z and σ_h, then:

$$\sigma_c = \tfrac{1}{3}(\sigma_z + 2\sigma_h)$$

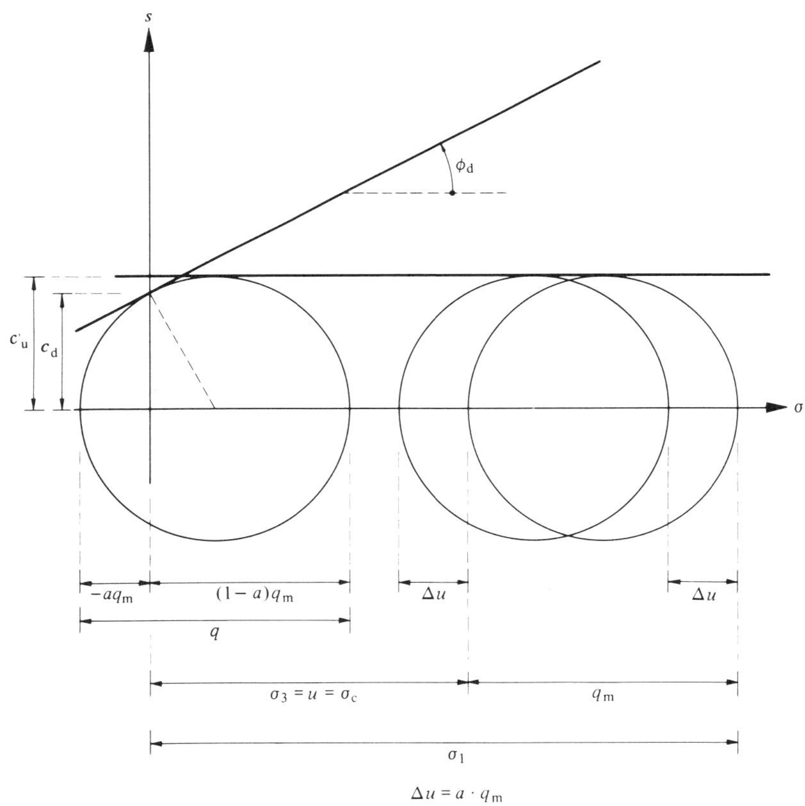

Fig. 7-II.4 Undrained shear strength in cohesive soils.

calling $K_0 = \sigma_h/\sigma_z$, we can write:

$$\sigma_c = \tfrac{1}{3}(1 + 2K_0)\,\sigma_z \qquad \text{22-II.4}$$

For each value of σ_c in nature a value of c_u may be obtained. The mechanics of the acquisition of cohesion in clay sediments will be discussed in further paragraphs.

4.4 Determination of the Average Shear Strength Parameters c and ϕ

Using the shear strength equation $s = c + \sigma_n \tan\phi$ and substituting into expressions 14 and 15-II.4 for the principal stresses at the instant of failure, we obtain

$$\frac{1}{2}(\sigma_1 - \sigma_3)\cos\phi = c + \left[\frac{\sigma_1 + \sigma_3}{2} - \frac{\sigma_1 - \sigma_3}{2}\sin\phi\right]\tan\phi$$

After ordering terms and algebraic operations we have

$$\tfrac{1}{2}(\sigma_1 - \sigma_3) = c \cdot \cos\phi + \tfrac{1}{2}(\sigma_1 + \sigma_3)\sin\phi \qquad \text{23-II.4}$$

Let us define

$$c' = c \cdot \cos\phi$$

$$\sin\phi = \tan\alpha$$

A straight line equation may be obtained in a coordinate system, plotting as ordinates $(1/2)(\sigma_1 - \sigma_3)$ and as abscissas $(1/2)(\sigma_1 + \sigma_3)$ (Fig. 8-II.4). The ordinate at the origin is c' and the slope of the line $\tan\alpha$. Therefore

$$\tfrac{1}{2}(\sigma_1 - \sigma_3) = c' + \tfrac{1}{2}(\sigma_1 + \sigma_3)\tan\alpha \qquad \text{24-II.4}$$

The above equation can be made to represent the average results of tests in a material with the same index properties. Thus, one can obtain the average angle of internal friction representative of the tests:

$$\phi = \sin^{-1}(\tan\alpha)$$

and the cohesion:

$$c = \frac{c'}{\cos\phi}$$

Notice from the geometry of Fig. 8(b)-II.4, where a maximum stress circle is drawn tangent to the shear strength line, that the equation of the straight line given by formula 24-II.4 may be also obtained.

4.5 Coulomb-Mohr's Plasticity Condition

At the instant the shear strength is totally mobilized, and represented by the maximum stress circles tangent to the Coulomb shear strength curve, the Coulomb-Mohr condition for plastic behavior may be obtained. The circles are characterized by their similarity on the approximate assumption that the shear strength law is a

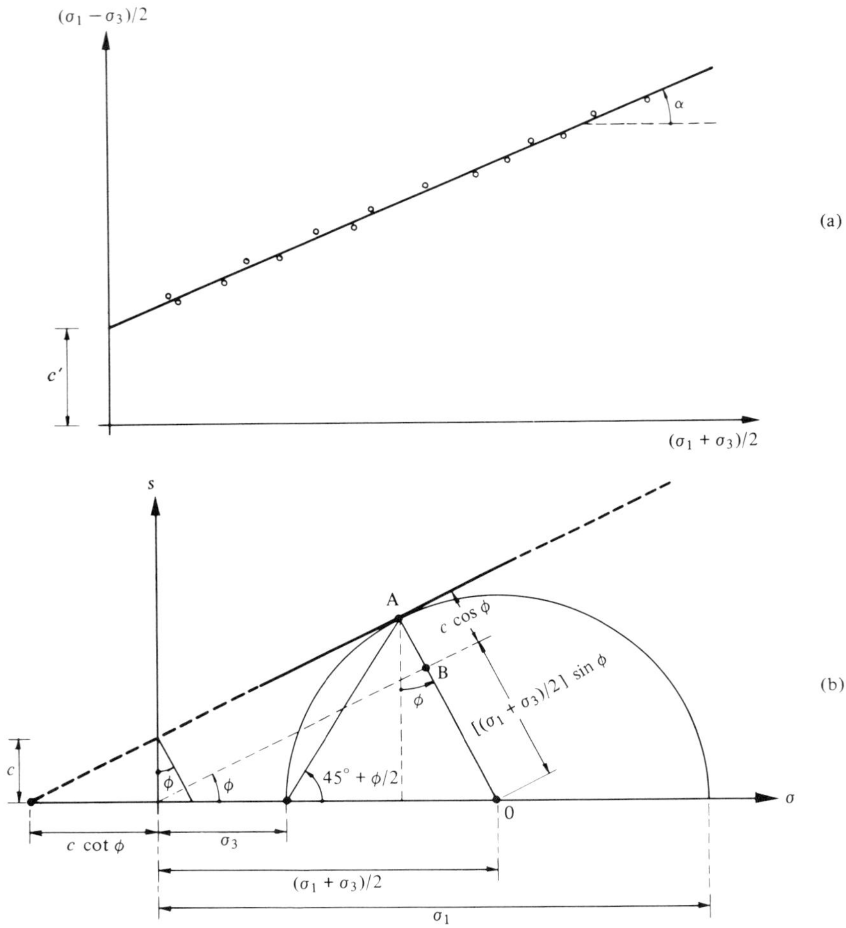

Fig. 8-II.4 Determination of average shear strength parameters.

straight line in a certain stresses level range. Therefore, we may establish the condition

$$\sin\phi = \frac{\frac{1}{2}(\sigma_1 - \sigma_3)}{c\cot\phi + \frac{1}{2}(\sigma_1 + \sigma_3)} \tag{25-II.4}$$

After transformations we obtain

$$\sigma_1 = \frac{2 \cdot \cos\phi}{1 - \sin\phi} \cdot c + \sigma_3 \frac{1 + \sin\phi}{1 - \sin\phi} \tag{26-II.4}$$

Using Terzaghi's notation, N_ϕ is defined as the flow number

$$N_\phi = \frac{1 + \sin\phi}{1 - \sin\phi}, \quad \text{or} \quad N_\phi = \tan^2\left(45° + \frac{\phi}{2}\right)$$

Then formula 26-II.4 reads as follows:

$$\sigma_1 = 2c\sqrt{N_\phi} + \sigma_3 N_\phi \qquad \text{27-II.4}$$

The above equation expresses the tangency of maximum Mohr's stress circles with Coulomb's shear strength law. The values σ_1 and σ_3 are the major and minor principal stresses, respectively, necessary to fulfill this condition for plane strain or plane stress problems.

Notice that in a material without cohesion, one obtains

$$\frac{\sigma_1}{\sigma_3} = N_\phi \qquad \text{28-II.4}$$

Furthermore, in an undrained test or constant volume test, one can assume that the apparent angle of internal friction is equal to zero. Therefore

$$\sigma_1 = 2c + \sigma_3 \qquad \text{29-II.4}$$

For any other intermediate condition, the apparent values of the shear parameters corresponding to c and ϕ may be substituted into formula 27-II.4 to obtain the failure condition or incipient state of plastic flow.

4.6 Rheological Considerations

It is important that the foundation engineer visualize the qualitative rheological properties of the soil to achieve a better understanding of the mechanical behavior of natural sediments in nature during shear failure, and from this knowledge, select properly the shear strength applying to the specific problem he has under consideration.

The rheological concepts in case of strain-stress-time behavior have been amply discussed in Section II.3. The conclusion was that the soil is an elasto-plastic-viscous material, and for this reason it is necessary to consider the soil to be formed of elements representing these three different mechanical properties. Furthermore, the soil is a two-phase material containing a solid and a liquid phase. The solid is formed by two essentially different types of elements: cohesionless materials giving shear strength by friction, and cohesive materials providing bond between the grains forming the skeleton structure of the soil. The cohesion in case of clay minerals shows a plasto-viscous shear strength behavior important to consider in foundation engineering. One has to stress the fact that static stability is only obtained because of the response of the elastic elements, which in a chain-like arrangement provide the soil grains with firm contacts. Upon partial breakage of the firm contacts, the material enters into a plasto-viscous state behavior. Theoretically, the shear strength behavior working hypotheses may be established as follows:

a) The elastic elements may be assumed to have a fragile failure, reached independently of the rate at which the effective stress is applied. Therefore, the true strength of these elements may be represented by $c_e = c_0 + s_e$ in which s_e is a func-

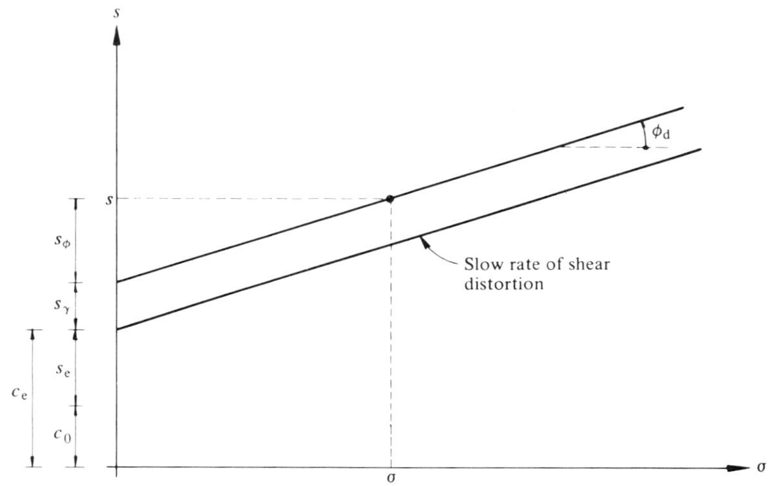

Fig. 9-II.4 Total shear strength for slow rate of shear distortion.

tion of the number of firm contacts between the grains, the state of compaction, and the cementing agent contained in the soil sediment.

b) The viscous elements will produce a virtual resistance that is a function of the rate at which the failure stress level is reached. Therefore, this property has the tendency to overshadow the shear strength of the material. The following statement may be visualized assuming linear viscosity during failure, applying Newton's law:

$$s_\gamma = \phi_f^{-1} \cdot \dot{\gamma} \qquad \text{30-II.4}$$

in which $\dot{\gamma}$ is the rate of shear deformation and ϕ_f is a coefficient representing the fluidity of the material.

c) The plastic phenomenon implies that the frictional shear strength due to relative displacement of soil grains will satisfy the Coulomb-Terzaghi equation

$$s_\phi = (p - u) \tan \phi_d = \sigma \tan \phi_d$$

Therefore, the total shear strength equation from the rheological point of view (Fig. 9-II.4) may be written as follows:

$$s = c_0 + s_e(\sigma_c) + s_\gamma(\phi_f, \dot{\gamma}) + s_\phi(p, u, \phi_d) \qquad \text{31-II.4}$$

The above-mentioned formula includes all the effects that may be expected to affect shear strength, for the case of a soil possessing cohesion and frictional resistance, and permitting the study of different types of shear strengths. In materials without cohesion, that is to say, without cementing material (Fig. 10-II.4), we have

$$s = s_\phi(\sigma, \phi_d) \qquad \text{32-II.4}$$

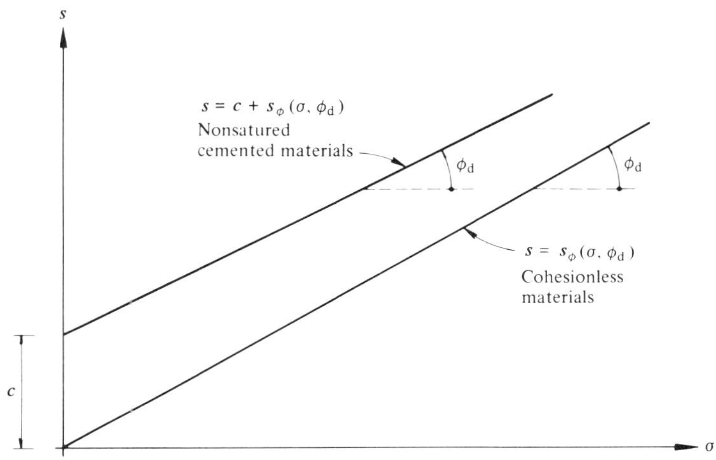

Fig. 10-II.4 Shear strength as determined from rheological considerations.

from which one can conclude that the apparent cohesion may be expressed by:

$$c_d = c_0 + s_e(\sigma_c) + s_\gamma(\phi_f, \dot{\gamma}) \qquad \text{33-II.4}$$

in which

c_0 = the true initial cohesion of the sediment;
s_e = the shear strength due to compaction of the material cementing agents, and is a function of its void ratio, water content and load history;
s_γ = the shear strength obtained due to the rate of angular distortion at which the material is stressed during failure. For a very slow rate of strain $s_\gamma = 0$.

For nonsaturated materials like sands and silts, with cementing agents like carbonates or silicates, one may write

$$s_d = c_d + s_\phi(\sigma, \phi_d) \qquad \text{34-II.4}$$

In materials for which the cohesion is given by clay minerals, the shear strength is represented by 31-II.4. In such cases, it may be observed that the viscosity plays a very important part. The terms s_γ and s_ϕ are affected by the loading rate at which the test is made. Consequently, the total shear strength of the material is a function of the rate of shear distortion during the test and, under field conditions, of the rate of load application.

From the above considerations, the different forms taken by the rheological shear strength equation may be studied in accordance with the application of stresses originating the failure:

a) Very slow rate of shear distortion and drained (Fig. 11-II.4):

$$s_1 = c_0 + s_e(\sigma_c) + s_\phi(\sigma, \phi_d) \qquad \text{35-II.4}$$

b) Rapid rate of shear distortion and constant volume $(p - u) = 0$ (Fig. 11-II.4):

$$s_{ro} = c_o + s_e(\sigma_c) + s_\gamma(\phi, \dot{\gamma}) \qquad 36\text{-II.4}$$

c) Very slow rate of shear distortion and constant volume $(p - u) = 0$ (Fig. 11-II.4):

$$s_{lo} = c_o + s_e(\sigma_c) \qquad 37\text{-II.4}$$

d) For an intermediate application of the rate of loading and volume change, expression 31-II.4 is valid (Fig. 9-II.4).

From the above discussion, one may judge that from the rheological point of view, different values of the shear strength may be expected, depending fundamentally on the following factors:

1. The rate of shear distortion upon reaching failure.
   a) By plasto-viscous effects.
   b) By hydrodynamic effects.
2. Loading history.
   a) Preconsolidation.
   b) Mineralogical cementing agents.

The first concept may be investigated by performing very slow tests in such a way as to insure that $s_\gamma \doteq 0$ and $u \doteq 0$. The second rheological concept concerning cohesion may be studied with the considerations described in the following.

Take a clayey material and add sufficient water to bring it to the liquid limit; the corresponding void ratio will be e_{LL}, obtaining under this condition a very

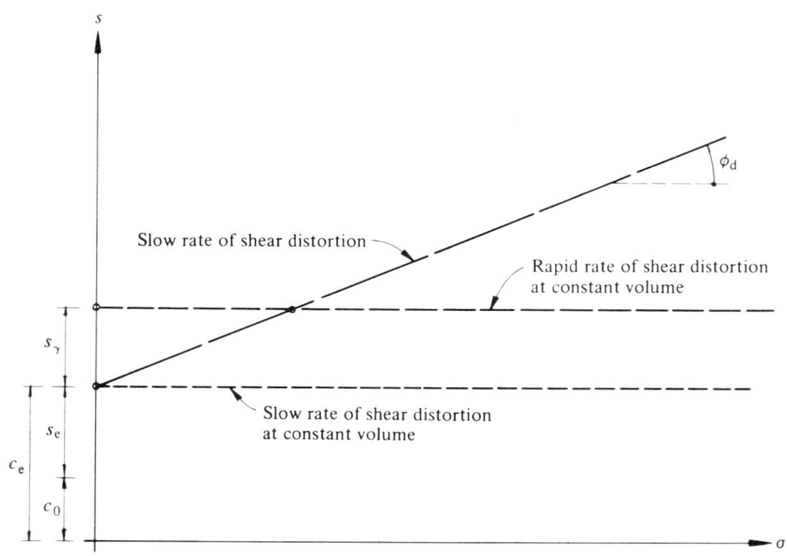

Fig. 11-II.4 Effect of rate of shear distortion on shear strength of clay.

small shear strength; call it $s_0 = c_0$. Let us take several test specimens of this material and consolidate them to the following volumetric effective stresses: $\sigma_{c1}, \sigma_{c2}, \ldots \sigma_{ci}, \ldots \sigma_{cn}$, obtaining for each, respectively, the void ratios $e_1, e_2, e_3, \ldots, e_i, \ldots e_n$. Let us plot these values in a graphical form e vs. σ_c, therefore obtaining a typical volumetric compressibility curve (Fig. 12-II.4). This test indicates that the natural consistency of the material has increased upon reduction of water content, and consequently the void ratio has also reduced owing to compaction induced by volumetric stress σ_c. The real cohesion gained may be measured in each one of the consolidated materials by means of a test under the conditions that $s_\gamma = 0$ and $s_\phi = 0$, implying very slow shear distortion at constant volume, that is to say, preserving the original void ratio of the test specimen in such a way that the effective stresses during the test are kept at any time equal to zero: $\sigma = p - u = 0$. The values so obtained represent the true cohesion gained for the different values of the preconsolidation stress σ_{ci}; hence:

$$c_e = c_0 + s_e(\sigma_c) \qquad \text{38-II.4}$$

The values are plotted in s, σ_c coordinates and a curve representing the gain in shear strength is obtained, according to formula 38-II.4. These values are then referred to their corresponding void ratios (Fig. 12-II.4), from which one obtains:

$$c_e = f_3(e) \qquad \text{39-II.4}$$

The compressibility curve is expressed by $e = f_1(\sigma_c)$; therefore, one can calculate

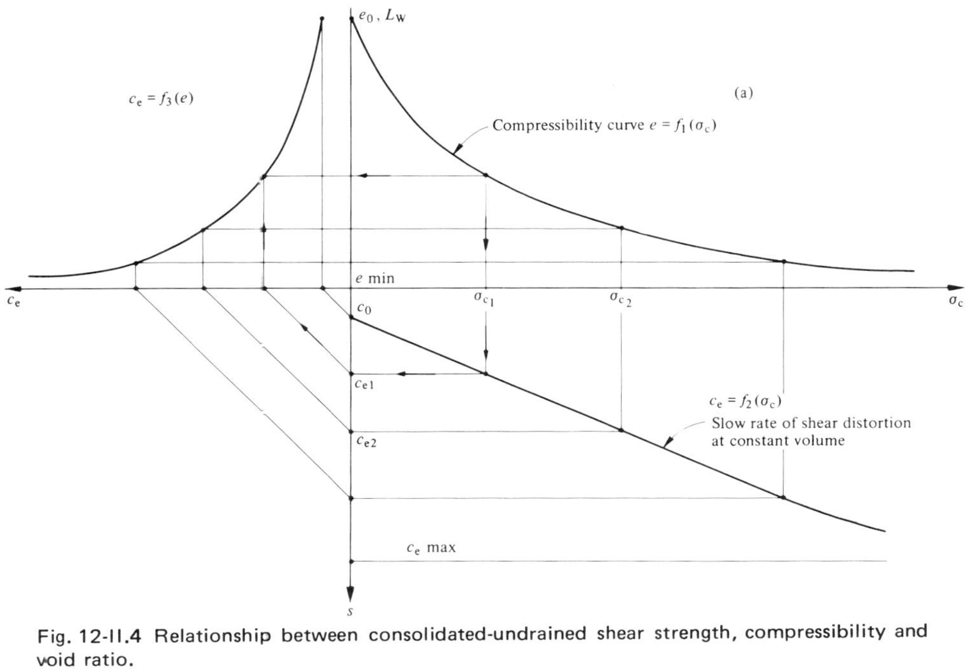

Fig. 12-II.4 Relationship between consolidated-undrained shear strength, compressibility and void ratio.

the value of c_e. As an example, assume that c_e is approximately a linear function of σ_c. Furthermore, we may write in the same range of stresses, from Chapter II.3:

$$e_2 = e_1 - C_c \log \frac{\sigma_{c2}}{\sigma_{c1}} \qquad \text{40-II.4}$$

or

$$\sigma_{c2} = \sigma_{c1} e_N^{2.3(e_1 - e_2)/C_c} \qquad \text{41-II.4}$$

Using the hypothesis $c_e = n \cdot \sigma_c$, we obtain the following expression for 39-II.4:

$$c_{e2} = n \cdot \sigma_{c1} \cdot e_N^{2.3(e_1 - e_2)/C_c} \qquad \text{42-II.4}$$

The value of n, however, has to be determined from tests.

From the rheological point of view and for practical application, it is important to investigate the dynamic or residual shear strength. This property is important to consider in problems where progressive shear failures may be expected. It has been stated before that the viscous intergranular phenomenon may be represented approximately by a Newtonian ideal fluid; however, soils do not behave according to such a simple law and investigations performed by Buisman (1936), Zeevaert (1947), Geuze (1948-60) have induced the author to conclude that the phenomenological law of distortion for the shear stress may be represented approximately by the following expression:

$$\gamma_t = \phi_f \cdot \log \frac{t_0 + t}{t_0} \cdot s_\gamma \qquad \text{43-II.4}$$

Therefore

$$\dot{\gamma}_t = \frac{\phi_f}{t_0 + t} \cdot s_\gamma \qquad \text{44-II.4}$$

The value t_0 represents the instant at which the phenomenon starts to take place after load application, and ϕ_f is a constant measuring the fluidity of the material. Therefore

$$\frac{s_\gamma}{\dot{\gamma}_t} = \frac{1}{\phi_f}(t_0 + t) \qquad \text{45-II.4}$$

represents a straight line with slope $1/\phi_f$ that measures the viscosity of the material (Fig. 13-II.4). Note that if the rate of angular distortion inducing the shear stress is small, the shear stress will be also small. Therefore, from the general shear strength formula 31-II.4, it may be noticed that upon reaching the shear failure with certain shear rate distortion, the strength $c_0 + s_e$ will be destroyed, and the terms s_γ and s_ϕ will subsist; however s_ϕ will have a smaller value because of the dynamic phenomenon after reaching the maximum shear strength and pore pressures at the instant of failure. The term s_γ should disappear, theoretically, when the rate of the angular distortion is very small; in that case $u \doteq 0$. Since s_γ is only a function of the intergranular viscosity, it may be neglected for noncohesive ma-

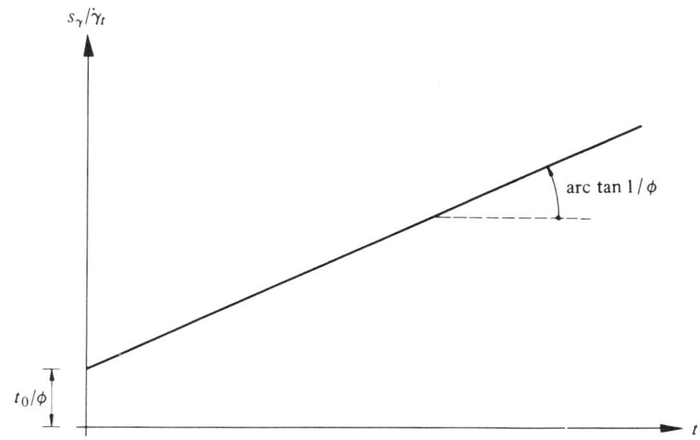

Fig. 13-II.4 Apparent viscosity behavior of soil.

terials. The phencmenon is shown by means of curve s vs. γ obtained in these types of tests (Fig. 14-II.4), from which one can judge that upon reaching the maximum shear strength, the visco-plastic phenomenon has effect and the shear strength has the value $s_{ru} = s_\gamma + s_{\phi K}$; hence:

$$s_{ru} = \frac{1}{\phi_f}(t_0 + t)\,\dot{\gamma}_t + (p - u)\tan\phi_K \qquad \text{46-II.4}$$

Therefore, the value s_{ru} is a function of the rate of angular distortion at which the phenomenon occurs, and of the pore water excess pressures in the laboratory and under field conditions. For very small rates $s_{ru} = p_n \tan\phi_K$.

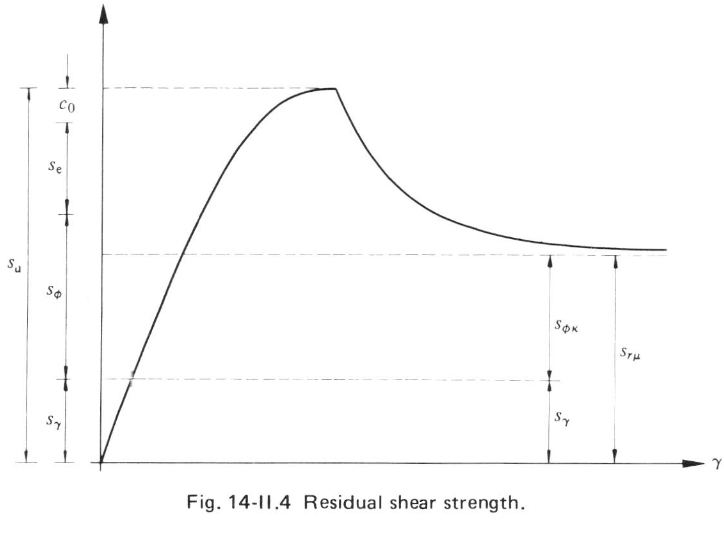

Fig. 14-II.4 Residual shear strength.

4.7 Shear Strength Applications

From the basic discussion of shear strength properties in preceding articles, it is concluded from theoretical and practical points of view that different values of this soil property may be obtained, depending on the physical and geometrical conditions developed at the instant of shear failure. Furthermore, in cohesive materials, the shear strength is an important function of the load history of the material, that is, of the state of stress at which the soil sediment was consolidated in the past, and of the bond between grains obtained by cementing substances or minerals.

Among the different shear strength values obtained, there are some more common in the practice of foundation engineering for typical stability problems. The general philosophy just discussed in previous articles may be useful to investigate the shear strength for the following materials:

a) Cohesionless
   1. Permeable: sands
   2. Semi-impermeable: fine sands and silty sands
b) Cemented (nonplastic)
   1. Permeable and semi-impermeable: sands and silts
c) Cohesive (plastic)
   1. Impermeable: clays and silts

The soil sediments should be studied in their natural state in connection with foundation engineering, with the exception of artificially compacted fills on which foundations will be supported. The determination of the shear strength property will depend on the characteristics of the specific problem: the stratigraphy, the soil sediment, the hydraulic conditions and the rate at which the state of stress is applied. It is also important to classify the different problems from the point of view of the load history.

Foundation engineering problems may be classified into two groups, namely: (A) Those for which the applied state of stress produces failure in natural deposits previously consolidated by the overburden pressure. For these studies, the engineer needs representative undisturbed soil samples and knowledge about the stratigraphical and hydraulic conditions at the site in question. (B) Those materials used for fills or embankments on which foundations will be supported.

GROUP A:

a) In foundations, the load is applied at a certain, usually low rate, over periods that vary from one to several months, and on occasion several years. To decide on the shear strength to be used, the stratigraphic conditions with respect to the drainage surfaces should be taken carefully into consideration, deciding if failure may occur approximately at constant volume. This estimate is made taking into account the permeability of the material, and estimating by means of the theory of consolidation the hydrostatic excess pressures induced during the application of the stresses or change in effective stresses, in addition to the existing hydrostatic pressures that should be investigated by means of piezometric water level observations.

b) In excavations, the state of stress in the soil mass is reduced. The rate of excavation will have an important influence in the selection of the shear strength. This problem, however, in the majority of the cases, does not increase the hydrostatic pressures. The hydraulic pressures may be investigated at the site by means of piezometric water level observations.

c) Fills and embankments, in general, increase the state of stress similar to the problem of loading the soil by construction of foundations. In this case, it is important to select the shear strength according to the conditions established for the construction problem of the embankment or fills. The loads in this type of construction are applied with a certain rate that is usually well known. Therefore, it is possible to estimate the hydrostatic excess pore pressure generated by the application of the load, and decide on the shear strength to be used.

d) In deposits subjected to load and found to be in a state of consolidation or for certain dynamic conditions of the subsoil water, the initial state of stress is estimated measuring piezometric water elevations by means of piezometers. On occasion, the hydraulic conditions may be estimated theoretically with the use of flow nets when the problem is well defined and the permeability of the subsoil well known.

e) The shear strength for long-term problems is defined by the hydraulic pressures in situ and its possible changes with time, as is the case for natural stability of slopes, the load capacity of point bearing piles, friction piles, and problems of negative skin friction (Chapters VIII to X).

GROUP B:

Group B is concerned with the stability of fills and embankments that support footings for buildings. They are usually constructed of selected materials with only sufficient cohesion to be able to obtain good compaction. In these cases the material with a low water content is usually placed very close to the plastic limit of the fine fraction; therefore, there is no problem of consolidation because of water excess pressures; however, if these materials get saturated after they are in place because of water seeping through them, the shear strength may be considerably reduced. Swelling of the material may be one of the causes. Under these conditions, it is important to investigate the future environment in which they will work, and to select properly the shear strength corresponding to the final conditions in the field. The granular materials, which are considered permeable, do not represent special shear strength problems.

The only means to study the shear strength in natural clay materials is using undisturbed soil samples. For this purpose, consider several test specimens of exactly the same cohesive material whose shear strength is to be determined. The material in question has been consolidated on a long-term basis in nature under the effective vertical stress σ_{0z}, and horizontal stress σ_{0h}, corresponding to a void rate e_0.

Therefore, the volumetric effective stress corresponding to these conditions is

$$\sigma_{co} = \tfrac{1}{3}(1 + 2K_0)\,\sigma_{z0} \qquad \text{47-II.4}$$

The magnitude of K_0 assumes a value on the order of 0.75 for clay and clayey silt deposits. The test specimen is placed in the triaxial chamber, and the capillary effect is eliminated; because of this action, the material will suffer some expansion when liberated from the capillary tension that approximately was holding constant the natural void ratio e_0. As the material expands, it will absorb water to reach void ratio e_0' and corresponding water content equal to w_0'. Under these conditions, a nondrained or constant volume test is performed and the apparent cohesion c_u obtained. Using other test specimens with equal index properties, reconsolidation is performed under volumetric pressures σ_c, and tested to failure at constant volume, obtaining shear strength curve A in Fig. 15-II.4.

In the range of stresses up to σ_{co}, the shear strength obtained may be expressed as follows:

$$s_{cu} = c_{cu} + \sigma_c \tan\,[\phi_{cu}, \sigma_c] \qquad \text{48-II.4}$$

Here ϕ_{cu} is variable and a function of σ_c. For the value σ_{co}, equivalent to the overburden stress, the shear strength will be almost identical to that obtained from unconfined compression tests, and corresponds approximately to the effective stress at which the material was precompressed in nature.

From the above discussion, we see that in clays, except in cases of shear strength on a long-term basis, the shear strength corresponding to the effective state of stress of confinement in the field is given by $s_{cu} \doteq q_u/2$, that is to say, it is approximately one-half the natural consistency of the material, implying an apparent angle of internal friction $\phi \doteq 0$. In case of remoulded and reconsolidated soils under a state of stress, the general concepts already explained may be applied; thus

$$s = \sigma_n \cdot \tan\phi_r \qquad \text{49-II.4}$$

This concept has application in estimating the negative friction in piles (see Chapter VIII). Furthermore, the natural consistency, as mentioned before, may be measured by unconfined compression strength tests. When the natural state of stress is removed, the material will undergo certain expansion. The test specimen, however, will still remain confined by capillary stresses induced in the soil skeleton structure: $\sigma_{cK} < \sigma_{co}$.

The value of the unconfined compression strength q_u under this condition is slightly smaller than would be obtained if the sediment is confined to its natural conditions σ_{co}, that is to say:

$$[q_u]_{\sigma_{cK}} < [q_u]_{\sigma_{co}} \qquad \text{50-II.4}$$

In order to investigate the capillary effect on the shear strength, it is necessary to perform three tests:

1) The test specimen is confined to σ_{co} in the compression triaxial chamber, and then stressed to failure under constant volume, obtaining $[q_u]_{\sigma_{co}}$.

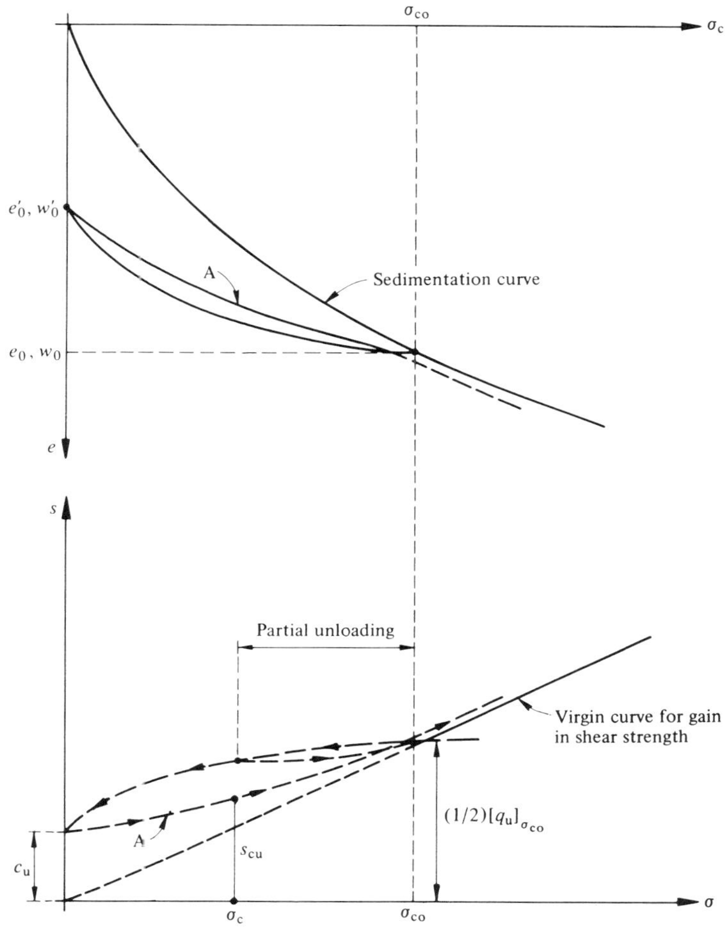

Fig. 15-II.4 Hysteresis shear strength properties of clay.

2) A test is performed under nonconfined condition without eliminating the capillary effect, therefore obtaining the conventional natural consistency $[q_u]_{\sigma_{cK}}$.
3) A nonconfined compression test is performed under water eliminating the capillary effect, obtaining $[q_u]_e$

The resulting limiting stress circles described above have been traced in Fig. 16-II.4 and related to the volume compressibility curve shown above, from which we conclude that

$$[q_u]_e < [q_u]_{cK} < [q_u]_{\sigma_{co}} \qquad \text{51-II.4}$$

Moreover, it may be noticed that the change in void ratio in the compressibility

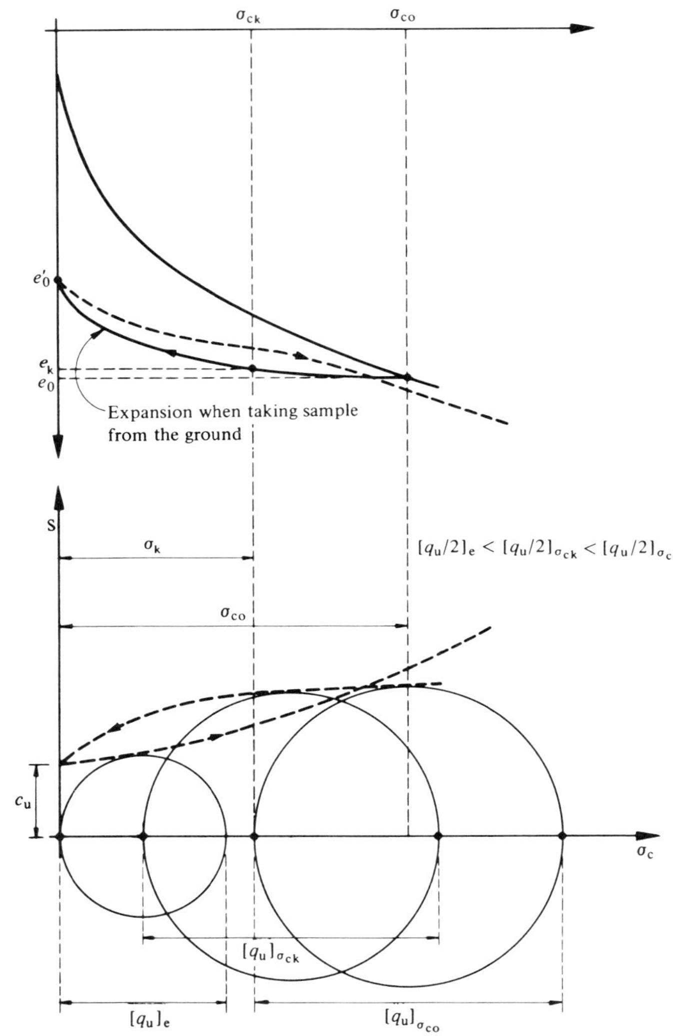

Fig. 16-II.4 Natural consistency of clay.

curve when passing from σ_{cO} to σ_{cK} is small; hence for practical purposes we have

$$[q_u]_{cK} \doteq [q_u]_{\sigma_{cO}} \qquad \text{52-II.4}$$

Therefore, one half of the natural consistency will be an approximate measure of the shear strength of the clay for cases in which failure takes place under consolidated natural conditions at constant volume. It is important in this type of test, however, to observe the rate of strain, since the viscous effect in the sediment increases the strength by approximately 20 to 25% when compared with tests performed under an extremely low rate of shear distortion.

The sediments found illuviated with cementing agents like carbonates or silicates enter into the category of residual soils, eolic sediments, soft volcanic tuff and soft sand stones. These soils may be considered with semi-impermeable hydraulic properties; the shear parameters c_d and ϕ_d may be obtained from drained tests at slow loading rates. In the soils previously mentioned, the term c_d may be seriously disturbed when samples are taken from the ground, or in the field, for instance when driving piles.

Residual Strength. The residual shear strength may be measured after reaching the maximum shear strength. The phenomenon implies that the shear strength experiences a relaxation of stress that may be maintained at a certain rate of shear distortion by the visco-plastic phenomenon. In slow tests, the residual shear strength may have the following values with respect to the ultimate shear strength: 0.4 in plastic clays, and 0.6 in clayey silts. The total shear strength along the sliding surface will depend on the homogeneity and isotropy of the material and the dynamic conditions of the problem. These conditions should be carefully investigated. The ultimate shear failure is the statistical average of the maximum and residual shear strengths, respectively, that may be present at the instant of failure. Therefore, it is important to determine the shear strength at the sliding surface from the point of view of angular distortion under ultimate conditions. When the problem is one of continuous relative movement at the sliding surface, such as in the case of negative friction in piles or any other similar problem, the shear strength used to analyze the specific problem is the ultimate residual shear strength obtained in remoulded soil specimens under rates of loading compatible with those to which the material will be subjected in the field.

4.8 Conclusions

From discussions presented in preceding articles concerning shear strength for different types of sediments and stability problems, one can reach the conclusion that in order to assign a proper value to the shear strength of the material in question, it is necessary to make first a proper diagnosis of the problem under consideration. This diagnosis should be performed knowing the following engineering properties of the soil deposits and the environmental circumstances.

1. Stratigraphy and load history.
2. Hydraulic conditions of the subsoil and drainage surfaces.
3. Permeability and compressibility properties of the sediments, consolidation taking place in the sediment.
4. Geometry and loading characteristics of the problem
5. Rate of load application.

Once the diagnosis is performed, shear strength tests may be performed closely representative to the conditions in the field. The time factor concerning the rate of application of the state of stress should not be overlooked. This information should be carefully considered in conjunction with the loading history of the sediments and drainage affecting the potential surface of sliding. Therefore, the shear strength has to be visualized for each one of the strata in consideration, which requires knowledge about the effective stress normal to the surface of sliding, when the real parameters c_d and ϕ_d are used. To find the value of σ_n at the potential sliding surface, it is necessary to estimate the pore water pressure in the soil existing at the onset of construction, and also the excess water pressure that may be generated upon load application and changes that may be expected to occur in the future.

A summary of the angles of internal friction expected in cohesionless materials is given in Fig. 17-II.4. We note that the form of the grains, dimensions and relative density D_r are important factors in determining the value ϕ_d. From Fig. 17-II.4 one can interpolate values for practical purposes. Moreover, from the practical point of view for shear strength estimates in semi-impervious soils, it has been found that the shear strength parameter ϕ under consolidated-undrained conditions assumes the following approximate values: in fine silty sands $\phi_{cu} = (2/3)\phi_d$ and in silts $(1/2)\phi_d$. When the mass is an impervious material like clays and clayey silts, and the load is applied without permitting important consolidation of the material,

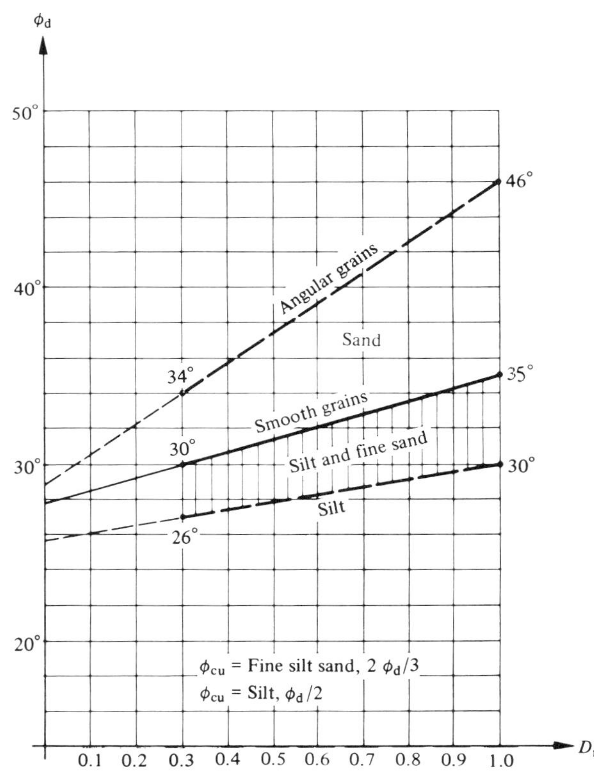

Fig. 17-II.4 Angle of internal friction.

then the shear strength may be measured by means of simple compression tests representing the natural consistency of the clay material. Table 1-II.4 gives the normal classification in such cases:

TABLE 1-II.4 Shear Strength in Clay

Natural Consistency	Shear strength in kg/cm^2
Very soft	< 0.125
Soft	0.125–0.25
Semirigid	0.25–0.50
Rigid	0.5–1.0
Very rigid	1.0–2.0
Hard	> 2.0

The natural consistency may be estimated in occasions for preliminary investigations by the tact or with a penetrometer; however, this requires much experience from the foundation engineer. When the residual shear strength is required to be known approximately, it may be estimated using the ratio q_r/q_u, with approximate limiting values of 0.4 for clays of high plasticity and 0.6 for clayey silts. Moreover, quick tests in clays and clayey silts show a shear strength which is on the order of 20% larger when compared with tests performed very slowly and at constant volume, that is to say, the shear strength determined at a high rate of load application gives higher values because of the resistance of the viscointergranular elements. In practice, however, this effect is not generally deducted from the test values because it is claimed that it compensates for partial disturbance taking place when obtaining the material from the ground. This is in part true; however, careful observation is required.

BIBLIOGRAPHY

ASCE (1960). *Research Conference on Shear Strength of Cohesive Soils*, Soil Mechanics and Foundation Division, ASCE. University of Colorado.

Barden, L. (1965). Consolidation of Clay with Non-linear Viscosity, *Geotechnique*, Vol. XV.

Barden, L. (1971). Sample Disturbance in the Investigation of Clay Structure, *Geotechnique* 21, No. 3, pp. 211–222.

Beles, A. A. and Staneulesco, I. (1961). Tassements des Constructions Fondées sur les Sols à Grand Porosité, *Proc. V ICOSOMEF*, Vol. I, p. 587, Paris.

Bjerrum, L., Jönson, W. and Ostenfeld, C. (1957). Settlement of a Bridge Abutment on Friction Piles, *Proc. IV ICOSOMEF*, Vol. II, p. 14, London.

Buisman, K. A. S. (1936, 1941). Results of Long Duration Settlement Tests, *Proc. I ICOSOMEF*, Vol. I, p. 103, and Grondmechanica 1941, Waltman, Delft.

Casagrande, A. and Fadum, R. E. (1940). *Notes on Soil Testing for Engineering Purposes*, Harvard University, Soil Mechanics Series No. 8, Graduate School of Engineering.

Casagrande, A. (1948). Classification and Identification of Soils, *Transactions ASCE*, Vol. 113.

Croce, Arrigo (1948). Secondary Time Effect in the Compression of Unconsolidated Sediments of Volcanic Origin, *Proc. II ICOSOMEF*, Vol. I, p. 166, Rotterdam.

DeBeer, E. (1950). The Cell Test, *Geotechnique*, Vol. II (1950–51).

DeBeer, E. (1948). Correlation between the Results of Cell-test and Compression Test, *Proc. II ICOSOMEF*, Vol. I, pp. 173-4.

De Mello, F. B. (1969). Foundations of Buildings in Clay, State-of-the-Art Report, *Proc. VII ICOSOMEF*, Vol. 3, p. 127, Mexico.

Egorov, K. E., Kuzmin, P. G. and Popov, B. P. (1957). The Observed Settlements of Buildings as Compared with Preliminary Calculation, *Proc. IV ICOSOMEF*, Vol. I, p. 291, London.

Freudenthal, Alfred M. (1950). *The Inelastic Behavior of Engineering Materials*, John Wiley and Sons, Inc. New York.

Gouze, E. and Tang, Tjong Kie (1950). The Shearing Properties of Soils, *Geotechnique*, Vol. II (1950-51).

Grim, R. E. (1948). Some Fundamental Factors Influencing the Properties of Soil Materials, *Proc. II ICOSOMEF*, Vol. III, p. 8, Rotterdam.

Holtz, W. G. and Hilf, J. W. (1961). Settlement of Soil Foundations due to Saturation, *Proc. V ICOSOMEF*, Vol. I, p. 673, Paris.

Hvorslev, Juul M. (1949). *Subsurface Exploration and Sampling of Soils for Civil Engineering Purposes*, Research Project ASCE. Waterways Experiment Station. Vicksburg, Miss.

Jennings, J. E. and Knight, K. (1957). The Additional Settlement of Foundations due to a Collapse of Sandy Subsoils on Wetting, *Proc. IV ICOSOMEF*, Vol. I, p. 316, London.

Jennings, J. E. (1961). Comparison Between Laboratory Prediction and Field Observation of Heave of Buildings on Desiccated Subsoils, *Proc. V ICOSOMEF*, Vol. I, pp. 689–692, Paris.

Koppejan, A. W. (1948). A Formula Combining the Terzaghi Load-Compression Relationship and the Buisman Secular Time Effect, *Proc. II ICOSOMEF*, Vol. III, p. 32, Rotterdam.

Lambe, T. W. (1951). *Soil Testing for Engineers*, John Wiley and Sons, Inc., New York.

Leonards, G. A. and Ramiah, B. K. (1959). *Time Effects in the Consolidation of Clays*, Prepared for Symposium on Time-Rate of Loading in Testing Soils. ASTM Annual Meeting, Atlantic City, N.J.

Machado, J. (1961). Settlement of Structures in the City of Santos, Brazil. *Proc. V ICOSOMEF*, Vol. I, pp. 719–725, Paris.

Moretto, O. (1948). Effect of Natural Hardening on the Unconfined Compression Strength of Remolded Clays, *Proc. II ICOSOMEF*, Vol. I, p. 137, Rotterdam.

Murayama, Sakuro and Shibata, Toru (1961). Rheological Properties of Clays. *Proc. V ICOSOMEF*, Vol. I, p. 269, Paris.

Peck, O. K. and Peck, R. B. (1948). Settlement Observations on a Large Water Tank, Salt Lake City, Utah, U.S.A. *Proc. II ICOSOMEF*, Vol. IV, p. 1, Rotterdam.

Rosenqvist, Th. (1963). *The Influence of Physico-Chemical Factors Upon the Mechanical Properties of Clays*, Norweigian Geotechnical Institute, Publication No. 54.

Rowe, P. W. (1957). Hypothesis for Normally Loaded Clays at Equilibrium, *Proc. IV ICOSOMEF*, p. 189, London.

Salas, J. A. and Serratosa, J. M. (1953). Compressibility of Clays, *Proc. III ICOSOMEF*, Vol. I, p. 192, Switzerland.

Salas, J. A. and Serratosa, J. M. (1957). Foundations on Swelling Clays, *Proc. IV ICOSOMEF*, Vol. I, p. 426, London.

Scott, R. and Hon-Yim, Ko. (1969). Stress-Deformation and Strength Characteristics, State-of-the-Art Report, *Proc. VII ICOSOMEF*, Vol. 3, p. 115, Mexico.

Schultze, Edgar and Coesfeld, Gunther (1961). Elastic Properties of Ballast, *Proc. V ICOSOMEF*, Vol. I, p. 323, Paris.

Schultze, Edgar and Kotzias, Panaghistis (1961). Geotechnical Properties of Lower Rhine Silt, *Proc. V ICOSOMEF*, Vol. I, p. 329, Paris.

Schultze, Edgar and Moussa, A. (1961). Factor Affecting the Compressibility of Sand. *Proc. V ICOSOMEF*, Vol. I, p. 335, Paris.

Seed, H. B., Woodward, J. R. and Lundgren, R. (1962). Prediction of Swelling Potential for Compacted Clays, *Journal of SMFE*, ASCE, Vol. 88, No. SM3.

Skempton, A. W. (1948). A Study of the Immediate Triaxial Test on Cohesive Soils, *Proc. II ICOSOMEF*, Vol. I, pp. 192-6.

Skempton, A. W. and Bishop, M. A. (1950). The Measurement of the Shear Strength of Soils, *Geotechnique*, Vol. II (1950–51).

Sowers, G. F. (1963). Engineering Properties of Residual Soils Derived from Igneous and Methamorphic Rocks, *Proceedings II Panamerican Conference on Soil Mechanics and Foundation Engineering*, Vol. I, p. 39, Brazil.

Taylor, D. W. and Merchant, Wilfred (1940). A Theory of Clay Consolidation Accounting for Secondary Compressions, *Journal of Mathematics and Physics*, Vol. XIX, No. 3.

Terzaghi, K. and Fröhlich, O. K. (1936). *Theorie der Setzung von Tonschichten*, Franz Deuticke, Vienna.

Terzaghi, K. (1941). Undisturbed Clay Samples and Undisturbed Clays, *Journal of the Boston Society of Civil Engineers*, No. 324.

Terzaghi, K. (1947). Recent Trends in Subsoil Exploration, *Proceedings VII Texas Conference on Soil Mechanics and Foundation Engineering*, January.

Trollope, D. H., Rosengren, K. J. and Brown, E. T. (1965). The Mechanics of Brown Coal, *Geotechnique*, Vol. XV.

Wahls, E. Harvey (1962). Analysis of Primary and Secondary Consolidation, *Journal of Soil Mechanics and Foundation Division*, ASCE, Vol. 88, No. SM6.

Zeitlen, J. G. and Komornik, A. (1961). Deformations and Moisture Movements in Expansive Clays. *Proc. V ICOSOMEF*, Vol. I, p. 813, Paris.

Zeevaert, L. (1945). Conceptos y Experimentos Fundamentales que se Aplican al Diseño de Cimentaciones en Arcillas Saturadas, *Revista Ingeniería y Arquitectura,* May.

Zeevaert, L. (1949). *An Investigation of the Engineering Characteristics of the Volcanic Lacustrine Clay Deposits in Mexico City*, Ph.D. Thesis, University of Illinois. Unpublished.

Zeevaert, L. (1952). Compresibilidad de la Arcilla Volcánica de la Ciudad de México. IV Centenario de la Universidad Nacional de México, *Revista Ingeniería*, Vol. XXV.

Zeevaert, L. (1953). Ratio of Horizontal to Vertical Pressure of Natural Unconsolidated Sedimentary Deposits. *Proc. III ICOSOMEF*, Vol. III, p. 113, Zürich, Switzerland.

Zeevaert, L. (1957). *Consolidation of Mexico City's Volcanic Clay*, Joint Meeting ASTM and SMMS, p. 28.

Zeevaert, L. (1962). Compresibilidad de los Sedimentos Eólicos Finos, *I Reunión Regional de la Sociedad Mexicana de Mecánica de Suelos.* Puebla.

Zeevaert, L. (1964). *Propiedades Esfuerzo-Deformación-Tiempo. Apuntes de Mecánica de Suelos*, Universidad Nacional Autónoma de México, 3rd Edition.

Zeevaert, L. (1967). Consolidation Theory for Materials Showing Intergranular Viscosity, *Proceedings III Panamerican Conference on Soil Mechanics and Foundation Engineering*, Vol. I, p. 89, Caracas, Venezuela.

▶III◀ SOLID AND LIQUID PHASES OF SOIL

III.1 BASIC CONCEPTS

Soil is a material composed of three elements: solid matter, water and air. It has been seen that the mechanical behavior of soil depends greatly on the proportions in which these elements are present in the soil. Solids are represented by the hard grains forming a more or less complicated skeleton structure. The grains may be only the product of disintegration of rock with the same lithological properties, in which case the soil may be considered cohesionless, or the material may be the product of decay of rocks, in which case the very fine fraction contains clay minerals giving cohesion to the soil. Organic material may be also present. The skeleton structure formed by the grains can be very complicated and sensitive to changes in effective stress levels from those to which the material is subjected in the field. The mechanical behavior of the solid phase has been studied in section II.3, where it has been shown that to calculate changes of strain in the solid phase, it is necessary to determine, in addition to the stress-strain-time and shear strength mechanical properties of the soil, the change in stresses in the soil mass induced by loads applied at the ground surface.

We may identify two groups of soil mechanics problems in which the foundation engineer is interested:

a) Stability problems falling in the category of ultimate shear strength analysis, investigated by means of simplified methods and assumptions of plastic behavior, such as the ultimate bearing capacity of footings, stability of slopes and earth pressure.
b) Deformation problems where the foundation engineer is interested in estimating displacements because of stress changes, making use of the stress-strain-time mechanical relationships for the solid phase.

The liquid phase is closely related with the mechanical behavior of the soil skeleton structure. Moreover, when the soil is saturated, permeability is important, and it is necessary to take into consideration the fundamental soil mechanics theorem: total pressure equals effective stress added to hydrostatic pressure. Therefore, at any point in a soil mass

$$p_i = \sigma_i + u_i \qquad \text{1-III.1}$$

Generally, the value of the intergranular stress may be found with reasonable accuracy from the engineering point of view from theory and assumptions. The hydraulic pressure, however, is difficult to determine when water is flowing through the soil mass. The only means will be by theoretical investigations or field determinations, depending on the characteristics of the problem and its importance in foundation engineering.

III.2 SOLID PHASE

2.1 Effective Stresses

To illustrate the importance of the state of stresses in the ground, let us consider the problem of determining the effective stress and hydraulic pressure at a point located in the subsoil at depth z. The subsoil is stratified as shown in Fig. 1-III.2. The average unit weight of each stratum of thickness Δz_i is γ_i. The mass is under unknown hydrodynamic conditions; therefore, the only means to determine the state of hydraulic pressure will be with the use of piezometers A, B, C and D

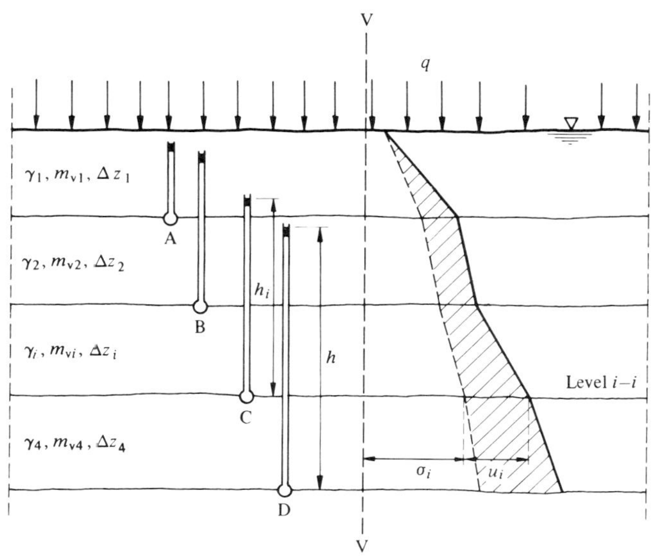

Fig. 1-III.2 Vertical effective stresses in the subsoil.

installed at different depths in the soil mass. The total pressure at the base of the i-th stratum is

$$p_i = q + \sum_1^i \gamma_i \Delta z_i \qquad \text{1-III.2}$$

The hydraulic pressure is measured in piezometer C, and the value found is $u_i = h_i \gamma_w$, where h_i is the piezometric head. Under this condition, the effective stress may be calculated by 1-III.1 at depth z_i; hence

$$\sigma_i = q + \sum_1^i \gamma_i \Delta z_i - h_i \gamma_w \qquad \text{2-III.2}$$

Thus we see that a change in effective stress takes place in the following two ways:

a) When the total pressure remains constant, a change in pressure in the liquid phase will produce an equal and opposite change of effective stress in the solid phase; therefore

$$\Delta \sigma_i = -\Delta u_i$$

b) When a change in the total pressure takes place, it immediately produces a change in pressure in the liquid phase and a change of effective stress in the solid phase; therefore

$$\Delta p_i = \Delta \sigma_i + \Delta u_i$$

The excess pore pressure, however, will be released as water flows toward the drainage surfaces encountered in the soil mass, that is to say, toward more pervious layers. A process of consolidation will take place in the impervious strata and ultimately the excess pore pressure $\Delta u_i \to 0$, and $\Delta p_i = \Delta \sigma_i$. The soil skeleton structure will take the total load. For static equilibrium, the soil structure works under elasto-plastic conditions; otherwise the soil fails and enters into visco-plastic state conditions.

The foundation engineer is interested in practical methods to estimate the change in stresses in the solid phase because of application of loads at the ground surface, and those to evaluate the hydraulic pressures in the liquid phase for certain hydrodynamic conditions to be encountered in the subsoil mass for a specific problem. The knowledge of these two conditions will enable the foundation engineer to visualize the solution of foundation problems. The change in effective stress at a point in the solid phase due to load applied at the ground surface may be estimated by the theory of elasticity, or in certain well defined cases, by means of simplified assumptions. The precision obtained depends on the magnitude of the change in effective stress. It was already noted in section II.3 that soil material is far from being elastic. Therefore, the relationship between ϵ and σ is not linear. The results obtained by the theory of elasticity, however, may be used satisfactorily if the following consideration is made:

$$\frac{\Delta \epsilon_i}{\Delta \sigma_i} = \overline{M}_i, \quad m_{vi} = \text{constant} \qquad \text{3-III.2}$$

The value of $\overline{M}_i$ represents the equivalent secant or average strain modulus (see equation 44-II.3) and m_{vi} the coefficient of unit volume compressibility, whatever the case may be within the stress levels σ_i and σ_{i+1}. Therefore, in case of a stratum of thickness Δz_i confined to zero lateral deformation

$$\Delta\delta_i = m_{vi} \cdot \Delta\sigma_i \cdot \Delta z_i \qquad \text{4-III.2}$$

and the total vertical displacement is the sum of the deformations of all the compressible strata (Fig. 1-III.2) and may be written as follows:

$$\delta_n = \sum_1^n m_{vi} \cdot \Delta\sigma_i \cdot \Delta z_i \qquad \text{5-III.2}$$

The value of m_{vi} is a function of confining stress level and time, and should be evaluated for each stratum as previously discussed in section II.3. To investigate the change in effective stresses due to either increase or removal of load it is necessary to investigate as a first step the initial state of stresses in the subsoil. The overburden stresses may be calculated from the index properties, the stratigraphical conditions at the place in question, and the hydraulic state of pressures determined by means of piezometric observations (equation 2-III.2). The horizontal stresses in the subsoil may be determined from simplified assumptions based on the theory of elasticity.

Assume a point within the soil mass has a vertical stress σ_z and a corresponding horizontal stress σ_h. Therefore,

$$\frac{\sigma_h}{\sigma_z} = K_0 \qquad \text{6-III.2}$$

where K_0 represents the ratio of the effective stresses at rest, and depends on the type of soil, stratigraphical conditions, and the stress history of the soil mass.

For normally loaded sediments, the value of K_0 varies from about 0.4 for sands to about 0.80 for clays, and for highly preconsolidated sediments, values on the order of 1.0 and larger may be encountered. Under field conditions, the equivalent confining overall stress at the point under consideration is

$$\sigma_c = \tfrac{1}{3}(1 + 2K_0)\,\sigma_z \qquad \text{7-III.2}$$

From the point of view of the theory of elasticity in a semi-infinite mass, $K_0 = \nu/(1 - \nu)$, in which ν is Poisson's ratio. For sands $\nu = 0.25$, and for clays $\nu = 0.45$; these represent extreme values that agree reasonably well with the values of K_0 reported before. For intermediate cases formula 7-III.2 may be used, if ν is previously determined in the laboratory in the range of static elasto-plastic conditions.

The value of σ_c at any depth in the subsoil is important in correlating the results of laboratory tests with field conditions, since the value of the strain modulus is a function of the confining stress level σ_c. In section II.3.5, it was found that K_0 also depends on the isotropy of the soil; hence

$$K_0 = \frac{\nu}{1 - \nu} \cdot \frac{M_z}{M_h}$$

Under confined conditions of zero lateral deformation the formula $\Delta \sigma_h = K_0 \Delta \sigma_z$ applies approximately when the load is incremented at the surface of a compressible soil mass of limited thickness.

Assume it is necessary to find the deformation of a fine sand stratum confined to zero lateral deformation, with a thickness of $2H$ and on which the load will be increased from vertical stress level σ_1 to σ_2. The volumetric strain change produced by the change in effective stress is given by the formula

$$\Delta \epsilon_v = \nu_c \cdot [M_{ep}]_{ave} (1 + \kappa_v)(\sigma_2 - \sigma_1) \qquad \text{50-II.3}$$

For sand the strain modulus is

$$M_{ep} = C_0 \cdot \sigma_c^{-n}$$

When the confining stress level passes from σ_{c1} to σ_{c2}, the average strain modulus (Fig. 2-III.2) is

$$[M_{ep}]_{ave} = \frac{C_0}{\sigma_{c1} - \sigma_{c2}} \cdot \int_{\sigma_{c1}}^{\sigma_{c2}} \sigma_c^{-n} \, d\sigma_c \qquad \text{8-III.2}$$

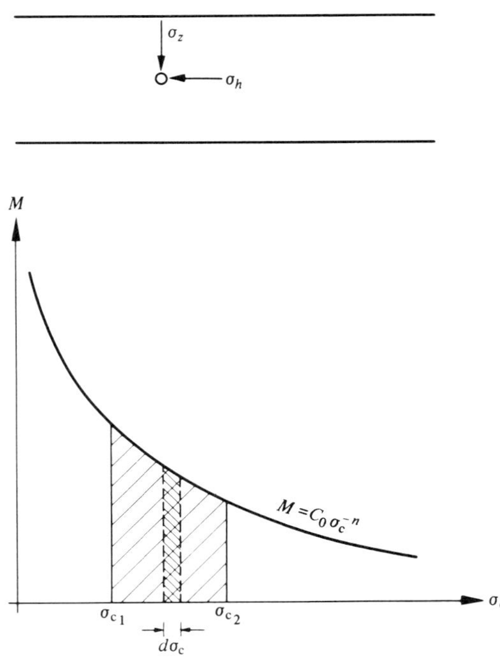

Fig. 2-III.2 Average strain modulus.

or after integration for $n \neq 1$

$$[M_{ep}]_{ave} = \frac{C_0}{1-n} \frac{\sigma_{c2}^{1-n} - \sigma_{c1}^{1-n}}{\sigma_{c2} - \sigma_{c1}} \qquad \text{9-III.2}$$

To illustrate the case with numerical values, assume the stratum thickness is $2H = 200$ cm. The material is a medium silty sand, and the stratum is subjected to an average vertical stress $\sigma_1 = 1.2 \text{ kg/cm}^2$. Because of excavation in a large area, the stress is reduced by 0.4 kg/cm²; thereafter, a load of 1.6 kg/cm² is applied in the same excavated area, considered large as compared with the thickness of the stratum. Therefore, the vertical stresses are:

a) Initial vertical stress: $\sigma_1 = 1.2 \text{ kg/cm}^2$
b) Initial vertical stress less excavation: $\sigma_1 - \Delta\sigma_{exc} = 1.2 - 0.4 = 0.8 \text{ kg/cm}^2$
c) Increment of stress: $\Delta\sigma_1 = 1.6 \text{ kg/cm}^2$
d) Final stress: $\sigma_2 = 2.4 \text{ kg/cm}^2$

From laboratory investigations the material has the following mechanical properties:

$$\nu = 0.3, \quad \nu_c = 0.742, \quad K_0 = 0.43, \quad \frac{1 + 2K_0}{3} = 0.62$$

$$C_0 = 0.004, \quad n = 0.5, \quad \kappa_v = 0.5$$

Hence:

$$[M_{ep}]_{ave} = 0.008 \frac{\sigma_{c2}^{1/2} - \sigma_{c1}^{1/2}}{\sigma_{c2} - \sigma_{c1}}, \quad \frac{\text{cm}^2}{\text{kg}}$$

Therefore:

Layer	σ_2	σ_1	σ_{c2}	σ_{c1}	$(\sigma_{c2} - \sigma_{c1})$	$\sigma_{c2}^{1/2}$	$\sigma_{c1}^{1/2}$	$\sigma_{c2}^{1/2} - \sigma_{c1}^{1/2}$	M_{ep}
1	2.4	0.8	1.485	0.495	0.990	1.219	0.704	0.515	0.00416

Using equation 50-II.3, we obtain:

$$\Delta\epsilon_v = 0.742 \cdot 0.00416(1 + 0.5)\,1.6, \text{ or } \Delta\epsilon_v = 0.0074$$

and the vertical displacement is $\delta = 0.0074 \times 200 = 1.48$ cm.

2.2 Stress Distribution in Soil Mass

The stress distribution in the subsoil may be calculated from the theory of elasticity. Boussinesq (1885) found the solution for the stresses at a point induced by a concentrated load applied at the surface of a semi-infinite isotropic elastic solid.

According to the geometry shown in Fig. 3-III.2, the stresses at a point are

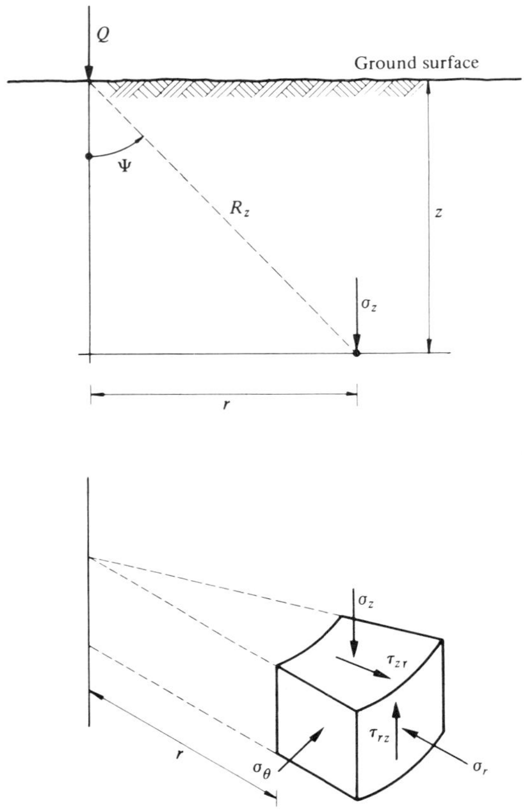

Fig. 3-III.2 Concentrated load applied at the surface of a semi-infinite solid.

calculated by means of the following expressions:

$$\sigma_z = \frac{3Q}{2\pi} \cdot \frac{z^3}{R_z^5} \tag{10-III.2}$$

$$\sigma_r = -\frac{Q}{2\pi}\left[\frac{1-2\nu}{r^2}\left(1-\frac{z}{R_z}\right)-\frac{3r^2z}{R_z^5}\right] \tag{11-III.2}$$

$$\sigma_\theta = -\frac{Q(1-2\nu)}{2\pi}\left[-\frac{1}{r^2}+\frac{z}{r^2R_z}+\frac{z}{R_z^3}\right] \tag{12-III.2}$$

$$\tau_{rz} = \frac{3Q}{2\pi} \cdot \frac{rz^2}{R_z^5} \tag{13-III.2}$$

To study the effect of the concentrated load in the isotropic mass, let us assume $\nu = 0.5$, that is to say, the soil is incompressible. This assumption will not affect the values of σ_z and τ_{rz}, since they are independent of ν. Furthermore, let us call

$$I_{Bq} = \frac{3}{2\pi}\left(\frac{z}{R_z}\right)^5$$

The value I_{Bq} is dimensionless and may be calculated and plotted against values of r/z. From Fig. 3-III.2, we find $R_z^2 = z^2 + r^2$; hence

$$I_{Bq} = \frac{3}{2\pi}\left[\frac{1}{1+\left(\frac{r}{z}\right)^2}\right]^{5/2} \qquad \text{14-III.2}$$

Therefore, the stresses at a point may be written as follows:

$$\sigma_z = \frac{Q}{z^2}\left[I_{Bq}\right] \qquad \text{15-III.2}$$

$$\sigma_r = \frac{Q}{z^2}\left[\left(\frac{r}{z}\right)^2 \cdot I_{Bq}\right] \text{ for } \nu = \tfrac{1}{2} \qquad \text{16-III.2}$$

$$\sigma_\theta = 0 \text{ for } \nu = \tfrac{1}{2} \qquad \text{17-III.2}$$

$$\tau_{rz} = \frac{Q}{z^2}\left[\frac{r}{z} I_{Bq}\right] \qquad \text{18-III.2}$$

The value I_{Bq} expressed by 14-III.2 is known as Boussinesq's influence value for vertical stresses. It is a function of the position (r/z) of the point at which the

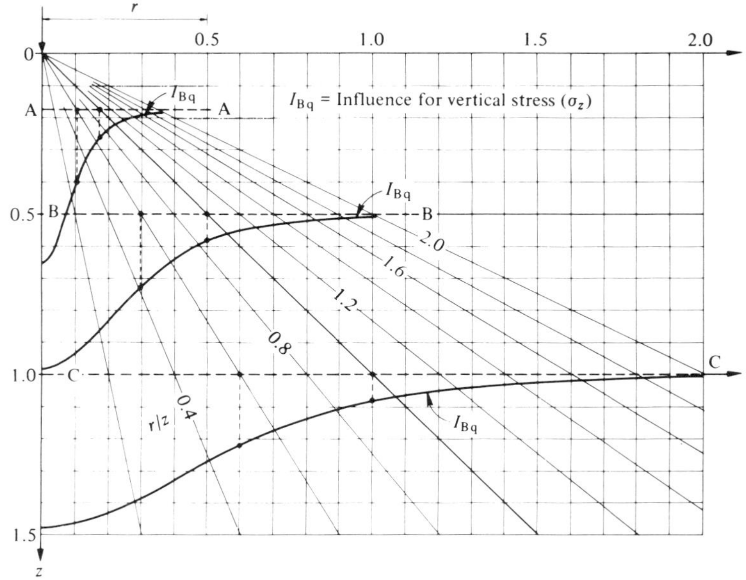

Fig. 4-III.2 Vertical stress distribution in the subsoil due to concentrated surface load.

vertical stress is required. The value of I_{Bq} is represented graphically in Fig. 4-III.2, where its variation in the horizontal radial direction for any depth z indicates the stress dissipation in the ground due to the concentrated load at the surface. The relative magnitude of the influence values for vertical radial and shear stresses, expressed by the brackets of formulas 16 and 18-III.2, respectively, are shown in Fig. 5-III.2. Moreover, it may be seen from expressions 15, 16 and 18-III.2 that the stresses decrease with the square of the depth. The ratio of radial to vertical stresses is $(r/z)^2$, and that of shear to vertical stresses is (r/z). These ratios remain constant with depth along conical surfaces with their vertexes at the point of application of the load. At the line of action of the vertical load these ratios are zero, and at an angle of 45° their value is unity.

From preceding discussion it may be concluded that the vertical stresses in the ground induced by a concentrated load at the surface dissipate rapidly in the soil mass. The shear stresses have a faster dissipation if compared with the vertical stresses, and the horizontal radial stresses dissipate considerably faster than the vertical stresses, as may be visualized from influence values given in Figs. 4 and 5-III.2. The above observations are interesting in foundation engineering in those problems in which assumptions are made concerning stress dissipation in the soil mass, for cases in which mathematical solutions cannot be found readily.

The Boussinesq formula for the vertical stress applies only for isotropic soil masses. When the soil is stratified, Westergaard (1938) worked out a limiting solution on the assumption that the soil is restricted from horizontal deformation but is free to deform in the vertical direction. In practice, this is partially true when the soil mass is reinforced by numerous rigid layers of cemented sand with a very low horizontal strain modulus. The vertical stress according to Westergaard is

$$\sigma_z = \frac{Q}{z_t^2}\left\{\frac{1}{2\pi}\left[\frac{1}{1+\left(\dfrac{r}{z_t}\right)^2}\right]^{3/2}\right\} \qquad \text{19-III.2}$$

In the deduction of this formula the vertical coordinate is modified by a linear transformation such that

$$z_t = k_w \cdot z$$

where

$$k_w = \sqrt{\frac{1-2\nu}{2(1-\nu)}} \qquad \text{20-III.2}$$

in which the value of Poisson's ratio refers to the soil layers between the rigid sheets.

Introducing the value of k_w into 19-III.2, we obtain

$$\sigma_z = \frac{Q}{z^2}\left\{\frac{1}{2k_w^2\pi}\left[\frac{1}{1+\left(\dfrac{r}{k_w \cdot z}\right)^2}\right]^{3/2}\right\} \qquad \text{21-III.2}$$

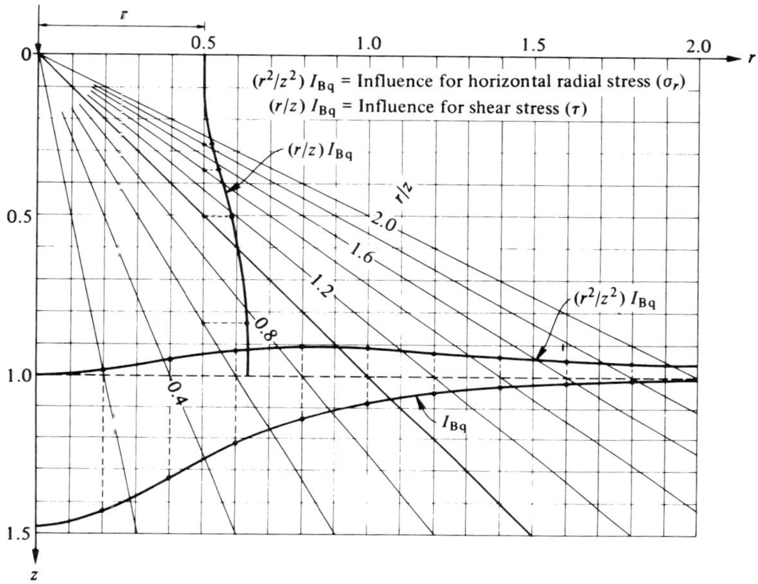

Fig. 5-III.2 Vertica , radial and shear stresses in the subsoil due to concentrated surface load.

and

$$I_{Wd} = \frac{1}{2k_W^2\pi}\left[\frac{1}{1 + \dfrac{1}{k_W^2}\left(\dfrac{r}{z}\right)^2}\right]^{3/2}$$

Under the assumption that the reinforcing sheets are close together and $\nu = 0$, then $k_W^2 = ½$, and the influence value is

$$I_{Wd} = \frac{1}{\pi}\left[\frac{1}{1 + 2\left(\dfrac{r}{z}\right)^2}\right]^{3/2} \qquad \text{22-III.2}$$

in which z is now the real vertical coordinate. Hence

$$\sigma_z = \frac{Q}{z^2} \cdot I_{Wd} \qquad \text{23-III.2}$$

The value of I_{Wd} may be found plotted in Fig. 6-III.2, against r/z values, where it may be seen that the stress under the load has smaller values than those given by the isotropic case. At the load axis, $r/z = 0$, the influence value $I_{Wd} = 1/\pi$; therefore, vertical stresses are 2/3 smaller than for the isotropic Bq-case. This solution gives limiting values under the extreme assumption $M_h/M_z = 0$; however, such conditions are rarely encountered in practice.

In natural deposits the strain modulus decreases with the confining pressure (see section II.3.7); hence

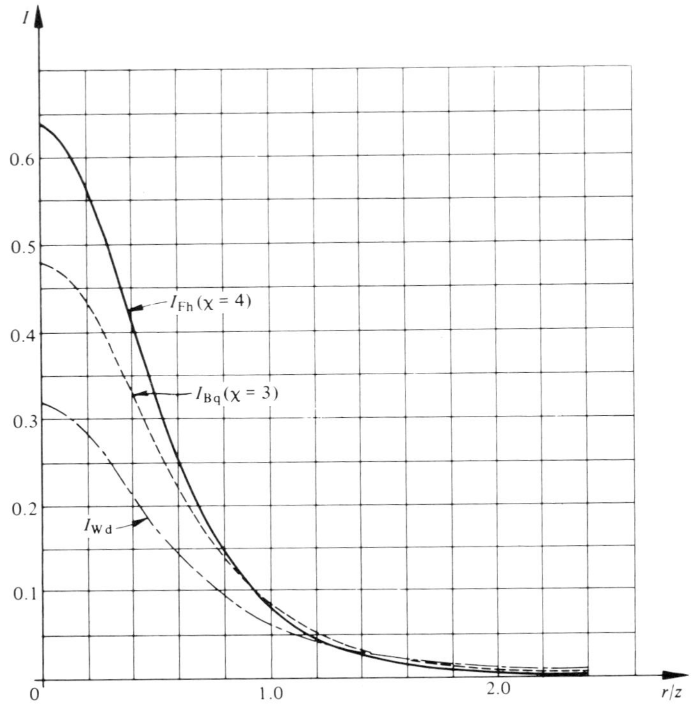

Fig. 6-III.2 Influence values for stress distribution due to concentrated load.

for cohesionless material: $M_z = C_0 \cdot \sigma_c^{-n}$
for cohesive materials: $M_z = M_{z0} e_N^{-n\sigma_c}$

The theory of elasticity as previously discussed, based on an homogeneous isotropic semi-infinite solid, does not apply rigorously in natural soil deposits; therefore, the soil should be considered anisotropic. The result is that the stresses do not dissipate with depth in the soil mass, as fast as given by the isotropic elastic theory. Fröhlich (1942) investigated the radial stress distribution for these cases, satisfying the static equilibrium conditions, and proposed a formula to calculate the vertical stresses induced by a concentrated load at the surface of a semi-infinite elastic anisotropic solid. Boussinesq's solution was modified as follows:

$$\sigma_z = \frac{\chi Q}{2\pi z^2} \cos^{\chi+2} \Psi \qquad \text{24-III.2}$$

Since $\cos^2 \Psi = z^2/(r^2 + z^2)$, Fröhlich's influence value is:

$$I_{Fh} = \frac{\chi}{2\pi} \left[\frac{1}{1 + \left(\frac{r}{z}\right)^2} \right]^{\frac{\chi+2}{2}} \qquad \text{25-III.2}$$

The factor χ is Fröhlich's stress distribution factor. Hence:

$$\sigma_z = \frac{Q}{z^2} I_{\text{Fh}} \qquad \text{26-III.2}$$

Here it may be noticed that when $\chi = 3$, the isotropic solution is obtained. A qualitative indication of the strain modulus varying with depth for equivalent confining pressures is shown in Fig. 7-III.2. The curve closer to a real variation of the strain modulus is marked with a heavy line, and corresponds to medium dense sand where a value of $n = 1/2$ may be assumed. Fröhlich proposed for sand $\chi = 4$, assuming a linear variation of $1/M_z$ with depth. Nevertheless, from Fig. 7-III.2 it may be seen this variation is stronger than real, and therefore that the stress distribution factor $\chi = 4$ is high. The conclusion reached from reported field measurements in sand deposits is that $3 < \chi < 4$. Values of I_{Fh} plotted against r/z are found in Fig. 6-III.2, where the difference may be observed as compared with the isotropic case $\chi = 3$. Under the line of action of the concentrated load,

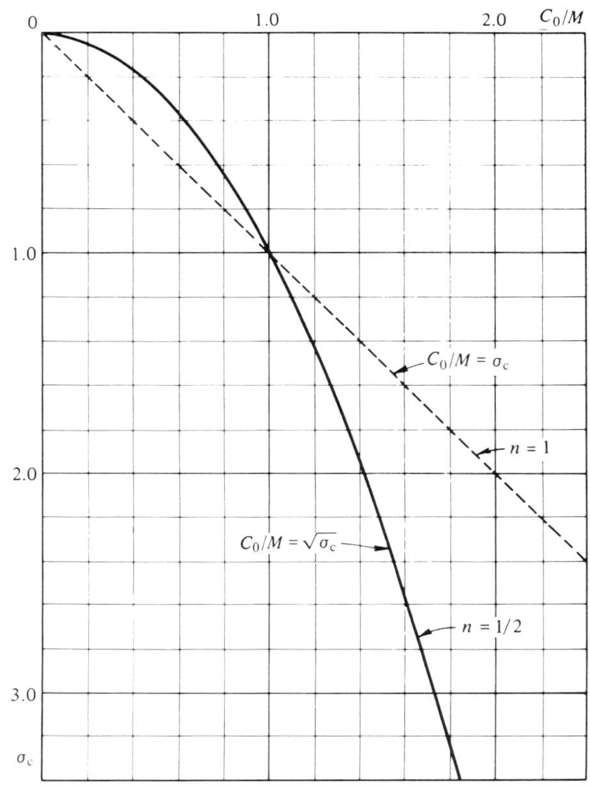

Fig. 7-III.2 Strain modulus variation with depth.

Fröhlich stress distribution for $\chi = 4$ gives a stress concentration 1/3 greater than Boussinesq's solution. Also it may be noticed that Boussinesq's solution is approximately intermediate between Westergaard and Fröhlich's solutions. In nature, a soil deposit has both properties, that is to say, it is stratified and the strain modulus decreases with depth; therefore, Boussinesq's solution represents empirically a closer approximation for most real subsoil conditions in the field. The general formula may be expressed as

$$\sigma_z = \frac{Q}{z^2} \cdot I_\chi \qquad \text{27-III.2}$$

and may be used to find the stress distribution with depth, under a loaded area.

There are cases when the vertical stress at depth z is required under the center of a uniformly loaded area of rectangular shape as shown in Fig. 8-III.2. Assume a small element at the surface with area $dx \cdot dy$ loaded with a uniform load per unit surface q. Therefore, the stress induced by this action at a point with coordinates x, z and $y = 0$ is

$$\Delta\sigma_z = \chi \frac{q \cdot dxdy}{2\pi z^2} \cos^{\chi+2} \overline{\Psi} \qquad \text{28-III.2}$$

From Fig. 8-III.2 we obtain

$$y = z \frac{\tan \alpha}{\cos \Psi}, \quad \text{and} \quad dy = \frac{z}{\cos \Psi} \sec^2 \alpha d\alpha$$

also

$$x = z \tan \Psi, \quad \text{and} \quad dx = z \sec^2 \Psi d\Psi$$

furthermore, $\cos \overline{\Psi} = \cos \Psi \cos \alpha$ by substitution of the above expressions in 28-III.2, and after algebraic operations we obtain the general formula for the stress at a depth z under the center of a uniformly loaded rectangular area

$$\sigma_z = \frac{\chi q}{2\pi} \int_{-\alpha_0}^{\alpha_0} \cos^\chi \alpha d\alpha \int_{\Psi_1}^{\Psi_2} \cos^{\chi-1} \Psi d\Psi \qquad \text{29-III.2}$$

The above expression may be integrated for different values of χ. Especially important is $\chi = 2$, intermediate between Westergaard and Boussinesq, that may represent a stratified soil mass with strain modulus practically constant with depth. The value $\chi = 3$ represents the Boussinesq solution for an isotropic soil mass. Another interesting value is the Fröhlich, $\chi = 4$, representing a uniform deposit of cohesionless soil where the compressibility or strain modulus decreases strongly with depth. Hence, from integration of 29-III.2 we obtain:

Case I: $\chi = 2$, $\sigma_z = q \cdot I_{\chi 2}$

$$I_{\chi 2} = \frac{1}{\pi} \left(\alpha_0 + \frac{1}{2} \sin \alpha_0\right) \sin \Psi_1 - \sin \Psi_2 \qquad \text{30-III.2}$$

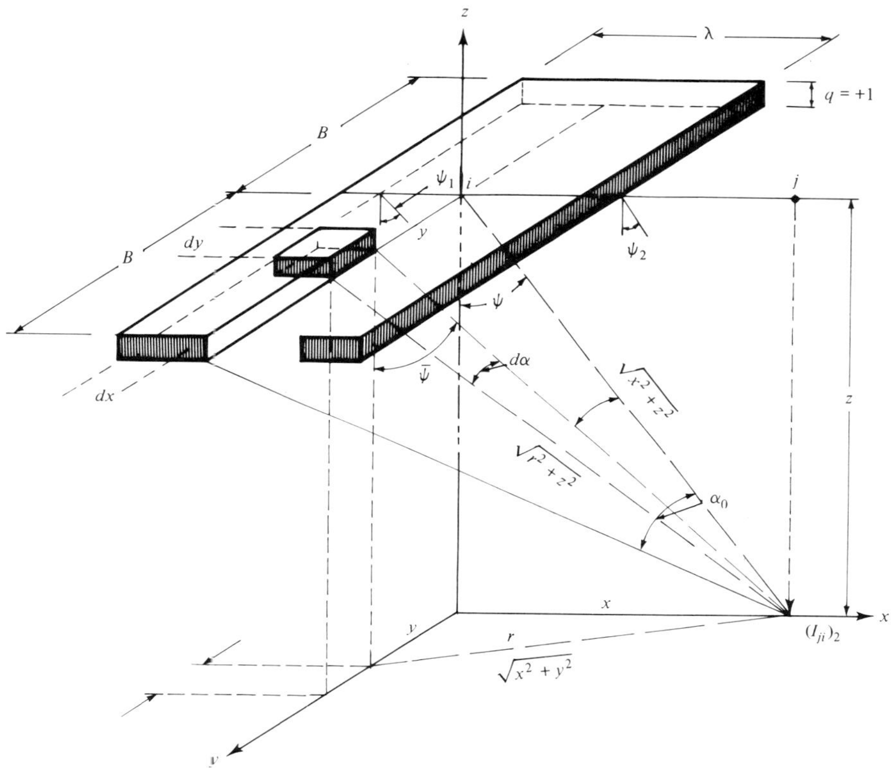

Fig. 8-III.2 Vertical stress at a point due to a uniformly loaded rectangular area.

Case II: $\chi = 3,\ \sigma_z = q \cdot I_{\chi 3}$

$$I_{\chi 3} = \frac{3}{2\pi}\left(\sin \alpha_0 - \frac{\sin^3 \alpha_0}{3}\right)\{(\Psi_1 - \Psi_2 + \sin(\Psi_1 - \Psi_2)\cos(\Psi_1 + \Psi_2)\} \quad \text{31-III.2}$$

Case III: $\chi = 4,\ \sigma_z = q \cdot I_{\chi 4}$

$$I_{\chi 4} = \frac{1}{\pi}\left(\frac{3}{2}\alpha_0 + \frac{3}{4}\sin 2\alpha_0 + \sin \alpha_0 \cos^3 \alpha_0\right)$$

$$\times \left\{(\sin \Psi_1 - \sin \Psi_2) - \frac{1}{3}(\sin^3 \Psi_1 - \sin^3 \Psi_2)\right\} \quad \text{32-III.2}$$

The angular arguments in the above formulas have the following values:

$$\alpha_0 = \tan^{-1}\frac{B}{\sqrt{x^2 + z^2}}, \quad \Psi_1 = \tan^{-1}\frac{x + \frac{\lambda}{2}}{2}, \quad \Psi_2 = \tan^{-1}\frac{x - \frac{\lambda}{2}}{2} \quad \text{33-III.2}$$

It is of particular importance to consider a uniformly loaded sector. According to Fig. 9-III.2, we can write

$$\sigma_z = \frac{q}{z^2} \int I_\chi \, da \qquad \text{34-III.2}$$

hence

$$\sigma_z = \frac{\chi}{2\pi} \frac{q}{z^2} \int_0^\theta d\theta \int_{R_1}^{R_2} \left[\frac{1}{1 + \frac{r^2}{z^2}} \right]^{(\chi+2)/2} r dr \qquad \text{35-III.2}$$

after integration we have

$$\sigma_z = q \cdot \frac{\theta}{2\pi} \left\{ \left[1 + \left(\frac{R_1}{z} \right)^2 \right]^{-\chi/2} - \left[1 + \left(\frac{R_2}{z} \right)^2 \right]^{-\chi/2} \right\}$$

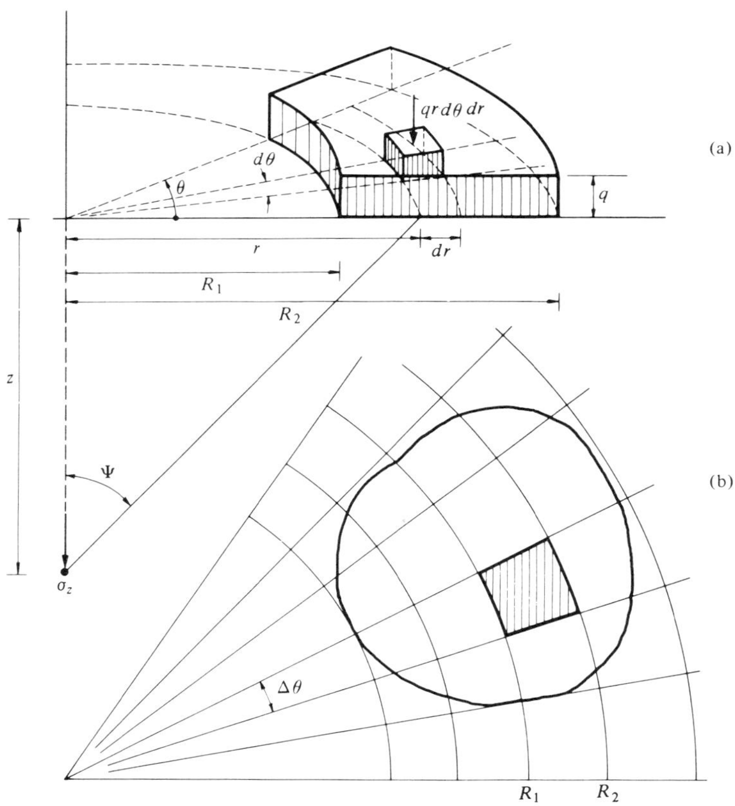

Fig. 9-III.2 Stress at a point for a uniformly loaded sector.

When $R_1 = 0$, and $R = R_2$, the total loaded sector of angle θ and radius R is obtained:

$$\sigma_z = q \cdot \frac{\theta}{2\pi} \left\{ 1 - \left[1 + \left(\frac{R}{z} \right)^2 \right]^{-\chi/2} \right\} \qquad \text{36-III.2}$$

When $\theta = 2\pi$, we obtain the stress under a circular uniformly loaded area of radius R_a:

$$\sigma_z = q \left\{ 1 - \left[1 + \left(\frac{R_a}{z} \right)^2 \right]^{-\chi/2} \right\} \qquad \text{37-III.2}$$

From the above formulas we can write the influence value as

$$I_{R\chi} = \left\{ 1 - \left[1 + \left(\frac{R_a}{z} \right)^2 \right]^{-\chi/2} \right\} \qquad \text{38-III.2}$$

or in terms of the central angle (Fig. 9-III.2)

$$I_{R\chi} = (1 - \cos^{\chi} \Psi) \qquad \text{39-III.2}$$

For the isotropic case, $\chi = 3$. The influence factor given by 35-III.2 may be found plotted for different values of χ in Fig. 10-III.2.

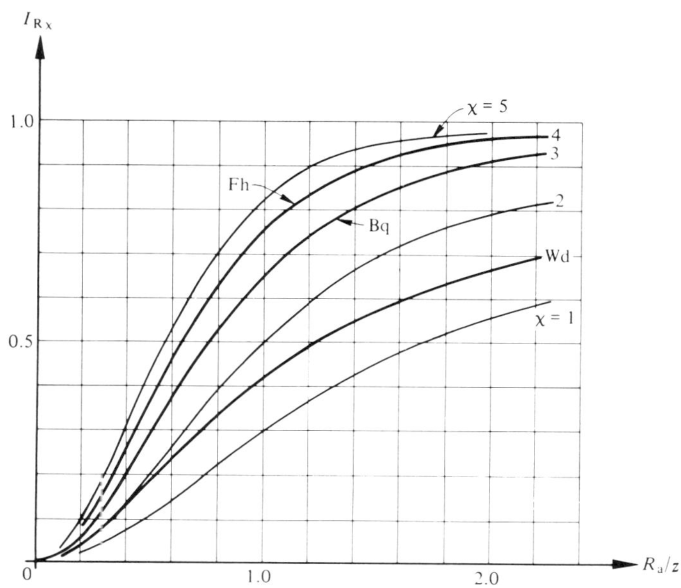

Fig. 10-III.2 Influence values at the center of a uniformly loaded surface of radius R_a.

2.3 The Stress Nets

In foundation engineering it is necessary as a practical procedure to determine the stress distribution in the soil mass, because of different magnitude of loads and loaded areas applied at the surface of the ground. For these purposes, let us continue operating with the influence factor $I_{R\chi}$. The increment of stress at depth z because of a uniformly loaded area in a sector limited by R_2 and R_1 is

$$\Delta\sigma_z = q\,\frac{\theta}{2\pi}\,[\Delta I_{R\chi}]_{R_1}^{R_2} \qquad \text{40-III.2}$$

From the above, it can be seen that any area is built up of small areas of a sector or fractions, as shown in Fig. 9(b)-III.2. By summing the small loaded areas, we obtain the total stress at a point. Let us call

$$N_\chi = \frac{\theta}{2\pi}\,[\Delta I_{R\chi}] \qquad \text{41-III.2}$$

The value N_χ representing the influence of one curvilinear rectangle is made constant. Therefore, the polar coordinates surface system will be formed by a net of equal influence curvilinear rectangles. This ingenious idea was introduced by N. M. Newmark (1942), and the stress net constructed in this manner for Boussinesq's solution is known as Newmark's Chart. Using the stress net, it is only necessary to count the number of rectangles and their fraction entering in the uniformly loaded area. The total stress at a point under the origin of the polar coordinates is

$$\sigma_z = (\text{no. of rectangles})\,q \cdot N_\chi \qquad \text{42-III.2}$$

The construction of the stress net is very simple using the graphical construction herein described. The influence value $I_{R\chi}$ is made to represent Westergaard, Boussinesq, Fröhlich or any other intermediate solution for a uniformly loaded circular area. The values of $I_{R\chi}$ are plotted against R/z in Fig. 10-III.2 Equation 40-III.2 may be written in the form

$$\frac{\Delta\sigma_z}{q} = \frac{\theta}{2\pi}\,\Delta I_{R\chi} \qquad \text{43-III.2}$$

The value of θ is conveniently selected dividing the polar coordinate system into any number of sectors between 20 to 60, depending on the size of the net and precision required; hence: $1/20 < \theta/2\pi < 1/60$. A constant value for $\Delta I_{R\chi}$ may be chosen, let us say, between 0.1 to 0.05. The graphical representation of $I_{R\chi}$ is divided into equal increments, and referred to the R/z axis; then circles are traced for the values of R/z found. Thereafter, the surface is divided into the number of sectors selected, and the net is ready for use.

As an example let us consider that the net for Fröhlich's stress distribution factor $\chi = 4$ is required. The values selected are $\Delta I_{R4} = 0.1$ and $\theta/2\pi = 1/40$.

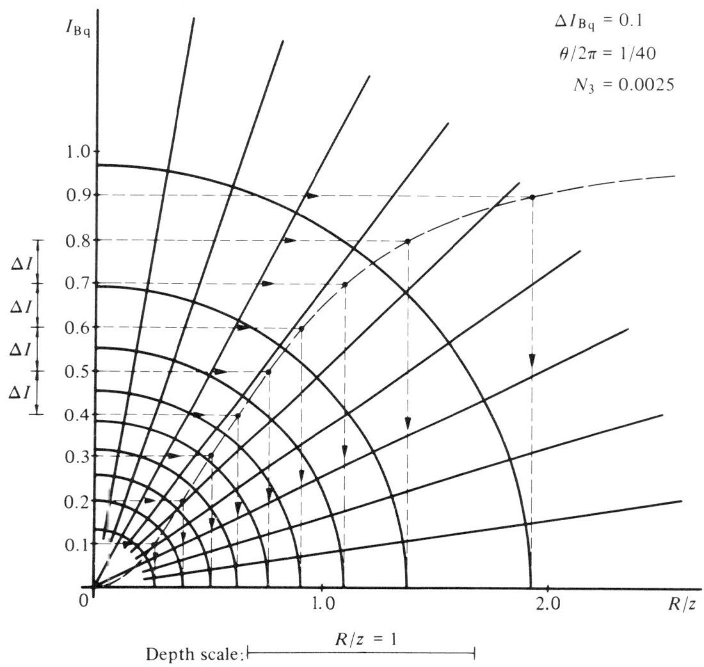

Fig. 11-III.2 Boussinesq stress net, $\chi = 3$.

Therefore, $N_4 = 0.00250$ will be the influence value of each one of the curvilinear rectangles obtained for $\chi = 4$, according to graphical construction shown in Fig. 12-III.2. The construction for stress nets for Boussinesq and Westergaard are shown in Figs. 11 and 13-III.2, respectively. When using the stress net to determine stresses at the edge of the loaded area where a discontinuity in the unit load occurs, it is convenient to determine the stress at about 5% of the side dimensions of the plot as shown in Fig. 14-III.2. In this way, as $z \to 0$, the total net is included at the edge point, and $\sigma_z = q$.

To illustrate the use of the stress net, assume it is required to determine the stress at a depth of 20 m under corner a of the L-shaped uniformly loaded area shown in Fig. 15-III.2. The area dimensions are divided by the depth $z = 20$ m, in order to make the plan dimensionless. The dimensionless ratios are shown in circles close to dimensions of the area. Then the loaded area is traced at the same scale of that of the stress net shown at the bottom of the drawing representing $R/z = 1$. Thereafter, point a is placed at the origin of the polar coordinate system, and the number of curvilinear rectangles is counted. In this particular case, $n = 60$, including fractions. Since $N = 0.0025$, then:

$$\frac{\sigma_z}{q} = 60 \times (0.0025) = 0.150$$

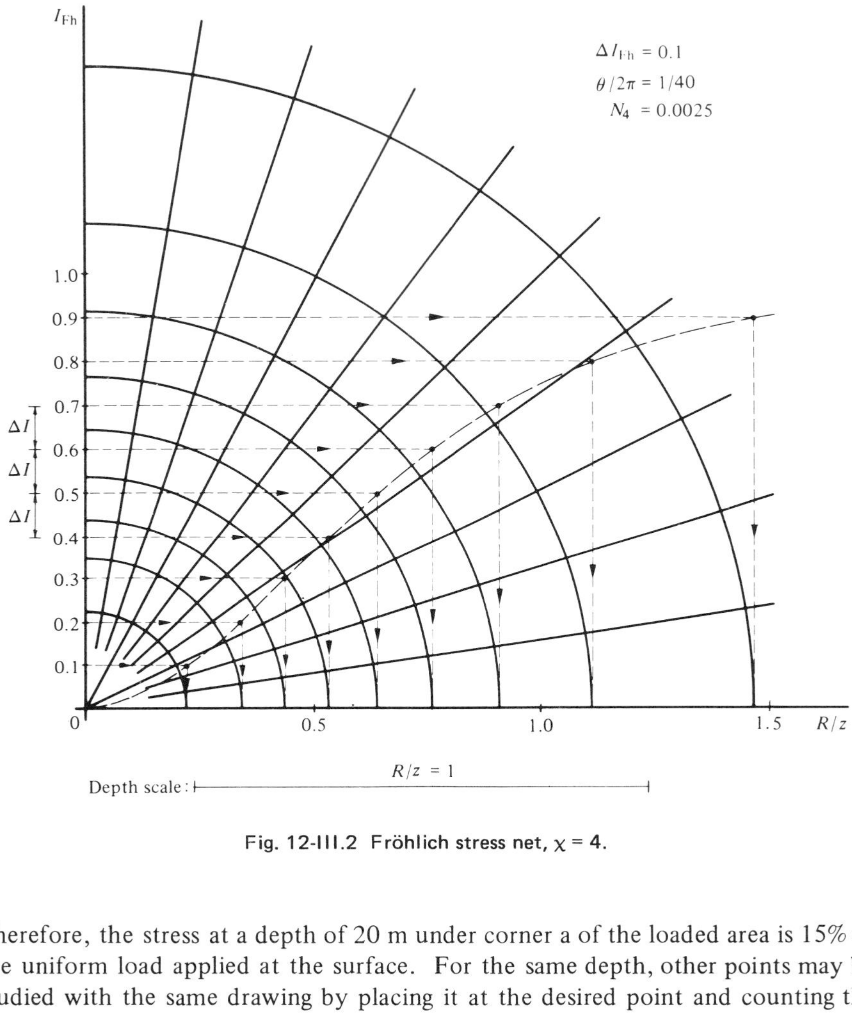

Fig. 12-III.2 Fröhlich stress net, $\chi = 4$.

Therefore, the stress at a depth of 20 m under corner a of the loaded area is 15% of the uniform load applied at the surface. For the same depth, other points may be studied with the same drawing by placing it at the desired point and counting the rectangles enclosed in the area for the new position. At any other depth, however, a new ratio R/z is determined and a new drawing to the scale of the net is made. For instance, if the vertical stress at the same corner is required for $z = 15$ m, then the scale of the net will now represent this depth. The new ratios are 4/3 on the long sides of the loaded area and 2/3 on the short sides. The new drawing will be drawn 33% larger than for $z = 20$ m. In this case, the number of rectangles in the area is $n = 75$, and

$$\frac{\sigma_z}{q} = 75\,(0.0025) = 0.187$$

Hence, the stress at 15 m depth is 18.7% of the uniform surface load q. Vertical

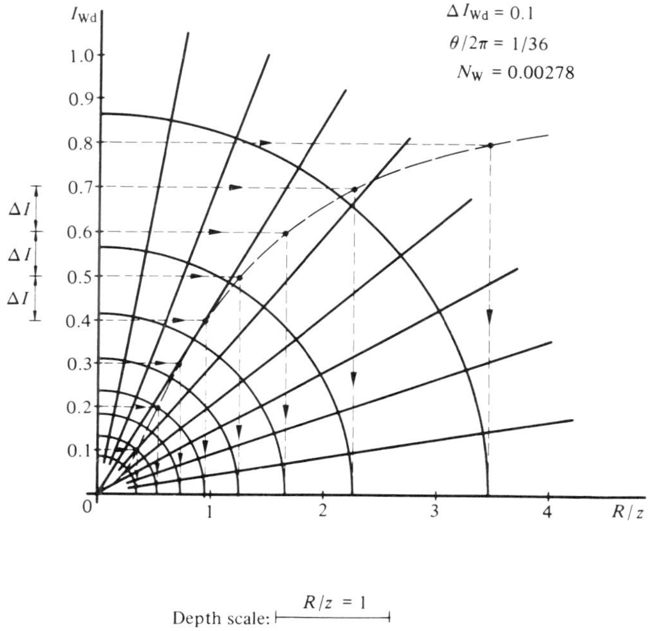

Fig. 13-III.2 Westergaard stress net.

stress nets for Westergaard, and for $\chi = 1, 2, 3, 4$ and 5 are found in Appendix B; they may be used in determining vertical stresses in the subsoil.

It has been stated that to estimate the strain modulus, it is necessary to know the confining volumetric stress with depth (Fig. 16-III.2):

$$\sigma_c = \frac{1 + 2K_0}{3}\,\sigma_{0z} + \Delta\sigma_{qc}$$

or

$$\sigma_c = \sigma_{0c} + \Delta\sigma_{qc} \qquad \text{44-III.2}$$

The value σ_{0c} may be estimated knowing K_0, and the value $\Delta\sigma_{qc}$, from the increment of stresses induced by the applied load q:

$$\Delta\sigma_{qc} = \tfrac{1}{3}\,[\Delta\sigma_x + \Delta\sigma_y + \Delta\sigma_z]$$

Using Bq-solution for the sector in polar coordinates

$$\Delta\sigma_z = q\,\frac{\theta}{2\pi}\,[1 - \cos^3\Psi]$$

and for $\nu = 0.5$, $\Delta\sigma_\theta = 0$. However

$$\Delta\sigma_r = q\,\frac{\theta}{2\pi}\,[3(1 - \cos\Psi) - (1 - \cos^3\Psi)]$$

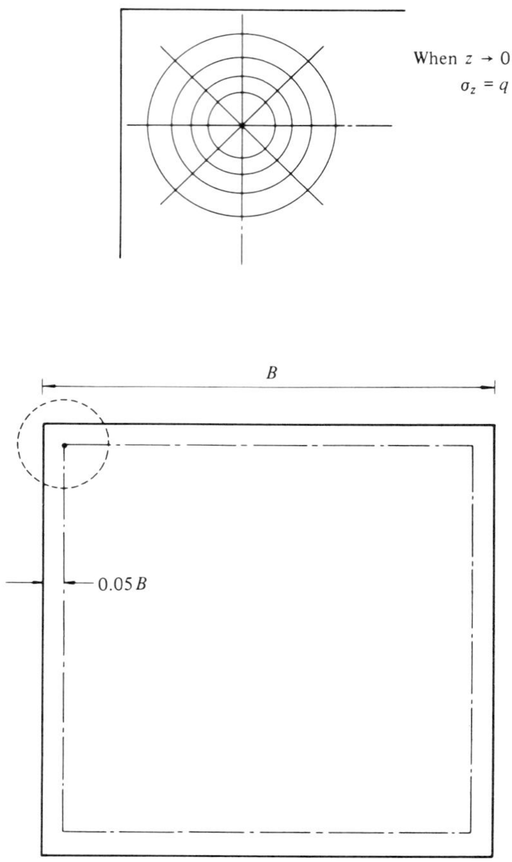

Fig. 14-III.2 Recommendation for determining stresses under the edge of a uniformly loaded surface.

Moreover

$$\Delta\sigma_x = \Delta\sigma_r \cos^2\theta + \Delta\sigma_\theta \sin^2\theta$$

$$\Delta\sigma_y = \Delta\sigma_r \sin^2\theta + \Delta\sigma_\theta \cos^2\theta$$

from which:

$$\Delta\sigma_{qc} = \tfrac{1}{3}\,(\Delta\sigma_z + \Delta\sigma_r) \qquad \text{45-III.2}$$

Substituting the values of $\Delta\sigma_z$ and $\Delta\sigma_r$ into 45-III.2, we obtain the value for $\Delta\sigma_{qc}$ for a sector with angle θ:

$$\Delta\sigma_{qc} = q\,\frac{\theta}{2\pi}\,(1 - \cos\Psi) \qquad \text{46-III.2}$$

The values of $I_{Rc} = (1 - \cos\Psi)$ corresponding to $\chi = 1$ are plotted in Fig. 10-III.2;

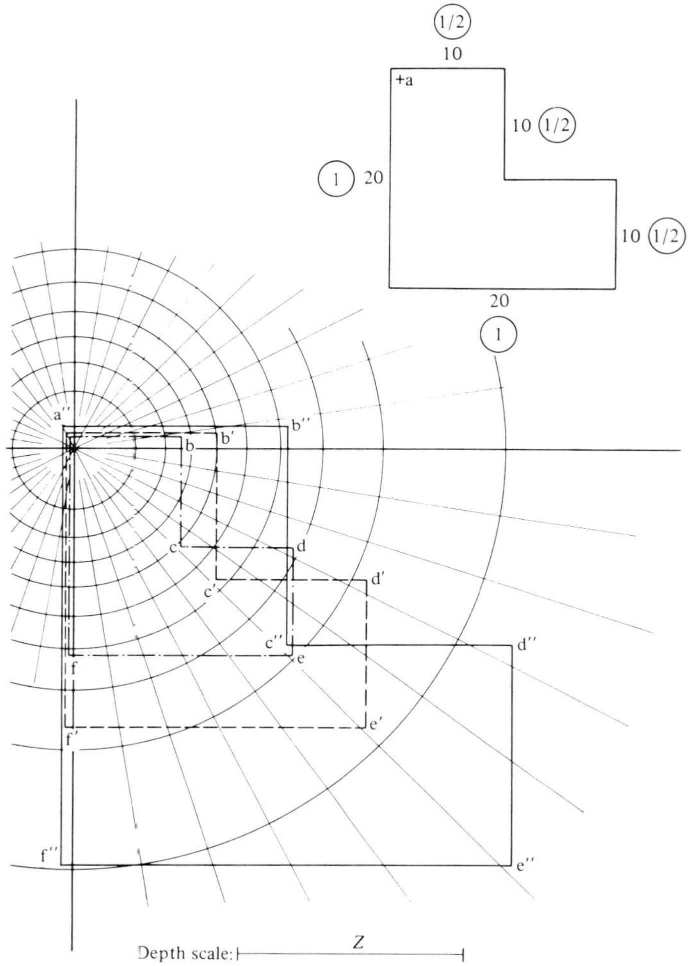

Fig. 15-III.2 Use of a stress net.

hence for a curvilinear rectangle in the sector θ uniformly loaded:

$$\Delta(\Delta\sigma_{qc}) = q\,\frac{\theta}{2\pi}\,\Delta I_{Rc}$$

from which we may call: $(\theta/2\pi)\;\Delta I_{Rc} = N_c$, the influence value for each rectangle in the stress net. Therefore:

$$\Delta\sigma_{qc} = (\text{no. of rectangles})\,q\cdot N_c \qquad \text{47-III.2}$$

The stress net for $\Delta\sigma_{qc}$ may be traced using the graphical procedure described before, which may be found in Appendix B. For soils where the value of ν is

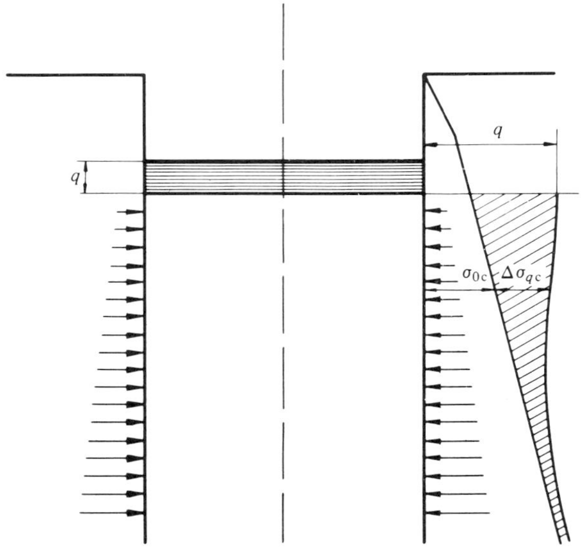

Fig. 16-III.2 Confining volumetric stresses.

different from 0.5, the values given by 47-III.2 are multiplied by (2/3) $(1+\nu)$. The values of $\Delta\sigma_{qc}$ so obtained are used in expression 44-III.2 to calculate the volumetric stress at the depth the strain modulus of the soil is required, as explained before (see formula 9-III.2 and section II.3.7).

In case of a rectangular uniformly loaded finite strip we can obtain a formula to calculate the octahedral or volumetric confining stress $\Delta\sigma_{qc}$ at any point on a perpendicular plane passing through the center of the loaded strip, according to:

$$\Delta\sigma_{qc} = \tfrac{1}{3}(\Delta\sigma_z + \Delta\sigma_y + \Delta\sigma_x)$$

for $\nu = 0.5$ we obtain in polar coordinates from 45-III.2

$$\Delta\sigma_{qc} = \tfrac{1}{3}(\Delta\sigma_z + \Delta\sigma_r) \qquad \text{45-III.2}$$

We calculate the value of $\Delta\sigma_{qc}$ for $\chi = 3$, for which the vertical stress at a point in the soil mass has been determined already, 29-III.2, hence:

$$\Delta\sigma_z = \frac{3q}{2\pi}\int_{-\alpha_0}^{\alpha_0} \cos^3 \alpha d\alpha \int_{\Psi_1}^{\Psi_2} \cos^2 \Psi d\Psi \qquad \text{48-III.2}$$

which after integration gives formula 31-III.2, Case II, for $\chi = 3$.

In order to find the value of $\Delta\sigma_{qc}$ we have to evaluate $\Delta\sigma_r$. From the point load formulas given early in this section, when $\nu = \frac{1}{2}$

$$\Delta\sigma_r = \frac{Q}{z^2}\left[\left(\frac{r}{z}\right)^2 \cdot I_{Bq}\right] \qquad \text{16-III.2}$$

$$\Delta\sigma_\theta = 0$$

here $Q = q \cdot dxdy$, and with the aid of Fig. 8-III.2 the following values have been previously obtained:

$$dy = \frac{z}{\cos \Psi} \sec^2 \alpha d\alpha$$

$$dx = z \sec^2 \Psi d\Psi$$

$$\cos \overline{\Psi} = \cos \Psi \cos \alpha$$

Furthermore,

$$\tan \overline{\Psi} = \frac{r}{z}$$

$$I_{Bq} = \frac{3}{2\pi} \cos^5 \overline{\Psi}$$

After proper substitutions and algebraic operations we find:

$$\Delta\sigma_r = \frac{3q}{2\pi} \int_{-\alpha_0}^{\alpha_0} \cos \alpha d\alpha \int_{\Psi_1}^{\Psi_2} d\Psi - \frac{3q}{2\pi} \int_{-\alpha_0}^{\alpha_0} \cos^3 \alpha d\alpha \int_{\Psi_1}^{\Psi_2} \cos^2 \Psi d\Psi$$

We notice that the second term in the second member has the value of formula 48-III.2, hence

$$\Delta\sigma_r = \frac{3q}{\pi} (\Psi_1 - \Psi_2) \sin \alpha_0 - \Delta\sigma_z$$

From the octahedral stress we obtain

$$\Delta\sigma_{qc} = \frac{q}{\pi} (\Psi_1 - \Psi_2) \sin \alpha_0 \qquad \text{49-III.2}$$

For values of $\nu \neq \frac{1}{2}$ we may use the following expression

$$\Delta\sigma_{qc} = \frac{2(1+\nu)}{3\pi} q(\Psi_1 - \Psi_2) \sin \alpha_0 \qquad \text{50-III.2}$$

The angular arguments in this case are the same as already given in 33-III.2. Calling the confining stress index value due to a rectangular loaded strip

$$I_{qc} = \frac{2(1+\nu)}{3\pi} (\Psi_1 - \Psi_2) \sin \alpha_0 \qquad \text{51-III.2}$$

we write

$$\Delta\sigma_{qc} = I_{qc} \cdot q \qquad \text{52-III.2}$$

The formula 51-III.2 is especially important when we use it in cohesionless soils to find the volumetric confinement state of stress in the soil mass under the foundation area when we wish to evaluate the strain modulus from $M_z = C_0 \sigma_c^{-n}$ (40-II.3; see Chapter VI).

2.4 Stratified Soil Masses

The elasticity solutions based on Boussinesq's problems may be used approximately to solve the stress distribution in stratified masses, if for the depth z we substitute $\eta \cdot z$, where η is a depth factor. For isotropic material $\eta = 1$; for highly stratified material $\eta = 1.5$; and for M_z decreasing very strongly with depth $\eta = 0.75$. The results obtained when using the depth factor for a circular uniformly loaded area are shown in Fig. 17-III.2. It may be demonstrated that curve C_1 for $\eta = 0.75$ lies close to Fröhlich's solution for $\chi = 5$, and curve C_2 for $\eta = 1.5$ approximates Westergaard's solution; the latter falls close to $\chi = 1.5$, giving a more reasonable stress distribution. For the isotropic case, the depth factor $\eta = 1$, for $\chi = 3$, falls

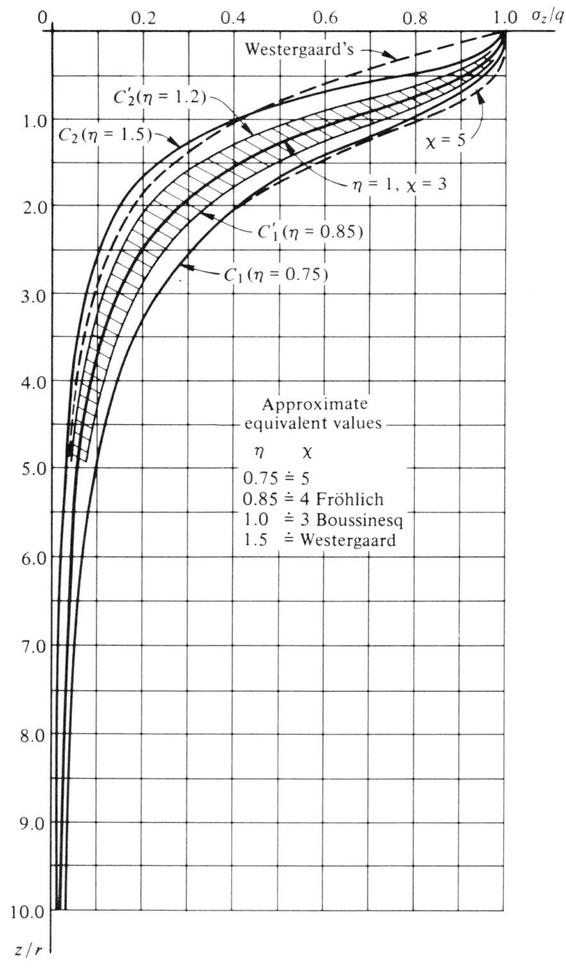

Fig. 17-III.2 Stress distribution under a circular loaded area for various nonisotropic cases.

approximately between both curves. Hence, it may be concluded that for most practical purposes $\chi = 3$ may be used. The depth factors proposed may be used in extreme cases; any other intermediate condition, however, can be interpolated to achieve a better guess on the stress distribution. Under field conditions, values may fall approximately on curve C_2' for stratified cohesive soils, and on curve C_1' for cohesionless soils Fig. 17-III.2. The depth factors are $\eta = 1.2$ and $\eta = 0.85$, respectively.

From the studies just made, it is concluded that in practice for quasi-isotropic semi-infinite masses, a solution other than Boussinesq's is seldom justified, since it has been shown that his solution empirically falls into the intermediate case found in nature in nonisotropic soil deposits. Since most of the cases in nature, however, do not deal with semi-infinite isotropic masses, and a stratum of very rigid or soft material is usually encountered interbedded in the soil mass, the following typical cases in practice should be investigated and may be solved approximately with the aid of the stress nets given in Appendix B:

A. Rigid stratum underlain by soft compressible soil deposit.
B. Soft and compressible soil stratum on firm base.
C. Compressible soil stratum interbedded between two rigid soil deposits.
D. Stratified soil deposit.
E. Nonstratified sand deposit, where compressibility decreases strongly with depth.

Case A. This case is illustrated in Fig. 18-III.2, and occurs when the load is applied on the surface of a rigid soil stratum underlain by a deposit of medium to high or very high compressibility. In this case, the rigid stratum has the tendency to produce a bridging action on the soft soil deposit; therefore, the stresses at the interface will be smaller than those given by Boussinesq. In the underlying soil mass, a corrected Boussinesq stress distribution may be used. In the rigid stratum, an approximation to Westergaard's solution or stress-net for $\chi = 1.5$ may be used. At the interface, the stresses will be: $[\sigma_d]_{1.5}$ and $[\sigma_d]_3$. In the underlying compressible soil deposit Boussinesq's solution is corrected in the proportion

$$\frac{[\sigma_d]_{1.5}}{[\sigma_d]_3} \qquad \text{53-III.2}$$

Figure 18-III.2 shows the configuration of stress distribution when this is the case.

Case B. The loaded area is resting on a compressible soil stratum on a firm base. In this case, the stresses have the tendency to concentrate at the rigid soil deposit because of restriction of the shear deformations on vertical planes, and at the horizontal interface plane $d - d$. The stresses at the rigid base may be calculated using Boussinesq's solution with a depth factor of $\eta = 0.75$ (Terzaghi (1943), Chapter XVIII), corresponding to stress net for $\chi = 5$. The case of a uniformly loaded circular area is illustrated in Fig. 19-III.2. Away from the rigid interface plane and into the soft stratum, the stresses have the tendency to dissipate rapidly into Boussinesq's

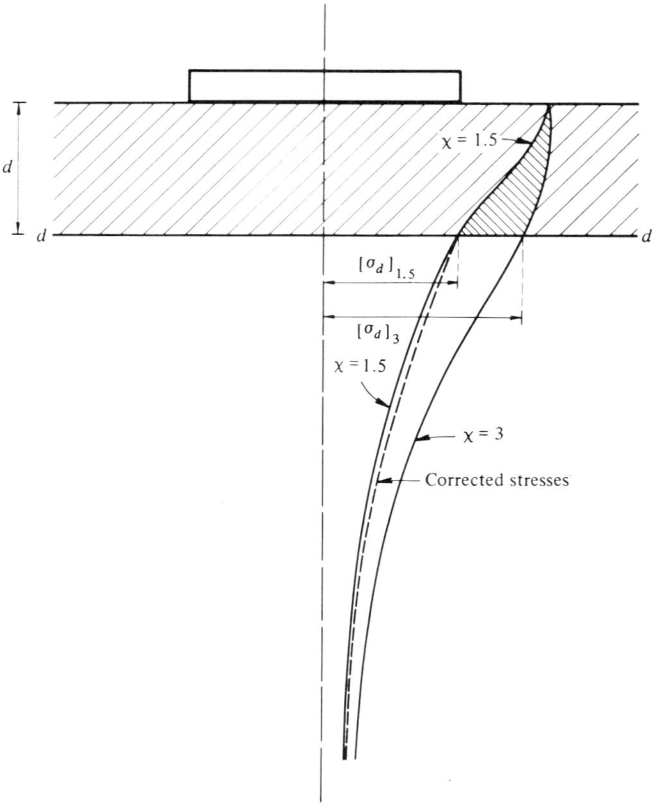

Fig. 18-III.2 Rigid stratum on compressible soil mass.

distribution. Toward the middle of the soft stratum, the concentration of stresses is very small; hence one can make the following assumptions:

for $z/d = 1$ the stress is equal to $[\sigma_d]_5$, stress net for $\chi = 5$
for $z/d = 0$ the stress is equal to $[\sigma_d]_3$, stress net for $\chi = 3$
for $z/d = \frac{1}{2}$ the stress is close to σ_{Bq}, stress net for $\chi = 3$

From $z = d/2$ to d the stress calculated with stress net $\chi = 3$ should be increased by factor f_z in such a way that:

$$\sigma_z' = [\sigma_{Bq}] \cdot f_z$$

The following correcting factor may be used:

$$f_z = 1 + \left(\frac{[\sigma_d]_5}{[\sigma_d]_3} - 1\right)\left(1 - \sin \pi \frac{d - z}{d}\right), \qquad \text{54-III.2}$$

from which it may be seen that when $z = d/2, f_z = 1$, and when $z = d, \sigma_z' = [\sigma_d]_5$.

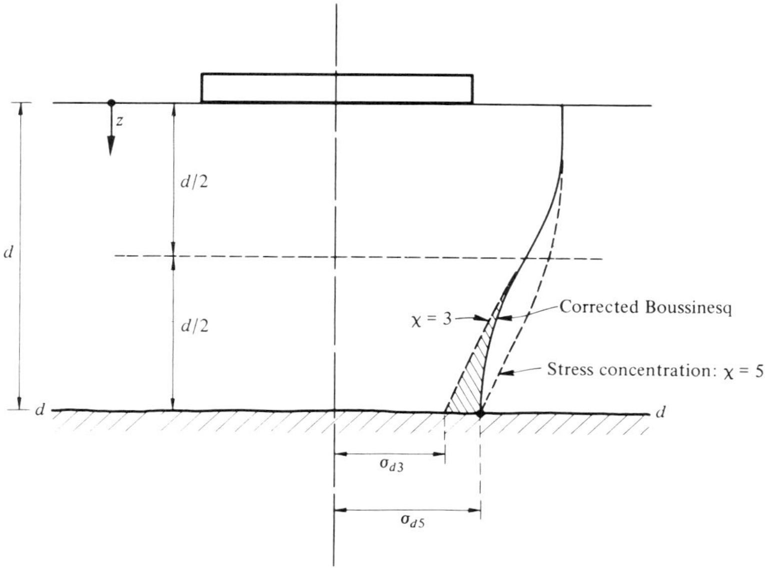

Fig. 19-III.2 Compressible soil mass on firm base.

Case C. In many practical cases in the field, it is customary to find the compressible soil stratum resting on firm ground, but also bounded at the upper part by a rigid soil stratum. Therefore, this case is the combination of Cases A and B. The upper rigid stratum produces on the soft compressible stratum a bridging effect; therefore, a reduction of stresses takes place. The stresses in the compressible stratum can be found with a corrected Boussinesq distribution as suggested for Case A, until reaching the firm stratum. With corrected σ'_{Bq} stresses, a second correction is performed as explained in Case B.

Case D. This condition is encountered when series of compact or semicompact sand strata are interbedded in soft compressible clay or silty clay strata and the compressibility decreases with depth (Fig. 20-III.2). In this case, an intermediate solution between $\chi = 1.5$ and $\chi = 4$ may be used. Hence for practical purposes, the stress net for $\chi = 2$ or 3 may be used, depending on the stratification encountered and variation with depth of m_{vh}/m_{vz} values.

Case E. This case is present in a thick, homogeneous nonstratified sand deposit when the compressibility decreases strongly with depth, in which case the solution using stress net for $\chi = 4$ may be applied.

In most practical cases, the precision of calculating vertical stresses using Westergaard, Boussinesq or Fröhlich solutions or any other intermediate semiempirical solution for the semi-infinite solid is far more accurate than the precision one obtains in determining the stress-strain-time characteristics of the soils. Therefore, the foundation engineer may use with confidence the solutions proposed based on the theory of elasticity. No matter how precise the foundation engineer's calculations

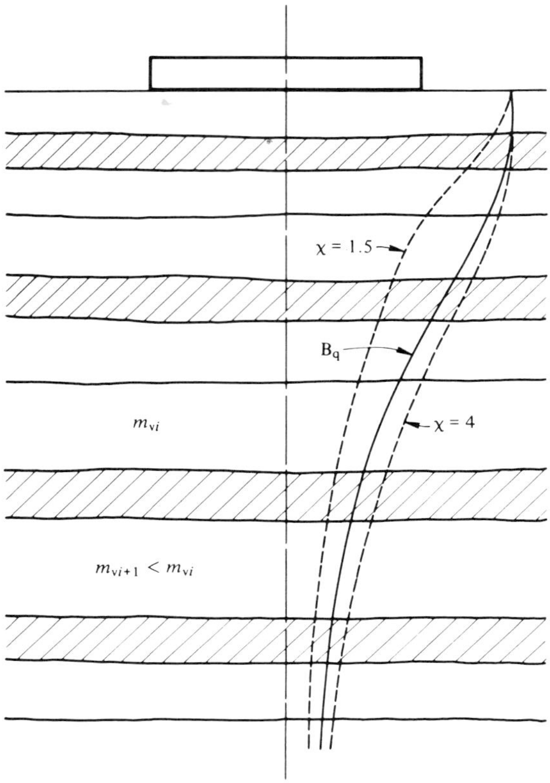

Fig. 20-III.2 Stratified soil mass.

of stresses in the ground and his knowledge of the stress-strain-time characteristics of the soil strata, he should always consider a factor of safety for settlement estimates. The foundation engineer should always be sure to remain under his reported settlement values. In general, the tendency is to overestimate the compressibility parameters determined in the laboratory with the real values in the field. The observed settlements show this fact, that is to say, settlements calculated with the use of semiempirical solutions of the theory of elasticity and the compressibility parameters determined in the laboratory are larger than the observed settlements.

Knowing the stress distribution under the loaded surface, the vertical displacements may be estimated at any location desired:

$$\delta_n = \Sigma\, M_i(\sigma_c, t)\, \Delta\sigma_i \cdot \Delta z_i \qquad \text{55-III.2}$$

in which M_i is the average strain modulus for a stratum of thickness Δz_i; it is a function of the confining stress level σ_c and time t (section II.3). The average increment of stress $\Delta\sigma_i$ in the strata due to load application is determined by means of the corresponding stress net solutions described above for the engineering characteristics of the subsoil encountered.

2.5 Vertical Displacements of Rigid Footings

The foundation engineer is interested in studying the vertical displacement of rigid footings and, in general, rigid foundations. When the foundation structure may be considered rigid in comparison with the compressibility of the subsoil, the footing will undergo a uniform vertical displacement when the loads and loaded area are symmetrical. This case may be investigated from the solution given by the theory of elasticity for rectangular rigid footings on a semi-infinite isotropic elastic solid:

$$\delta_a = \alpha_e (2B)(1 - \nu^2) M_e \cdot q_a \qquad \text{56-III.2}$$

in which $2B$ is the width and $2L$ the length of the footing; the factor α_e is a function of L/B; M_e is the elastic strain modulus; ν is Poisson's ratio, and q_a is the average unit contact stress.

From the values of α_e given in Fig. 21-III.2, we see that for a square footing α_e = 0.95, and this value increases as a linear function of the logarithm of L/B. In prac-

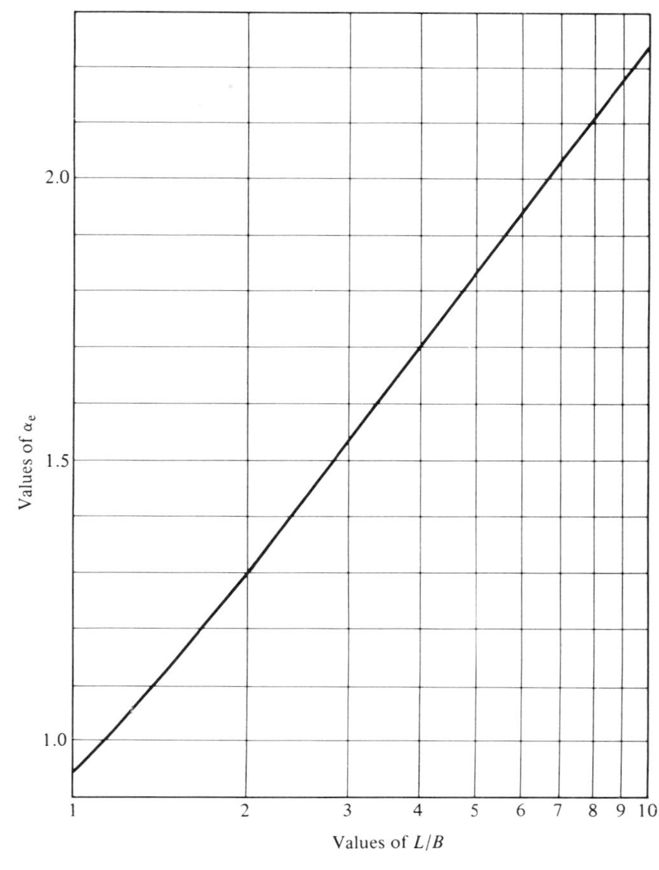

Fig. 21-III.2 The factor α_e for rigid rectangular footings on semi-infinite isotropic soil mass.

tice the foundation engineer, however, seldom has to consider rectangular rigid footings larger than $L/B = 6$, for which $\alpha_e = 2$, in which case the settlement is on the order of two times greater than for a square footing. Furthermore, from the theory of elasticity, the settlement of a square uniformly loaded area on the semi-infinite isotropic elastic solid is:

at the center:

$$\delta_c = 1.12(2B)(1 - \nu^2)M_e q_a$$

at the corner: 57-III.2

$$\delta_e = \tfrac{1}{2}\,\delta_c$$

The square loaded area has an average settlement of $\delta_a = 0.85\,\delta_c$.

In nature, however, and in large loaded areas, the stratigraphical conditions of the subsoil are important to consider. The compressibility usually decreases with depth and eventually at a certain depth, a soil deposit of low or very low compressibility is encountered, where the vertical displacements may be neglected if compared with those obtained in the overlying compressible soil strata. Therefore, the solutions given by the theory of elasticity based on a semi-infinite isotropic solid are only of academic value and of general guidance. The only means to perform more accurate settlement estimates is to calculate the increment of stresses at each one of the strata by means of the stress nets discussed in preceding paragraphs, and from them calculate the vertical displacements corresponding to the compressibility properties of each one of the strata encountered.

The settlement analysis is performed for several points in the loaded surface considering no foundation stiffness. The total loaded surface is divided into small areas a_n called tributary areas and the vertical displacement δ_n is calculated for each one of them (Fig. 22-III.2). If the footing is rigid, then the average settlement is obtained from the vertical displacement volume of the loaded area; hence

$$\delta_a = \frac{\Sigma \delta_n a_n}{\Sigma a_n} \qquad \text{58-III.2}$$

Moreover, if the footing surface is not loaded uniformly, the rigid footing will rotate. The magnitude of the calculated settlements will not be symmetrical with respect to the principal axis of the loaded surface. When such is the case, the value of δ_a given by 58-III.2 will only represent the average vertical displacement at the center of gravity of the vertical displacement volume. The rotation of the footing in a plane will increase the vertical deflection at point x_n, y_n, because of rotation along the XX principal axis in $\theta_x \cdot x_n$, and along the YY principal axis in $\theta_y \cdot y_n$, respectively (Fig. 22(b)-III.2). Taking moments of the volume displacements for the tributary areas a_n, we obtain for the XX axis

$$\Sigma(\theta_x \cdot x_n)a_n \cdot x_n = \Sigma a_n \delta_n x_n$$

Therefore, in the XX axis direction

$$\theta_x = \frac{\Sigma a_n \delta_n x_n}{\Sigma a_n x_n^2} \qquad \text{59-III.2}$$

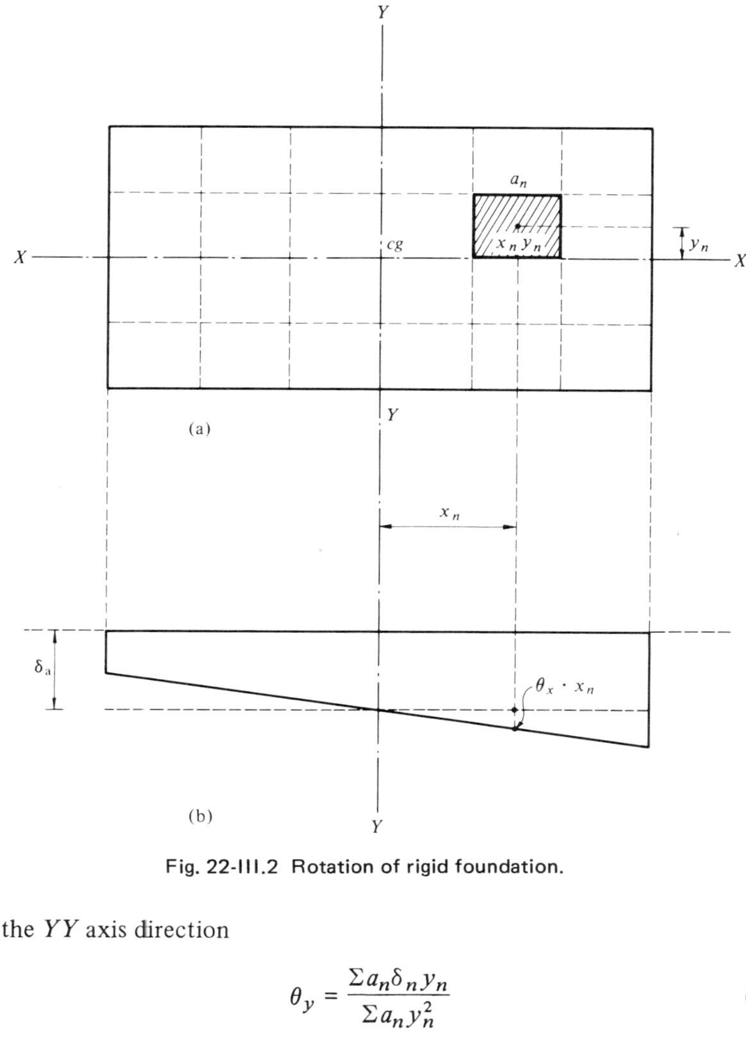

Fig. 22-III.2 Rotation of rigid foundation.

and in the YY axis direction

$$\theta_y = \frac{\Sigma a_n \delta_n y_n}{\Sigma a_n y_n^2} \qquad \text{60-III.2}$$

The total deflection of the rigid foundation at points with coordinates x_n, y_n is

$$\delta_n = \delta_a \pm \theta_x \cdot x_n \pm \theta_y \cdot y_n \qquad \text{61-III.2}$$

The procedure given to calculate the average vertical displacement and rotations of a foundation on stratified soil deposits of high and very high compressibility is particularly useful to investigate the behavior of monolithic rigid foundation structures. Furthermore, this procedure is useful when it is required to investigate approximately the effect of overturning moments in tall buildings subjected to earthquake or heavy wind horizontal forces (see Chapter XII). The bending moments and shears, however, depend on the contact stresses at the foundation grade elevation imposed by the stiffness of the foundation structure. This problem is discussed in Chapters IV, VI and XII on soil-foundation structure interaction.

III.3 HYDRAULIC PRESSURES: LIQUID PHASE

3.1 Water Flow Components

The hydraulic state of pressure in the soil mass must be determined, since the total stress and the hydraulic pressure are related in all the problems of subsoil stability. The hydraulic pressure may be measured in the field by means of piezometers (Fig. 1-III.3). The porous point is installed at the depth at which it is desired to measure the hydraulic pressure. The installation is made in a drill hole, and seals of commercial bentonite are placed immediately underneath and above the porous point, to confine the depth at which the recording is made. Between the two seals, a filter of sand is placed to permit the water easy access into a piezometric pipe, where the water can reach the elevation necessary to balance the hydraulic pressure. Different types of piezometers are available commercially, from elec-

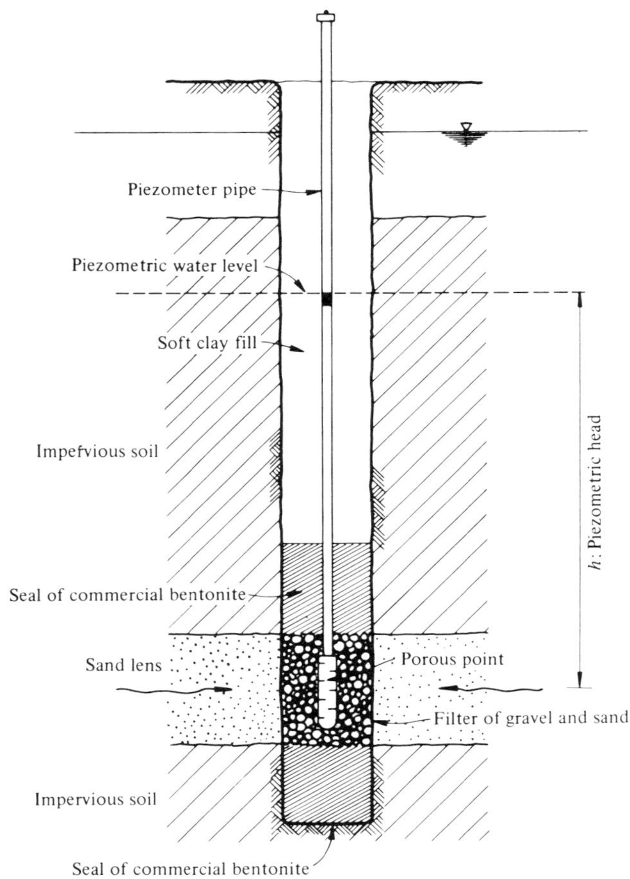

Fig. 1-III.3 Piezometer.

tronic to simple head-pipe piezometers. The description of these instruments is beyond the scope of this book; however, the foundation engineer should be acquainted with the instrumentation necessary to measure pore pressures in the ground and in the laboratory. The piezometric height is called the piezometric head, and the level the water reaches into the piezometer is called the piezometric water level. In design problems, usually the piezometric water levels may be estimated theoretically by means of flow nets.

Consider several piezometers installed at different elevations into the subsoil; the stratigraphy and unit weight of different soil strata are known (Fig. 2-III.3). The vertical stresses have been calculated for different conditions, that is, for static conditions when all the piezometric water levels rise at the water surface elevation. If the piezometric water levels suffer a drawdown, the water enters into hydrodynamic conditions with a component of vertical descending flow; the effective stress in the soil mass may be calculated for any depth by means of the following expression:

$$\sigma_z = p_z - h\gamma_w \qquad \text{1-III.3}$$

in which h is equal to the piezometric head at the point under consideration. Notice that the increase in the effective stress is equivalent to the reduction in the hydraulic pressure $\lambda\gamma_w$.

Thus one can conclude that a flow of water can increase or diminish the effective pressure, depending on the direction of the flow, and one can visualize its importance in connection with intergranular stresses. In order to generalize the problem and be able to relate it with further studies, let us study the components of the water flow. Consider a small element, through which the vertical component of a

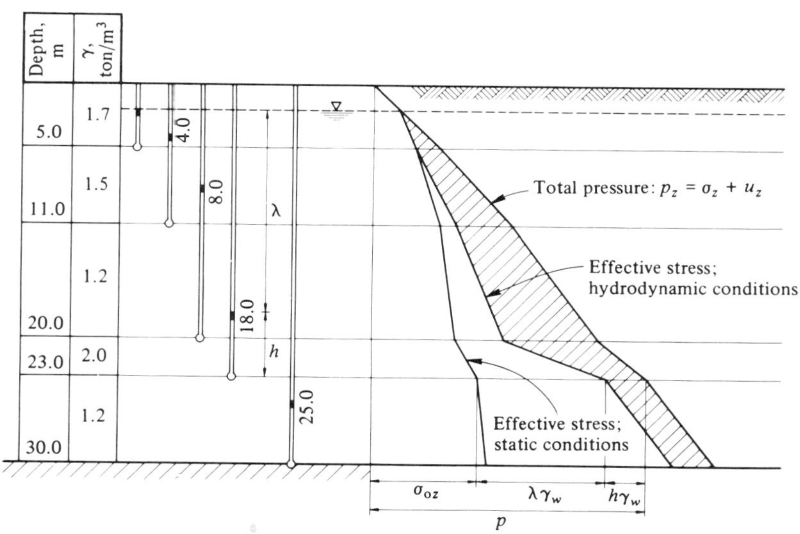

Fig. 2-III.3 Effective stresses under vertical water flow.

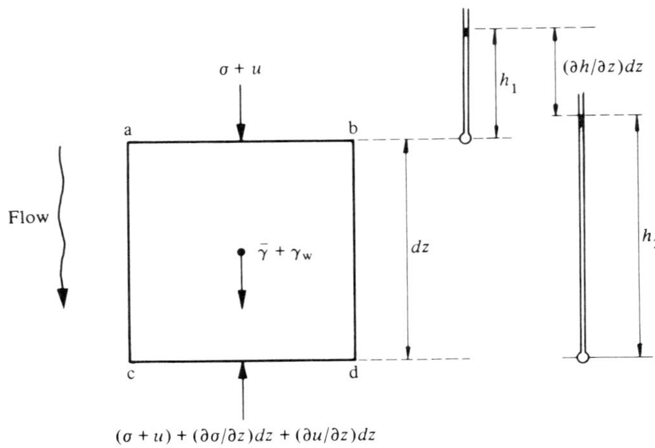

Fig. 3-III.3 Vertical flow component.

water current is passing; the pressure conditions on the upper and lower sides are indicated in Fig. 3-III.3. The element with submerged unit weight $\overline{\gamma}$ is in static equilibrium; therefore

$$\frac{\partial \sigma_z}{\partial z} dz + \frac{\partial u}{\partial z} dz = (\overline{\gamma} + \gamma_w) dz \qquad \text{2-III.3}$$

but the increment of hydraulic pressure in the face *cd* of the element is

$$\frac{\partial u}{\partial z} dz = -\gamma_w \frac{\partial h}{\partial z} dz + \gamma_w dz$$

Therefore, from 2-III.3

$$\frac{\partial \sigma_z}{\partial z} - \gamma_w \frac{\partial h}{\partial z} = \overline{\gamma}$$

Notice that the ratio $\frac{\partial h}{\partial z}$ represents the hydraulic gradient i_z for the considered element, in the direction of the hydrodynamic flow, and therefore

$$\frac{\partial \sigma_z}{\partial z} = \gamma_w \cdot i_z + \overline{\gamma} \qquad \text{3-III.3}$$

The gradient may be considered positive when a downward flow is present and negative when the flow is upward, that is to say, the gradient will be positive when in the direction of the gradient there is an increase in the effective stress, and negative when there is a reduction. The horizontal component (Fig. 4-III.3) may

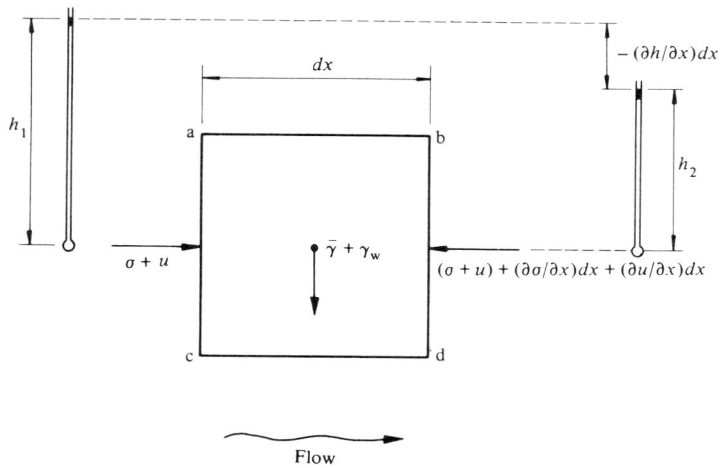

Fig. 4-III.3 Horizontal flow component.

be treated in the same form, and for equilibrium we obtain

$$\frac{\partial \sigma_x}{\partial x} + \frac{\partial u}{\partial x} = 0 \tag{4-III.3}$$

However, $u_x = h_x \gamma_w$, and for any depth z we obtain

$$\frac{\partial \sigma_x}{\partial x} = -\gamma_w \frac{\partial h_x}{\partial x}$$

Therefore:

$$\frac{\partial \sigma_x}{\partial x} = -\gamma_w \cdot i_x \tag{5-III.3}$$

By means of 3 and 5-III.3 important problems related with intergranular pressures in the soil mass may be studied. In order to illustrate one of the most important problems, let us imagine an isotropic mass where the water is in hydrodynamic conditions.

3.1a Downward Flow. In this case, as depth increases the piezometric levels diminish and the effective stresses increase; therefore, the hydraulic gradient is positive. Furthermore, if we assume that the gradient is constant with depth, we can integrate expression 3-III.3:

$$\sigma_z = (\gamma_w \cdot i_z + \bar{\gamma})z + C \tag{6-III.3}$$

and for $z = 0$ and $\sigma_z = 0$, we have $C = 0$. Hence, the intergranular pressure may be calculated by means of the following expression:

$$\sigma_z = (\gamma_w \cdot i_z + \bar{\gamma})z \tag{7-III.3}$$

from which one can conclude that if $i_z = 0$, the conditions in the water are static and the intergranular stress is equivalent to

$$\sigma_z = \overline{\gamma} z$$

This formula expresses the submerged weight of a soil column with depth z. On the other hand, if the piezometers indicate a piezometric water level close to the elevation at which the porous point is installed, then the hydraulic gradient is equal to unity, and the maximum value of the effective stress will be

$$\sigma_z = (\gamma_w + \overline{\gamma})z$$

This expresses the total weight of a soil column from the surface to a depth z.

3.1b Upward Flow. In this case, the hydraulic gradient may be considered negative. The piezometric elevations increase, reducing the effective stresses in the soil mass. Under these conditions

$$\sigma_z = (\overline{\gamma} - \gamma_w \cdot i_z)z + C$$

for $z = 0$ and $\sigma_z = 0$, we obtain $C = 0$, hence:

$$\sigma_z = (\overline{\gamma} - \gamma_w \cdot i_z) \cdot z \qquad \text{8-III.3}$$

Therefore, an upward water flow reduces the initial effective stresses. The following special condition may be encountered if $\sigma_z = 0$:

$$\overline{\gamma} - \gamma_w \cdot i_z = 0 \qquad \text{9-III.3}$$

When the material is noncohesive, this condition produces a state of unstable equilibrium; since the intergranular stress is zero the friction forces between grains is zero. The material enters into a state of liquefaction known as quicksand. The critical hydraulic gradient for this phenomenon is the following:

$$i_c \geqslant \frac{\overline{\gamma}}{\gamma_w} \qquad \text{10-III.3}$$

The phenomenon takes place at the ground surface where the water leaves the soil when the hydraulic gradient is larger than the ratio of weights shown above. The water velocity is high and produces a state of flotation of the material carrying grains of sand and silt, producing the phenomenon known as boiling, usually observed at the toe of the structure. The boiling is followed by erosion of the soil extending into the soil mass, endangering the stability of the bottom of deep excavations. Therefore, special precautions should be taken to avoid a failure. Because of this phenomenon, it is necessary mainly to reduce the hydraulic gradient to values considerably smaller than critical, or to protect the soil surface with filters to stop the escape of fine grains from the subsoil material.

3.2 The Flow Net

From the above discussion, the foundation engineer can realize the importance of knowing the state of hydraulic pressures in the ground for a particular problem in the zones where seepage forces may be expected. In order to solve this problem theoretically, it is necessary to know the position of seepage lines and the variation of the hydraulic potential making the water flow along the current lines. In case of two-dimensional problems, it is relatively simple to obtain a reasonable solution from the practical foundation engineering point of view, giving satisfactory results to estimate the problem under study.

Let us consider two current lines limiting a flow channel in the system. One can establish that the discharge in this conduit is constant, and assuming that the problem under study is a bidimensional flow, then the current lines may be represented by the following general equation:

$$\Phi = f_1(x, y) \qquad \text{11-III.3}$$

here x and y are the coordinates of reference; a curve is obtained for each constant value of Φ representing a flow line. On the other hand, an equipotential surface or surface of equal piezometric level may be defined by the following relation:

$$\Psi = h + z \qquad \text{12-III.3}$$

Here z represents the height of the point over the horizontal reference plane, and h is the piezometric head for that point with reference to the equipotential surface. Therefore, along an equipotential line the value of Ψ is constant. Piezometers can be installed along an equipotential line, and the water will reach the same elevation; therefore in a bidimensional flow an equipotential surface is represented by a curve in the vertical plane. From these considerations, one can construct the following theorem: the geometrical location of points of equal piezometric level represent an equipotential line. The equation of the family of equipotential lines may be expressed by

$$\Psi = f_2(x, y) \qquad \text{13-III.3}$$

in which for each constant value of Ψ, a curve representing an equipotential line or line of equal piezometric level is obtained at the considered cross section. The system formed by the family of flow lines and that of the equipotential lines defines two families of curves that are known as the flow net, and each problem according to the boundary conditions and certain specific geometry has one particular solution.

The object is then to study the mathematical solutions of seepage problems according to 11 and 13-III.3. Consider a small element of the soil mass subjected to seepage. The water velocity that enters and leaves the element may be represented vectorially as indicated in Fig. 5-III.3; however, if there is no loss of water, that is to say, if there is continuity of flow, the water discharge entering sides *ab*

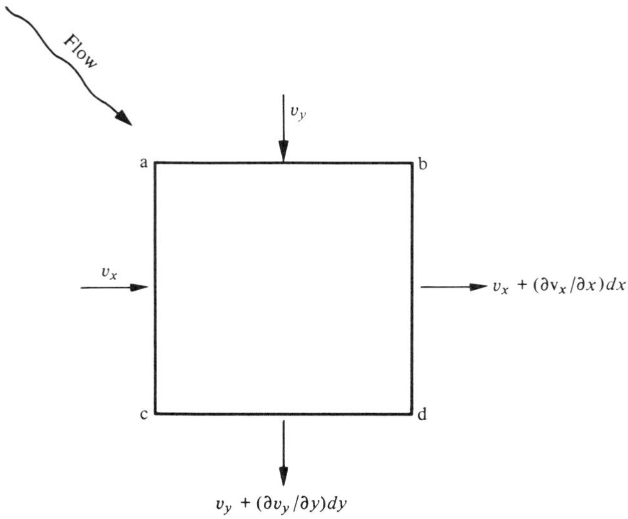

Fig. 5-III.3 Water flow through a soil element.

and *ac* will be the same as that leaving sides *bd* and *dc*. Therefore, one can establish the condition of flow continuity:

$$\frac{\partial v_x}{\partial x} + \frac{\partial v_y}{\partial y} = 0 \qquad \text{14-III.3}$$

Assuming laminar flow according to Darcy's law, and calling the coefficients of permeability in both directions parallel to the coordinate axis k_x and k_y, respectively, then we have

$$k_x \frac{\partial i_x}{\partial x} + k_y \frac{\partial i_y}{\partial y} = 0$$

The hydraulic gradient $i = d\Psi/ds$ may be expressed by the formula

$$\frac{d\Psi}{ds} = \frac{\partial \Psi}{\partial x} \cdot \frac{dx}{ds} + \frac{\partial \Psi}{\partial y} \cdot \frac{dy}{ds} \qquad \text{15-III.3}$$

Therefore, the components of the gradient along the coordinate axis are

$$i_x = \frac{\partial \Psi}{\partial x}, \quad \text{and} \quad i_y = \frac{\partial \Psi}{\partial y}$$

and substituting we obtain

$$k_x \frac{\partial^2 \Psi}{\partial x^2} + k_y \frac{\partial^2 \Psi}{\partial y^2} = 0 \qquad \text{16-III.3}$$

3.2a Isotropic Soil Mass. In case the material may be considered isotropic, then $k_x = k_y = k$, therefore:

$$\frac{\partial^2 \Psi}{\partial x^2} + \frac{\partial^2 \Psi}{\partial y^2} = 0 \qquad \text{17-III.3}$$

This differential equation expresses the condition of the potential function for continuous and laminar flow. The family of the equipotential lines will be $\Psi = f_2(x, y)$ and should satisfy 17-III.3, known in mathematics as Laplace's equation.

From mathematical considerations one can also establish that the flow lines expressed by $\Phi = f_1(x, y)$ satisfy the equation

$$\frac{\partial^2 \Phi}{\partial x^2} + \frac{\partial^2 \Phi}{\partial y^2} = 0 \qquad \text{18-III.3}$$

It may be shown also by physicomathematical considerations that the functions f_1 and f_2 are perpendicular to each other; in other words, the family of flow lines are orthogonal to the family of equipotential lines or lines of equal piezometric level. This is only true when Laplace's equations are satisfied by the above-mentioned functions and the soil mass is isotropic.

In accordance with the above discussion, the solution would consist in finding two algebraic functions Ψ and Φ, satisfying Laplace's equations and the boundary conditions of the problem. This procedure is practically impossible to achieve for all problems found in foundation engineering. The solution of Laplace's equations, however, may be obtained either solving the differential equations by finite differences or by a graphical method that in the majority of the cases gives sufficient accuracy from the practical point of view.

To find the graphical solution to trace flow nets, consider Fig. 6-III.3, a family of flow lines perpendicular to the equipotential lines in an isotropic material. This may be arranged graphically, in such a way that the drop Δh in piezometric head between two equipotential lines is constant, and that all the flow channels formed by pairs of flow lines have the same discharge. The flow net traced under these conditions will be formed by curvilinear rectangles, as in Fig. 6-III.3. In fact, the magnitude of the discharge in each one of the conduits just described, when the material is isotropic, may be expressed by:

$$\Delta q = k \cdot \frac{\Delta h}{l_1} s_1 = k \cdot \frac{\Delta h}{l_2} s_2 = \cdots\cdot k \cdot \frac{\Delta h}{l_n} s_n \qquad \text{19-III.3}$$

hence:

$$\frac{s_1}{l_1} = \frac{s_2}{l_2} = \cdots\cdots \frac{s_n}{l_n}$$

In order to facilitate the tracing of the flow net one can write $s_n/l_n = 1$; hence the flow net will be formed by curvilinear squares with dimension of $l_n = s_n = a_n$.

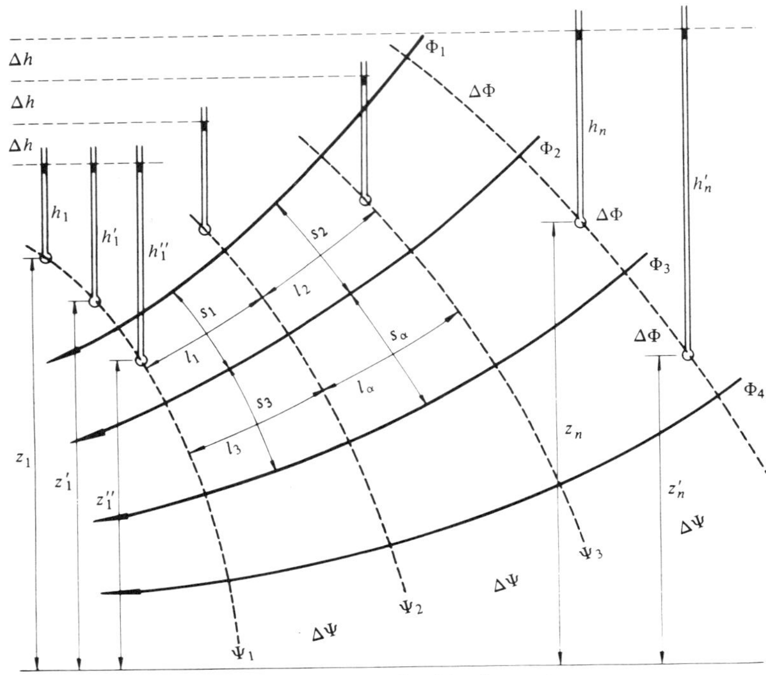

Fig. 6-III.3 Water flow net.

Therefore, if the flow net is drawn with curvilinear squares satisfying the boundary conditions of the problem, a solution is obtained for the hydrodynamic conditions satisfying Laplace's equations; at the same time, the following theorems will be satisfied:

1. The material is considered homogeneous and isotropic and Darcy's law is valid.
2. The flow lines and those representing equal piezometric water elevations are orthogonal to each other.
3. Flow channels limited by two flow lines have the same discharge: $\Delta q = k \cdot \Delta h$, in which Δh is the drop in head between two successive equipotential lines.
4. The drop Δh between equipotentials is constant.
5. The drop in head Δh between two successive equipotentials is equal to the total drop divided by the number N_p of equipotential drops $\Delta h = H_t/N_p$.
6. The total discharge of the flow net is equal to the discharge of one flow channel Δq multiplied by the number N_f of channels: $Q = \Delta q \cdot N_f$.

Knowing the flow net, it is possible to determine at each point of the flow zone the following properties:

a) hydraulic pressure: $u = \gamma_w \cdot h$;

b) the hydraulic gradient: $i = H_t/a \cdot N_p$;
c) the Darcy's velocity of the current in the conduits: $v = k \cdot H_t/a \cdot N_p$;
d) finally, the discharge:

$$Q = kH_t \frac{N_f}{N_p} \qquad \text{20-III.3}$$

The relation N_f/N_p, defines the characteristic geometrical solution of the flow net. For each problem there will be one and only one characteristic value:

$$R_c = \frac{N_f}{N_p} \qquad \text{21-III.3}$$

Using Fig. 7-III.3, we illustrate the application of the above-discussed graphical procedure. We seek the discharge per lineal meter for an excavation limited by an impermeable sheet-pile wall and the hydraulic pressures on the sheet-pile wall along plane (*c-d*) and in the soil mass on a horizontal plane (*a-b*). In this problem, the perimeter of the sheet-pile wall defines a boundary line of flow and the impermeable base another. The horizontal surfaces on each side of the sheet pile respectively correspond to equipotential lines. From the figure, the solution found for the flow net is 8 flow channels and 9 potential drops; therefore, the characteristic value of the net is $R_c = 8/9$. If the coefficient of permeability is $k = 10^{-3}$ cm/sec, the discharge per lineal meter will be

$$\begin{aligned} Q &= k \cdot H_t \cdot R_c \\ &= 10^{-3} \cdot 600 \cdot 0.89 = 53 \text{ cm}^3/\text{sec per lineal meter} \end{aligned}$$

The hydraulic pressure distribution against the sheet-pile is shown in the figure; also shown is the uplift pressure on plane (*a-b*), determined from the equipotential

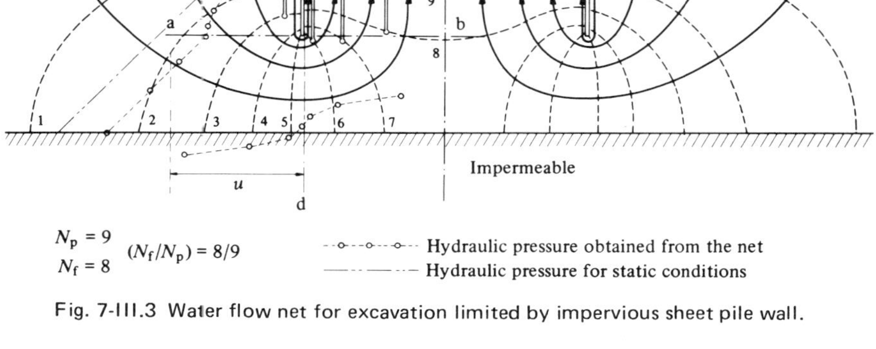

Fig. 7-III.3 Water flow net for excavation limited by impervious sheet pile wall.

lines using the expression $\Psi = h + z$. The hydraulic pressure at any point is $u = \gamma_w \cdot h$. Notice that the effect of the vertical flow diminishes considerably the static hydraulic pressures against the sheet pile. On the other hand, if the hydraulic pressure on horizontal plane (*a-b*) is larger than the effective weight of the material on this plane, then the bottom of the excavation will rise and failure will occur. This phenomenon is observed only in cohesive materials. In noncohesive materials, it is necessary to study the hydraulic gradient when the water leaves the soil to investigate the stability of the bottom of the excavation, because of the boiling effect. Notice that the largest hydraulic gradient is found close to the sheet pile. In this particular case, its value is 1/2 (Fig. 7-III.3); therefore, the bottom of the excavation may be considered stable.

3.2b Stratified Soil Mass The case of stratified soil masses may be represented by equation 16-III.3, where k_x and k_y are the average coefficients of permeability of the soil in the horizontal and vertical directions, respectively. To solve the problems by means of the graphical procedure just explained, a linear transformation of one of the coordinate axes is made, writing equation 16-III.3 as follows:

$$\frac{\partial^2 \Psi}{\partial x_t^2} + \frac{\partial^2 \Psi}{\partial y^2} = 0 \qquad \text{22-III.3}$$

Again Laplace's equation is obtained in an x_t, y coordinate system. In the new system of coordinates, one can solve the problem by the same procedure described before. The scale of transformation is found analyzing the first term of expression 22-III.3:

$$\frac{\partial^2 \Psi}{\partial x_t^2} = \frac{\partial^2 \Psi}{\partial x^2} \left(\frac{\partial x}{\partial x_t}\right)^2$$

However $\dfrac{\partial^2 \Psi}{\partial x_t^2} = \dfrac{k_x}{k_y} \cdot \dfrac{\partial^2 \Psi}{\partial x^2}$, and therefore the scale of transformation $\lambda = x_t / x$ is given by

$$\lambda = \sqrt{\frac{k_y}{k_x}} \qquad \text{23-III.3}$$

The equivalent permeability in the transformed system may be obtained if one compares the discharge of the two systems. Considering the discharge through a differential element, in the real system:

$$\Delta q = k_x \frac{\partial \Psi}{\partial x} dy + k_y \frac{\partial \Psi}{\partial y} dx$$

but in the transformed system $\dfrac{\partial \Psi}{\partial x} \cdot \lambda = \dfrac{\partial \Psi}{\partial x_t}$ and $dx = \dfrac{1}{\lambda} dx_t$; therefore, making proper substitutions we obtain

$$\Delta q = \sqrt{k_x k_y}\, \frac{\partial \Psi}{\partial x_t} dy + \sqrt{k_x k_y}\, \frac{\partial \Psi}{\partial y} dx_t$$

from which the equivalent coefficient of permeability in the transformed system is found to be

$$k_t = \sqrt{k_x k_y} \qquad \text{24-III.3}$$

Therefore, the discharge in the stratified soil with average permeabilities k_x and k_y will be:

$$Q = k_t \cdot H \cdot \frac{N_f}{N_p} \qquad \text{25-III.3}$$

The procedure to solve the flow net is simple. We have only to reduce the geometry, multiplying all horizontal dimensions by λ, and trace the flow net in the new geometry. The above-described theorems for graphic solution of Laplace's equations are valid in the transformed system x_t, y. The real geometry of the flow net is determined in the real system multiplying all the abscissas of the transformed system by $1/\lambda$, in which the flow lines are no longer orthogonal to the equipotential lines. Notice, however, that the characteristic value R_c of the flow net remains the same in both systems. Since the ordinates do not change, the hydraulic pressures will also remain the same in both systems for corresponding points; however, the hydraulic gradient has to be obtained in the real system. The following theorems are also valid:

1. The drop between equipotential lines is constant.
2. All the flow channels have the same discharge.

3.3 Average Coefficients of Permeability

In the case of stratified soils and when the permeability does not change considerably from one layer to another, it is necessary to find the average coefficient of permeability in the horizontal and vertical directions, respectively, since in practice the soil deposits are encountered stratified. Let Fig. 8-III.3 represent a stratified soil deposit for which one wishes to evaluate the average coefficient of permeability in the horizontal direction. Assume the coefficients of permeability in the horizontal direction for the different strata are $k_1, k_2, k_3, \ldots, k_n$, and the thicknesses of these strata, respectively, are $d_1, d_2, d_3, \ldots, d_n$. If k_x is the average coefficient of permeability in the horizontal direction, the discharge is:

$$k_x \cdot i \sum_1^n d_i = \sum_1^n k_i \cdot i d_i$$

Under equal gradient conditions for all the strata, we obtain

$$k_x = \frac{\sum_1^n k_i d_i}{\sum_1^n d_i} \qquad \text{26-III.3}$$

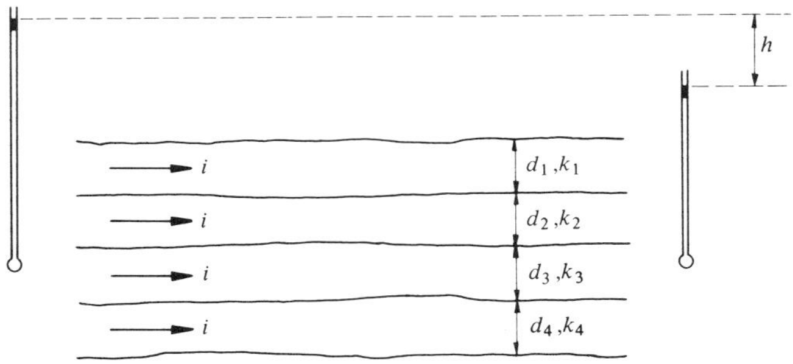

Fig. 8-III.3 Average coefficient of permeability for horizontal flow.

If the flow takes place in the vertical direction, the discharge per unit horizontal surface imposes the following conditions of flow continuity (Fig. 9-III.3):

$$v_1 = v_2 = \cdots = v_i = v_n$$

therefore:

$$k_1\frac{h_1}{d_1} = k_2\frac{h_2}{d_2} = \cdots = k_i\frac{h_i}{d_i} = \cdots = k_n\frac{h_n}{d_n}$$

On the other hand, we can write

$$H = h_1 + h_2 + \cdots + h_i \cdots + h_n$$

in which H is the total loss of potential for the seepage through the strata. After making proper substitutions, one obtains for the average coefficient of permeabil-

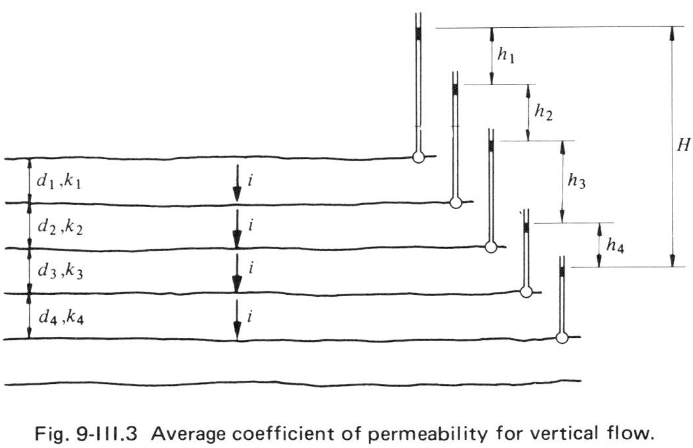

Fig. 9-III.3 Average coefficient of permeability for vertical flow.

ity in the vertical direction the following formula:

$$k_y = \frac{\sum_1^n d_i}{\sum_1^n \frac{d_i}{k_i}} \qquad \text{27-III.3}$$

Notice that this formula may be also used to calculate the average coefficients of consolidation, since this is proportional to the coefficients of permeability (see section II.1). When the problem takes place in series of sand deposits, the coefficient of permeability may be estimated by means of Hazen's approximate formula $k = 100D_{10}^2$ cm/sec, where D_{10} is the effective diameter in centimeters limiting 10% of the finer material contained in the sample. Formulas 26 and 27-III.3 are therefore valid for estimating the average coefficients of permeability

3.4 Vertical Flow in Stratified Soil Deposits

There are cases in practice when it is necessary to estimate the reduction of piezometric pressures in stratified soil deposits when only the change in the piezometric water elevations at the boundaries of the deposit is known.

Let us consider a stratified deposit limited by pervious strata, Fig. 10-III.3. The piezometric water elevations are known by means of piezometers at the upper and lower boundaries and at the interphase of the soil strata that may be considered as having approximately the same permeability characteristics. The conditions shown imply a vertical downward flow of water from the upper boundary toward the lower coarse sand deposit boundary. It is foreseen that there will be a change in the piezometric water elevations at the upper boundary λ_0 and at the lower boundary λ_b. These changes will induce water elevation changes of the intermediate piezometers at the interphase of the strata. The changes in the piezometric water elevations will increase the stresses in the subsoil and will produce the compression of the strata. Calling $(m_v d)_i = \alpha_i$ the compression of a stratum N, the change in thickness will be:

$$\Delta\delta_i^N = \alpha_i^N \gamma_w \frac{\lambda_{i-1} + \lambda_i}{2} \qquad \text{28-III.3}$$

Therefore, the compression for the total thickness of the deposit may be calculated if the values of λ_i are known for all the interphases, hence

$$\delta_i = \gamma_w \sum_A^N \alpha_i \frac{\lambda_{i-1} + \lambda_i}{2} \qquad \text{29-III.3}$$

From the above reasoning, it is shown to be important to develop a computational method to forecast the changes λ_i, when a drop in the piezometric elevations takes place at the boundary conditions. From continuity conditions the discharge per

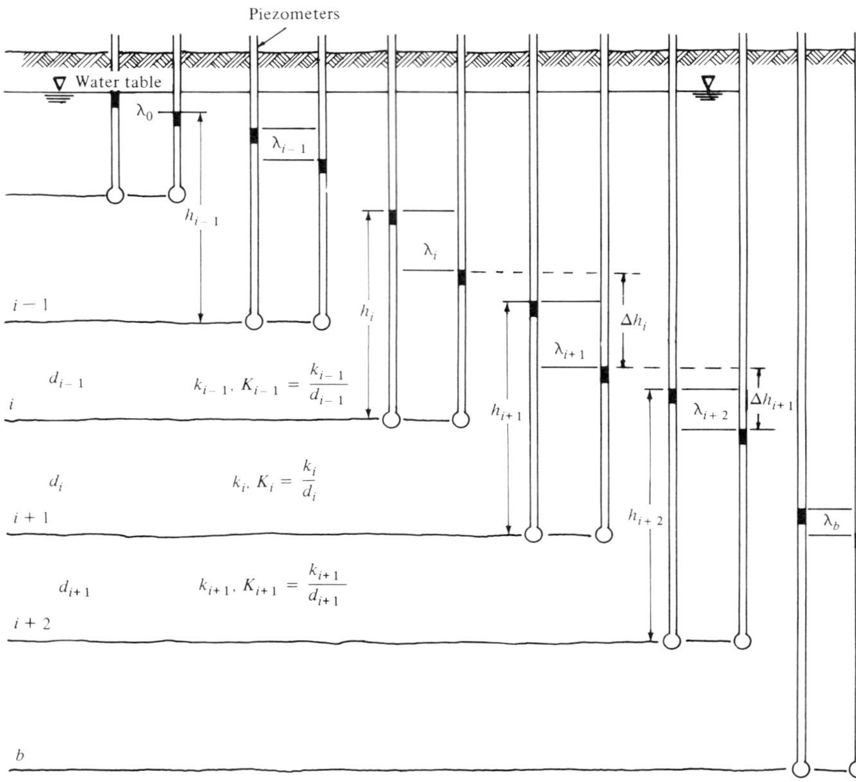

Fig. 10-III.3 Piezometric water elevation change in subsoil vertical component.

unit surface

$$k_1 \frac{\Delta h_1}{d_1} = k_2 \frac{\Delta h_2}{d_2} = \cdots \text{constant} \qquad \text{30-III.3}$$

From Fig. 10-III.3 we can write

$$\Delta h_i = (h_{i-1} + d_i - \lambda_{i-1}) - (h_i - \lambda_i)$$
$$\Delta h_{i+1} = (h_i + d_{i+1} - \lambda_i) - (h_{i+1} - \lambda_{i+1})$$

arranging terms

$$\Delta h_i = (h_{i-1} - h_i + d_i) - (\lambda_{i-1} - \lambda_i)$$
$$\Delta h_{i+1} = (h_i - h_{i+1} + d_{i+1}) - (\lambda_i - \lambda_{i+1}) \qquad \text{31-III.3}$$

considering that

$$k_i \cdot \frac{\Delta h_i}{d_i} = k_{i+1} \cdot \frac{\Delta h_{i+1}}{d_{i+1}}$$

we obtain

$$\frac{k_i}{d_i}(h_{i-1} - h_i + d_i) - \frac{k_i}{d_i}(\lambda_{i-1} - \lambda_1) = \frac{k_{i+1}}{d_{i+1}}(h_i - h_{i+1} + d_{i+1}) - \frac{k_{i+1}}{d_{i+1}}(\lambda_1 - \lambda_{i+1})$$

Calling

$$\Delta_i = (h_{i-1} - h_i + d_i), \qquad K_i = \frac{k_i}{d_i}$$

$$\Delta_{i+1} = (h_i - h_{i+1} + d_{i+1}), \qquad K_{i+1} = \frac{k_{i+1}}{d_{i+1}} \qquad \text{32-III.3}$$

and substituting these values and arranging terms we obtain

$$K_i\lambda_{i-1} - (K_i + K_{i+1})\lambda + K_{i+1}\lambda_{i+1} = K_i\Delta_i - K_{i+1}\Delta_{i+1} \qquad \text{33-III.3}$$

Using equation 33-III.3 the unknown values of λ may be determined. Assume a deposit is divided in four strata with thickness and known coefficients of permeability. At the boundaries a known piezometric drawdown takes place with values λ_0 and λ_b respectively. Therefore, the unknown values to be determined are λ_1, λ_2, and λ_3. From equation 33-III.3 we obtain

$$K_1\lambda_0 - (K_1 + K_2)\lambda_1 + K_2\lambda_2 = K_1\Delta_1 - K_2\Delta_2$$

$$K_2\lambda_1 - (K_2 + K_3)\lambda_2 + K_3\lambda_3 = K_2\Delta_2 - K_3\Delta_3$$

$$K_3\lambda_2 - (K_3 + K_4)\lambda_3 + K_4\lambda_4 = K_3\Delta_3 - K_4\Delta_4 \qquad \text{34-III.3}$$

Since $K_1\lambda_0$ and $K_4\lambda_4$ are known the set of simultaneous equations may be put in the following form:

$$\begin{bmatrix} -(K_1 + K_2) & K_2 & 0 \\ K_2 & -(K_2 + K_3) & K_3 \\ 0 & K_3 & -(K_3 + K_4) \end{bmatrix} \times \begin{vmatrix} \lambda_1 \\ \lambda_2 \\ \lambda_3 \end{vmatrix} = \begin{vmatrix} K_1\Delta_1 - K_2\Delta_2 - K_1\lambda_0 \\ K_2\Delta_2 - K_3\Delta_3 \\ K_3\Delta_3 - K_4\Delta_4 - K_4\lambda_4 \end{vmatrix}$$

35-III.3

It may be noticed that the matrix of the coefficients K is a three band diagonal matrix. Calling

$$T_1 = (K_1\Delta_1 - K_2\Delta_2) - K_1\lambda_0$$

$$T_2 = (K_2\Delta_2 - K_3\Delta_3)$$

$$T_3 = (K_3\Delta_3 - K_4\Delta_4) - K_4\lambda_b \qquad \text{36-III.3}$$

we write in matrix form

$$[K_i] \cdot |\lambda_i| = |T_i| \qquad \text{37.III.3}$$

or

$$|\lambda_i| = [K_i]^{-1} \cdot |T_i| \qquad \text{38-III.3}$$

Example:

Assume the initial conditions given in Fig. 11-III.3, and the change in boundary conditions, upper boundary $\lambda_0 = 3.0$ m, bottom boundary $\lambda_b = 2.0$ m; then

$$\Delta_1 = (3 - 4 + 3) = 2\text{m}, \quad K_1 = 0.66 \times 10^{-6}, \quad K_1\Delta_1 = 1.32$$

$$\Delta_2 = (4 - 8 + 5) = 1\text{m}, \quad K_2 = 1.60 \times 10^{-6}, \quad K_2\Delta_2 = 1.60$$

$$\Delta_3 = (8 - 7 + 6) = 7\text{m}, \quad K_3 = 1.00 \times 10^{-6}, \quad K_3\Delta_3 = 7.0$$

$$\Delta_4 = (7 - 6 + 2) = 3\text{m}, \quad K_4 = 0.2 \times 10^{-6}, \quad K_4\Delta_4 = 0.6$$

From which the relative values for the independent terms from 36-III.3 are:

$$T_1 = 1.32 - 1.6 - 1.98 = -2.26$$

$$T_2 = 1.60 - 7.0 = -5.40$$

$$T_3 = 7.0 - 0.6 - 0.4 = 6.00$$

The simultaneous equations to solve are

λ_1	λ_2	λ_3	T
-2.26	1.60	0	-2.26
1.60	-2.60	1.0	-5.40
0	1.0	1.2	6.00

from which $\lambda_1 = 4.69$ m, $\lambda_2 = 5.22$ m, and $\lambda_3 = 0.65$ m. Considering the coefficients of volume compressibility given in Fig. 11-III.3, the total compression in the deposit will be, on a long time basis, the sum of the compression of all the strata affected by the piezometric elevations drawdown

$$\delta = \sum (m_v d)\Delta\sigma$$

or

$$\delta = \sum \gamma_w (m_v d) \frac{\lambda_i + \lambda_{i+1}}{2}$$

hence

$$\delta = 0.03 \frac{0 + 3.0}{2} + 0.09 \frac{3 + 4.69}{2} + 0.01 \frac{4.69 + 5.22}{2} + 0.006 \frac{5.22 + 0.65}{2} + 0.001 \frac{0.65 + 2.0}{2}$$

and, the settlement at the ground surface due to the piezometric changes indicated is $\delta = 0.46$ m.

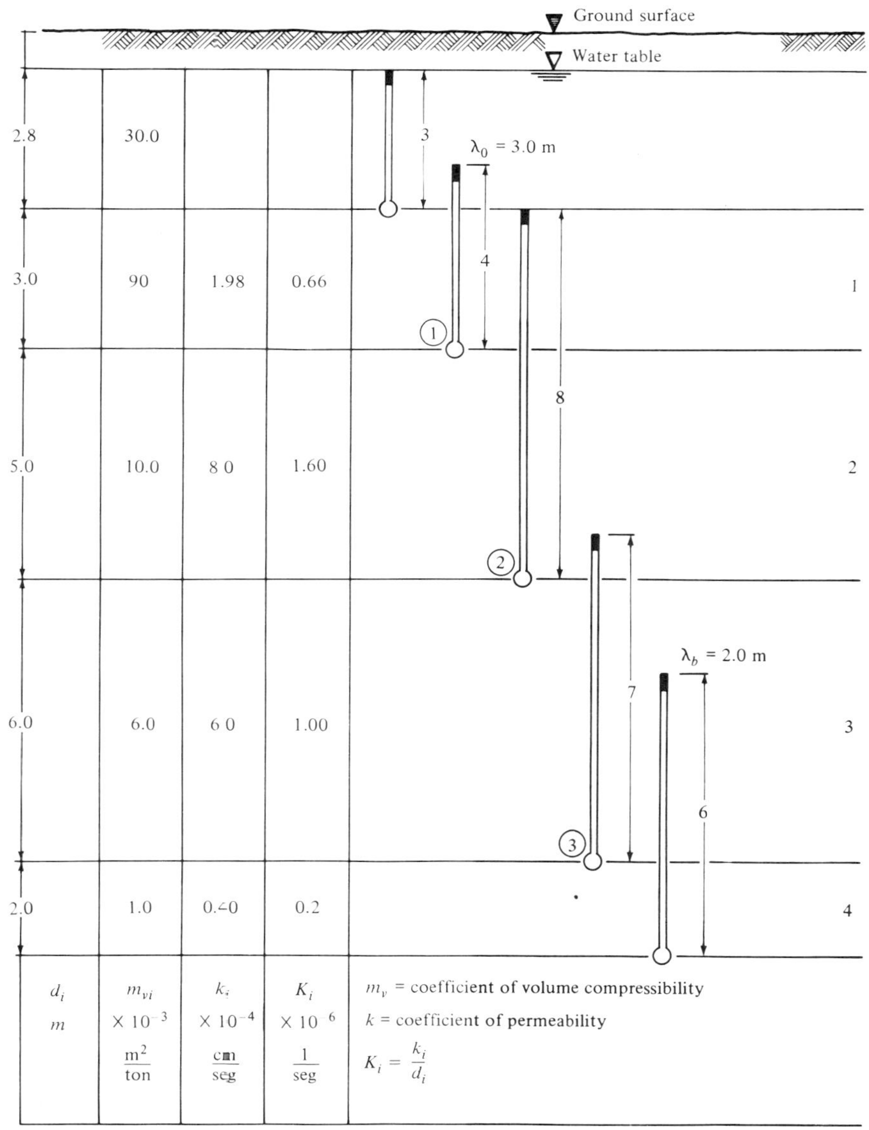

Fig. 11-III.3 Piezometric initial conditions and soil properties.

3.5 Dewatering by Wells

The study of seepage of water toward wells is of importance for different problems in foundation engineering, mainly when it is necessary to lower the water table or water pressure to perform excavations for foundation construction of buildings, bridges and other engineering works. Furthermore, it is necessary to foresee the hydraulic pressures and hydraulic gradients during pumping, and to study the total discharge of water from the wells to recommend the pumping system for the problem under consideration.

3.5a Study of a Single Well. Consider a well of diameter $2r_0$ which has been driven to an impermeable stratum. The well cuts different strata with thicknesses d_n and coefficients of permeability k_n. Suppose that during pumping the water table inside the well suffers a depression S_0. This depression does not exceed the thickness of the first deposit with coefficient of permeability k_0. Applying Dupuit's hypothesis, one may establish that the hydraulic gradient at the considered section is equal to the slope of the free water surface. Therefore, establishing the condition of discharge and applying Darcy's law at section YY (Fig. 12-III.3), at the radial distance r we obtain

$$q = 2\pi r \cdot y k_0 \frac{dy}{dr} + 2\pi r \cdot d_1 k_1 \frac{dy}{dr} + \cdots + 2\pi r d_n k_n \frac{dy}{dr} \qquad 39\text{-III.3}$$

Performing the integration and after the water flow regime is established for $r = R_0$ and $y = d_0$, the following expression for the discharge is found:

$$q = \frac{\pi k_0 (d_0^2 - y_0^2) + 2\pi (d_0 - y_0) \sum_1^n k_i d_i}{\operatorname{Log} \dfrac{R_0}{r_0}} \qquad 40\text{-III.3}$$

If the equation is written in terms of the water depression S_0 in the well, then the following formula may be written:

$$q = \left(\frac{\pi k_0 (d_0 + y_0) + 2\pi \sum_1^n k_i d_i}{\operatorname{Log} \dfrac{R_0}{r_0}} \right) \cdot S_0 \qquad 41\text{-III.3}$$

From above formula, two particular cases may be observed:

First Case. The surface deposit of thickness d_0 is much more permeable than the underlying deposits (Fig. 13-III.3):

$$\frac{k_n}{k_0} \doteq 0$$

therefore:

$$q = \frac{\pi k_h (d_0 + y_0)}{\operatorname{Log} \dfrac{R_0}{r_0}} \cdot S_0 \qquad 42\text{-III.3}$$

in which k_h is the average horizontal coefficient of permeability.

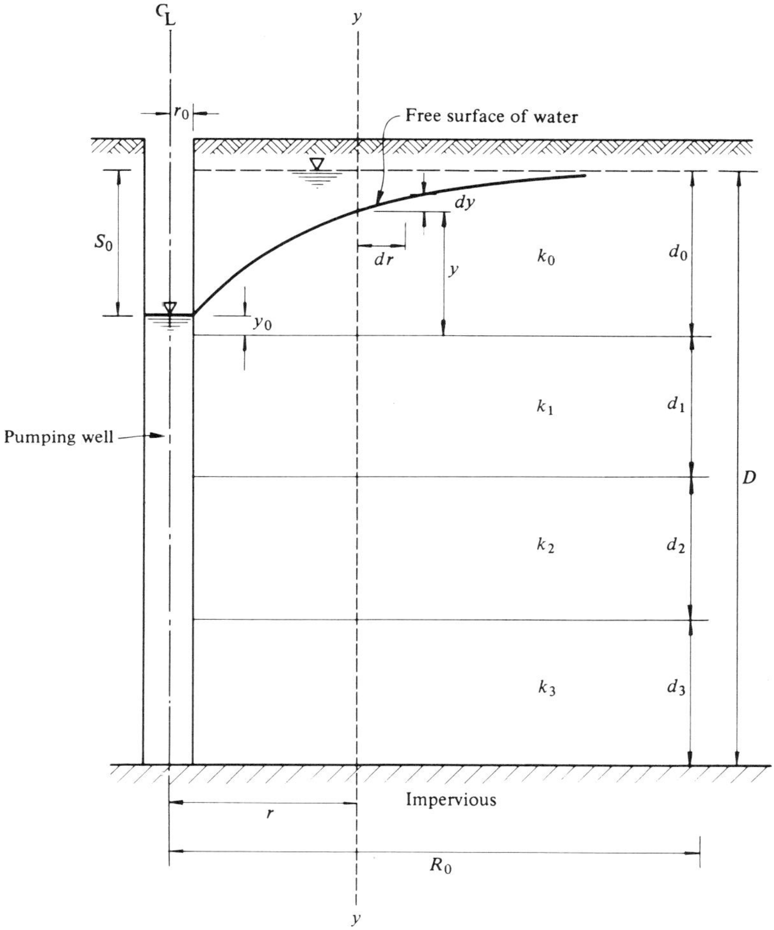

Fig. 12-III.3 Water flow toward a well.

Second Case. The superficial deposit of thickness d_0 is impermeable (Fig. 14-III.3). In this case the underlying strata of thicknesses d_n are confined aquifers, and the discharge is expressed by the following formula:

$$q = \frac{2\pi \sum_{1}^{n} k_i d_i}{\text{Log} \dfrac{R_0}{r_0}} \cdot S_0 \qquad \text{43-III.3}$$

The problem of water seepage toward the well may be considered in this case as a typical problem of horizontal flow, such that if the subsoil is formed of different strata with permeabilities $k_1, k_2, k_3, \ldots, k_n$ and respective thicknesses $d_1, d_2, d_3, \ldots, d_n$, the solution of the problem may be solved satisfactorily using

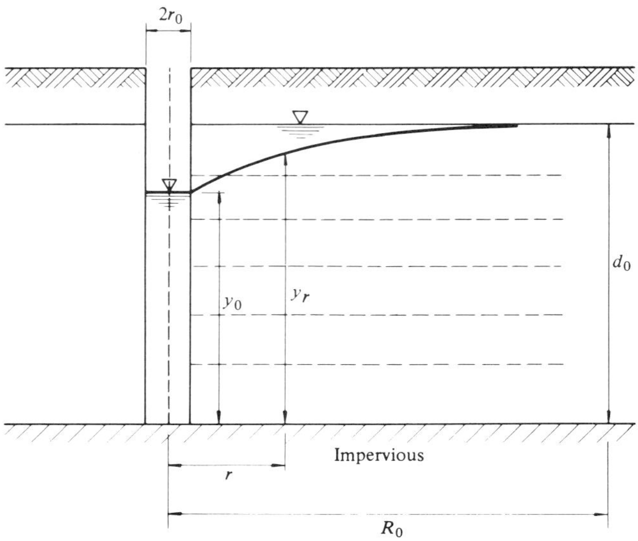

Fig. 13-III.3 Water flow into a well in stratified nonconfined subsoil strata.

the average permeability

$$k_h \cdot D = \sum_{1}^{n} k_i d_i$$

The well formula established here indicates that the discharge diminishes as R_0 increases. From the practical point of view, R_0 cannot increase indefinitely.

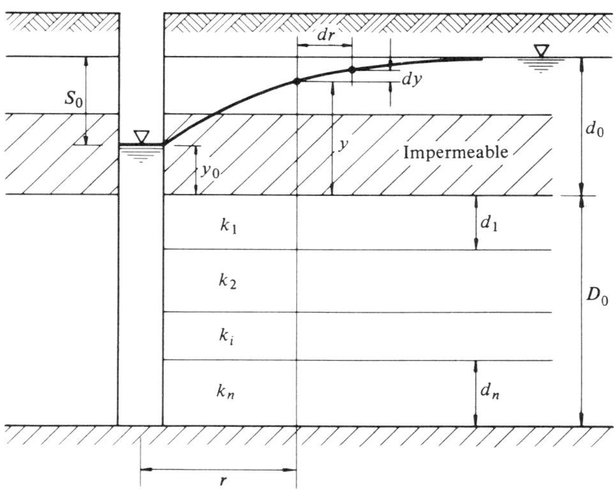

Fig. 14-III.3 Water flow into a well in stratified confined subsoil strata.

When the aquifers are of unlimited capacity or properly supplied, R_0 will be limited by the distance of supply. The empirical formula of Sichardt may be used for wells to estimate the value of $R_0 = R_s$, when the water supply is at a greater distance than the value R_s:

$$R_s = C_s \cdot S_0 \sqrt{k} \qquad \text{44-III.3}$$

where S_0 is the dynamic water table inside the well in centimeters, k is the coefficient of permeability in cm/sec and the value of C_s (sec/cm)$^{1/2}$ may be taken as 300 for wells and 200 for trenches or well lines. The value of R_s is obtained in centimeters.

For example, a well in sand where S_0 = 1000 cm and $k = 10^{-2}$ cm/sec yields

$$R_s = 300 \times 1000 \times \frac{1}{10} = 30000 \text{ cm} = 300 \text{ m}$$

and for a trench in silt with $k = 10^{-4}$ cm/sec and S_0 = 200 cm, the following value for R_s is obtained:

$$R_s = 200 \times 200 \times \frac{1}{100} = 400 \text{ cm} = 4.0 \text{ m}$$

3.5b Study of Well Groups. In the problem of dewatering of excavations we need to estimate the discharge of each one of the wells in the group in order to foresee the pumping problem. It is also important to determine approximately the water drawdown within the area enclosed by the wells. From 15-III.3, the drawdown produced by well i in the group at a distance r_{mi} is calculated with the following expression:

$$S_{mi} = D - y_{mi} = \frac{q_i}{2\pi \sum_1^n k_i d_i} \text{Log} \frac{R}{r_{mi}} \qquad \text{45-III.3}$$

in which q_i is the discharge of the ith well in the group. Call $2\pi \Sigma_1^n k_i d_i = M$. Moreover, assume the total drawdown at point m is equal to the sum of the drawdowns of all the wells in the group, hence

$$S_m = S_{m1} + S_{m2} + S_{m3} + \cdots + S_{mi} \qquad \text{46-III.3}$$

therefore

$$S_m = \frac{1}{M} \sum_1^n q_i \text{ Log} \frac{R}{r_{mi}} \qquad \text{47-III.3}$$

In order to calculate the values of S_m for different points within the group of wells, it is necessary to estimate the discharge q_i at each one of the wells. For this purpose we calculate the drawdown at each one of the wells. Calling S_{0i} the draw-

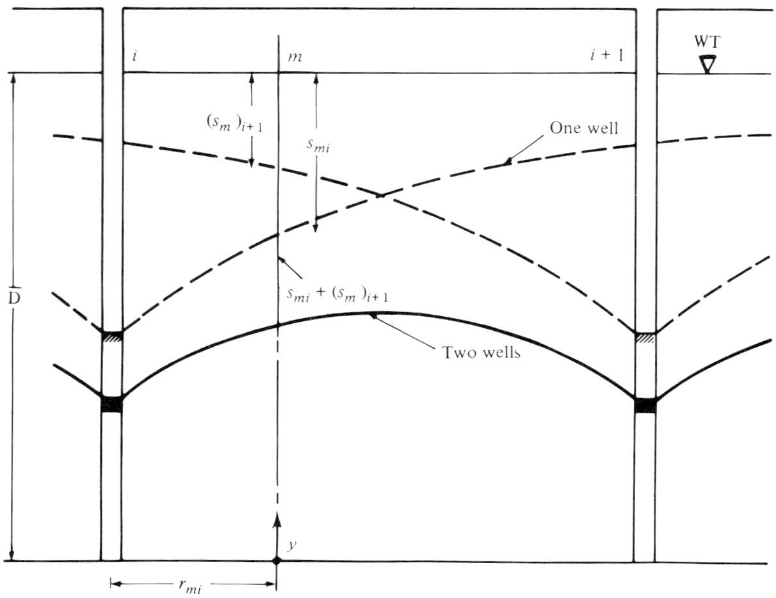

Fig. 15-III.3 Water drawdown due to pumping from well groups.

down at well i we can write the following expressions:

For well i

$$S_{0i} = \frac{1}{M} q_i \operatorname{Log} \frac{R}{r_{ii}} + \frac{1}{M} q_j \operatorname{Log} \frac{R}{r_{ij}} + \cdots$$

For well j

$$S_{0j} = \frac{1}{M} q_i \operatorname{Log} \frac{R}{r_{ji}} + \frac{1}{M} q_i \operatorname{Log} \frac{R}{r_{jj}} + \cdots$$

where r_{ji} is the distance from well j to well i. Arranging terms we can write the following system of simultaneous equations:

$$\left(\operatorname{Log} \frac{R}{r_{ii}}\right) q_i + \left(\operatorname{Log} \frac{R}{r_{ij}}\right) q_j = S_{0i} \cdot M$$

$$\left(\operatorname{Log} \frac{R}{r_{ji}}\right) q_i + \left(\operatorname{Log} \frac{R}{r_{jj}}\right) q_j = S_{0j} \cdot M \qquad \text{48-III.3}$$

The above array of terms may be written in matrix form:

$$\left[\operatorname{Log} \frac{R}{r_{ij}}\right] \cdot |q_i| = |S_{0i} M| \qquad \text{49-III.3}$$

Assigning a constant water drawdown at each one of the wells S_0 and solving the simultaneous equations given by 48-III.3, we find the q_i values. Using expression 47-III.3 the drawdown at any point in the area of the well group may be estimated.

The study of this problem may also be undertaken from the practical engineering point of view, by means of already established concepts which are applicable to all flow net theorems in the case of a horizontal flow through a permeable stratum confined by impermeable strata. In order to establish the validity of the flow net theorems, consider Fig. 16-III.3, which represents the conditions of horizontal flow under confined conditions. Assume that the hydraulic gradient is constant at any point along a vertical section, and equal to the slope of the free surface determined by the piezometric water levels of the confined strata. The discharge going through sides *ab* and *ac* of the differential horizontal element shown in Fig. 16-III.3 flows out at sides *bd* and *cd* of the same element, respectively.

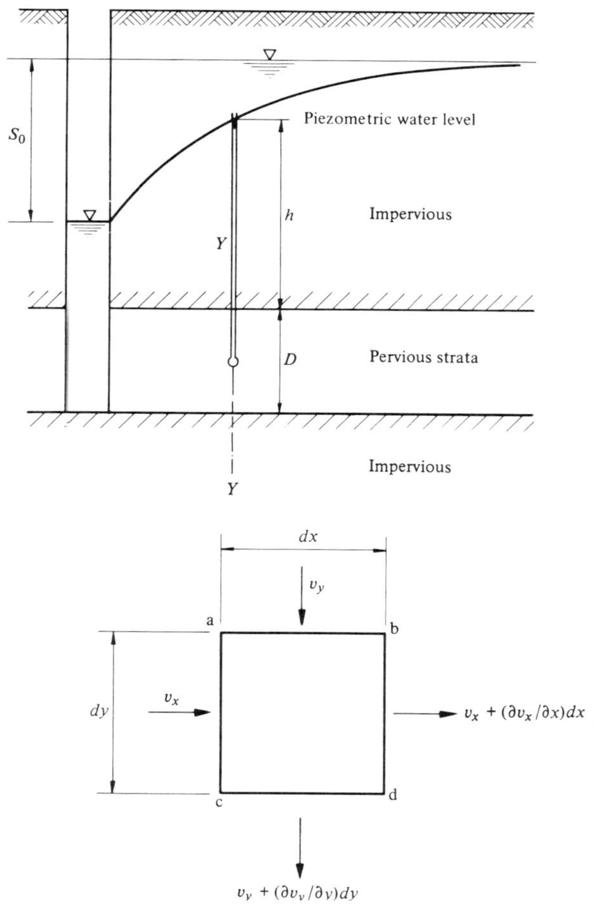

Fig. 16-III.3 Horizontal water flow conditions in confined strata.

The condition for continuity of flow establishes that

$$\frac{\partial v_x}{\partial x} + \frac{\partial v_y}{\partial y} = 0 \qquad \text{50-III.3}$$

If the potential or piezometric head at a certain section may be represented by

$$\Psi = h + d$$

with respect to a horizontal base of the pervious stratum, then according to Darcy's law and assuming that the average coefficient of permeability is equal for any direction of the horizontal flow, we have

$$\frac{\partial^2 \Psi}{\partial x^2} + \frac{\partial^2 \Psi}{\partial y^2} = 0$$

Therefore, Laplace's equation is obtained and is satisfied by a family of equipotential lines or of equal piezometric level $\Psi = f_1(x, y)$. Following the above reasoning, one can demonstrate that the flow lines represented by $\Phi = f_2(x, y)$ satisfy

$$\frac{\partial^2 \Phi}{\partial x^2} + \frac{\partial^2 \Phi}{\partial y^2} = 0$$

Therefore, the system of equipotential curves of equal piezometric water level and the flow lines are orthogonal to each other, and the hydrodynamic flow problem of a group of wells may be solved by means of the same theorems used for the flow nets, as already discussed in previous paragraphs. In this case, the total discharge of the group system of wells is

$$Q = k_h \cdot D \cdot S_0 \frac{N_f}{N_p} \qquad \text{51-III.3}$$

in which k_h is the average coefficient of permeability in the horizontal direction, D is the thickness of the permeable strata, S_0 is the dynamic water level in the wells, N_f is the number of flow channels approaching the well, and N_p is the number of potential drops in the system. The characteristic value of the flow net is

$$R_c = \frac{N_f}{N_p} \qquad \text{52-III.3}$$

To illustrate the procedure, take as an example a single well of radius r_0 going through a permeable stratum of thickness D_0 with coefficient of permeability k. The flow net shown traced in Fig. 17-III.3 has the characteristic value of 1.6 to the fifth equipotential line. Hence, the discharge is equal to:

$$Q = 1.6(k_h \cdot D_0 \cdot S_0) \qquad \text{53-III.3}$$

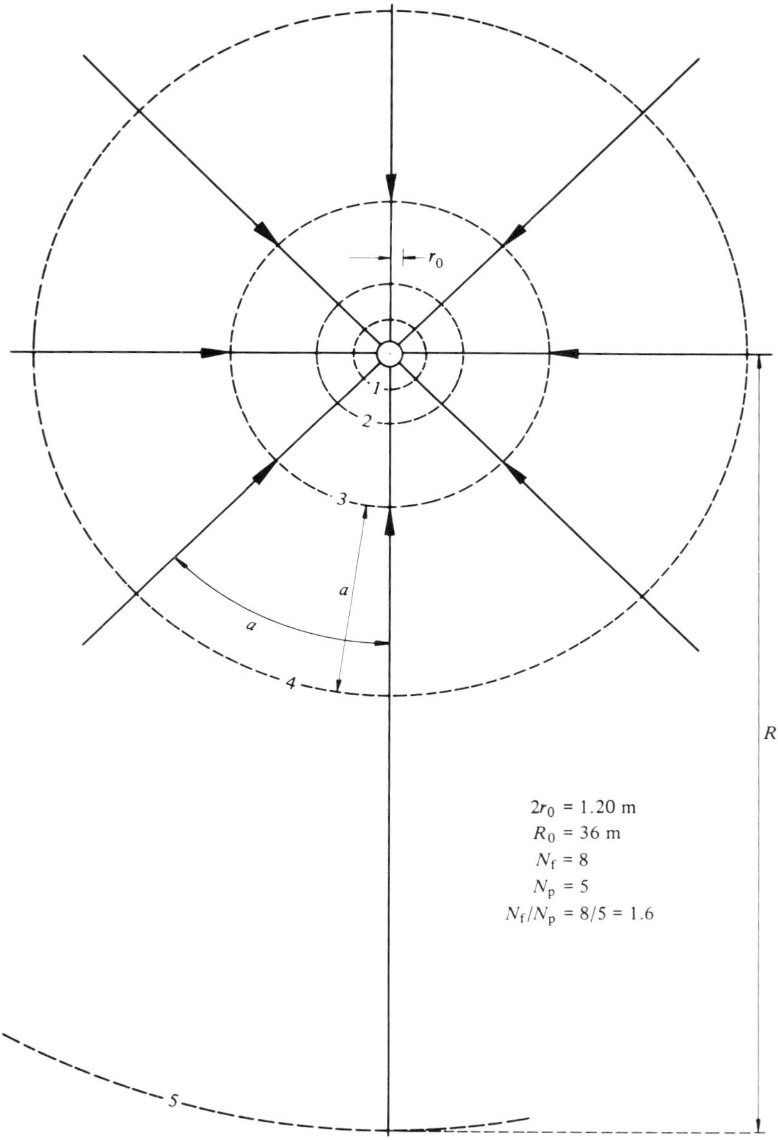

Fig. 17-III.3 Water flow net for one well.

This may be checked using formula 43-III.3 deduced directly from Dupuit's hypothesis for a unique confined stratum, which reads:

$$Q = \frac{2.73}{\log \dfrac{R_0}{r_0}} (k_h \cdot D_0 \cdot S_0) \tag{54-III.3}$$

Comparing both formulas, it may be observed that the characteristic value R_c is a function of the ratio R_0/r_0, that is to say:

$$R_c = \frac{2.73}{\log \dfrac{R_0}{r_0}} \tag{55-III.3}$$

From Fig. 17-III.3 for five equipotential drops, $R_0/r_0 = 60$, from which we have

$$\frac{2.73}{\log \dfrac{R_0}{r_0}} = 1.53$$

Therefore, the results of applying the flow net method to one well and those obtained by means of the theoretical formula may be considered equivalent. Before the flow net is traced, one has to estimate the value of R_0, by means of Sichardt's empirical formula, in order to establish the equipotential boundary of the flow net at a shorter distance from the water supply. The concept expressed above will serve to study groups of wells subjected to horizontal water flow in confined permeable strata. For the case of strata that are not confined, one can obtain a good approximation by means of formula 42-III.3, and Fig. 13-III.3:

$$Q = 2\pi k_h \frac{\frac{1}{2}(d_0 + y_0) \cdot S_0}{\mathrm{Log} \dfrac{R_0}{r_0}}$$

By means of the average thickness or equivalent cross section D_a

$$D_a = \frac{1}{2}(d_0 + y_0)$$

we obtain

$$Q = (k_h D_a S_0) \frac{2.73}{\log \dfrac{R_0}{r_0}} \tag{56-III.3}$$

It may be seen that formula 56-III.3 is structurally equivalent to formula 54-III.3, and the characteristic value R_c of the flow net expressed by 55-III.3 is the same as that for confined strata. The values of d_0 and y_0 should be taken as shown in Fig. 13-III.3.

In the following paragraphs, various systems of well groups are studied with use of the procedures proposed.

A. The flow net for the case of two wells separated by a distance L is shown in Fig. 18-III.3, where we assume that the fifth equipotential line represents the exterior boundary of the flow net. From the figure one obtains:

a). The number of flow channels $N_f = 24$;

b). The number of equipotential drops up to the fifth equipotential line $N_p = 8$;

c). The total discharge is obtained:

$$Q = 3.0(k_h \cdot D_0 \cdot S_0)$$

The lines of equal piezometric level in the case of horizontal flow are also equipressional or isobaric lines, and they can be used to estimate the state of hydraulic pressures at any point on the horizontal plane x, y.

B. Let us consider now the case of eight wells located in an area supplied at the boundary $ABCD$, as shown in Fig. 19-III.3, in such a way that the perimeter of the

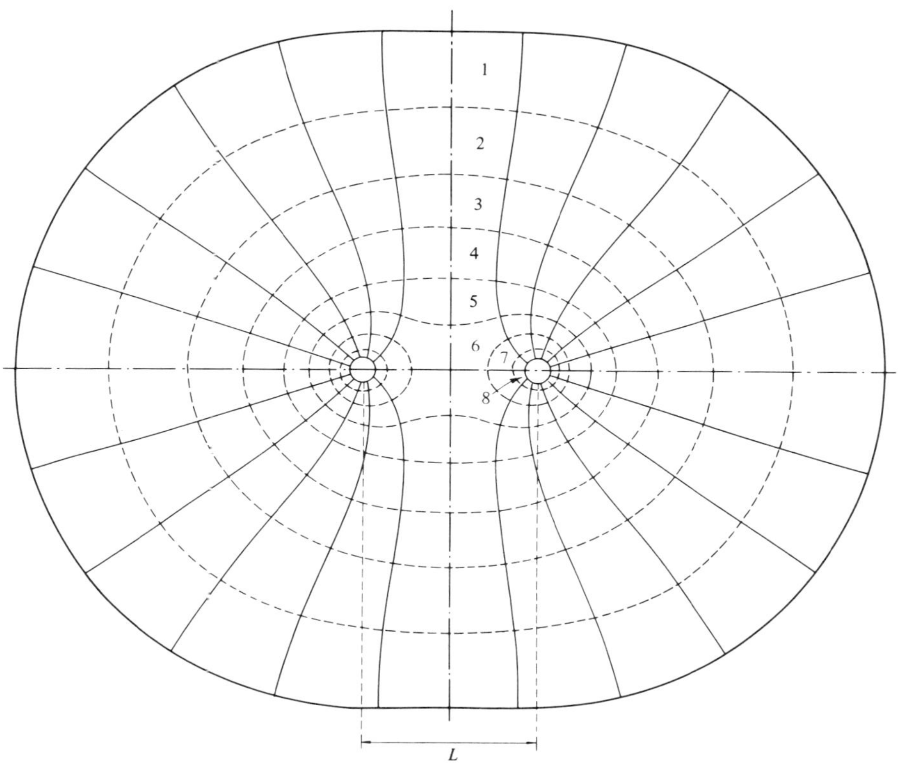

Fig. 18-III.3 Water flow net for two wells.

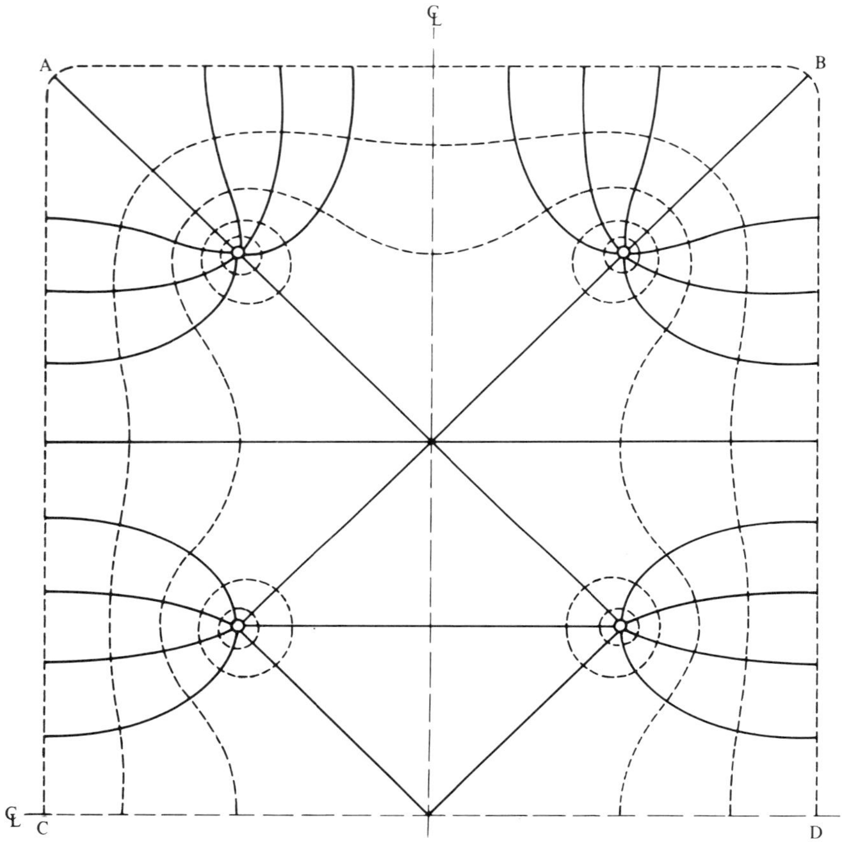

Fig. 19-III.3 Water flow net for eight wells.

area represents the boundary equipotential line, that is to say, the water has at the perimeter the same piezometric level. The eight wells of diameter $2r_0$ are distributed as shown. The dynamic water level in all wells is the same. The thickness of the permeable stratum is D_0 and the average coefficient of permeability k_h.

From the flow net one obtains for each one of the wells at the corner:

$$q = \frac{8}{5}(k_h D_0 S_0)$$

and for each one of the intermediate wells:

$$q = \frac{5}{5}(k_h D_0 S_0)$$

Therefore, the total discharge will be:

$$Q = 10.4(k_h D_0 S_0)$$

From the above examples it is concluded that by means of the flow nets one can solve complicated well groups. Nevertheless, the distance R_0 corresponding to the boundary equipotential line, which is considered unaffected during pumping from the wells, should first be determined.

3.6 Ratio of the Discharge in One Well and in a System of Wells

To obtain information about the hydraulic properties of a region, it is necessary to investigate the discharge and average coefficient of permeability of the area by means of one or several test wells. The following formula permits the calculation of the average coefficient of permeability:

$$q_1 = (k_h \cdot D_a \cdot S_{01})R_{c1} \qquad \text{57-III.3}$$

in which R_{c1} is the characteristic of the flow net for one well investigated; D_a is the equivalent section of water flow, depending on the stratigraphy of the subsoil in the area investigated; S_{01} is the dynamic water level in the test well; and k_h is the average coefficient of permeability.

On the other hand, if the characteristic value of the system of wells is R_{cs}, then the total discharge

$$Q_s = (k_h \cdot D_a \cdot S_{os})R_{cs} \qquad \text{58-III.3}$$

If the stratigraphy for one well and the group of wells may be considered approximately the same, one may assume that the average coefficient of permeability has the same value for one well as for the group of wells. Hence, we may write the following relation:

$$\frac{Q_s}{q_1} = \frac{S_{os}}{S_{01}} \cdot \frac{R_{cs}}{R_{01}} \qquad \text{59-III.3}$$

If the drawdown in the test well is preserved equal to that in the group of wells ($S_{os} = S_{01}$), we can obtain the discharge for the total system by means of the following expression:

$$Q_s = \frac{R_{cs}}{R_{c1}} \cdot q_1 \qquad \text{60-III.3}$$

As an example, suppose that from investigation of one well we obtain a discharge $q_1 = 0.010\ \text{m}^3/\text{sec}$, and the characteristic value is $R_{c1} = 2.0$. A group of eight wells located as shown in Fig. 19-III.3 has the characteristic value $R_{cs} = 10.4$; hence the discharge for the system is:

$$Q_s = \frac{10.4}{2.0} \cdot 0.010 = 0.052\ \text{m}^3/\text{sec}$$

III.4 SHEAR STRENGTH BEHAVIOR IN SOIL MASS

4.1 Basic Considerations

Stability problems in soil mechanics call for the understanding of the combined physical action of the solid and liquid phases, in accordance with the fundamental equation of effective stresses and hydraulic pressures at a point. They are estimated by ultimate shear strength methods using the simplified working assumptions of plasticity theory, which give results on the safe side when using the proper factor of safety selected for each case in particular, based on the importance of the problem in consideration. Ultimate shear strength problems in foundation engineering are those in connection with footings bearing capacity, ultimate load capacity of pile and pier foundations, earth pressure, and stability of excavations and slopes.

Current methods of solving shear strength problems is the scope of elementary soil mechanics. The discussion is presented only to establish a judgment for application to foundation engineering used in the following chapters. From the point of view of solid and liquid phase behavior, however, it is important for the foundation engineer to study the limitations of these methods and be aware of the factors of safety to be used in different occasions, to classify and make a diagnosis of the problem, and to select the data required concerning the physical characteristics of the problem under study. A precise evaluation of the ultimate stability condition in such problems is practically impossible to obtain for the following reasons:

1. Even if the ultimate shear strength properties of the material may be determined with great precision in the laboratory, the soil engineering conditions in nature change considerably. Therefore, the best information the foundation engineer can obtain only allows him to make a good average guess of what these mechanical properties may be under field conditions.
2. The stratigraphy conditions in any stability problem, no matter the effort made to obtain precision in most cases at a very high cost, is no more than an average condition of the general geometry of the site.
3. The prevailing hydraulic conditions are known only approximately and they have to be averaged. These conditions may change with time, and in this case, the ultimate shear strength characteristics may change accordingly.
4. The assumption on the position and form of the potential surface of sliding in stability analysis is not correct. The best one can obtain is a crude approximation of what may take place in the field under ideal conditions. The assumed position of the potential surface of sliding is greatly affected by the conditions of stratification of the soil mass. This fact should be taken carefully into consideration, according to the experience of the foundation engineer.

The uncertainty connected with these factors will obviously affect the precision of the calculations of stability problems, and the foundation engineer has to be aware that his best guess may be wrong by 50% to 100% or more. One must consider the use of factors of safety of not less than 2 to 3 when one has a good knowledge of all the environmental factors that may affect the stability of the problem in question, and when the theories used have been deduced from hypotheses on the safe side. The first important decision in analysis of stability problems concerns the shear strength parameters for the conditions encountered in the field. This may be achieved as already discussed in section II.4. Once this is done, the stability analysis on ultimate shear strength conditions may be performed by current methods.

4.2 Bearing Capacity

4.2a. Shallow Footings. The ultimate bearing capacity of footings may be estimated by the following formula proposed by Terzaghi (Figs. 1 and 2-III.4):

$$q_d = \alpha_1 c N_c + \alpha_1' \sigma_z N_q + \alpha_2 \gamma \cdot B \cdot N_\gamma \qquad \text{1-III.4}$$

in which N_c, N_q and N_γ are bearing capacity factors which depend exclusively on the real or apparent angle of internal friction and the geometry of the assumed potential surface of sliding. Several authors have investigated these values under different theoretical assumptions. The result of the different theories exposed may be found in the literature. The value c is the apparent cohesion assigned to the material in question; $2B$ is the minimum width of the footing and α_1 and α_2 are shape factors of the contact surface of the foundation with the soil. They have been found to be function of the angle of internal friction; however, for practical purposes, the following approximate values for flat foundation bases may be assumed:

Footing	Width	α_1	α_1'	α_2
Continuous	$2B$	1	1	1
Square and circular	$2B$	1.3–1.2	1.2	0.8–0.6

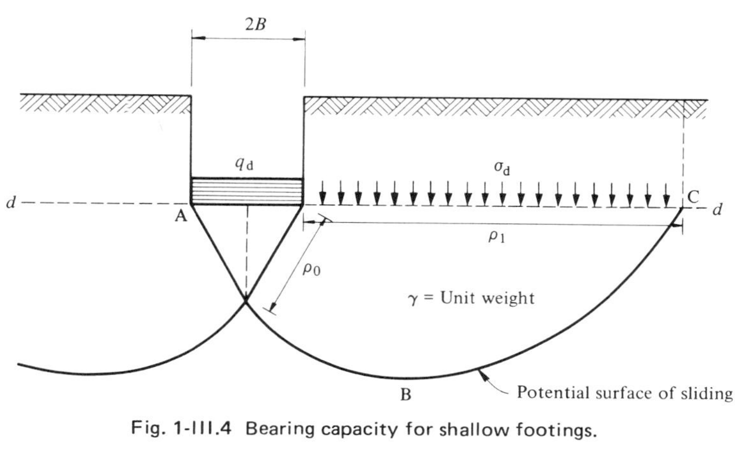

Fig. 1-III.4 Bearing capacity for shallow footings.

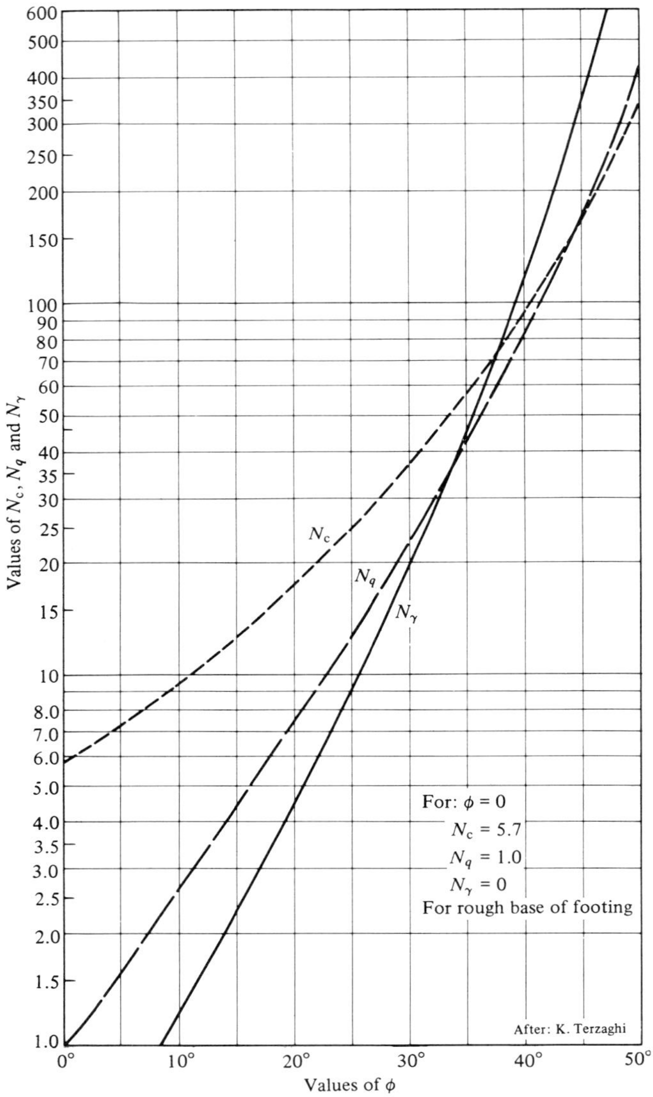

Fig. 2-III.4 Bearing capacity factors for shallow foundations.

From formula 1-III.4, it may be recognized that the ultimate bearing capacity is obtained by the following three independent terms: the first term is a function only of the soil shear strength parameters c and ϕ, and of the assumed shape of the surface of sliding; the second term is a function of the vertical confining effective stress at the footing grade elevation, and of the assumed shape of the surface of sliding; the third term is a function of the width of the footing on the average weight of the soil mass under the footing, and of the assumed shape of the surface of sliding. The position of the water table affects only second and third terms, as they are functions of the unit weight of the soil.

The full value of formula 1-III.4 is based on the assumption that the footing breaks into the ground when the ultimate shear strength is fully mobilized along the entire potential surface of sliding, and this can only happen if the material may be considered incompressible, as otherwise the footing may break into the ground without necessarily mobilizing all the shear strength along curve ABC, Fig. 1-III.4. Therefore, one encounters a very important factor to consider in the application of this formula, namely, the compressibility of the material. To compensate for it we assume from the practical foundation engineering point of view that there is a relation between the relative density of the soil and the compressibility. The simplest way to take this into consideration for the time being is to assume that the relative density of the soil D_r is approximately proportional to q_d. The author has used the following correction based on the relative density of the material:

$$\bar{q}_d = q_d(D_r + 0.1) \qquad \text{2-III.4}$$

Therefore, the bearing capacity in a loose sand with an angle of internal friction of 33° will be only about 1/3 of that calculated with the same angle by formula 1-III.4, as if the material were almost incompressible. On the other hand, the relative density of a well graded material is related with the angle of internal friction approximately as follows:

$$\phi = 20°D_r + 26° \qquad \text{3-III.4}$$

Note here that there is a twofold effect which affects the ultimate bearing capacity: first, the low angle of internal friction; and second, the inability of the loose compressible material to mobilize the ultimate shear strength along the total length of the potential surface of sliding. It is reasonable to think that in case of saturated clay materials and upon rapid load application the shear strength is mobilized along the entire sliding surface; in such cases, the value of ϕ is taken equal to zero. Upon consolidation of the clay, the bearing capacity may be investigated replacing the value of D_r by the relative consistency of the soil:

$$C_r = \frac{L_w - w_n}{I_w} \qquad \text{4-III.4}$$

The values of N_c, N_q and N_γ may be used as given in Fig. 2-III.4.

4.2b Deep Footings. In case of deep footings foundations, the bearing capacity may be estimated omitting the third term for all practical purposes. The geometry of the surface of sliding for a fairly isotropic soil mass may be assumed as shown in Fig. 3-III.4, that is to say, a logarithmic spiral starting under the foundation and ending with a vertical tangent. The confining stress σ_z may be assumed applied as shown in Fig. 3-III.4. The limiting plastic equilibrium conditions may

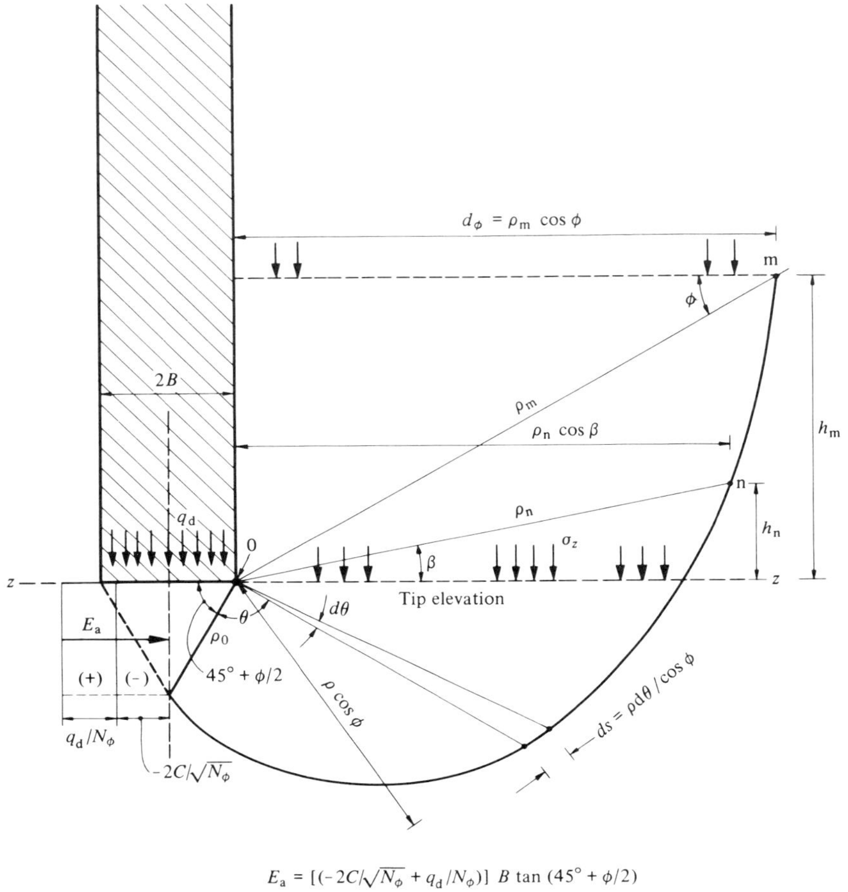

Fig. 3-III.4 Bearing capacity for a deep strip foundation.

be obtained under the conservative assumption that the mass of soil enclosed in the logarithmic spiral is weightless. Taking moments through point O at the edge of the strip footing, the following value is obtained for the moment of the active forces:

$$M_a = [q_d - c\sqrt{N_\phi}]\, \rho_0^2 \cos^2\left(45° + \frac{\phi}{2}\right) \qquad \text{5-III.4}$$

and for the passive forces due to mobilization of the total ultimate shear strength along the logarithmic spiral under the confining stress σ_z, one obtains:

$$M_p = \frac{1}{2}\sigma_z \rho_n^2 \cos^2\beta + \frac{c\rho_0^2}{2\tan\phi}\left[e_N^{2\theta\tan\phi} - 1\right] \qquad \text{6-III.4}$$

For limiting plastic equilibrium $M_a = M_p$; therefore, the ultimate unit bearing capacity for a deep seated strip footing is given by

$$q_d = c\left[\sqrt{N_\phi} + \frac{e_N^{2\theta \tan\phi} - 1}{2\cos^2\left(45^\circ + \dfrac{\phi}{2}\right)\tan\phi}\right] + \left[\frac{\cos^2\beta}{2\cos^2\left(45^\circ + \dfrac{\phi}{2}\right)} e_N^{2\theta \tan\phi}\right]\sigma_z \qquad \text{7-III.4}$$

and considering that $\theta = (3/4)\pi + \phi/2$ and $\beta = \phi$ the values of N_c and N_q may be found:

$$N_c = \tan\left(45^\circ + \frac{\phi}{2}\right) + \frac{e_N^{(\frac{3}{2}\pi + \phi)\tan\phi} - 1}{2\cos^2\left(45^\circ + \dfrac{\phi}{2}\right)\tan\phi} \qquad \text{8-III.4}$$

$$N_q = \frac{(\cos^2\phi)e_N^{(\frac{3}{2}\pi + \phi)\tan\phi}}{2\cos^2\left(45^\circ + \dfrac{\phi}{2}\right)} \qquad \text{9-III.4}$$

Taking into consideration the base shape factor and the relative density D_r, then the base ultimate bearing capacity is given by

$$\overline{q}_d \geqslant \alpha_1 \left[cN_c + \sigma_z N_q\right](D_r + 0.1) \qquad \text{10-III.4}$$

The values of N_c and N_q are plotted against ϕ in Fig. 4-III.4. In case of clays on a long-term basis, the ultimate point bearing capacity may be estimated by substitution of D_r for C_r.

The confining stress σ_z may be taken as the effective stress at the horizontal level of the deep seated footing. Furthermore, it may be important for the application of the ultimate bearing capacity formula to establish an assumption for square or circular bases. The potential surface of sliding may be considered as a vertical channel, shown in Fig. 5-III.4, with an average width on the order of 1.3 in cohesive materials. This working hypothesis is important when considering interference in the ultimate load carrying capacity of the ground, when supporting elements are close together as is the case of pile groups or piers. The height h_m at which the surface of sliding would be fully mobilized for very dense soil may be obtained from the ratio h_m/B plotted against ϕ in Fig. 6-III.4.

To illustrate the procedure just described, assume it is required to find the unit bearing capacity for a circular deep footing. The material has a relative density of 0.4. The cohesion is 0.4 kg/cm^2 and the apparent angle of internal friction is 35°. Assuming that $\alpha_1 = 1.2$ we have from Fig. 4-III.4 that $N_c = 138$ and $N_q = 66$. But these factors are reduced by the relative density, and thus $N_c' = 69$ and $N_q' = 33$. The reduced bearing capacity factors correspond to an equivalent angle of in-

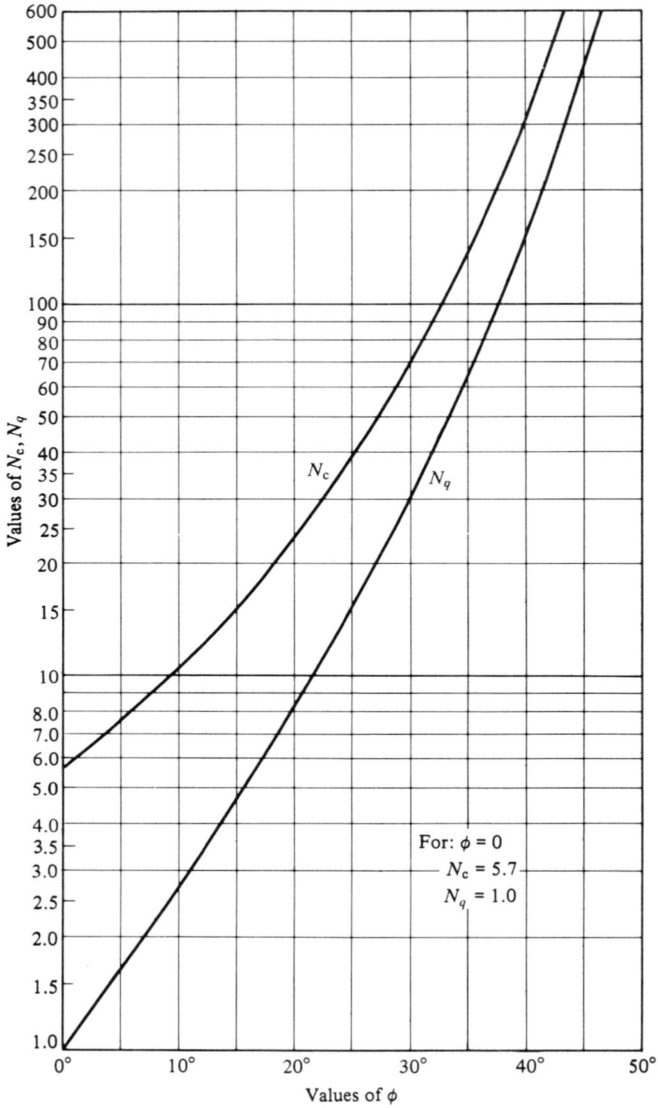

Fig. 4-III.4 Bearing capacity factors for a deep strip foundation.

ternal friction of $\phi = 30°$. Therefore, the ultimate base bearing capacity:

$$q_d = 1.2[0.4 \times 69 + 33\sigma_z]$$

and if

$$\sigma_z = 1.0 \text{ kg/cm}^2, \text{ then } q_d \geqslant 73.0 \text{ kg/cm}^2$$

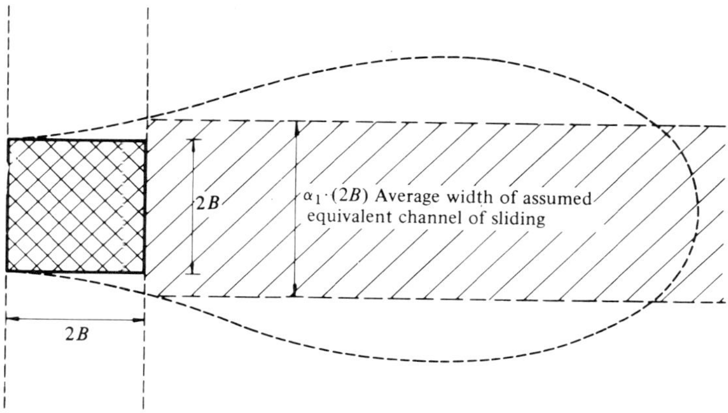

Fig. 5-III.4 Cross section of assumed failure channel at tip elevation of pile or pier.

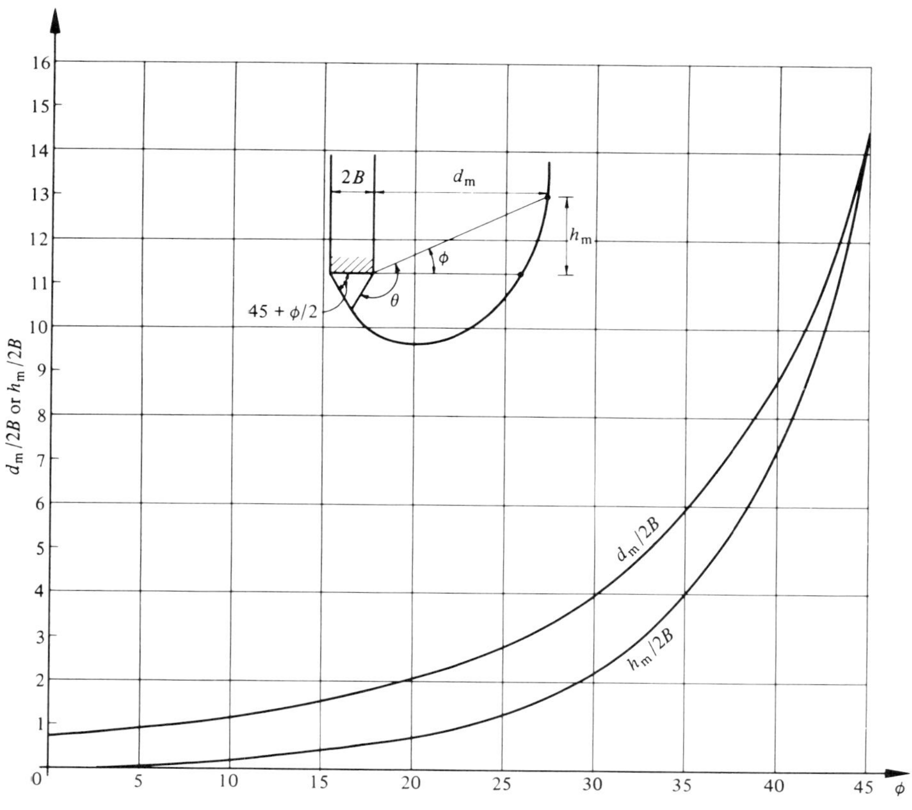

Fig. 6-III.4 Geometrical characteristics of assumed surface of sliding logarithmic spiral.

The factor of safety, when all the causes affecting the ultimate value have been taken into account, should not be less than 2. Therefore, the allowable unit bearing capacity in this case will be taken as $q_a = 36.5$ kg/cm^2.

The values of the unit ultimate bearing capacity computed by means of formula 10-III.4 are on the safe side; hence the q_d-value should be expected somewhat higher because the following physical actions have been omitted:

a). The third term in equation 1-III.4.
b). The side friction on the lateral surfaces limiting the potential surface of sliding in case of deep seated square and circular bases has been neglected; however, this effect is taken into consideration to a certain extent, approximately, by the shape factor.
c). The compaction the pile produces in the ground has the tendency to increase by a few degrees the angle of internal friction.

In case of deep foundations like piles and piers, it is necessary to consider other important factors like skin friction along the shaft that affect the base ultimate load capacity. The discussion of this phenomenon is out of the scope of this chapter, and is considered in Chapter VIII. The ultimate load capacity may be also affected by the hydraulic conditions prevailing at the site in question, or changes that may be expected in the future life of the structure on such foundations. The construction procedure of deep supporting load elements, like piers, may also affect the ultimate load carrying capacity of the ground. Therefore, the foundation engineer should not overlook the effects of the hydraulic conditions when excavating shafts under water to cast in place these elements.

BIBLIOGRAPHY

Berezantzev, V. G. and Yaroshenko, V. A. (1957). The Bearing Capacity of Sands under Deep Foundations, *Proc. IV ICOSOMEF*, Vol. 1, p. 283, London.

Berezantzev, V. G. Khristoforov, V. S. and Golubkov, V. N. (1961). Load Bearing Capacity and Deformation of Pile Foundations, *Proc. V ICOSOMEF*, Vol. II, p. 11, Paris.

Brinch, Hansen and Lundgren, H. (1965). *Hauptprobleme der Bodenmechanik*, Springer Verlag, pp. 223–246, Berlin.

Casagrande, Arthur (1937). *Seepage Through Dams*, Publication from the Graduate School of Engineering, Harvard University, No. 209, Soil Mechanics Series No. 5.

De Beer, E. E. (1970). Experimental Determination of the Shape Factors and the Bearing Capacity Factors of Sand, *Geotechnique* 20, No. 4, pp. 387–411.

Fröhlich, O. K. (1934). *La Repartición de Presiones*, translated from first German edition, Tipografía Artística, Alameda 12, Madrid.

Jumikis, Alfred S. (1956). Rupture Surfaces in Sand under Oblique Loads, Journal Soil Mechanics and Foundations Division, *Proceedings ASCE*, SM1, paper 861.

Kögler, F. (1936). Stress Distribution in Soils, *Proc. I ICOSOMEF*, Vol. III, pp. 66–70.

Kozeny, Josef (1953). *Hydraulik*, Springer-Verlag.

Manzur, I. and Kaufman, R. (1962). *Dewatering, Foundation Engineering*, G. A. Leonards, editor, McGraw-Hill Book Company, Inc. New York.

Meyerhof, G. G. (1950). The Ultimate Bearing Capacity of Foundations, *Geotechnique*, Vol. II, No. 4, p. 301.

Meyerhof, G. G. (1953). The Bearing Capacity of Foundations under Eccentric and Inclined Loads, *Proc. III ICOSOMEF*, Vol. I, pp. 440–445, Zurich.

Meyerhof, G. G. (1955). Influence of Roughness of Base and Ground Water Conditions on the Ultimate Bearing Capacity of Foundations, *Geotechnique*, Vol. V, No. 3, pp. 227–242.

Meyerhof, G. G. and Asce, F. (1961). The Ultimate Bearing Capacity of Wedge-Shaped Foundations, *Proc. V ICOSOMEF*, Vol. II, p. 105, Paris.

Muskat, M. (1937). *Flow of Homogeneous Fluids Through Porous Media*, McGraw-Hill Book Company, Inc. New York.

Newmark, M. N. (1942). *Simplified Computation of Vertical Pressures in Elastic Foundations*, Circular No. 24, Engineering Experiment Station, University of Illinois, Urbana.

Skempton, A. W. (1951). *The Bearing Capacity of Clays*, Building Research Congress 1951, Div. I, pp. 180–189, London.

Skempton A. W., Yassin, A. G. and Gibson, R. E. (1952). Theorie de la Force Portante des Pieux dans le Sable, *Journées de Mechanique des Sols*, Juillet.

Timoshenko, S. and Goodier, J. N. (1951). *Theory of Elasticity*, McGraw-Hill Book Company, Inc., New York.

Todd, D. K. (1967). *Ground Water Hydrology*, John Wiley and Sons, Inc., New York,

Vesić, S. A. (1967). *A Study of Bearing Capacity of Deep Foundations. Report Project B-189*, Georgia Institute of Technology, Atlanta, Georgia.

Westergaard, H. M. (1938). *A Problem of Elasticity Suggested by a Problem in Soil Mechanics: Soft Material Reinforced by Numerous Strong Horizontal Sheets*, Stephen Timoshenko 60th Anniversary Volume, MacMillan Co.

Zeevaert, L. (1942). Flujo de los Materiales de Tierra Anisótropos, *Revista de Ingeniería*, July.

Zeevaert, L. (1964). *General Considerations on Problems Related with Pile and Pier Foundations*, Congreso Cimientos Profundos, Mexico, D. F.

Zeevaert, L. (1969). *Propiedades Hidráulicas y Comportamiento Hidrodinámico del Agua del Subsuelo*, Apuntes de Mecánica de Suelos. Universidad Nacional Autónoma de México. Third Edition.

Zeevaert, L. (1980). *ISE-Interaccion Suelo-Estructura de Cimentaciones Superficiales y Profundas Sujetas a Cargas Estaticas y Sismicas*, Editorial LIMUSA, Mexico, D.F.

►IV◄ SUBGRADE REACTIONS

IV.1 GENERAL CONSIDERATIONS

To calculate shear forces and bending moments for structural foundation design purposes, it is necessary to investigate the subgrade reaction distribution against the foundation slab. Foundation structures are usually constructed of beams and slabs of reinforced concrete, and it is therefore necessary to determine the amount of bending and shear reinforcement. A subgrade reaction with uniform distribution is usually assumed because of the difficulty in determining the variation in contact stresses between foundation structure and soil, mainly due to lack of knowledge of subsoil conditions and mechanical properties of the soil sediments. This assumption may be approximately true when the foundation structure is of high stiffness and is supported on soils of high and very high compressibility and low shear strength. The vertical average displacements may be obtained from a settlement analysis considering the application of a uniform load.

When the stiffness of the foundation structure is introduced, the contact stresses are redistributed and the vertical displacement configuration of the foundation will show less curvature because bridging action takes place. Normally, the contact stresses increase toward the edges of the foundation structure and decrease toward the center. The magnitude of the bridging action is a function of the relative stiffness of the foundation structure and soil compressibility. Obviously, the largest bridging action will take place when the foundation structure is very stiff. On the other hand, if the subsoil has very low compressibility and the foundation structure is considered very flexible, a small or nonbridging action takes place and the load is taken by soil reaction in a narrow area directly under the columns. The following paragraphs will discuss cases with which the foundation engineer should be acquainted when determining contact stresses, and approximate methods of approach are recommended to obtain the most practical possible results to estimate subgrade reactions in stratified soil deposits.

IV.2 FOUNDATION MODULUS

The contact stresses between the foundation slab and soil should be such that the distribution of stress is compatible with the deflections of the foundation structure and soil displacements, since, when loading the soil, the foundation structure and ground surface should undergo the same vertical displacements. Therefore, depending on the stress-strain-time characteristics of the soil, the contact stresses will change under the foundation slab. The subgrade reactions are determined using the concept of the unit foundation modulus defined by:

$$k = \frac{q}{\delta} \qquad \text{1-IV.2}$$

in which q is the unit load and δ is the vertical displacement.

In problems concerning beams on elastic foundations the value of k is usually assumed to be a constant soil property. Nevertheless, the foundation modulus is not only a function of the soil properties, but depends also on the geometry and load distribution on the loaded surface; therefore, the name foundation modulus implies that its magnitude is a function of the foundation characteristics, namely, the load distribution, the geometry of the foundation surface, stratigraphy and compressibility characteristics of the soil strata, which also depend on the stress level induced in the subsoil because of the applied load at the ground surface.

Calling q_n the average unit load on a certain finite small area a_n, such that the total load on this area is $\Delta Q_n = q_n a_n$, then we find

$$k_n a_n = \frac{\Delta Q_n}{\delta_n}, \text{ or } K_n = \frac{\Delta Q_n}{\delta_n} \qquad \text{2-IV.2}$$

The value of K_n will be called the concentrated foundation modulus or simply the foundation modulus for tributary area a_n.

For a strip-footing

$$k(2B) \cdot 1 = \frac{q(2B) \cdot 1}{\delta}, \text{ or } K_l = \frac{Q_l}{\delta} \qquad \text{3-IV.2}$$

in which Q_l is the load per unit length of the strip-footing and K_l may be called the foundation modulus per unit length of footing or linear foundation modulus. The values of K, K_l and k may be determined when we know the vertical displacement δ produced by the loads ΔQ_n, Q_l or q_n, respectively.

IV.3 RIGID FOUNDATIONS

Consider a very rigid disk of radius R loading the ground surface. The average soil reaction is q_a; hence the total load on the disk will be:

$$Q = q_a \cdot \pi \cdot R^2 \qquad \text{1-IV.3}$$

If the geometrical and mechanical properties of the subsoil strata are known, and assuming a uniform stress distribution on the loaded area, we may calculate the vertical displacement δ_c at the center from the formula

$$\delta_c = \sum_1^n M_{vi} \cdot \Delta\sigma_i \cdot \Delta z_i \qquad \text{2-IV.3}$$

in which M_{vi} is the strain modulus at depth level i (see section II.3). From considerations of large foundation stiffness treated in section III.2, the average vertical displacement is $\delta_a = 0.85\delta_c$, and the foundation modulus for the rigid footing with loaded surface A is approximately

$$K = \frac{q_a \cdot A}{0.85\delta_c} \qquad \text{3-IV.3}$$

From formula 3-IV.3, it may be recognized that the foundation modulus is a function of the following factors:

1. Size of the loaded area.
2. Stratigraphy of the subsoil.
3. Stress distribution in the subsoil, which depends on the stiffness of the foundation structure.
4. Stress-strain properties of the soil strata encountered.

Moreover, it may be concluded that the value of K cannot be considered a constant mechanical property of the subsoil. This fact may be visualized in a large uniformly loaded area considered flexible. In such case, the settlement is not uniform (Fig. 1-IV.3). The settlement may be computed approximately at any point (x, y) of the loaded area with formula 2-IV.3; hence according to definition 1-IV.2 the value of k will vary accordingly:

$$k_{xy} = \frac{q_{xy}}{\sum_1^n M_{vi} \cdot \Delta\sigma_i \cdot \Delta z_i} \qquad \text{4-IV.3}$$

Therefore, the unit foundation modulus is a function of the position of the point in question within the loaded area. The values at any point of the loaded surface, in case of deep compressible stratified subsoil conditions, may be calculated practically up to a stress of ±5% of the applied load at the foundation grade elevation.

The average settlement of a rigid footing of width $2B$ and length $2L$ is placed on the surface of an isotropic cohesive soil mass of semi-infinite extent may be estimated (see section III.2.5) by

$$\delta_{ave} = \alpha_e(2B)(1 - \nu^2)M_e \cdot q_a \qquad \text{5-IV.3}$$

where the factor α_e is a function of the ratio L/B; for $L/B = 10$, it takes a value on the order of $\alpha_e = 2.3$. The strain modulus of the soil M_e is considered constant

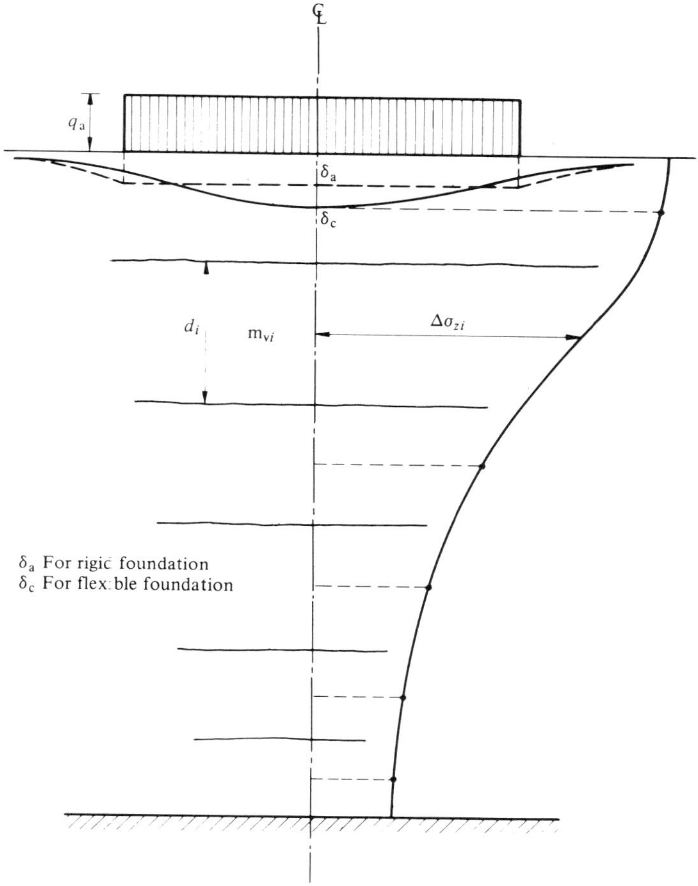

Fig. 1-IV.3 Vertical displacements under loaded area.

with depth, and q_a is the average unit load. From this expression, the average unit foundation modulus may be determined for a footing on an elastic semi-infinite isotropic subsoil:

$$k_a = \frac{1}{\alpha_e(2B)(1 - \nu^2)M_e} \tag{6-IV.3}$$

For a square footing $\alpha_e = 0.95$, and we have

$$k_a = \frac{1}{1.90\,B \cdot (1 - \nu^2)M_e} \tag{7-IV.3}$$

When the subsoil may be considered as having the same mechanical properties in an area where other footings are placed, we can establish the following relation between the k-value for a footing of area $(2B_1)(2L_1)$ and that for one of area

$(2B_2)(2L_2)$:

$$\frac{k_2}{k_1} = \frac{\alpha_1 B_1}{\alpha_2 B_2} \qquad \text{8-IV.3}$$

The values of α_1 and α_2 may be obtained from Fig. 23-III.2.

It was stated before that the unit foundation modulus cannot be considered a constant mechanical property of the soil, since it is a function also of the geometry and load distribution of the loaded surface. In order to illustrate further this fact, consider a circular area loaded with a uniform unit load q_a placed on a soil mass of semi-infinite extent with elastic and isotropic properties. The elastic strain modulus M_e is constant with depth. Under these conditions, from the theory of elasticity we find the vertical displacements and from them the k-values of the loaded area at the center and edge, respectively:

$$\text{center:} \quad \delta_c = 2M_e(1 - \nu^2)R \cdot q_a, \quad k_c = \frac{q_a}{\delta_c}$$

$$\text{edge:} \quad \delta_e = \frac{4}{\pi} M_e(1 - \nu^2)R \cdot q_a, \quad k_e = \frac{q_a}{\delta_e} \qquad \text{9-IV.3}$$

Hence

$$\frac{k_e}{k_c} = \frac{\pi}{2}$$

From the preceding discussion, we conclude that the k-values assume different magnitudes under a uniformly loaded area, and the equivalent unit foundation modulus is $\pi/2$ times larger at the edge than at the center of the loaded surface. Now consider a rigid disk of radius R on a compressible soil mass. The material is assumed, as before, elastic and isotropic with depth. When the load is applied, the disk suffers a vertical displacement equal at every point at the ground surface. Under these conditions, the theory of elasticity gives the following values:

a) the vertical displacement:

$$\delta_a = \frac{\pi}{2} M_e(1 - \nu^2)Rq_a \qquad \text{10-IV.3}$$

b) the distribution of contact stresses:

$$q_r = \frac{q_a}{2\sqrt{1 - \frac{r^2}{R^2}}} \qquad \text{11-IV.3}$$

From formula 11-IV.3, it may be noticed that at the center of the footing the unit reaction is $q_c = (1/2)q_a$, and the unit foundation modulus will be $k_c = q_a/2\delta_a$. For any other point on the loaded area at a distance r from the center

$$k_r = \frac{q_r}{\delta_a} \qquad \text{12-IV.3}$$

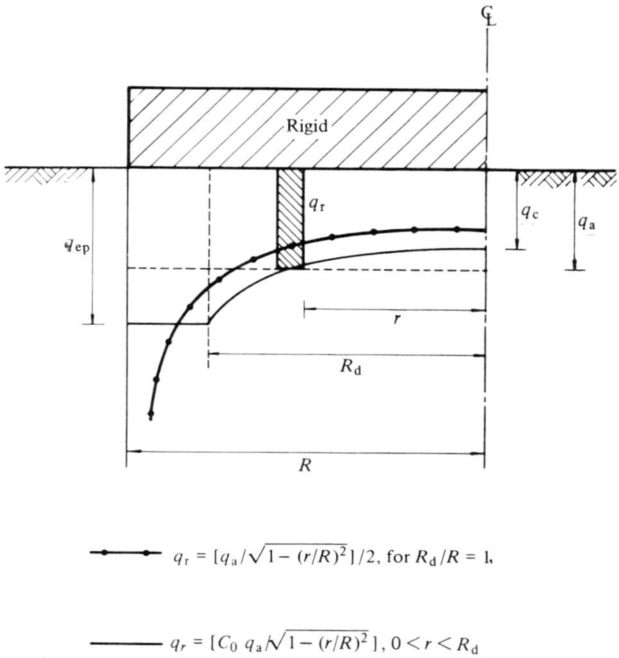

Fig. 2-IV.3 Contact stresses under rigid circular foundation.

The average value is

$$k_a = \frac{q_a}{\delta_a}$$

At the edge of the rigid disk ($r = R$) we find that the unit stress q_e assumes theoretically very high values; hence $k_e \to \infty$. Nevertheless, soils yield because of plasto-viscous behavior at the edge of the foundation. Therefore, the contact stresses cannot reach, at this place, very high values. The importance of this phenomenon may be investigated assuming that in the elastic range (Fig. 2-IV.3) the contract stress distribution is represented by the following formula:

$$q_r = C_0 \frac{q_a}{\sqrt{1 - \frac{r^2}{R^2}}} \qquad \text{13-IV.3}$$

in which R is the radius of the rigid disk, C_0 is a constant to be determined, and q_a is the average unit soil reaction. Furthermore, as stated before, because of the imposed vertical displacement at the edge of the foundation, the stress is limited in magnitude because of the visco-plastic properties and relaxation of the soil. Therefore, calling the limiting stress at the edge q_{ep} (Fig. 2-IV.3), we can write the following working hypothesis proposed by the author (1945) for elasto-plastic equilibrium:

$$\pi R^2 q_a = \int_{R_d}^{R} q_{ep}(2\pi r)dr + \int_0^{R_d} q_r(2\pi r)dr \qquad \text{14-IV.3}$$

After integration and calling $R_d/R = a$, we obtain

$$q_a = q_{ep}(1 - a^2) + 2C_0 q_a[1 - \sqrt{1 - a^2}] \qquad \text{15-IV.3}$$

From 13-IV.3 and 15-IV.3, after performing algebraic operations we find:

$$C_0 = \frac{1}{2 - \sqrt{1 - a^2}} \qquad \text{16-IV.3}$$

and

$$\frac{q_a}{q_{ep}} = 2\sqrt{1 - a^2} - (1 - a^2) \qquad \text{17-IV.3}$$

The values of q_a/q_{ep} and C_0 vs. a may be found plotted in Figs. 4 and 5-IV.3, respectively.

In the case of a rigid strip foundation on a semi-infinite isotropic soil mass, the contact stresses may be estimated with the same working hypothesis as before.

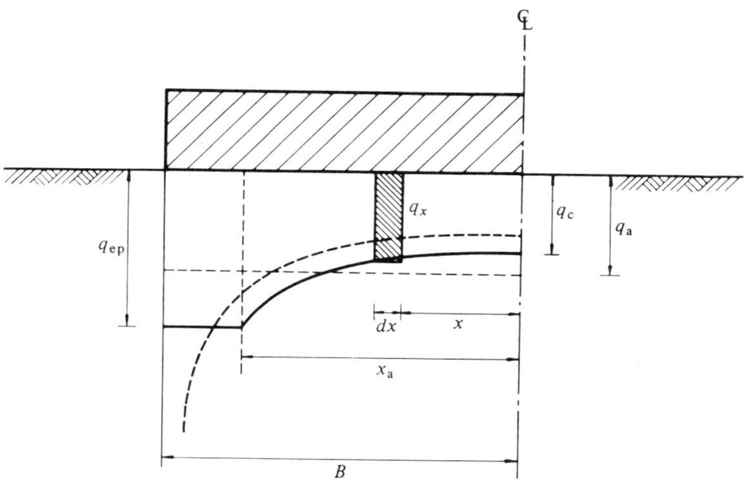

---- $q_x = [2q_a/\pi\sqrt{1-(x/B)^2}]$, for $x_a/B = 1$

—— $q_x = [C_s q_a/\sqrt{1-(x/B)^2}]$, $x \leqslant x_a$

Fig. 3-IV.3 Contact stresses under rigid strip foundation.

According to Fig. 3-IV.3, we write:

$$(2B)q_a = 2\int_{x_a}^{B} q_{ep}\,dx + 2\int_{0}^{x_a} q_x\,dx \qquad \text{18-IV.3}$$

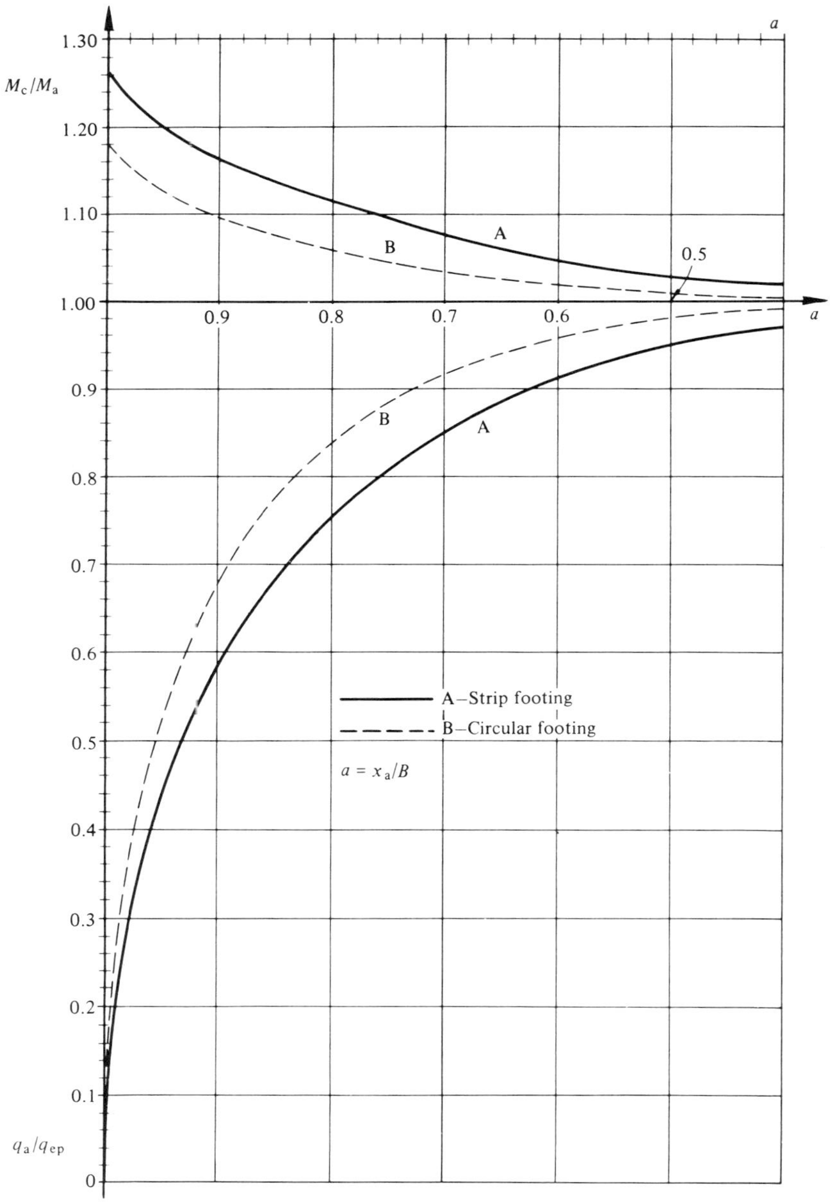

Fig. 4-IV.3 Values of q_a/q_{ep} vs. a for calculating contact stresses in strip and circular rigid foundations.

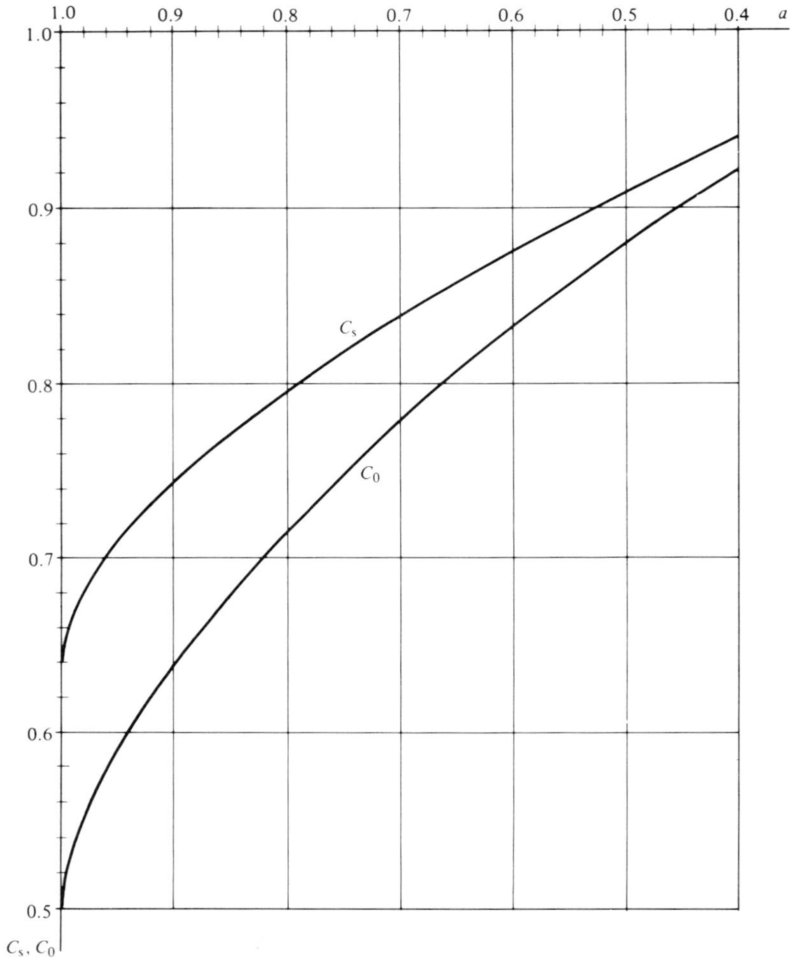

Fig. 5-IV.3 Values of C_s and C_0 vs. a for circular and strip rigid foundations.

In the elastic range we assume again

$$q_x = C_s \frac{q_a}{\sqrt{1 - \frac{x^2}{B^2}}} \qquad \text{19-IV.3}$$

(see Fig. 3-IV.3). In the plastic range $q_x = q_{ep}$ is considered constant, and so for $x_a < x < B$

$$B \cdot q_a = q_{ep}(B - x_a) + C_s \cdot q_a \cdot \int_0^{x_a} \frac{dx}{\sqrt{1 - \frac{x^2}{B^2}}} \qquad \text{20-IV.3}$$

Calling $x_a/B = a$, we integrate and obtain for the strip foundation

$$q_a = q_{ep}(1 - a) + C_s q_a \cdot \sin^{-1} a$$

with this equation and 19-IV.3, we find

$$C_s = \frac{\sqrt{1 - a^2}}{(1 - a) + \sqrt{1 - a^2} \cdot \sin^{-1} a} \qquad \text{21-IV.3}$$

and

$$\frac{q_a}{q_{ep}} = (1 - a) + \sqrt{1 - a^2} \cdot \sin^{-1} a \qquad \text{22-IV.3}$$

At the center of the footing, for $x = 0$

$$\frac{q_c}{q_a} = C_s, \text{ and } k_c = \frac{q_c}{\delta_a}$$

The values of 21 and 22-IV.3 are plotted in Figs. 4 and 5-IV.3, respectively.

From the foundation engineering point of view, it is important to study the moments and shears induced in the rigid foundation because of the variable soil reaction as compared to the uniform reaction currently assumed. The moment about an axis through the center of the strip footing may be investigated according to 19-IV.3; hence

$$M_c = \int_{x_a}^{B} q_{ep} \cdot xdx + \int_0^{x_a} q_x \cdot xdx$$

and if M_a is the moment for a uniform reaction distribution q_a, then we find:

$$\frac{M_c}{M_a} = \frac{2\sqrt{1 - a^2} - (1 - a^2)}{(1 - a) + \sqrt{1 - a^2} \cdot \sin^{-1} a} \qquad \text{23-IV.3}$$

When the soil reaches full plasticity under the rigid footing $a = 0$, then $M_c = M_a$. If the soil works under full elastic conditions, then $a = 1$ and

$$\lim_{a \to 1} \frac{M_c}{M_a} = 1.27 \qquad \text{24-IV.3}$$

The values of the ratio M_c/M_a for a strip footing and also for a circular rigid disk are plotted, in Fig. 4-IV.3, against the distance of the plasticity action at the edge of the footing represented by the parameter a. We observe that the largest bridging action occurring in a rigid strip footing is 27% of the moment M_a for a uniform reaction distribution, and in case of a rigid circular disk only 18%. These values will decrease considerably if the soil yields at the edge. Thus we see that in the particular case of isolated footings with concentrated load at the center, the error involved when the moment is calculated assuming a uniform contact stress is unimportant, since the factor of safety used in structural design of footings is ample to include any uncertainty in the contact stress distribution. Nevertheless, in case of a wide strip foundation subjected to several concentrated loadings, the difference in shears and bending moments may be important for different assumptions of the contact stresses, and therefore, the foundation structural analysis and economy may be affected.

Moreover, to select the value of q_{ep} at the edge of the foundation structure in order to find the approximate stress distribution as discussed in previous paragraphs, it is necessary to consider the plasto-viscous flow behavior and relaxation of the soil, because of the imposed fixed vertical displacement by the rigid foundation structure. From rheological considerations, it may be found that relaxation under fixed vertical displacement is represented by the following expression for one rheological unit:

$$p_t = p_e + (p_0 - p_e) e_N^{-\frac{\phi_a}{\alpha_e} \cdot t} \qquad \text{25-IV.3}$$

in which p_0 is the applied load, p_e is the elastic response load, and ϕ_a represents viscous fluidity and α_e elastic properties (see section II.3). Therefore, for a large time t, the relaxation phenomenon approaches the elastic response load p_e under fixed deformation.

The relaxation phenomenon takes place in a small zone at the foundation edge until the elastic soil elements stop the plasto-viscous flow. Tests performed to investigate the characteristics of this phenomenon in clays and silty clays verify formula 25-IV.3 and give elastic response on the order of 1/2 of the ultimate shear strength. Furthermore, at the edge of the strip foundation, the maximum shear stress according to the theory of elasticity is reached when the load imposed by the foundation slab is $\sigma_q = \pi \cdot s$. Hence, the relaxation stress may be taken approximately as $\sigma_{qr} = (3/2)s$, where s is the shear strength of the soil. If the effective confining stress at the foundation grade elevation is called σ_{0d}, then we find

$$q_{ep} \doteq \frac{3}{2} s + \sigma_{0d} \qquad \text{26-IV.3}$$

Nevertheless, in case of sensitive sediments of the preconsolidated type (section II.3) it is necessary in selecting the q_{ep}-value to consider the critical confined compressive stress σ_b, at which the compressibility curve representing the soil in question shows a break, and the material becomes more compressible. Therefore, the minimum value between that given by 26-IV.3 and σ_b should be selected for q_{ep}.

To illustrate the problem, assume it is required to know the approximate contact stress distribution for a rigid strip foundation placed at a depth $D = 2.0$ m on a semi-infinite isotropic subsoil of unit weight 1.5 ton/m^3. The rigid foundation structure is subjected to the loads shown (Fig. 6-IV.3). The foundation width is 12 m, and therefore, the average unit load is $q_a = 12$ ton/m^2. The soil is of the preconsolidated type and has high sensitivity and medium compressibility, the critical stress $\sigma_b = 26.4$ ton/m^2 and the shear strength of the soil is $s = 13.6$ ton/m^2. The

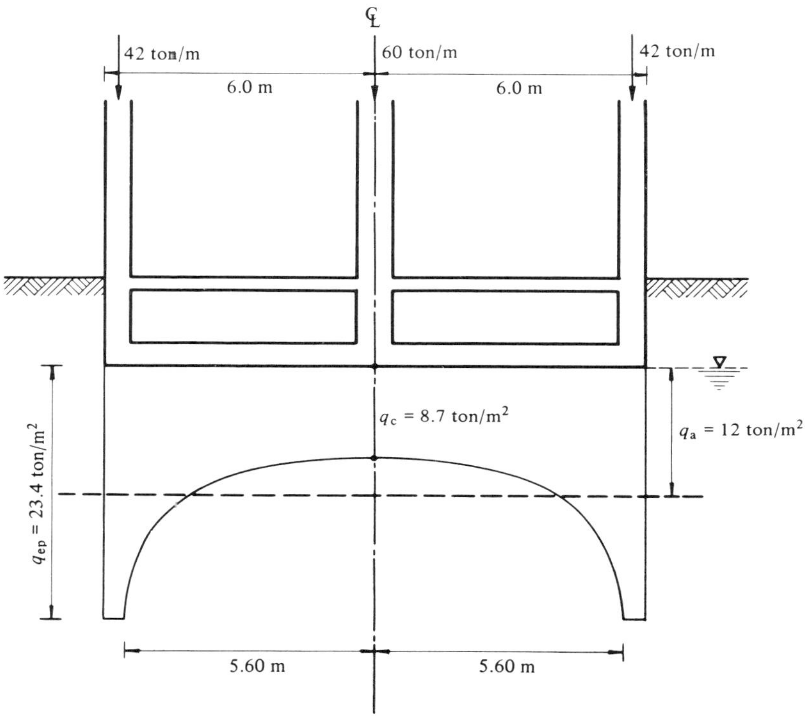

Fig. 6-IV.3 Contact stress under rigid strip foundation.

value of K_0 is 0.8. The allowable bearing capacity of the footing for a factor of safety of three is

$$q_d = \frac{5.7 \times 13.6}{3} + 2 \times 1.5 = 28.8 \text{ ton/m}^2$$

Therefore, the average load $q_a = 12$ ton/m^2 is satisfactory. At the edge because of plastic yield and relaxation

$$q_{ep} = \frac{3}{2} \times 13.6 + 3.0 = 23.4 \text{ ton/m}^2$$

The value of $q_{ep} = 23.4$ ton/m^2 is smaller than σ_b thus should be used; hence

$$\frac{q_a}{q_{ep}} = 0.512$$

From Figs. 4 and 5-IV.3, we find $a = 0.93$ and $C_s = 0.725$; hence the contact stress at the center of the footing is $q_c = 8.7$ ton/m^2, and $M_c/M_a = 1.18$. The bending moment at the center of the strip foundation, under the assumption that all the loads are linearly concentrated at the center line, is 18% higher than for the assumption of uniform soil stress reaction. Therefore:

for $B > x > x_a$

$$q_{ep} = 23.4 \text{ ton/m}^2$$

for $x_a > x > 0$

$$q_x = \frac{8.7}{\sqrt{1 - \dfrac{x^2}{B^2}}} \text{ ton/m}^2$$

The contact stress distribution is plotted in Fig. 6-IV.3. To facilitate the use of formula 19-IV.3 for the contact stress distribution of rigid strip foundations under

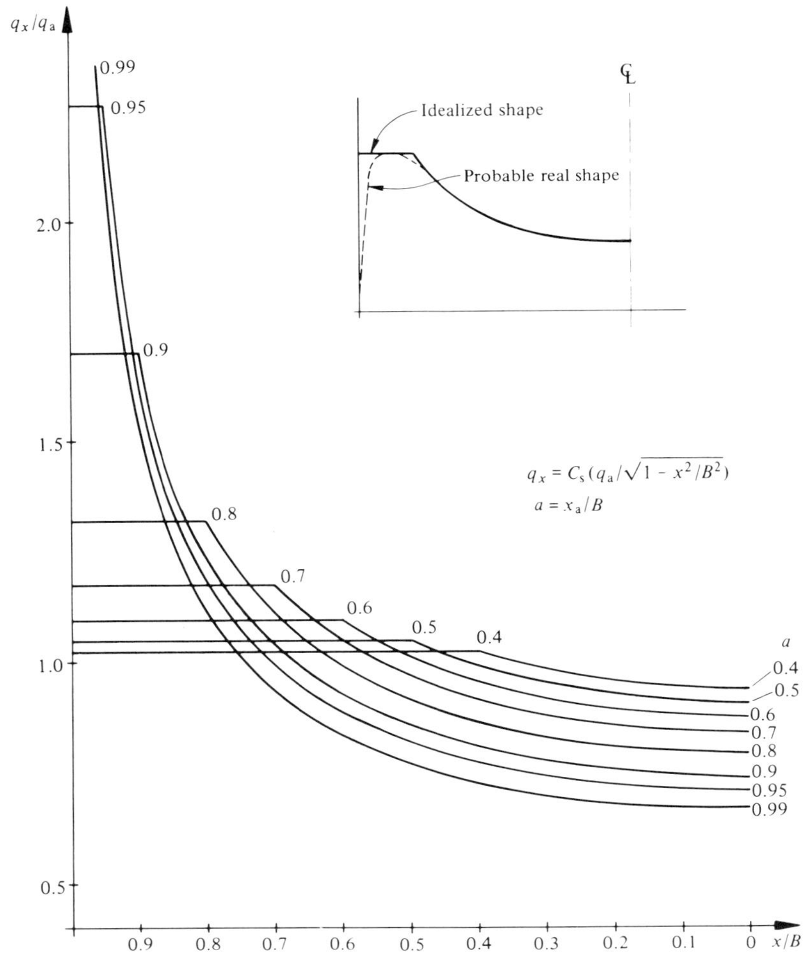

Fig. 7-IV.3 Contact stress distribution for various elasto-plastic conditions under a rigid strip foundation.

elasto-plastic conditions, in Fig. 7-IV.3 values of q_x/q_a are plotted against x/B for various conditions of the plastic range.

IV.4 BENDING MOMENTS AND SHEARS IN RIGID FOUNDATION

The relative deflections of the foundation structure may be small in comparison with the total vertical displacements because of the compressibility of the subsoil strata under the foundation grade elevation. In those cases, the subgrade reaction will assume a magnitude close to that given by the formulas obtained in section IV.3 based on a semi-infinite isotropic subsoil. The study of the shears and bending moments is useful for the first stage of calculations and when conditions in the field may be considered approximately isotropic to great depth.

For the rigid strip foundations, the shears and bending moments per unit length due to soil reaction can be calculated introducing a correction ΔV_x and ΔM_x, respectively, to the shears V_a and bending moments M_a based on uniform subgrade reaction q_a (Fig. 1-IV.4). Therefore:

$$V_x = V_{ax} + \Delta V_x$$

$$M_x = M_{ax} + \Delta M_x \qquad \text{1-IV.4}$$

4.1 Shear Correction

The shear correction is expressed as follows:
for $B > x > x_a$

$$\Delta V_x = q_{ep}(B - x) - q_a(B - x)$$

Hence:

$$\Delta V_x = q_a B\left(\frac{q_{ep}}{q_a} - 1\right)\left(1 - \frac{x}{B}\right) \qquad \text{2-IV.4}$$

for $x_a > x > 0$

$$\Delta V_x = q_{ep}(B - x_a) + \int_x^{x_a} q_x\, dx - \int_x^B q_a\, dx \qquad \text{3-IV.4}$$

where

$$q_x = \frac{\sqrt{1 - a^2}}{(1 - a) + \sqrt{1 - a^2}\sin^{-1} a} \cdot \frac{q_a}{\sqrt{1 - \dfrac{x^2}{B^2}}} \qquad \text{4-IV.4}$$

and for $x = x_a$

$$q_{ep} = \frac{1}{(1 - a) + \sqrt{1 - a^2}\sin^{-1} a} \cdot q_a \qquad \text{5-IV.4}$$

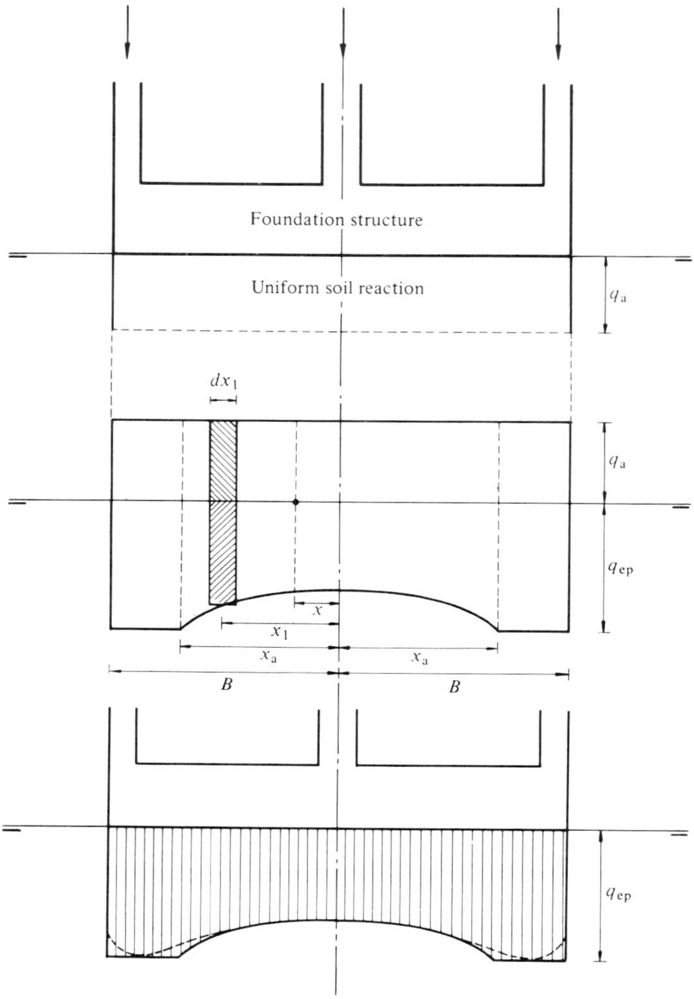

Fig. 1-IV.4 Subgrade reaction.

After substituting 4 and 5-IV.4 into 3-IV.4 and performing integration and algebraic operations, we obtain

$$\Delta V_x = q_a B \left[\frac{(1 - a) + \left(\sin^{-1} a - \sin^{-1} \dfrac{x}{B} \right) \sqrt{1 - a^2}}{(1 - a) + \sqrt{1 - a^2} \sin^{-1} a} - \left(1 - \frac{x}{B} \right) \right] \qquad \text{6-IV.4}$$

which can be written in the form

$$\Delta V_x = q_a B \cdot V_1 \left(a, \frac{x}{B} \right) \qquad \text{7-IV.4}$$

The function $V_1 \left(a, \frac{x}{B} \right)$ is plotted in Fig. 2-IV.4.

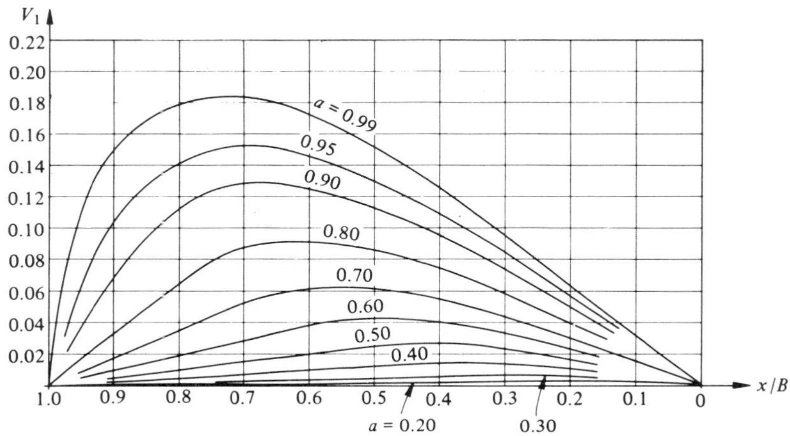

Fig. 2-IV.4 Function $V_1(a, x/B)$.

4.2 Moment Correction

The moment correction according to Fig. 1-IV.4 is:

$$\Delta M_x = \frac{1}{2} q_a B^2 \left(\frac{q_{ep}}{q_a} - 1\right)\left(1 - \frac{x}{B}\right)^2 \qquad \text{8-IV.4}$$

for $B > x > x_a$, and

$$\Delta M_x = q_{ep}(B - x_a)\left[(B - x) - \frac{1}{2}(B - x_a)\right] + \int_x^{x_a} q_x(x_1 - x)\,dx_1 - \int_x^B q_a(x_1 - x)\,dx_1 \qquad \text{9-IV.4}$$

for $x_a > x > 0$.

Using expressions 4 and 5-IV.4 for q_x and q_{ep}, respectively, and after integration and algebraic arrangements, we obtain

$$\Delta M_x = \frac{1}{2} q_a \cdot B^2 \cdot \left\{ \frac{2\sqrt{\left(1 - \frac{x^2}{B^2}\right)(1 - a^2)} - \frac{2x}{B}\left[\sqrt{1 - a^2}\left(\sin^{-1} a - \sin^{-1}\frac{x}{B}\right) + (1 - a)\right] - (1 - a^2)}{(1 - a) + \sqrt{(1 - a^2)}\,\sin^{-1} a} - \left(1 - \frac{x}{B}\right)^2 \right\} \qquad \text{10-IV.4}$$

from which we have

$$\Delta M_x = \frac{1}{2} q_a B^2 \cdot M_1\left(a, \frac{x}{B}\right) \qquad \text{11-IV.4}$$

The values of the function $M_1(a, x/B)$ are plotted in Fig. 3-IV.4.

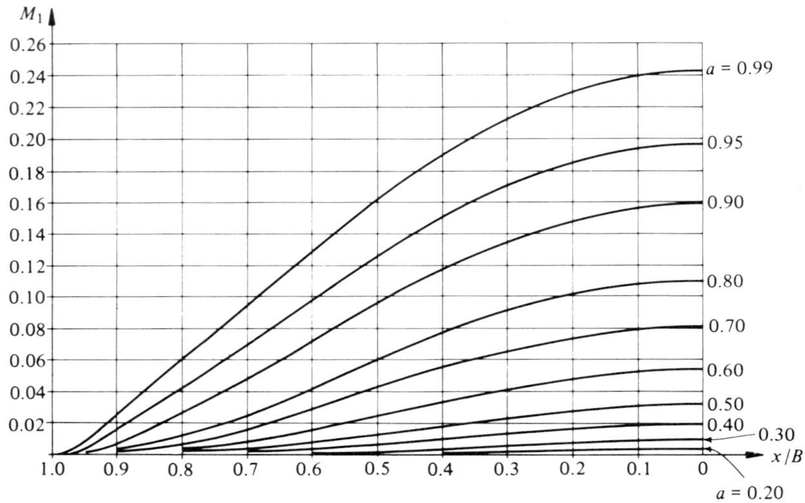

Fig. 3-IV.4 Function $M_1(a, x/B)$.

By means of formulas 7 and 11-IV.4, the example shown in Fig. 6-IV.3 has been completed. The shears and bending moments are plotted in Fig. 4-IV.4 for two conditions: a) a uniform subgrade reaction, and b) a variable subgrade reaction for rigid strip foundation. From this calculation we see the importance of taking into consideration the rigidity of the foundation structure. The results obtained above, however, are considered extreme conditions for the following reasons:

1. The soil deposits are usually encountered stratified and limited in depth by firm ground.
2. The strain modulus or compressibility decreases with depth.
3. The foundation structure is not of infinite stiffness in comparison to the soil compressibility, except in soils exhibiting high or very high compressibility.

These conditions reduce the bridging effect of the foundation, and on many occasions this is translated into economy in the design of the foundation structure. The functions $V_1(a, x/B)$ and $M_1(a, x/B)$ to simplify calculations are useful to the foundation engineer in performing estimates of the shear forces and bending moments to be expected under approximate isotropic conditions, as those assumed in the preceding discussion.

IV.5 RECOMMENDED *k*-VALUES

Precise evaluation of the *k*-values for a particular problem has always been the concern of the foundation engineer. Only approximate values can be determined; therefore, mathematical refinements are not justified. The easiest and least time-consuming methods giving results on the safe side should be chosen by the founda-

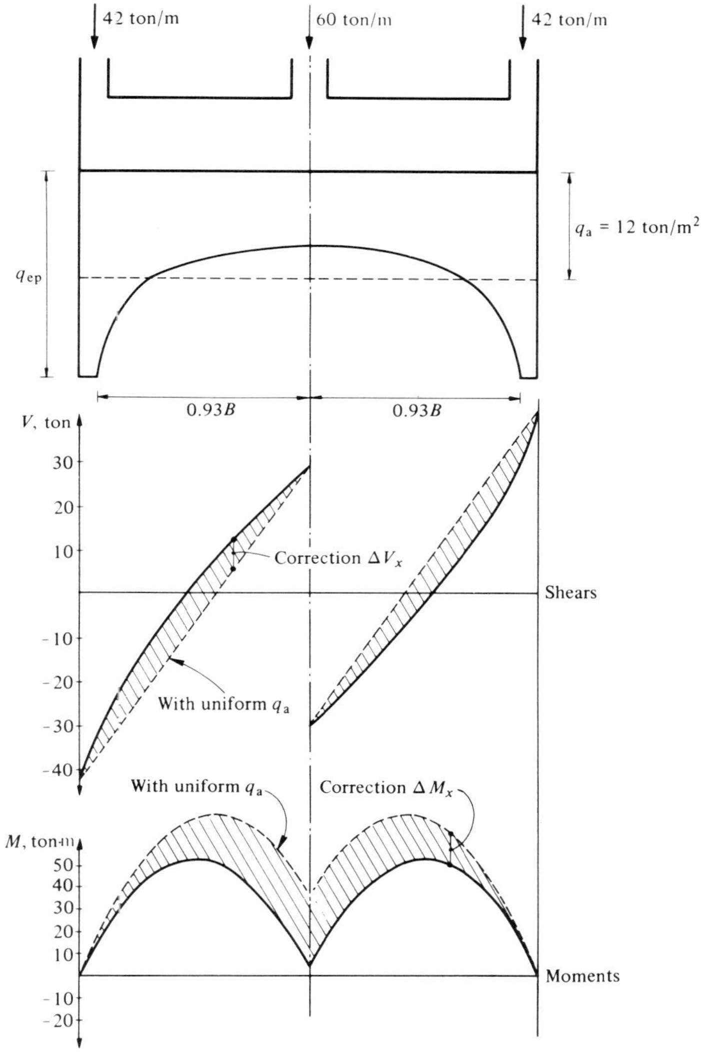

Fig. 4-IV.4 Example of shear and moment correction.

tion engineer. Terzaghi has given empirical rules to orient the foundation engineer in the use of k-values. The recommendations are based on tests made in the field with 1-ft × 1-ft rigid plates. This method, however, should be used with care. It may be applied only approximately in cases in which the depth of the soil affected by the width of the footing may be considered isotropic, and one is able to extrapolate the results of the test plates to larger areas. If $\overline{k}_{s1}$ is the average unit foundation modulus determined for the 1-ft × 1-ft test plate given in ton/m³, then

TABLE 1-IV.5 Proposed average values of $\overline{k}_{s1}$ for 1-ft × 1-ft square plates and long 1-ft wide strips, after Karl Terzaghi (1955)

		Average $\overline{k}_{s1}$ values		Range of $\overline{k}_{s1}$ values	
		tons/ft^3	kg/cm^3	tons/ft^3	kg/cm^3
Sand:	loose	40	1.29	20–60	0.64–1.92
	medium	130	4.17	60–300	1.92–9.62
	dense	500	16.10	300–1000	9.62–32.1
Clay:	stiff	75	2.41	50–100	1.6–3.21
	very stiff	150	4.82	100–200	3.21–6.42
	hard	300	9.64	300	9.60

For dry sand multiply by 1.5 and for submerged sand by 0.6. Here 1 ton = 2000 lb.

Terzaghi suggests the use of the following formula for sand:

$$k_s = \overline{k}_{s1}\left(\frac{B + 0.15}{2B}\right)^2 \qquad \text{1-IV.5}$$

where $2B$ is the width in meters of the footing of any length equal or greater than $2B$, or in a rigid mat foundation it represents the distance in meters between line loads or the width of the tributary area of that particular column row. In case of clays

$$k_s = \overline{k}_{s1}\,\frac{n + 0.5}{1.5n} \qquad \text{2-IV.5}$$

in which the contact area has the shape of a rectangle of width B and length $L = nB$. Values of $\overline{k}_{s1}$ collected by Terzaghi may be found in Table 1-IV.5; they may serve as a guidance for estimates. The k-values, however, should be investigated for the specific case of subsoil conditions and loaded areas by means of formulas 3 and 4-IV.3, with the use of the mechanical properties of compressibility of the material investigated by means of good undisturbed samples.

BIBLIOGRAPHY

Barden, L. (1962). Distribution of Contact Pressure under Foundations, *Geotechnique*, Vol. XII, p. 181.

Borowicka, H. (1936). Influence of Rigidity of a Circular Foundation Slab on the Distribution of Pressures over the Contact Surface, *Proc. I ICOSOMEF*, Vol. II, pp. 144–149, Cambridge, Mass.

Schultze, Edgar (1961). Distribution of Stress Beneath a Rigid Foundation, *Proc. V ICOSOMEF*, Vol. I, p. 807, Dunod, Paris.

Zeevaert, L. (1953). Equilibrio Elástico-Plástico en la Superficie de Contacto entre un Suelo Arcilloso y una Placa Rígida, *Revista Ingeniería*, Vol. XXX, September–October.

Sunken area at Long Beach, California, due to oil extraction from subsoil.

▸V◂ GROUND SURFACE SUBSIDENCE

V.1. INTRODUCTION

Settlement of the ground surface is observed in places where heavy pumping of ground water is taking place, as is the case in numerous cities and regions like the San Joaquin Valley, Tokyo, Mexico City, and so many others. Ground surface subsidence takes place also because of oil extraction, such as in Long Beach, California, and Maracaibo, Venezuela, where the ground surface is found below sea level, which poses severe problems of protection against ocean water flooding the sunken areas. In these cases, there is always the danger that the dikes restraining the water may break with catastrophic consequences. One of the most effective remedies, though expensive, has been to fill up the depressions.

The problems encountered because of deep seated pumping of water and of lowering the water table are important to consider in foundation engineering. In places where the water table has been lowered in highly compressible soil sediments, cracks develop at the ground surface, running deep into the ground. Before foundation construction in these areas, the cracks must first be treated to restore continuity in the ground surface. The lowering of the water table produces shrinkage of the lateral supporting soil in point bearing piles, inducing them to buckle. In case of friction piles, the adhesion of the soil is considerably reduced, and uneven settlements take place because of reduction in the load capacity of the piles.

The occurrence of large differential vertical displacements of the ground surface produces damages to public utilities such as the sewage system, streets, railroads, and highways. The maintenance of these services becomes a matter of primary importance for the city services. To achieve a proper design of foundations of buildings or other engineering projects founded on the subsiding ground surface, the foundation engineer has to take into account the rate of settlement of the ground surface. It is important to investigate the local characteristics of this phenomenon in the area where the project is proposed, since in general, it is found that the ground settlement is not uniform from one place to the other. Pile foundations supported on deeper strata are very sensitive to ground surface subsidence. The relative displacement between piles and ground develops large dragging forces on the piles that increase considerably their load and decrease their bearing capacity.

When designing foundations in areas where ground surface subsidence is taking place, the foundation engineer must know the stratigraphical and hydrodynamic conditions under the area affected and the mechanical properties of the subsoil materials.

V.2 MECHANICS OF GROUND SURFACE SUBSIDENCE

The change in vertical intergranular stress with depth in the subsoil when subjected to hydrodynamic conditions (section III.3) may be expressed as follows:

$$\frac{\partial \Delta\sigma_z}{\partial z} = \gamma_w \cdot i_z \qquad \text{1-V.2}$$

in which i_z is the vertical component of the hydraulic gradient of the water current, taken as positive when the flow is downward, increasing the effective intergranular stresses. Where there is no water current, $i_z = 0$, and the subsoil gravitational water conditions are static. Under this condition, piezometers installed at different elevations within the subsoil mass will show the same piezometric water level, corresponding to the free water surface elevation. The effective stress at a depth z under static conditions may be calculated from the formula

$$\sigma_{0z} = \sum_{1}^{n} \gamma_i \Delta z_i \qquad \text{2-V.2}$$

where γ_i represents the unit weight of the soil, either the submerged unit weight or that above the free water surface. These conditions are changed as soon as the water enters into motion. A downward water current will increase the stress at a depth z, according to formula 1-V.2, in the following amount:

$$\Delta\sigma_z = \int_0^z \gamma_w \cdot i_z \cdot dz \qquad \text{3-V.2}$$

If the drop between piezometric water levels at distance dz is called $d\lambda$, then the induced hydraulic gradient is $i_z = \dfrac{\partial\lambda}{\partial z}$, and by substitution in 1-V.2, we obtain

$$\frac{\partial(\Delta\sigma_z)}{\partial z} = \gamma_w \frac{\partial\lambda}{\partial z}$$

Hence:

$$\Delta\sigma_z = \gamma_w \cdot \lambda_z \qquad \text{4-V.2}$$

Therefore, the increment in effective stresses at depth z caused by a downward water current is equal to the total loss of piezometric water pressure induced by such current. This case is illustrated in Fig. 1-V.2.

The increment of effective stress induced by the water current will produce at depth z a strain

$$\frac{\partial\delta}{\partial z} = m_{vz} \int_0^z \gamma_w \cdot i_z \cdot dz \qquad \text{5-V.2}$$

Thus the ground surface subsidence between firm ground and depth z of the compressible soil deposit of thickness H is

$$\delta_z = \int_z^H \left[m_{vz} \int_0^z \gamma_w \cdot i_z \cdot dz \right] dz \qquad \text{6-V.2}$$

where m_{vi} is the coefficient of unit volume compressibility and i_z the hydraulic gradient at depth z, respectively.

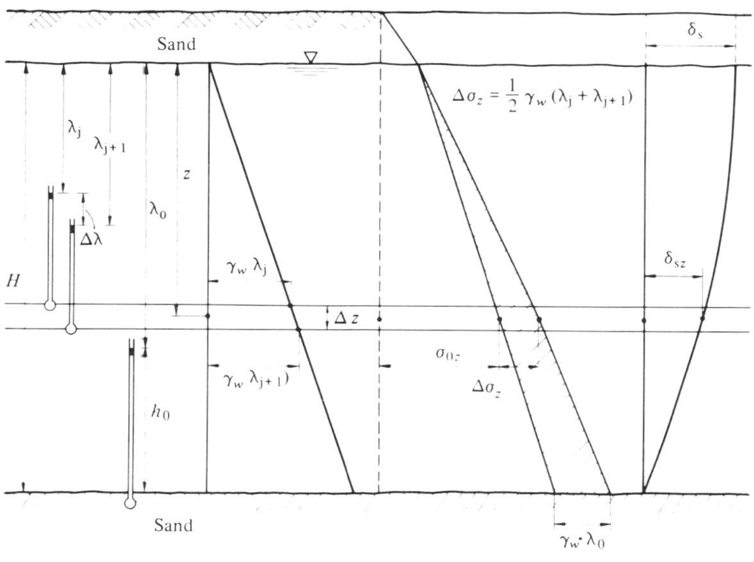

σ_{0z} Initial overburden effective stress

$\Delta\sigma_z$ Stress increments because of downward water flow

δ_{sz} Ground subsidence at depth z

δ_s Ground subsidence at surface ($z = 0$)

Fig. 1-V.2 Increment in effective stresses due to reduction in piezometric water elevations.

When the soil deposit may be considered isotropic, the value of m_{vz} is constant with depth. The hydraulic gradient because of total piezometric water level reduction at the bottom of the deposit is λ_0/H (Fig. 1-V.2). Therefore, in this particular case after integration of 6-V.2, we obtain:

$$\delta_{sz} = m_v \gamma_w \cdot \tfrac{1}{2}\, i_z\, (H^2 - z^2) \qquad \text{7-V.2}$$

but $i_z = \lambda_0/H$, and the ultimate ground surface subsidence at $z = 0$ is given by

$$\delta_s = m_v \left(\tfrac{1}{2}\, \gamma_w \lambda_0\right) H \qquad \text{8-V.2}$$

Furthermore, the drop in the piezometric water levels usually takes place slowly with time; hence, it is important to investigate the rate of subsidence or downward velocity of the ground surface; hence:

$$\frac{\partial \delta_s}{\partial t} = m_v \cdot H \cdot \left(\frac{1}{2}\, \gamma_w \frac{\partial \lambda_0}{\partial t}\right) \qquad \text{9-V.2}$$

Therefore, from 7-V.2, the rate of subsidence at any depth z is given by

$$\frac{\partial \delta_{sz}}{\partial t} = \frac{\partial \delta_s}{\partial t}\left[1 - \frac{z^2}{H^2}\right] \qquad \text{10-V.2}$$

In case of an isotropic soil deposit of thickness H, the above formula applies. The ground surface subsidence with depth has a parabolic variation as shown graphically in Fig. 1-V.2.

In practice, however, soil deposits are stratified and the mechanical properties of compressibility are not constant with depth. Also the hydraulic gradient may change from one stratum to another (Fig. 2-V.2). In such cases, it is necessary to install piezometers at the interface of the different strata forming the soil deposit where more permeability may be encountered. From the piezometric water levels one can calculate the hydraulic gradient for the stratum in question; this may be considered constant throughout the thickness of each stratum. Under these conditions, the increment of effective stress at the bottom of each stratum j of thickness d_j because of the downward water current is

$$\Delta\,(\Delta \sigma_{zj}) = \gamma_w\, i_j \cdot d_j$$

and the total stress increment induced by the water flow to the base of the n-th stratum is

$$\Delta \sigma_z = \sum_1^n \gamma_w \cdot i_j \cdot d_j \qquad \text{11-V.2}$$

but $i_j = \Delta\lambda_j/d_j$; hence from 11-V.2 we have

$$\Delta \sigma_z = \gamma_w \sum_1^n \Delta\lambda_j, \text{ or } \Delta \sigma_z = \gamma_w \cdot \lambda_z \qquad \text{12-V.2}$$

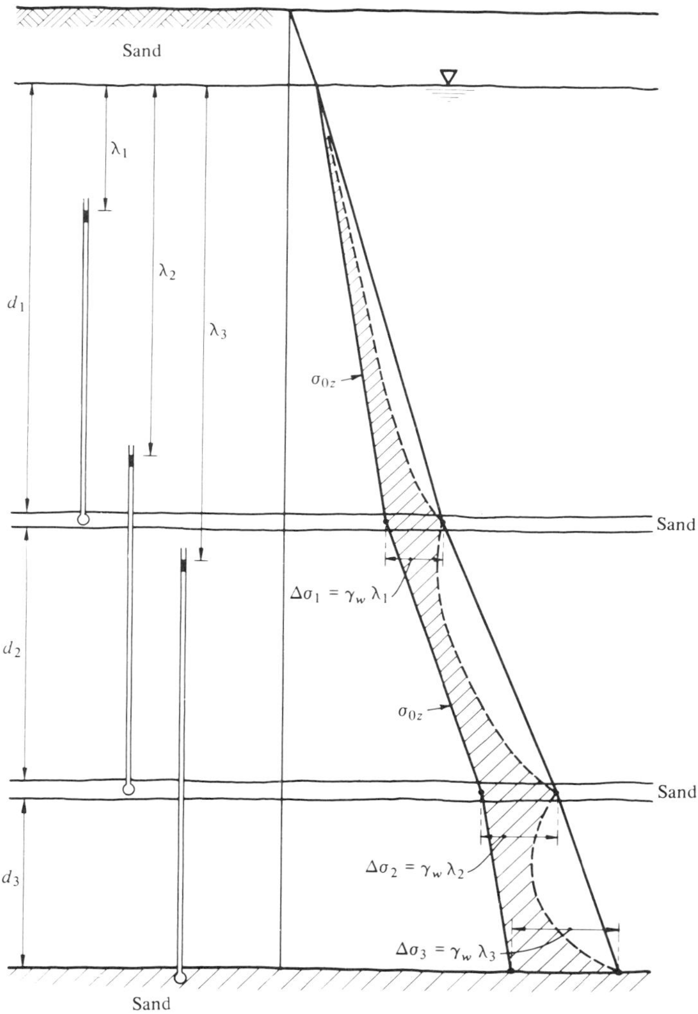

Fig. 2-V.2 Reduction of piezometric water elevations in stratified soil mass.

where λ_z is the total change in piezometric water level at the bottom of the n-th stratum, and $\Delta\sigma_z$ is the effective stress caused by the piezometric water level change in the strata from ground surface to the same depth. The deformation of each stratum may be calculated with the formula

$$\Delta\delta_j = m_{vj} \frac{\Delta\sigma_j + \Delta\sigma_{j+1}}{2} d_j \qquad \text{13-V.2}$$

and the total settlement at the ground surface will be the summation of all the compressible strata, hence

$$\delta_s = \tfrac{1}{2} \sum_{1}^{n} m_{vj} (\Delta\sigma_j + \Delta\sigma_{j+1}) d_j \qquad 14\text{-V}.2$$

This may be written in the form

$$\delta_s = \tfrac{1}{2} \gamma_w \sum_{1}^{n} m_{vj} (\lambda_j + \lambda_{j+1}) d_j \qquad 15\text{-V}.2$$

From formula 15-V.2, the rate of subsidence may be calculated in terms of the rate of piezometric water level reduction; using the notation $\frac{\partial \delta_s}{\partial t} = \dot{\delta}_s$ and $\frac{\partial \lambda}{\partial t} = \dot{\lambda}$ we obtain

$$\dot{\delta}_s = \tfrac{1}{2} \gamma_w \sum_{1}^{n} m_{vj} (\dot{\lambda}_j + \dot{\lambda}_{j+1}) d_j \qquad 16\text{-V}.2$$

where $(1/2)(\dot{\lambda}_j + \dot{\lambda}_{j+1})$ is the average rate of piezometric water level reduction for such strata.

In practice, fine sediments of medium to high and very high compressibility show elasto-plastic-viscous behavior that should be considered in the calculation of the ground surface subsidence. The rheological mechanical properties of these soils have been discussed in section II.3.13, where it was found that for linearly increasing stresses, the strain-time behavior may be expressed by

$$\Delta\epsilon_v = m_{ep} \Delta\dot{\sigma} \left[\phi(T_v) + \frac{\beta A_1}{2.3} \right] \cdot t \qquad 17\text{-V}.2$$

in which m_{ep} is the elasto-plastic coefficient of unit volume compressibility, $\phi(T_v)$ is a dimensionless function derived from Terzaghi's theory of consolidation as described in section II.3, the value $T_v = (c_v/H^2)t$ and $\beta = m_t/m_{ep}$, where m_t is the coefficient of unit volume compressibility for intergranular viscosity. The strain for one stratum is

$$\Delta\epsilon_j = \frac{\Delta\delta_j}{d_j}$$

hence for stratum j

$$\Delta\delta_j = \left\{ m_{ep} \left[\phi(T_v) + \frac{\beta A_1}{2.3} \right] \right\}_j \left[\tfrac{1}{2} \gamma_w (\dot{\lambda}_j + \dot{\lambda}_{j+1}) \right] d_j \cdot t \qquad 18\text{-V}.2$$

Moreover, if the rate of piezometric water level reduction is limited to a period t_c, then $\dot{\lambda} = \lambda/t_c$, and we can write

$$\Delta\delta_j = \tfrac{1}{2}(m_{ep} \cdot d)_j \cdot \gamma_w \cdot (\lambda_j + \lambda_{j+1}) \cdot \frac{T_v}{T_{vc}} \left[\phi(T_v) + \frac{\beta A_1}{2.3} \right]$$

in which:

$$T_v \left[\phi(T_v) + \frac{\beta A_1}{2.3}\right] = f_1$$

The values of A_1 are plotted in Fig. 50-II.3. Hence:

$$\Delta\delta_j = (m_{ep} \cdot d)_j \cdot (\lambda_j + \lambda_{j+1}) \cdot \gamma_w \cdot \frac{f_1}{2T_{vc}} \qquad \text{19-V.2}$$

and the ground surface subsidence is obtained adding up all strata for equal elapsed times. Therefore

$$\delta_s = \sum_1^n \Delta\delta_j \qquad \text{20-V.2}$$

In order to illustrate the use of formula 19-V.2, assume that Fig. 2-V.2 represents a silty clay soil deposit formed of three compressible strata with mechanical characteristics given in Table 1-V.2. The rate of piezometric water level drawdown is 50 cm/year at the bottom of each one of the strata and remains constant for a period of $t_c = 20$ years. The calculations are given in Table 1-V.2 and are self-explanatory. The compression of the layers after 20 years shows a smaller rate because the effective stresses have reached their maximum value. Therefore, at the end of $t_c = 20$ years, the deformation of each one of the strata is given by:

$$\Delta\delta_{jt_c} = \left\{m_{ep}\left[\phi(T_{vc}) + \frac{\beta A_1}{2.3}\right]\right\}_j \left[\frac{1}{2}\gamma_w\,(\lambda_j + \lambda_{j+1})\right] d_j \qquad \text{21-V.2}$$

The strata, however, will continue compressing because of further intergranular viscosity consolidation with an average stress of $\Delta\sigma_c = 1$ kg/cm^2, and according to the expression found in section II.3.13, formula 121-II.3, we have

$$\Delta\epsilon_v = \left\{F(T_v - T_{01}) + \beta \log\left[1 + \frac{4.62}{\beta}(T_v - T_{vc} \cdot A_2)\right]\right\} m_{ep}\Delta\sigma_c \qquad \text{22-V.2}$$

In this particular case, T_{vc} is calculated for $t_c = 20$ years.

The compression for each one of the strata drained on both sides has been calculated up to a period of 60 years, and plotted in Fig. 3-V.2. The sum of the compressions for equal times represented by curves A, B and C, respectively, gives the total subsidence; its value is given by curve D. This simple example shows how the foundation engineer can estimate the settlement of the ground surface as a function of time, if he knows the stratigraphy, the hydraulic conditions varying with time, and the mechanical properties of consolidation of the different strata that may be affected by the reduction of the piezometric water levels. Moreover, it may be observed that for large periods of time, the consolidation function $\phi(T_v) \doteq 1$, and thus from 18-V.2

$$\delta_s = t \cdot \sum_1^n \left[m_{ep}\left(1 + \frac{\beta A_1}{2.3}\right)\right]_j \left[\tfrac{1}{2}\gamma_w\,(\dot\lambda_j + \dot\lambda_{j+1})\right] d_j \qquad \text{23-V.2}$$

TABLE 1-V.2 $\Delta\sigma_c = \gamma_w \cdot \frac{1}{2}(\lambda_j + \lambda_{j+1}) = 1\ kg/cm^2$

1	2	3	4	5	6	7	8	9	10	11	12	13	14	15	16	17	18	19	20
t, years	d cm	m_{ep} cm^2/kg	β	c_v cm^2/sec	$\frac{4c_v}{d^2}$	T_v	$\phi(T_v)$	$F(T_v)$	T_v/β	$A_1 = f\left(\frac{T_v}{\beta}\right)$	f_1	$\frac{\gamma_w}{2}(\lambda_j + \lambda_{j+1})$ kg/cm^2	$(T_v - T_{01})$	$F(T_v - T_{01})$	$\frac{T_{vc}}{\beta}$	$A_2 = f\left(\frac{T_{vc}}{\beta}\right)$	$T_v - T_{vc} \cdot A_2$	f_2	δ cm
5						2.52	0.87	1.0	2.10	0.76	3.19	0.25							3.99
10						5.04	0.93	1.0	4.20	0.85	6.92	0.50							17.31
20	1000	0.05	1.2	0.004	0.504	10.00	0.97	1.0	8.33	0.91		1.00	1.30	0.96	8.33	0.96	0.4	1.45	72.29
40						20.16	0.99	1.0	16.80	0.94		1.0	11.46	1.0			10.56	2.95	147.50
60						30.24	1.00	1.0	25.20	0.96		1.0	21.54	1.0			20.64	3.29	164.50
$T_{vc} = 10$ $m_{ep} \cdot d = 50.0$ $T_{01} = f(T_{vc}) = 8.70$																			
5						3.5	0.90	1.0	3.5	0.82	4.40	0.25							1.41
10						7.0	0.95	1.0	7.0	0.89	9.36	0.50							6.03
20	600	0.03	1.0	0.002	0.700	14.0	0.97	1.0	14.0	0.93		1.0	1.44	0.975	14.0	0.975	0.35	1.39	25.02
40						28.0	1.0	1.0	28.0	0.96		1.0	15.44	1.0			14.35	2.83	50.90
60						42.0	1.0	1.0	42.0	0.97		1.0	29.44	1.0			28.35	3.12	56.11
$T_{vc} = 14.0$ $m_{ep} \cdot d = 18.0$ $T_{01} = 12.56$																			
5						3.94	0.91	1.0	6.57	0.89	4.50	0.25							0.29
10						7.87	0.95	1.0	13.12	0.92	9.37	0.50							1.19
20	400	0.01	0.6	0.001	0.787	15.74	0.97	1.0	26.23	0.96		1.00	1.48	0.98	26.23	0.985	0.24	1.25	5.01
40						31.48	1.0	1.0	52.47	0.98		1.00	17.22	1.0			15.98	2.26	9.04
60						47.22	1.0	1.0	78.70	0.99		1.00	32.96	1.0			31.72	2.43	9.72
$T_{vc} = 15.74$ $m_{ep} \cdot d = 4.00$ $T_{01} = 14.26$																			

FORMULAS:

$T_v \leqslant T_{vc}$, $\quad \Delta\delta_j = (m_{ep}d)\,\frac{1}{2}\,\gamma_w(\lambda_j + \lambda_{j+1})\,\frac{f_1}{T_{vc}}$; $\quad f_1 = T_v[\phi(T_v) + \beta A_1/2.3]$

$T_v \geqslant T_{vc}$, $\quad \Delta\delta_j = (m_{ep}d)\,\Delta\sigma_c \cdot f_2$, $\quad f_2 = F(T_v - T_{01}) + \beta \log\left[1 + \frac{4.62}{\beta}(T_v - T_{vc} \cdot A_2)\right]$

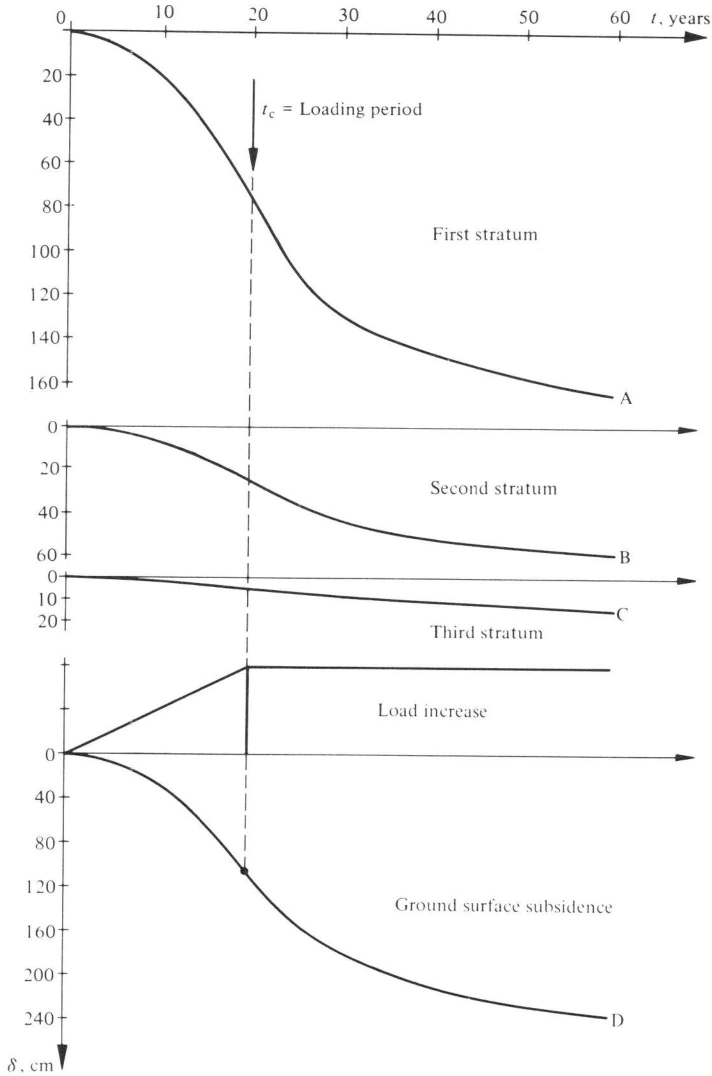

Fig. 3-V.2 Ground surface subsidence.

Therefore, the rate of ground surface subsidence $\dot{\delta}_s$ may be calculated, assuming $A_1 = 1$, from the formula

$$\dot{\delta}_s = \sum_1^n \left[m_{ep} \left(1 + \frac{\beta}{2.3} \right) \right]_j \left[\tfrac{1}{2} \gamma_w \left(\dot{\lambda}_j + \dot{\lambda}_{j+1} \right) \right] d_j + t \sum_1^n \left[m_{ep} \left(1 + \frac{\beta}{2.3} \right) \right]_j \left[\frac{1}{2} \gamma_w \frac{\partial (\dot{\lambda}_j + \dot{\lambda}_{j+1})}{\partial t} \right] d_j \quad \text{24-V.2}$$

When the ground surface subsidence is known over certain time range and the rate of piezometric drop may be considered constant, the average compressibility of the entire deposit under consolidation can be calculated, and setting $A_1 = 1$;

$$m_{va} = m_{ep}\left(1 + \frac{\beta}{2.3}\right) \qquad \text{25-V.2}$$

The value of m_{va} is found from

$$m_{va} = \frac{\dot{\delta}_s}{\frac{1}{2}\gamma_w \sum_1^n (\dot{\lambda}_j + \dot{\lambda}_{j+1})\, d_j} \qquad \text{26-V.2}$$

Here the denominator represents the rate of increment of stress area for the entire deposit under compression, because of a constant rate of reduction in the piezometer water pressures.

To illustrate the application of 26-V.2, let us use Fig. 12-V.3 for Mexico City, where a constant rate of ground surface subsidence was observed, on the order of 24 cm in the second clayey silt stratum Tarango Clay I of thickness 900 cm, in the 4-yr period from 1953 through 1956. The total piezometric water reduction corresponding to the same period is 4.2 m at the bottom and at the top of this stratum, respectively. Hence:

$$m_{va} = \frac{24}{420 \times 0.9} = 0.0635 \text{ cm}^2/\text{kg}$$

Therefore, the average unit volume compressibility of Tarango Clay I including the intergranular viscosity phenomenon is $m_{va} = 0.0635$ cm^2/kg.

When the reduction of piezometric water pressures has reached its maximum value, deformation due to intergranular viscosity still continues. Formula 22-V.2 may be written, assuming $A_2 = 1$, in the form

$$\delta_s = \sum_1^n (m_{ep} d)_j\, \Delta\sigma_c \left\{1 + \beta_j \log\left[1 + \left(\frac{18.5\, c_v}{\beta d^2}\right)_j (t - t_c)\right]\right\} \qquad \text{27-V.2}$$

where d_j represents the thickness of one stratum for which $(m_{ep})_j$, c_v and β_j are considered average values. If t_2 and t_1 are total times in the range in which the phenomenon is observed under constant piezometric water pressure, and calling $\tau_j = \left(\frac{\beta d^2}{18.5\, c_v}\right)_j$ we can write:

$$\Delta\delta_s = \sum_1^n (m_{ep}\beta)_j\, d_j \Delta\sigma_c \log \frac{\tau_j + (t_2 - t_c)}{\tau_j + (t_1 - t_c)} \qquad \text{28-V.2}$$

For large values of $(t_2 - t_c)$ and $(t_1 - t_c)$ and considering that Tarango Clay I in this case is a clayey silt deposit interbedded with numerous sand lenses permitting fast drainage of the layers, τ_j is small and we can write

$$\Delta\delta_s = \Delta\sigma_c \log \frac{t_2 - t_c}{t_1 - t_c} \sum_1^n (m_{ep}\beta)_j\, d_j \qquad \text{29-V.2}$$

from which, calling $(m_{ep}\beta)_a$ an average value, we obtain

$$(m_{ep}\beta)_a = \frac{\Delta\delta_s}{d \cdot \Delta\sigma_c \log \dfrac{t_2 - t_c}{t_1 - t_c}} \qquad \text{30-V.2}$$

To find the average value of $(m_{ep}\beta)$ take again as an example, the ground surface subsidence in Mexico City shown in Fig. 12-V.3, where it may be observed that the piezometric water pressures have remained practically unaltered with time, starting about January 1, 1957, at depths 34 and 49 m (top and bottom respectively) of the Tarango Clay I deposit of 900 cm thickness. Moreover, if we take t_2 as January 1, 1963, we have $t_2 - t_c = 72$ months, and we take t_1 to be January 1, 1960, obtaining $t_1 - t_c = 36$ months. In the period $t_2 - t_1$, a compression of $\Delta\delta_s = 5.0$ cm is observed. The corresponding increment of effective stresses increased since September 1949 to $\Delta\sigma_{34} = 0.70$ kg/cm^2 at the top of the clayey silt deposit and $\Delta\sigma_{49} = 0.80$ kg/cm^2 at the bottom. The average increase is $\Delta\sigma_a = 0.75$ kg/cm^2. Therefore, applying these numerical values in 30-V.2, we find

$$(m_{ep}\beta)_a = \frac{5}{900 \times 0.75 \log \dfrac{72}{36}} = 0.0246 \text{ cm}^2/\text{kg}$$

The value of $(m_{ep}\beta)_a$ represents the average coefficient of unit volumetric intergranular viscosity m_t. Hence from 25-V.2

$$m_{va} = m_{ep} + \frac{1}{2.3} m_t$$

Using the value $m_t = 0.0246$ cm^2/kg and the previous value found for m_{va} = 0.0635 cm^2/kg, then with the above equation we can compute the values of m_{ep} = 0.0528 cm^2/kg and $\beta = 0.465$.

Using formula 30-V.2, one can also make a forecast for the compression in future years should the piezometric water pressures remain unaltered. In fact, assume that we wish to forecast the subsequent subsidence after January 1, 1963, and up to January 1, 1970, that is to say, in the next 7 years or $\Delta t = 84$ months. In this period the average piezometric pressures are assumed to remain unaltered and equal to $\Delta\sigma_c = 0.75$ kg/cm^2:

$$\Delta\delta_s = 0.0246 \times 900 \times 0.75 \log \frac{72 + 84}{72}, \quad \Delta\delta_s = 5.57 \text{ cm}$$

The forecast is on the order of 5.57 cm for future compression up to January 1, 1970. Observing the records given in Fig. 12-V.3, we find that $\Delta\delta_s$ was on the order of 4.5 cm, and the average increment of effective stresses in this period has been of $\Delta\sigma_c = 0.62$ kg/cm^2, instead of the assumed value of 0.75 kg/cm^2. Therefore, the correction will give $0.62 \times 5.57/0.75 = 4.61$ cm, which is approximately equal to the observed value.

From the preceding example, it is recognized that the foundation engineer may be able to calculate approximate forecasts for ground surface subsidence, if he is acquainted with settlement and piezometric water pressure observations, and has the knowledge of the stratigraphy and mechanical properties of the subsoil materials. Moreover, the environmental conditions of the area in question should be known, in connection with future loading at the ground surface and expected reduction of the piezometric water levels because of pumping from deep water bearing sand strata.

V.3 GROUND SURFACE SUBSIDENCE IN MEXICO CITY

3.1 General and Historical Considerations

The importance of the ground surface settlement owing to continuous reduction in the piezometric water levels can be better visualized from the study of a case history, which will give the reader a better understanding of the significance of the general trends of this important phenomenon. The author will discuss the ground surface subsidence of Mexico City, where he has had experience throughout his professional practice. The mechanics of the phenomenon observed is similar in general theoretical characteristics to any other subsidence problem, as just described in previous paragraphs. The analysis of the subsidence taking place in Mexico City can make concrete the general philosophy governing the behavior of the subsoil and related problems caused by this phenomenon, and will illustrate the particular information and engineering characteristics required for competent analysis of such problems.

The ground settlement assumes different values throughout the city, depending upon the rate of reduction of the piezometric water pressures in the subsoil and different thicknesses and compressibilities of the clay deposits. The problems encountered are many and varied. In places where the water table has been lowered in highly compressible volcanic clay, strong shrinkage takes place and large fissures and cracks develop at the ground surface and into the clay deposits. The occurrence of large and uneven differential displacements at the ground surface produces damage in private and public structures and utilities, such as the city sewage system, streets, highways, underpasses, bridges, buildings and in the conduction of the polluted and pluvial waters out of the valley of Mexico. Maintenance of these services is expensive and becomes a matter of primary importance.

The first credited levelings of Mexico City were performed by Professor Cavallari, Director of the National Academy of Fine Arts of San Carlos in 1860-61. Fifteen years later, new levelings showed settlements of the ground surface that varied from about 0.07 to 0.40 m. Studies of rates of ground surface subsidence were first performed at the end of the last century by A. T. Pizarro, and in the second and third decades of the present century by J. L. Cuevas. Unfortunately, the latter did not leave published time records of his observations. The maximum ground surface subsidence observed by the author took place in Mexico City around the year 1949, when the rate of settlement of the ground surface amounted to about 35 cm/year in the central part of Mexico City with respect to the second sand stratum at a depth of 48 m. This stratum is taken as a reference datum. Water pumping within the city area has been forbidden since the year 1955, when the Mayor of the City of Mexico passed a decree forbidding the drilling of new water wells, and stopping pumping in those that were operating at that time within the city area. Therefore,

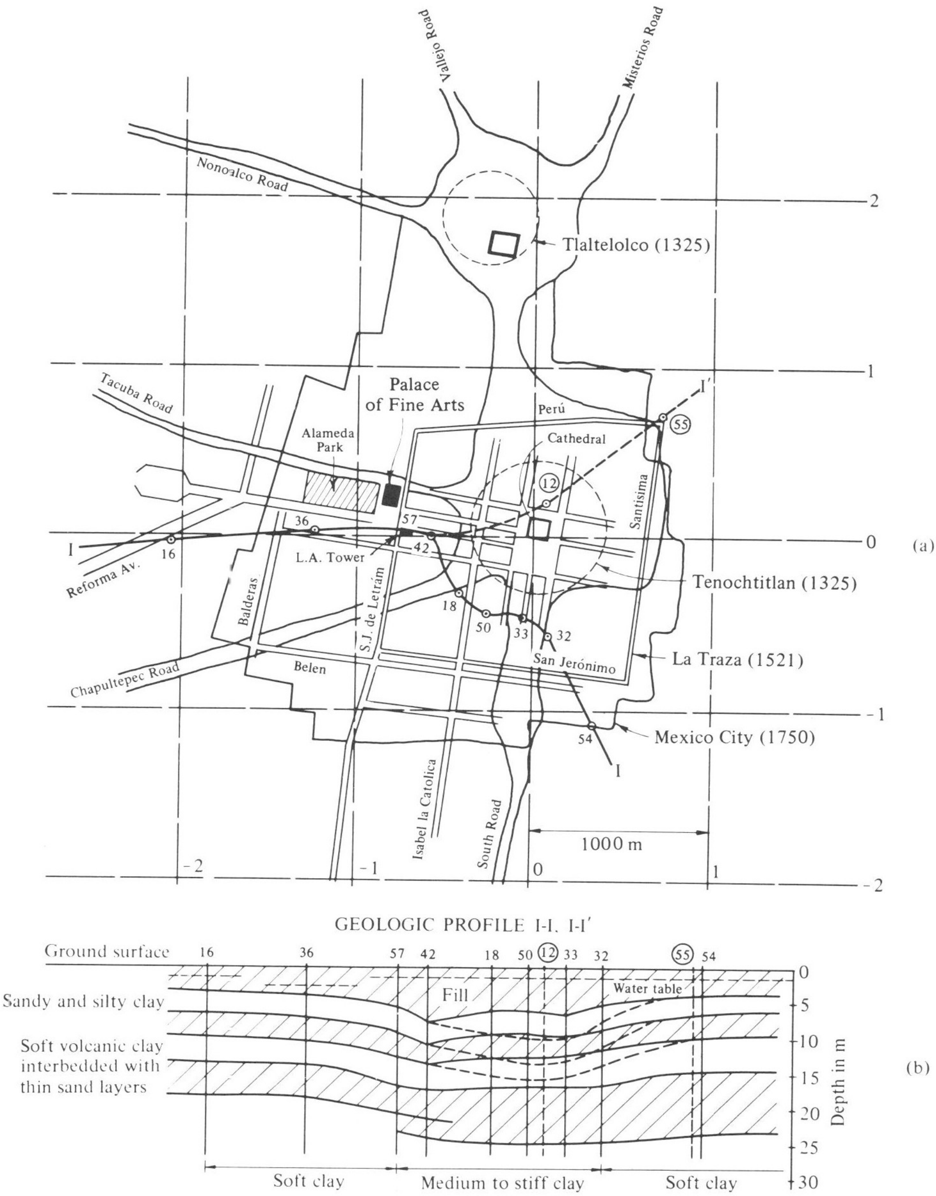

Fig. 1-V.3 Geologic profile of Mexico City correlated with the construction of the Old City.

the rate of reduction of the piezometric water levels in the subsoil in the following years decreased considerably, and there was a corresponding reduction in ground surface settlement (Fig. 12-V.3).

Since the middle of the last century until about 1955, a large number of wells drilled between 50 and 500 m deep had been in use for the extraction of water from the subsoil in central Mexico City, in order to aid the water supply. In the city area, the large pumping of water created a strong reduction of the piezometric water levels in the pervious strata, particularly in depths greater than 28 m. The surface water table, however, remained unaltered; first, because of the imperviousness of the thick clay deposit overlain by the water bearing top deposits; and second, because the water table in the area known as La Traza is supplied by many pervious fields interbedded in the upper crust of the subsoil. The pre-Spanish canals and the large-section early sewers built in Spanish times are filled with coarse materials (Fig. 1-V.3), and the rupture of the present sewage system also contributes to the preservation of the surface water table. A hydraulic gradient originates in the vertical direction because of the difference of piezometric water pressures between the ground surface and the water bearing layers at greater depth. The descending water flow across the highly compressible silty clay deposits increases the effective stresses producing their consolidation, thus originating the ground surface subsidence.

In order to investigate the effective stresses in the subsoil and the rate of subsidence of the ground surface, the author installed, since 1949, piezometers and benchmarks at different depths for the design of large foundations. One of the most complete observations is that performed for the foundation problem of La Latino Americana, the 43-story tower owned by the Life Insurance Company. In addition to the investigations made at the site of the tower to solve this particular problem, a similar set of piezometers and benchmarks were installed at a distance of about 260 m in the City Park, Alameda Central. The object was to measure, in this unloaded area of the city, the effects of the seepage stresses in the compressible clay deposits. The observations have been made over a period of 22 years on benchmarks and piezometers at depth of 3, 34 and 49 m below ground surface. They have yielded very significant information for the understanding of the seat of settlement.

3.2 Subsoil Conditions

The understanding of ground surface subsidence requires a brief discussion of the environment in which the subsoil in the lacustrine area of the basin of the valley was performed. The Valley of Mexico is situated at the south end of the highest part of the Mexican Plateau (Fig. 2-V.3). It has the form of a closed basin extending in a north-south direction; it is bounded in the east by the Sierra Nevada with snow peaks Iztaccihuatl and Popocatepetl, about 5,300 m above sea level. On the west and northwest, the valley is bounded by the Sierra Madre Occidental. On the north it is limited by the Pachuca range and on the south by the Ajusco range with an altitude of 4,000 m. The mountain range of Guadalupe extending eastward

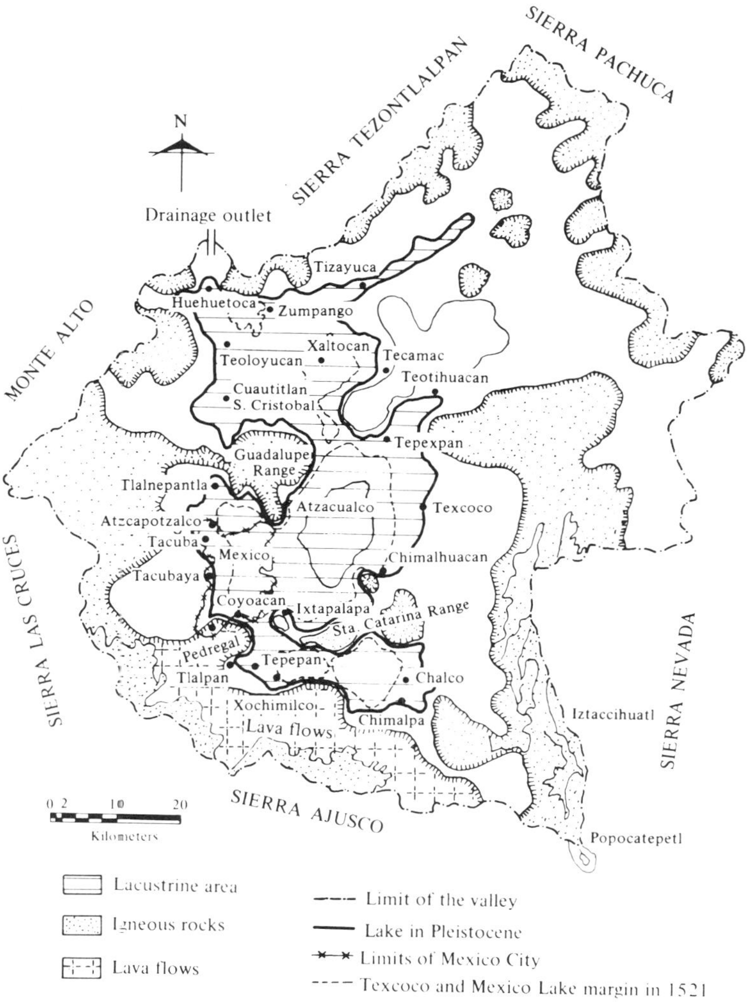

Fig. 2-V.3 Basin of the Valley of Mexico.

toward the center of the basin forms the north protection of Mexico City. The lowest part of the basin has a conventional altitude of 2,236 m above sea level. Many small rivers flow into the valley. They bring water to the lowest part of the closed basin forming a series of shallow lakes; nowadays, they are considered lake plains. From north to south, these are Zumpango, Xaltocan, San Cristobal, Texcoco, Xochimilco and Chalco; Texcoco being the lowest. At present, only a reduced area of Zumpango, Texcoco and Xochimilco has water during the dry season.

At the close of the Pliocene when the great volcanic activity was about to end, the high peaks attained their maximum elevation and formed the large closed basin of the Valley of Mexico. At the north, a low pass remained in water-laid volcanic materials underlain by cretaceous limestones. The large accumulation of fossils and water-laid sediments, found during excavations made in the seventeenth century to open a drainage outlet for the basin of Valley of Mexico, suggested that this pass may have served as one outlet in the early Pleistocene into another valley to the north. Thereafter, the natural outlet was blocked and eventually a large lake occupied practically the entire basin. At the south along the Sierra Nevada, another pass was blocked by the volcanic detritus mainly from the Popocatepetl. The water level in the lake early in the Pleistocene must have attained high elevations. Late findings show that this water level was as high as 26 m above the present level of Texcoco Lake.

During the Pleistocene, the deepest part of the large closed basin was filled by water-transported materials. The decomposed rocks, the residual clays, pyroclastic materials and the gravels and sands representing the disintegration products of andesitic rocks of the surrounding hills were gradually eroded and transported. All these materials accumulated in the basin and formed a series of deposits of gravel, sand and silty clays several hundred meters thick. In the center of the city, the upper surface of these deposits is encountered at depths greater than about 35 m. The soft fine-grained lake sediments shown in geologic profile (Figs. 3 and 4-V.3) date from late in the Pleistocene. These deposits are the products of volcanic effusions of fine and ultrafine fractions of basaltic lava and very fine water-transported materials. The effusions were accompanied by steam explosions forming dense clouds containing fine volcanic ash and other pyroclastic materials. The fine pyroclastics carried by the clouds were deposited as a rain on the waters of the lake covering the basin. The fine volcanic ash decomposed into bentonitic clay with the mineral montmorillonite comprising about 20%, containing a large percentage of diatoms and ostracods. The clay fraction amounts to about 40%. The glass particles of sand size, spreading over the lake after steam explosions, formed thin clean layers that were covered immediately by the ultrafine and fine materials already in the process of sedimentation. These well defined thin sand layers are good markers for stratigraphical studies. The winds that entered the basin from the northeast dropped fine materials that helped to fill up the central part of the lake deposits. The currents of water running toward the center of the lake aided the transportation of the fine material, and copious rains washed the fine dust and ashes from the air.

From the end of the volcanic effusions to recent time, a fill was formed in the lake region. It consists largely of coarser pyroclastic materials from decomposed rocks and residual clays eroded from the hills and mountains surrounding the basin. Minor volcanic activity during this time is recorded by lenses of wind-deposited volcanic clean ash interbedded in the top water-laid sediments. The geological sequence and stratigraphy of the materials encountered in the subsoil beneath Mexico City, the water content profile and the relative consistency and compressibility of these sediments, are shown in Fig. 3-V.3. From the surface of the ground to a depth of about 6 m, archaeological deposits may be encountered in the central part of the city, underlain by fine alluvial sediments corresponding to the closure of the Pleistocene.

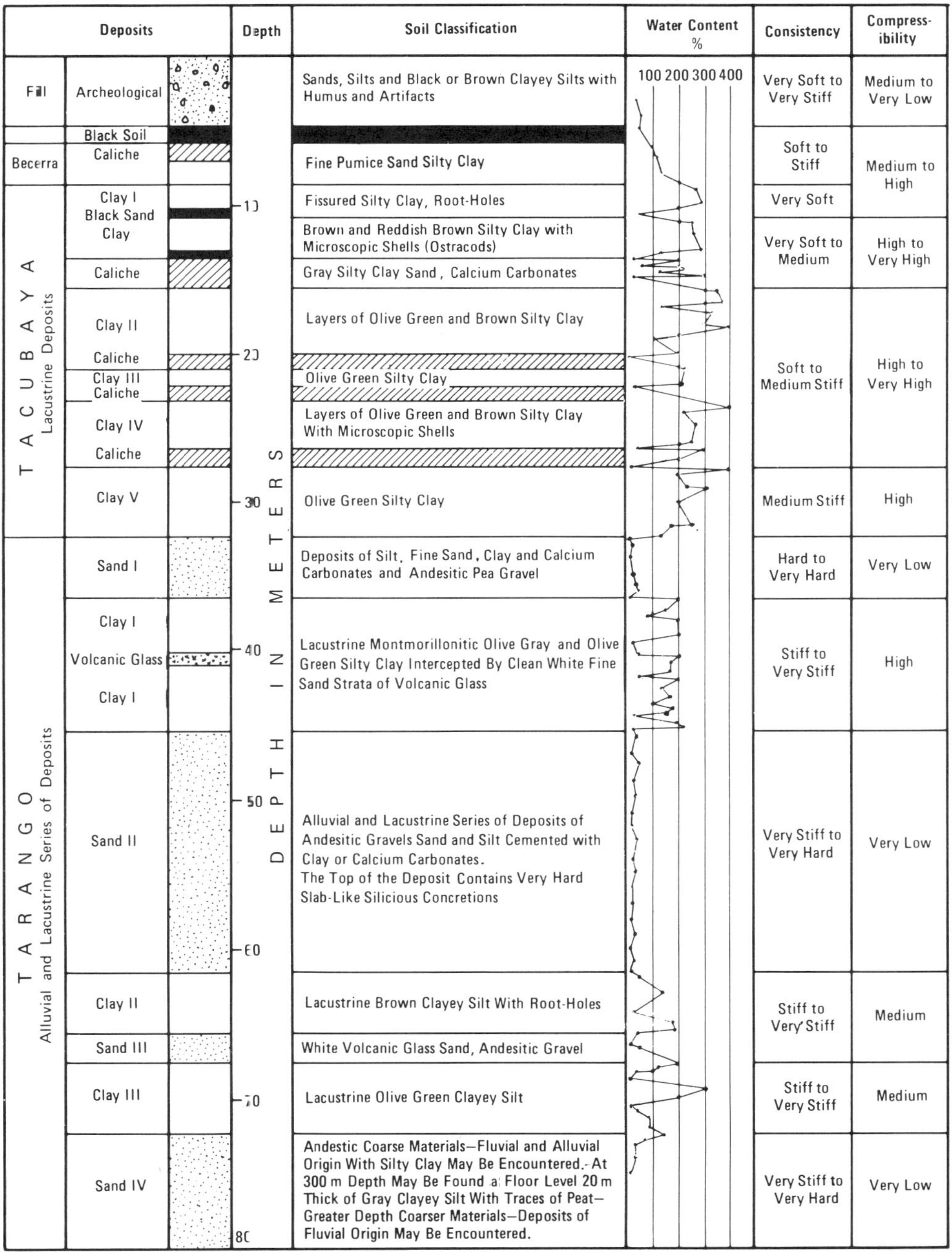

Fig. 3-V.3 Subsoil profile, Mexico City.

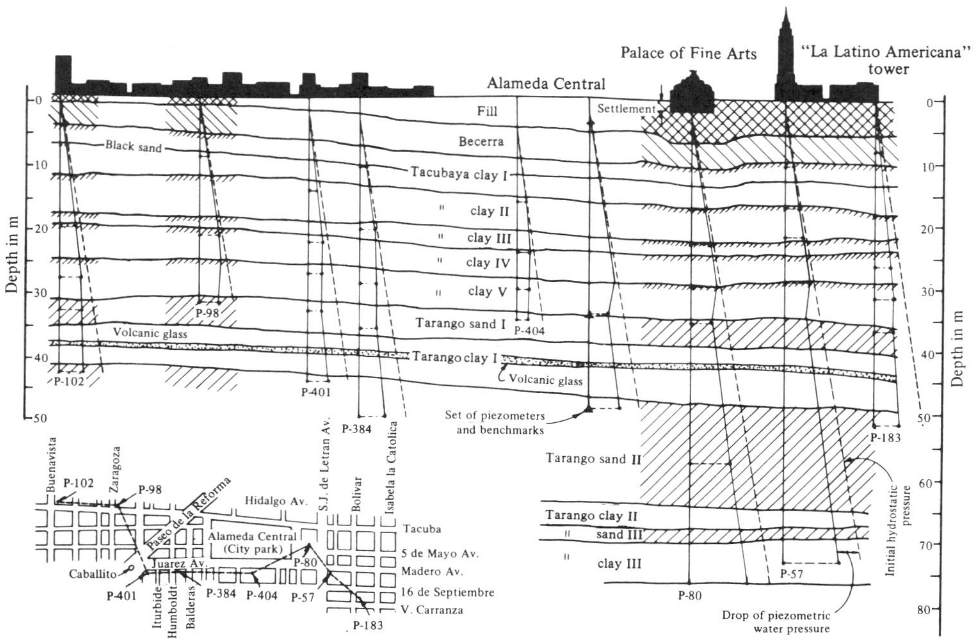

Fig. 4-V.3 Stratigraphy, Mexico City.

From about 9 m to 33 m depth, we find a silty clay deposit of very high water content and soft to semirigid consistency interbedded with numerous sand layers, products of the rains of pyroclastic materials. The silty clay developed high bond between grains under the overburden effective stresses of the past. Another particular observation during periods of minor activity is the formation of five typical clay layers representing, respectively, different periods of sedimentation. Between 33 m and 38 m depth, we encounter a stratum formed of a series of thin deposits of sand and silt with a variable cementation given by calcium carbonates and clay. Piles for building foundations are resting on this stratum in the central part of the city. From about 38 m to 48 m depth, a lacustrine deposit of olive green high-water-content clayey silt may be found, intercepted at the middle by a well defined white volcanic glass stratum, Fig. 3-V.3. At depths greater than 48 m follow series of deposits of gravel, sand and silt. To much greater depth, cycles of alluvial sediments may be found, becoming coarser with depth.

The sediments called Becerra, Tacubaya and Tarango, respectively, identify the different layers and may be correlated with the geological studies under way in the upper ridges of the lacustrine area (Fig. 4-V.3). For building foundations in the city area, the hydrostatic and mechanical properties of these deposits are important. About 15 to 20% of the total ground surface subsidence with respect to the mountains may be attributed to the compressibility of the sediments encountered at depths greater than 50 m.

The piezometric levels are measured with a series of piezometers installed at different elevations in the pervious materials, at depths of 8, 14, 28, 34, 48 and 75 m, respectively. The geological section presented shows the piezometric pressure existing in the places marked (Fig. 4-V.3). The pervious strata contain volcanic ashes, pumice and ostracod shell sand in Tacubaya sediments, and andesitic sands and gravel in the Tarango sediments. The layer found at 27–28 m in the Tacubaya is particularly pervious and contains a high percentage of pumice sand and ostracods. These series of thin sand strata are water bearing, having maintained the piezometric water levels in the upper layers practically unaltered in the central part of the city, in spite of pumping from the deep water bearing strata. At a depth of 33 m, however, we find that a considerable reduction of the piezometric water level has taken place. Equally important is the reduction that may be observed in the more pervious deposits at depths of 48 m and 75 m, respectively.

3.3 General Soil Properties

The upper, highly compressible deposit of fine sediments is responsible for the behavior of foundations. Two distinctive types of strength may be found in Mexico City's highly sensitive volcanic clay: the shear strength measured by the unconfined compression test, and the critical compressive strength measured by means of the confined compression test in the oedometer. The shear strength is determined in undisturbed clay specimens by the conventional unconfined compression tests performed rapidly at constant water content. The material exhibits high elasticity in spite of its very large water content. The stress-strain curve shows an almost perfectly rectilinear relationship up to about 75% of the ultimate failure, reached at a strain level ranging usually from 2 to 3%. A sudden failure with a well defined shear plane is usually obtained. The unconfined compressive strength varies from soft to stiff consistency for different areas of the city. If the material is perfectly remoulded, its strength decreases considerably; it becomes a viscous mass flowing easily with small pressure. The sensitivity measured by the ratio of the unconfined compressive strength of the material in the undisturbed and remoulded states, respectively, reaches values as large as 20. Considering the very high water content, it should be recognized that the shear strength and rigidity of this material is unique. This may be demonstrated when compared with other clays, with water contents as low as 30% (Fig. 5-V.3). The shear strength equation obtained in drained slow shear strength tests has a break at the critical compressive stress. This stress was defined as the effective stress corresponding to the break in the compressibility curve obtained in oedometer tests (Figs. 7 and 8-V.3). This definition is differentiated from the ultimate bearing capacity calculated from the ultimate shear strength. The flat portion of the shear strength curve shows regular slopes of 4% to

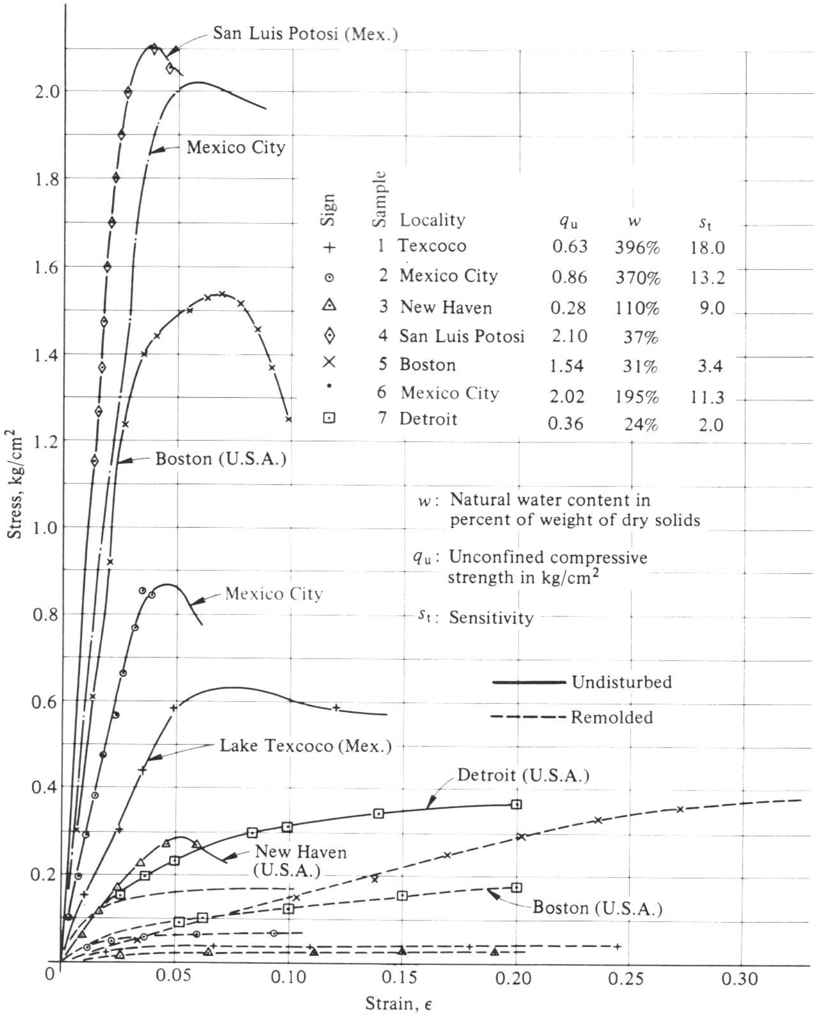

Fig. 5-V.3 Consistency characteristics for sensitive clays.

8% and up to 12%, and the inclined portion after passing the critical compression stress shows a slope on the order of 30° (Fig. 6-V.3).

After the break in the compressibility curve is reached, the material undergoes substantial consolidation, and a new structural arrangement is formed, creating higher strengths (see section II.4). The compressibility is high, and becomes many times higher after the critical stress has been passed. Before the break is reached, the compressibility is comparable with the highest compressibilities encountered in other clays (Fig. 8-V.3). In the flat portion of the curve, the compressibility

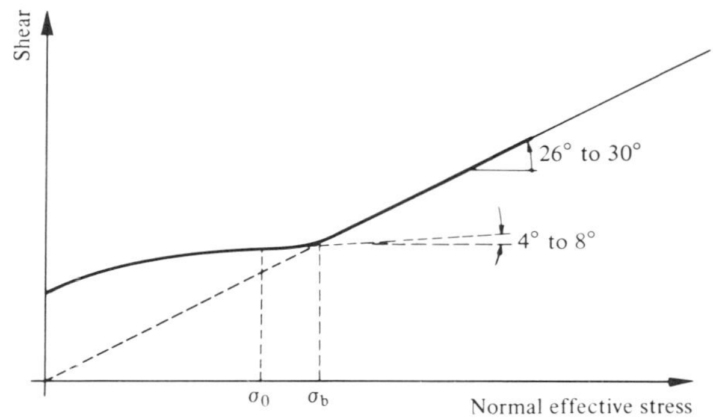

Fig. 6-V.3 Drained shear strength of undisturbed silty clay.

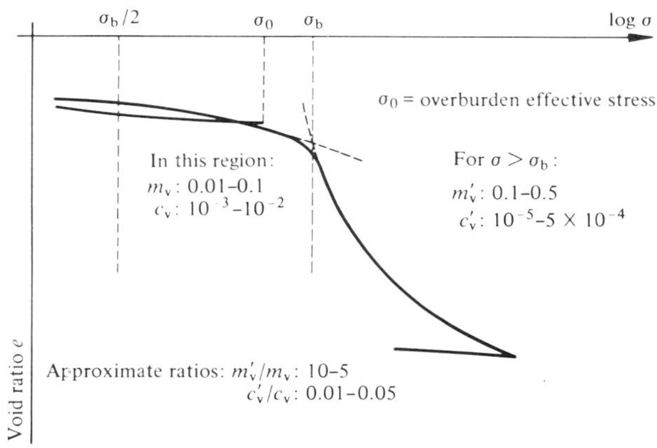

Fig. 7-V.3 Compressibility curve for undisturbed Mexico City clay.

due to recompression is on the order of five times smaller than that obtained in the steep branch of the curve in the neighborhood of the break. Therefore, in this stress level range, the effective stresses produce very significant displacements in the vertical and horizontal directions. The gain in strength in the clay was obtained by the solidification processes (see section II.3.6). This process implies that the gain in rigidity and strength is due to physicochemical action, aided by effective stresses, in the adsorbed water films at the contact surface between clay particles. For example, in the heart of Mexico City, in places where the overburden pressure may be considered the same as it was 100 years ago, the difference between the critical stress and the total overburden effective stress may reach values as high as 0.6 to 0.8 kg/cm^2. The difference of stress defined above represents the gain in strength or the hardening phenomenon that took place in the clay under an overburden effective stress smaller in the past than in the present (section II.3.6a).

In designing building foundations in Mexico City, attention should be paid to keep

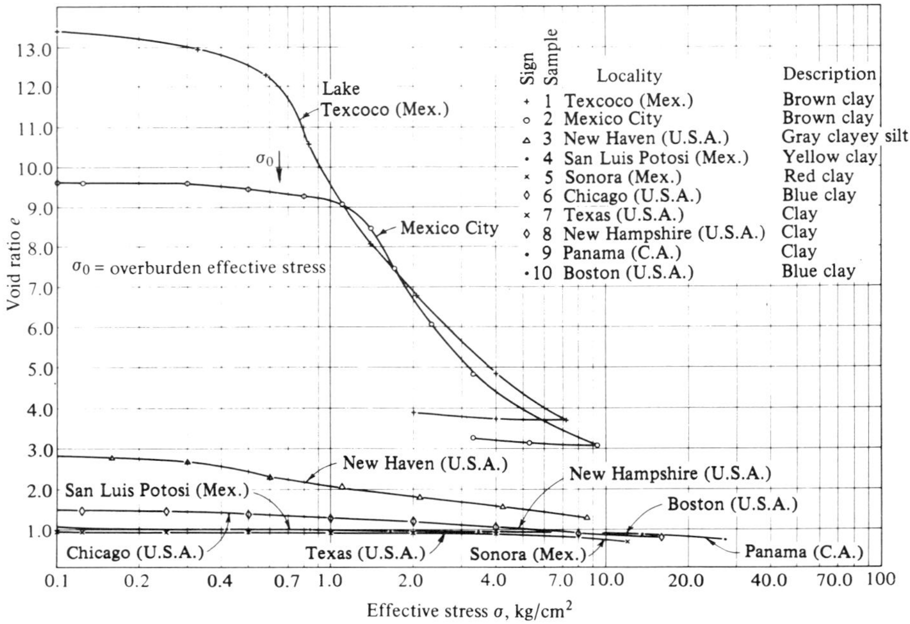

Fig. 8-V.3 Compressibility curves for various localities.

safe the critical compression stress of the clay. In places where the overburden effective stress has been reached because of heavy buildings on raft foundations, the settlements in a few years are at least 5 times larger than the largest settlements that are observed if the increment of stress does not reach the above-mentioned strength in the natural clay deposit. To illustrate the importance of the critical compressive stress, a statistical relationship is presented in Fig. 9-V.3 including in the existing overburden effective stress, the increment of effective stresses due to loads at the ground surface. The statistical relationship demonstrates that the critical compressive stress is above the present total effective stress, except for those places where very recently the overburden effective stresses have been strongly increased. Therefore, Mexico City's typical silty clay may be considered as of the preconsolidated type, having the property of hardening under a sustained load on a long-term basis.

Touring the city, one observes remarkable large vertical displacements at the surface of the ground, due to excavation, construction of buildings and ground surface subsidence. Large settlements have always been a great problem ever since the foundation of the Aztec city in 1325, when cedar posts driven close together were used to build mats to spread the load and stabilize heavy structures on the silty bottom of the lake (Fig. 10-V.3). At this time the lake occupied the entire city area, leaving only two small shallow islands, known as Tenochtitlan and Tlaltelolco where the Aztecs settled (see Fig. 1-V.3). After the Spanish Conquest, the heavily loaded Aztec city was completely demolished. The channels forming the outskirts

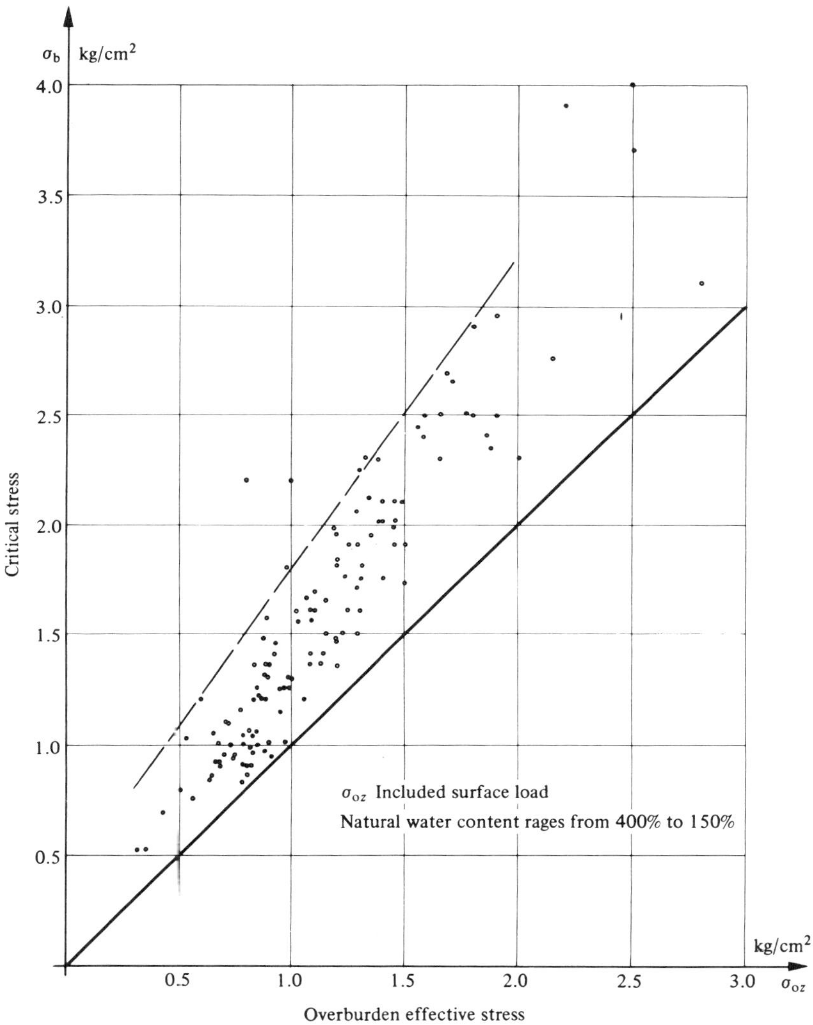

Fig. 9-V.3 Relationship between the critical and the overburden stress.

of the Venice-like Aztec city, similar to the present gardens of Xochimilco, were filled completely with refuse, leaving just a few main canals that were rectified and used for navigation. The Spanish constructed a larger and heavier city limited by what they called La Traza. Old buildings inside this area have undergone large differential settlements, and the very old ones, mostly from the seventeenth and eighteenth centuries, are in their first floor partially buried under the actual street level. Excavations performed in the center of the city show patios and street levels, as well as old foundations of buildings constructed in the sixteenth century, well

Fig. 10-V.3 Old pile foundation in Mexico City.

below the actual street elevation. The differential deformation of the primarily horizontal lacustrine volcanic clay deposit, because of the heavy weight imposed by the construction that followed the Conquest up to our days, is shown at bottom of Fig. 1-V.3. The large consolidation that took place in the La Traza area changed the consistency of the clay in this area from a very soft clay to a medium and stiff clay, and was primarily due to overloading the surface of the clay deposit in that area.

The secondary consolidation representing the intergranular viscosity in the volcanic clay is important. This phenomenon is taken into consideration in settlement analysis. Secondary consolidation takes place also when reloading the silty clay deposits after load relief because of excavation without passing the critical stress or even the overburden effective stress (Fig. 11-V.3). This phenomenon demonstrates that during unloading and reloading of the silty clay structure, the grains suffer a reorientation resulting in creep effects even when reloading the material with the same stress level it had before. The compressibility properties of preconsolidated type soil materials are given in section II.3.

3.4 Piezometric Pressure and Surface Subsidence Measurements

The recording of the behavior of the subsoil deposits because of pumping from deep water bearing strata is of vital importance in Mexico City. The piezometers and

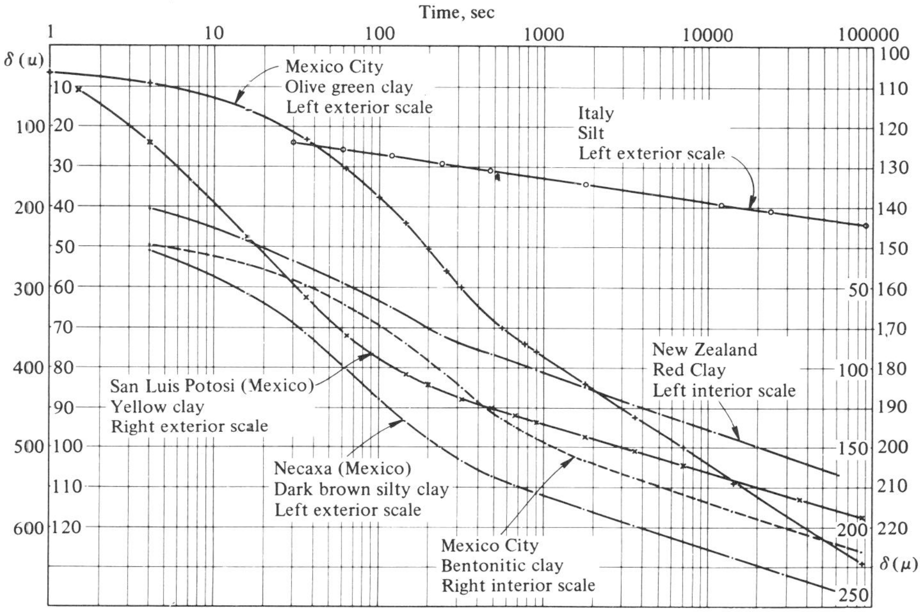

Fig. 11-V.3 Typical consolidation curves of undisturbed material showing secondary consolidation.

benchmarks installed at depths of 3, 34 and 49 m in the city park and observed for a good number of years have aided in understanding the phenomenon in this particular area. At the park area, the subsidence is taking place only because of pumping water from the ground. The piezometric measurements show that from June 1949 to November 1950, a practically constant drop in pressure was taking place at depths 34 and 49 m, with an average of about 1.9 m per year. The surface water table, however, remained practically unaltered during this period of observation. In correlation with the drop of piezometric water levels, the surface subsidence was taking place at a practically uniform rate over the same period (Fig. 12-V.3). The rate of settlement decreased toward July 1951 as the reduction in the piezometric levels slowed.

With the completion of the new aqueduct bringing water from the Valley of Toluca to add to the general water supply of Mexico City, a certain amount of wells was cancelled. The effect was instantaneously observed, as the rate of drawdown in the piezometric water levels decreased considerably. In fact, from January 1957 to the present, the piezometric water levels have remained practically unchanged. This effect was immediately registered in the settlement measurements. Nevertheless, the settlements of the ground surface continue to take place, though at a rate decreasing with time. The hydrostatic pressure distribution with depth obtained from piezometric observations shows that the upper part of the soft silty clay deposit, at a depth of about 28 m below ground surface, has not suffered an appreciable reduction because of pumping in the deep water bearing strata. The main drawdown is

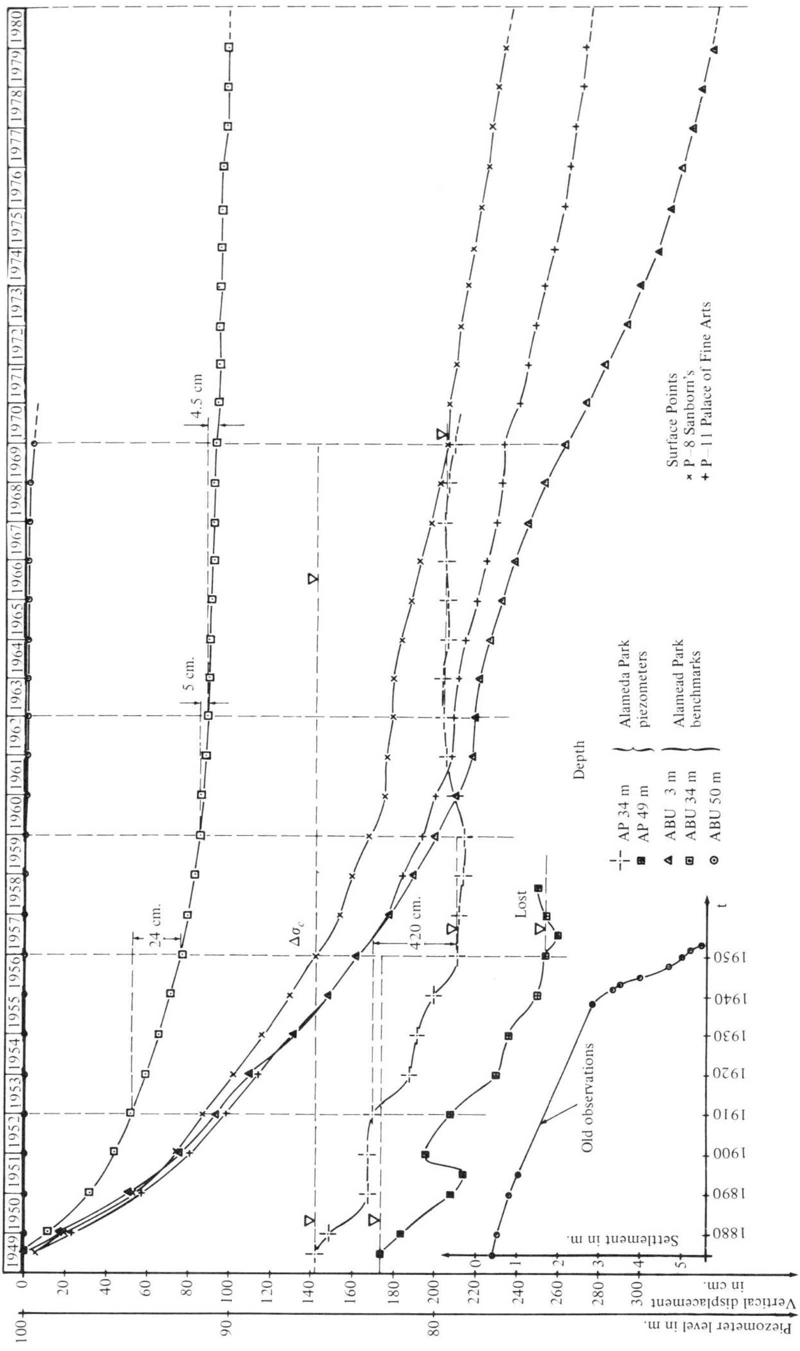

Fig. 12-V.3 Observed settlements with respect to ABN-49 m benchmark in Alameda Park, Mexico City.

observed in the first hard stratum at 33 m depth, and in the sand strata at greater depths (Figs. 3 and 4-V.3). Therefore, if in the future the water bearing sand strata in the upper part of the Tacubaya sediments connect with the lower sand strata, the upper part of the highly compressible deposit will experience a reduction in the piezometric water pressures, and consequently, a strong increase in the ground surface subsidence may take place. The seat of compression of the upper clay deposit Tacubaya, extending to a depth of 33 m, is taking place in the present at the lower part of this deposit between 28 and 33 m depth, corresponding to Tacubaya Clay V (Fig. 4-V.3). Therefore, it is concluded that the horizon at 27–28 m depth formed of series of pervious deposits of volcanic ashes, pumice and large amount of ostracods sand is supplying water maintaining the normal hydrostatic pressure in the upper deposits, thus avoiding large seepage stresses in the uppermost part of the silty clay sediments of high to very high compressibility.

The silty clay deposit called Tarango Clay I is responsible for a large share in the surface subsidence; its consolidation reached as much as 22 cm per year in 1949 (Fig. 12-V.3). The hard stratum at 33 m depth has settled 93 cm from June 1949 to January 1970. The upper clay deposit, Tacubaya, has compressed 172 cm in the same period, for a total of 265 cm for the total surface subsidence in the last 20 years with respect to the benchmark in the Central Park installed at 50 m depth.

Recent observations in the periods 1970–1980 of the reference points shown in Fig. 12-V.3 with respect to ABN-48 benchmark in the Alameda Park, Mexico City, have shown that the subsidence in that area continues at a practically constant rate with the following values:

1) △, ABN 3 m	7.33 cm/year
2) P8, (Sanborn's)	2.33 cm/year
3) P11, Palace of Fine Arts	4.30 cm/year
4) ⊡, ABN 34 m	1.00 cm/year

The piezometers have been damaged and no further records can be obtained from them.

3.5 Foundation Problems

3.5a General Considerations. The hydraulic conditions in the subsoil are available throughout the city. It was stated, however, that the large drop in the piezometric water levels occurs in the first and second hard strata. The lower clay deposit, Tarango Clay I, and the lower part of the upper clay deposits, Tacubaya V, have compressed greatly because of continuous change in the piezometric water pressures. In spite of the fact that at present water extraction from wells is prohibited in the central part of the city, it may be noticed that subsidence still continues because of the intergranular viscosity phenomenon. The compression of upper Tacubaya deposits, however, has almost doubled its rate of compression if compared with the one it had in 1965–1967. This is mainly taking place because of reduction of piezometric water levels within this deposit, from the surface to a depth of 28 m. The cause of this important increase remains to be investigated.

It is obvious that the engineering characteristics of the nonconsolidated sedimentary deposits of the basin of the Valley of Mexico, and in particular those that may be encountered under the city area in conjunction with the hydrostatic conditions created by the strong extraction of water from the deep aquifers, produce important problems to be taken into consideration in the design of engineering works constructed in the city. The most important case problems observed readily are the following:

Case I: Local subsidence because of pumping from deep water wells.
Case II: Shrinkage problem produced by the drawdown of the surface water table.
Case III: Strong settlement caused by heavy buildings at the ground surface.
Case IV: Damage produced because of construction of buildings on point bearing piles.

3.5b Case I: Water Wells. The water wells drilled in the past in the central zone of Mexico City extracted large quantities of water from the subsoil. When not well cemented they collected water from the upper layers, originating a strong depression of the hydrostatic water levels around the wells, and thus increasing the effective pressures consolidating strongly the compressible soil. The result was the formation of large bowl-like depressions around the water wells, producing damage and tilting of the nearby constructions, and in some cases to such an extent that they had to be abandoned and demolished (Fig. 13-V.3). The large local depressions caused by the water wells also produce damage to public utilities, particularly those conducting the sewage waters and water supply. The large distortions of the ground surface in the neighborhood of the water wells demonstrated the need, in Mexico City, to obtain water by other means and not from the extraction of large quantities of this fluid with wells concentrated in the city area.

3.5c Case II: Shrinkage Problem. Mexico City has grown to be a very large and extended city, and away from the heart of the city, the foundation problems are different, as the top fill and the clay strata decrease in thickness. In the outskirts of the city, the drawdown of the water table has taken place because the upper pervious deposits are not supplied rapidly with water, as is the case in the central part of Mexico City, where as mentioned before, pervious materials and coarse refuse of historical times are encountered. The reduction of the surface water levels in the outskirts of the city is also caused by the depression of the piezometric water level in the deep aquifers because of the strong extraction of this liquid by means of wells for industrial and domestic water supply purposes. The drawdown of the water table reaches deep into the highly compressible and soft bentonitic clays, producing strong shrinkage in the clay by desiccation. The result is the development of large shrinkage cracks at the ground surface (Fig. 14-V.3). Some of these cracks are rather extensive and when they are not treated properly and protected from evaporation, they open at the ground surface to about 1 m and reach a depth over

Fig. 13-V.3 Ground subsidence caused by pumping from water wells under a heavy building.

15 m. If these cracks open in empty fields, they do not produce sensitive damage; however, in constructed districts the damage to the private and public interests can assume large proportions. The behavior of the subsoil under these conditions is complex. When engineering works are contemplated in zones exposed to the shrinkage phenomenon, it is necessary to investigate carefully the prevailing conditions at these places, with the object of foreseeing in the foundation designs–mainly housing and industrial developments–the possible effects of this phenomenon.

To counteract further shrinkage, the surface of the clay sediments should be maintained well irrigated with water by recirculation; however, before this is done, the cracks should be perfectly sealed using the same clay or commercial bentonite mixed with cementing agents, in order to give the soil material the same consistency it has in its natural conditions. The mix has to be stirred well, as water is added, to produce a clay slurry that may be injected into the cracks. When the clay shrinks away from piles, the lack of lateral support may induce the piles to buckle with corresponding damage to the superstructure. In surface foundations, the treatment of the cracks and irrigation of the surface of the clay deposit may prove sufficient to avoid distortion of the ground surface, but in case of any doubt, the best solution, though expensive, will be to use piers whenever a hard stratum may be reached at a reasonable economical depth.

Fig. 14-V.3 Investigation of surface shrinkage cracks affecting nearby construction.

Another important problem induced by nonuniform ground surface subsidence and induced cracks is the distortion of the sewage systems (Fig. 15-V.3). This creates a loss in the slope of the sewage pipes and rupture. The solution used in the city area has been to establish pumping stations where large depressions are observed, pumping the sewage water to higher levels into undamaged sections of the system. The water supply system is also damaged and sometimes the pipes burst with considerable loss of water. When the pipes break in the ground, it is difficult to detect the damage and much potable water may be lost.

Needless to say, the expense of maintenance of sidewalks, streets, tracks for the trainways and sewage and water supply conduits are items of considerable importance for the city authorities. Mexico City, however, has learned to live with these problems, and they have become the everyday problem of the foundation engineer.

Fig. 15-V.3 Old aqueduct on piles distorted because of ground surface subsidence.

Moreover, it can be said that no building can be considered perfectly vertical. A maximum rotation allowance of 0.0025 is generally accepted, in order to be able to decide on the type of foundation and building construction (section I.3).

3.5d Case III: Buildings on Surface Foundations. When a building is founded at the ground surface, and the shear strength and compressibility of the Tacubaya clay deposit is not properly taken into consideration, the buildings suffer very strong total and differential settlements; they actually sink into the ground. This usually happens when the critical compressive stress of the soil is passed. The consequence is that light construction adjacent to heavy buildings is forced to settle, and damage takes place, sometimes without repair and the necessity of demolishing the affected building (Fig 16-V.3). As a matter of course, the public utilities going into the settling building are also strongly distorted.

Fig. 16-V.3 Light construction damaged by an adjacent heavy building.

Heavy buildings on raft foundations covering the complete area of construction can damage the clay structure and give rise to very large settlements, on the order of 1.0 to 1.5 m in a few years. Nevertheless, if an adequate factor of safety is used against a structure breakdown, the settlement may amount only to several inches. The large settlements may be avoided by reducing the stresses at the top of the volcanic clay deposit. This is achieved by the excavation of sufficient soil to compensate partially or totally for the weight of the building. This procedure, when properly applied, results in small and tolerable settlements, and marked economy in the foundation structure may be gained. A foundation of this type is called compensated. The weight of the building may be completely compensated by an equal weight of earth removal, providing the building with one or two basement levels (see Chapter VII). When the loads imposed by the building are high and do not permit within economical limits the use of a compensated foundation, it is necessary to use piles.

The only way large settlements may be avoided is to investigate carefully the compressibility properties of the subsoil materials at the site in question, and select the depth of the foundation and the weight of the building so as not to exceed the critical compressive stress of the subsoil materials. Therefore, the foundation is compensated in such a way as to reduce the settlements to an allowable range, protecting from damage adjacent structures and public utilities (section I.3). Allowance should be made, however, for the increment of effective stresses in deep seated soft compressible soil layers because of the reduction in the piezometric water levels.

Problems of surface foundations are encountered when long buildings are founded on isolated and continuous footings that undergo significant differential settlements damaging the superstructure because of the uneven ground surface subsidence. In those cases, one should connect columns with stiff beams to minimize the differential vertical displacements as stated in Chapter VI. There are, however, cases in which even if these precautions are taken, the column supports may yield. The best technical and economical practice is to design, in conjunction with the stiffening beams, numerous joints to avoid distortion in the superstructure. Another important problem encountered in areas for storage occurs when, owing to distortion of the ground surface, the handling of merchandise becomes difficult, in which case a continuous releveling of the floors has to be performed. As a summary, it may be said that owing to its nonuniform characteristics the ground surface subsidence often introduces cases of maintenance that should have been considered in the design of the project.

Fixed benchmarks installed on the so-called hard deposit found at a depth of 48 m below ground surface in the central part of the city show differential vertical displacements; therefore, precise levelings should be referred to the mountains. The settlement of buildings is usually judged from the point of view of differential vertical displacements in an area with a radius of about 100 m. Sometimes large deformations are accepted, if the stability of private and public properties is not affected.

3.5e Case IV: Buildings on Piles. Point bearing piles resting on the first hard stratum found at 33 m depth in the central part of Mexico City are subjected to downward dragging forces because of the relative rate of ground subsidence between the supporting stratum where the piles are bearing, and the strata pierced by the piles. The shear strength of the clay along the pile shaft is fully mobilized, and therefore large negative skin friction forces are induced on the piles in this type of foundation. On the other hand, if the problem concerns friction piles embedded in the Tacubaya silty clay deposit, the piles are carried down by the subsidence. The negative friction force acting on point bearing piles may be, on occasion, so large that the reduction of the confining stress on the bearing strata reduces the bearing capacity of the piles to such an extent that they penetrate into the bearing stratum.

This problem becomes more critical if the bearing stratum at the place in question has very low or no cohesion, causing the piles to penetrate the fine sand stratum continuously forced by the negative skin friction force. This phenomenon may induce a nonuniform rate of penetration of the piles, and differential settlements may occur in the foundation, damaging or tilting the buildings. The problem may be of sufficient significance to require the injection of cementing substances into the sand stratum, in order to achieve higher bearing capacities; alternatively, the number of piles may be increased to reduce the negative friction load on them, and give the piles better chance to support the building. When the point bearing piles do not penetrate the sand stratum, owing to their good bearing capacity, the building emerges gradually from the ground surface at a rate equal to the differential rates of settlement of the ground surface and the bearing stratum. If this phenomenon is not properly considered, other important problems may arise with respect to adjacent buildings and the street level.

A building on a surface foundation adjacent to the one on piles usually tilts away, as illustrated by Fig. 17-V.3. To diminish this danger a deep timber sheet pile coated with repellent agents may be driven into the clay enclosing the building on piles to cut the upper stronger strata. A slip surface develops between sheet pile

Fig. 17-V.3 House damaged due to ground surface subsidence close to heavy buildings on point bearing piles.

and soil, minimizing the phenomenon mentioned above. It is also important to protect the connections of public utilities entering the building by means of flexible joints designed in such a way that they may be enlarged as the ground surface settles away from the building. In buildings where these precautions have been taken, damage is very limited. Certain maintenance, however, should be observed. The problem of the ground floor emerging from the ground surface may be minimized if it is designed in such a way that it can be easily lowered periodically as the sidewalk elevation settles. An estimate of the ground surface subsidence is made for a certain number of years to determine the space that should be allowed in lowering the ground floor, and the rate at which these corrections should be performed.

In order to reduce the inconveniences produced by buildings emerging from the ground surface, the author developed during his professional practice the so-called friction pile compensated foundation. This type of foundation may be used for intermediate weight buildings and is somewhere between a compensated foundation and a point bearing pile foundation. The friction pile group, however, will be carried down by the soil mass as a unit. The rate of vertical displacement will fall between that of the ground surface and that of the pile points. Due to this condition, there is always a differential velocity between piles and soil, and the shear strength of the soil is mobilized to the full length of the piles. The lower part of the piles will work under ultimate point resistance and positive friction, and the upper portion of the piles under negative friction. Under these conditions, there will be a small emerging of the building from the ground surface. The design of friction pile foundations is explained further in this book in Chapters VIII and IX.

In the case of ground surface subsidence, the friction piles may be used in the two following ways:

1. The piles may be allowed to settle at an intermediate rate, thus reducing considerably the differential vertical displacements problem as compared with that of a point bearing pile foundation.
2. The friction piles may be designed in such a way as to eliminate negative friction and the differential settlement problem at the ground surface, by allowing the friction piles to penetrate the clay deposit at a rate equal to that of the ground surface. They are used in conjunction with a compensated foundation.

In the first case, the piles may be designed with the ultimate residual friction load accounting for small negative friction. In the second case, the problem of negative friction and the building emerging from the ground surface is eliminated, designing the piles with their ultimate positive friction load and point bearing capacity. In either case of this type of foundation design, however, the ultimate load for each pile may differ within the foundation area. The probable differential load should be estimated in order to be able to design the foundation structure with sufficient stiffness to absorb the pile load differences. The foundation structure may be given mechanical means to control differential settlements, and should be constructed deep into the ground, waterproof, and sufficiently stiff to control tilting of the building. In places where very soft clay is found extending to great depth, the differential pile loads are not an important problem, since at the ultimate load range, small load differences equalize pile velocities. During his professional practice the author has designed several foundations of this type in Mexico City with satisfactory results.

The friction pile compensated foundation may be used advantageously in designing projects with units of different weights. When a light and a heavy unit are both designed with a compensated foundation, the heavier one may suffer settlements several times larger than the lighter unit, thus producing damage and creating problems of communication between the two units. To avoid this undesirable situation, the heavier unit is founded on a compensated friction pile foundation, in such a way as to compensate for the rate of differential settlements including the ground surface subsidence. Therefore, the length and spacing of the friction piles are designed to produce a rate of settlement approximately equal to that of the light unit on a surface foundation (Fig. 18-V.3).

In conclusion, one should recognize that for proper design of pile and compensated foundations in Mexico City, one must consider the ground surface subsidence,

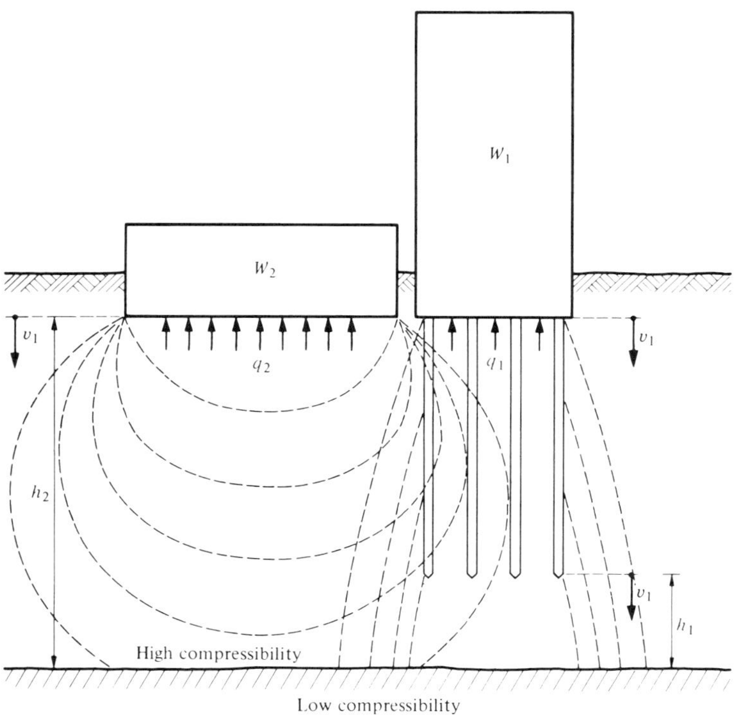

Fig. 18-V.3 Equalizing rate of settlement of units with different weights.

in conjunction with the engineering characteristics of the subsoil deposits and environmental conditions of the site at which the building is to be constructed.

BIBLIOGRAPHY

Arellano, A. R. V. (1951). *Estratigrafía de la Cuenca de México*, Congreso Científico Mexicano, IV Centenario de la Universidad de México.

Bryan, K. (1948). Los Suelos Complejos y Fósiles de la Altiplanicie de México en Relación a los Cambios Climáticos, *Boletín Sociedad Geológica Mexicana*, Tomo XIII.

Carrillo Flores, N. (1948). Influence of Artesian Wells in the Sinking of Mexico City, *Proc. II ICOSOMEF*, Vol. 7, p. 157, Rotterdam.

Cuevas, J. A. (1936). Foundation Conditions in Mexico City, *Proc. I ICOSOMEF*, Vol. III, pp. 233–237, Cambridge, Mass.

Cummings. A. E. (1947). *The Foundation Problem in Mexico City*, 7th Texas Conference on Soil Mechanics and Foundation Engineering. January.

Dawson, F. R. (1963). Land Subsidence Problems, *ASCE*, *Journal Surveying and Mapping Division*, SU2.

Fang, H. Y. and Cleary, F. T. (eds.) (1976). Subsidence, Analysis and Design of Building Foundations, Envo Publ. Co.

Geers, F. J. and Van Mierlo, W. C. (1948). Estimation of Future Settlements of the Dikes of the North-East Polder, based on Observations of Past Settlements, *Proc. II ICOSOMEF*, Vol. II, p. 122, Rotterdam.

Grim, R. (1949). *An Investigation of the Engineering Characteristics of the Volcanic Lacustrine Clay Deposit Beneath Mexico City*, Ph.D. Thesis by L. Zeevaert, University of Illinois, pp. 221–227.

Harris, F. R. and Harlow, H. (1948). Subsidence of the Terminal Island-Long Beach Area, California, *Trans ASCE*, Vol. 113, p. 375–395.

Lee, K. C. and Shen C. K. (1969). Horizontal Movements Related to Subsidence, Soil Mechanics and Foundation Division, Vol. 95, SM1, p. 139.

Marsal, R., Hiriart, F. and Sandoval, R. (1951). *Hundimiento de la Ciudad de México. Observaciones y Estudios Analíticos*, Ingenieros Civiles Asociados, S. A. de C. V. Serie B Ingeniería Experimental, No. 3.

Marsal, R., Mazari, M. and Hiriart, F. (1953). *Comparación de Propiedades Mecánicas en Muestras Remoldeadas e Inalteradas de las Arcillas del Valle de México*, Ingenieros Civiles Asociados, S. A. de C. V. Publicación No. 15 de la Serie B.

Palerm, L. A. and Brown, P. B. (1957). Settlement Analysis for Areas of Continuing Subsidence. *Proc. IV ICOSOMEF*, Vol. I, p. 395, London.

Pizarro, T. Adrian (1900). *Apuntes Acerca de los Cimientos de los Edificios de la Ciudad de México.*

Zeevaert, L. (1949). *Present Building Foundation Problems in Mexico City*, Meeting ASCE, Mexico City, July.

Zeevaert, L. (1949). *An Investigation of the Engineering Characteristics of the Volcanic Lacustrine Clay Deposit Beneath Mexico City*, University of Illinois.

Zeevaert, L. (1952). Estratigrafía y Problemas de la Ingeniería de los Depósitos de Arcilla Lacustre de la Ciudad de México, *Revista de Ingeniería*, Vol. XXV, Julio-Agosto.

Zeevaert, L. (1952). Compresibilidad de la Arcilla Volcánica de la Ciudad de México, *Revista de Ingeniería*, Vol. XXV.

Zeevaert, L. (1953). Outline of the Stratigraphical and Mechanical Characteristics of the Unconsolidated Sedimentary Deposits in the Basin of the Valley of Mexico. IV Congress INQUIA, Rome-Pisa, Italy.

Zeevaert, L. (1953). Pore Pressure Measurements to Investigate the Main Source of Surface Subsidence in Mexico City. *Proc., III ICOSOMEF*, Vol. II, p. 299, Switzerland.

Zeevaert, L. (1956). *Heavy and Tall Building Problems in Mexico City,* Proc. ASCE, paper No. 917.

Zeevaert, L. (1957). Foundation Design and Behavior of Tower Latino Americana in Mexico City, *Geotechnique*, Vol. VII, pp. 115-133, London.

Zeevaert, L. (1958). Consolidation of Mexico City Volcanic Clay, *Proc. Joint Meeting ASTM and SMMS*, p. 28.

Zeevaert, L. (1962). Foundation Problems Related to Ground Surface Subsidence in Mexico City, *Proc. IV Pacific Area National Meeting, ASTM*, Los Angeles, California, Symposium on Field Testing of Soils. Special Technical Publication No. 322.

VI: SOIL-FOUNDATION STRUCTURE INTERACTION

VI.1 INTRODUCTION

In Chapter IV a procedure has been presented for estimating the contact stresses for a rigid foundation supported on a homogeneous and isotropic soil mass of infinite extent. The procedure gives an approximate idea of the contact stresses when the above mentioned soil assumptions are fulfilled. In practice, however, the soil mass is found stratified and limited in depth to a firm soil deposit with low or very low compressibility. Furthermore, the foundation structure in many cases has certain flexibility measured by EI, in which E is the modulus of deformation of the foundation structure material, and I is the moment of inertia of the cross section. Hence, for the same load layout, the contact stresses may show a different configuration from that obtained with the theoretical procedure given in Chapter IV, since they are a function of EI, the subsoil stratigraphy and the stress-strain-time properties of the subsoil strata. To find the soil-structure interaction under these conditions, the compatibility of displacements at the interphase of foundation slab and soil should be established.

To illustrate the procedure theoretically, we assume a raft foundation of stiffness EI, Fig. 1-VI.1. Columns are distributed as shown. For simplicity of illustration assume that the column loads are symmetrical, that the compressible subsoil strata are parallel to the ground surface, and the limiting depth to firm ground is known. Furthermore, the foundation structure does not suffer rotation, although the method given here is valid for nonsymmetrical conditions.

The foundation structure is to be built of a system of short beams and long beams. The short beams are required to take soil reactions and column loads considering the longitudinal beams as a statically determinate system. Obviously, if the differential settlements with this assumption are fulfilled, the longitudinal beams are not

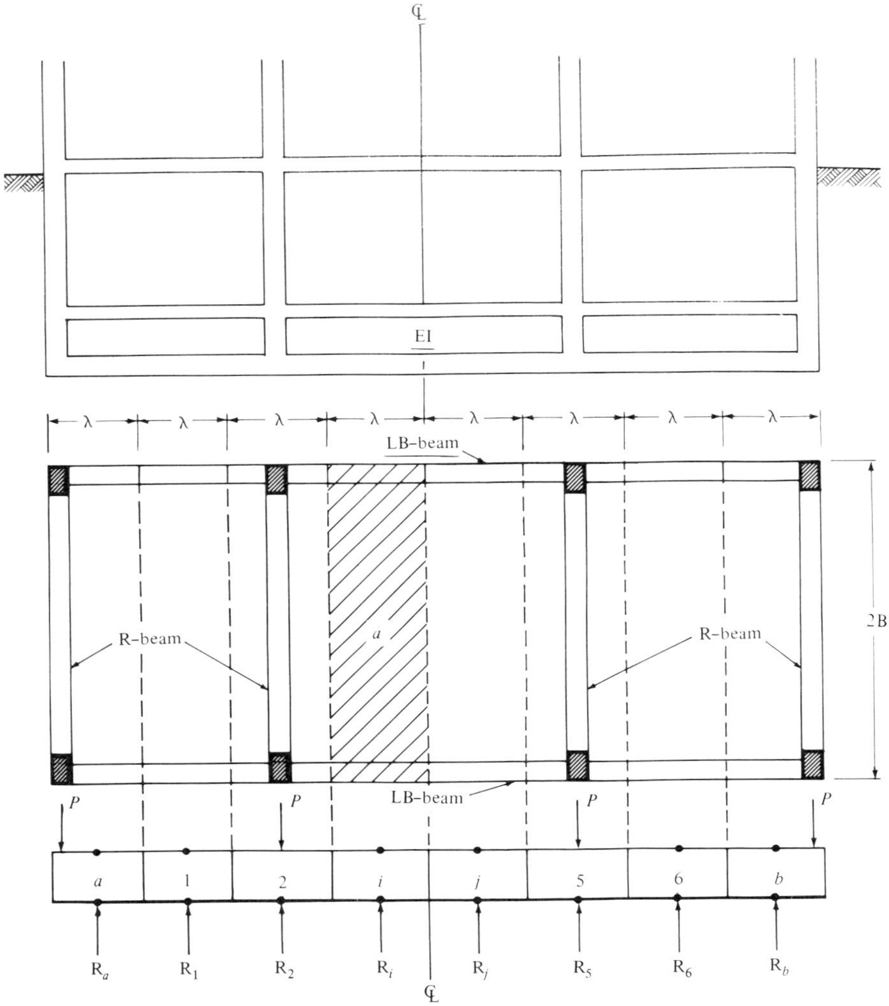

Fig. 1-VI.1 Foundation plant and elevation.

required to have any special stiffness except that necessary between spans to put in static equilibrium the reactions of the short beams. Any torsion in the foundation structure will be taken by the foundation box formed by the reaction slab, longitudinal beams and the top slab. The longitudinal beams, however, may be designed with the required stiffness necessary to reduce differential settlements to allowable values or less. In this case, the longitudinal beams are used to distribute the unbalanced loads of the short beams required to obtain continuity in the subgrade reactions. Therefore, the longitudinal beams will be called load balancing beams, LB-beams, to distinguish them from the reaction beams, R-beams.

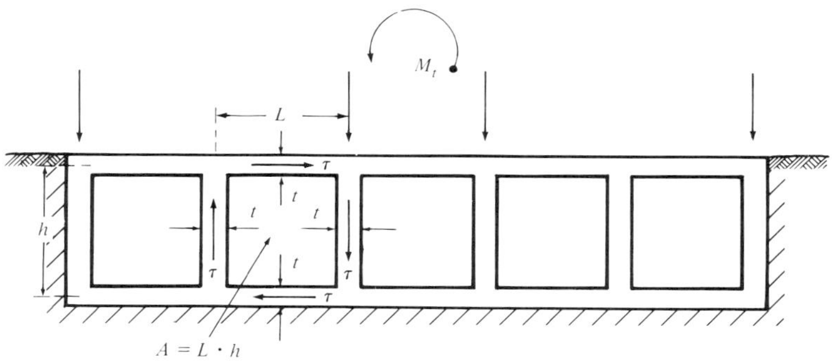

Fig. 2-VI.1 Rigid box type foundation.

Owing to nonsymmetrical column loads, the foundation structure may be subjected to high torsion that cannot be taken properly with subgrade reactions when the soil is of high compressibility. In this case a monolithic rigid box type foundation structure may be used, as shown in Fig. 2-VI.1. The box type structure will take the torsion and, therefore, the soil reactions will be those obtained as if the foundation structure had only flexural stiffness. The torsion taken by the cellular monolithic foundation structural frame may be investigated by usual methods of shear flow described in text books on strength of materials. Hence

$$M_t = 2 \sum_{1}^{n} A_i \cdot q_i \qquad \text{1-VI.1}$$

where A_i is the area of one cell in the foundation structure, and q_i is the shear flow per unit length along the element forming the cell considered.

The settlement of the mat foundation, however, should be investigated accounting for the deformation of all the compressible strata underlying the foundation slab affected by the final distribution of reactions found from the analysis discussed below. Since this type of foundation is usually constructed on soils having medium to high compressibility and viscous-intergranular viscosity mechanical properties, the foundation analysis should be extended to include calculation of secondary settlements (see Section II.3).

Mat foundations in difficult subsoil conditions, however, enter into the limiting design of compensated foundations described in Chapter VII. The determination of subgrade reactions follows the same philosophy as previously described, except that in the latter case, the structural foundation frame used for tall buildings has greater stiffness because heavier loads have to be supported.

The foundation engineer seldom encounters in nature what he may call an homogeneous and isotropic soil mass of semi-infinite extent. The usual field conditions are stratified subsoil deposits where the mechanical properties of stress-strain-time assume different values for the strata encountered at the volumetric stress level at which the soil sediment is confined. Usually the compressibility decreases with depth, and eventually the deposit is limited by firm ground or materials of low to very low compressibility. The vertical displacement varies in accordance with its location in the loaded surface, and the K-values also have different magnitudes (Chapter IV). Hence, to investigate these values it will be necessary to perform a settlement analysis of the loaded surface, determining sufficient points to obtain reasonable accuracy for the foundation structure calculations. The first approximation is obtained using subgrade reactions calculated by means of 19-IV.3, represented graphically in Fig. 7-IV.3, and by the method given in section IV.3, that is to say, assuming a rigid foundation. The K-values are determined, and a conventional computation is performed as for a beam on an elastic foundation to determine the vertical displacements including the stiffness of the foundation structure.

From the new set of vertical displacements and with the K-values previously found, new reactions are determined at corresponding points. These reactions are used to perform an improved settlement analysis, correcting the compressibility parameters for the new stress levels found (see Section II.3). The result is a set of new vertical displacements from which improved K-values can be determined. The cycle is repeated until the required precision is obtained.

VI.2 SOIL-STRUCTURE INTERACTION

The foundation structure surface is divided into tributary band areas, where the load is considered uniform. The subsoil is assumed to have the same mechanical properties under the area covered by the foundation structure. Nevertheless, the method explained below may be applied when the mechanical properties of the subsoil show variations in the area of the foundation structure. The vertical stresses within the soil mass may be determined by the theory of elasticity (Chapter III.2.2).

The average properties of each stratum N are found for the corresponding stress level $(\sigma_{0i} + \Delta\sigma_i/2)^N$ under point i, where $\Delta\sigma_i^N$ is the average increment of stress at the Nth stratum of thickness d_i induced by the load applied at the ground surface. Hence,

$$\delta_i = \sum (M_z \cdot d \cdot \Delta\sigma)_i^N \qquad \text{1-VI.2}$$

Call $(M_z \cdot d)_i^N = \alpha_{ci}^N$, the unit compression of the stratum N for which $\Delta\sigma_i^N$ is the average increment of effective stress and M_z^N is the average or secant unit strain modulus of the considered stratum. The value of M_z is determined as discussed in Chapter II. Therefore, we may write

$$\delta_i = \sum_{A}^{N} (\alpha_c \Delta\sigma)_i^N \qquad \text{2-VI.2}$$

Assume band a, Fig. 1-VI.2 is loaded with unit surface load of $q = +1$ the influence value under the center line of any other band j and for stratum N will be called

+1

STR	z_i	d_i	a	1	2	i	j	5	6	b	α_C^N
A	z_A	d_A	I_{aa}^A	I_{1a}^A	I_{2a}^A	I_{ia}^A	I_{ja}^A	I_{5a}^A	I_{6a}^A	I_{ba}^A	α_C^A
B	z_B	d_B	I_{aa}^B	I_{1a}^B	I_{2a}^B	I_{ia}^B	I_{ja}^B	I_{5a}^B	I_{6a}^B	I_{ba}^B	α_C^B
C	z_C	d_C	I_{aa}^C	I_{1a}^C	I_{2a}^C	I_{ia}^C	I_{ja}^C	I_{5a}^C	I_{6a}^C	I_{ba}^C	α_C^C
N	z_N	d_N	I_{aa}^N	I_{1a}^N	I_{2a}^N	I_{ia}^N	I_{ja}^N	I_{5a}^N	I_{6a}^N	I_{ba}^N	α_C^N
M	z_M	d_M	I_{aa}^M	I_{1a}^M	I_{2a}^M	I_{ia}^M	I_{ja}^M	I_{5a}^M	I_{6a}^M	I_{ba}^M	α_C^M

Fig. 1-VI.2 Unit influence values and strata unit compression values.

I_{ja}^N. Hence, the influence settlement at point j due to the unit load applied at any band a is:

$$\bar{\delta}_{ja} = \sum_{A}^{N} (I_{ja}\alpha_c)^N \qquad \text{3-VI.2}$$

The other influence settlements under bands i, j and b may be calculated in the same manner. Therefore, in matrix form:

$$|\bar{\delta}_{ji}| = [I_{ji}^N]^T \, |\alpha_c^N| \qquad \text{4-VI.2}$$

in which

- $|\bar{\delta}_{ji}|$ is the column matrix of vertical displacements at any point j due to the unit load at band i,
- $[I_{ji}^N]^T$ is the square transposed matrix of the influence coefficients due to the unit load $q_i = +1$ at tributary area i, and
- $|\alpha_c^N|$ is the column matrix of the unit compression of the strata from A to N.

Assume now that bands a, i, j, and b are loaded with q_a, q_i, q_j, and q_b, then the settlements at the center of the bands will be:

Band a $\qquad \delta_a = |\bar{\delta}_{ja}|^T \cdot |q_i|$

Band i $\qquad \delta_i = |\bar{\delta}_{ji}|^T \cdot |q_i|$

Band j $\qquad \delta_j = |\bar{\delta}_{jj}|^T \cdot |q_i|$

Band b $\qquad \delta_b = |\bar{\delta}_{jb}|^T \cdot |q_i| \qquad$ 5-VI.2

The above expressions may be written in matrix form as follows:

$$|\delta_i| = [\bar{\delta}_{ji}]^T \cdot |q_i| \qquad \text{6-VI.2}$$

Hence,

$$\begin{bmatrix} \bar{\delta}_{aa} & \bar{\delta}_{ai} & \bar{\delta}_{aj} & \bar{\delta}_{ab} \\ \bar{\delta}_{ia} & \bar{\delta}_{ii} & \bar{\delta}_{ij} & \bar{\delta}_{ib} \\ \bar{\delta}_{ja} & \bar{\delta}_{ji} & \bar{\delta}_{jj} & \bar{\delta}_{jb} \\ \bar{\delta}_{ba} & \bar{\delta}_{bi} & \bar{\delta}_{bj} & \bar{\delta}_{bb} \end{bmatrix} \cdot \begin{vmatrix} q_a \\ q_i \\ q_j \\ q_b \end{vmatrix} = \begin{vmatrix} \delta_a \\ \delta_i \\ \delta_j \\ \delta_b \end{vmatrix}$$

in which $[\overline{\delta}_{ij}]$ is the transposed matrix of the unit vertical displacements. The expression 6-VI.2 will be called EMA, matrix equation of vertical displacements. Therefore, knowing the q_i values in each tributary area, the vertical displacements may be calculated. When the tributary areas $\overline{a}$ are selected to be equal, then $\overline{\delta}_{ji} = \overline{\delta}_{ij}$, and the matrix $[\overline{\delta}_{ji}]^T = [\overline{\delta}_{ij}]$ becomes a symmetrical square matrix.

The influence coefficients I_{ji}^N for certain stratigraphical and mechanical conditions of the subsoil may be determined from stress nets (Section II.2.3), and from the M_z values and the strata thicknesses; the values of α^N for each stratum are calculated. From the vertical displacements δ_i obtained with EMA, the foundation moduli K_a, K_i, K_j, K_b per band tributary area $\overline{a}$ may be obtained, hence

$$K_i = \frac{q_i \overline{a}}{\delta_i} \qquad \text{7-VI.2}$$

When the foundation structure is considered of high rigidity in comparison with the soil displacements, then EMA may be solved as a system of simultaneous equations, considering a uniform displacement: $\delta_i = \delta_a$ = constant.

The approximate value of δ_a may be obtained by solving EMA for δ_i, assuming q_i constant, hence,

$$\delta_a' = \frac{1}{n} \sum_1^n \delta_i \qquad \text{8-VI.2}$$

For the solution of the contact stresses q_i, EMA is solved in the form

$$|q_i'| = [\overline{\delta}_{ij}]^{-1} \cdot |\delta_a'| \qquad \text{9-VI.2}$$

The configuration of reactions q_i' so obtained should be corrected to adjust them proportionally to the weight of the foundation structure, and using EMA again the real values of q_i and δ_a may be checked for compatibility.

When the foundation structure has a rigidity EI that may influence the reactions configuration in the longitudinal direction of the foundation structure, the problem will be to determine the q_i values establishing the compatibility of vertical displacements at the interphase of soil and foundation structure. Assume now that the solution of Fig. 2a-VI.2 requires knowledge of the reactions $X_i = q_i \cdot \overline{a}$. If we assume $X_i = X_j = 0$, a statically determinate foundation structure is obtained with reactions R_{ao} and R_{bo} at the edge bands a and b, respectively. This condition will be called condition $X_i = 0$, Fig. 2b-VI.2. The vertical displacements at points i will be called Δ_{io}, and include the vertical displacements produced by the loads and those produced by the rotation of the foundation structure due to yielding of the supports a and b caused by the soil deformation. Hence

$$\Delta_{io} = \Delta_{io}^P + \Delta_{io}^A$$

With the same support conditions we apply a unit load $X_i = +1$ at point i on the structure and soil, Fig. 2c-VI.2. This condition will be called condition $X_i = +1$. The displacements of the foundation structure at point i will be called $\overline{S}_{ii}$, and the displacement at point j due to the load applied at point i is $\overline{S}_{ji}$. Due to the unit load

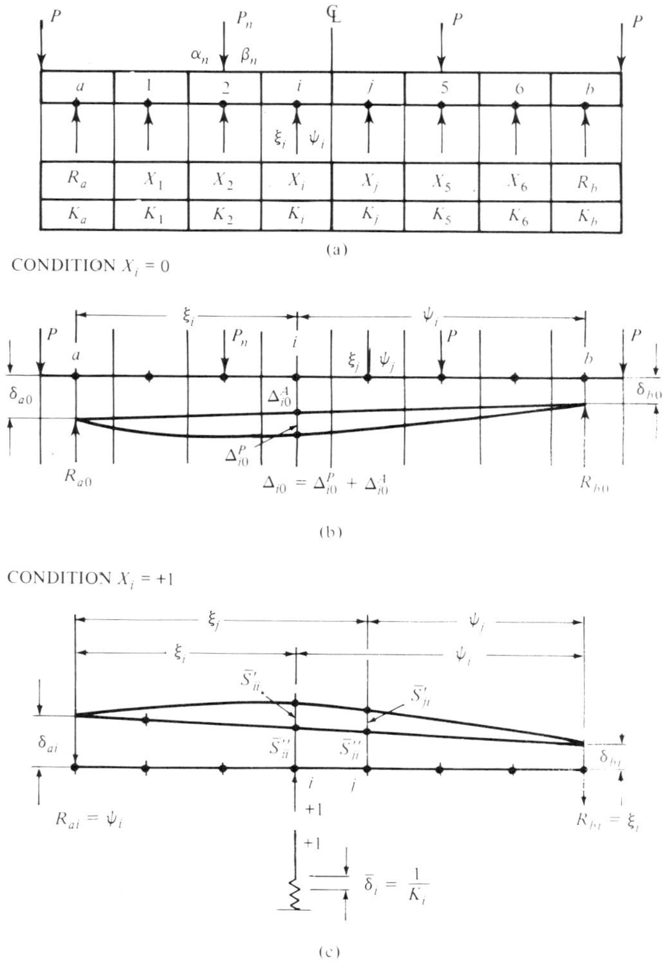

Fig. 2-VI.2 Displacement conditions.

applied at point i, the soil will be displaced by the amount $1/K_i$, in which K_i is the foundation modulus for tributary area $\bar{a}$ as defined before (7-VI.2). The displacement $\bar{S}_{ii}^{1}$ represents the deflection of the foundation structure because of its flexural rigidity EI, and $\bar{S}_{ji}^{11}$ represents, the vertical displacement and rotation of the foundation structure as a rigid body when the supports a and b yield ψ_i/K_a and ξ_i/K_b, respectively. Therefore, $\bar{S}_{ji} = \bar{S}_{ji}^{1} + \bar{S}_{ji}^{11}$ is the total displacement of the foundation structure at point i. If δ_i represents the yielding of the soil because of the unknown reactions, and increment of stress imposed to the soil, we obtain $\delta_i = X_i/K_i$. From the above discussion and using the principle of superposition, we may establish a condition of compatibility for point i, adding up all the effects induced by the unknown reactions on the foundation structure and soil displacements δ_i. Hence,

$$\bar{S}_{ii}X_i + \bar{S}_{ij}X_j = \Delta_{io} - \delta_i \qquad \text{9-VI.2}$$

When there is no yielding at the points i, $\delta_i = 0$, and we obtain a foundation structure firmly supported on these points. When the soil yields under points i, j, due to the reactions X_i, then $\delta_i = X_i/K_i$, and equation 9-VI.2 may be written as follows:

$$\left(\bar{S}_{ii} + \frac{1}{K_i}\right) X_i + \bar{S}_{ij}X = \Delta_{io} \qquad \text{10-VI.2}$$

Similar equations of interaction may be written for all the bands considered. Finally, we obtain the following system of simultaneous equations

$$\left(\bar{S}_{ii} + \frac{1}{K_i}\right) X_i + \bar{S}_{ij}X_j = \Delta_{io}$$

$$\bar{S}_{ji}X_i + \left(\bar{S}_{jj} + \frac{1}{K_j}\right) X_j = \Delta_{jo} \qquad \text{11-VI.2}$$

This system of equations may be written in matrix form as follows:

$$\{[\bar{S}_{ij}] + [1/K_i]_D\} \, |X_i| = |\Delta_{io}| \qquad \text{12-VI.2}$$

from which the unknown values of the reactions may be determined.

The matrix equation 12-VI.2 will be called EMI, Interaction Matrix Equation. The problem of soil-interaction may be solved assuming as a first approximation any logical set of reaction values X_i. Using EMA, the values of the displacements and $q_i = X_i/\bar{a}$ are determined. With the first values of X_i and δ_i the foundation moduli are calculated, $K_i = X_i/\delta_i$, and used in EMI (12-VI.2) for first approximation of X_i^1; thus the next unit reactions are $q_i^1 = X_i^1/\bar{a}$, and EMA is solved again for improved values of δ_i and new values of $K_i^1 = X_i^1/\delta_i^1$, to be used in EMI for the next iteration. This process of iteration is repeated until the values of X_i do not change substantially and satisfy EMA and EMI. From X_i values, the reactions R_a, R_b at the end bands a and b are calculated and the shears and moments in the foundation structure determined. Nevertheless, it should be observed that the high contact stresses taking place in the soil at the edge of the foundation structure are limited, especially in sensitive soils, by relaxation and the plasto-viscous phenomenon. The edge stress q_{ep} is determined by the critical compressive stress of the soil sediment or by expression (26-IV.3), whichever is smaller. Hence, the edge reactions may reach the limited condition $R_{ep} = q_{ep} \cdot \bar{a}$. The edge reactions may be considered in the interaction calculation with fixed values R_{ep} at the edge bands (Fig. 3-VI.2). If W_e is the effective weight of the foundation and building, then:

$$W_e = 2R_{ep} + R_a + R_b + \sum_1^n X_i \qquad \text{13-VI.2}$$

or

$$W_e - 2R_{ep} = R_a + R_b + \sum_1^n X_i \qquad \text{14-VI.2}$$

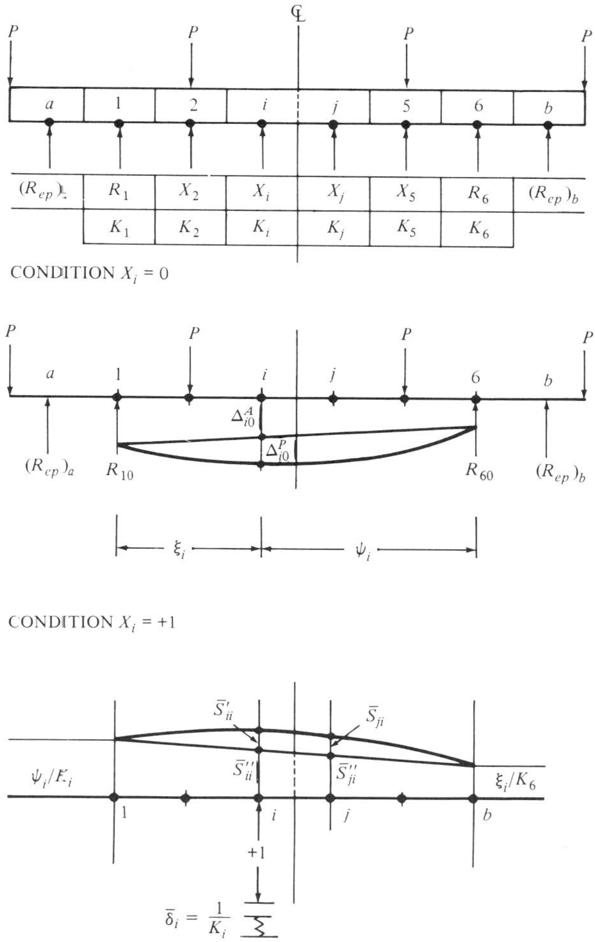

Fig. 3-VI.2 Displacement conditions when limiting reaction is developed at the edges of the foundation.

The EMA settlement matrix equation reads

$$\begin{bmatrix} \bar{\delta}_{aa} & \bar{\delta}_{a1} & \bar{\delta}_{ai} & \bar{\delta}_{aj} & \bar{\delta}_{an} & \bar{\delta}_{ab} \\ \bar{\delta}_{1a} & \bar{\delta}_{11} & \bar{\delta}_{1i} & \bar{\delta}_{1j} & \bar{\delta}_{1n} & \bar{\delta}_{1b} \\ \bar{\delta}_{ia} & \bar{\delta}_{i1} & \bar{\delta}_{ii} & \bar{\delta}_{ij} & \bar{\delta}_{in} & \bar{\delta}_{ib} \\ \bar{\delta}_{ja} & \bar{\delta}_{j1} & \bar{\delta}_{ji} & \bar{\delta}_{jj} & \bar{\delta}_{jn} & \bar{\delta}_{jb} \\ \bar{\delta}_{na} & \bar{\delta}_{n1} & \bar{\delta}_{ni} & \bar{\delta}_{nj} & \bar{\delta}_{nn} & \bar{\delta}_{nb} \\ \bar{\delta}_{ba} & \bar{\delta}_{b1} & \bar{\delta}_{bi} & \bar{\delta}_{bj} & \bar{\delta}_{bn} & \bar{\delta}_{bb} \end{bmatrix} \cdot \begin{vmatrix} R_{ep} \\ R_1 \\ X_i \\ X_j \\ R_n \\ R_{ep} \end{vmatrix} = \begin{vmatrix} \delta_a \\ \delta_1 \\ \delta_i \\ \delta_j \\ \delta_n \\ \delta_b \end{vmatrix} \qquad \text{15-VI.2}$$

Since R_{ep} is known the above matrix equation (15-VI.2) may be reduced to

$$\begin{bmatrix} \bar{\delta}_{11} & \bar{\delta}_{1i} & \bar{\delta}_{1j} & \bar{\delta}_{1n} \\ \bar{\delta}_{i1} & \bar{\delta}_{ii} & \bar{\delta}_{ij} & \bar{\delta}_{in} \\ \bar{\delta}_{j1} & \bar{\delta}_{ji} & \bar{\delta}_{jj} & \bar{\delta}_{jn} \\ \bar{\delta}_{n1} & \bar{\delta}_{ni} & \bar{\delta}_{jn} & \bar{\delta}_{nn} \end{bmatrix} \cdot \begin{vmatrix} R_1 \\ X_i \\ X_j \\ R_n \end{vmatrix} = \begin{vmatrix} \delta_1 - (\bar{\delta}_{aa} + \bar{\delta}_{ab})R_{ep} \\ \delta_i - (\bar{\delta}_{ia} + \bar{\delta}_{ib})R_{ep} \\ \delta_j - (\bar{\delta}_{ja} + \bar{\delta}_{jb})R_{ep} \\ \delta_n - (\bar{\delta}_{na} + \bar{\delta}_{nb})R_{ep} \end{vmatrix} \qquad \text{16-VI.2}$$

Accordingly, the EMI matrix equation becomes

$$\begin{bmatrix} \bar{S}_{ii} + \dfrac{1}{K_i} & \bar{S}_{ij} \\ \bar{S}_{ji} & \bar{S}_{jj} + \dfrac{1}{K_j} \end{bmatrix} \cdot \begin{vmatrix} X_i \\ X_j \end{vmatrix} = \begin{vmatrix} \Delta_{io} \\ \Delta_{jo} \end{vmatrix} \qquad \text{17-VI.2}$$

The system of the above simultaneous equations may be solved for X_i, and the process of iterations performed as explained previously.

To calculate the approximate reactions distribution in the transverse direction of the foundation structure, it is necessary to analyze the configuration of the deflection of the foundation structure based on a unit average load $\bar{q}_{mi}$ applied on tributary band areas, considering the stiffness of the foundation structure EI in the transverse direction. With the values so encountered, the real reactions are calculated proportional to the average reactions $\bar{q}_{mi}$ found from the analysis in the longitudinal direction. Assume $\bar{q}_{\alpha i}$ is the unit reaction at any point α due to the configuration obtained in the transverse direction due to a unit average reaction $\bar{q}_{mi}$ in the transverse band i. Therefore, calling q_{mi} the average reaction in transverse band i obtained from the analysis in the longitudinal direction, the configuration of reactions in the transverse bands may be obtained by $q_{\alpha i} = \bar{q}_{\alpha i} \cdot q_{mi}/\bar{q}_{mi}$. (See Zeevaert (1980). ISE-Interacción Suelo-Estructura de Cimentación.)

The time factor in calculating shears and bending moments in the foundation structure with the method just described has an important bearing on the results obtained, since the settlements are a function of time. Therefore, to determine the limiting values for design of the foundation structure, it will be necessary to perform the calculations for two limiting conditions, namely, one at the end of the construction period, and the other for a number of years for which the subsequent settlements may be considered negligible. The creep of concrete also has an important effect in the results, since with time we observe a decrease in the stiffness of the concrete foundation structure. For this purpose, it is recommended that one perform the first analysis at the end of the construction period with a modulus of elasticity of concrete equal to one-half that of the nominal modulus, and for long term with one-third of nominal. The foundation structure is designed to cover the maximum stress conditions obtained in the investigation.

VI.3 SOIL-STRUCTURE INTERACTION MATRIX EQUATION

So far it has been found that to solve the soil-foundation structure interaction problem, it will be necessary to use EMA and EMI with the K-values as a convenient interacting variable, and perform iterations until the values of the unknown reactions do not change significantly, say ±3%. However, it is possible to establish the unknown reaction coefficients and independent terms to find a unique matrix equation to solve the problem without the iteration work and to obtain more accurate results. The new matrix equation will be called EMISE, Soil-Structure Interaction Matrix Equation, (Zeevaert (1980). ISE).

To solve the problem we have to introduce EMA into EMI, hence,

$$\text{EMA into EMI} = \text{EMISE} \qquad \text{1-VI.3}$$

Let us operate with EMA (6-VI.2) in the following form:

$$\begin{bmatrix} \bar{\delta}_{aa} & \bar{\delta}_{ai} & \bar{\delta}_{aj} & \bar{\delta}_{ab} \\ \bar{\delta}_{ia} & \bar{\delta}_{ii} & \bar{\delta}_{ij} & \bar{\delta}_{ib} \\ \bar{\delta}_{ja} & \bar{\delta}_{ji} & \bar{\delta}_{jj} & \bar{\delta}_{jb} \\ \bar{\delta}_{ba} & \bar{\delta}_{bi} & \bar{\delta}_{bj} & \bar{\delta}_{bb} \end{bmatrix} \cdot \begin{vmatrix} R_a \\ X_i \\ X_j \\ R_b \end{vmatrix} = \begin{vmatrix} \delta_a \\ \delta_i \\ \delta_j \\ \delta_b \end{vmatrix} \qquad \text{2-VI.3}$$

Notice that the values of $\bar{\delta}_{ij}$ should be divided by the tributary band areas $\bar{a}_i$, as we apply to them a unit load equal to $1/\bar{a}_i$ or $X_i = +1$. Calculate δ_a and δ_b from (2-VI-.3),

$$\delta_a = \bar{\delta}_{aa} R_a + \sum \bar{\delta}_{ai} X_i + \bar{\delta}_{ab} R_b$$

$$\delta_b = \bar{\delta}_{ba} R_a + \sum \bar{\delta}_{bi} X_i + \bar{\delta}_{bb} R_b \qquad \text{3-VI.3}$$

Now, we examine EMI,

$$\{[\bar{S}_{ji}] + [1/K_i]_D\} |X_i| = |\Delta_{io}| \qquad \text{4-VI.3}$$

Here,

$$\bar{S}_{ji} = \bar{S}_{ji}^{1} + \bar{S}_{ji}^{11}$$
$$\Delta_{io} = \Delta_{io}^{P} + \Delta_{io}^{A} \qquad \text{5-VI.3}$$

where

- $\bar{S}_{ji}^{1}$ are the unit deflections due only to the flexibility of the foundation structure for conditions $X_i = +1$,
- $\bar{S}_{ji}^{11}$ are the unit deflections produced in the foundation structure considered as a rigid body because of yielding of the supports with K_a and K_b values respectively for conditions $X_i = +1$,
- Δ_{io}^{P} is the deflection due to the loads applied to the foundation structure in condition $X_i = 0$, and

Δ_{io}^{A} is the deflection of the foundation structure as a rigid body in condition $X_i = 0$ because of yielding of the supports a and b respectively with K_a and K_b values.

Therefore, we can write EMI as follows:

$$\{[\bar{S}_{ji}^{1}] + [\bar{S}_{ji}^{11}] + [1/K_i]_D\}\ |X_i| = |\Delta_{io}^{P} + \Delta_{io}^{A}| \qquad \text{6-VI.3}$$

We notice that matrices $[\bar{S}_{ji}^{1}]$ and $|\Delta_{io}^{P}|$ are independent of the soil displacements. The investigation, therefore, is confined to finding the significance of the following matrix equation

$$\{[\bar{S}_{ji}^{11}] + [1/K_i]_D\}\ |X_i| = |\Delta_{io}^{A}|, \qquad \text{7-VI.3}$$

for which the forces acting will be as shown in Fig. 1a-VI.3, Hence, taking moments we find reactions R_a and R_b as functions of R_{ao} and R_{bo} for condition $X_i = 0$ in EMI, and of the unknown reactions X_i, X_j,

$$R_a = R_{ao} - \sum \psi_i X_i$$
$$R_b = R_{bo} - \sum \xi_i X_i \qquad \text{8-VI.3}$$

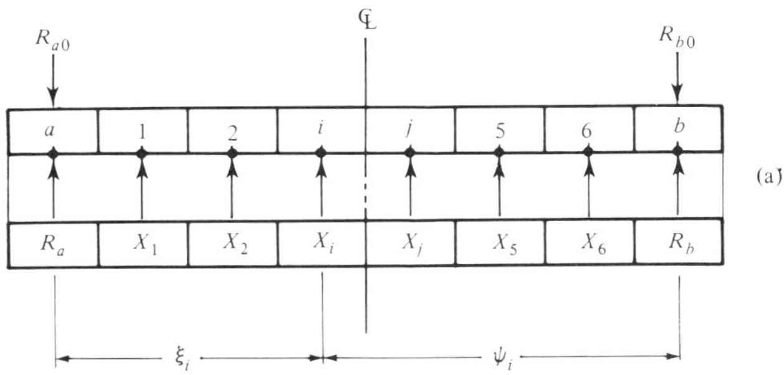

Rotation of beam as a rigid body.

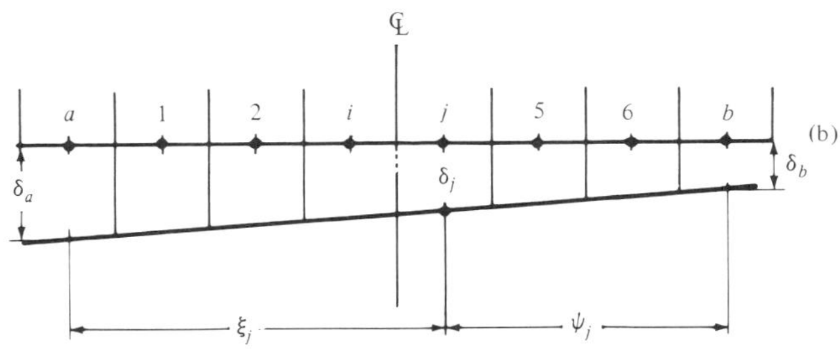

$$\delta_j = \delta_a \psi_j + \delta_b \xi_j$$

Fig. 1-VI.3 Condition of beam as a rigid body.

Substituting R_a and R_b in 3-VI.3 we obtain

$$\delta_a = \bar{\delta}_{aa} R_{ao} + \bar{\delta}_{ab} R_{bo} + \sum (-\bar{\delta}_{aa} \psi_i + \bar{\delta}_{ai} - \bar{\delta}_{ab} \xi_i) X_i$$

$$\delta_b = \bar{\delta}_{ba} R_{ao} + \bar{\delta}_{bb} R_{bo} + \sum (-\bar{\delta}_{bb} \xi_i + \bar{\delta}_{bi} - \bar{\delta}_{ba} \psi_i) X_i \qquad \text{9-VI.3}$$

On the other hand, the rotation of the foundation structure as a rigid body, Fig. 1a-VI.3, for any point j due to the unit load applied at i will be:

$$\delta_j = \delta_a \psi_j + \delta_b \xi_j \qquad \text{10-VI.3}$$

Substituting the values of δ_a and δ_b from 10-VI.3 we obtain

$$\delta_j = (\bar{\delta}_{aa} \psi_j + \bar{\delta}_{ba} \xi_j) R_{ao} + (\bar{\delta}_{bb} \xi_j + \bar{\delta}_{ab} \psi_j) R_{bo} + \sum (-\bar{\delta}_{aa} \psi_j \psi_i - \bar{\delta}_{bb} \xi_j \xi_i$$

$$+ \bar{\delta}_{ai} \psi_j + \bar{\delta}_{bi} \xi_j - \bar{\delta}_{ab} \psi_j \xi_i - \bar{\delta}_{ba} \psi_i \xi_j) X_i \qquad \text{11-VI.3}$$

Now, from EMA (2-VI.3) for the j row substituting the values obtained for R_a and R_b from 8-VI.3 we have

$$\delta_j = \bar{\delta}_{ja} R_{ao} + \bar{\delta}_{jb} R_{bo} - \sum (\bar{\delta}_{ja} \psi_i + \bar{\delta}_{jb} \xi_i) X_i + \sum \bar{\delta}_{ji} X_i \qquad \text{12-VI.3}$$

Hence, in order to make EMA and EMI compatible, expression 11-VI.3 must be equal to 12-VI.3. Equating these equations and rearranging terms we find

$$\sum \{\bar{\delta}_{aa} \psi_j \psi_i + \bar{\delta}_{bb} \xi_j \xi_i - \bar{\delta}_{ai} \psi_j - \bar{\delta}_{bi} \xi_j - \bar{\delta}_{ja} \psi_i$$

$$- \bar{\delta}_{jb} \xi_i + \bar{\delta}_{ab} \psi_j \xi_i + \bar{\delta}_{ba} \psi_i \xi_j\} X_i + \sum \bar{\delta}_{ji} X_i$$

$$= (\bar{\delta}_{aa} \psi_j + \bar{\delta}_{ba} \xi_j - \bar{\delta}_{ja}) R_{ao} + (\bar{\delta}_{bb} \xi_j + \bar{\delta}_{ab} \psi_j - \bar{\delta}_{jb}) R_{bo} \qquad \text{13-VI.3}$$

The above expression shows the coefficients of the unknown reactions for position j when the unit load is at position i. Comparing with EMI (7-VI.3) we may write:

$$[\bar{S}_{ji}^{11}] = [\bar{\delta}_{aa} \psi_j \psi_i + \bar{\delta}_{bb} \xi_j \xi_i - \bar{\delta}_{ai} \psi_j - \bar{\delta}_{bi} \xi_j - \bar{\delta}_{ja} \psi_i - \bar{\delta}_{jb} \xi_i + \bar{\delta}_{ab} \psi_j \xi_i + \bar{\delta}_{ba} \psi_i \xi_j] \qquad \text{14-VI.3}$$

$$[1/K] = [\bar{\delta}_{ji}] \qquad \text{15-VI.3}$$

$$|\Delta_{io}^{A}| = |(\bar{\delta}_{aa} \psi_j - \bar{\delta}_{ba} \xi_j - \bar{\delta}_{ja}) R_{ao} + (\bar{\delta}_{bb} \xi_j + \bar{\delta}_{ba} \psi_j - \bar{\delta}_{jb}) R_{bo}| \qquad \text{16-VI.3}$$

The complete EMISE will read as follows:

$$[\bar{S}_{ji}^{1} + \bar{S}_{ji}^{11} + \bar{\delta}_{ji}] \cdot |X_i| = |\Delta_{io}^{P} + \Delta_{io}^{A}| \qquad \text{17-VI.3}$$

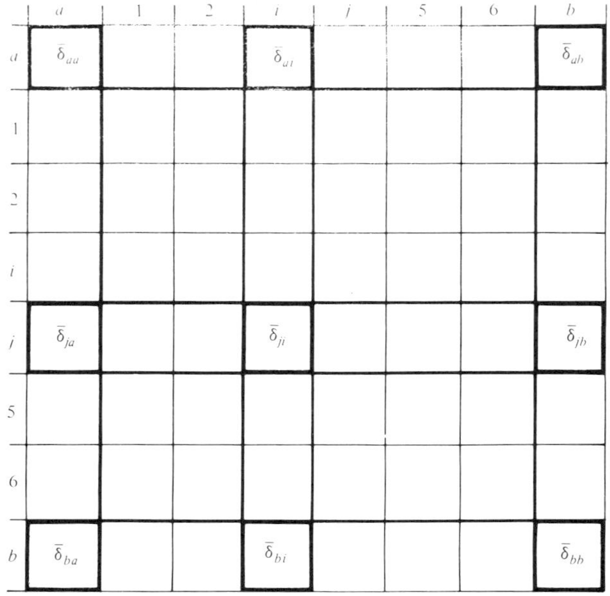

Fig. 2-VI.3 Format from EMA to obtain coefficients and independent terms in EMISE.

The values of $\bar{S}_{ji}^{1}$ and Δ_{io}^{P} are calculated from the deflection of the foundation structure supported on points a and b without yielding, for conditions $X_i = +1$ and $X_i = 0$, respectively.

In order to facilitate the calculation of the coefficients for the unknown reactions X_i (14 and 15-VI.3) and the independent terms (16-VI.3), the format for the EMA square matrix may be used as shown in Fig. 2-VI.3. The values of $(\bar{S}_{ji}^{11} + \bar{\delta}_{ji})$ and Δ_{io}^{A} may be obtained by columns.

The reaction distribution in the transverse direction may be estimated by computing by the same method the reaction configuration based on a unit average reaction in the band considered, and then adjusting proportionally to the average band reaction values obtained from the soil-foundation structure interaction in the longitudinal direction.

A numerical example of a strip foundation similar to the one shown in Fig. 1-VI.1 may be found in Appendix E for Chapter VI, where numerical values of real soil properties have been used for illustration.

BIBLIOGRAPHY

Baker, A. L. L. (1965). *Raft Foundations*, Third Edition, Concrete Publications, Ltd. London.

Chamecki, S. (1956). Structural Rigidity in Calculating Settlements, Journal Soil Mechanics and Foundations Division, *Proceedings ASCE*, Vol. 82, No. SM1, paper 865.

Chamecki, S. (1969). Calcul des Tassements Progressifs des Fondations, *Annales de L'Institut Technique du Batiment et des Travaux Publics,* No. 261, Septembre.

De Beer, E. E. (1957). The Influence of the Width of a Foundation Raft on the Longitudinal Distribution of the Soil Reactions. *Proc. IV ICOSOMEF*, Vol. I, p. 269, London.

De Beer, E. E., Grasshof, H. and Kany, M. (1966). Die Berechnung elastischer Gründungsbalken auf nachgrebigem Untergrund. Westdeutscherverlag Köln und Opladen.

Heil, H. (1969). Studies on the Structural Rigidity of Reinforced Concrete Building Frames on Clay, *Proc. VII ICOSOMEF*, Vol. II, p. 115, Mexico City.

Hetenyi, M. (1964). *Beams on Elastic Foundations.* University of Michigan Press. 7th Printing.

Janbu, N. (1957). Foundations of Structures, *Proc. IV ICOSOMEF*, Vol. III, pp. 155-156, London.

Jennings, J. E. and Evans, G. A. (1962). Practical Procedures for Buildings in Expansive Soil Areas, *The South African Builder*, October.

Popov, E. P. (1950). Successive Approximations for Beams on Elastic Foundations, *Trans. ASCE*, paper 2457.

Sowers, G. F. (1962). *Shallow Foundations*, *Foundation Engineering*, G. A. Leonards, editor, McGraw-Hill Book Company, Inc.

Sowers, G. G. (1963). Foundations, Discussion, *II Panamerican Conference on Soil Mechanics and Foundation Engineering*, Vol. II, p. 607, Brazil.

Tsytovich, N. A. (1957). Foundation of Structures. *Proc. IV ICOSOMEF*, Vol. III, pp. 155-156, London.

Zeevaert, L. (1974). *Concrete Structure-Soil Mass Interaction*, IMCYC and ACI, IMCYC Vol. 111, p. 73.

Zeevaert, L. (1975). The Role of Soil Mechanics in Foundation Structure-Soil Interaction, in: W. J. Hall (ed.), *Structural and Geotechnical Mechanics*, Vol. honoring N. M. Newmark, Prentice-Hall, Inc.

Zeevaert, L. (1980). *ISE-Interacción Suelo-Estructura de Cimentaciones Superficiales y Profundas*, Publ. LIMUSA, Mexico, D. F.

▸VII◂ COMPENSATED FOUNDATIONS

VII.1 BASIC CONCEPTS

Compensated foundations, commonly known as floating foundations, are used in soil deposits of low shear strength and medium to very high compressibility extending to great depth. The materials showing these engineering characteristics contain a large percentage of fine and very fine soil fractions, as seen in section II.3. The foundation structure is designed of the rigid or semirigid box type mat foundation and is placed deep in the ground.

To analyze the real meaning of a compensated foundation from the physical engineering point of view, and to be able to recognize the problem in relation to design and construction, it is necessary to consider the soil a two-phase material, in which water is the liquid phase and the skeleton structure the solid phase with well defined mechanical properties of shear strength and compressibility (see Chapter II). When an excavation is performed and its weight is thereafter replaced by an equivalent building weight, one replaces the water pressure and the effective stresses the ground has at the foundation grade elevation (Fig. 1-VII.1), that is to say, the counteraction of hydraulic pressure and effective stress removed by excavation is compensated for or balanced by the weight of the building. The first application of this by the author was in 1943 for compensated footing foundations housing a factory building in the highly compressible silty clay of the Valley of Mexico (Fig. 2-VII.1).

A compensated foundation is governed by the following equation:

$$q_d = \sigma_{0d} + u_d \qquad \text{1-VII.1}$$

where q_d is the average unit weight of the building to be applied on the soil at depth d, at which σ_{0d} is the effective overburden stress, and u_d is the uplift water

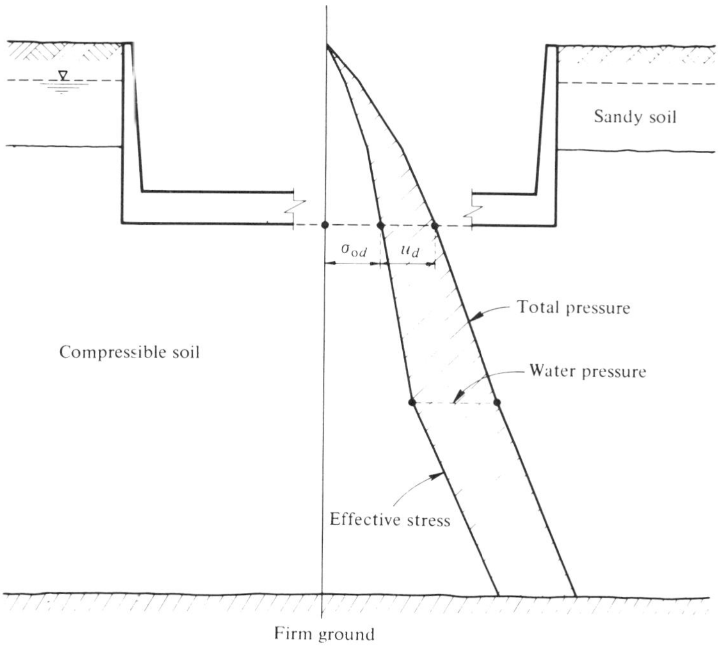

Fig. 1-VII.1 Effective stress and water pressure at the foundation grade elevation.

pressure under the foundation slab. For a definite building weight, the behavior of the foundation with time is governed by the relation

$$\frac{\partial \sigma_{od}}{\partial t} = -\frac{\partial u_d}{\partial t} \qquad \text{2-VII.1}$$

Hence, a change in the uplift water pressure will produce an equal and contrary change in the effective stress under the foundation slab. Consequently, the solid phase will govern the behavior of the foundation according to the stress-strain-time characteristics of the soil skeleton structure.

The contact stress σ_d under the foundation slab is not uniform (Fig. 3-VII.1); however, the uplift water pressure u_d may be considered uniform. Calling A the total area in contact, and W_a the total weight of the building, the compensation of the building load is given by

$$W_a = \int_A \sigma_d \cdot da + u_d A \qquad \text{3-VII.1}$$

In cases when the soil is classified of the preconsolidated type, the foundation may be designed over-compensated, and equation 3-VII.1 reads as follows:

$$W_a = \int_A \sigma_d da + u_d \cdot A + \int_A \Delta \sigma_d da \qquad \text{4-VII.1}$$

Fig. 2-VII.1 Compensated rigid box type footing.

The design of a compensated foundation should be made with caution recalling that its behavior depends only on the change taking place in the effective stress in the soil mass; the smaller these changes are, the smaller the deformational behavior of the foundation will be. Therefore, a compensated foundation is designed to achieve the following:

(a) The building load is compensated with soil effective stresses and water uplift pressures at the foundation grade elevation. The foundation structure is placed deep into the ground obtaining more stability and bearing capacity. The change in effective stress because of excavation should be kept to a minimum, to reduce heave and subsequent settlement.
(b) The foundation design is made to satisfy the allowable total and differential settlements for the building project itself and adjoining construction.

The total settlement can be reduced if during excavation the change in effective stress is kept to a minimum. If it were possible to place the foundation structure at the depth required in such a way as to eliminate changes in the over-burden effective stresses, and without alteration of the existing hydraulic pressures, then no settlement would take place.

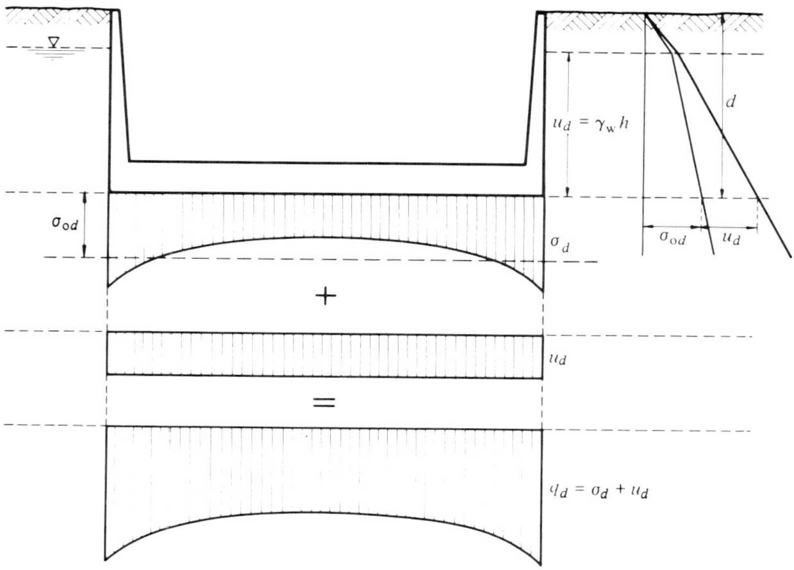

Fig. 3-VII.1 Contact stresses under foundation slab.

When clay is found underlying the foundation grade elevation, the allowable soil pressure q_a with factor of safety G_s may be taken as:

$$q_a = 5.7 \frac{s_{cu}}{G_s} + \sum_{1}^{d} \gamma \Delta z \qquad \text{5-VII.1}$$

in which s_{cu} is the average shear strength along the potential surface of sliding, determined by one-half of the unconfined compressive strength for each one of the subsoil strata. The second term represents the soil effective stress σ_{0d} at the foundation grade elevation.

The behavior of compensated foundations constructed in soil deposits of medium to high compressibility are governed by settlement. In the case of totally compensated or over-compensated foundations, no significant excess shear forces are induced in the ground for static conditions. Under dynamic loading, however, the overturning moments in case of slender buildings may induce increments in the subgrade reactions at the foundation edges greater than the bearing capacity of the soil, and permanent rotational deformations may take place. A compensated foundation is designed, as stated before, with a watertight rigid box type foundation structure, to be constructed deep into the ground in an excavation. Usually, the foundation structure is considerably more rigid if compared with the soil compressibility. Therefore, when calculating subgrade reactions, the foundation structure is currently assumed infinitely stiff (see Chapters IV and VI).

The foundation engineer should understand the problems related with stability and deformation occurring during deep excavations and reapplication of the load in compressible soil deposits, as discussed in Chapters II and XI. To design properly a compensated foundation, it is necessary to know the stratigraphy of the subsoil to a depth at which firm ground may be encountered. The hydraulic properties in the soil deposit should also be investigated by means of piezometers, to establish whether water conditions are static or dynamic. Special considerations should be given for foundation designs for which piezometric pressures are decreasing with time, as is the case in areas where ground surface subsidence is taking place. The effective stresses in the soil mass are continuously increasing, and therefore the foundation should be designed slightly over-compensated to allow for the change in effective stresses to take place and to avoid exceeding the critical load of any of the compressible subsoil strata at the site in question. The foundation engineer should be well acquainted with the properties of compressibility and elasticity of the soil, as well as shear strength of the different soil strata encountered. These mechanical properties may be estimated using the best possible undisturbed soil samples and techniques available in the laboratory.

A compensated foundation is constructed in deep excavations made to a depth of one or several basements. During the excavations, adjacent property should not be damaged and the soil not disturbed, otherwise the mechanical properties may change considerably. This is especially important in cases of highly sensitive soils. The phenomenon of heave at the bottom of the excavation has a very important bearing on the future behavior of a compensated foundation. The lateral deformation of the sides of the excavation should be carefully considered. Before analyzing the important phenomena mentioned above, and discussing the procedures to minimize their effects, it is necessary to point out, from the practical engineering point of view, the properties of the soil sediments that should be considered in this type of foundation design.

VII.2 SHEAR STRENGTH

The materials on which compensated foundations are supported are usually saturated clayey silts and clays. The shear strength may be evaluated by the relations

$$s_{cu} = \tfrac{1}{2}\, q_u \text{ in clay and } s_{cu} = c_{cu} + \sigma \tan \phi_{cu} \text{ in silts} \qquad \text{1-VII.2}$$

Thus, we must determine one-half of the unconfined compressive strength for clays, and in the case of clayey silts and silts, the consolidated undrained parameters c_{cu} and ϕ_{cu}. In both cases, these properties are determined in the laboratory on undisturbed soil samples. The specimens are subjected to a volumetric confining effective stress equivalent to the one the material had in the ground at the depth the soil specimen was taken:

$$\sigma_{ci} = \tfrac{1}{3}(1 + 2K_0)\sigma_{zi}$$

where σ_{zi} is the vertical overburden effective stress, and $K_0 = \sigma_z/\sigma_h$ is the coefficient of earth pressure at rest. Hence, if $K_0 = 0.70$, then $\sigma_{ci} = 0.8\sigma_{zi}$ will be the average volumetric confining stress level to be used in the laboratory, at which the shear strength is determined in the triaxial compression chamber, as described in section II.4.

VII.3 COMPRESSIBILITY AND CRITICAL STRESS

Soft sensitive materials with medium to very high compressibility are usually of a complicated internal skeleton structure, which may be easily altered because of changes in stresses, swelling or overloading. Therefore, it is absolutely necessary to determine the compressibility characteristics of the soil on good undisturbed samples. First of all, it is necessary to determine whether the soil is of the preconsolidated type or not. To illustrate the importance of this problem, let 0 be a point representing the effective vertical stress σ_0 and the void ratio e_0 in the field at a certain depth from the ground surface (Fig. 1-VII.3). Starting from this condition, let us load the material in nature obtaining either one of two different phenomenological laws of compression (section II.3):

(a) Compression takes place along curve A up to a stress $\sigma_{bz} > \sigma_{0z}$, where σ_{0z} is the present overburden effective stress. At stress σ_{bz}, a break in the compressibility curve may be observed. At this point, a collapse of the in-

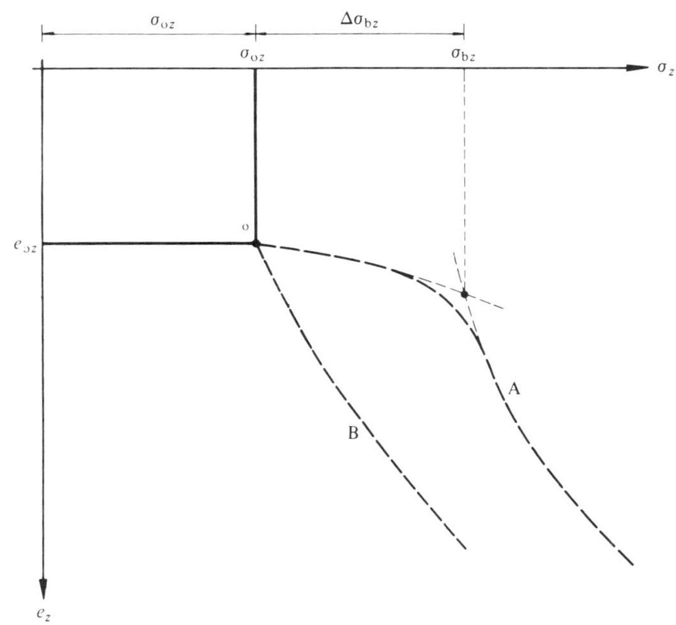

Fig. 1-VII.3 Preconsolidated and normally consolidated types of soils.

ternal soil skeleton structure takes place and thereafter, large compression follows for small stress increments. The soil material showing these characteristics is known to be of the preconsolidated type, and σ_{bz} is the confined compressive critical stress (section II.3).

(b) The compression follows along curve B with no break. Small increments of stresses over σ_{0z} produce large void-ratio changes. This soil material is known as a normally loaded or consolidated type soil.

From the above-mentioned behavior one reaches the conclusion that neither of the steep branches of the curves just explained—either curve A or B—should be used in designing compensated foundations. The value σ_{bz}, in the case of soils of the preconsolidated type, measures a confined compressive strength or critical stress representing the load necessary to produce, in the soil skeleton structure under laterally confined conditions, an internal failure, upon which the soil skeleton structure suffers damage, thus showing considerably larger compressibility. In normally loaded soil, however, the process of compaction may be considered a continuous process, and the concept of critical stress thus has no significance. In the case of preconsolidated type soils, the extra strength of the skeleton structure $\Delta\sigma_{bz} = \sigma_{bz} - \sigma_{0z}$ may have been given in nature by either one or the combination of the following principal causes (section II.3.6a):

(a) Larger effective stresses of the past induced by loading.
(b) Effective stresses of the past induced by capillary forces because of drying.
(c) Bond developed by active clay minerals or other cementing materials.

In this type of soil after reaching the critical stress σ_b corresponding to the break in the compressibility curve, the bond between particles is damaged and the material is forced to compress strongly to build up a more compact and stable structure; the toll is a large volumetric deformation.

The increment of allowable stress $\Delta\sigma_{ba}$ that may be added to the in situ effective overburden stress at any depth z may be taken in this case as

$$(\Delta\sigma_{ba})_z = \frac{1}{G_s}(\sigma_b - \sigma_0)_z \qquad \text{1-VII.3.}$$

where G_s is a factor of safety on the order of 1.5–2.0.

In soils of high compressibility, no increment of stress should be allowed over the existing critical compressive stress if large deformations are to be eliminated. The importance of the critical compressive stress may be illustrated from buildings in Mexico City (Fig. 2-VII.3) where we see an example of how overloading the soil mass can lead to settlements on the order of several feet instead of several inches.

Thus, we may conclude that in compressible soil deposits the design of compensated foundations for settlement should be made only with mechanical properties representing the recompression or flat branch of the compressibility curve (Figs. 3 and 4-VII.3). For preconsolidated type soil between stresses

$$\left[\sigma_0 + \frac{1}{G_s}(\sigma_b - \sigma_0)\right]_z$$

Fig. 2-VII.3 Total settlement of 125 cm.

and

$$(\sigma_0 - \Delta\sigma_{ex})_z \qquad \text{2-VII.3}$$

in which $(\Delta\sigma_{ex})_z$ is the relief in effective stress at depth z because of excavation, the following secant coefficients of unit volume compressibility for recompression of each stratum are used:

A. For recompression due to excavation up to the overburden effective stress,

$$m_{rc} = \frac{\Delta e'_{ex}}{(1 + e'_0)\ \Delta\sigma_{ex}} \qquad \text{3-VII.3}$$

B. For compression over the effective overburden stress in preconsolidated type soils, for an increment of stress $\Delta\sigma_{ba} = \Delta\sigma_b/G_s$:

$$m_{vb} = \frac{G_s \cdot \Delta e'_b}{(1 + e'_0)\ \Delta\sigma_b} \qquad \text{4-VII.3}$$

With the values of m_{rc} and m_{vb}, the settlement of the building after excavation relief may be calculated for applied stress of $\Delta\sigma_{ex}$ and $\Delta\sigma_{bd}$; hence:

$$\delta_c = \sum_1^n (m_{rc}\Delta\sigma_{ex} \cdot d)_i + \sum_1^n (m_{vb}\Delta\sigma_{ba} \cdot d)_i \qquad \text{5-VII.3}$$

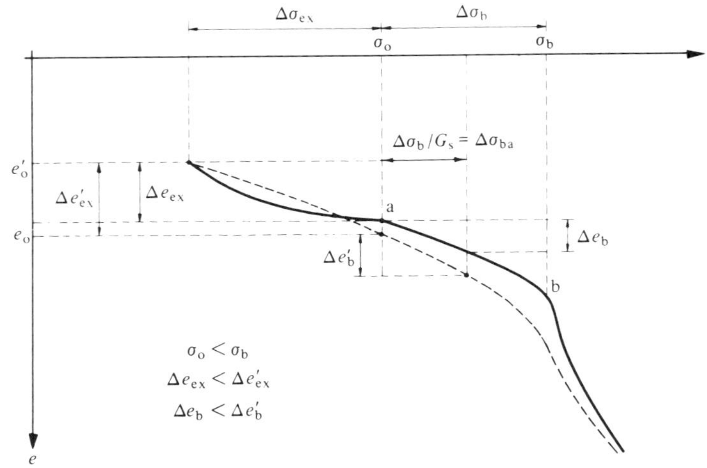

Fig. 3-VII.3 Preconsolidated material.

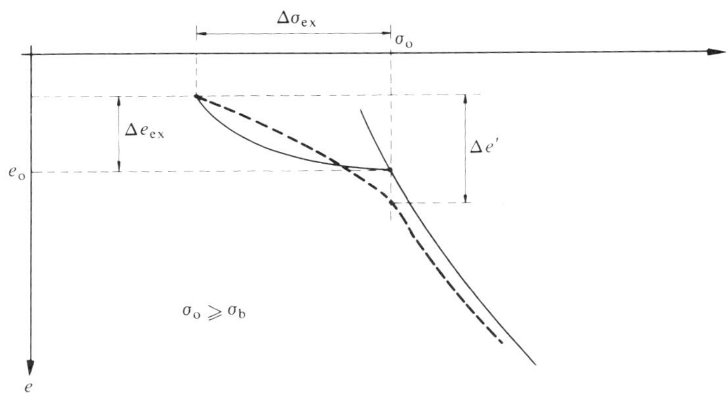

Fig. 4-VII.3 Normally loaded material.

where the subscript i stands for average values of one layer of thickness d_i in the deposit under consideration, between depths z_i and z_{i+1}.

In the laboratory one performs a recompression of the material, obtaining the compressibility curves shown in Figs. 3 and 4-VII.3 by a dashed line. Therefore, if $\sigma_0 \geqslant \sigma_b$, the soil may be considered as normally consolidated; when $\sigma_0 < \sigma_b$, the soil may be recognized as of the preconsolidated type and the flat branch of the compressibility curve representing the recompression will closely follow line a–b, but will have a steeper slope (Fig. 3-VII.3).

The most important problems related with the shear strength and stress-strain-time properties of the soil in the design and construction of compensated foundations requiring deep excavations are:

(a) Plastic flow into the excavation.
(b) Elastic heave and subsequent settlement.
(c) Lateral contraction and settlement outside the excavation.

VII.4 PLASTIC FLOW

The danger of plastic flow at the bottom of the excavation can be minimized by careful study of the probable potential surface of sliding, and by allowing shear stresses not larger than half of the shear strength values. If the soil deposit is stratified, the potential surface of sliding may be assumed as shown in Fig. 1-VII.4. The problem may be treated like a bearing capacity problem; hence:

$$\sigma_d = \frac{1}{G_s}(5.7 \cdot s_{cu})$$

where $\sigma_d = \sigma_{0d} + q_0$ is the total effective stress at the foundation grade elevation and G_s a factor of safety. Therefore, the safe average shear strength is given by

$$s_{cu} \geqslant \frac{G_s \sigma_d}{5.7} \qquad \text{1-VII.4}$$

In case of stratified soil deposits, the potential surface of sliding may be found closer to the bottom of the excavation as shown in Fig. 1-VII.4, surface of sliding labeled B. In this case, however, the factor of safety is larger provided that $B \leqslant d_a$,

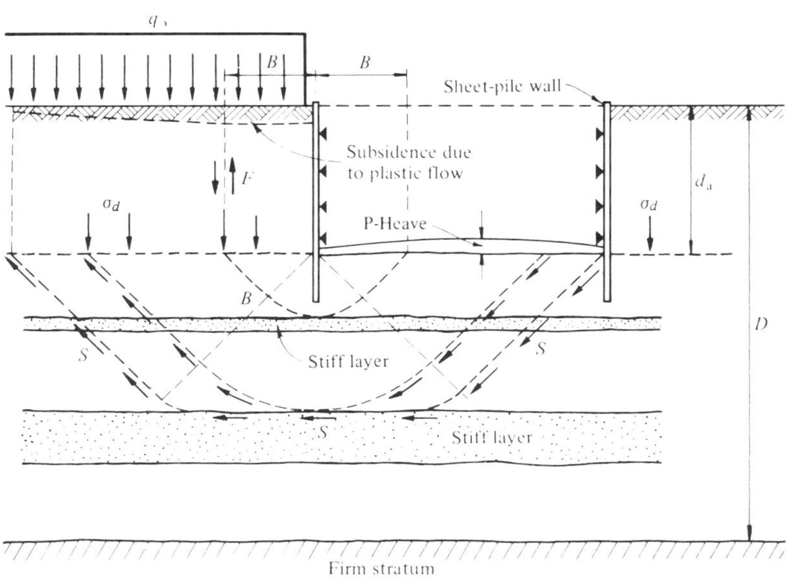

Fig. 1-VII.4 Potential surface of failure in stratified cohesive soils.

where d is the depth of the excavation, since the shear force F mobilized along the vertical plane tends to reduce the value of σ_d in the amount F/B. Hence, it may be seen that the depth of excavation is limited if one wishes to avoid P-heave because of plastic flow and important subsidence at the edges of the excavation where buildings may be encountered. The allowable depth is obtained from:

$$s_{cu} = \frac{G_s}{5.7}\left(\sum_{1}^{d} \gamma \Delta z + q_0\right) \tag{2-VII.4}$$

Assuming for illustration a constant unit weight of the soil through depth D, then the allowable depth d_a of excavation is

$$d_a = \frac{5.7 \dfrac{s_{cu}}{G_s} - q_0}{\gamma} \tag{3-VII.4}$$

Therefore, assuming $s_{cu} = 4$ ton/m^2, $q_0 = 3.0$ ton/m^2 and the unit weight of the soil $\gamma = 1.5$ ton/m^3, the allowable depth of excavation with a factor of safety of 2 to avoid P-heave will be $d_a \leqslant 5.60$ m.

VII.5 ELASTIC HEAVE AND SUBSEQUENT SETTLEMENT

In spite of the fact that plastic flow toward the bottom of the excavation is eliminated, there is always the E-heave problem because of response of the elastic elements of the material upon load relief. This phenomenon may be accompanied by swelling of the clay minerals when the soil is released from the load and water molecules enter into the clay mineral structure. The additional heave because of swelling, S-heave, may be strongly present if the excavation is kept without reloading for a long time, or if it is permitted to be flooded for several days (see Chapter XI). The stress relief is usually increased because of the upward water seepage forces taking place in the subsoil close to the bottom of the excavation, and at the time it is dewatered. The sides of the excavation are protected by an impervious sheet-pile wall, and therefore, a downward water seepage takes place outside of the excavation increasing the effective stresses and producing settlement of the adjacent structures. The corresponding E-heave and subsequent settlement after load application and the settlement outside the excavation may be estimated knowing the elastic response and compressibility properties of the subsoil materials, as determined in undisturbed samples. The E-heave and subsequent settlement is illustrated in Fig. 1-VII.5 with vertical displacement records of a compensated foundation having an excavation 4.75 m deep, performed in Mexico City's silty clay deposits. We note that heave due to response of elastic soil elements takes place rapidly during the excavation and foundation construction period.

Calling M_{ei} the expansion secant strain modulus of elasticity for an average effective stress relief $\Delta\sigma_{ei}$ of any stratum N with thickness d_i, the E-heave δ_e at a point

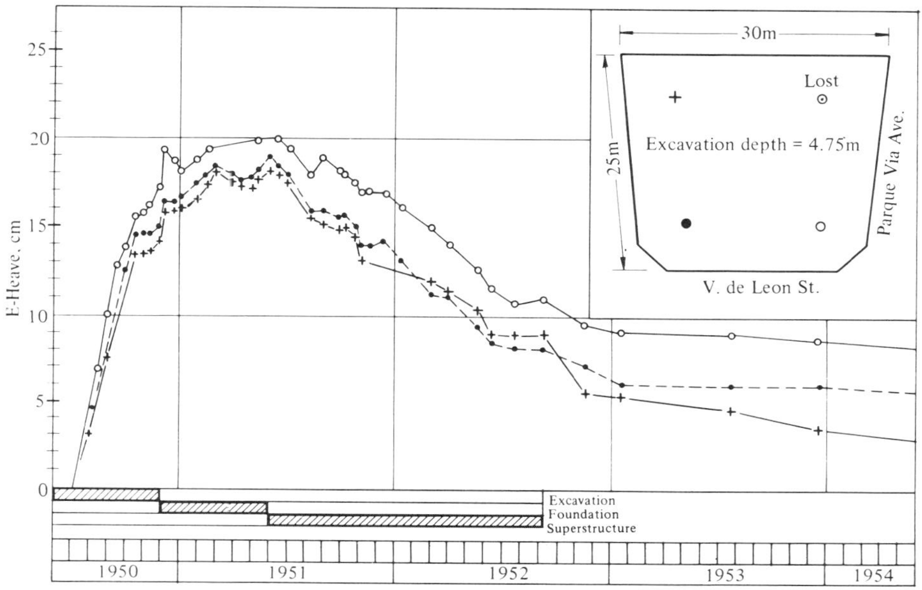

Fig. 1-VII.5 E-heave and subsequent settlement.

i on the bottom of the excavation is

$$\delta_{ei} = \sum_{A}^{N} (M_{ei} d_i)^N \cdot \Delta\sigma_{ei}^N \qquad \text{1-VII.5}$$

Here $(M_{ei} \cdot d_i)^N = \alpha_e^N$ is the unit expansion of the stratum N (CHAPTER VI); for any other points we write in matrix form

$$|\delta_{ei}| = [\Delta\sigma_{ei}^N] \cdot |\alpha_e^N| \qquad \text{2-VII.5}$$

The state of stress relief $[\Delta\sigma_{ei}^N]$ in the subsoil under the foundation area must be determined taking into consideration the change in hydraulic conditions because of pumping from the excavation. Consider the initial conditions before excavation is performed, $p_{0i} = (\sigma_{0i} + u_{0i})$, and the final conditions after the excavation is made and pumping is stabilized, $p_i = (\sigma_i + u_i)$. Therefore, the change in total pressure is

$$p_i - p_{0i} = (\sigma_i - \sigma_{0i}) + (u_i - u_{0i})$$

and, calling the average change in effective stresses in stratum N, $\Delta\sigma_{ei}^N = (\sigma_i - \sigma_{0i})^N$, we have

$$\Delta\sigma_{ei}^N = (u_{0i} - u_i)^N - (p_{0i} - p_i)^N \qquad \text{3-VII.5}$$

where $(u_{0i} - u_i)^N = \Delta u_i^N$, represents the change in hydraulic pressures. If λ_i^N is the drop in piezometric water elevation at the center of stratum N, then $\Delta u_i^N = \lambda_i^N \gamma_w$. The change in total pressure is represented at the same point by $\Delta p_i^N = (p_{0i} - p_i)^N$.

It is a function of the total pressure p_{0d} removed by excavation up to the foundation grade elevation. Substituting 3-VII.5 in 2-VII.5, we obtain:

$$|\delta_{ei}| = [\Delta u_i^N]^T \cdot |\alpha_e^N| - [\Delta p_i^N]^T \cdot |\alpha_e^N| \qquad \text{4-VII.5}$$

or

$$|\delta_{ei}| = |\delta_{ei}^\lambda| - |\delta_{ei}^p| \qquad \text{5-VII.5}$$

From Chapter VI we notice that the state of stress relief in the subsoil is expressed by:

$$[\Delta p_i^N] = \sum [I_{ji}] \cdot |p_{0d}|. \qquad \text{6-VII.5}$$

Also, the expansion due to the total stress relief is

$$|\delta_{ei}^p| = [\bar{\delta}_{ji}]^T \cdot |p_{0d}| \qquad \text{7-VII.5}$$

Finally, we obtain the following expressions to calculate the E-heave:

1) The state of effective stress relief

$$[\Delta\sigma_{ei}^N] = [\Delta u_i^N] - [\Delta p_i^N] \qquad \text{8-VII.5}$$

2) The E-heave

$$|\delta_{ei}| = [\Delta u_i^N]^T \cdot |\alpha_e^N| - [\bar{\delta}_{ji}]^T \cdot |p_{0d}| \qquad \text{9-VII.5}$$

The average total stress relief Δp_i^N for each stratum N and under point i is calculated from the theory of elasticity as discussed in Section III.2 and Chapter VI. (See L. Zeevaert, 1980, ISE). The value of Δu_i is the change in hydraulic pressure estimated from the hydrodynamic conditions in the soil, created because of pumping from the excavation; it may be estimated tracing a flow net (see section III.3). An example is shown in Fig. 2-VII.5, where a water bearing firm stratum is underlying an isotropic silty clay deposit. The water flow takes place upward toward the bottom of the excavation. The water table is encountered stable in the pervious fill material overlying the silty clay deposit. The excavation is protected by an impervious sheet-pile wall. The water table is reduced by pumping to the excavation grade elevation. From the flow net, the values of $u_{0i} = \gamma_w h_{0i}$ may be found from the geometrical position of each equipotential line in the soil mass under the bottom of the excavation. The water pressure change for this example is shown graphically for a vertical plane located at the center of the excavation, it may be expressed at any point in the soil mass by

$$\Delta u_i = \lambda_i \gamma_w \qquad \text{10-VII.5}$$

where λ_i is the drop in piezometric water elevation at depth $z = i$ from the ground surface, it may be estimated from the flow net and d_w is the depth at which the water table has been reduced because of pumping from the excavation. When the subsoil is stratified containing water bearing sand strata under high pressure, pumping from deep water wells will be necessary to reduce the uplift pressures and avoid

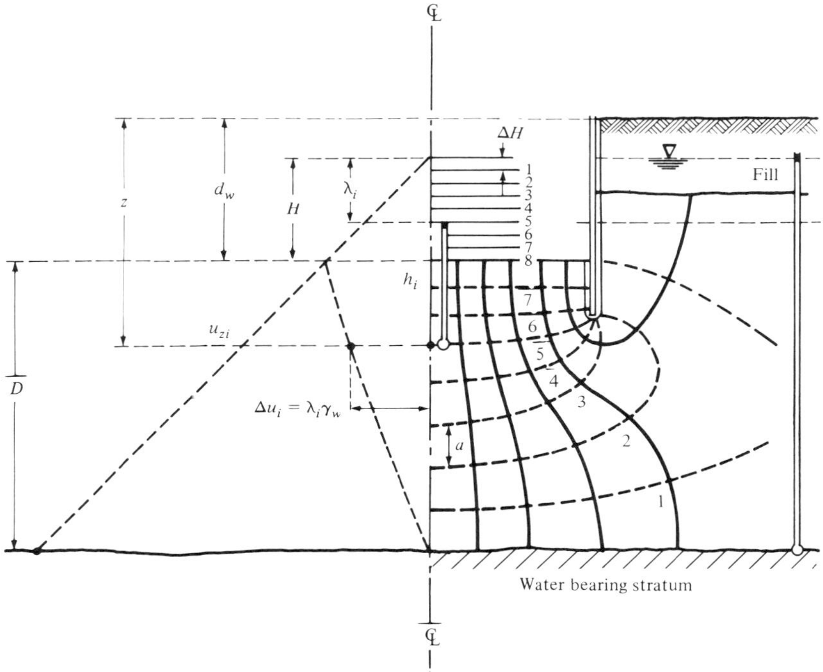

STRATUM	1	2	3	4	5	6	α_e^N
A	Δu_1^A	Δu_2^A	Δu_3^A	Δu_4^A	Δu_5^A	Δu_6^A	α_e^A
B	Δu_1^B	Δu_2^B	Δu_3^B	Δu_4^B	Δu_5^B	Δu_6^B	α_e^B
C	Δu_1^C	Δu_2^C	Δu_3^C	Δu_4^C	Δu_5^C	Δu_6^C	α_e^C
D	Δu_1^D	Δu_2^D	Δu_3^D	Δu_4^D	Δu_5^D	Δu_6^D	α_e^D
N	Δu_1^N	Δu_2^N	Δu_3^N	Δu_4^N	Δu_5^N	Δu_6^N	α_e^N

Fig. 2-VII.5 Flow net to estimate the change in effective stresses.

a blow out or high expansion of the bottom of the excavation (see Chapters III and XI).

The matrix $[\Delta u_i^N]$ for the change in hydraulic pressure obtained from the study of the flow net for every stratum N under point i of the excavation area is shown graphically in Fig. 2-VII.5. The subsequent settlement due to reapplication of p_{od} is calculated by means of

$$|\delta_{ci}| = [\bar{\delta}_{ji}]^T \cdot |p_{od}| - [\Delta u_i^N]^T \cdot |\alpha_c^N| \qquad \text{11-VII.5}$$

However, it may be noticed that now $[\bar{\delta}_{ji}]^T$ should be calculated using the unit compression of the strata α_c^N (Chapter VI, or L. Zeevaert, 1980, ISE).

Once the effective stress relief is estimated, the E-heave and subsequent settlement of the foundation may be calculated. The relief of effective stresses due to the excavation varies with depth. Therefore, the process of unloading and reloading the ground in determining the soil compressibility characteristics should be fully understood. A compression test is shown in Fig. 3-VII.5 where several unloadings and reloadings have been performed for stresses $\sigma_i < \sigma_{0z}$. The value σ_{0z} is considered the overburden effective stress. From the test it may be noticed that all the hysteresis loops are parallel to each other when total unloading is performed. However, for partial unloading, that is to say, for small hysteresis loops, the expansion and recompression characteristics are found smaller (Fig. 4-VII.5). Moreover, it may be noticed that upon recompression, the material undergoes a larger strain ϵ_{0c}, if compared with the expansion strain ϵ_0 owing to total stress relief σ_0. The ratio of these strains may be expressed as follows:

for full stress relief

$$a_0 = \frac{\epsilon_{0c}}{\epsilon_0}$$

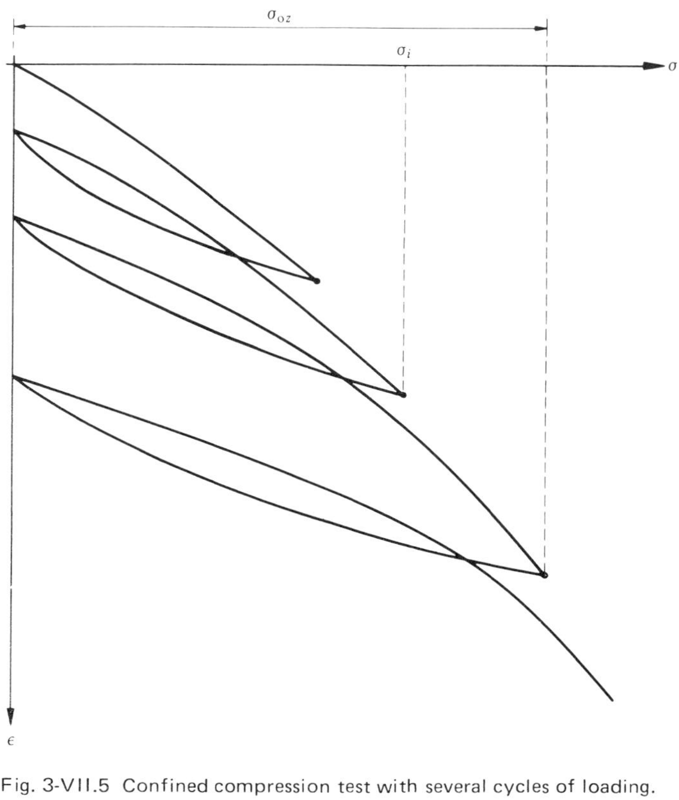

Fig. 3-VII.5 Confined compression test with several cycles of loading.

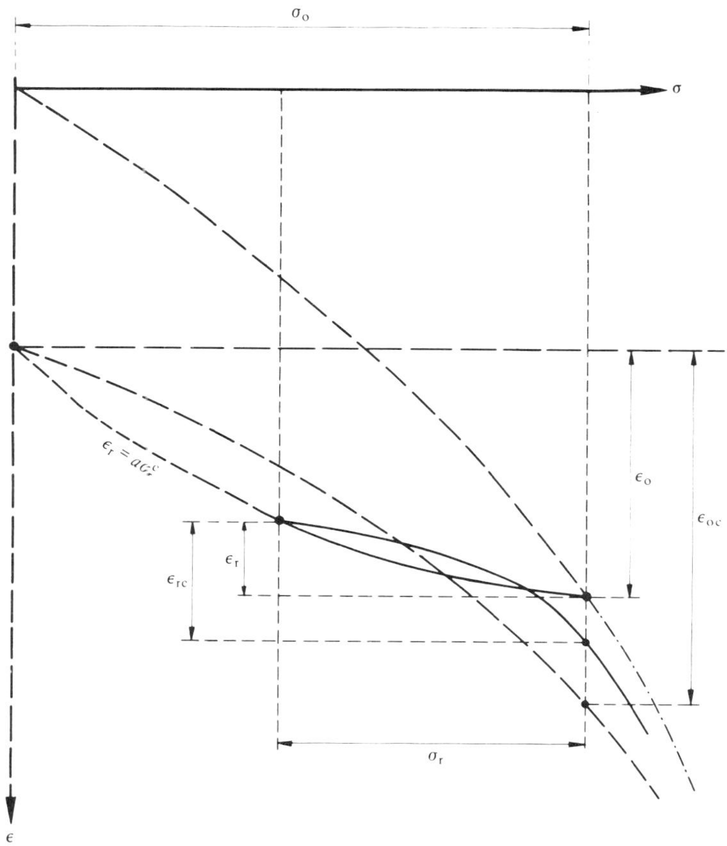

Fig. 4-VII.5 Expansion and recompression characteristics.

For partial stress relief

$$a_r = \frac{\epsilon_{rc}}{\epsilon_r} \qquad \text{12-VII.5}$$

The ratio a_r is a function of the magnitude of stress relief. For total stress relief $a_r = a_0$, and for very small values of stress relief $a_r \to 1$ (Fig. 4-VII.5).

Tests show that the phenomenological law governing the relation between expansion strain vs. stress relief may be represented as a straight line in logarithmic scales (Fig. 5-VII.5). Therefore:

$$\epsilon_r = a\,\sigma_r^c \qquad \text{13-VII.5}$$

in which coefficients a and c are functions of the soil characteristics. For Mexico City's typical silty clay, c assumes an average value of 1.5.

The slope ϵ_r/σ_r represents the secant strain modulus of expansion for stress relief

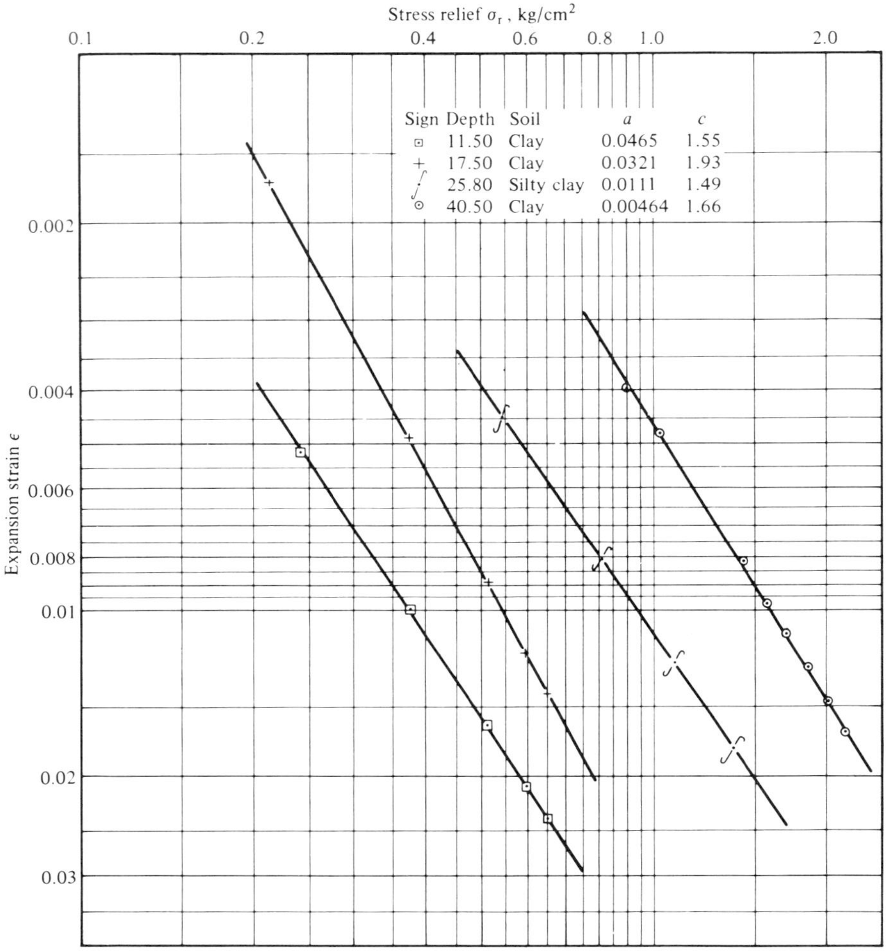

Fig. 5-VII.5 Expansion strain vs. stress relief.

σ_r; hence

$$M_{er} = \frac{\epsilon_r}{\sigma_r} \quad \text{or} \quad M_{er} = a \cdot \sigma_r^{c-1} \qquad \text{14-VII.5}$$

Now call M_{e0} the expansion strain modulus obtained from the hysteresis loop for complete unloading; $\sigma_r = \sigma_0$, as determined in the laboratory in unconfined and undisturbed soil specimens. Therefore, the M-ratio of intermediate to full stress relief will be:

$$\rho_e = \left(\frac{\sigma_r}{\sigma_0}\right)^{c-1} \qquad \text{15-VII.5}$$

The value ρ_e is the stress relief expansion factor.

From laboratory tests in unconfined soil specimens by means of full hysteresis loops, one can determine the value of M_{eo}. Therefore, expression 15-VII.5 may be used for E-heave calculations:

$$M_{er} = \rho_e M_{eo} \qquad \text{16-VII.5}$$

To find the M-ratio for recompression, we observe that:

$$\frac{\epsilon_{rc}}{\epsilon_{oc}} = \frac{\epsilon_r}{\epsilon_o} \cdot \frac{a_r}{a_o} \qquad \text{17-VII.5}$$

and multiplying this expression by σ_o/σ_r, we obtain:

$$\frac{M_{rc}}{M_{oc}} = \left(\frac{\sigma_r}{\sigma_o}\right)^{c-1} \cdot \frac{a_r}{a_o} \qquad \text{18-VII.5}$$

Here M_{rc} represents the recompression strain modulus for partial stress relief, and M_{oc} for total stress relief. The value M_{oc} is determined in the laboratory in undisturbed soil samples, and represents the secant unit strain modulus for certain increments of stress (Section 3.7, Chapter II).

In case of recompression of the soil, however, the visco-plastic strain is added to the elastic strain, hence

$$\frac{\epsilon_{rc}}{\epsilon_{oc}} = \frac{\epsilon_{er} + \epsilon_{vpr}}{\epsilon_o + \epsilon_{vp}}$$

Calling

$$\kappa_{vpr} = \frac{\epsilon_{vpr}}{\epsilon_{er}} \quad \text{and} \quad \kappa_{vpo} = \frac{\epsilon_{vpo}}{\epsilon_o}$$

we can write

$$\frac{\epsilon_{rc}}{\epsilon_{co}} = \frac{\epsilon_{er}(1 + \kappa_{vpr})}{\epsilon_o(1 + \kappa_{vpo})}$$

and, according to 9-VII.5 and 6-VII.5, we obtain

$$\frac{M_{rc}}{M_{oc}} = \rho_e \frac{1 + \kappa_{vpr}}{1 + \kappa_{vpo}} \qquad \text{19-VII.5}$$

Assuming that κ_{vp} varies approximately linearly with the stress level,

$$\frac{\kappa_{vpr}}{\kappa_{vpo}} \approx \frac{\sigma_r}{\sigma_o}$$

However, this assumption should be investigated further in the laboratory for different soils. For the present we assume

$$\frac{a_r}{a_o} = \frac{1 + \kappa_{vpo} \cdot \sigma_r/\sigma_o}{1 + \kappa_{vpo}} \qquad \text{20-VII.5}$$

where

$$\frac{1 + \kappa_{vpo} \cdot \sigma_r/\sigma_o}{1 + \kappa_{vpo}} = f(\kappa_{vp}, \sigma_r/\sigma_o)$$

Therefore, calling ρ_c the recompression factor,

$$\frac{M_{rc}}{M_{co}} = \rho_c$$

and

$$\rho_c = \rho_e \cdot f(\kappa_{vp}, \sigma_r/\sigma_o) \quad \text{21-VII.5}$$

or

$$\frac{\rho_c}{\rho_e} = f(\kappa_{vp}, \sigma_r/\sigma_o)$$

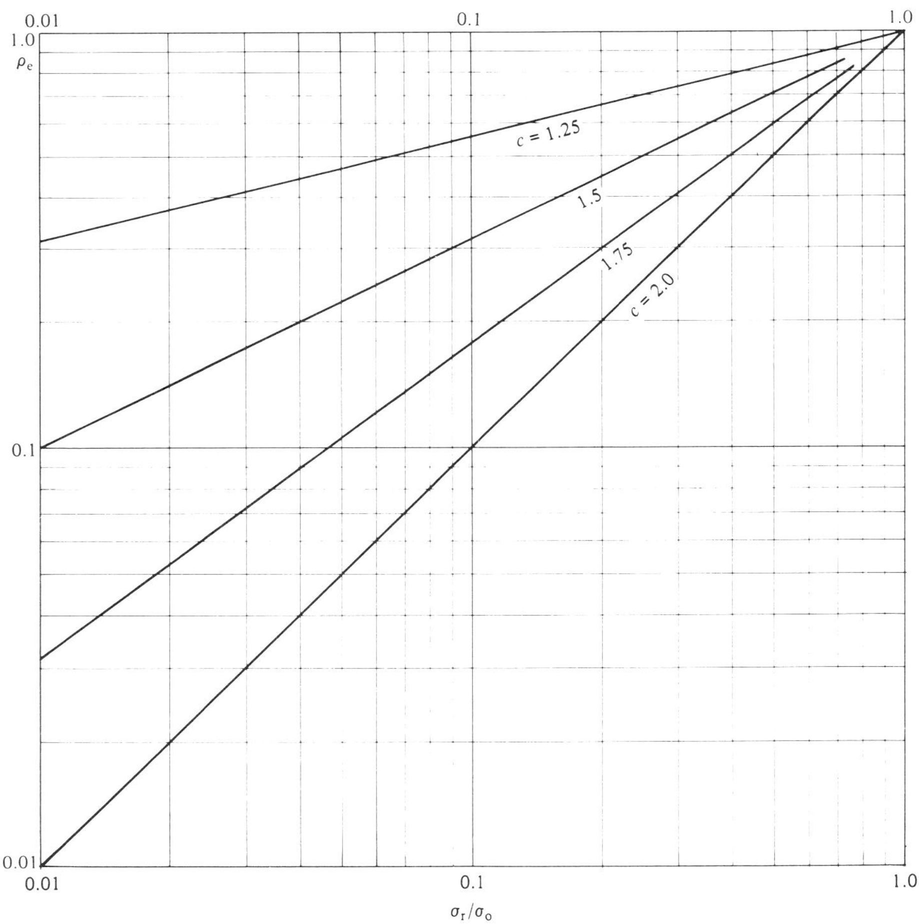

Fig. 6-VII.5 Expansion factor ρ_e.

The recompression factor expressed by 21-VII.5 is used to perform corrections for recompressions smaller than the one taking place in the laboratory, and is used to correct the value of m_{vc} in the oedometer or M_{0c} in the triaxial stress-strain test.

To aid the calculations of E-heave and subsequent settlements, the factors ρ_e and ρ_c/ρ_e have been plotted against σ_r/σ_0 for different values of the coefficient c and of κ_{vp} in Figs. 6 and 7-VII.5, respectively. Furthermore, calling $\sigma_r = (\Delta\sigma_{ex})$ the average relief of effective stress due to excavation including the change in the hydraulic conditions due to pumping, and σ_{0i} the average overburden effective stress

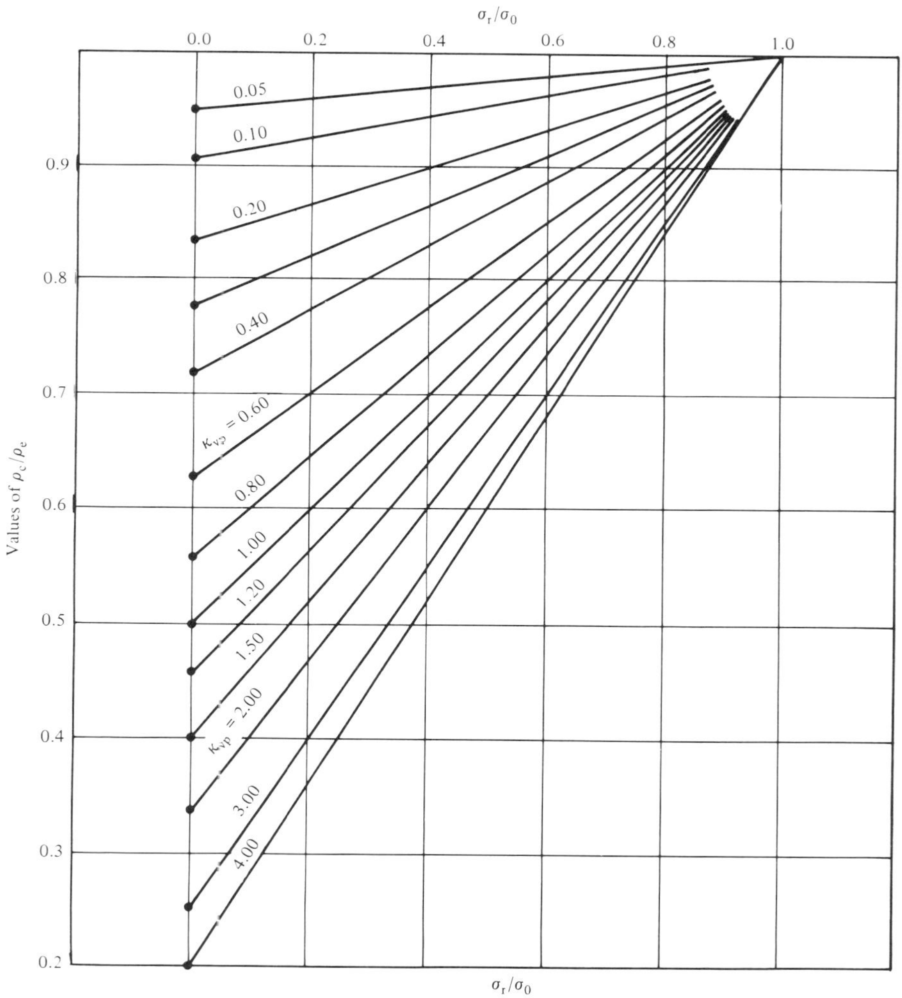

Fig. 7-VII.5 Recompression factor.

for a layer of thickness d_i at depth $(z_i + z_{i+1})/2$, we can write for the heave

$$\rho_{ei} = \left(\frac{\Delta\sigma_{ex}}{\sigma_0}\right)_i^{c-1}$$

Hence, in the case of laterally confined strata to zero horizontal displacement we obtain

$$\delta_e = \sum_1^n \rho_{ei} \nu_c (M_{0c})_i (\Delta\sigma_{ex})_i \, d_i \qquad \text{22-VII.5}$$

and, similarly, for subsequent settlement we have

$$\delta_c = \sum_1^n \rho_{ci} (m_{vc})_i (\Delta\sigma_{ex})_i \, d_i + \sum_1^n (m_{vb})(\Delta\sigma_{ba})_i \, d_i \qquad \text{23-VII.5}$$

The m_v values represent the nominal average recompression coefficient of unit volume compressibility for each stratum in the range of the stress change expected; the coefficient is also a time function as discussed in section II.3, where its value is found for impermeable soils exhibiting intergranular viscosity.

The first term in 23-VII.5 represents the recompression of the soil because of stress reapplication. For a totally compensated foundation $(\Delta\sigma_{ba})_i = 0$; hence, the second term is not used. When the subsoil material is of the preconsolidated type, one may allow the soil to take, in addition to the overburden effective stress σ_{0i}, an extra increment of stress $(1/G_s)\Delta\sigma_b = \Delta\sigma_{ba}$, according to expression 1-VII.3. In such cases, the additional ultimate settlement is calculated with the second term in expression 23-VII.5.

VII.6 LATERAL CONTRACTION AND SETTLEMENT OUTSIDE THE EXCAVATION

The other important problem to consider during excavations is the horizontal compression of the soil at the bottom of the excavation. This phenomenon is interrelated with E-heave since, when the soil mass is relieved of effective vertical stress, horizontal stress relief also occurs (Fig. 1-VII.6). The effective stress change may be calculated at any depth from the theory of elasticity, by means of the stress nets for confining volumetric stress and vertical stresses, respectively (section III.2). When the problem is considered a plane strain problem, the following expressions taken from the theory of elasticity may be used to estimate the relief of the vertical and horizontal stresses.

For vertical stress relief

$$\Delta\sigma_z = -\frac{\Delta\sigma_{ex}}{\pi}\left[\psi + \sin\psi \cos\psi\right]_{\psi_1}^{\psi_2}, \quad \Delta\sigma_z = -\Delta\sigma_{ex} \cdot I_z \qquad \text{1-VII.6}$$

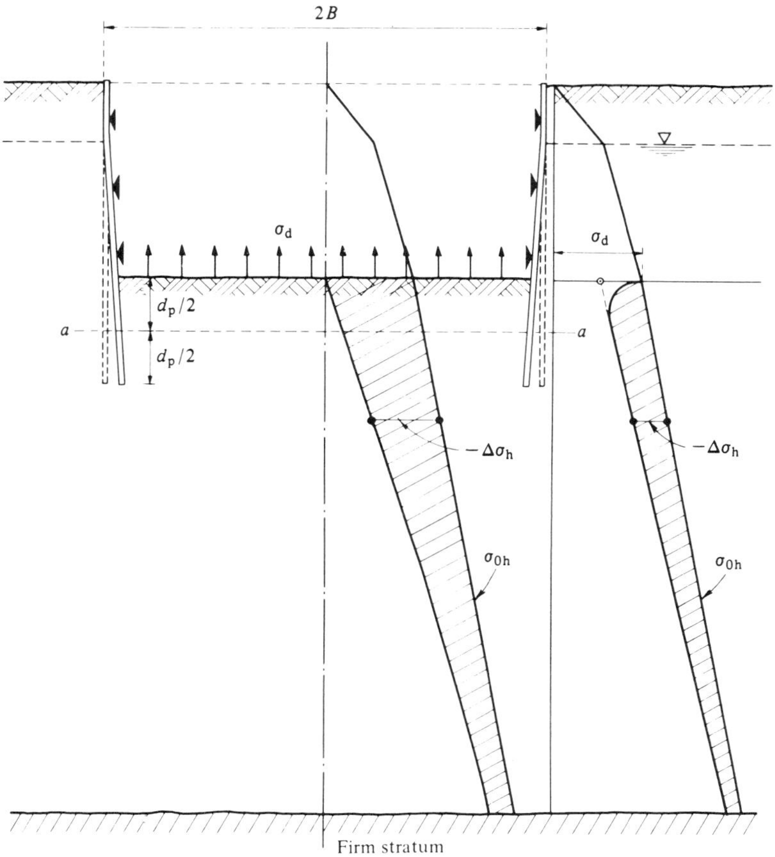

Fig. 1-VII.6 Horizontal stress relief.

for horizontal stress relief

$$\Delta\sigma_h = -\frac{\Delta\sigma_{ex}}{\pi}\,[\psi - \sin\psi\cos\psi]_{\psi_1}^{\psi_2},\quad \Delta\sigma_h = -\Delta\sigma_{ex}\cdot I_h \qquad \text{2-VII.6}$$

and for shear

$$\Delta\tau = -\frac{\Delta\sigma_{ex}}{\pi}\,[\sin^2\psi]_{\psi_1}^{\psi_2} \qquad \text{3-VII.6}$$

Notice from 3-VII.6 that the shear stress assumes a maximum at points located along a circle with diameter $2B$ and center at the foundation grade elevation center; its value is $\Delta\tau = -\Delta\sigma_{ex}/\pi$. This is an important observation in foundation engineering.

In Fig. 2-VII.6 the values of the influence coefficients I_z and I_h are plotted against z/B for vertical planes located at the center and edge of the excavation, respectively.

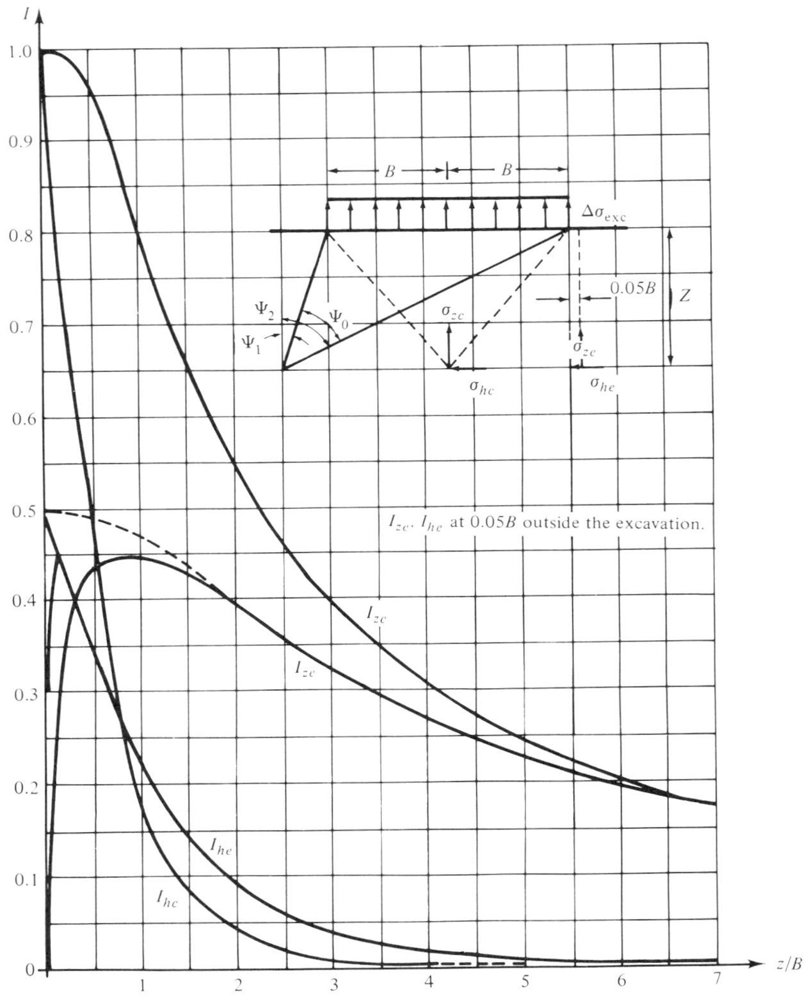

Fig. 2-VII.6 Influence coefficients for horizontal and vertical stress relief for a strip excavation.

Let us assume we have calculated the horizontal stress relief at any depth, at the center $\Delta\sigma_{hc}$ and edge $\Delta\sigma_{he}$, respectively. Therefore, the average horizontal stress relief at that depth is

$$\Delta\sigma_{ha} = \frac{1}{3}\,(\Delta\sigma_{he} + 2\Delta\sigma_{hc})_{hz} \qquad \text{4-VII.6}$$

and for the same depth calling M_{ha} the average horizontal linear strain modulus, then the total horizontal displacement at each side of the excavated area of width

$2B$ at any depth z from the bottom:

$$\delta_{hz} = [\rho_e M_{ha} \Delta\sigma_{ha}]_z \cdot B \qquad \text{5-VII.6}$$

To estimate the average horizontal displacement of the sheet pile wall at the bottom of the excavation, we compute the average horizontal stress relief at a depth corresponding to the middle depth of the buried portion of the sheet piles (level a–a, Fig. 1-VII.6). When the subsoil of compressible material is stratified with less compressible soils, one can estimate the average horizontal strain modulus from values determined in the laboratory for each one of the layers; calling M_{hi} the horizontal linear strain modulus of one stratum with thickness d_i, then the average horizontal strain modulus is given by

$$M_{ha} = \frac{\sum_1^n d_i}{\sum_1^n \frac{d_i}{M_{hi}}} \qquad \text{6-VII.6}$$

Fig. 3-VII.6 Lateral displacement of timber sheet pile.

Hence, the horizontal displacement of the sheet pile wall toward the excavation is

$$\delta_h = \rho_{ea} M_{ha} \cdot \Delta\sigma_{ha} \cdot B \qquad \text{7-VII.6}$$

The induced settlement taking place on the sides of the excavation may damage adjoining construction; therefore it should be carefully investigated. The downward water flow induced by pumping from the bottom of excavation increases the effective stresses in the soil outside the sheet pile wall. This change should also be included in the calculation of the horizontal compression, and may be estimated from studies of flow nets, by calculation of the increment in effective stresses due to the change in piezometric water pressures (see Chapter XI).

The lateral displacement of the sheet pile wall may be reduced if shoring is placed carefully under pressure, as excavation proceeds. A case is shown in Fig. 3-VII.6 where a very large lateral displacement of the sheet pile took place owing to lack of proper bracing across the width of the excavation. The sheet piles are driven to a depth necessary to compensate for the downward friction forces exerted on them by the earth pressure. When subsidence outside of the excavation takes place, the adjacent construction may be damaged. The foundation engineer should investigate the tolerances in distortion that may be permitted in adjacent buildings, according to their weight and foundation structure. In stratified subsoils with sand strata, water under pressure is injected in them outside of the excavation to counteract the downward seepage forces, thus preserving the original effective stress, and minimizing the vertical displacements outside of the excavation.

VII.7 METHODS TO REDUCE HEAVE

The phenomenon of heave and horizontal contraction can be minimized if the change in effective stress relief is made as small as possible. If it were possible to place the foundation structure by total substitution replacing step by step the weight of the building for the weight of soil excavated, then δ_e, δ_h and also the subsequent settlement δ_c would be very small. The excavation, however, can be performed to final depth D in two stages, that is to say, the overburden effective stress can be removed in two parts:

$$\Delta\sigma_{ex} = (\Delta\sigma_{ex})_1 + (\Delta\sigma_{ex})_2 \qquad \text{1-VII.7}$$

in which $(\Delta\sigma_{ex})_1$ is the effective stress removal to depth D_1, corresponding to the first stage of the excavation, and $(\Delta\sigma_{ex})_2$ the effective stress removal corresponding to the second stage of excavation D_2; however, the latter is performed substituting step by step the weight of the foundation structure and building as excavation proceeds.

Hence, for total pressure compensation we may write

$$q_a = [(\Delta\sigma_{ex})_1 + \Delta u_1] + (\Delta\sigma_{ex})_2 + (u - \Delta u_1) \qquad \text{2-VII.7}$$

where u is the uplift water pressure at the foundation grade elevation and Δu_1 is

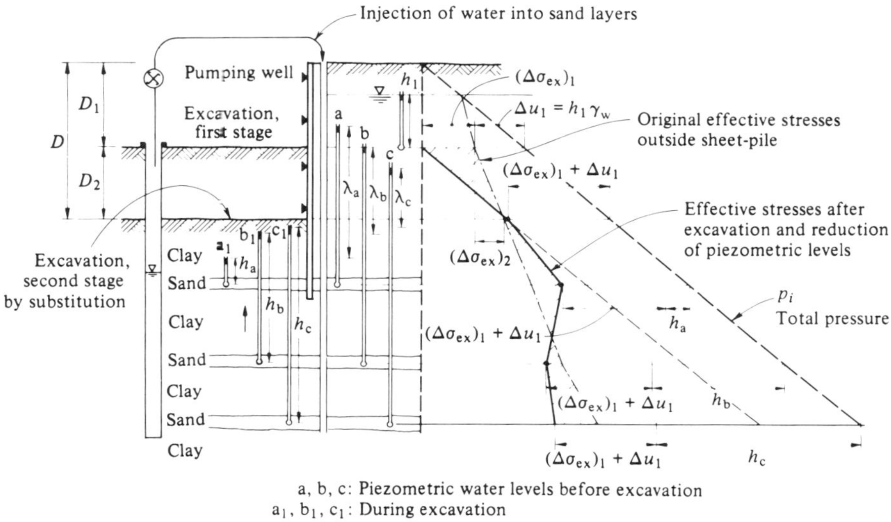

Fig. 1-VII.7 Injection of water into sand layers outside the excavation.

that corresponding to depth D_1. Thus we see that the term in the brackets is the one producing E-heave and correspondingly horizontal soil displacements, as studied before. Therefore, this term should be made as small as possible. When the subsoil is stratified, this may be easily achieved reducing the water pressure below the bottom of the excavation; for this purpose, a sand layer may be used (Fig. 1-VII.7). The piezometric water pressure is reduced to a minimum in the following proportion:

$$\lambda_a \gamma_w = (\Delta\sigma_{ex})_1 + h_1 \gamma_w \tag{3-VII.7}$$

where λ_a is the necessary drop of piezometric head in sand layer a; see section XI.4 on dewatering of excavations.

To gain economy, in deep excavations certain allowable and controlled changes of effective stresses and displacements are permitted during construction. To reduce the change in effective stresses to a minimum, the author (1948) introduced a procedure to perform excavations in soft silty clay to a depth of 13 m for the foundation construction of the Latino Americana Tower in Mexico City. In this case, the piezometric water pressures were strongly reduced by pumping from deep water wells located in the excavated area inside the sheet pile wall, and the water thus obtained injected into the sand layers outside of the excavation to preserve the existing effective stresses, avoiding settlement of adjacent property (Figs. 1 and 2-VII.7).

The increase in effective pressure in the foundation area using a water pressure drawdown may be used to larger depths when stratigraphy of the subsoil is favorable and when the foundation design calls for it, achieving in this manner a reduc-

Fig. 2-VII.7 Adjacent building on north side of 13-m-deep excavation in soft clay for LAT foundation.

tion of heave during excavation and subsequent large settlements of the building. After the second part of the excavation, that is, after the step by step substitution of the weight of the structure for that of the soil removed, the water piezometric level is permitted to rise gradually at the same time the building weight increases, and until full building load is applied, and final equilibrium is obtained when the water table and subsoil piezometric levels are fully restored. See the example of a heave calculation in Appendix E to Chapter VII.

The pictures in Figs. 3 to 7-VII.7 illustrate several stages of construction for a large and deep compensated foundation. The process of shoring the sheet pile, and installation of bracing levels corresponding to first stage of excavation is shown in Figs. 3 and 4-VII.7. The water pressure at this stage is reduced to final position. The excavation in trenches to achieve the weight substitution of concrete for soil is shown in Fig. 5-VII.7. The beams may be seen already poured in the trenches in Fig. 6-VII.7. Thereafter, the panels between beams are progressively excavated and the construction of the foundation slab to final grade finished. The panels are filled up immediately with water to load them, and substitute for the excavated soil weight, Fig. 7-VII.7.

Fig. 3-VII.7 First and second shoring levels for a deep excavation.

VII.8 OVERTURNING MOMENT AND BASE SHEAR

8.1 Introduction

Compensated foundations may be used for slender buildings provided the stresses induced by overturning moments do not produce intolerable permanent tilting of the building. Overturning moments may be induced in the foundation in a fraction of a second by earthquakes or wind forces, that is to say, as transient loadings. Under this action the deformation because of intergranular viscosity does not play an important role, the problem becomes only one of elasto-plastic deformation based on the elastic response of the soil skeleton structure. For this purpose, the ratio of elasticity to plasticity dynamic behavior should be investigated in soil samples subjected to periodic loading, applied at stress levels from 1/4 to 1/3 of the ultimate shear strength, and in periods of about 0.3 to about 3.0 seconds, thus covering the fundamental periods of vibration in buildings (Chapter XII.5.2).

8.2 Elastic Response

Assume it is required to estimate the stresses induced by an overturning moment (Fig. 1-VII.8). The resultant force producing this condition is applied at height *H*

Fig. 4-VII.7 Shoring of a wakefield sheet-pile wall for a deep excavation.

from the foundation slab. At the foundation grade elevation this force produces a base shear V_b and an overturning moment O_T. Under these circumstances, one may analyze the mechanics of the problem under the following working assumptions:

1. The foundation box is rigid.
2. The base shear is resisted by the soil at the foundation slab elevation with an ample factor of safety. Hence, no slippage takes place between soil and foundation slab.
3. The soil response is considered elastic.
4. The rotation of the box foundation takes place around the middle point, at the elevation of the foundation slab.
5. The soil is compressed on one side of the retaining wall with a uniform stress distribution.
6. The increment in vertical soil reaction induced by the overturning moment at the base of the foundation structure has a straight line variation, and may be added to the static subgrade reaction distribution.

According to these assumptions, the overturning moment at the foundation grade

Fig. 5-VII.7 Excavation in trenches to cast in place foundation beams.

elevation may be expressed as follows in terms of reactions R_v and R_h:

$$O_T = \tfrac{1}{2} dR_h + \tfrac{4}{3} BR_v \qquad \text{1-VII.8}$$

From geometry (Fig. 1-VII.8) we have

$$\delta_h = \tfrac{1}{2} d\Delta\theta, \quad \text{and} \quad \delta_v = \tfrac{2}{3} B \cdot \Delta\theta \qquad \text{2-VII.8}$$

Let us call K_{hl} and K_{vl} the linear foundation moduli for the load distribution imposed in the horizontal and vertical directions, respectively; then

$$R_h = K_{hl} \cdot \delta_h, \quad \text{and} \quad R_v = K_{vl}\delta_v \qquad \text{3-VII.8}$$

Fig. 6-VII.7 Foundation beams already constructed in the trenches.

Hence

$$\frac{R_h}{R_v} = \frac{K_{hl}}{K_{vl}} \cdot \frac{\delta_h}{\delta_v} \qquad \text{4-VII.8}$$

Substituting values given by 2-VII.8, we get

$$\frac{R_h}{R_v} = \frac{3}{4}\frac{K_{hl}}{K_{vl}} \cdot \frac{d}{B} \qquad \text{5-VII.8}$$

and using equilibrium equation 1-VII.8 and calling $\lambda_{Kl} = K_{hl}/K_{vl}$ we obtain

$$R_h = \frac{2}{d\left[1 + \dfrac{32}{9\lambda_{Kl}} \cdot \dfrac{B^2}{d^2}\right]} O_T, \quad R_v = \frac{3}{4B\left[1 + \dfrac{9}{32}\lambda_{Kl}\dfrac{d^2}{B^2}\right]} O_T \qquad \text{6-VII.8}$$

From equation 6-VII.8, one should recognize the importance of designing the foundation deep into the ground to achieve good stability of the building. The in-

Fig. 7-VII.7 Panels filled with water to replace weight of excavated material.

crement in the edge stresses may be determined by

$$\Delta\sigma_{\text{h}} = \frac{R_{\text{h}}}{d}, \quad \text{and} \quad \Delta\sigma_{\text{v}} = \pm\frac{2R_{\text{v}}}{B} \qquad \text{7-VII.8}$$

From 2 and 3-VII.8 the angle of rotation is given by

$$\Delta\theta = \frac{3}{2}\frac{R_{\text{v}}}{K_{\text{vl}}\cdot B} \qquad \text{8-VII.8}$$

Moreover calling O_{TB} the overturning moment taken by the foundation slab, then the rotational linear foundation modulus is:

$$K_{\theta B} = \frac{O_{\text{TB}}}{\Delta\theta}, \quad \text{or} \quad K_{\theta B} = \frac{8}{9}K_{\text{vl}}B^2 \qquad \text{9-VII.8}$$

In the same way for the wall:

$$K_{\theta\text{W}} = \frac{O_{T\text{W}}}{\Delta\theta}, \quad \text{or} \quad K_{\theta\text{W}} = \frac{1}{4}K_{\text{hl}}d^2 \qquad \text{10-VII.8}$$

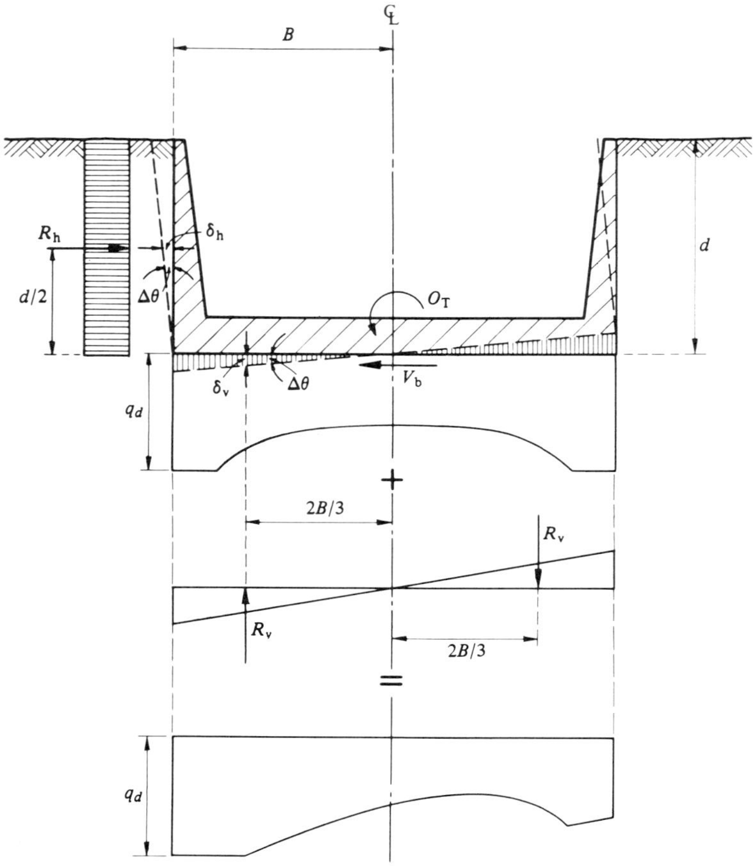

Fig. 1-VII.8 Stresses induced by an overturning moment.

Therefore:

$$\frac{K_{\theta W}}{K_{\theta B}} = \frac{9}{32}\frac{K_{hl}}{K_{vl}}\frac{d^2}{B^2} \qquad \text{11-VII.8}$$

and calling $\lambda_{K\theta} = K_{\theta W}/K_{\theta B}$, we may write formula 6-VII.8 in the form

$$R_h = \frac{2\lambda_{K\theta}}{d(1+\lambda_{K\theta})}O_T, \quad \text{and} \quad R_v = \frac{3}{4B(1+\lambda_{K\theta})}O_T \qquad \text{12-VII.8}$$

and the elastic rotation of the foundation from equations 8 to 12-VII.8 is

$$\Delta\theta_e = \frac{O_T}{K_{\theta B}(1+\lambda_{K\theta})} \qquad \text{13-VII.8}$$

The values of $K_{\theta B}$ and $K_{\theta W}$ per unit length of foundation may be investigated by calculating the vertical and horizontal displacements produced, respectively, by moments O_{TB} and O_{TW}, applied to the soil deposit, taking into consideration the elastic response mechanical properties of the subsoil strata, and with the aid of formulas 51 to 53-III.2, given in Chapter III, from which the value of $\lambda_{K\theta}$ may be determined. For shallow foundations $\lambda_{K\theta} = 0$.

In the case of an homogeneous and isotropic soil mass of semi-infinite extent, Fröhlich (1953) has found, based on the theory of elasticity, that the rotation of a shallow rigid strip footing of width $2B$ subjected to a bending moment $\overline{M}$ per unit length of footing is given by

$$\tan \nu = \frac{4}{\pi}(1 - \nu^2)\frac{\overline{M}}{EB^2} \qquad \text{14-VII.8}$$

Using the preceding nomenclature we have $\tan \nu = \Delta\theta_e$, the response elastic strain modulus $M_{ev} = 1/E$ and calling $\overline{M} = O_{TB}$, we can write for a saturated impervious soil where $\nu = 0.5$:

$$\Delta\theta_e = \frac{3}{\pi}\frac{O_{TB}}{B^2} \cdot M_{ev} \qquad \text{15-VII.8}$$

from which the linear rotational foundation modulus of the footing:

$$K_{\theta B} = \frac{\pi B^2}{3M_{ev}} \qquad \text{16-VII.8}$$

Formula 16-VII.8 may be used when the soil is approximately isotropic and of unlimited extent; however, its limitations should be taken into consideration if the subsoil is stratified and M_e cannot be considered constant with depth (Chapter XII.5.2).

The horizontal displacement at mid-height because of rotation of the wall may be calculated approximately assuming:

$$\delta_h = M_{eh}(\tfrac{1}{2}d)\Delta\sigma_h \qquad \text{17-VII.8}$$

Therefore $K_{hl} = 2/M_{eh}$, where M_{eh} is the average elastic response strain modulus of the soil in the horizontal direction. The rotational linear wall modulus is accordingly

$$K_{\theta W} = \frac{1}{2}\frac{d^2}{M_{eh}} \qquad \text{18-VII.8}$$

Using formula 16-VII.8, we find:

$$\lambda_{K\theta} = \frac{3}{2\pi}\frac{d^2}{B^2}\left(\frac{M_{ev}}{M_{eh}}\right) \qquad \text{19-VII.8}$$

If one assumes $M_{ev}/M_{eh} = 1$, then $\lambda_{K\theta} = (3/2\pi)(d^2/B^2)$. Formulas 15 and 19-VII.8 may be used for estimates. Overturning moments in connection with seismic engineering problems are treated in Chapter XII.

8.3 Permanent Tilt

The permanent tilt of buildings subjected to earthquakes and wind should be investigated for the first cycles of vibration. Since the soil behavior is elasto-plastic, the permanent rotation of the foundation is estimated from the following expression:

$$\Delta(\Delta\theta)_d = \Delta\theta_{ep} - \Delta\theta_e \qquad \text{20-VII.8}$$

in which $\Delta\theta_{ep}$ is the rotation calculated with elasto-plastic parameters based on the first cycles of loading applied to the soil, as in the case when the building is subjected to several strong oscillations during earthquakes.

Using for illustration the case of a strip foundation on a semi-infinite isotropic mass, where

$$\Delta(\Delta\theta)_d = \frac{3}{\pi}\frac{O_{TB}}{B^2}(M_{ep} - M_{ev})$$

and calling $\kappa_{ep} = M_{ep}/M_{ev}$ we can write

$$\Delta(\Delta\theta)_d = \Delta\theta_e[\kappa_{ep} - 1] \qquad \text{21-VII.8}$$

The investigation of κ_{ep} may be performed by means of dynamic unconfined compression tests, studying the elastic and plastic deformation produced by hysteresis loop cycles for stresses below one-half of the ultimate shear strength, or in the range of stress increments compatible with those expected in the ground (Fig. 2-VII.8). The value of κ_{ep} is, however, a function of the stress level, up to 1/2 of the unconfined compressive strength. In Mexico City a maximum permanent tilt corresponding to a rotation of 0.0025 is allowed.

For numerical illustration assume a building with a compensated foundation placed on a compressible isotropic soil deposit extending to great depth. The foundation is 6 m below ground surface and has a width of 20 m. The height of the building is 60 m from the foundation grade elevation. The maximum overturning moment per unit length of the building during a strong earthquake is estimated in 500 ton · m. From dynamic unconfined compression tests performed at a stress level $\Delta\tau \leqslant (1/4)q_u$, the average elastic strain modulus is found to be $M_{ev} = 0.01\ \text{cm}^2/\text{kg}$, and $\kappa_{ep} = 1.20$. Using 15-VII.8, the elastic rotation is calculated not considering the foundation depth:

$$\Delta\theta_e = \frac{3}{\pi}\frac{500}{(10)^2}\,0.001 = 0.0048$$

The value of $\lambda_{K\theta}$ can be computed from 19-VII.8. Assuming that $M_{ev}/M_{eh} = 2$ we have

$$\lambda_{K\theta} = \frac{3}{2\pi}\left(\frac{6}{10}\right)^2 \cdot 2 = 0.345$$

Hence, the elastic rotation considering the foundation depth is obtained from 13-VII.8: $\Delta\theta_e = 0.0048/1.345 = 0.00356$, from which the elastic vertical displace-

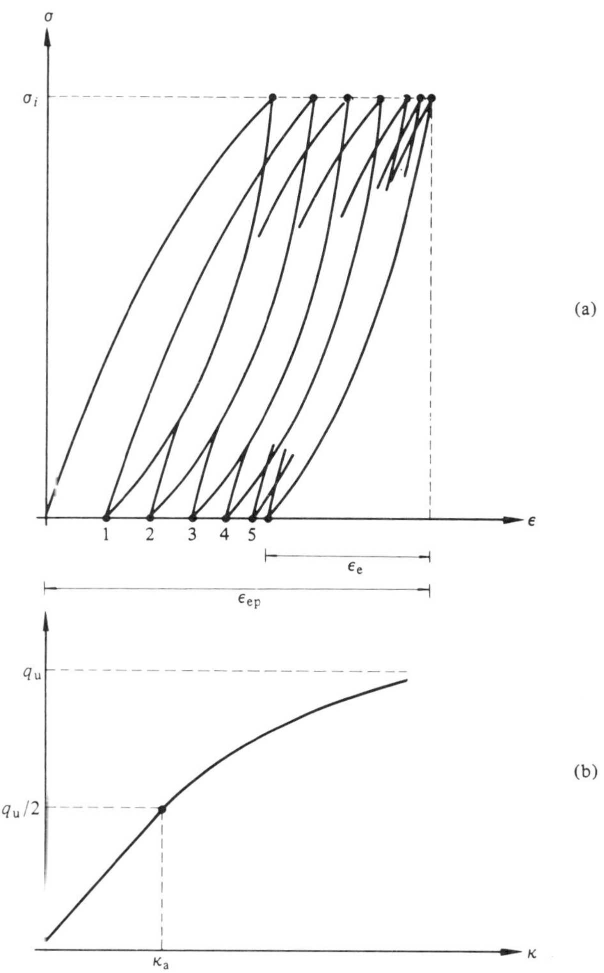

Fig. 2-VII.8 Hysteresis cycles to determine κ_{ep} values.

ment at the edge is 3.56 cm. By means of formula 21-VII.8 the permanent rotation of the foundation structure may be estimated: $\Delta(\Delta\theta)_d = 0.00356(1.20 - 1) = 0.00071$. Therefore, the permanent tilt at the top of the building after the strong motion has taken place will be on the order of 4.3 cm. This is acceptable, since it is considerably less than the allowable of 1/4 of 1% or 15 cm.

The value obtained for the permanent tilt may be considered on the safe side, since usually the subsoil is stratified and limited in depth. Therefore, to obtain a better approximation the foundation modulus should be determined under the real subsoil conditions encountered and the stress levels at which the soil will be subjected (Chapter XII.5.2).

8.4 Base Shear

According to point (2), section VII.8.2, of the assumptions, the shear strength parameters should be determined by means of consolidated undrained tests, from which the average shear strength at the foundation grade elevation is:

$$s_{cu} = c_{cu} + \frac{W}{A} \tan \phi_{cu} \qquad \text{22-VII.8}$$

in which A is the foundation area and W is the total weight of the building. The total base shear strength is $S = s_{cu} \cdot A$. Calling V_b the base shear induced by earthquake or wind, then the safety factor is

$$G_s = \frac{S + E_p}{V_b} \qquad \text{23-VII.8}$$

in which E_p is the passive earth pressure against the retaining wall. If 23-VII.8 is satisfied, then the rest of the working assumptions so far discussed in this article apply.

VII.9 BENDING MOMENTS AND SHEARS IN THE FOUNDATION STRUCTURE

Compensated foundations placed deep into the ground to accommodate two or three basements for heavy buildings are usually designed as rigid box type raft foundations forming one or several units. The deflection of the foundation structure is usually small if compared with the vertical displacements produced in highly compressible soil deposits. Therefore, for practical purposes, one may estimate the bending moments and shears in the foundation structure on the assumption that the vertical displacement is constant all over the loaded area (sections VI.4 and VI.2). In long buildings, however, the bending moments are found to be very large because of this assumption. Therefore, it will be necessary to investigate the shears and bending moments owing to foundation structure stiffness. To reduce high bending moments, one can design a semiflexible foundation using joints of the type described below. The foundation structure is formed by a series of rigid pontoon-like sections joined together to fulfill the conditions that have been imposed by the superstructure design and allowable differential settlements.

The computation of bending moments and shears in the foundation structure is beyond the scope of this book, and falls within structural engineering design procedures. It is important, however, for the foundation engineer to know the general philosophy applied to estimate the changes in contact stresses when joints are introduced in a rigid foundation. From the final corrected subgrade reactions due to pontoon action, it is possible to calculate the mechanical elements for design of the foundation structural frame. Calling, at a specific joint, M the bending moment, S the shear and T the torsional moment produced by the reactions and loads acting on the foundation structure, then we may use the joints shown in Table 1-VII.9 in the design of a pontoon-like compensated foundation.

TABLE 1-VII.9

Joint Type	M	S	T
I	0	0	0
II	0	0	$T \neq 0$
III	0	$S \neq 0$	$T \neq 0$
IV	M_p	$S \neq 0$	$T \neq 0$

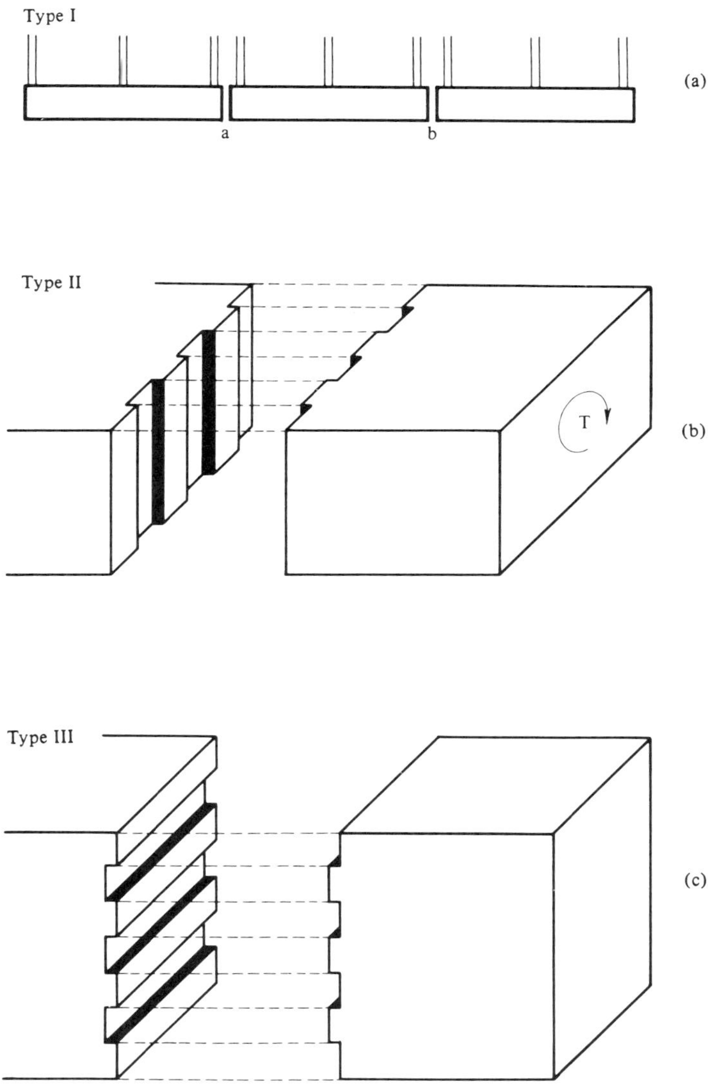

Fig. 1-VII.9 Types of joints used in foundation design.

The type I joint is a free point, and no forces are transmitted from one pontoon unit to the other. Therefore, the units are free to settle and rotate away from each other (Fig. 1a-VII.9).

The type II joint is able to transmit only torsion from one pontoon unit to the other, however, permitting differential vertical displacements between pontoons (Figs. 1b and 2-VII.9).

The type III joint is designed in such a way that shear and torsion may be transmitted, but the moment is maintained zero (Fig. 1c-VII.9). This joint permits only rotation between the pontoon rigid foundations.

The type IV joint is designed to take shear, torsion and a fixed plastic bending moment. The joint is designed structurally to fulfill this purpose. It is mainly used when it is necessary to adjust differential settlements to allowable values, which is not possible with the other types of joints. Using the joints described, especially types III and IV, the foundation engineer is able to reduce bending moments to gain economy in the foundation structural frame. The allowable differential settlements, however, should not be overlooked.

When designing a long compensated foundation structure, first the general bending moments, shears and torsion in the structural foundation frame are calculated in the longitudinal direction, estimating the contact stresses against the foundation slab. The procedures recommended in section IV.4 may be used. In long and deep

Fig. 2-VII.9 Foundation joint, type II.

compensated foundations, the stiffness of the retaining walls may be used to advantage in some specific cases, but in general, this may result in being uneconomical because of the large amount of longitudinal reinforcing steel required. In many cases this procedure is unnecessary, when the allowable differential vertical displacements are fulfilled by the foundation structure. In such cases, the walls are cut, providing them with numerous waterproof type I joints.

Consider a long foundation structure acted upon by loads and corresponding subgrade soil reactions. The bending moments have been calculated as if the structure had an infinite stiffness (Fig. 3a-VII.9). To reduce the high bending moments, several joints of type III or IV are introduced, dividing the long foundation structure into small and rigid sections or pontoons, thus giving to the system more flexibility (Fig. 3b-VII.9). The bending moments at the joints between pontoons are permitted to relax to fixed known values, and accordingly corrections are made to shears and subgrade reaction at the joints. The change in subgrade reactions in each pontoon is assumed as rectilinear variation. Considering the n-th pontoon, the change in reaction at the i-th and $(i + 1)$-st joints is, respectively

$$(\Delta q_i)_n = \frac{\Delta S_i - \Delta S_{i+1}}{A_n} - \frac{\Delta M_i + \Delta M_{i+1}}{S_n} + \frac{\Delta S_i + \Delta S_{i+1}}{S_n} l_n$$

1-VII.9

$$(\Delta q_{i+1})_n = \frac{\Delta S_i - \Delta S_{i+1}}{A_n} + \frac{\Delta M_i + \Delta M_{i+1}}{S_n} - \frac{\Delta S_i + \Delta S_{i+1}}{S_n} l_n$$

in which ΔS_i and ΔS_{i+1} are the shears necessary, at joints i and $i + 1$, respectively, to bring the n-th pontoon into compatibility with the change in subgrade reactions. Similar expressions may be written for each one of the pontoons in the system. The following conditions are established for the change in reactions: for pontoons $n - 1$ and n, at the i-th joint:

$$[\Delta q_i]_{n-1} = [\Delta q_i]_n \qquad \text{2-VII.9}$$

for pontoons n and $n + 1$, at the $(i + 1)$-st joint:

$$[\Delta q_{i+1}]_n = [\Delta q_{i+1}]_{n+1} \qquad \text{3-VII.9}$$

For the n-th pontoon we define

$$a_n = \left[\frac{l_n}{S_n} - \frac{1}{A_n}\right] = \frac{2}{A_n}$$

$$b_n = \left[\frac{l_n}{S_n} + \frac{1}{A_n}\right] = \frac{4}{A_n}$$

$$c_n = \frac{\Delta M_i + \Delta M_{i+1}}{S_n} \qquad \text{4-VII.9}$$

$$S_n = \tfrac{1}{3} A_n l_n$$

$$A_n = 2B(2l_n)$$

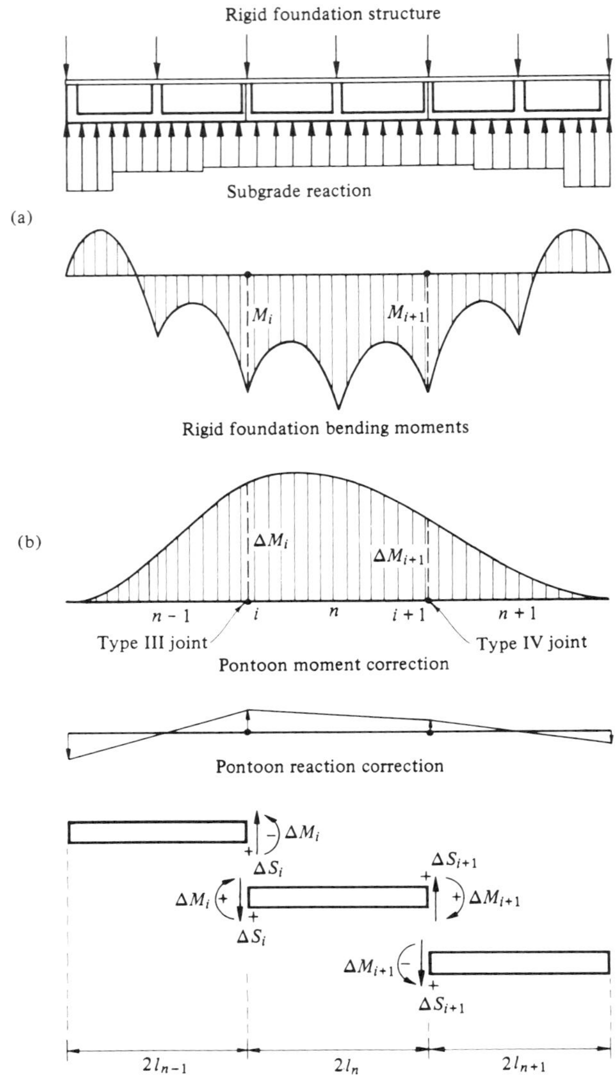

Fig. 3-VII.9 Pontoon foundation.

The expression for the i-th joint takes the form

$$a_{n-1}\Delta S_{i-1} + (b_{n-1} + b_n)\Delta S_i + a_n \Delta S_{i+1} = (c_{n-1} + c_n) \qquad \text{5-VII.9}$$

and similarly for all joints; thus we obtain a set of equations equal in number to the number of joints in the system. These equations are solved for the values of ΔS_{i-1},

ΔS_i, ΔS_{i+1}, ... The Δq_i values for each pontoon joint are calculated from 1-VII.9, and correspondingly we obtain the shears and bending moments. In this equation, ΔM_i is the reduction in bending moment required, and is deducted from the total bending moment curve at the i-th joint location (Fig. 3a-VII.9). If one wishes to retain the plastic moment M_{pi}, then the correction will be $\Delta M_i = M_i - M_{pi}$.

To illustrate the procedure to reduce bending moments when several pontoon-like foundations are used to form a semiflexible compensated foundation, let us calculate the case of a foundation with width 12 m and length 60 m.

The large bending moments may be reduced introducing two joints, one of type III and the other of type IV. The long foundation is divided into three equal sections 20 m long. The calculation is shown in Appendix E to Ch. VII, and is self-explanatory under the following stages:

First stage. The subgrade reactions, shears and bending moments shown were obtained with account for the subsoil engineering characteristics by means of the unit foundation moduli method described in section IV.4 and Chapter VI. The foundation structure at this stage is considered of infinite stiffness.

Second stage. The joint selected at point a is of type III; hence $\Delta M_a = M_a$. The moment at this point is reduced to zero. At point b a joint of type IV will be used to preserve the plastic moment M_{pb} required to retain certain stiffness between pontoons 2 and 3, and therefore, $\Delta M_b = M_b - M_{pb}$. To solve this problem, equations such as 5-VII.9 have been established and a set of two equations is obtained satisfying conditions of equal change in reactions at joints a and b, respectively. After solving for ΔS_a and ΔS_b, the change in subgrade reactions is computed from equation 1-VII.9; thereafter the correction for shears and bending moments for each one of the pontoons are calculated, and they are deducted from the first stage of calculation when considering the entire foundation structure rigid. The vertical displacements are analyzed with the corrected soil reactions to verify the design for allowable differential settlements. When the differential settlements are larger than those permitted, larger plastic moments should be retained at joints a and b and the calculations repeated until the proper selection of the plastic moments is made and the allowable differential settlements are satisfied.

VII.10 PRACTICAL CONSIDERATIONS

The foundation design methods that have been described are applicable in practical foundation engineering. It must not be overlooked, however, that for this purpose we have assumed continuity in the strata under the foundation grade elevation. The foundation engineer should be aware that, in nature, nonhomogeneity in the ground is often present. The soil deposits may contain either rigid or softer spots that should be bridged with the foundation structure. Therefore, the stratigraphical characteristics of the sediments under the proposed site should be carefully investigated.

BIBLIOGRAPHY

Brinch Hansen, J. (1966). Some Stress-Strain Relationships of Soils, *Proc. VI ICOSOMEF*, Vol. I, pp. 231–234, Montreal.

Casagrande, A. and Fadum, R. E. (1944). Application of Soil Mechanics in Designing Building Foundations, *Trans. ASCE*, Vol. 109, p. 383.

Cuevas, J. A. (1936). The Floating Foundation of the New Building of the National Lotery of Mexico, *Proc. I ICOSOMEF*, Vol. I, pp. 294–301, Cambridge, Mass.

Fröhlich, O. K. (1953). On the Settling of Buildings Combined with Deviation from their Originally Vertical Position. *Proc. III ICOSOMEF*, Vol. I, p. 362–365, Zurich.

Glick, G. W. (1936). Foundations on the New Telephone Building at Albany, *Proc. I ICOSOMEF*, Vol I, p. 278, Cambridge, Mass.

Golder, Hugh Q. (1963). Floating Foundations for a Factory Building, *The Canadian Consulting Engineer*, Vol. 5, No. 10, pp. 65–69, Toronto, Canada, October.

Golder, Hugh Q. (1965). State of-the-Art of Floating Foundations, *Journal of Soil Mechanics and Foundations Division*, Proceedings ASCE, p. 543.

Habib, P. and Puyo A. (1970). Stabilité des Fondations des Constructions de Grande Hauteur, *Annales de l'Institute Technique du Batiment et des Trauvaux Public*, No. 275, November.

Skempton, A. W. (1971). The Albion Mill Foundations, *Geotechnique* 21, No. 3, pp. 203–210.

Terzaghi, K. (1944). Discussion of Application of Soil Mechanics in Designing Building Foundations by A. Casagrande and R. E. Fadum, *Trans. ASCE*, Vol. 109, p. 427.

Tettinek, W. (1953). A Contribution to Calculating Inclination and Eccentrically Loaded Foundations, *Proc. III ICOSOMEF*, Vol. I, pp. 461–465, Switzerland.

Zeevaert, L. (1944). Discussion to: Application of Soil Mechanics in Designing Building Foundations by A. Casagrande and R. E. Fadum, *Trans. ASCE*, Vol. 109, p. 383.

Zeevaert, L. (1945). Conceptos y Experimentos Fundamentales que se Aplican al Diseño de Cimentaciones en Arcillas Saturadas, *Revista Ingeniería y Arquitectura*, p. 335.

Zeevaert, L. (1943). *The Outline of a Mat Foundation Design on Mexico City Clay*, Proceedings VII Texas Conference on Soil Mechanics and Foundations. January.

Zeevaert, L. (1959). Compensated Foundations, *Proceedings I Panamerican Conference on Soil Mechanics and Foundation Engineering*, Vol. III, pp. 1109–1126, Mexico City.

Zeevaert, L. (1961). Grobe Ausgleichsgründungen auf Vulkanischem Ton mit hoher Zusammendrückbarkeit in Mexico City, *Internationaler Baugrundkursus*, Essen, Germany, July 12–13.

Zeevaert, L. (1972). *Design of Compensated Foundations. State of-the-Art Report*, Joint Committee ASCE and International Association for Bridge and Structural Engineering, Committee No. 11 Planning and Design of Tall Buildings, Lehigh University, Bethlehem, Pa.

Zeevaert, L. (1976), in: H. V. Fang (ed.), *Compensated Foundations Heave and Subsequent Settlement*, pp. 389–416, Envo Publ. Co., Inc.

Zeevaert, L. (1980). *ISE-Interacción Suelo-Estructura de Cimentuciones Superficiales y Profundas.* Publ. LIMUSA, México, D.F.

▸VIII◂ ULTIMATE LOAD CAPACITY OF PILES AND PIERS

VIII.1 INTRODUCTION

When the surface soil deposits exhibit low bearing capacity and the compressibility of the soil is found high for the required load to be placed on the ground surface, the foundation engineer has to investigate deeper strata to support the proposed loads with smaller vertical displacements, therefore making use of a deep foundation with piles or piers. These elements may be constructed of masonry, concrete, steel or timber. The design and behavior of pile and pier foundations have always been the concern of the foundation engineer and may be considered one of the less investigated fields in soil mechanics, because of the difficulty in performing expensive field measurements and tests, and expensive subsoil and laboratory investigations. The design of pile foundations, in most cases, is achieved by means of empirical rules and experience supplemented by simple pile tests and explorations in the field with the use of sounding penetration tools.

In this chapter simplified working hypotheses are introduced on which the foundation engineer can support his decisions in designing deep foundations. The procedures herein presented have been used by the author during many years of professional practice in Mexico City with satisfactory results. Subsoil conditions are far from being homogeneous and isotropic, and therefore, the theories and working hypotheses are no better than the knowledge the foundation engineer obtains from the engineering characteristics of the subsoil materials.

A pile may be defined as a rigid and strong structural member driven by means of hammers or pushed by jacks into the ground causing compaction of the soil in the neighborhood of the pile. In contrast, a pier is an element carrying load to the supporting deposit located at a certain depth from the ground surface. This element is usually cast inside of a prebored well to the depth of the hard stratum. The subsoil usually encountered in pile and pier problems is complex, since these elements are driven through stratified subsoil conditions where the layers may have different mechanical properties. The theories based on assumptions of homogeneity and isotropy of the subsoil are only of academic value in these cases. Each subsoil stratum has engineering characteristics that must be considered in the behavior of pile and pier foundations. Therefore, piles and piers are structural members that have to be analyzed with account for the ultimate load capacity and ability to carry an allowable load under certain total vertical displacement.

Structures resting on pile and pier foundations are, in general, sensitive to settlements because, owing to the assumption that total and differential settlements of structures on these foundations will be of a small order of magnitude, the foundation structure and structural frame is usually not designed to take differential settlements.

Piles should be designed according to their ability to carry load under an allowable settlement in group action, and not by the individual ultimate bearing capacity and vertical displacement obtained in customary single unit tests, since settlement of one pile is by no means representative of the behavior of the group. The mechanical behavior of a pile group has not been investigated thoroughly in the field, owing to the fact that it is less difficult and more economical to investigate the behavior of one pile than it is to investigate the action of a group of piles. In this regard, the only suitable method is to observe the behavior of pile foundations already constructed, and to attempt theoretical correlations with single pile behavior in the area under consideration. Numerous theories to compute ultimate load capacity and settlement of single piles have been proposed in the literature on this subject, with the aim of finding better and more precise means of design and interpretation of single pile tests. Behavior results for a single pile are not to be extrapolated to explain the behavior of large groups of piles. In these cases, it is far more satisfactory to establish simple theoretical working hypotheses taking into consideration the statics of the group of piles in conjunction with the load applied, and then use these hypotheses to make proper decisions on the probable mechanical behavior of the group.

The use of scale model tests has indicated, under ideal laboratory conditions, the behavior of one pile and that of small pile groups, and for correlation efficiency coefficients have been proposed. The results, however, are not satisfactory for practical application, when compared with field observations, except in those particular cases when the dynamic similitude between field conditions and model piles is in general agreement. It is very rare to find natural deposits with ideal isotropic and homogeneous conditions. Many uncertainties arise concerning the use of mechanical properties of the subsoil material in calculations of a single pile and pile groups even considering ideal conditions. In nature, the subsoil deposits are stratified and the mechanical properties assume different values in the vertical and horizontal directions. It is therefore practically impossible, in the case of deep foundations, to set forth design rules to obtain accurate results, since they are highly dependent on the local mechanical properties and stratigraphy of the subsoil materials, especially when difficult subsoil conditions are encountered.

The foundation engineer should be able to recognize the important aspects of pile and pier behavior in conjunction with the mechanical properties of the materials acting on these structural members. The engineer should recognize the difference between the general behavior of a single pile and that of a group of piles, taking into account the role of the ultimate load capacity and vertical displacement in the two cases. When the phenomenon is properly visualized, in our present state of knowledge, simplified working hypotheses may be made on the safe side to establish, in general, the mechanics of the problem giving satisfactory results in foundation engineering practice. The working assumptions established for every job should be tested. The behavior of existing foundations should be studied where these assumptions have been put into practice.

A pile or pier foundation is a group of structural members whose load is supported by the reaction these elements receive from the soil, which takes the pressure at the tip by means of shear stresses and by skin friction along the shaft. Therefore, one should investigate these two ways of behavior of a pile separately, and learn how they act together.

To decide if a pile or pier will develop point resistance and either positive or negative friction along the shaft, it is first necessary to learn if either one of these effects or both of them will be present. This decision depends largely on the hydraulic conditions at the site under consideration, as well as on the stratigraphical and mechanical properties of the subsoil materials. It is well known, in the case of consolidating compressible layers overlying a firm stratum where the tip of the piles is supported, that these elements will be subjected to negative friction (see Chapter V). When this phenomenon is present, special considerations are required to estimate the point resistance. Negative friction in group action cannot be determined by conventional pile tests. It can only be estimated by learning how to apply the properties of the subsoil materials to the mechanics of these structural elements, in conjunction with the dynamic conditions found in the subsoil, and knowing the stratigraphical and mechanical properties of the site in question. Then, by means of theoretical calculations representing the phenomenon in simplified form, the problem may be solved satisfactorily from the viewpoint of practical accuracy.

Many conventional pile foundations are calculated as point bearing piles with the assumption of positive friction on the pile shaft, which leads to the inclusion into the foundation design of a positive force adding to the total capacity of the pile. However, when a group of piles is driven into the ground, the soil is displaced by the piles in the horizontal and vertical directions, and significant ground surface heave takes place which cannot be ignored in the design of the pile groups, especially in those cases where impervious soil deposits are present. Upon dissipation of the excess pore water pressures, the soil tends to resume its initial elevation, and necessarily, the point bearing piles will be working from the very beginning with considerable negative frictional forces. Usually, these forces are not considered in the design, or even worse, the not-present positive friction forces are included in addition to the point bearing capacity calculations.

The difference in considering positive friction instead of negative friction in a point bearing pile is translated in general into subsequent settlements, until the point bearing pile has penetrated sufficiently to change the negative friction into positive friction. In contrast, piles considered as point bearing piles, owing to high compressibility at the point supporting layer, work like positive friction piles. Therefore, it is important also to be able to evaluate the point vertical displacement because of the compressibility under the pile point to decide if this is included or omitted.

To investigate the lateral friction, it is important to study the horizontal effective stresses acting against the pile shaft to be able to estimate the shear strength along the potential surface of sliding. This is an important problem that needs careful investigation in sand, silt and clay materials, since the state of stress induced because of negative friction also affects the point bearing capacity of the pile very strongly, as will be demonstrated in further sections of this chapter.

In case of difficult subsoil conditions, the foundation engineer should be careful when using the results of conventional single pile tests to design group of piles by means of empirical rules. This line of thinking in many instances may be detrimental. Frequent cases of defective work of pile foundations where this criterion has been used are reported in the engineering profession. The author's experience has been that in most cases it is best to visualize the behavior of pile groups by means of representative simple mechanical models, and set forth simple working hypotheses of calculation made compatible with the statics of the group as a whole. The results of single pile tests should be only the guidance and means of learning the expected mechanical properties of the subsoil under different testing conditions, that in conjunction with sounding tests and the direct determination of the mechanical properties of the materials on well taken undisturbed samples, will give the adequate means for proper interpretation of the pile tests, that may be used later for foundation design within the accuracy of practical engineering problems.

The great number of pile forms and empirical systems found in the literature of the past appears to be the result of poor understanding of the mechanical action of rigid elements introduced into the soil, as well as of groups of these elements. Pile forms have developed empirically, as have the dynamic formulas used until the last decades in the attempt to improve piles and methods in the competitive struggle of a pile market.

The author has concluded through his professional experience, as shown in further paragraphs of this chapter, that with good judgment in soil mechanics the foundation engineer is able to design his own pile usually better than any proposed complicated pile offered in the market at the moment, but which may not be suitable for a certain specific problem. In general, it is found that the required pile will be the simplest one compatible with the mechanics of behavior of the pile group and the preservation or improvement of the soil properties where the piles are driven. Usually, the best sections are circular, orthogonal, hexagonal or square. Other shapes of pile sections are not justified. The same may be said for pile points, which will be discussed in further articles concerning point bearing capacity.

From the above discussion, certain general conclusions may be drawn before entering into the calculation of deep foundations. Working hypotheses in the development of theories should be, however, in general agreement with field observations on the mechanical behavior of different types of materials under the pile tips and in the neighborhood of the pile shafts. General agreement has to be investigated by the observation of actual large pile foundations, where the behavior is well known, and where the working assumptions used for calculations are also known and may be compared with observational facts.

The pile foundation problem will be divided into two parts, the stability problem and the elasticity problem. The first concerns the pile or pile group breaking into the ground because it exceeds the shear strength of the soil material where it is supported. The second problem concerns total and differential settlements of a pile or pile group, or pile fields; the engineer must be able to prescribe the foundation structure rigidity necessary to support differential vertical displacements, to avoid damage to the superstructure. The second problem is far more common in practice, since many foundations have failed to work as foreseen because of excessive differential settlement of the pile groups. Therefore, it is necessary to predict and control the vertical displacements, making them compatible with the precision required in foundation engineering and structural design.

The stability problem is understood by the ultimate load capacity of a pile or pier: a) by the point bearing capacity, and b) by the ultimate friction load capacity.

a) The ultimate point bearing capacity is known to be a function of the geometrical dimensions, shape and roughness of the point, confining pressure at the point, shear strength, and the density and compressibility of the soil under and around the point of the pile after driving. In the case of piers or bored piles, the possible alteration of the natural mechanical properties of the material during excavations should not be overlooked.

b) The ultimate friction load capacity is a function of the shear strength of the soil along the pile shaft. In clays the time element is important. The horizontal effective stresses remaining after pile driving because of volume displacements are important, as well as the compaction or consolidation of the soil around the pile.

In clayey soils the remolded material, because of pile driving, consolidates against the pile shaft regaining shear strength. This is easily demonstrated in practical cases when the pile is pushed into silty clay deposits with small effort, and after a certain period, the same pile can take a large load. The surface of sliding in most cases, is not at the interface between soil and pile. The shear surface may be at a short distance from the pile shaft. In clays, the sensitivity is an important factor. Highly sensitive clays never regain their total strength lost during pile driving, and therefore, a minimum strength is encountered at certain distance from the pile shaft. In some cohesive soils, the cohesion may be entirely broken during pile driving, leaving only the friction term. Therefore, it is imperative to investigate the real horizontal effective stresses remaining after pile driving for different subsoil materials. For this purpose, it is also important to know how the excess pore pressures interfere with the phenomenon.

There are empirical methods based on penetration devices discussed in elementary soil mechanics, which in some cases are the only means to estimate the load capacity of piles; however, these methods are highly dependent on the size of the device and characteristics of the subsoil materials. Therefore, large deviations may be expected in practice. On the other hand, the penetration test devices like the standard and cone penetration tests are valuable tools in areas where experience has been gained empirically by their application during a great number of years.

The use of rational methods based on approximate working hypotheses derived from statics to estimate the load capacity and vertical displacement behavior of pile groups is, in the majority of the cases, the only means to obtain sufficient accuracy and proper behavior of pile foundations, and achieve the best possible economy without sacrificing safety. It is obvious that in many cases of difficult subsoil conditions the empirical methods based on penetration devices are not powerful enough, and that careful investigation of the properties of shear strength and compressibility of the materials is necessary. The hydraulic conditions should be well known, especially in areas where ground surface subsidence is taking place.

The negative friction because of such conditions may be recognized by the fact that the vertical effective stresses in the soil mass have the tendency to decrease. On the other hand, the positive friction phenomenon increases the vertical stresses. The frictional forces under these two conditions have different meaning and magnitude. The importance of this action has to be carefully investigated, and properly taken into account when performing the foundation design.

From the discussions given in the preceding paragraphs and taking into consideration our present state of knowledge, it is desirable to mention some general conclusions and recommendations to be considered by the foundation engineer in the attempt to understand more thoroughly the behavior of this type of foundation.

First, it must be kept in mind that the computation of the load capacity of piles or piers and the behavior of large groups will be no more precise than the accuracy that may be obtained in the investigation of the stratigraphical, mechanical and hydraulic conditions of the subsoil materials applied to the problem in question. Careful investigations on actual-size model piles are recommended to gain experience in the application of the parameters of shear strength and compressibility of the soil. The point bearing capacity and lateral skin friction will be investigated separately and the results checked with the working hypotheses on the subject. From careful investigations, one will discover better methods in the application of the factors affecting the bearing capacity parameters. The time element when testing actual-size piles should be carefully considered. It is well known that when using piers, the shear strength is damaged to a certain extent close to the shaft and base of the pier, depending on the sensitivity of the soil.

Single pile loading tests should be planned carefully and special devices used to perform the tests. The best device, in the author's opinion, is one similar to the Dutch cone penetrometer with which the point bearing for actual-size piles could be determined separately from the skin friction. It is important in loading tests on piles to define the critical load, and the meaning of this value may be compared with the so-called ultimate load, so that for the different methods of testing, one may be able to interpret properly the load-settlement curves. It would be highly desirable in the foundation engineering profession to specify standardization, within certain flexibility, of loading tests on piles to gain consistent information. A procedure is proposed in section 6 of this chapter.

Under earthquake conditions, the excess pore water pressure should be measured in the soil mass close to the pile shaft during the time the strong ground motion takes place. The behavior of existing pile foundations in seismic areas that have been subjected to strong ground motions must be investigated after earthquakes have taken place. This problem is treated in Chapter XII concerning seismic behavior in connection with foundation design.

Finally, it may be concluded that the art of successfully designing pile or pier foundations, especially in difficult subsoil conditions, is to learn to apply full knowledge about all environmental engineering and geological factors encountered at the site under consideration, and how these factors are brought together. They have to be investigated one by one, studying their action separately. There is no law of similitude that can represent all the factors in one. Therefore, knowledge should be gained by careful study of the stratigraphical and hydraulic conditions, and the mechanical properties of the subsoil materials in conjunction with actual-size pile tests. The working hypotheses used in design should be tested with observations of the statics and forces involved, and last but not least, with behavior histories for already constructed foundations that have been exposed to the time element and strong ground motions.

VIII.2 POINT BEARING CAPACITY

2.1 Theory Review

Theories on point bearing capacity of piles and piers are still not satisfactory. However, they may be used conveniently if properly adjusted to subsoil conditions and mechanical properties of the site in question with the understanding that a factor of safety should be selected to account for possible inaccuracy of the theories, variation of the soil and stratigraphical properties of the ground under consideration, and changes in the hydraulic conditions.

A single pile is shown in Fig. 1-VIII.2, where only the point resistance will be measured in order to separate the action of point bearing from skin friction. Let us assume the pile has been driven into a bearing stratum separately from the shaft of the pile. The point of the pile has the tendency to compact the material around and under the pile point to about one diameter. When the pile is tested to failure, the potential surface of sliding may be anywhere between surfaces A or B in Fig. 1-VIII.2, according to different theoretical studies and observation on model single pile tests in uniform sand, performed by several investigators and reported in the soil mechanics literature. The potential surface of sliding is also affected by the stratigraphy in the neighborhood of the pile point.

To illustrate the problem, let us start with Terzaghi's approximate proposition on

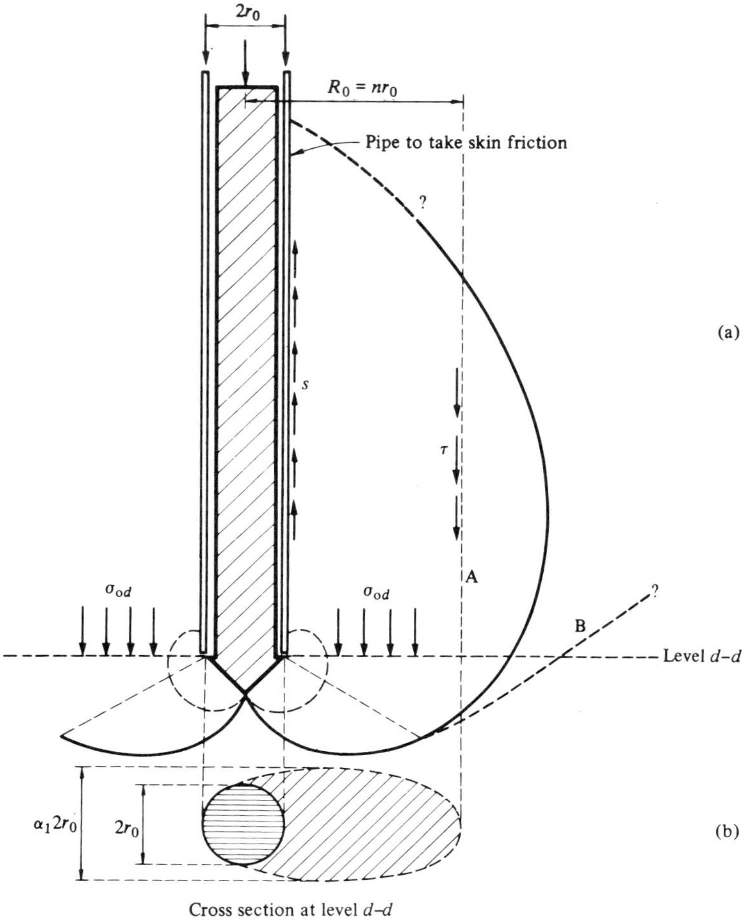

Fig. 1-VIII.2 Ultimate bearing capacity.

the ultimate bearing load capacity for a pier or deep footing:

$$Q_{pu} = \pi r_0^2 \left[1.3cN_c + \gamma_1 D_f N_q + 0.6\gamma_2 N_\gamma r_0\right] \qquad \text{1-VIII.2}$$

$$\gamma_1 = \gamma + 2\frac{f_s + n\tau}{(n^2 - 1)r_0} \qquad \text{2-VIII.2}$$

where $2r_0$ is the diameter of the pier, γ_2 is the unit weight of the soil under the tip of the pier and γ_1 represents the unit weight of the soil above the toe affected by the mobilization of the shear strength along the vertical cylindrical surface, limited by the shaft and by a cylinder of radius nr_0 measured from the center of the pier. The quantity f_s represents the skin friction along the shaft of the pile or the pier, and τ is the shear stress developed on an outer cylindrical surface of radius R_0. The mobilization of f_s and τ depends, however, on the compressibility of the soil above the base of the pier. Moreover, the value of τ does not obtain at failure in the outer total soil cylinder of radius R_0, since from the horizontal plane d-d at the tip of the pile, the channel formed by the potential surface of sliding has the tendency to spread only in a segment, according to the conventional ultimate bearing capacity formula (Fig. 1-VIII.2). Therefore, the value of $n\tau$ in formula 2-VIII.2 is very much smaller and of unknown value, and has little influence on the total bearing capacity of the pile or pier and may be safely neglected. Hence, 2-VIII.2 may be written in the form

$$\gamma_1 = \gamma + \frac{2f_s}{(n^2 - 1)r_0}$$

The theory parameters N_c and N_q are functions of the angle of internal friction and shape and length of the surface of sliding. They are sensitive to the compressibility of the material. The way these parameters vary with the compressibility has not yet been investigated to an extent that definite conclusions may be reached. Therefore, in order to take into account approximately in the values of N_c and N_q the compressibility of the material, they may be multiplied by the relative density D_r, as suggested in section III.4.

Formula 1-VIII.2 may be used also for piles. The effect of the shape and dimensions of the point of the pile have been recently investigated theoretically and experimentally in model tests. Therefore, the ultimate point bearing capacity formula on a semiempirical basis for piles may be written as follows:

$$Q_{pu} = \pi r_0^2 \alpha_1 \; [cN_c + \sigma_{0d} N_q] \; (D_r + 0.1) \qquad \text{3-VIII.2}$$

The factor D_r has the effect of reducing the theoretical values of N_c and N_q because of the compressibility of the soil. When the soil under the pile point is in a dense state, $D_r + 0.1 = 1$, for a semidense state the reduction may be to as low as 0.6, and for a very loose or a very compressible material D_r may reach values as low as 0.30. The value of σ_{0d} is the effective confining vertical stress at the elevation of the horizontal plane d-d corresponding to the pile point. Notice that the first and second terms are affected by the shape factor $\alpha_1 = 1.2$, in contrast with the case of a plane strain problem where $\alpha_1 = 1$. Therefore, the factor α_1 is an approximate measure of the width of the failure channel (Fig. 1b-VIII.2).

By means of equation 3-VIII.2 and well conducted pile tests made in specific soil conditions associated with soil investigations of the mechanical properties of the material and stratigraphy of the site in question, the theoretical values of N_c and N_q may be investigated and calibrated to be used in the design of pile groups.

2.2 Pile Groups

The mechanics of a pile within a group may be visualized differently from that of a single pile (Fig. 2-VIII.2). Consider a pile group to be tested for ultimate load bear-

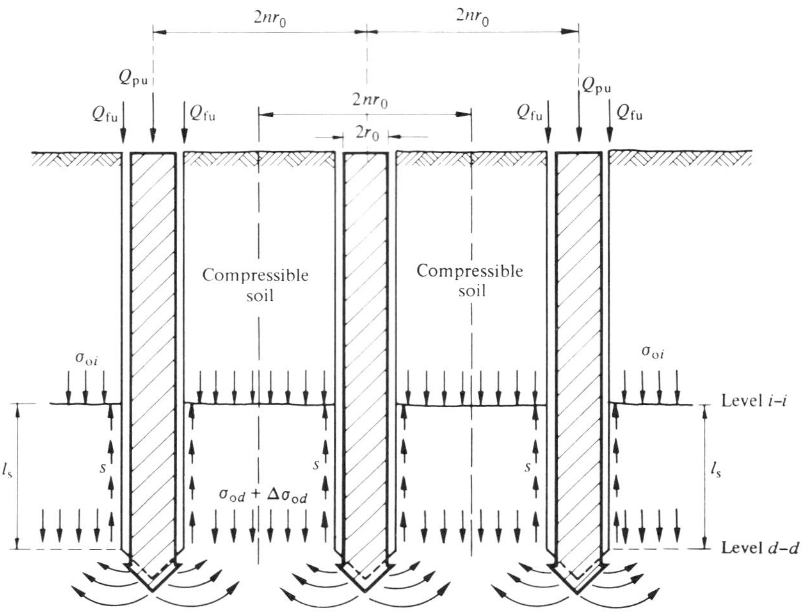

Fig. 2-VIII.2 Pile field.

ing capacity by the same process used for a single pile. The assumption can be made that any pile in the group will show a minimum strength failure channel of approximate width $\alpha_1(2r_0)$ located within the pile group, as shown in Fig. 3-VIII.2, which even with a spacing as small as three diameters will not be sensibly affected by the presence of the other point bearing piles. Under these considerations, formula 3-VIII.2 for ultimate point bearing capacity applies for a pile within the group. The positive frictional forces at the shaft of the pile, close to the point, increase the confining vertical effective stresses and more bearing capacity of the piles is obtained.

In order to investigate this important problem of group action, assume the piles are driven to point bearing with a penetration of l_s into a sand stratum (Fig. 2-VIII.2). When the piles are driven into the proposed sand stratum, the soil is remolded and reconsolidated close to the pile shaft. The cohesion gained after driving may be assumed small. Hence, the shear strength along the pile shaft is reduced to:

$$s_{0z} = c_r + \sigma_{hz} \tan \phi_r \qquad \text{4-VIII.2}$$

where σ_{hz} is the horizontal effective stress, and the shear strength parameters c_r and ϕ_r are the cohesion and angle of internal friction of the remolded and reconsolidated soil close to the pile shaft. The angle ϕ_r assumes the value of a material in a semidense state.

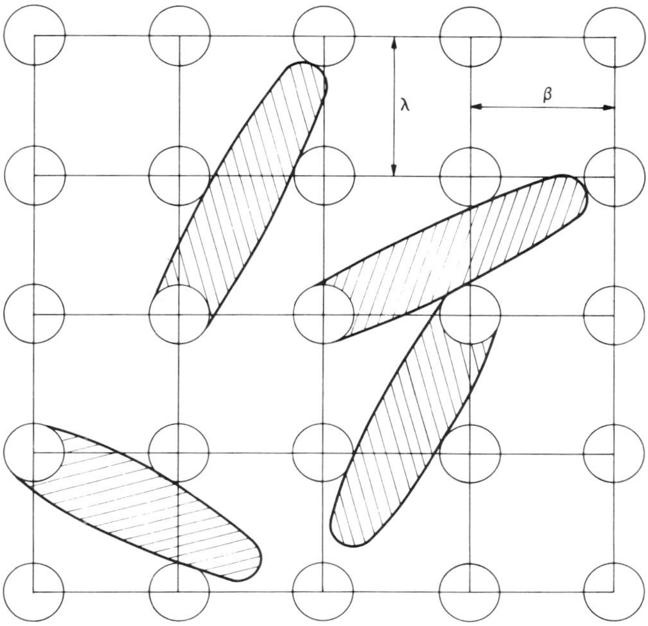

Fig. 3-VIII.2 Possible channels of failure of a pile in a group of piles.

The skin friction force developed in the embedded part of the point of the pile is

$$Q_{fu} = \overline{\omega} \int_0^{l_s} s_{0z}\, dz \qquad \text{5-VIII.2}$$

and if $\overline{a}$ is the tributary area between piles of spacing λ and β; then the average increment in effective vertical stress at the pile tip is given by

$$\Delta\sigma_{0d} = \frac{Q_{fu}}{\overline{a}} \qquad \text{6-VIII.2}$$

Therefore, the ultimate point bearing capacity of a pile in a group driven into a sand stratum to a length l_s of penetration is

$$Q_{pu} = a_p \left\{1.2\left[cN_c + (\sigma_{0d} + \Delta\sigma_{0d})N_q\right]\right\}(D_r + 0.1) \qquad \text{7-VIII.2}$$

and the total end bearing capacity of the pile is

$$Q_u = Q_{pu} + Q_{fu}$$

To calculate the value of $\Delta\sigma_{0d}$ required in formulas 6 and 7-VIII.2, one has to establish the mechanical behavior of friction along the pile shaft embedded in the sand bearing stratum, under simplified working hypotheses. In this particular case, assume that the shear parameters along the shaft are constant over the total em-

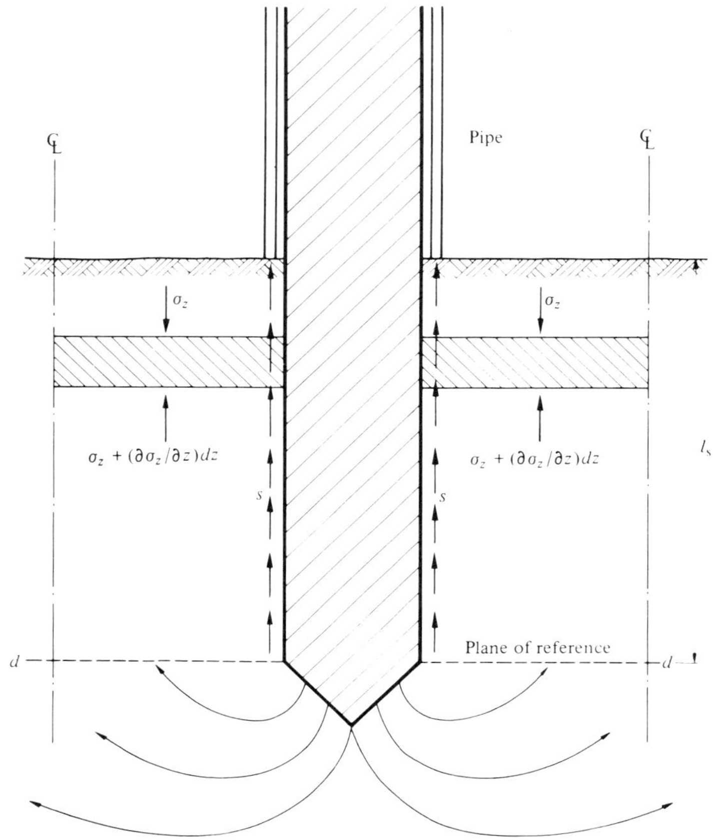

Fig. 4-VIII.2 Plastic equilibrium conditions at the pile shaft.

bedded length l_s. The pile perimeter is defined as $\overline{\omega} = 2\pi r_0$. Furthermore, assume that under ultimate penetration of the pile point in the bearing stratum, the shear strength at the pile shaft in the section of length l_s is fully mobilized. The mobilization of the shear strength in the upper strata of compressible soil is not considered. Therefore, from Fig. 4-VIII.2, we can establish under plastic equilibrium conditions that

$$\overline{a} \cdot \frac{\partial \Delta\sigma_z}{\partial z} dz = \overline{\omega} \cdot s_{0z} \cdot dz \qquad \text{8-VIII.2}$$

and using the shear strength given by 5-VIII.2, for $c_r = 0$, we have

$$\frac{\partial \Delta\sigma_z}{\partial z} = \frac{\overline{\omega}}{\overline{a}} \left[\frac{\tan \phi_r}{\overline{N}_\phi} (\sigma_{0z} + \Delta\sigma_z) \right] \qquad \text{9-VIII.2}$$

The value of $\overline{N}_\phi$ is the ratio of horizontal to vertical effective stresses close to the pile shaft under plastic conditions:

$$\frac{1}{\overline{N}_\phi} = \frac{\sigma_{hz}}{\sigma_{0z} + \Delta\sigma_z}$$

and has the following value (see section 3 of this chapter):

$$\overline{N}_\phi = \frac{1 + \sin^2 \phi_r}{1 - \sin^2 \phi_r} \qquad \text{10-VIII.2}$$

The value of $\tan \phi_r/\overline{N}_\phi = K_\phi$ assumes a maximum of 0.353, for values of $\phi_r = 35°$, and has very little sensitivity to variations of the angle of internal friction in this range of angles. The differential equation may be written as follows:

$$\frac{\partial \Delta\sigma_z}{\partial z} - \mu\Delta\sigma_z = \mu\sigma_{0z} \qquad \text{11-VIII.2}$$

in which $\mu = (\overline{\omega}/\overline{a})K_\phi$.

The integration of this equation yields the following value:

$$\Delta\sigma_z \cdot e_N^{-\mu z} = \mu \int_0^{l_s} \sigma_{0z} \cdot e_N^{-\mu z}\, dz + C \qquad \text{12-VIII.2}$$

The problem may be solved with sufficient accuracy, from the foundation engineering point of view, by means of simplifying assumptions. Using differential equation 9-VIII.2, the cohesion term may be included as follows:

$$\frac{\partial \Delta\sigma_z}{\partial z} dz = \frac{\overline{\omega}}{\overline{a}} c_r\, dz + \frac{\overline{\omega}}{\overline{a}} K_\phi \sigma_{0z} dz + \frac{\overline{\omega}}{\overline{a}} K_\phi \Delta\sigma_z dz \qquad \text{13-VIII.2}$$

Integrating this equation from 0 to l_s, we obtain

$$\Delta\sigma_{0d} = \frac{\overline{\omega}}{\overline{a}} \int_0^{l_s} c_r\, dz + \frac{\overline{\omega}}{\overline{a}} K_\phi \int_0^{l_s} \sigma_{0z} dz + \frac{\overline{\omega}}{\overline{a}} K_\phi \int_0^{l_s} \Delta\sigma_z dz \qquad \text{14-VIII.2}$$

Using Fig. 5-VIII.2, we can make the following assumptions when l_s is not very large:

1. $$\int_0^{l_s} c_r\, dz = c_{ra}\, l_s$$

2. $$\int_0^{l_s} \sigma_{0z} dz = \frac{1}{2}(\sigma_{0i} + \sigma_{0d}) l_s \qquad \text{15-VIII.2}$$

3. $$\int_0^{l_s} \Delta\sigma_z dz = \frac{1}{2}\Delta\sigma_{0d}\, l_s$$

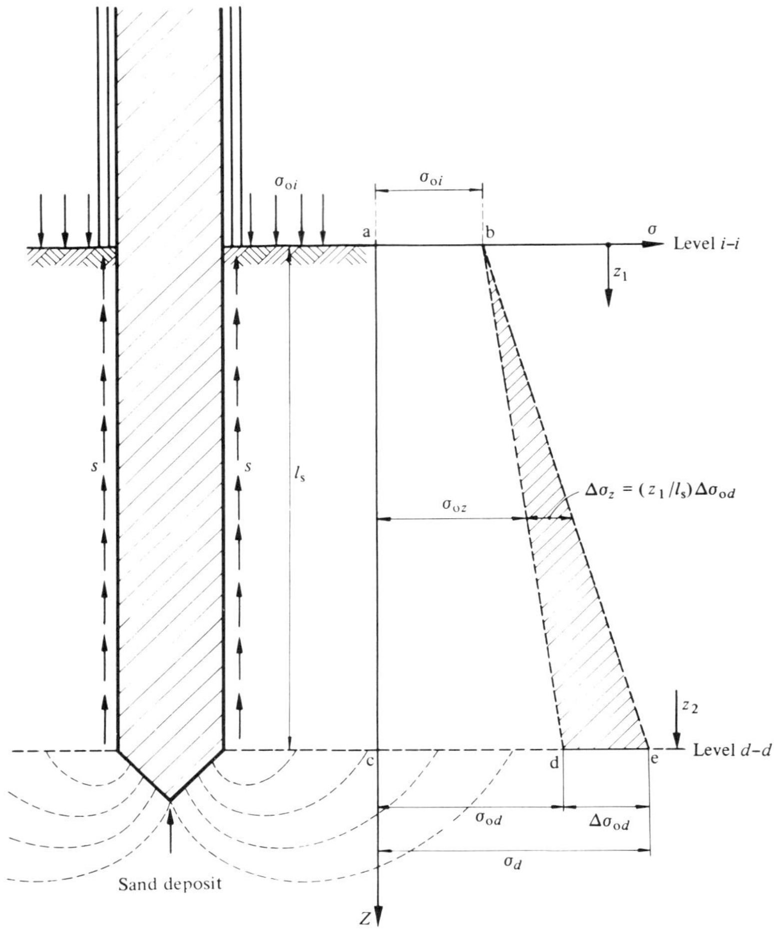

Fig. 5-VIII.2 Load capacity of one pile in a group.

Therefore, substituting these assumptions in 14-VIII.2, we obtain

$$\Delta\sigma_{0d} = \frac{\overline{\omega}}{\overline{a}} K_\phi \frac{l_s}{2}(\sigma_{0i} + \sigma_{0d}) + \frac{\overline{\omega}}{\overline{a}} K_\phi \frac{l_s}{2}\Delta\sigma_{0d} + \frac{\overline{\omega}}{\overline{a}} c_{ra}\, l_s$$

and calling $(l_s/2)(\overline{\omega}/\overline{a})K_\phi = \overline{m}$, and $(\overline{\omega}/\overline{a})l_s = \overline{n}$, we can write

$$\Delta\sigma_{0d} = \frac{\overline{m}}{1-\overline{m}}(\sigma_{0i} + \sigma_{0d}) + \frac{\overline{n}}{1-\overline{m}} c_{ra} \qquad \text{16-VIII.2}$$

and for the frictional ultimate load we have

$$Q_{fu} = \overline{a}\cdot\Delta\sigma_{0d} \qquad \text{17-VIII.2}$$

Hence the total load capacity of one pile in a group may be approximately the sum of Q_{pu} and Q_{fu}. Therefore, the total allowable load each pile in the group can carry with a factor of safety of G_s will be:

$$Q_{pa} = \frac{1}{G_s}(Q_{pu} + Q_{fu}) \qquad \text{18-VIII.2}$$

When the length l_s embedded in the resisting stratum is large and the strata show different shear strength properties and unit weights, then the assumptions 15-VIII.2 are not valid. Furthermore, the tributary area $\bar{a}$ assumed constant may also change with depth. If such is the case, a step by step integration procedure is used to calculate Q_{fu}, as described in sections 3 and 4 of this chapter.

A simple example will show the application of 16 and 17-VIII.2. Assume a field of piles 0.5 m in diameter is driven into a sand deposit. The penetration in the sand is l_s = 3 m. The tributary area of one pile in the group is $\bar{a}$ = 4 m^2. Hence $\overline{m}$ = 0.21, and if the values of σ_{0i} = 20 ton/m^2 and σ_{0d} = 21.5 ton/m^2, then the value for the increment of stress at the pile points is obtained because of positive friction developed in the pile shaft for the embedded length l_s. Hence:

$$\Delta\sigma_{0d} = \frac{0.21}{0.79}(41.5) = 11 \text{ ton/m}^2$$

Thus, the confining stress at the pile tips will be increased by 11 ton/m^2, and therefore, the point bearing capacity should be calculated with a confining vertical stress of 32.5 ton/m^2 instead of 21.5 ton/m^2 corresponding to the overburden effective stress. The friction load to be added to the point bearing capacity is Q_{fu} = 11 × 4 = 44 ton. If the sand under the pile tip has an angle of internal friction of $\phi = 35°$ and the material is encountered under the pile tip in a semicompact state (D_r = 0.4), then using formula 7-VIII.2 we obtain

$$Q_{pu} = a_p(1.2)(32.5\,N_q)(0.4 + 0.1)$$

From Fig. 4-III.4 in section III.4, we obtain N_q = 63, and since the pile area a_p = 0.196 m^2, then Q_{pu} = 240 ton, and the total ultimate capacity of the pile is Q_{pa} = 284 ton. Having taken all physical factors into consideration, we may apply a safety factor of 3, and assign for the pile an allowable point bearing load of 95 ton.

2.3 Point Vertical Displacements

The vertical displacement of the pile point after it has been driven into the bearing stratum can be estimated approximately from the deformations taking place in the zone affected by the stresses applied on the soil by the pile tip: $q_a = Q_{pa}/a_p$ where a_p is the cross-section of the pile. The zone to a depth $2r_0$ under the pile tip may be assumed incompressible because during pile driving the soil to this depth has been compacted very strongly; therefore, only zones 1, 2 and 3 may be considered

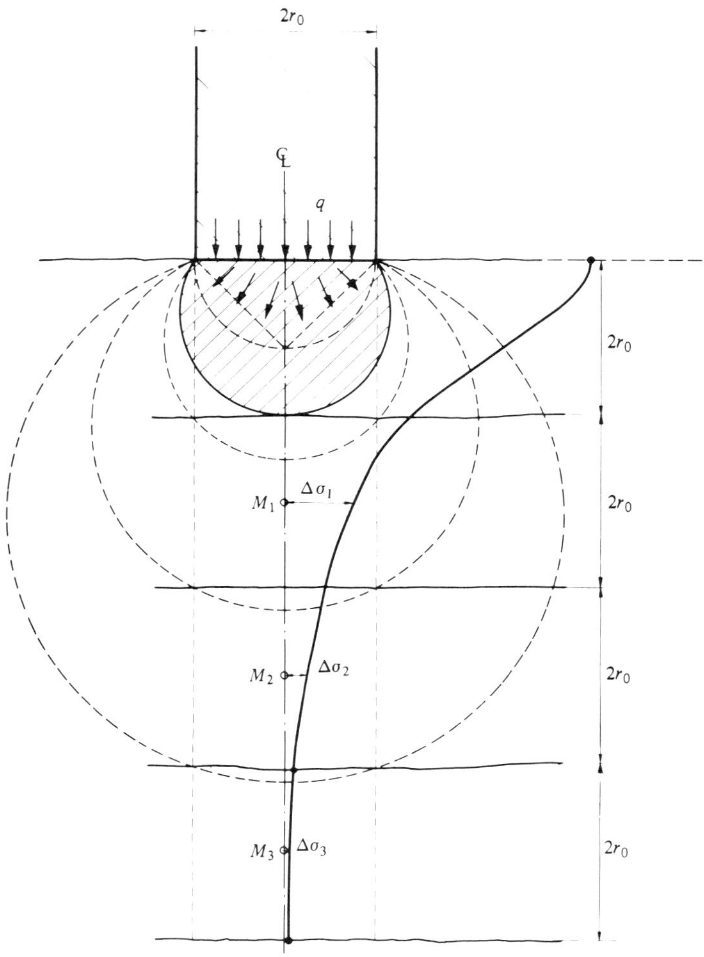

Fig. 6-VIII.2 Zones considered for settlement calculations.

for settlement calculations (Fig. 6-VIII.2). Hence

$$\delta = (M_1 \Delta\sigma_1 + M_2 \Delta\sigma_2 + M_3 \Delta\sigma_3) 2r_0 \qquad \text{19-VIII.2}$$

The M-values are the strain moduli of the soil and may be estimated in the laboratory, if possible on undisturbed samples subjected to volumetric confining effective stresses compatible with those found in the ground (see section II.3.7). The $\Delta\sigma_z$-values under the pile point are determined by the theory of elasticity as described in Chapter III. Fröhlich stress charts may be used for this purpose if M decreases strongly with depth. To illustrate this case, assume a pile 45 cm in diameter at the point, where the stress discharged by the pile tip to the soil is q_a = 30 kg/cm^2 under working conditions. The increment of stresses $\Delta\sigma_z$ and the average M-values found in the three zones under the corresponding confining state of stress σ_c (sections II.3.5-3.7 and III.2) are as follows:

TABLE 1-VIII.2

Zone	M, cm^2/kg	$\Delta\sigma$, kg/cm^2	$M\Delta\sigma(2r_0)$, cm
1	0.003	6	0.81
2	0.002	2	0.18
3	0.005	0.5	0.11
			δ = 1.10 cm

Therefore, the value of $\delta = 1.10$ cm represents the approximate penetration that should be expected of each one of the piles of the group in the bearing stratum. This vertical displacement will be added to the total settlement of the group that depends on how the strata underlying the pile tips are stressed and on their compressibility properties.

2.4 Pile Group Bearing Capacity

After the load bearing capacity of the piles in the group has been estimated, and the allowable load against plastic penetration of the pile tips into the bearing stratum selected, it is necessary to investigate the group load capacity, considering it as a large pier. The load capacity when the bearing stratum extends to great depth may be estimated by means of formula 7-VIII.2. In many cases, however, the stratum bearing the pile point is underlain by lower shear strength strata; under these conditions, the group ultimate load capacity may be smaller than the problem analyzed for penetration of each individual pile. The group local failure may be investigated by:

$$Q'_{gp} = 4BL\alpha_1 \left[c_a N_c + (\sigma_{0d} - \Delta\sigma_{0d})N_q\right](D_r + 0.1) \qquad \text{20-VIII.2}$$

where: c_a, N_c and N_q are average values determined from the soil shear strength properties; σ_{0d} is the confining overburden stress and $\Delta\sigma_{0d}$ may be a reduction in the confining vertical average stress due to the effect of negative friction caused by consolidation of the upper compressible soil deposits.

More important, however, is the case shown in Fig. 7-VIII.2 in which the bearing stratum has a limited thickness and is overlain and underlain by clay or silty clay deposits. Therefore, in this case, it may happen that even though the individual piles do not penetrate the hard stratum, a subsidence of the group may take place if the base load bearing capacity of the entire pile group exceeds the shear strength of the underlying impervious soil deposits and of the hard stratum on which they are bearing. The load capacity of the underlying soil strata may be calculated approximately using expression 20-VIII.2 in the form

$$Q'_{gu} = (2B + D)(2L + D)\left[\alpha_1 (5.7\, c_a)\right] \qquad \text{21-VIII.2}$$

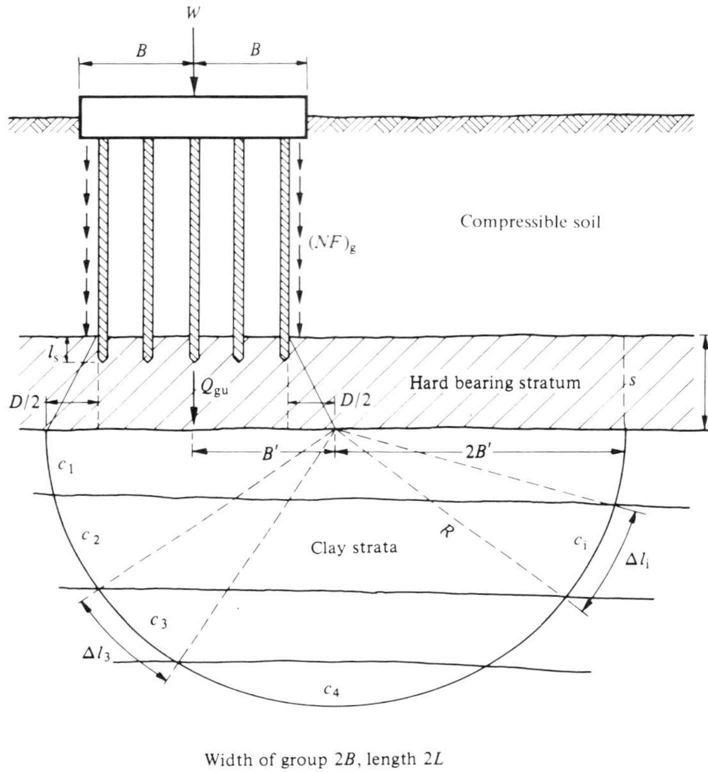

Fig. 7-VIII.2 Group of piles supported by a hard bearing stratum underlain by clay strata.

where c_a is the average unit cohesion obtained from one half of the unconfined compressive strength representative of each stratum cut by the potential surface of sliding. Assuming in this case an arc of a circle, the average unit shear strength may be calculated by means of

$$c_a = \frac{\Sigma c_i \Delta l_i}{\pi(2B')}$$

But $2B' = (2B + D)$, and therefore

$$c_a = \frac{\Sigma c_i \Delta l_i}{\pi(2B + D)} \qquad \text{22-VIII.2}$$

On the other hand, the punching shear capacity of the hard stratum may be estimated knowing its unit shear strength s_D, hence

$$Q''_{gu} = 4(B + L) \cdot \sum_{0}^{D} s_D \cdot \Delta z$$

and the total base load capacity of the pile group under these conditions is

$$Q_{gu} = 1.82\alpha_1(2L + D) \sum_{1}^{n} c_i \Delta l_i + 4(B + L) \sum_{0}^{D} s_D \Delta z \qquad 23\text{-VIII.2}$$

Once the values of the load capacity Q_{gu} and the group negative friction $(NF)_g$ corresponding to the total downward dragging force acting in the entire group are estimated, the factor of safety may be calculated with respect to the total building load W_a, applied to the group of piles; hence

$$W_a + (NF)_g = \frac{1}{G_s} Q_{gu}$$

from which we obtain the factor of safety

$$G_s = \frac{Q_{gu}}{W_a + (NF)_g} \qquad 24\text{-VIII.2}$$

Moreover, the settlement of the group under working load conditions should be investigated, since in most cases it is a decisive factor in the behavior of the pile foundation, because of the compressibility properties of the underlying soil strata.

VIII.3 NEGATIVE FRICTION

3.1 Basic Concepts

Point bearing piles or piers are often supported on a deep firm stratum overlaid by compressible soil deposits containing sediments like silt and clay encountered under consolidation owing to downward seepage forces or loads placed at the ground surface. In this case a continuous relative movement between pile and soil takes place, and the ultimate shear strength close to the pile shaft is fully mobilized (Fig. 1-VIII.3). Therefore, part of the weight of the surrounding soil will be transferred to the piles because of the skin friction developed at the shaft.

The phenomenon mentioned above is known as negative skin friction; it introduces serious problems into the load capacity of piles or piers, and is responsible for the behavior of the foundation. It is recognized from the fact that the vertical effective stresses decrease in the soil mass in proportion to the reduction in the soil mass weight because of load transfer to the piles, therefore introducing a twofold action, namely:

1) A load increase on the pile or pier, and
2) a reduction of the confining effective stresses at the tip elevation with a corresponding reduction in the ultimate point bearing capacity.

In order to investigate this phenomenon, it is necessary to learn about the horizontal effective stresses developed during the process, and from them evaluate the shear strength of the soil close to the pile shaft. Since there is a continuous relative

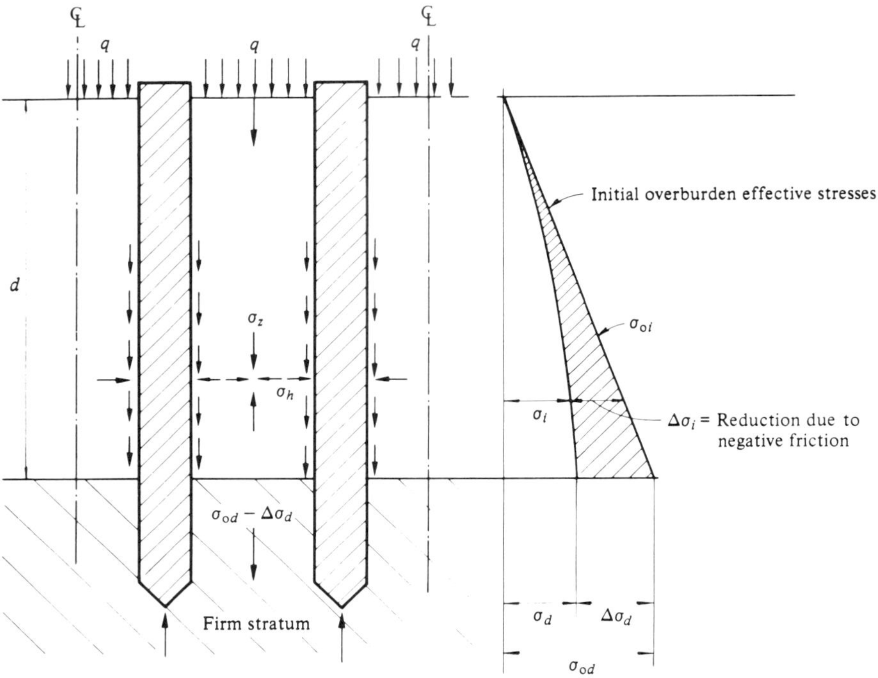

Fig. 1-VIII.3 Negative friction developed in piles resting on a firm stratum.

movement between pile and soil, the shear strength parameters of the remolded material close to the pile shaft are investigated in remolded soil specimens under sustained load after reaching the ultimate soil strength. The load on the sample is applied at very small rates of shear strain, compatible with those that may be observed in the field related with the relative movement between soil and pile.

The piles are driven through impervious soil deposits to reach the hard stratum on which they will be supported; therefore, a volume of soil approximately equal to that of the pile is displaced during driving. The soil material will be disturbed and displaced like a viscous liquid under the tip of the pile (Fig. 2-VIII.3), forming an annular ring around the pile shaft. If r_0 is the radius of the pile, then theoretically the disturbed soil ring will have a radius of $r_0' = \sqrt{2}\, r_0$. During this process, the soil loses its natural strength and high hydrostatic excess water pore pressures are developed that, however, dissipate as soon as the water flows into the surrounding soil and through the pile itself when the pile material is timber or concrete. A thin ring of perfectly disturbed soil around the pile shaft will consolidate and the soil will regain shear strength according to the formula

$$s_{0z} = \sigma_{hz} \tan \phi_r \qquad \text{1-VIII.3}$$

in which σ_{hz} is the effective horizontal stress at depth z from the ground surface

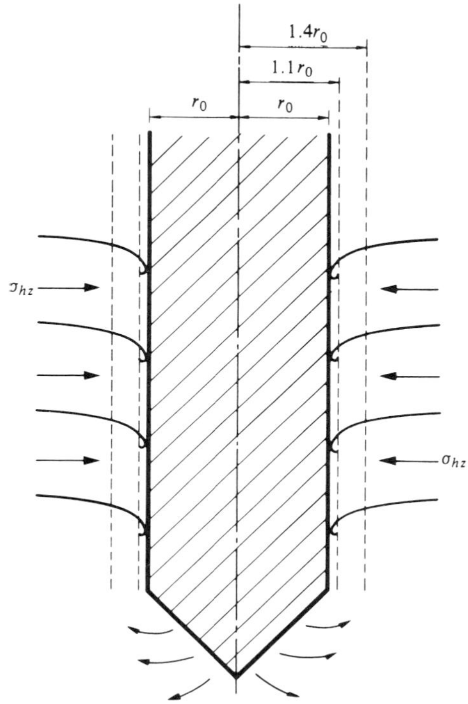

Fig. 2-VIII.3 Disturbed material around the pile shaft.

producing the consolidation; tan ϕ_r is a proportionality factor measuring the rate of gain in shear strength as the soil consolidates under effective stresses σ_{hz}. The gain in shear strength is a maximum at the interface of the pile shaft and soil where adhesion takes place. This is demonstrated from the material found clogged to the shaft of piles when performing pulling tests. The shear strength, however, has a minimum value at a distance of approximately $\bar{r}_0 = 1.05r$ from the pile center. Therefore, the potential surface of sliding should be taken located at distance $\bar{r}_0$ for negative skin friction calculations.

The value of tan ϕ_r may be determined approximately in the laboratory in soil samples remolded at constant water content and then consolidated by increasing the effective stresses and sheared off under zero hydrostatic pore pressure. If any pore pressure is present at the instant of failure, this should be deducted from the applied stress on the sample (Fig. 3-VIII.3). The shear strength tests should be made under small rates of shear strain compatible with the relative movement that may be observed or expected in the field between pile and soil (see section III.3.5c). The Dutch cell may be adapted to perform this type of shear strength tests by means of model piles (Fig. 4-VIII.3).

The gain in shear strength with time close to the pile shaft, according to the ulti-

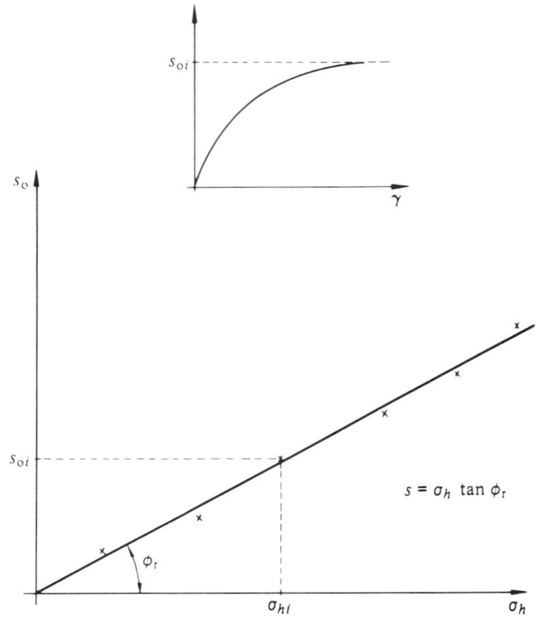

Fig. 3-VIII.3 Rate of gain of shear strength under effective stresses.

mate value expressed by formula 1-VIII.3, should be investigated. The average degree of consolidation of the remolded soil against the pile shaft is computed with the coefficient of consolidation c_{vr}, determined by means of consolidation tests on remoulded soil samples at constant water content. According to section II.3.9, we can write

$$s_{0z} = \sigma_{hz} F(T_v) \tan \phi_r \qquad \text{2-VIII.3}$$

or

$$(s_{0z})_t = \sigma_{hz} \tan \phi_r'$$

where $\tan \phi_r' = F(T_v) \tan \phi_r$, and T_v is the primary consolidation time factor:

$$T_v = \frac{c_v}{H^2}\, t$$

The soil disturbance takes place from the pile shaft in a thickness of about $0.4r_0$ (Fig. 2-VIII.4) and is drained by the surrounding soil of more permeability, because the coefficient of consolidation of remolded sensitive soils assumes values of 0.1 to 0.01 of that in the undisturbed state. Hence, the average time factor for the remolded layer drained on both sides is

$$T_v = \frac{(c_{vr})_a}{(0.2r_0)^2} \cdot t$$

where $(c_{vr})_a$ stands for the average coefficient of consolidation in the zone of the disturbed soil.

Fig. 4-VIII.3 Dutch cell for investigation of skin friction on piles.

To illustrate the role of the time element in these problems, assume a steel pipe pile with diameter: $2r_0 = 40$ cm. The pile is driven into a sensitive impermeable soil. Consolidation tests indicate a remolded coefficient of consolidation on the order of $c_{vr} = 0.00002$ cm²/sec, and in the undisturbed state $c_v = 0.001$ cm²/sec. The value of $\tan \phi_r = 0.5$. Furthermore, assume that the average coefficient of consolidation at the disturbed zone close to the pile shaft with thickness $0.40 \cdot 20 = 8.0$ cm, is from Chapter III.3, formula 27-III.3 and Fig. 2-VIII.3:

$$(c_{vr})_a = \frac{\sum_1^n H_i}{\sum_1^n \dfrac{H_i}{c_{vi}}} = \frac{8}{\dfrac{4}{0.00002} + \dfrac{4}{0.001}} = 3.93 \times 10^{-5} \text{ cm}^2/\text{sec}$$

Therefore, the time factor

$$T_v = \frac{3.93 \times 10^{-5}}{64} t = 6.15 \times 10^{-7}\, t$$

in terms of days $t = 86{,}400\, t_d$, and we obtain $T_v = 5.3 \times 10^{-2}\, t_d$. Using Terzaghi's function (Fig. 34-II.3) we find the following values:

TABLE 1-VIII.3

t_d	T_v	$U\%$	$F(T_v) \tan \phi_r$	ϕ_r'	K_ϕ'
5	0.265	58	0.29	16.2°	0.245
10	0.53	78	0.39	21.2°	0.295
20	1.06	92	0.46	24.7°	0.322
30	1.59	99	0.49	26.2°	0.330

Therefore, when making interpretations on field pile tests in impermeable soils, we consider the time element very important. The piles shall be provided with pore pressure measuring devices along their shafts, in order to estimate properly the effective soil stresses acting at the pile shaft.

In case of bored piles or cast-in-place piles or piers, the ultimate shear strength may be calculated by means of the following expression:

$$s_d = c_d + \sigma_h \tan \phi_d \qquad \text{3-VIII.3}$$

in which the drained shear strength parameters c_d and ϕ_d are the cohesion and angle of internal friction determined on undisturbed representative samples for each of the strata pierced by the shaft of the pier or pile.

The horizontal effective stress under plastic conditions close to the pile shaft is related to the vertical effective stress. Using Fig. 5-VIII.3 and the Coulomb-Mohr ultimate state of stress condition of section II.4.3, we may write the ratio of the horizontal to vertical effective stresses as a function of the principal stresses:

$$\frac{\sigma_h}{\sigma_z} = \frac{(\sigma_1 + \sigma_3) - (\sigma_1 - \sigma_3)\cos 2\alpha}{(\sigma_1 + \sigma_3) + (\sigma_1 - \sigma_3)\cos 2\alpha}$$

From the ultimate stress circle we find $2\alpha = \phi_r + (\pi/2)$, and since $\sin \phi_r = \cos 2\alpha$ and $\sin \phi_r = (\sigma_1 - \sigma_3)/(\sigma_1 + \sigma_3)$, after performing substitutions and algebraic operations on 3-VIII.3, we find

$$\frac{\sigma_h}{\sigma_z} = \frac{1 - \sin^2 \phi_r}{1 + \sin^2 \phi_r} = \frac{1}{\overline{N}_\phi} \qquad \text{4-VIII.3}$$

The flow value for this condition will be called $\overline{N}_\phi$. From tests in the remolded material at constant water content the value of ϕ_r may be investigated. Calling $\tan \phi_r$ the coefficient of skin friction between pile and soil, then the skin friction

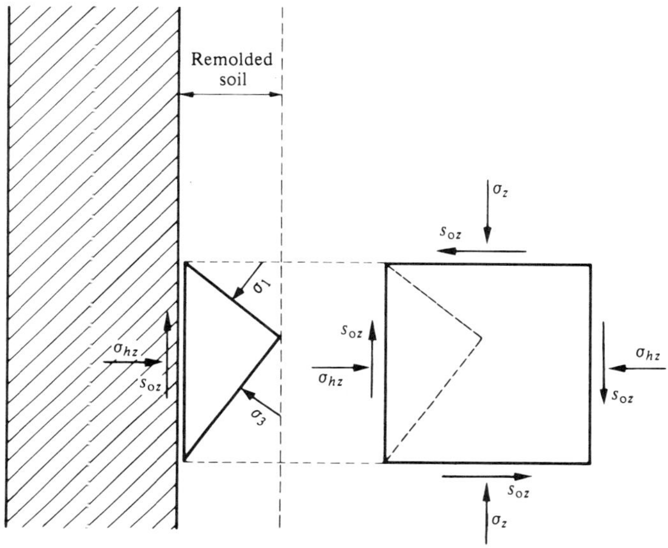

Fig. 5-VIII.3 State of stress at pile shaft.

at the pile shaft will be approximately

$$s_{0z} = \left(\frac{1 - \sin^2 \phi_r}{1 + \sin^2 \phi_r} \cdot \tan \phi_r\right) \cdot \sigma_z$$

or

$$s_{0z} = K_{\phi z} \cdot \sigma_z \qquad \text{5-VIII.3}$$

where $K_\phi = \tan \phi_r / \overline{N}_\phi$. The skin friction against the pile shaft may be diminished using water-soil repellents. Different products may be investigated as paints, bitumen or bentonite admixtures, and the value of $\tan \phi_p$ determined. When this is the case $1/\overline{N}_\phi$ maybe replaced by $(1 - \sin \phi_r)$, [Jáky (1944)] and $K_{\phi p} = (1 - \sin \phi_r) \tan \phi_p$. The value of K_ϕ is plotted in Fig. 6-VIII.3 against ϕ_r, for different values of $\tan \phi_p$. For uncoated piles when $\phi_r = \phi_p$ it may be noticed that in the range of $\phi_r = 25\text{–}30°$, the value of K_ϕ does not change much. Most soils of the impervious type disturbed at constant water content give angles for the gain in shear strength against the pile shaft material that are close to those mentioned above. Thus, an average value of $K_\phi = \frac{1}{3}$ may be usually assumed for calculations of negative friction in uncoated piles. Drained tests performed in the Dutch cell on remolded Mexico City silty clay show values on the order of $\phi_r = 26°$. The value of K_ϕ may be reduced when the pile is coated with a repellent agent, in fact, if $\tan \phi_p = 0.3$ and $\phi_r = 26°$, then $K_{\phi p} = 0.17$ against $K_\phi = 0.33$ for an uncoated pile. Therefore, the negative skin friction may be diminished if a strong and durable coating is used.

Let us assume now we wish to calculate approximately the negative skin friction on a pier or pile of constant diameter $2r_0$. According to Fig. 7-VIII.3, we may

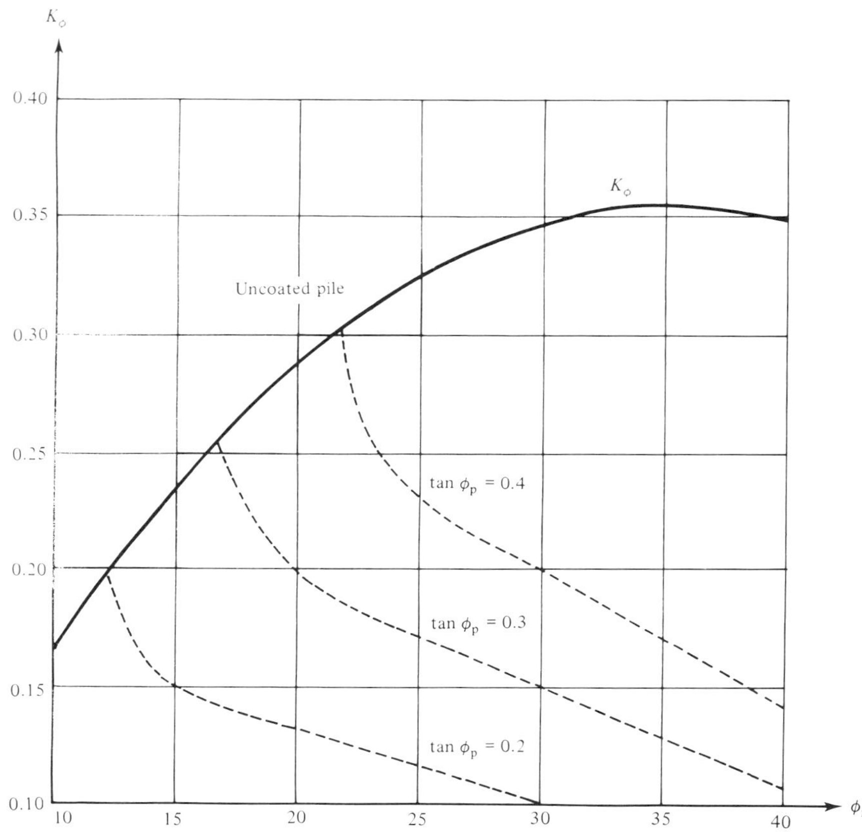

ϕ_p = Angle of skin friction of coated pile
ϕ_r = Angle of internal friction of remolded soil

Fig. 6-VIII.3 Values of K_ϕ for uncoated and coated piles.

write

$$\frac{\partial (NF)}{\partial z}\, dz = 2\pi r_0\, s_{0z}\, dz \qquad \text{6-VIII.3}$$

Letting $2\pi\bar{r}_0 = \bar{\omega}$, and integrating, we obtain

$$(NF)_d = \bar{\omega} \int_0^d s_{0z}\, dz \qquad \text{7-VIII.3}$$

The ultimate shear strength of the reconsolidated material at the interface of soil and pile shaft at a depth z is

$$s_{0z} = K_\phi (\sigma_{0z} - \Delta\sigma_z) \qquad \text{8-VIII.3}$$

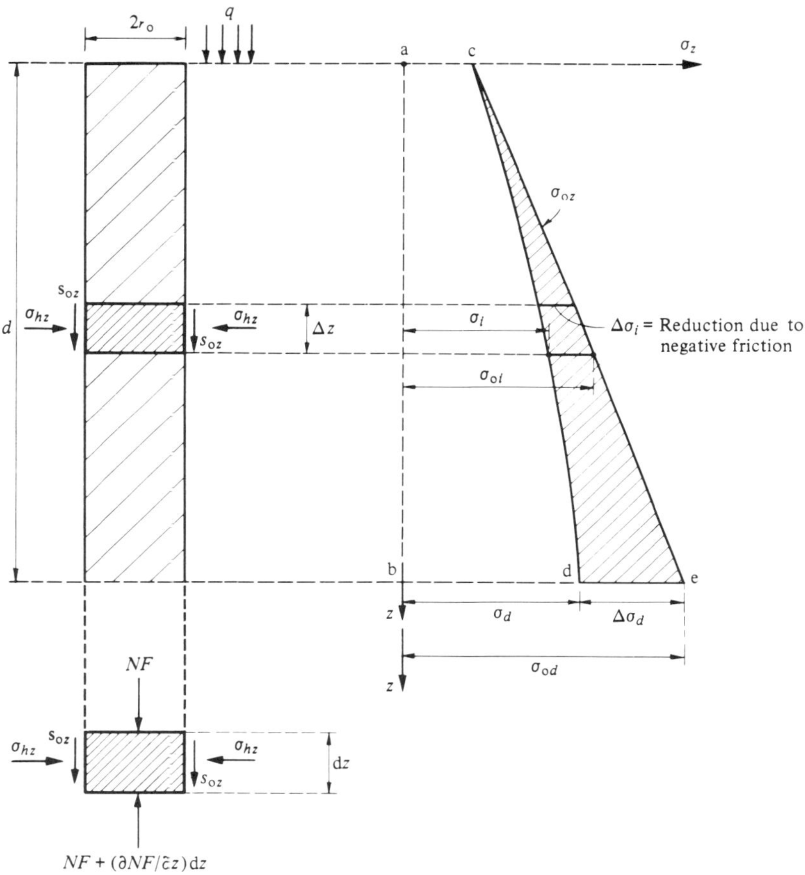

Fig. 7-VIII.3 Negative friction acting on pile.

Therefore:

$$(NF)_d = \overline{\omega}K_\phi \int_0^d \sigma_{0z}\,dz - \overline{\omega}K_\phi \int_0^d \Delta\sigma_z\,dz \qquad \text{9-VIII.3}$$

In this expression using Fig. 7-VIII.3, we find:

1) $\int_0^d \sigma_{0z}\,dz$ – area (abec) of initial stresses

10-VIII.3

2) $\int_0^d \Delta\sigma_z\,dz$ – area (cde) of vertical relief of stresses because of *NF* on pile or pier; it may be assumed approximately equal to $(\frac{1}{3})\Delta\sigma_d \cdot d$

Hence:

$$(NF)_d = \bar{\omega} K_\phi \int_0^d \sigma_{0z}\, dz - \bar{\omega} K_\phi \left(\frac{1}{3} \Delta\sigma_d \cdot d\right) \qquad \text{11-VIII.3}$$

On the other hand, calling $\bar{a}$ the effective tributary area and assuming it is constant with depth, $\bar{a} = \pi(nr_0)^2$, then:

$$\Delta\sigma_d = \frac{(NF)_d}{\bar{a}} \qquad \text{12-VIII.3}$$

Substituting this value into 11-VIII.3 and solving for NF, we obtain:

$$(NF)_d = \frac{\bar{\omega} K_\phi}{1 + \dfrac{\bar{\omega} K_\phi}{3} \dfrac{d}{\bar{a}}} \int_0^d \sigma_{0z} \cdot dz \qquad \text{13-VIII.3}$$

For practical purposes, formula 13-VIII.3 gives usually satisfactory estimates.

The value of $\bar{a}$ may be taken as $\pi(12r_0)^2$ for isolated piles or piers. In case of pile fields $\bar{a}$ is the nominal tributary area for each pile in the group. The load capacity of the pile or pier may be calculated taking into consideration the average confining vertical stress at the base, $\sigma_d = \sigma_{0d} - \Delta\sigma_d$. In problems representing real field conditions, the tributary area is not constant with depth. Furthermore, the assumption of $\Delta\sigma_z$ cannot be represented by a simple mathematical function of z. If the equivalent tributary area $\bar{a}_{ez}$ and the total initial or overburden effective stresses are known with depth, the integration to calculate the negative skin friction is obtained by a numerical step by step method that, however, can be also performed with a computer program in accordance with the algorithms given in further paragraphs.

Let Fig. 8-VIII.3 represent the effective stresses induced by the negative skin friction in an element Δz_i of the soil-pile system at a depth z. The equilibrium equation under plasticity conditions reads as follows:

$$(\sigma_{0i} - \sigma_i)\bar{a}_i - (\sigma_{0i-1} - \sigma_{i-1})\bar{a}_{i-1} = \bar{\omega} s_{0i} \Delta z_i \qquad \text{14-VIII.3}$$

but the average value of the shear strength at the pile shaft for any depth $z = i$ is:

$$s_{0i} = \tfrac{1}{2} K_{\phi i}(\sigma_i + \sigma_{i-1}) \qquad \text{15-VIII.3}$$

On the other hand we have

$$(NF)_i = (\sigma_{0i} - \sigma_i)\bar{a}_i \qquad \text{16-VIII.3}$$

and

$$(NF)_{i-1} = (\sigma_{0i-1} - \sigma_{i-1})\bar{a}_{i-1} \qquad \text{17-VIII.3}$$

Defining $\pi \bar{r}_0 K_\phi = m_i$ and assuming that the integration will start at the ground surface where $(NF)_{i-1} = 0$ and $\sigma_{i-1} = \sigma_{0i-1} = q$, and since the values of σ_{0i}, m_i and $\bar{a}_i$ are known with depth, then we have

$$\sigma_{0i}\bar{a}_i - \sigma_i\bar{a}_i - (NF)_{i-1} = m_i\sigma_i\Delta z_i + m_i\sigma_{i-1}\Delta z_i \qquad \text{18-VIII.3}$$

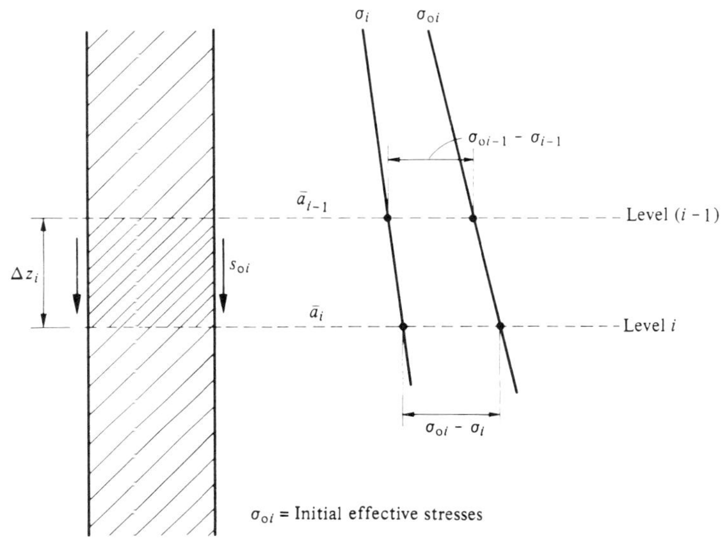

Fig. 8-VIII.3 Vertical stress decrease due to negative friction.

Solving the above for σ_i, we obtain

$$\sigma_i = \frac{\sigma_{0i}\bar{a}_i - [(NF)_{i-1} + m_i \sigma_{i-1} \Delta z_i]}{\bar{a}_i + m_i \Delta z_i} \qquad \text{19-VIII.3}$$

The negative friction load at the i-th level may be calculated for each step by means of formula 16-VIII.3, with the value of σ_i previously computed from 19-VIII.3. The process is started at the ground surface and repeated step by step until reaching any depth z required. An example to illustrate the numerical procedure may be found in Appendix E to Chapter VIII.

The average confining stress is used to calculate the point bearing load capacity of the piles at the elevation of the pile points:

$$\sigma_d = \sigma_{0d} - \frac{(NF)_d}{\bar{a}_0} \qquad \text{20-VIII.3}$$

where $\bar{a}_0$ is the nominal tributary area of the pile considered in the pile field or group.

3.2 Effective Tributary Area

Piles in use range from small groups to large pile fields, and therefore the problem arises concerning the effective tributary area, which is a function of depth and pile layout. The tributary area is used in formulas 16 to 19-VIII.3 to obtain more accuracy in calculating pile skin friction problems in corner, center and intermediate piles, or partially coated piles. The relief in overburden effective stress $\Delta\sigma_z$ because of negative skin friction is larger at the pile shaft than the average value, and its magnitude is a function of the pile spacing and depth. For a constant tributary area

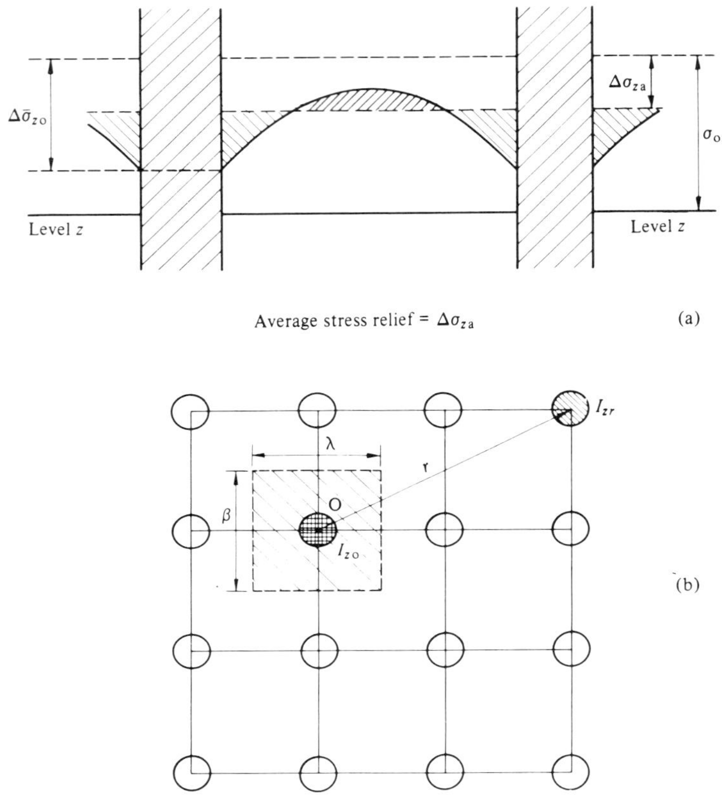

Fig. 9-VIII.3 Influence of pile group on a pile O of the group.

$\bar{a}_0 = \lambda \cdot \beta$, where λ and β define the pile spacing, one obtains from calculations an average vertical stress relief $\Delta\sigma_{za} = \sigma_{0z} - \sigma_z$ defined as $\Delta\sigma_{za} = (NF)_d/\bar{a}_0$ (Fig. 9-VIII.3). The stress at the pile shaft, however, is a function of the influence of each pile in the field on the other piles and should be considered in the calculations. From the figure it may be established that if $\Sigma_0^n [\Delta\sigma_{zo}] = \Delta\bar{\sigma}_{zo}$ represents at any depth z the total change in vertical effective overburden stress at the shaft of pile O, because of the effect of all the piles in the group including pile O under study, then we may write:

$$\Delta\bar{\sigma}_{zo} = \alpha_{zo} \cdot \Delta\sigma_{za} \qquad \text{21-VIII.3}$$

here α_{zo} is a correction factor that is a function of the depth applied to pile O in the group, and is used to obtain from the average stress relief $\Delta\sigma_{za}$, the relief of stress concentration at the shaft of pile O (Fig. 9a-VIII.3). Moreover, calling $\bar{a}_{ez}$ the effective tributary area, the following condition may be established:

$$\bar{a}_{ez} \cdot \sum_1^n [\Delta\sigma_{zo}] = \Delta\sigma_{za} \cdot \bar{a}_0 \qquad \text{22-VIII.3}$$

Hence

$$\frac{\bar{a}_0}{\bar{a}_{ez}} = \alpha_{z0} \qquad \text{23-VIII.3}$$

Therefore, if the relative value $\Delta\bar{\sigma}_{z0}/\Delta\sigma_{za}$ may be estimated, then an approximate value of $\bar{a}_{ez}$ may be obtained at any depth z, to be used in formulas 16 and 19-VIII.3 for skin friction calculations.

For this purpose, first it will be necessary to calculate the approximate value of $\Delta\sigma_{z0}$ at the pile shaft for one isolated pile. Consider a pile or pier subjected to negative skin friction *NF* (Fig. 10-VIII.3). The effective stress reduction $d(\Delta\sigma_{zr})$ at a distance r from the pile center and on a horizontal plane at depth z because of the

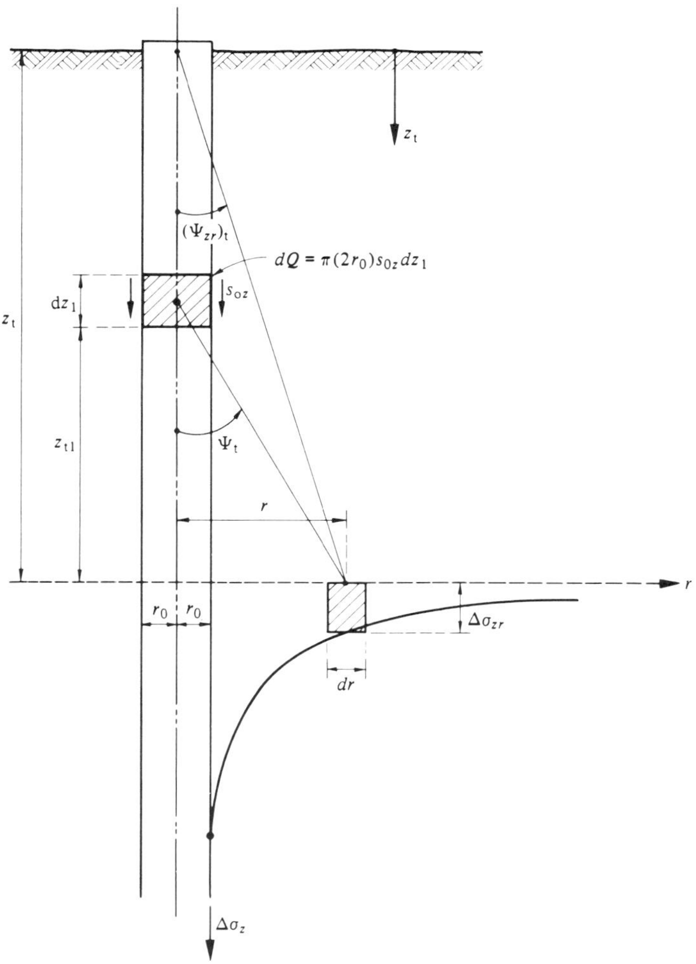

Fig. 10-VIII.3 Vertical stress reduction due to friction on pile shaft.

elementary load $\bar{\omega}s_{0z}dz$ may be calculated approximately if we assume Westergaard's stress distribution in the subsoil (section III.2, formula 19-III.2) and use the method proposed by Terzaghi (1943), Art. 143:

$$d(\Delta\sigma_{zr}) = -\frac{1}{2\pi}\frac{\bar{\omega}s_{0z}\cdot dz_t}{z_t^2}\cos^3\psi_t \qquad \text{24-VIII.3}$$

where $z_t = k_w \cdot z$ and ψ_t correspond to a linear transformation of the vertical coordinate where the factor is $k_w = 1/\sqrt{2}$. Therefore, the total stress at depth z_t in the transformed coordinate system, at a radial distance r from the pile center, is

$$\Delta\sigma_{zr} = -r_0\int_0^z s_{0z}\frac{\cos^3\psi_t}{z_t^2}dz_t \qquad \text{25-VIII.3}$$

To integrate the above expression, it is necessary to know approximately the skin friction distribution with depth. This is, of course, the problem that should be solved with formulae 16 and 19-VIII.3, if the approximate value of the equivalent tributary area $\bar{a}_{ez}$ is known with depth. From previous investigations on the subject, however, the author has found that for practical engineering purposes, the value of $\bar{a}_{ez}$ is not very sensitive to the real skin friction distribution in the calculation of the *NF*-load, if this is assumed directly proportional to depth. Nevertheless, it is important to study also the value of $\bar{a}_{ez}$ in case the skin friction may be assumed constant along the pile shaft. Therefore, the integration of equation 25-VIII.3 is achieved studying the following cases: Case I for $s_{0z} = kz$, Case II for $s_{0z} = s_e$, and Case III for $s_{0z} = s_e + kz$.

CASE I. From Fig. 10-VIII.3 we have $s_{0z} = k(z_t - z_{t1})$, and therefore

$$\Delta\sigma_{zr} = -r_0 k\int_0^z \frac{z_t - z_{t1}}{z_{t1}^2}\cos^3\psi_t dz_{t1} \qquad \text{26-VIII.3}$$

but $z_{t1} = r\cot\psi_t$, and $dz_{t1} = -r\csc^2\psi_t d\psi_t$, and substituting in 26-VIII.3 we get

$$\Delta\sigma_{zr} = r_0 k\left[\frac{z_t}{r}\int_{\pi/2}^{(\psi_{zr})_t}\cos\psi_t d\psi_t - \int_{\pi/2}^{(\psi_{zr})_t}\frac{\cos^2\psi_t}{\sin\psi_t}d\psi_t\right] \qquad \text{27-VIII.3}$$

After integration and algebraic arrangements,

$$\Delta\sigma_{zr} = -r_0 k\left\{\frac{z_t}{r}(1 - \sin(\psi_{zr})_t) + \left[\cos(\psi_{zr})_t + \mathrm{Log}\left(\tan\frac{1}{2}(\psi_{zr})_t\right)\right]\right\} \qquad \text{28-VIII.3}$$

in which the angular arguments for

$$\frac{z_t}{r} = \frac{z}{r\sqrt{2}}$$

are

$$\sin(\psi_{zr})_t = \frac{1}{\sqrt{(z_t/r)^2 + 1}}, \quad \cos(\psi_{zr})_t = \frac{z_t}{r}\sin(\psi_{zr})_t$$

and

$$\tan\frac{1}{2}(\psi_{zr})_t = \frac{\sin(\psi_{zr})_t}{1+\cos(\psi_{zr})_t}$$

The values of $\Delta\sigma_{zr}$ calculated for the transformed system are used for the depth $z = \sqrt{2}\, z_t$ of the real system, and the distance r is the same. From correlation on pile tests subjected to negative and positive friction, respectively, the author has concluded that due to stratification encountered in practice in nearly all soil sediments and because of the presence of the piles, the vertical effective stresses induced in the subsoil are well represented by the Westergaard stress distribution.

The dimensionless term in the brackets of 28-VIII.3 may be expressed as a function of z/r and represents an influence value I_{zr}. At the pile shaft the value of I_{r0} is obtained for $r = r_0$, also $\psi_{zr} = \psi_{z0}$; hence:

$$\Delta\sigma_{z0} = r_0 k \cdot I_{z0} \tag{29-VIII.3}$$

The values of I_{zr} for pile O as a function of z/r in the real system may be obtained from Fig. 11-VIII.3, curve A_k.

In case of pile groups r is considered to be the distance from the studied pile O to other piles in the group. The influence values on pile O may be determined for each pile in the group from the influence curve I_{zr} vs. z/r. The total influence $\Sigma_0^n [I_{zr}]_k$ on pile O may be written as follows:

$$\bar{I}_{zk} = \sum_0^n [I_{zr}]_k = I_{z0} + \sum_1^n I_{zr} \tag{30-VIII.3}$$

Therefore, the total stress relief at the shaft of pile O is

$$\sum_1^n \Delta\sigma_{z0} = r_0 k \bar{I}_{zk} \tag{31-VIII.3}$$

On the other hand, since for this case we assumed $s_{0z} = kz$, the total skin friction load is

$$\Delta\sigma_{za} \cdot \bar{a}_0 = \pi r_0 k z^2 \tag{32-VIII.3}$$

and from 22-VIII.3, the effective tributary area for Case I is

$$\bar{a}_{ez} = \frac{\pi z^2}{\bar{I}_{zk}} \tag{33-VIII.3}$$

Formula 33-VIII.3 may be used to calculate the value of $\bar{a}_{ez}$ to any depth and pile spacing to be used in the integration of algorithms 16 and 19-VIII.3 for skin friction calculations. Calling R_{ez} the radius of the effective area at any depth z from the ground surface, neglecting the pile cross section, then $\bar{a}_{ez} = \pi R_{ez}^2$, and

$$R_{ez} = \sqrt{\frac{\bar{a}_{ez}}{\pi}} \tag{34-VIII.3}$$

CASE II. When the shear strength is assumed approximately constant with depth along the pile shaft, formula 25-VIII.3 may be integrated; for $s_{0z} = s_e = \text{const}$ we obtain

$$\Delta\sigma_{zr} = \frac{r_0 s_e}{z_t}\,[I_{zr}]_{tc} \qquad \text{35-VIII.3}$$

in which the influence value as a function of z/r is

$$I_{zr} = \frac{z_t}{r}\,[1 - \sin(\psi_{zr})_t] \qquad \text{36-VIII.3}$$

The value given by 36-VIII.3 is plotted in Fig. 11-VIII.3 in terms of z/r, curve A_c already affected by the depth factor. Furthermore, from the assumption of constant shear strength along the pile shaft in the real system we have

$$\Delta\sigma_{za} \cdot \bar{a}_0 = 2\pi r_0 s_e z \qquad \text{37-VIII.3}$$

Therefore, from 22-VIII.3 and 35-VIII.3, we obtain:

$$\bar{a}_{ez} = \frac{2\pi z^2}{\sqrt{2}\,\bar{I}_{zc}} \qquad \text{38-VIII.3}$$

The sum of all influence values $\Sigma_0^n\,[I_{zr}] = \bar{I}_{zc}$ has the same meaning given by 30-VIII.3: it represents the sum of the influence values I_{z0} for pile O and those I_{zr} from all other piles at a distance r from pile O.

CASE III. This case may be found when an initial skin friction s_e obtains at the head of the pile located at depth h_e from the ground surface. Following the above assumptions, at any depth z from the pile head $s_{0z} = s_e + kz$. Moreover, if it is considered that $s_e = kh_e$, then we may use Cases *I* and *II* to solve the problem. The induced vertical stress in the subsoil close to the pile shaft as function of z/r is

from Case *I*: considering kz,

$$[\Delta\bar{\sigma}_{z0}]_k = r_0 k \bar{I}_{zk}$$

from Case *II*: considering $k \cdot h_e$ constant with depth

$$[\Delta\bar{\sigma}_{z0}]_c = r_0 k \frac{h_e}{z_t} \bar{I}_{zc}$$

The sum of these two expressions may be equated to the total pile skin friction with linear variation up to depth z, that is, $2\pi r_0[kh_e z + \frac{1}{2}kz^2]$, and we obtain

$$2\pi r_0 \left[kh_e z + \frac{1}{2}kz^2\right] = \left\{r_0 k \cdot \bar{I}_{zk} + r_0 k \frac{h_e}{z_t} \bar{I}_{zc}\right\} \bar{a}_{ez}$$

from which we get the following expression for the equivalent tributary area:

$$\bar{a}_{ez} = \frac{(2h_e + z)\pi z^2}{z\bar{I}_{zk} + h_e \bar{I}_{zc}\sqrt{2}} \qquad \text{39-VIII.3}$$

The influence values $\bar{I}_{zk}$ and $\bar{I}_{zc}$ are obtained from Fig. 11-VIII.3, or from influence charts in Appendix B. The value of h_e may be estimated by $h_e = s_e/k$, where s_e is the cohesion at the pile head.

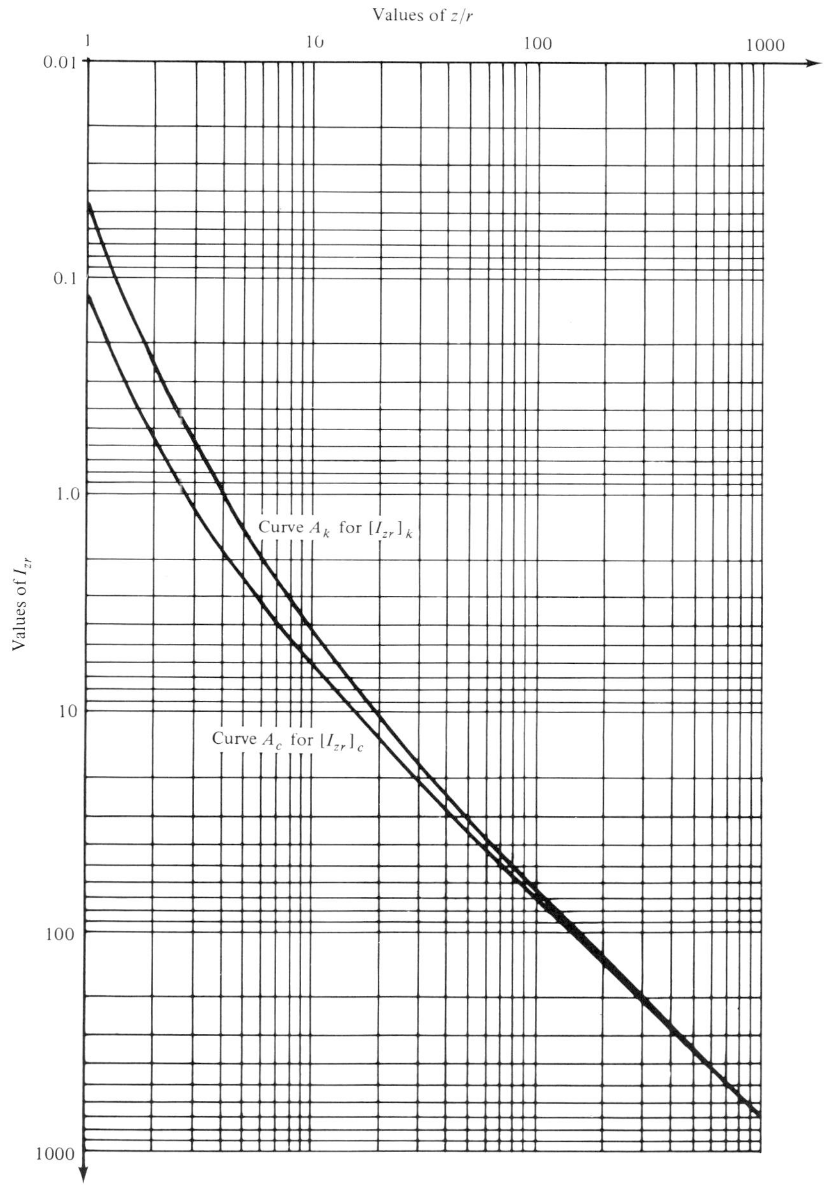

Fig. 11-VIII.3 Influence values for calculation of effective tributary areas.

3.3 Use of Influence Charts

Influence charts have been prepared to estimate easily the effective tributary area $\bar{a}_{ez}$ for corner and edge piles and for interior piles when $\bar{a}_{ez} < \bar{a} = \beta\lambda$. They are used in a similar way as the stress net charts discussed in section III.1. The charts are drawn on dimensionless parameters at a convenient scale λ_k given in the charts, Fig. 12-VIII.3, and in Appendix B, Figs. 8 and 9B.

The geometrical layout of the piles is divided by the required depth z at which the equivalent tributary area is required, and then it is drawn to the scale of the corresponding chart. An example will illustrate the use of influence charts.

Assume that the value of $\bar{a}_{ez}$ is required at a depth of 10 m for the condition $s_{0z} = kz$, and for a group of 16 piles of 0.4 m in diameter. The piles spacing is 2×1.5 m; therefore, the nominal tributary area $\bar{a}_0 = 3.0$ m^2. The drawing is prepared dividing by depth $z = 10$ m, and using the scale of the influence chart. Assume that the corner pile is placed at the center of the chart. The influence

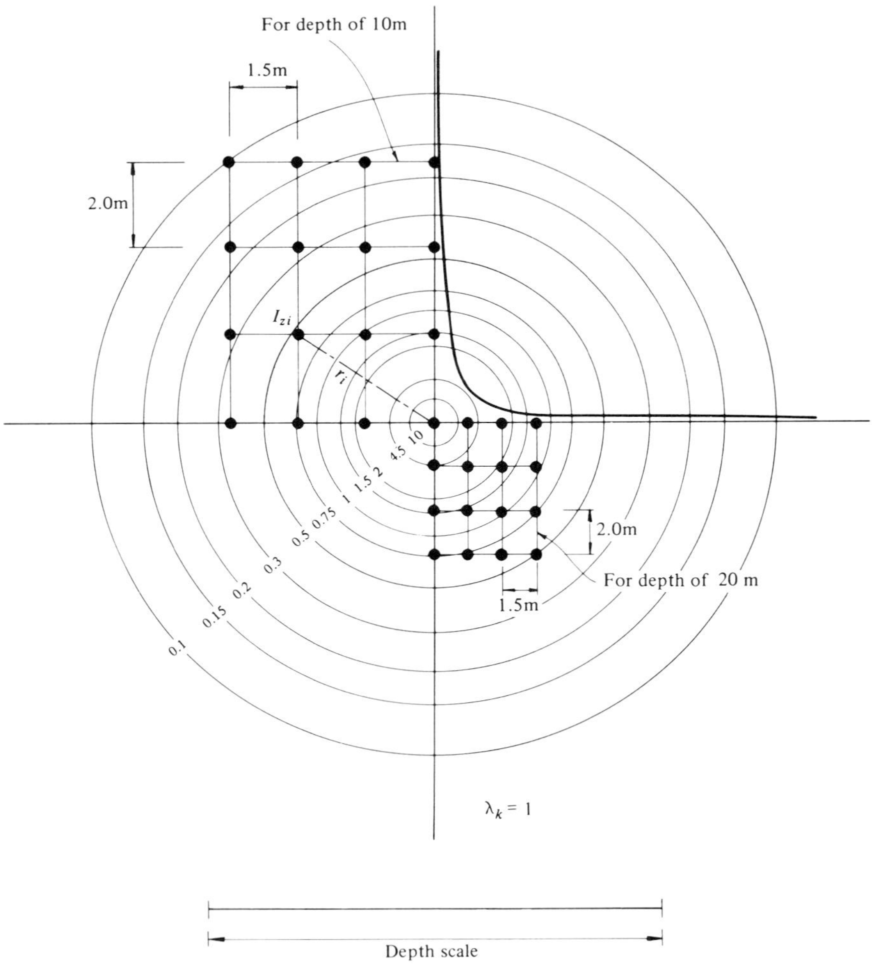

Fig. 12-VIII.3 Friction pile influence chart for determining the equivalent tributary area for $s = kz$.

values for all piles in the group are counted and added, which gives

$$\sum_{1}^{n} [I_{zr}]_k = 8.8$$

From Fig. 11-VIII.3, curve A_k for $z/r_0 = 50$, the value $I_{z0} = 32$ is found; therefore

$$\sum_{0}^{n} [I_{zr}]_k = 32 + 8.8 = 40.8$$

The equivalent tributary areas at a depth of 10 m according to 33-VIII.3, will be:

for one isolated pile $\bar{a}_e = \dfrac{\pi \cdot 10^2}{32} = 9.82 \text{ m}^2$

for the corner pile $\bar{a}_e = \dfrac{\pi \cdot 10^2}{40.8} = 7.7 \text{ m}^2$

The equivalent radius $R_{e1} = 1.56$ m.

In the same way, the values for the other piles may be calculated:

for an edge pile (short spacing) $\bar{a}_{e2} = 7.15 \text{ m}^2$, and $R_{e2} = 1.51$ m

for an interior pile $\bar{a}_{e4} = 6.63 \text{ m}^2$, and $R_{e4} = 1.45$ m

The tributary areas so obtained, however, are limited by the pile spacing. Therefore, the minimum values of the tributary areas should be investigated according to Fig. 13-VIII.3, using the following formulas applied for uniform spacing:

Corner pile

$$\bar{a}_{e1} = \frac{\pi}{4} R_{e1}^2 + \frac{R_{e1}}{2} [F_{A1} \cdot \beta + F_{B1} \cdot \lambda] + \frac{\lambda\beta}{4} - a_p$$

Edge pile on β spacing

$$\bar{a}_{e2} = F_{A2}(R_{e2}\beta) + \frac{\lambda\beta}{2} - a_p$$

Edge pile on λ spacing

$$\bar{a}_{e3} = F_{B3}(R_{e3}\lambda) + \frac{\lambda\beta}{2} - a_p$$

Interior pile

$$\bar{a}_{e4} = \lambda\beta - a_p$$

40-VIII.3

The factors F_A and F_B given in formulas 40-VIII.3 have the following values according to the geometry shown in Fig. 13-VIII.3:

$$F_{A2} = \frac{\pi A}{180^\circ} \frac{R_{e2}}{\beta} + \frac{1}{2} \cos A$$

$$F_{B3} = \frac{\pi B}{180^\circ} \frac{R_{e3}}{\lambda} + \frac{1}{2} \cos B \qquad \text{41-VIII.3}$$

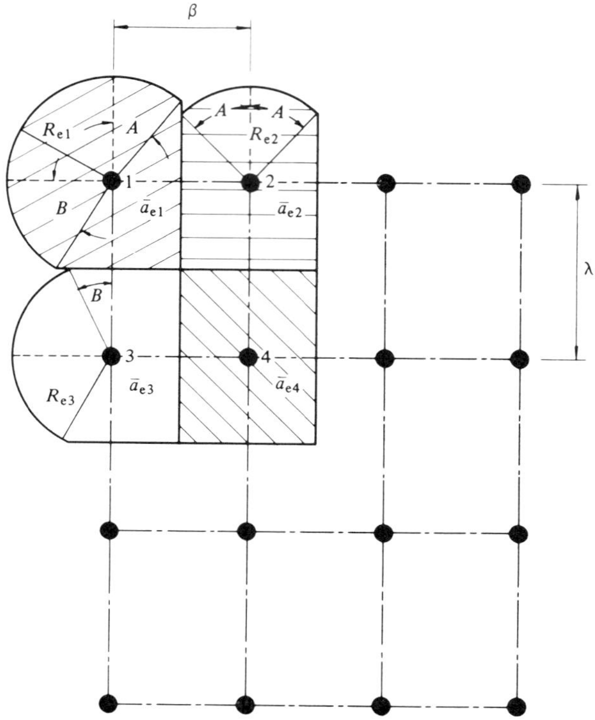

Fig. 13-VIII.3 Tributary areas in pile groups.

Applying numerical values in expressions 40-VIII.3, the tributary areas are, respectively,

$$a_{e1} = 5.1 \text{ m}^2, \quad a_{e2} = 3.5 \text{ m}^2, \quad a_{e3} = 4.1 \text{ m}^2, \quad \text{and} \quad a_{e4} = 2.87 \text{ m}^2$$

Notice that for this particular depth and pile spacing, the values given above based on Fig. 13-VIII.3 offer the minimum values for the effective tributary areas. Therefore, they may be used in algorithms given before to calculate skin friction pile problems.

3.4 Confining Stress at the Pile Point Elevation

To calculate the ultimate bearing capacity of the piles subjected to negative skin friction, it is necessary to know the average change in effective stress $\Delta\sigma_{da}$ at depth d corresponding to the elevation of the pile points. This value may be found after $(NF)_d$ has been calculated to full depth from formulas 16 and 19-VIII.3. Hence, for one pile in the group

$$\Delta\sigma_{d0} \cdot \bar{a}_{ed} = \Delta\sigma_{da} \cdot \bar{a}_0 = (NF)_d \qquad \text{42-VIII.3}$$

Therefore, the vertical average confining stress is given by

$$\sigma_d = \sigma_{0d} - \Delta\sigma_{da} \quad \text{or} \quad \sigma_d = \sigma_{0d} - \frac{(NF)_d}{\bar{a}_0} \qquad \text{43-VIII.3}$$

From the point bearing capacity formula 3-VIII.2, one obtains:

$$Q_{\text{pu}} = \alpha_1 \left[cN_c + \left(\sigma_{0d} - \frac{(NF)_d}{\bar{a}_0} \right) N_q \right] (D_r + 0.1)(\pi r_0^2)$$

Thus, we see that the phenomenon of negative friction on point bearing piles decreases the ultimate point capacity of the piles supported on the bearing sand stratum.

When calculating the compression of the sediments it should be noticed, however, that when using the tributary effective area $\bar{a}_e$ in algorithms for negative or positive friction, the value σ_i corresponds to the stress close to the pile shaft, and $(\sigma_{0i} - \sigma_i)$ or $(\sigma_i - \sigma_{0i})$, respectively, is the change in stress at the same place. Therefore, to compute the average compression between piles one should use the average change in stress; hence, at any elevation

$$\Delta\sigma_i \cdot \bar{a}_{ei} = (\Delta\sigma_i)_{\text{ave}} \cdot \bar{a}_0$$

where $\bar{a}_0$ is the nominal tributary area $(\beta\lambda - a_p)$. (see Section 3.2). Hence, if the change in stress $\Delta\sigma_i$ at depth $z = i$ is known, then the average change in stress in the tributary area is

$$(\Delta\sigma_i)_a = \Delta\sigma_i \frac{\bar{a}_{ei}}{\bar{a}_0}$$

3.5 Allowable Point Bearing Load

After the *NF* and ultimate point bearing load capacity of the pile have been calculated, it is necessary to select the allowable point bearing load at which the piles will work safely with a factor of safety G_s; hence

$$Q_{\text{pa}} + (NF)_d = \frac{1}{G_s} Q_{\text{pu}} \qquad \text{44-VIII.3}$$

from which the allowable point load is determined according to the formula

$$Q_{\text{pa}} = \frac{Q_{\text{pu}}}{G_s} - (NF)_d \qquad \text{45-VIII.3}$$

When the soil properties are known, a value of $G_s = 2$ may be used in practice. When it is necessary to secure piles working under load Q_{pa}, the factor of safety is

$$G_s = \frac{Q_{\text{pu}}}{Q_{\text{pa}} + (NF)_d} \qquad \text{46-VIII.3}$$

Thus we see that when a very low point bearing resistance is encountered, the piles will penetrate continuously into the bearing stratum forced by the negative friction without necessarily being loaded at their heads.

In Mexico City, cases are observed in which the occurrence of this phenomenon leads to corresponding structural damage to buildings supported on this type of foundations because of large differential vertical displacements. To overcome this problem, the foundation should be designed with sufficient rigidity to overcome the differential penetration of the piles. The piles at the corner and edges showing larger penetration induce the buildings to tilt because of uneven settlement of the piles. Building foundations, however, may be designed under these conditions where ground surface subsidence is taking place. The methods and hypotheses used in design of such foundations will be treated in the following sections and in Chapter IX on compensated friction pile foundations. In Appendix E to Chapter VIII, skin friction calculations on a single pile are given based on information reported in the literature from actual-size negative friction tests.

VIII.4 POSITIVE FRICTION ON PILES

4.1 Basic Concepts

The problem of correlating the behavior of single friction piles with friction pile groups represents more difficulties than the case of point bearing piles supported on a thick sand stratum overlaid by soft materials. In the case of skin friction pile tests, it is customary to load the pile to failure and then use a fraction of this load to design the pile group. This empirical practice should be abandoned in case of difficult subsoil conditions, in favor of more technical methods which permit the engineer to visualize the behavior of the pile group. In compressible materials pile groups may settle unevenly, as in the case of deep clay and silt deposits. The behavior of one pile in relation to that of a group will be discussed under simplified working assumptions for the case of compressible soil deposits.

When testing single friction piles to ultimate load, important information may be obtained on the skin friction value when this is measured along the pile shaft. Consider a pile tested in impermeable subsoil conditions where layers of different strength, compressibility and permeability are present. Experience has shown in such cases that the time element is of major importance to the behavior of the test pile (see section VIII.3). When a test is made at full capacity, the skin friction is mobilized completely in all the layers pierced by the pile, independently of their relative rigidity. The skin friction along the pile shaft depends on the loss of shear strength in the soil due to pile driving, and thereafter the regaining of shear strength due to consolidation of the material against the pile shaft. It has been found that complete remolding of the material takes place to a distance of about $0.1r_0$ from the pile shaft and the disturbance follows to a lesser degree up to about one diameter. Moreover, the spacing of the friction piles has much influence in the load carrying capacity of the groups.

From tests on single piles loaded at less than ultimate values, it has been shown that the shear stress along the pile shaft assumes a value corresponding to a fraction of the ultimate shear strength in proportion to the rigidity of the soil strata. Therefore, it is difficult, in stratified subsoil conditions, to determine the working shear stress distribution along the pile shaft for loads below the ultimate. Nevertheless, when the problem is considered as one of ultimate shear strength, the distribution given by the ultimate shear strength or skin friction along the pile shaft may be estimated.

The ultimate load capacity of a positive friction pile is calculated with simplified working assumptions similar to those established for the negative skin friction pile problems. When a *PF*-pile is loaded, the soil mass will support the pile by skin friction; at the same time the friction developed will induce an increment in the vertical stresses in the surrounding soil. Therefore, a *PF*-pile is recognized because the overburden initial soil stresses increase. The forces and stresses involved are shown in Fig. 1-VIII.4. Under ultimate load capacity of the pile, the equilibrium and shear strength equations may be established.

The average shear strength on the element Δz at depth z has the value

$$s_{0i} = c_i' + K_{\phi i} \frac{\sigma_i + \sigma_{i-1}}{2} \qquad \text{1-VIII.4}$$

The ultimate shear transferred into the soil mass is

$$(\sigma_i - \sigma_{0i})\bar{a}_i - (\sigma_{i-1} - \sigma_{0i-1})\bar{a}_{i-1} = \bar{\omega} s_{0i} \Delta z_i \qquad \text{2-VIII.4}$$

and the positive friction and vertical stresses interaction given by

$$(PF)_{i-1} = (\sigma_{i-1} - \sigma_{0i-1})\bar{a}_{i-1} \qquad \text{3-VIII.4}$$

and

$$(PF)_i = (\sigma_i - \sigma_{0i})\bar{a}_i \qquad \text{4-VIII.4}$$

In order to solve for σ_i as performed in case of *NF*-pile problems, the integration may be started at $z = 0$, where $(PF)_{i-1} = 0$, $\sigma_{i-1} = \sigma_{0i-1}$, and $\sigma_{0i-1} = q$ is known. Therefore, substituting expressions 1 and 3-VIII.4 into 2-VIII.4 and solving for σ_i, we obtain the following expression:

$$\sigma_i = \frac{(PF)_{i-1} + \sigma_{0i}\bar{a}_i + (\bar{\omega} c_i' + m_i \sigma_{i-1})\Delta z_i}{\bar{a}_i - m_i \Delta z_i} \qquad \text{5-VIII.4}$$

After the value of σ_i is known, expression 4-VIII.4 is used to find $(PF)_i$, and its value is substituted into 5-VIII.4 to calculate the next values of σ_{i+1} and $(PF)_{i+1}$, respectively. The process is repeated until full numerical integration step by step is performed to depth d, where $(PF)_d = Q_{\text{fu}}$ is obtained. The values at depth $z = i$ of $\bar{a}_i$ and m_i have the same significance as in the case of the *NF*-problem previously treated. The values of the equivalent tributary areas may be obtained from Fig. 11-VIII.3, and influence charts like the one shown in Fig. 12-VIII.3 can be found in Appendix B, Figs. 8 and 9-B.

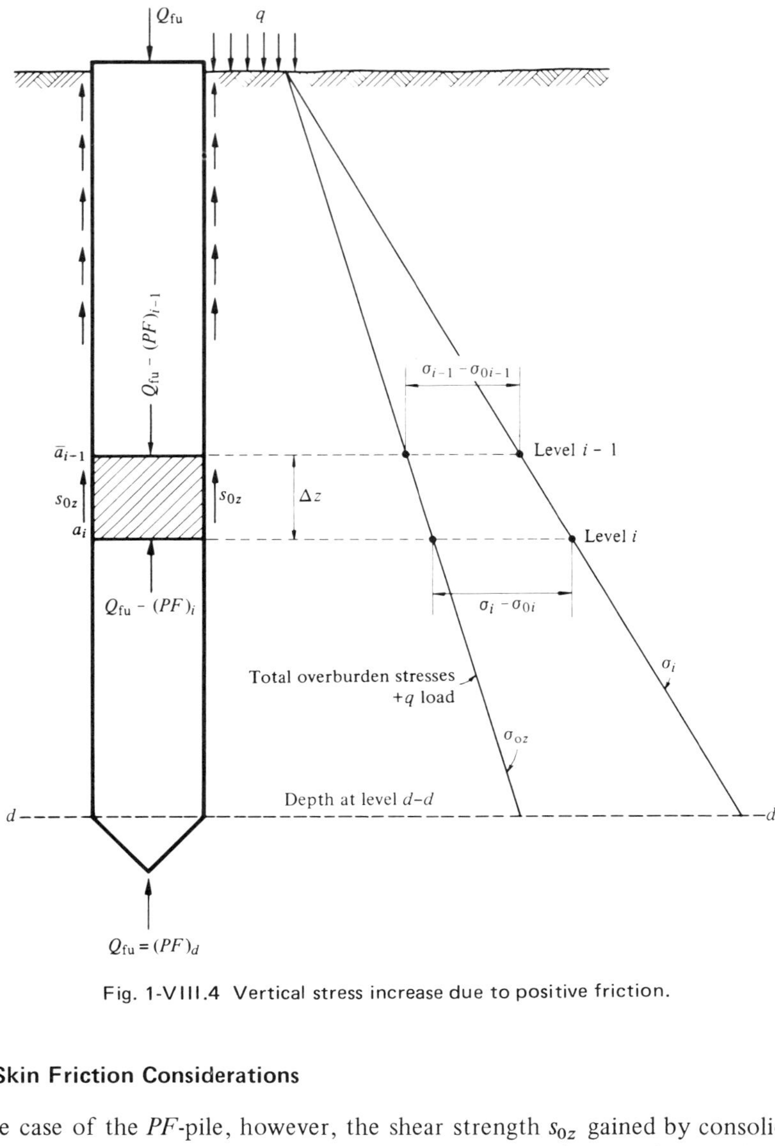

Fig. 1-VIII.4 Vertical stress increase due to positive friction.

4.2 Skin Friction Considerations

In the case of the *PF*-pile, however, the shear strength s_{0z} gained by consolidation may reach at a certain depth a larger value than that of the undisturbed soil surrounding the pile. To find the value of the natural shear strength representing the soil at different depths, undisturbed samples are required to investigate this mechanical property in the laboratory. The test is preferably performed in the triaxial apparatus. The sample is confined with equivalent all around effective stress it had in the field at the depth it was taken:

$$\sigma_c = \frac{1}{3}(1 + 2K_0)\sigma_z$$

The sample is permitted to dissipate all excess pore water pressure because of the application of σ_c and is then sheared under constant volume, that is to say, under undrained conditions. The shear strength may be approximately determined, for practical purposes, by means of unconfined compressive strength tests. The values so determined are slightly lower than if the samples are reconsolidated to the effective stress mentioned above (see section II.4). Therefore, the results obtained by this procedure are on the safe side.

The natural shear strength along the shaft of piles has been investigated by the author (1957) to understand better the behavior of positive friction piles used in foundation design. Investigations have been performed in undisturbed clay samples obtained close to the shaft of old driven piles. The samples were recovered in open pit deep excavations and tested in the laboratory. A set of results is shown in Figs. 2 and 3-VIII.4, from which the following conclusions may be drawn:

(a) A thin skin of perfectly remolded and reconsolidated soil was observed close to the pile shaft, not exceeding a distance of about 5% of the radius of the pile, thus giving rise to shear strengths according to the equation:

$$s_{0z} = K_\phi \sigma_{0z} \qquad \text{6-VIII.4}$$

(b) The water content (Fig. 2-VIII.4) at a distance greater than $1.05r_0$ was observed to remain practically unchanged, showing that there is only a small volume

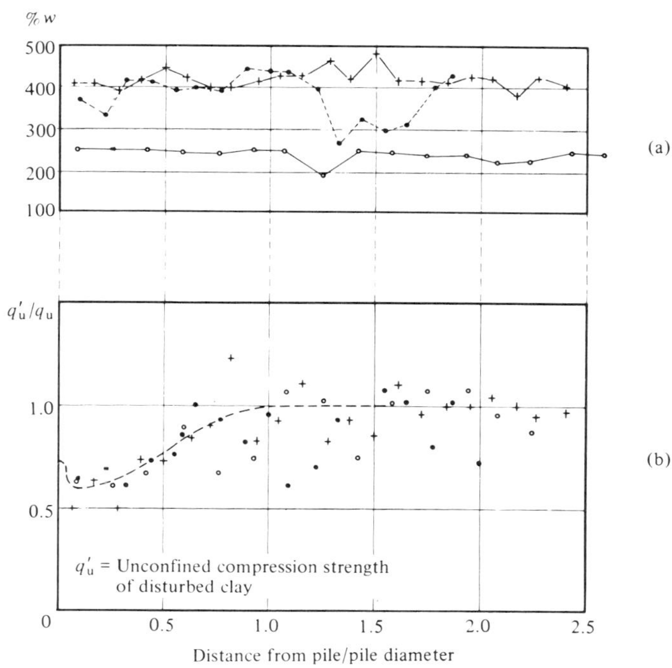

Fig. 2-VIII.4 Shear strength ratio and water content of disturbed clay close to the pile shaft.

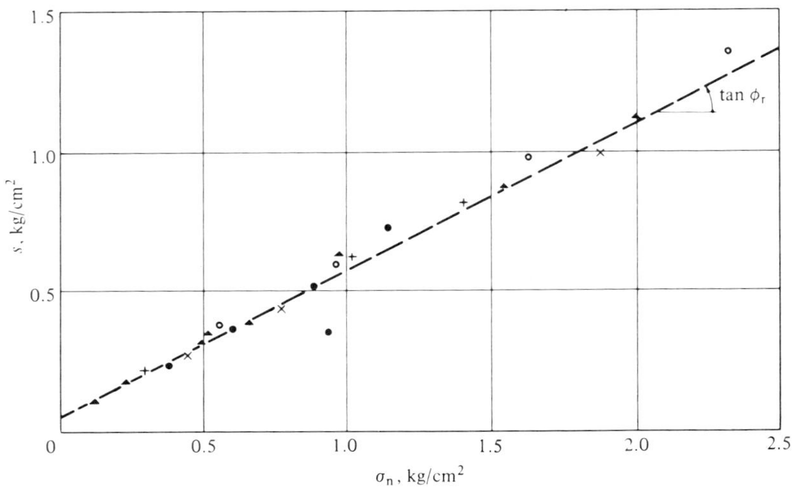

Fig. 3-VIII.4 Skin friction of concrete against remolded volcanic silty clay, obtained by means of slow shear strength tests.

change in the soil after pile driving. This fact may be explained if it is considered that the large excess pore water pressures induced during pile driving at constant volume produce a momentary reduction in the effective stresses and a corresponding expansion of the clay mass in the vertical direction. Upon dissipation of the excess pore water pressure and reconsolidation of the soil, the vertical effective stresses have the tendency to come back to its initial value close to the pile shaft. The water content in the soil zone close to the pile shaft has also the tendency to preserve its original value. This phenomenon is aided by the high thixotropic properties of the clay. Even though there is no appreciable water content change in the soil under these circumstances, a thin layer of remolded material is consolidated close to the pile shaft.

(c) The natural shear strength of the clay is affected by pile driving to a distance of one diameter away from the pile shaft. Close to the pile shaft, the shear strength of the soil is found reduced to about 60% of its natural undisturbed value (Fig. 2-VIII.4). The minimum shear strength of the semidisturbed clay extends to a distance of about $\bar{r}_0 = 1.1r_0$ from the center of the pile. The shear strength at this zone is approximately

$$s_{0z} = 0.3q_{uz} \qquad \text{7-VIII.4}$$

in which q_{uz} is the undisturbed natural consistency of the soil at depth z.

To investigate the ultimate frictional load at which a *PF*-pile starts to yield, it is necessary to investigate the minimum values of the shear strength close to the pile shaft, by means of formulas 6 and 7-VIII.4, respectively. The time element should be also considered. The values of $s_{0z} = c_z = 0.3q_{uz}$, representative of the shear strength of the strata close to the pile shaft in homogeneous saturated clay, may be used for the calculation described above. Therefore, in this case K_ϕ may be assumed equal to zero and equation 5-VIII.4 takes the following form:

$$\sigma_i = \sigma_{0i} + \frac{(PF)_{i-1} + \bar{\omega} c_i \Delta z_i}{\bar{a}_i} \qquad \text{8-VIII.4}$$

from which we obtain

$$(PF)_i = \bar{\omega} c_i \Delta z_i + (PF)_{i-1}$$

and finally the positive friction at level *i-i* is

$$(PF)_i = \bar{\omega} \sum_0^d c_i \Delta z_i \qquad \text{9-VIII.4}$$

Formula 9-VIII.4 is used when, at depth z_0 and below,

$$0.3 q_{\mathrm{u}z_0} \bar{\omega}_2 \leqslant \left[c_i + K_\phi \frac{\sigma_i + \sigma_{i-1}}{2} \right]_{z_0} \cdot \bar{\omega}_1 \qquad \text{10-VIII.4}$$

Therefore, the total ultimate positive skin friction of the pile from the ground surface to full depth d is

$$Q_{\mathrm{fu}} = [(PF)_K]_0^{z_0} + \bar{\omega}_2 \sum_{z_0}^d c_i \Delta z_i \qquad \text{11-VIII.4}$$

The shear strength distribution along the pile shaft for a pile under these circumstances is shown in Fig. 4-VIII.4. In the upper part it corresponds to the shear strength gained by reconsolidation of the soil after pile driving according to the concept given by 6-VIII.4; in the lower part of the pile the soil is considered with its natural partially disturbed shear strength according to the concept expressed by 7-VIII.4. A practical example showing this case is found in the appendix to Chapter VIII.

In the case of pile groups, the *PF*-load capacity of each pile in the group may be computed with the formulas just described. The tributary area at the depth required is determined by the same method used for negative skin friction calculations. The algorithms established for either *NF* or *PF* may be used readily to form a computer program with which the calculations may be performed speedily.

4.3 Pile Group as a Single Unit

When a pile group is not very large, it is necessary to investigate the group action as a whole. This situation may be encountered for pile spacings of 3 to 6 diameters—the piles and soil work like a rigid unit, and the problem becomes one of base ultimate bearing capacity of the group working like a large pier (Fig. 5-VIII.4). The ultimate load capacity of the group is the sum of the positive friction group capac-

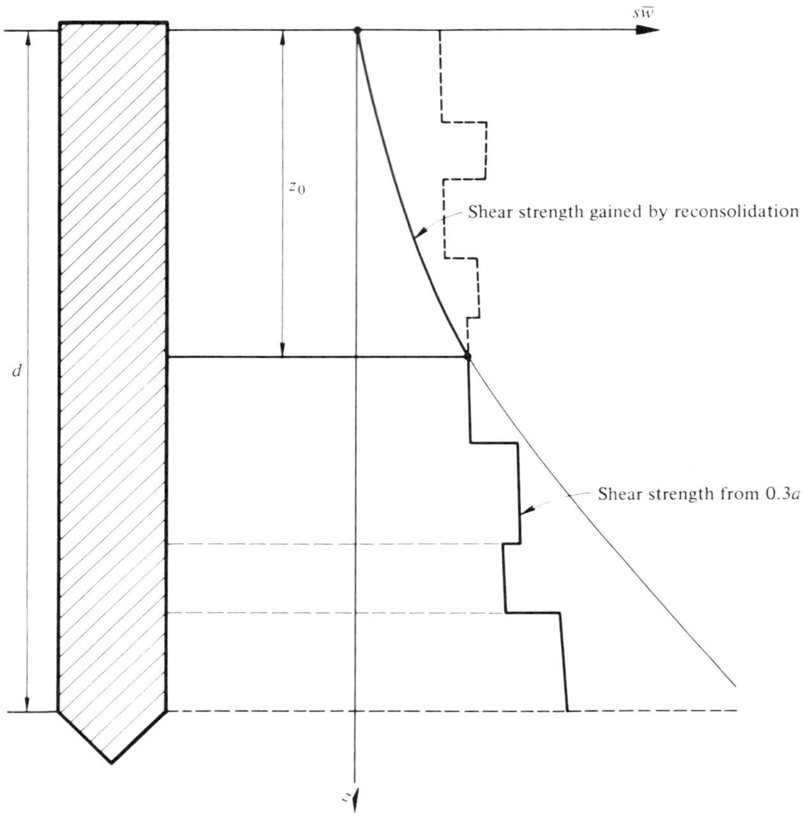

Fig. 4-VIII.4 Shear strength distribution in clay along pile shaft.

ity $(PF)_g$ and the ultimate base group capacity Q_{gb}. Hence

$$Q_{gu} = Q_{gb} + (PF)_g \qquad \text{12-VIII.4}$$

When the group breaks into the ground, the $(PF)_g$ is mobilized at the perimeter of the pile group. The potential surface of sliding will be located along the natural soil at the perimeter $4(L + B)$ in which $2L$ is the length and $2B$ is the width of the group. Let us call c_i the average unit shear strength measured by one half of the unconfined compressive strength of the impervious material representative of the layer with thickness d_i, overlying the pile points; the group positive friction is then

$$(PF)_g = 4(L + B) \sum_1^n c_i d_i \qquad \text{13-VIII.4}$$

On the other hand, the base of the group of piles rests on a deep deposit of impervious soils. The ultimate load capacity of the group may be obtained from

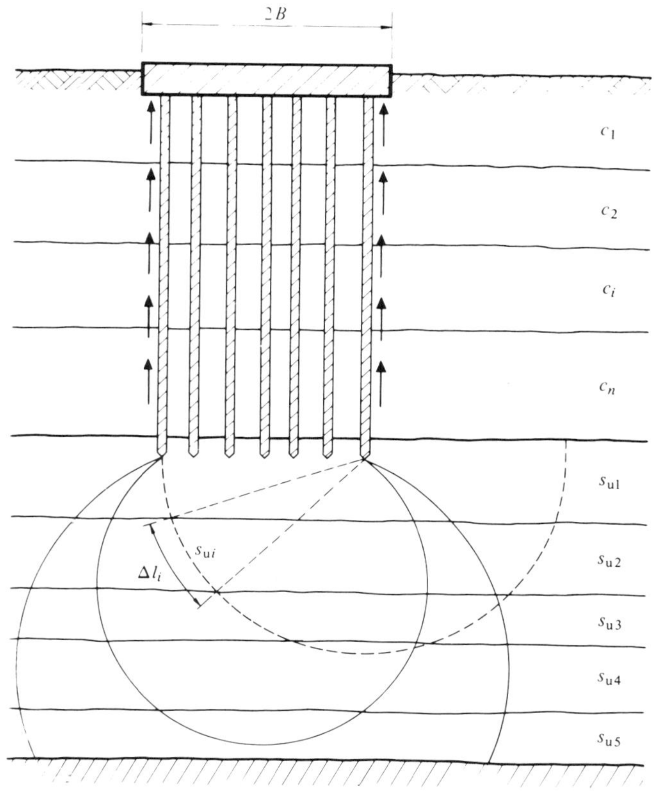

Fig. 5-VIII.4 Pile group base load capacity.

formula 23-VIII.2 for $D = 0$; hence

$$Q_{gb} = 3.64\alpha_1 L \sum_1^n s_{ui}\Delta l_i \qquad \text{14-VIII.4}$$

Here, the quantity s_{ui} is equal to half of the confined compressive strength determined in a consolidated-undrained test performed at the volumetric stress representative of the stratum cut by the potential surface of sliding corresponding to the arc segment Δl_i. This is determined by current laboratory methods in well taken undisturbed samples of the clay strata (see section II.4). The factor of safety of the group breaking into the ground is

$$G_s = \frac{Q_{gb} + (PF)_g}{[W_a]} \qquad \text{15-VIII.4}$$

where W_a is the allowable load that may be applied to the pile group under this condition with factor of safety G_s.

Nevertheless, it should be considered that under normal working conditions with load W_a, the group may be subjected to negative friction because of the following two reasons:

1. The upper soil sediments are compressible and may be found under consolidation.
2. When the piles are driven into the ground, a volume of soil equal to that of the piles is displaced and heave of the surrounding ground surface takes place. Upon dissipation of high pore water excess pressures, the heave is followed by a downward vertical displacement.

The group negative friction $(NF)_g$ may be calculated by means of expression 13-VIII.4. Notice that in case of group action, theoretically $(PF)_g = (NF)_g$. Therefore, the negative friction on the full group of piles is developed because the base of the group shows a smaller rate of subsidence than the ground surface. The base ultimate load capacity will be approximately the same as previously analyzed, in formula 14-VIII.4; therefore, the factor of safety

$$G_s = \frac{Q_{gb}}{W_a + (NF)_g} \qquad \text{16-VIII.4}$$

In this case, the average vertical displacement at the center of the group because of the compressibility of the sediments under the base area is calculated using an average unit load:

$$q_a = \frac{0.25\,[W_a + (NF)_g]}{L \cdot B} \qquad \text{17-VIII.4}$$

The increment of stresses in the subsoil due to the unit load applied at the elevation of the pile points is estimated from stress nets (section III.1). If the soil compressibility is represented by m_{vi} for layer d_i, then the vertical displacement of the group of piles is

$$\delta = \sum_{1}^{n} m_{vi} \cdot \Delta\sigma_i \cdot d_i \qquad \text{18-VIII.4}$$

It may be noticed that for the same factor of safety the value obtained for $[W_a]_{NF}$ is considerably smaller than the one obtained for $[W_a]_{PF}$; for $G_s = 1$, we obtain

$$\frac{[W_a]_{NF}}{[W_a]_{PF}} = \frac{Q_{gb} - (NF)_g}{Q_{gb} + (PF)_g} \qquad \text{19-VIII.4}$$

Therefore, W_a should be carefully investigated.

When the pile points are not designed to penetrate continuously, the compressible bearing soil deposits and the settlement of the group are the governing factors; the phenomenon of negative friction takes place, if not to the full length of the pile group, at least in the upper part; hence in this case, the allowable load of the building is

$$W_a = \frac{Q_{gb} + [(PF)_g]_d^i}{G_s} - [(NF)_g]_0^i G_{NF} \qquad \text{20-VIII.4}$$

The quantification of the positive and negative friction taking place should be made after study of the velocity of the ground surface subsidence and the probable rate $\partial\delta/\partial t$ of the base of the pile group. The time element here plays a very important role in the settlement of the base of the group in the application of equations 17 and 20-VIII.4 (see Chapter X).

In case of ground surface subsidence, it is concluded that groups of piles do not work under positive friction if their points do not penetrate continuously into the soil underlying the level of the pile tips. The group works as a rigid unit or a pier-like unit, subjected to negative friction, whereupon the behavior of the group depends on the load bearing capacity at the base, measured by the average shear strength and compressibility characteristics of the materials underlying that base. Therefore, it should be investigated by means of equations 18 and 20-VIII.4. The value of the factor of safety G_{NF} affecting the group negative friction term may be selected according to the reliability of the mechanical and stratigraphical properties of the subsoil, and to the experience of the foundation engineer in these types of foundation problems.

VIII.5 BEHAVIOR OF PILE FIELDS BASED ON MECHANICAL MODELS

5.1 Basic Considerations

The mechanics and methods of calculation for point bearing capacity and positive and negative friction of piles have been discussed in sections 2, 3 and 4. Using these methods, one is able to design under practical simplified working hypotheses pile groups for various conditions, and for different purposes in foundation engineering, using one or the combined action of *PB*, *PF* or *NF*. To visualize pile-soil interaction and to understand the separate types of behavior mentioned above or their combination from the practical foundation engineering point of view, it is convenient to base our reasoning on a mechanical model representing the pile-soil system. The mechanical model may be idealized as shown in Fig. 1-VIII.5.

The soil in the model is represented by a series of porous rigid plates that can slide along the piles, exerting on them Coulomb friction. Between the rigid thin plates, resisting elements take the interacting effective forces. The spaces between plates represent the soil pores, which may be filled with water or partially saturated. Therefore, the soil is properly represented by the porosity of the plates standing for the porosity of the soil, controlling permeability; the resisting elements represent the skeleton structure of the soil with stress-strain-time properties, as discussed in section II.3. The frictional strength along the piles is given by a mechanical brake-

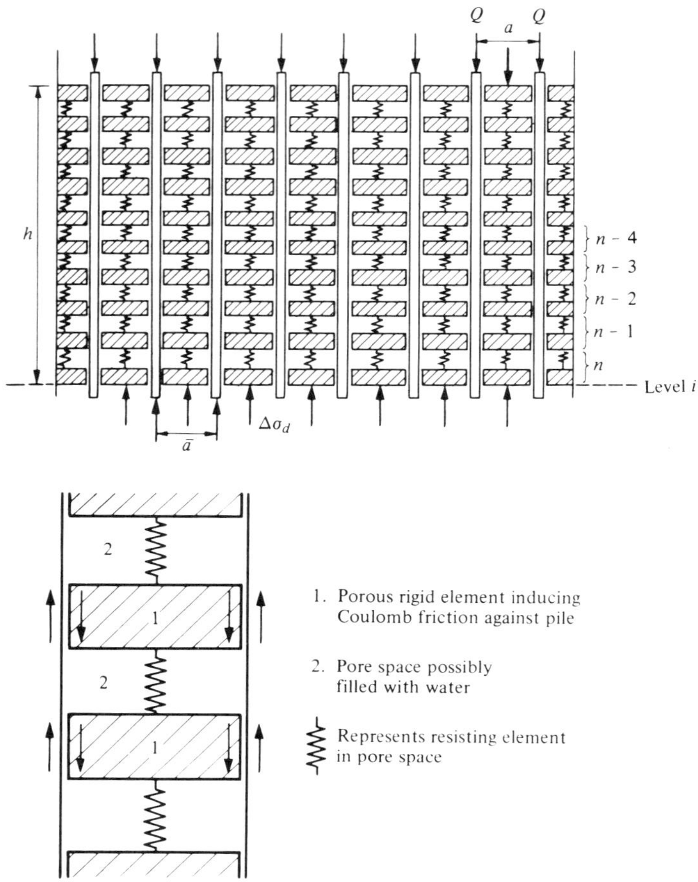

Fig. 1-VIII.5 Mechanical model for study of friction piles.

like element attached to the plates and transmitting frictional resistance along the pile shaft.

Assuming that the piles do not pierce the rigid plates, the model works to represent the soil mass, which may be assumed to have different mechanical properties in the horizontal and vertical directions, respectively. Therefore, the pile-soil interaction in the soil mass may be assumed with the desired anisotropy or compressibility varying with depth. Various practical cases will be studied using this model under the following working conditions:

1) The piles are considered incompressible with respect to the soil mass.
2) The soil at the pile shaft is working under shear strength ultimate conditions.
3) The shear stresses due to elastic action are disregarded.
4) Negative or positive friction calculations are performed as described in sections 3 and 4, and the hypotheses therein mentioned apply.

5) The results from analysis of the pile-soil interaction mechanical model have sufficient accuracy for foundation engineering purposes. Their limitations, however, should be understood and a proper factor of safety assigned.

With the model and working hypotheses mentioned, the foundation engineer has a powerful tool for the interpretation of the mechanical behavior of pile groups or pile fields, and is in the position to make a proper diagnosis of similitude of the pile problem using simple statics. Several cases commonly encountered in practice are described to illustrate the proposed method.

5.2 Case I: Positive Friction, No Point Resistance

This case is to illustrate the action of the soil-pile interaction model for a *PF*-pile, where the positive friction is only partially mobilized. Assume an equal load Q is applied on the piles (Figs. 1 and 2-VIII.5). The ground surface remains unloaded, and the load is applied gradually in all the piles at the same time. Furthermore, we assume that the piles are without point bearing. The load on the pile field has to be taken entirely by the soil tributary area for each pile at the level of the pile points. If $\bar{a}$ is called the tributary area of one pile, then $Q_f = \bar{a}\Delta\sigma_d$, where $\Delta\sigma_d$ is the average vertical increment of effective stress because of the load Q_f per pile at depth d. When the load Q_f is small, the n-th plate will take all the load by friction against the

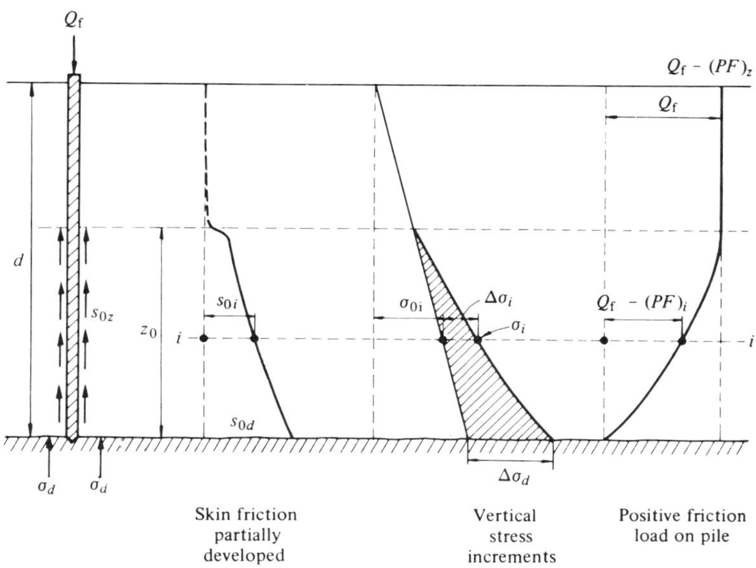

$s_{0i} = K_{\phi i}\sigma_i$ = Shear strength at level i–i

$\Delta\sigma_i$ = Average effective stress increment due to PF at level i–i

Fig. 2-VIII.5 Case Ia—Positive friction partially mobilized, no point resistance.

pile shaft, without relative movement between piles and soil (Fig. 1-VIII.5). The next plate (n - 1) will not receive stresses close to the pile shaft, since the n-th spring is not deformed. At the central part, however, the upper plates overlying the n-th plate may deform if the plates are considered flexible, but in the model they are assumed to be rigid for simplicity. Therefore, the upper plates will not receive any stresses through the springs until the load is increased and the underlying plate has mobilized its ultimate strength against the pile shaft. The implications of this assumption have been discussed using the concept of equivalent effective tributary area in section VIII.3.2.

Now the load is further increased and the ultimate shear strength of the n-th plate against the pile is mobilized. Obviously, upon consolidation of the n-th pore space, the n-th spring will transmit the excess load to plate (n - 1). Plate (n - 2) will not mobilize its shear strength along the pile shaft until the load is sufficiently large so that spring (n - 1) will be stressed by the excess load and transmit load to plate (n - 2). Increasing progressively the load on the piles, and after consolidation of the pore spaces has taken place, the shear strength will be also mobilized progressively in the rigid plates along the pile shaft up to the point at which plastic equilibrium is obtained for an applied load Q_{z_0} on the piles. Assuming plastic equilibrium is reached at a distance z_0 from the tip of the piles

$$Q_{z_0} = \bar{\omega} \int_0^{z_0} s_{0z}\, dz, \text{ and } (PF)_{z_0} = Q_{z_0} \qquad \text{1-VIII.5}$$

The overburden effective stresses are increased, since the load of the pile has to be gradually supported by vertical stresses in the tributary area (Fig. 2-VIII.5). At any level i-i

$$\Delta\sigma_i = \frac{(PF)_i}{\bar{a}} \qquad \text{2-VIII.5}$$

and the load the pile is carrying at the same level is

$$Q_{fi} = Q_{z_0} - (PF)_i \qquad \text{3-VIII.5}$$

Notice that at the upper part $d - z_0$ of the piles, the shear strength is not mobilized because the plates are considered rigid, and the piles are not under their full skin frictional capacity; any elastic stresses developed in this section of the pile are disregarded.

To find the total skin friction load capacity of the piles (Fig. 3-VIII.5) the load Q_f should be further increased until the shear strength of the soil along the total length of the pile shaft has been fully mobilized; then

$$Q_{fu} = \bar{\omega} \int_0^{d} s_{0z}\, dz \qquad \text{4-VIII.5}$$

The shear strength distribution along the pile shaft, the vertical increment of stresses along the soil within the piles and the pile load distribution are represented in a qualitative way for this case in Fig. 3-VIII.5.

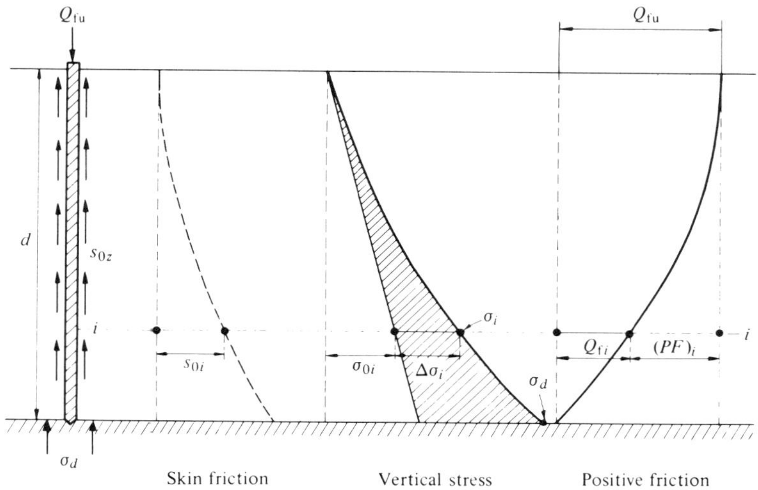

Fig. 3-VIII.5 Case Ib—Positive friction totally mobilized, no point resistance.

5.3 Case II: Positive Friction and Point Resistance

When the points of the piles are bearing on a loose sand stratum, the ultimate total load is reached when the point of the pile is forced to penetrate the bearing stratum (Fig. 4-VIII.5), hence:

$$Q_d = Q_{pu} + (PF)_d \qquad \text{5-VIII.5}$$

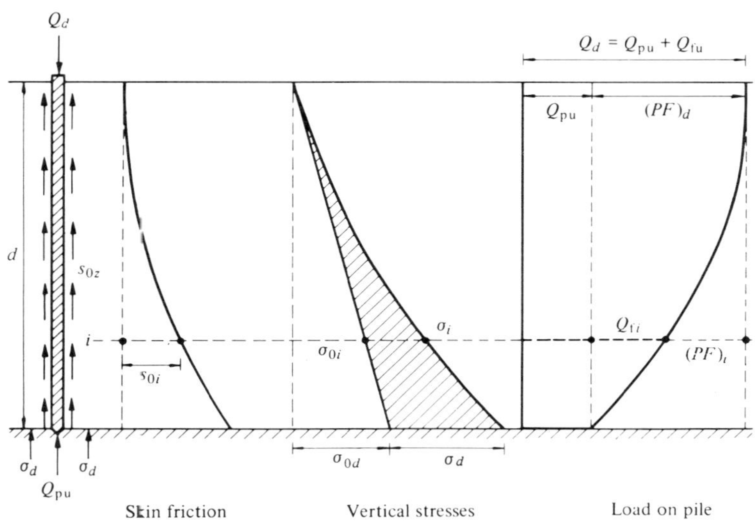

Fig. 4-VIII.5 Case II—Positive friction and point resistance.

The shear strength distribution, the increment of vertical stresses in the soil, and load distribution of the pile are shown in a qualitative way in Fig. 4-VIII.5. Here it may be noticed that the overburden effective stress σ_{0d} at the level of the pile points is increased because of the load transfer of the *PF* into the soil mass. This effect has the tendency to increase the point bearing capacity at the pile points.

5.4 Case III: Negative Friction

This case illustrates the mechanics of the soil-pile interaction for negative friction due to a consolidating soil deposit of thickness d, already discussed in section VIII.3. The piles are point bearing into a firm stratum. Their points do not penetrate because of the load imposed. The soil deposit above the pile tips has a continuous vertical relative displacement with respect to the piles; therefore, the shear strength of the soil against the pile shaft is fully mobilized, represented by the downward movement of the rigid plates. Therefore, the total *NF* on the pile is

$$(NF)_d = \bar{\omega} \int_0^d s_{0z}\, dz \qquad \text{6-VIII.5}$$

The phenomenon produced in the soil mass is qualitatively represented in Fig. 5-VIII.5. The shear strength distribution along the pile shaft is shown. In this case, part of the weight of the rigid plates in the model have the tendency to hang from the piles which are firmly supported on the hard bearing stratum, and by the same token, a reduction of the overburden initial stresses in the soil mass takes place. The transfer of the soil mass weight into the piles expressed at level *i-i* is $(\sigma_{0i} - \sigma_i)\bar{a}$,

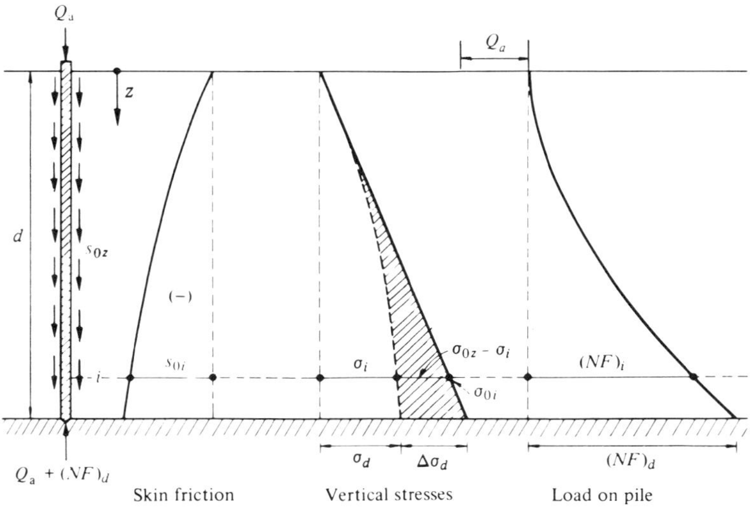

Fig. 5-VIII.5 Case III—Negative friction.

and the total load on the pile point due to negative friction at level d-d is

$$Q_p = Q_a + (NF)_d \qquad \text{7-VIII.5}$$

The average reduction in the overburden effective stresses taking place at level d-d is equal to $\Delta\sigma_d = (NF)_d/\bar{a}$; therefore, a reduction in the ultimate point bearing capacity of the piles takes place. This phenomenon cannot be overlooked in foundation engineering practice.

When negative friction is high and the bearing stratum is weak, penetration of the pile points takes place and damage of the structure supported on the pile field may be expected. This is especially so when the stratum bearing the piles has nonuniform shear strength characteristics over the pile field. Many cases of buildings, in practice, where this has happened may be encountered. It is common to find buildings with rigid foundations tilting; they have to be underpinned with more piles to absorb the NF and give more point bearing capacity to the pile field in those places where yielding of the point bearing piles is observed.

5.5 Case IV: Negative Friction Used for Building Support

The negative friction on piles has been used empirically to support foundations in Mexico City (González Flores, 1950). The piles are driven to firm ground and allowed to pierce through the foundation slab, therefore permitting the load of the building to rest at the ground surface (Fig. 6-VIII.5). To analyze this particular problem, first let us assume that no load is applied on the head of the piles, the total unit load of the building being applied on the ground surface between the piles. Under these circumstances, the unit load that may be applied at the ground surface is

$$q_a = \frac{\bar{\omega}}{\bar{a}} \int_0^d s_{0z}\, dz \qquad \text{8-VIII.5}$$

Here, the shear strength along the pile shaft is

$$s_{0z} = K_\phi(\sigma_{0z} + q_a - \Delta\sigma_z)$$

Notice that the overburden effective stress σ_{0z} is increased because of the load applied at the ground surface. This action is beneficial since a larger friction on the pile shaft is achieved in the upper part of the deposit to hold the load applied at the surface. Nevertheless, consolidation should be expected in the compressible soil between the piles in the upper part of the soil mass, where the material usually has higher compressibility, and is subjected to the largest vertical stresses induced by the surface load. Furthermore, it should be expected that the ultimate negative friction load $(NF)_d$ along the pile shaft is not uniform for each pile in the field. Therefore, a factor of safety should be added. This is achieved by mechanical means placing in each one of the piles a mechanical control device, with which one can apply a fraction of the working load to the pile head. This device may be adjusted periodically in order to eliminate possible differential set-

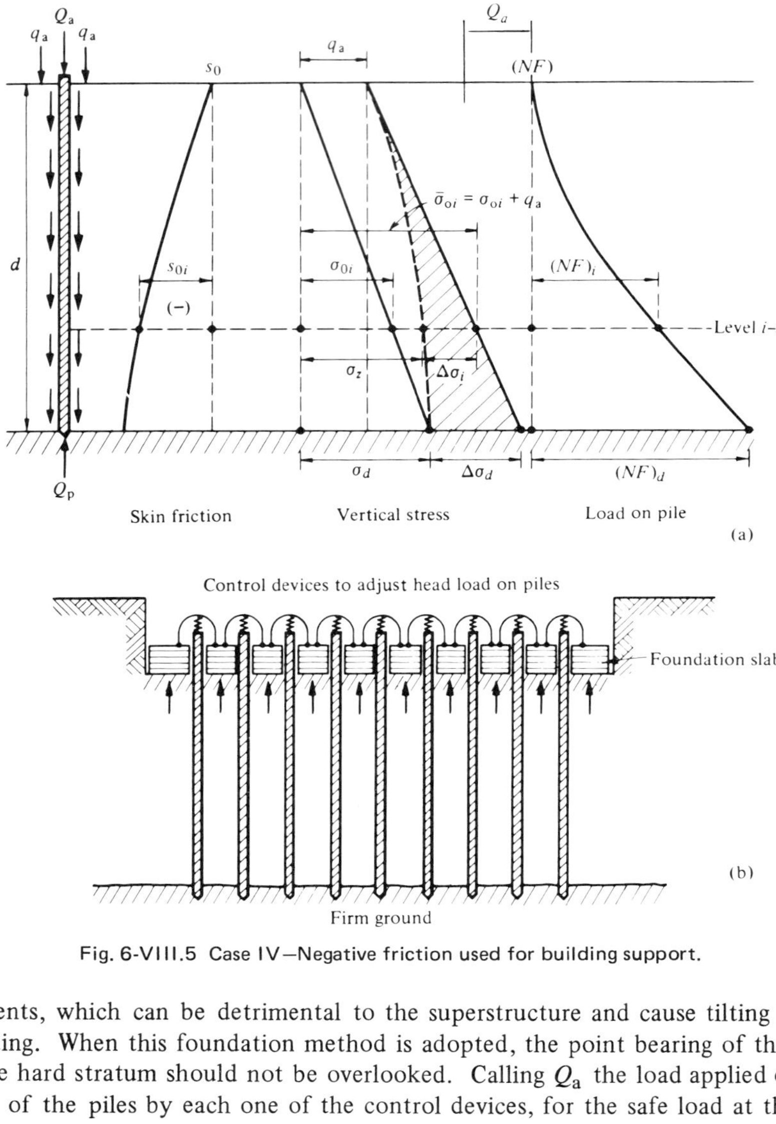

Fig. 6-VIII.5 Case IV—Negative friction used for building support.

tlements, which can be detrimental to the superstructure and cause tilting of the building. When this foundation method is adopted, the point bearing of the piles in the hard stratum should not be overlooked. Calling Q_a the load applied on the head of the piles by each one of the control devices, for the safe load at the pile points we have

$$Q_{pa} = Q_a + (NF)_d \qquad \text{9-VIII.5}$$

where $(NF)_d$ is the negative skin friction load produced on the piles by the surface load q_a including the effect of ground surface subsidence. The factor of safety of such a foundation is controlled by its point bearing capacity; hence

$$G_s = \frac{Q_{pu}}{Q_{pa}} \qquad \text{10-VIII.5}$$

The ultimate point bearing load should be investigated, as discussed in previous cases, considering a possible reduction in the confining stress at the pile tips. See Chapter X, Case III.

5.6 Case V: Restriction of Stress Relief in Soil Mass

This case is shown in Fig. 7-VIII.5, and takes place when friction piles are placed in the ground to depth d. The piles are considered without point bearing and are not loaded at their heads. Sometime after the pile field has remained driven in the ground, excavation takes place to a certain depth inducing a vertical stress relief q_{ex}. Upon stress relief, response is obtained from swelling or from the elastic elements of the soil represented in the model by the springs. The expansion will cause a mobilization of the shear strength in the upper part of the pile producing

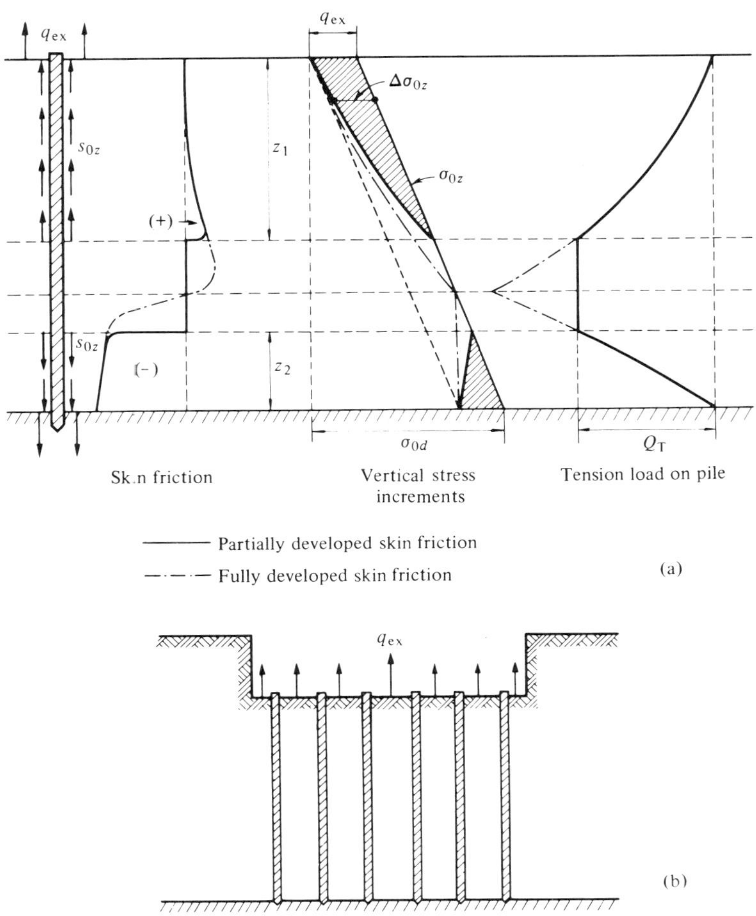

Fig. 7-VIII.5 Case V—Stress relief in soil mass, no point resistance, no load on pile.

a friction force to a depth z_1 until:

$$q_{ex} \cdot \bar{a} = \int_0^{z_1} s_{0z}\,dz \qquad \text{11-VIII.5}$$

This action will introduce a tension load in the pile equal to the value given by expression 11-VIII.5. The tension load is shown qualitatively in Fig. 7-VIII.5, and will be taken by friction in the lower portion of the pile to a distance z_2 from the tip of the piles, where the subsoil material may have a larger shear strength and less compressibility. Hence, for equilibrium we should write

$$\int_0^{z_1} s_{0z}\,dz = \int_0^{z_2} s_{0z}\,dz \qquad \text{12-VIII.5}$$

From expressions 11 and 12-VIII.5, the values of z_1 and z_2 may be obtained. In this case, we observe that the piles can be used to preserve the precompressed condition of the soil, thereby minimizing heave due to the excavation load relief. In effect, the elastic response without piles is

$$\delta_e = \int_0^{d} M_{ez}(\Delta\sigma_{ex})_z dz \qquad \text{13-VIII.5}$$

If the piles restrict the vertical upward displacement, the elastic heave will be only that represented by the shaded areas shown in Fig. 7-VIII.5, corresponding to the vertical effective stress relief diagram:

$$\delta_{ep} = \int_0^{z_1} M_{ez}(\Delta\sigma_{ex})_z dz + \int_0^{z_2} M_{ez}(\Delta\sigma_{ex})_z dz \qquad \text{14-VIII.5}$$

To illustrate the importance of this phenomenon, we assume M_{ez} constant throughout the deposit. Then

$$\delta_e = M_e \cdot q_{ex} \cdot d$$

and

$$\delta_{ep} \doteq \tfrac{1}{2} M_e q_{ex} z_1 + \tfrac{1}{2} M_e q_{ex} z_2$$

Hence

$$\frac{\delta_{ep}}{\delta_e} = \frac{z_1 + z_2}{2d} \qquad \text{15-VIII.5}$$

When friction is fully mobilized, $z_1 + z_2 = d$, and therefore the minimum restriction will be 1/2; the heave will be reduced in the same proportion. This action is extremely important in design of compensated friction pile foundations (Chapter IX). The subsequent settlement will also be reduced since the elastic response is restricted; the soil preserves part of its original precompressed conditions, owing to the presence of the friction piles.

Piles driven into a hard stratum may be able to take the full tension $Q_T = q_{ex} \cdot \bar{a}$. In this particular case, the restriction obtained may be larger since $z_2 = 0$. When the skin friction is mobilized to the total length of the piles, the restriction again has the tendency to approach 1/2, but if by the number and spacing of piles the skin friction is only mobilized to 1/2 of the pile length the restriction will be 3/4. This procedure thus considerably reduces the E-heave.

5.7 Case VI: Friction to Reduce Consolidation of Soil Mass

There are cases in which it is necessary to restrict the deformation of a consolidating soil mass when at the ground surface a unit load q is applied. This case is represented by the model shown in Fig. 8-VIII.5. The pile field under this condition will absorb friction in the upper and the lower parts, respectively, such that $(NF)_{z_1} = (PF)_{z_2}$, or

$$q\bar{a} = \int_0^{z_1} s_{0z} dz = \int_0^{z_2} s_{0z} dz \qquad \text{16-VIII.5}$$

This case is the reverse of Case V; the vertical displacement of the ground surface is caused by the compression of the shaded areas shown in the stress diagram of Fig. 8-VIII.5. Assuming that the compressibility of the soil is constant on the full length and that the stress variation is rectilinear, the vertical displacement in the zone of the pile field is given by

$$\delta_{ep} = \frac{1}{2} m_v q(z_1 + z_2)$$

against the settlement of the stratum without piles,

$$\delta_c = m_v q \cdot d$$

Therefore, the settlement in the pile zone is restricted in the same proportion as the expansion of the soil for Case V; hence

$$\frac{\delta_{ep}}{\delta_c} = \frac{z_1 + z_2}{2d}$$

The total vertical displacement, however, is $\delta_{ep} + \delta_2$, where δ_2 represents the settlement of the compressible soil strata under the pile points. In practice, however, the points of the pile field are always taken to the interface of materials of lower compressibility as compared with the sediments pierced by the piles.

When piles are bearing on a firm stratum, this case is similar to Case IV. The consolidation of the soil mass because of the applied load q at the ground surface may be restricted approximately in accordance to $z_1/2d$, depending on the spacing of the piles. This problem has applications in practice for large storage areas

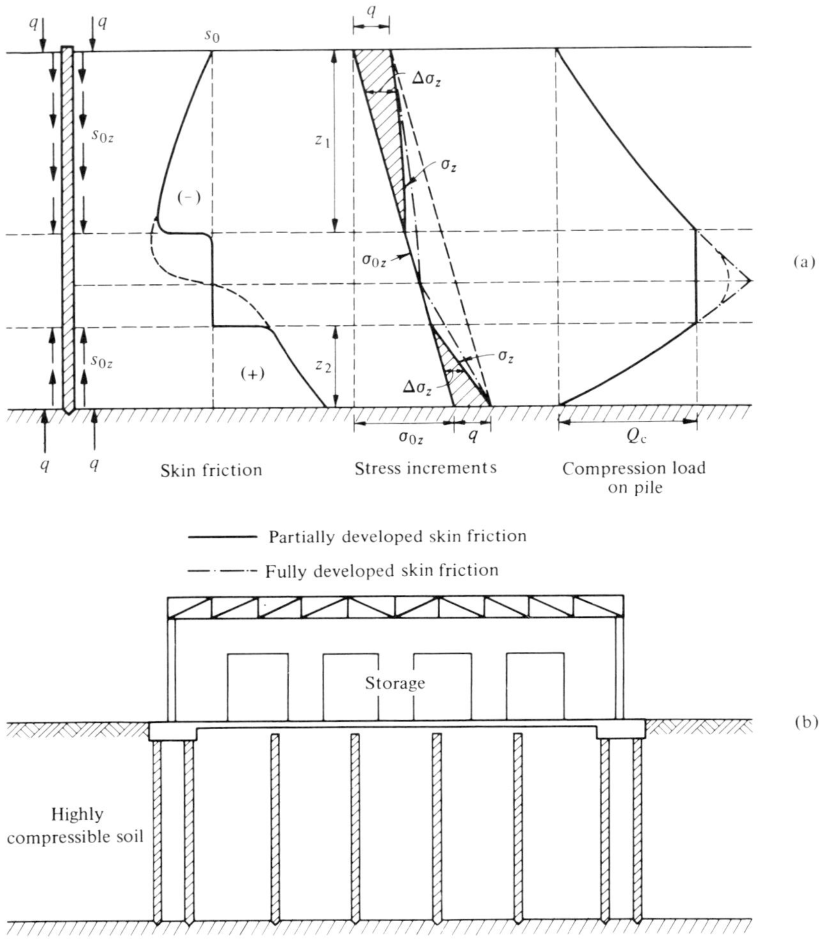

Fig. 8-VIII.5 Case VI—Friction to reduce consolidation in soil mass.

located in highly compressible soil materials and when it is desired to reduce subsidence due to the load applied at the ground surface (Fig. 8b-VIII.5). Therefore, the allowable point bearing load should be

$$[NF]_0^{z_1} = \frac{Q_{pu}}{G_s} \qquad \text{17-VIII.5}$$

where the load on the piles $[NF]_0^{z_1}$ is the negative friction mobilized to depth z_1. The important settlement of the ground surface because of compression of the soil within the piles is

$$\delta_c = \int_0^{z_1} m_{vz}\, \Delta\sigma_z dz \qquad \text{18-VIII.5}$$

5.8 Case VII: Friction Piles in Consolidating Mass

Another case of skin friction problem is that found in pile foundations driven in a consolidating soil mass. Let us assume that the rate of ground surface subsidence is known throughout the depth of the deposit (Chapter V). At the surface let it be V_1 and at the pile points V_2 (Fig. 9-VIII.5). Moreover, assume for simplicity a rectilinear variation of the subsidence velocity with depth, such that $V_1 > V_2$. The pile field under these conditions will be dragged down; however, since the piles are rigid, they cannot follow either the velocity of the ground surface or that of the level of the pile tips. Therefore, the pile field will have an intermediate downward motion V_n, $V_1 > V_n > V_2$. Furthermore, assume that the pile tips are resting on a clayey silt stratum where the piles penetrate under a point load Q_{pu}. Under these conditions, the positive friction $(PF)_n$ is mobilized to a certain distance from the points $(d - z_n)$, and the $(NF)_n$ to depth z_n. Under these conditions, the initial overburden stress is reduced in the upper part of the pile field because of $(NF)_n$; thereafter an increase in the vertical stresses takes place because of $(PF)_n$, such that for limiting plastic equilibrium

$$Q_a + (NF)_n = Q_{pu} + (PF)_n \qquad \text{19-VIII.5}$$

from which one obtains

$$Q_a = Q_{pu} + (PF)_n - (NF)_n \qquad \text{20-VIII.5}$$

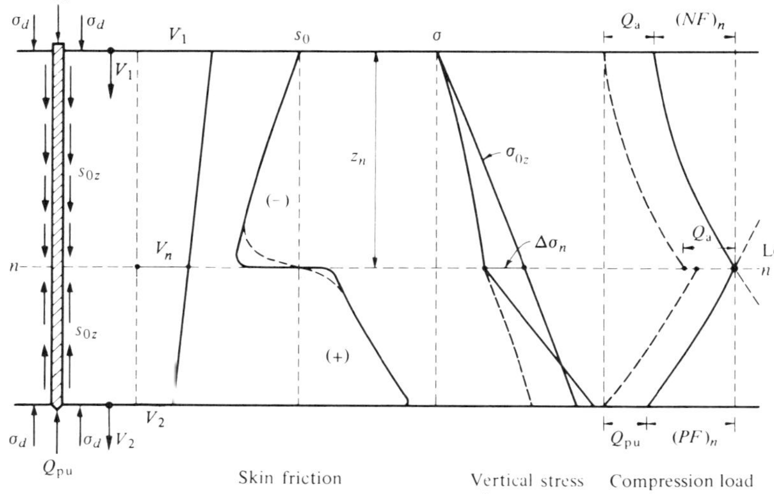

V_1 = Ground surface velocity
V_2 = Soil velocity at pile tip
V_n = Pile velocity

Fig. 9-VIII.5 Case VII—Consolidating soil mass.

The load on the pile given by 19-VIII.5 reaches a maximum value at level *n-n* corresponding to the neutral point of zero skin friction stresses. This elevation at the same time represents the velocity V_n of the pile group because of ground surface subsidence. The pile points penetrate continuously the silt stratum with load Q_{pu}, and are subjected to ultimate skin friction. The head of the piles will emerge from the ground surface at a velocity $(V_1 - V_n)$, and their points will penetrate the loose soil with $(V_n - V_2)$. The elevation at which the value of V_n obtains is determined by means of the self-explanatory load diagram on the right-hand side of Fig. 9-VIII.5.

This condition is important in compensated friction pile foundations; it was first used by the author in 1950 and reported in the literature (Chapter IX). When designing this type of foundation, however, it is necessary to allow the piles to penetrate in the subsoil with the same velocity as the ground surface, in order to prevent the building from emerging from the ground surface. Under working conditions, a fraction of the load is taken by the ground surface and the other fraction used to force the piles to work under their full ultimate $(PF)_d$ load capacity; hence $(NF) = 0$, and

$$Q_a = Q_{pu} + (PF)_d \qquad \text{21-VIII.5}$$

The value of $(PF)_d$ should be calculated including the unit load q_a, taken at the ground surface. The value of Q_{pu} is estimated with the confining pressure at the pile tips corresponding to $(\sigma_{0d} + \Delta\sigma_d)$. Therefore, if q_B is the unit load of the building and $\bar{a}$ the tributary area per pile, then:

$$q_B \cdot \bar{a} = Q_a + q_a \cdot \bar{a}$$

Hence

$$q_B = q_a \left[1 + \frac{Q_a}{\bar{a} q_a}\right] \qquad \text{22-VIII.5}$$

The value of Q_a is the load allowed on the piles and q_a that on the soil. The ratio $Q_a/\bar{a}q_a = R_q$ is called the loading pile-soil ratio. In practice for compensated friction pile foundations, the best results may be obtained if $1 < R_q < 4$, provided the stresses in the soil mass are fully compensated. For a material of the preconsolidated type, the total effective stresses at any depth should be observed not to pass the following limiting value:

$$\sigma_z \leqslant \sigma_{0z} + \tfrac{1}{2}(\sigma_{bz} - \sigma_{0z}) \qquad \text{23-VIII.5}$$

where σ_{0z} is the effective stress due to the overburden weight of the sediments, and σ_{bz} is the critical stress corresponding to the break in the compressibility curve representative of the stratum in question (see section II.3.6a). However, the settlement of the foundation should be investigated to learn if the loading pile-soil ratio R_q has been properly selected. The design of friction pile compensated foundations is discussed in Chapter IX.

Conclusions

Using the soil-interaction mechanical model in difficult subsoil conditions, the foundation engineer is able to study separately the action of point bearing, positive and negative friction phenomena taking place in the pile fields, and therefore has a powerful practical tool to analyze pile foundations on a rational basis. The cases discussed above are considered the ones most commonly encountered in foundation engineering practice; however, other complicated problems may be studied with the use of the mechanical model. Nevertheless, the geometrical and mechanical properties of the subsoil materials should be determined with sufficient precision to make them compatible with the working assumptions used to build up the theory herein explained.

VIII.6 FIELD LOADING TESTS ON PILES AND THEIR LIMITATIONS

6.1 Basic Concepts

A pile test is performed in practice, jacking down the test pile against a dead load. The dead load should be sufficiently large to take the expected ultimate load of the pile. The arrangement for a pile test is shown in Fig. 1-VIII.6. The dead load is bridged over the pile by means of I-beams supported on timbers. At the head of the pile, a concrete block is constructed on which one or several coupled hydraulic

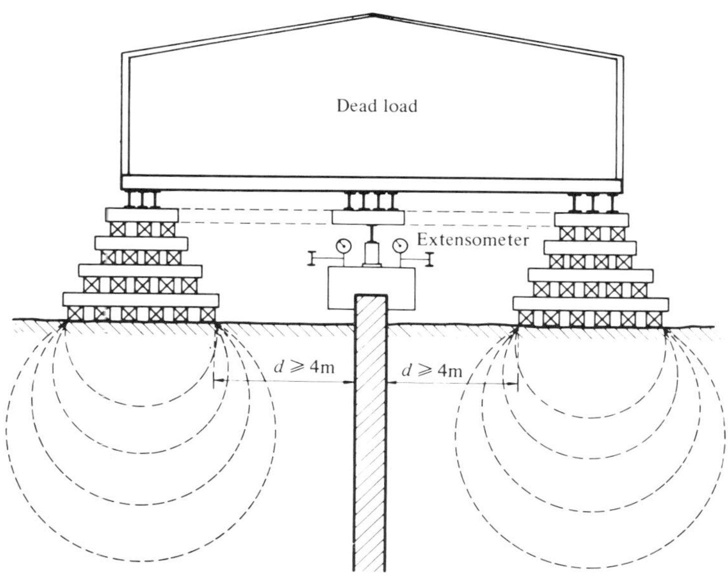

Fig. 1-VIII.6 Pile loading test.

jacks may be installed. The pumping system is provided with a manometer properly calibrated beyond the expected ultimate load of the pile. The vertical displacements of the pile are measured by means of three extensometers placed at an angle of 120° one from each other, and at equal distance from the pile center. The readings of the extensometers are averaged to obtain the pile axis displacement. These values are plotted against the load applied to the pile.

The extensometers are attached to long I-beams supported as far as possible from the test pile, not less than 4 m on each side of the pile. The I-beams supporting the measuring devices are wrapped with an insulating material to avoid large changes in temperature during the 24 hours of the day. Before starting the test, the extensometers are read for a minimum of 24 hours to determine the vertical displacements due to changes in temperature in the I-beams, to enable the engineer to make temperature corrections.

At least six load increments are applied to the pile, chosen from the expected ultimate load. Each load increment is applied with the hydraulic jack, and sustained until sufficient readings are made to be able to plot in semilog scales the observed displacements as a function of time. In the first part of the test, the vertical displacements show a logarithmic rectilinear law behavior with time for each load increment, similar to the one representative of plasto-viscous consolidation in which the rate of vertical displacement decreases strongly with time (Fig. 2-VIII.6). In the last part of the test for higher increments of load, the rate of vertical displacements has the tendency to remain constant or increase with time; this behavior may be interpreted as a visco-plastic flow caused by soil structure breakdown with time. The increment of vertical displacement $\Delta\delta$ is determined for each load increment

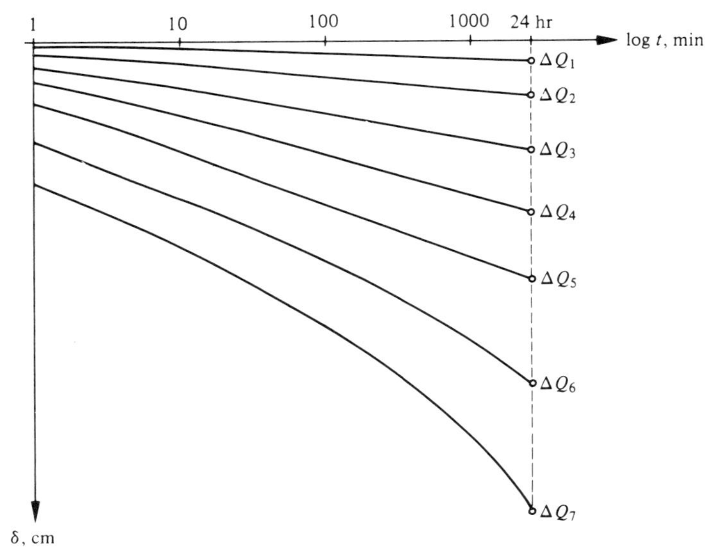

Fig. 2-VIII.6 Time-vertical displacements of pile.

ΔQ_i for a specific time. The values are plotted as shown in the load-displacement graph, Fig. 3-VIII.6, from which curve A is obtained for equal elapsed time.

Therefore, for the single pile loading test, we find two stages of behavior: the first, in which the pile deforms the soil under normal stable deformation laws, as described in section II.2—the phenomenological law of this behavior is usually represented by a rectilinear relationship in semilog scales; and the second stage, when the potential surface of sliding starts to be mobilized and plasto-viscous flow takes place; a final law may be observed with the tendency of vertical displacement showing a steady creep phenomenon.

Every pile has the tendency to work under point bearing and positive friction loads. The ratio

$$\frac{Q_f}{Q_p} \qquad \text{1-VIII.6}$$

changes from values below the ultimate to values for which $Q_u = Q_{fu} + Q_{pu}$ is reached. The knowledge of the ratio given by 1-VIII.6 is important to be able to perform more accurate settlement calculations of pile groups in difficult subsoil conditions. Therefore, if one wishes to investigate separately the positive friction taken by the pile and on the point bearing load, it is necessary to perform the tests with a device similar to the Dutch cone penetrometer (Fig. 4-VIII.6). This special test pile is given the same diameter as the proposed piles in the project. The tests are arranged in such a way as to be able to measure separately the point bearing resistance up to the ultimate load capacity, and afterwards the ultimate skin friction load. The results can be used to study the relative behavior of Q_f and Q_p for intermediate values and up to the ultimate load capacity. From this information,

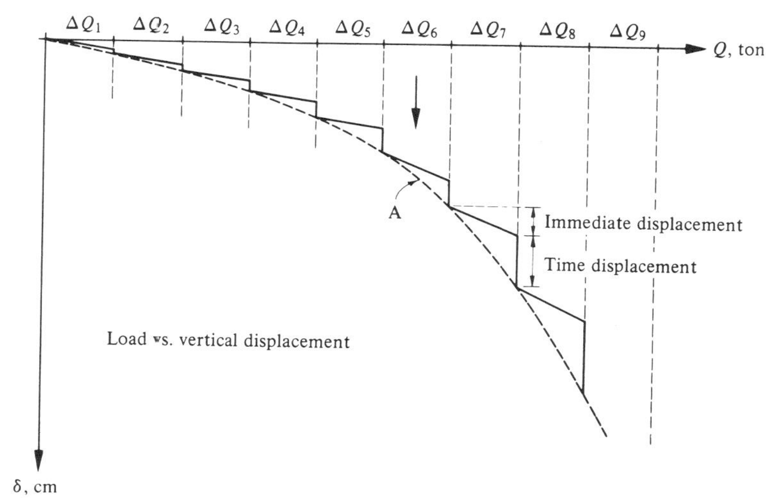

Fig. 3-VIII.6 Pile loading record.

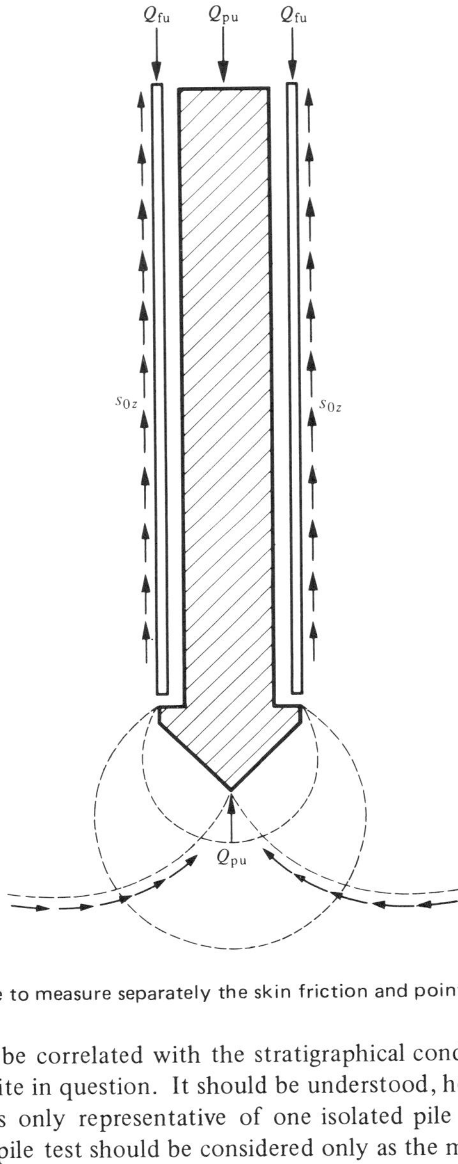

Fig. 4-VIII.6 Device to measure separately the skin friction and point bearing load capacity.

the pile tests may be correlated with the stratigraphical conditions and mechanical properties of the site in question. It should be understood, however, that the information obtained is only representative of one isolated pile and never of a group. Furthermore, the pile test should be considered only as the means to determine the average mechanical properties of the soil related with the stratigraphy. A pile test is shown in Fig. 5-VIII.6.

6.2 Friction Pile in Cohesive Soil

In deep soil deposits of clay or clayey silt sediments, one is often compelled to use friction piles, and therefore tests are desirable to check on the skin friction average

Fig. 5-VIII.6 Pile test arrangements.

properties and the theoretical calculations based on the stratigraphy and mechanical properties of the material previously determined in the laboratory. The point resistance, however, may be important at the ultimate load. The ratio Q_f/Q_p depends on the length and diameter of the pile placed in the ground. For instance, in Mexico City for a friction pile 0.40 m diameter and 20 m long driven in the silty clay deposit, this ratio reaches ultimate values on the order of 4 to 8. In case of pile groups, however, the ultimate point resistance Q_{pu} may be assumed fully developed as the group is loaded and much before the Q_{fu} of the piles in the group is reached (section VIII.5.2).

Field loading tests on friction piles in clay, where Q_{pu} is considered small, reveal a specific configuration, shown in Fig. 6-VIII.6. After the pile reaches the ultimate carrying load capacity Q_{fu}, the vertical displacement increases indefinitely, under a sustained residual load $Q_{fr} < Q_{fu}$. Assume a friction pile in the field is supported

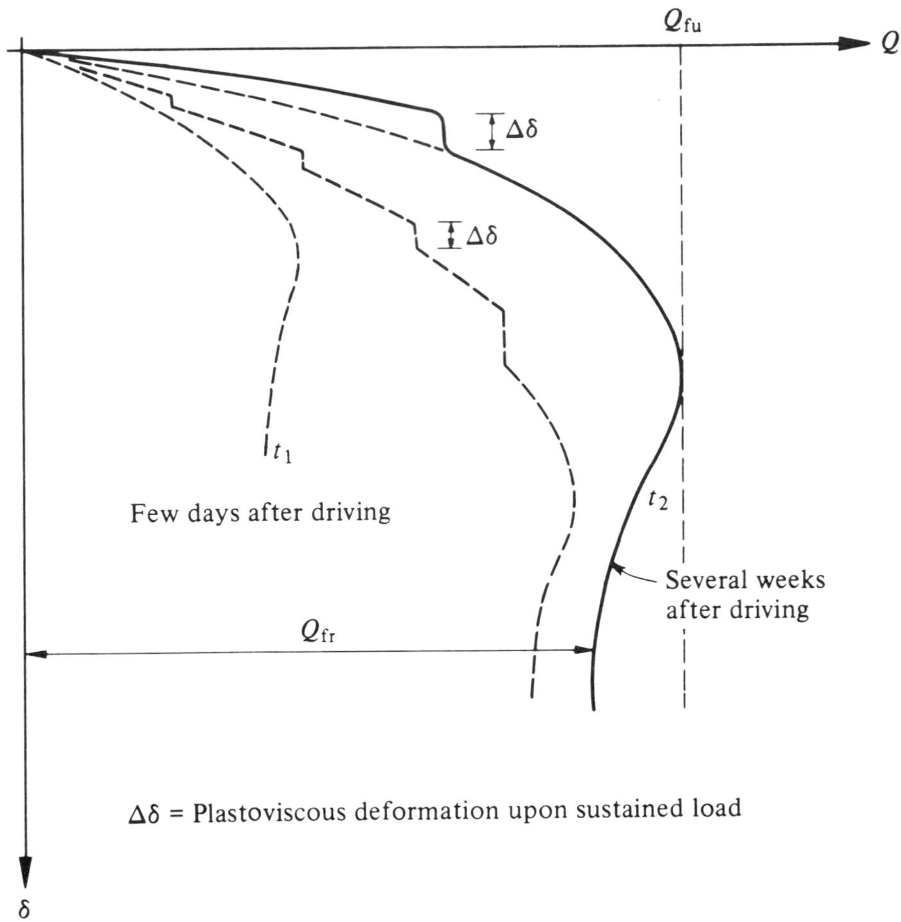

Fig. 6-VIII.6 Friction pile test in clay.

by several soil strata of thickness d_i. At the ultimate load, the skin friction is fully mobilized at the pile shaft according to basic concepts described in preceding articles:

for maximum friction load

$$Q_{fu} = \bar{\omega} \sum s_i d_i$$

and for residual friction load under sustained kinematic conditions

$$Q_{fr} = \bar{\omega} \sum s_{ri} d_i \qquad \text{2-VIII.6}$$

If l is the embedded length of the pile, the average skin friction per unit length of pile has the following two values:

$$(s_m)_a = \frac{Q_{fu}}{\bar{\omega} l}$$

and

$$(s_r)_a = \frac{Q_{fr}}{\overline{\omega} l} \qquad \text{3-VIII.6}$$

The average value of the skin friction obtained from the field tests may be compared with the shear strength calculated from 2-VIII.6, by means of the properties of the soil determined from laboratory tests for each stratum. The interpretation of the results of this test may be applied to calculate pile groups as described in preceding articles.

When the point bearing is determined in clay by means of the device described in Fig. 4-VIII.6, then:

$$Q_{pu} = \alpha_1 [c_a N_c' + \sigma_d N_q'] a_p \qquad \text{4-VIII.6}$$

from which the value of N_c may be investigated; however, the value of the average cohesion c_a under the pile point should be estimated from laboratory tests in undisturbed samples:

$$N_c' = \frac{Q_{pu} - \alpha_1 \alpha_p \sigma_d N_q'}{\alpha_1 c_a a_p} \qquad \text{5-VIII.6}$$

6.3 Point Bearing Piles in Sand

In the case of point bearing piles in sand strata, the load displacement curve configuration has the approximate shape shown in Fig. 7-VIII.6. No definite break may be observed in the configuration curve. As the load increases, the vertical displacement increases without reaching a definite ultimate value. Moreover, it may be observed that the curve in the first stages of the test has a flat configuration that merges gradually into a steep branch. This phenomenon may also be observed when the skin friction load is implicit in the test (Fig. 8-VIII.6). The two ultimate loads Q_{fu} and Q_{pu} may be measured separately with the device of Fig. 4-VIII.6. The ultimate point bearing capacity in case of cohesive sand materials is given by

$$Q_{pu} = a_p \{\alpha_1 [cN_c + \sigma_d N_q]\}(D_r + 0.1) \qquad \text{6-VIII.6}$$

In these cases, when c is large, it is not possible to find the values of N_c and N_q directly from the pile tests information. The foundation engineer can only compare field test results with ultimate bearing capacity calculations performed with the average shear strength properties of the soil obtained in the laboratory. Nevertheless, if the sand may be considered cohesionless, then the value of N_q may be easily investigated with formula 6-VIII.6.

The determination of the point bearing capacity by means of field tests in firm soil strata is useful, since this may be applied to pile groups and pile fields, to assure the point penetration of the individual piles. Nevertheless, the problem arises when deciding the Q_{pu}-load on which to apply a factor of safety. For this purpose, the relation between load and vertical displacement shown on natural scales in Fig. 7-VIII.6 is plotted in log-log scales in Fig. 9-VIII.6, where a definite break in the

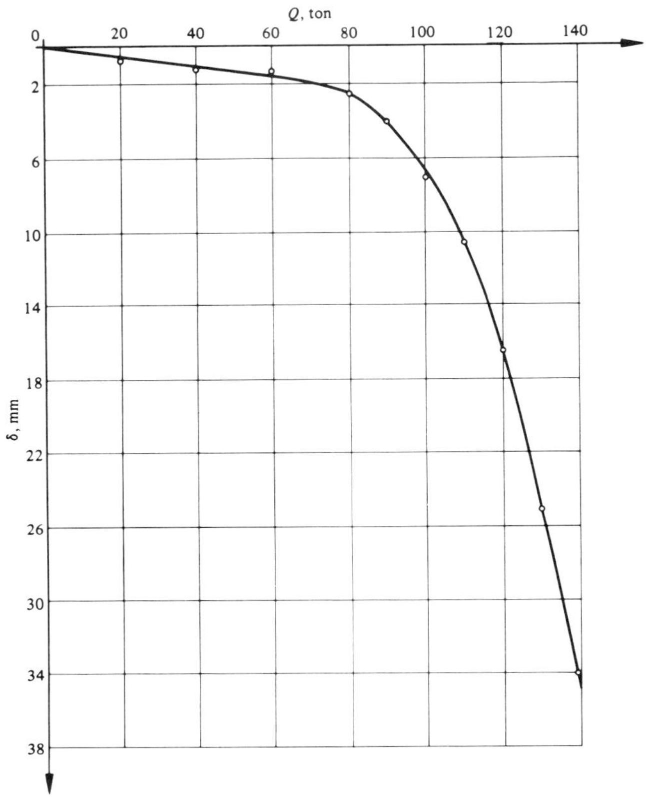

Fig. 7-VIII.6 Point bearing pile in sand strata.

curve may be observed, thus differentiating the two stages of behavior discussed before: a) the flat branch representing an elasto-plastic viscous behavior in which stability may be reached; and b) the plasto-viscous behavior that may be interpreted as the pile point breaking into the ground. Therefore, the load at the break of the curve may be considered to be the ultimate point bearing capacity of the pile Q_{pu}. The factor of safety of the entire group, however, should be investigated against breaking into the ground (Figs. 10 and 11-VIII.6). The results of loading tests on one pile cannot be applied to the calculation of the bearing capacity of the group since the shear properties of the strata not entering into the single pile tests have to be considered in the group. The procedures explained in sections 2 and 4 may be used for this purpose.

When the average cohesion is small and known from laboratory tests, we determine the values of N_q and N_c from the pile tests, using the point bearing capacity in the following form:

$$N_q = \frac{Q_{pu}}{1.2(D_r + 0.1)\,a_p\sigma_d} - \frac{c_a N_c}{\sigma_d} \tag{7-VIII.6}$$

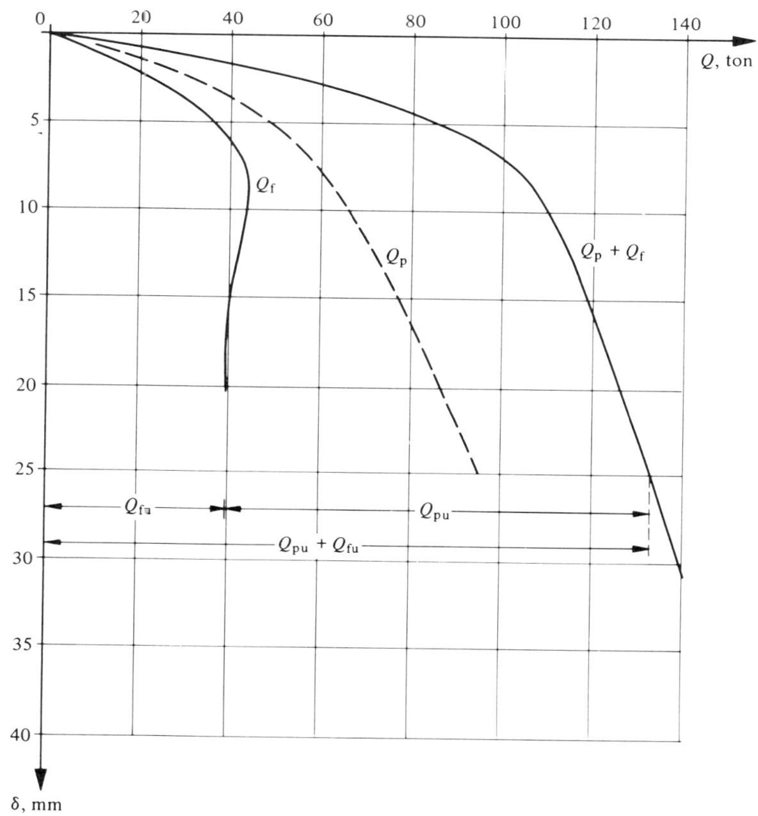

Fig. 8-VIII.6 Point bearing and positive friction ultimate loads.

where the value of σ_d is calculated from the stratigraphy, hydraulic conditions and index properties of the sediments encountered from ground surface to the pile tip. From shear strength tests, the values of c_a and ϕ_d, as well as the relative density of the bearing stratum, may be estimated.

The following example will illustrate the procedure used to investigate the values of N_c and N_q. A point bearing pile test gave $Q_{pu} = 250$ ton. At the tip elevation the pile has an area $a_p = 0.16\ m^2$. From laboratory tests on undisturbed samples of the bearing stratum, the following average values for the shear strength are obtained: $c_a = 2\ ton/m^2$ and $\phi_d = 40°$. The relative density of the soil is estimated from tests in $D_r = 0.5$. From the stratigraphy, hydraulic conditions and index properties of the soil strata overlying the hard stratum an overburden effective stress $\sigma_d = 20\ ton/m^2$ is found at the pile tip elevation. Hence, as a first approximation, using $\phi_d = 40°$ from the N factors graph (Fig. 4-III.4), we find $N_c = 315$. Using formula 7-VIII.6 and substituting numerical values, we have $N_q = 109 - 31.5 = 77.5$. For the value $N_q = 77.5$, we find from Fig. 4-III.4 $N_c = 165$, corresponding to an angle of in-

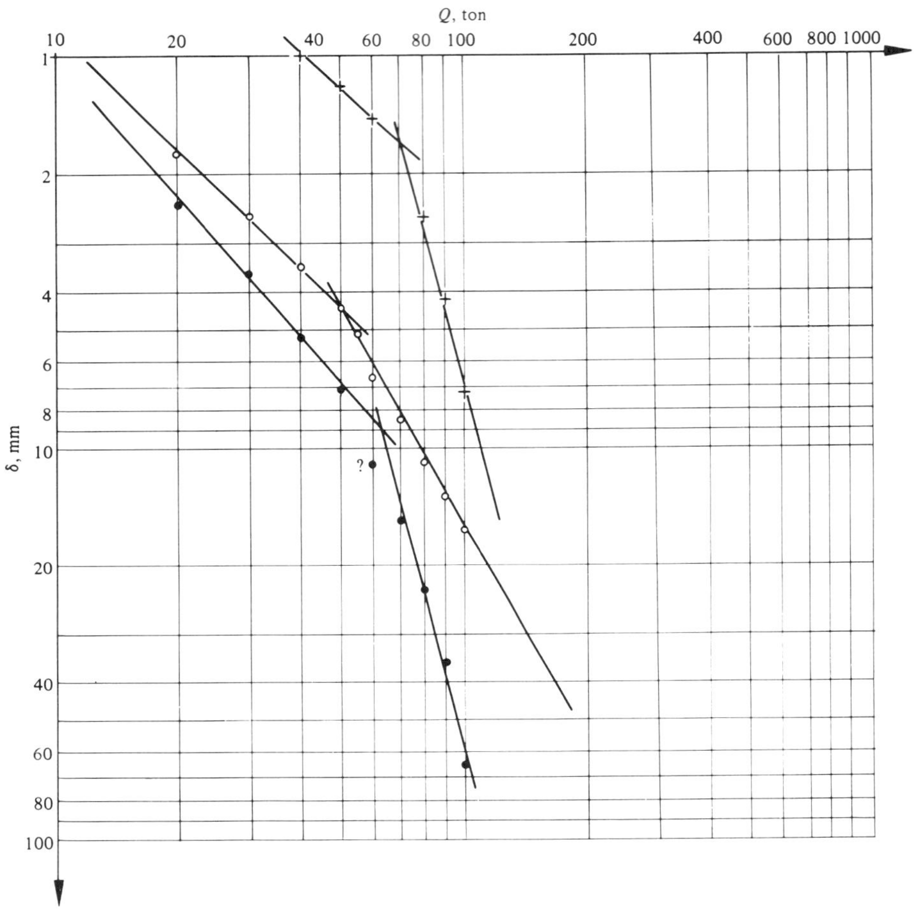

Fig. 9-VIII.6 Field tests, point bearing piles.

ternal friction ϕ_d = 36°. A second approximation yields N_q = 109 - 16.5 = 92.5. Again for N_q = 93, we find N_c = 190, and ϕ_d = 37°. On further cycles of calculation we find finally N_q = 90, N_c = 190 and ϕ_d = 37°.

In this way, the theory is calibrated to the pile test results, and N_q and N_c may be used to investigate other pile conditions for the same bearing stratum. Notice that the resulting average angle of internal friction obtained from the analysis of the pile test was ϕ_d = 37°, instead of 40° as determined in the laboratory.

6.4 Vertical Displacement of Single Pile Tests, and Pile Groups

The vertical displacements measured in single pile tests cannot be applied, in the case of pile groups, to estimate the settlement of the groups, since the subsoil stresses caused by a single pile are quantitatively different from those caused by a pile group

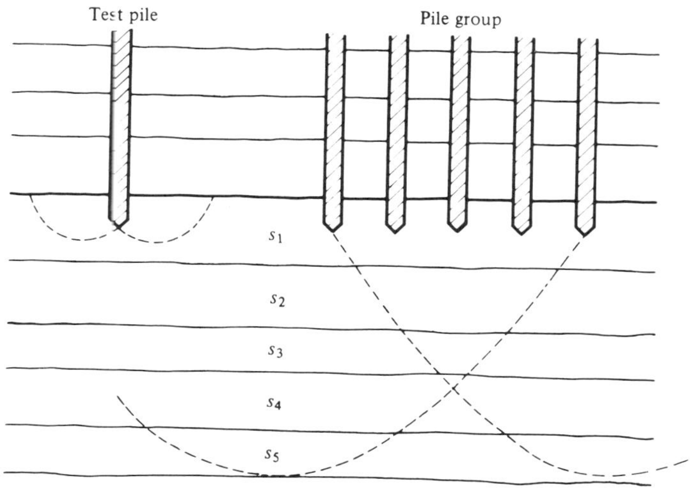

Fig. 10-VIII.6 Base bearing capacity.

or pile field (Figs. 12 and 13-VIII.6). The only means to estimate the settlement of pile groups is to calculate them from the stratigraphy, hydraulic conditions and compressibility properties of the soil materials, as well as from the increment of stresses induced in the subsoil by the pile group. The methods of computation have been explained in previous sections.

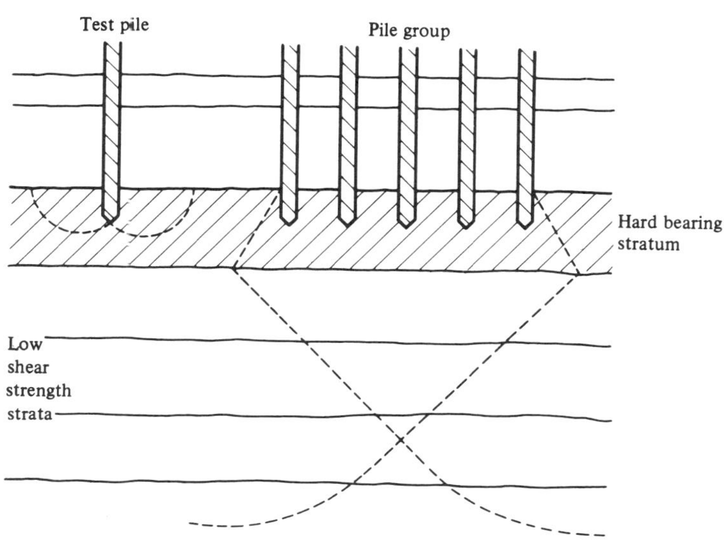

Fig. 11-VIII.6 Base bearing capacity.

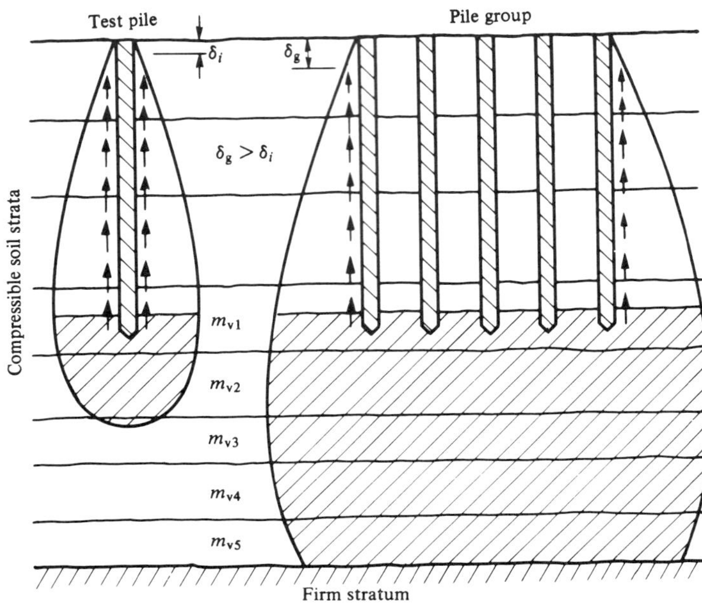

Fig. 12-VIII.6 Vertical displacement of one pile and group.

VIII.7 REVIEW ON PILE SELECTION AND DRIVING

7.1 Project Requirements

The selection of the type of pile and driving equipment for a specific pile job should be carefully considered by the foundation engineer, in conjunction with the environmental conditions at the site and the project requirements. Most important

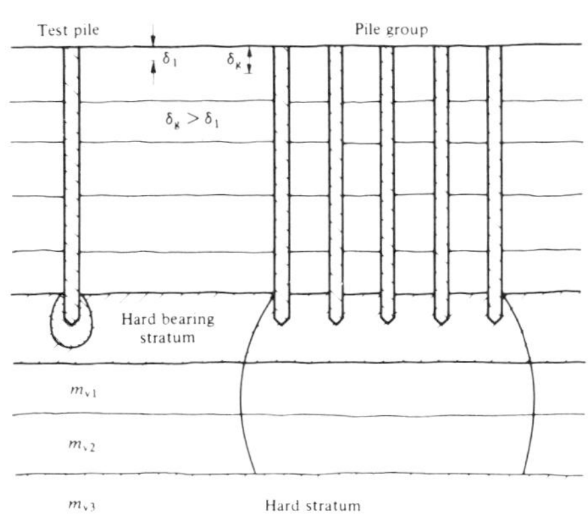

Fig. 13-VIII.6 Vertical displacement of one pile and group.

are the total and differential allowable settlements between different loaded areas of the building (Chapter I). Towers with low wings call for special attention. Another special case is that of heavy buildings to be constructed adjacent to expensive or historic constructions, where damage absolutely must be avoided. Special problems arise in the case of slender buildings in regions of strong earthquakes and wind storms, where these transient forces may impair the stability of the structure; this possibility should be taken into account with the foundation (Chapter XII).

In practice, there are a great number of pile types and pile driving equipment and procedures, from which the foundation engineer may choose the most suitable and economic type, compatible with the problem he has under consideration. The techniques in pile driving and construction, however, are mainly a function of the manufacturers and pile driving companies. A thorough discussion of them is beyond the scope of this book. The foundation engineer will find in the literature and from the pile contracting organizations information related with this topic. Nevertheless, it is important to mention that when the pile market is restricted in a country where the foundation problem is encountered, it is always possible to import techniques, patents and pile driving equipment. The foundation engineer should be acquainted with the international market of piles, equipment and pile driving. In this article, the author will only make a panoramic review of pile types, selection and driving, to introduce the foundation design engineer to the practical branch of piles and pile driving in engineering. The sequence is briefly as follows:

PROJECT CHARACTERISTICS

STRUCTURE LOADS → | ← PRE-SUBSOIL EXPLORATION

↓

FINAL SUBSOIL INVESTIGATION

PILE FOUNDATION → | ← SELECTION OF PILE TYPE

↓

DRIVING, FIELD TESTING AND CONTROL

The foundation engineer should first investigate the architectural and engineering characteristics of the project, its physical requirements, and the layout of loadings and their probable magnitude. At this stage, the preliminary subsoil engineering characteristics should also be included. From this information, the foundation engineer should be able to decide if the problem is a pile job, and perform a preliminary pile foundation design, finding the best and most economical solution for the problem under consideration.

Once the tentative solution is accepted, the foundation engineer has to decide on the extent and quality of the final subsoil investigation necessary to design the foundation, and make the selection of the pile type on a rational and economical basis. During the final design, the foundation engineer will proceed to write the construction procedures and specifications for the piles and pile driving. The pile driving organization, however, may have standardized procedures and specifications for the selected pile; when this is the case, the foundation engineer must revise and adjust them to the problem under his responsibility, specifying pile length and diameter, number of pile sections, verticality, material, hammer, etc. The supervision and control during construction of the piles and driving procedure should be clearly specified. Finally, the foundation engineer should specify with detail the type of field loading tests that should be performed before pile driving, and those that may be found necessary during and after pile driving. Furthermore, the foundation engineer is responsible for the interpretation of the pile loading tests and their implication in the proposed foundation design.

7.2 Structural Loads—Subsoil Exploration

The foundation engineer should know all about the structural characteristics of the project, since the design of pile foundations greatly depends on the magnitude of the loads and their spacing. When the structure calls for very heavy loads at large spans, the foundation engineer has to solve a foundation for large concentrations of load. The pile should be carefully selected to support this type of loading economically. The decision, however, must be based on preliminary subsoil explorations, performed by a physiographical and geological reconnaissance of the site environments, aided by sounding devices. The tool recommended for subsurface explorations is the standard penetration test, with an open split spoon from which representative samples of the soil sediments may be obtained to study their index properties. When compressible soil deposits are encountered, it is always important to establish deep benchmarks and surface reference points to determine whether a problem of ground surface subsidence is present. For this purpose, it is also necessary to determine the water table and the piezometric water levels at different elevations within the soil mass (sections III.3 and V.3).

Building foundations, however, are performed in cities where a previous knowledge of the geology is known. In these cases, the preliminary exploration is confined to the capability of the ground to take loads by means of piles or piers. After this is performed, and it is found that the subsoil in question may be able to take the structural loads proposed, a detailed subsoil investigation is made to determine at the proposed site the stratigraphy, hydraulic conditions and mechanical properties of the materials, taking for this purpose undisturbed samples that will permit laboratory tests of shear strength and compressibility; in addition the permeability of silt and sand strata found in the subsoil must be determined (section V.3). The final investigation program, however, depends on the foundation problem and project requirements. When deep excavations are not required, and favorable subsoil conditions are encountered, careful preliminary explorations may be sufficient to decide on pile foundations. From the standard penetration tests or cone sounding devices, one can determine the relative properties of the firm ground where piles may be supported by point bearing.

TABLE 1-VIII.7

State of compaction of sand	D_r	N	q_c	ϕ_d
Very loose	< 0.2	< 4	< 20	< 30°
Loose	0.2–0.4	4–10	20–40	30°–35°
Semicompact	0.4–0.6	10–30	40–120	35°–40°
Compact	0.6–0.8	30–50	120–200	40°–45°
Very compact	> 0.8	> 50	> 200	> 45°

D_r–relative density
N–number of blows in standard penetration test
q_c–unit pressure in penetration cone test in kg/cm^2. After G. G. Meyerhof (1956).

After the subsoil conditions are studied and the mechanical properties of the materials are known, the foundation engineer will be in the position to start a rational pile foundation design based on theoretical considerations. The first problem is to investigate the firm stratum for support of point bearing piles, and to investigate the negative skin friction the piles have to carry because of ground surface subsidence. Other problems should be considered in the foundation design, especially the possibility of sidewalks settling away from the buildings. Problems of differential levels are caused between public utilities and the entrance to parking garages and ground floor of the buildings. This problem, however, may be solved if the first floor is designed mobile, that is to say, permitting the first floor to be lowered periodically as the building emerges from the ground surface because of the ground surface subsidence. This procedure has been used by the author in Mexico City. On the other hand, if the building is not very heavy, this phenomenon can be overcome by means of a friction pile compensated foundation or a foundation on pile controls. For this purpose, the foundation engineer should study the positive friction capacity of the piles in the first case, and the negative friction capacity in the second case, as discussed in previous sections. References may be found in the bibliography for the use of pile mechanical control devices. In the design of the foundation, the foundation engineer must be sure to remain within the allowable settlements requirements, which, of course, will be different for each type of construction under consideration, depending mainly on whether the building is isolated from other buildings or adjacent to important buildings, or if the problem is a factory building, a warehouse, or another type of construction as discussed in section I.3.

7.3 Pile Foundation—Selection of Pile Type

It is important in selecting the type of pile to consider pile length and spacing and the probable loads the piles will carry in order to establish pile groups and the type of foundation structure that should be used to distribute the building load on the piles. It is also important to consider if batter piles are necessary. When this is the case, the angle of inclination and loads taken by the piles should be carefully studied. Batter piles, however, should not be used in areas where ground surface subsidence is taking place. In the selection of the pile type, it is important to consider the time element in the deterioration of the piles.

The factors affecting pile life in the case of timber piles should be carefully considered. Fungus is developed in timber piles due to wetting and drying processes. In ocean water, boring mollusks and crustaceans are often found to drill into the pile, producing very severe damage. Some of the many boring marine species also attack concrete piles. The borers do not operate in polluted waters. Also, in tropical countries especially, timber piles may be attacked by insects such as termites and ants, which feed on the cellulose.

Steel piles exposed to aerated areas corrode to differing degrees, depending on the acidity or salinity of the water and soils. The average corrosion rate in ocean waters is on the order of 0.001 inches per year. Therefore, in the design of steel piles exposed to corrosive waters, an allowance should be provided. To prevent corrosion and decay and increase the life of the piles, they can usually be treated with oil emulsions. Timber piles are usually treated with creosote.

Concrete piles crack because of the following causes: handling, tension stresses during driving, early driving, rich mixtures, improper storing and curing, crushing during driving. The reinforcing steel is attacked through the cracks by water containing nitric acid, sulfuric acid, hydrochloride acid, sulfates, nitrates, ammonias, and tannic acid. This may be observed in piles mainly used for trestles in structures at the water front. The iron oxide expands and makes the concrete pop away from the pile, exposing the reinforcing steel and damaging the pile very rapidly, sometimes beyond repair. The use of epoxys are sometimes convenient to repair piles that show this type of damage. The premature cracks in concrete piles are overcome if constructed of prestressed concrete sections.

The method of placing piles in the ground should be carefully considered in conjunction with the stratigraphy and subsoil properties, and the pile is selected to be a point bearing pile, a friction pile or a combination. They may be driven by means of hammers or jacks, or may be cast in place in the ground. The most common method of driving piles is by means of hammers: single steam-acting, differential or high-frequency vibrating hammers, depending on the type of soil and resistance offered when driving the pile, and on the bearing capacity assigned to the pile by the penetration required in the supporting stratum. In case of point bearing piles, the point is usually required to penetrate a certain distance into the firm layer; therefore the weight and type of hammer should be selected for this purpose. It should be borne in mind, however, that the weight of the hammer has to be compatible with the weight of the pile, otherwise the pile is damaged by overdriving, or else poor penetration in the bearing stratum is obtained. Many cases in practice have shown that the points or even the shafts of the piles have been seriously damaged because of overdriving. Therefore, one has to consider from the practical point of view, the type of hammer and the method to be used. For this purpose, dynamic formulas may be used and calibrated at the site in question, as described further in this article. In the case of impervious soils, driving is usually easy, and small hammers may be used. For cased cast-in-place piles, one has to drive first the point of the pile, and then cast the concrete into a shell. Other piles are cast in a hole in the ground; a certain volume of concrete is highly compacted at the point of the pile, and the hole is filled with concrete and vibrated to build up the pile shaft. Different procedures and patents may be found in the engineering profession for type of piles, equipment and pile driving.

It should be considered, however, that when piles are driven into the ground, heave takes place. This has to be carefully considered, since as mentioned in previous articles, ground surface heave will cause negative friction on point bearing piles after the ground surface reassumes its original position. In case of timber piles, hard driving may produce rupture or brooming (Figs. 1 and 2-VIII.7). The timber piles have to be provided at their points with steel shoes, and strong joints should be

Fig. 1-VIII.7 Brooming in point bearing timber pile due to over-driving.

Fig. 2-VIII.7 Joint failure in timber pile due to over-driving.

used to splice sections. When driving piles with hammers, attention should be given to strong vibrations and noise. The vibration may cause damage to adjacent construction, especially when piles are driven very close to them. In very deep soil deposits, the selection of piles is important from the point of view of the number of sections necessary to reach the point bearing. In cases of loose or medium loose sand deposits, jetting the piles may be necessary to make them penetrate into the sand, to obtain the frictional and point bearing carrying capacity required. The implication of jetting should be carefully considered for other constructions surrounding the site in question. The jetting process may be combined with driving by means of a hammer. Partial jetting or pre-excavation may be used in cohesive sandy soils around the head of the pile. The stream of water lubricates the pile, reducing friction and allowing more driving energy into the lower strata.

The material of the pile has to be carefully decided, taking into consideration the energy of driving. In very hard driving, it would be necessary to use steel piles, either a pipe or H-section. In soft driving, either concrete or timber piles may be used. The material of the pile is important under the environmental conditions to which the piles are going to be driven. At the ocean front or in corrosive soils, and when piles extend above the ground surface, part of the upper section of the pile will be exposed to weathering, corrosion, and marine borers and termites, and to overcome this problem composite piles may be used, that is to say, timber below the ground water level and prestressed concrete above. The steel pile may be treated in the same way, that is to say, a steel section under the water table and the upper section of concrete.

In the selection of the pile material and driving equipment, one has to take into consideration, from the design point of view, the stresses to which the piles may be subjected during driving, being sure to protect the point of the piles and the joints if different sections are driven. The head of the pile should be doubly protected. To assure full energy to drive the pile in the bearing stratum, it is important to obtain the proper number of strokes per minute, specified for single or differential acting hammers.

The foundation engineer must select the type of pile that is most suitable and economical for his foundation problem. For this purpose, he has to be acquainted with the possibilities of each one of the piles he can obtain in his district, and the equipment available to drive the piles.

7.4 Pile Types Most Commonly Used

Timber piles are used as friction or point bearing piles for allowable loads of 15 to 25 ton and depths not exceeding about 15 m. The pile tip should be protected with a shoe of sheet metal. Timber piles have a small diameter at the tip, and are therefore especially useful as friction piles. The use of timber piles, however, requires large quantities of timber from a pine forest, and therefore in countries low in timber, their use is forbidden.

Concrete piles with circular, square or orthogonal section may be used to advantage, and are sometimes more economical than timber piles. Concrete piles are precast in sections, and prestressed concrete may also be used when it proves convenient and economical. The sections are firmly spliced together; the joints may differ in types with specific details given by the manufacturer. The piles may be straight or tapered, and can be used with water jets to facilitate driving under certain subsoil conditions. The piles in this case are provided with a small diameter pipe in the center of the section, or with lateral attached pipes, or both. The head and tip of the piles are usually strongly reinforced. The points may be designed with different shapes, enlarging them to obtain greater bearing capacity in the firm stratum (Figs. 3 and 4-VIII.7). Here again, special details are offered by the different manufacturers and constructing firms for certain established prices. Concrete piles may be designed for large loads. Usually, the load as a column is not a problem in a concrete pile. The problem may be present, however, when very hard driving is required and the concrete pile may be damaged. In such cases, it is preferable to use steel pipe piles or H-sections with specially designed points.

Cast-in-place concrete piles may be of two types: those used with a steel shell and those cast in the ground without the shell. Cast-in-place piles with shell are generally constructed by driving a thin steel shell by means of a mandrel. At the same time, the mandrel pushes down the point of the pile to the desired elevation and

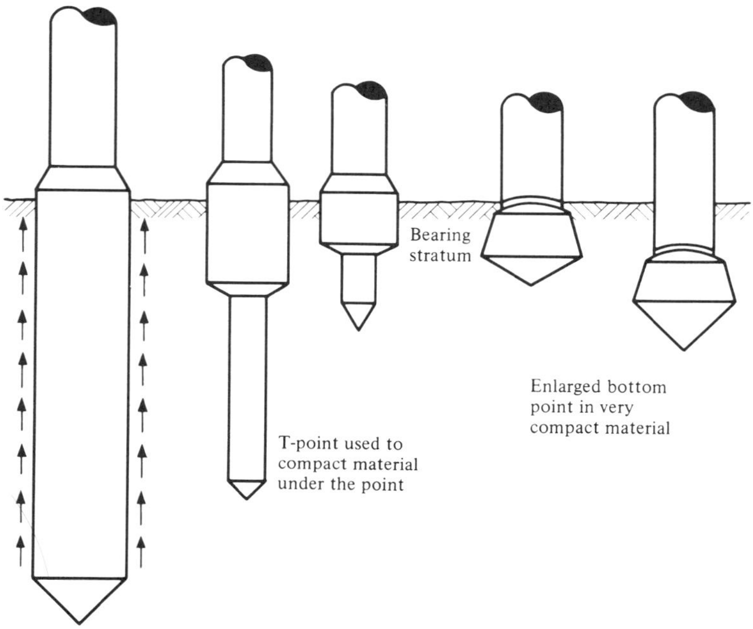

Fig. 3-VIII.7 Precast concrete pile points.

specified penetration in the bearing stratum. The point is prefabricated usually of reinforced concrete. After driving, the steel mandrel is withdrawn and the thin steel shell is filled with concrete. However, in these cases, precautions have to be taken. The concrete should be vibrated carefully or injected from the bottom of the shell, to avoid trapped air; otherwise, the concrete column may be interrupted with cavities.

Another type of cast-in-place piles is performed driving a steel shell or pipe to the desired bearing stratum. Afterwards, the mandrel is withdrawn and a small amount of concrete is poured to the bottom of the pile, and a pedestal is formed by ramming the concrete strongly into the bearing stratum. The thin shell forming the shaft of the pile is then filled with concrete. This type of pile is used only when the supporting capacity of the bearing stratum should be improved. If the consistency of the soil above the hard stratum is sufficient to maintain the hole open, then a concrete cast-in-place pile without shell may be used. The shaft of the pile is formed by placing concrete in the hole at the same time the casing pipe is withdrawn. This type of piles requires special driving techniques, like the McArthur type compressed concrete pedestal pile and the Franki pile. They are used when a good bearing stratum is found at shallow depth and heavy loads are to be supported.

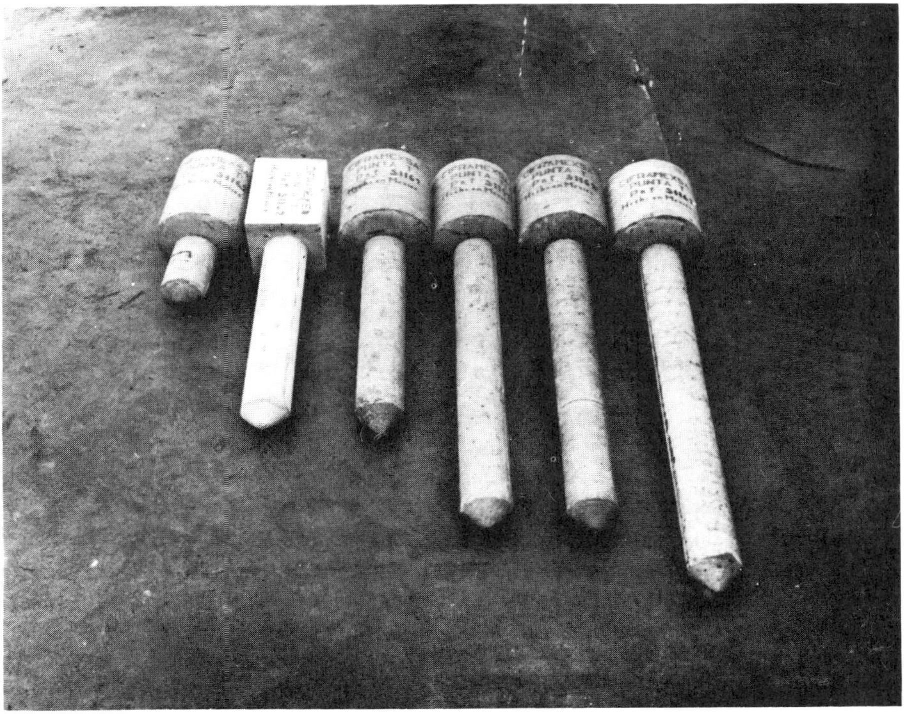

Fig. 4-VIII.7 T points specially designed to guide the pile vertically and compact the bearing stratum under the pile point.

A composite pile may be formed by driving first a section of timber or steel to the bearing stratum, and then finishing the pile with a cast-in-place or prefabricated concrete section. Composite piles are used, as stated before, with the sole purpose of protecting against corrosion or decay the upper part of the pile exposed to weathering or water level fluctuations. Precautions should be taken at the joint between the sections of different materials. Therefore, composite piles are only justified when because of scarcity of materials to fabricate a totally precast or prestressed concrete pile, the composite pile may represent economy in the foundation.

Precast concrete piles in short sections about one meter long may be used for underpinning buildings. They usually have a square or circular section and contain a 10 cm diameter hole at the center, which permits inspection of verticality during driving. This type of pile is known as the Mega pile. They are used extensively in Mexico City, as friction piles, point bearing piles and for underpinning jobs, in conjunction with a T-point to assure verticality during driving. The reinforcing steel is introduced in the center hole to the total length of the pile and cement grout injected. Hence, all the short sections will be strongly tied together. In some instances, post-tensioning may be used. Steel short sections may be used also in underpinning jobs, jacking them against the weight of the structure to the desired hard stratum or depth if used as friction piles. The pile is filled afterwards with concrete properly placed to avoid cavities. These types of piles are more expensive than concrete sections; they are used only in special cases when concrete cannot be used, mainly because their weight makes it difficult to handle the sections under the foundation to be underpinned.

Other types of piles mentioned in the literature may be used in some cases with advantage, such as sand piles, drill-in caissons, screw piles, disk piles and other piles that are proposed by different manufacturing and pile driving organizations. The foundation engineer is advised to consult the bibliography recommended in this chapter on the topic of pile types and pile driving, especially the book *Pile Foundations* by R. D. Chellis (1961).

7.5 Pile Driving and Control

The driving of friction piles in soft soil deposits offers no special problem. The need for careful pile driving and control arises when driving the piles through soft strata to point bearing strata, since in this case it is necessary to assure the carrying load capacity for each one of the piles, and be able to detect any irregularities of the bearing stratum during driving. The pile may be considered in itself to be an element like a sounding device, with which one can learn indirectly subsoil conditions. Therefore, the foundation engineer obtains from driving records a two-fold information: first, the characteristics of driving through the upper soil strata, and the energy absorption by the total number of blows required to reach the bearing stratum, and second, from the last number of blows of penetration if the piles have met the specific driving requirements in the firm stratum. For this purpose, the dynamic conditions during driving may be reviewed in order later to use a semiempirical formula that, upon calibration with pile loading tests performed at the site, can be used for pile driving control.

Assume a point bearing pile will be driven with a hammer of weight W_H falling at a height h from the pile head. The blow on the pile head will produce an effective penetration δ_e of the pile in the bearing stratum. Part of the total potential energy $W_H \cdot h$ is used to perform the work $\delta_e Q_{dy}$, where Q_{dy} is the dynamic load on the pile during penetration; the rest of the energy is used to overcome energy losses; hence

$$W_H \cdot h = \delta_e Q_{dy} + \text{energy losses} \qquad \text{1-VIII.7}$$

The most important energy losses are those due to friction in the hammer system and to impact at the head of the pile, loss of energy during the blow because of elasto-plastic deformation of soil and pile, and the loss due to friction of the soil along the pile shaft. An approximate value of the load a pile can carry may be obtained if the static load is assumed proportional to the potential energy and the energy losses. Therefore, according to 1-VIII.7 we obtain the following semiempirical formula:

$$W_H \cdot h = \delta_c Q_{pu} + \delta_0 Q_{pu} \qquad \text{2-VIII.7}$$

The quantity δ_0 is a coefficient with units of length which represents the virtual penetration lost to overcome all the energy dissipation during pile driving, except the work performed by the pile load on the real penetration δ_c. Therefore, the ultimate pile load capacity is

$$Q_{pu} = \frac{W_H \cdot h}{\delta_c + \delta_0} \qquad \text{3-VIII.7}$$

The value of δ_0 is determined from test results and for a specific ratio of weight of hammer to that of the pile:

$$\delta_0 = \frac{W_H}{Q_{pu}} \cdot h - \delta_c \qquad \text{4-VIII.7}$$

Formula 4-VIII.7 may be used for calibration at a specific site where static point bearing pile loading tests have been performed and their results are known in connection with the driving characteristics. The value of δ_0 thus obtained is used in formula 3-VIII.7 to make final adjustments in the last blows of penetration, δ_c, and consequently on the load Q_{pu}. A factor of safety on the order of 1.5 to 2 may be used, depending on the accuracy in the determination of δ_0 and consistency of the pile tests. A minimum of three pile tests is required before a final decision is made on the working load of the piles.

A practical example will illustrate the use of 4-VIII.7. The largest final penetration of three point bearing piles driven through soft soil was 24 blows in the last 5 cm, hence $\delta_c = 0.21$ cm per blow. A single-acting steam hammer was used with weight of 2.27 ton, 60 cpm and fall $h = 91.4$ cm. The weight ratio of hammer to pile was in this case 0.3. From loading tests a low average value on the order of $Q_{pu} = 200$ ton was found. Substituting the numerical values in 4-VIII.7 we obtain

$$\delta_0 = \frac{2.27}{200} 91.4 - 0.21 = 0.827 \text{ cm} \qquad \text{5-VIII.7}$$

from which the control driving formula will be:

$$Q_{pu} = \frac{2.27 \times 91.4}{\delta_c + 0.827} = \frac{207.5}{\delta_c + 0.827} \text{ ton} \qquad \text{6-VIII.7}$$

It may be noticed that formula 3-VIII.7 has the same structure as proposed by Wellington (1888) better known as the *Engineering News* formula, used extensively to calculate the allowable load on point bearing piles driven through soft sediments. It reads as follows:

$$Q_a = \frac{2W_H \cdot h}{\delta_c + c} \qquad \text{7-VIII.7}$$

Here W_H is given in pounds and h in feet; the values of δ_c and c are given in inches. A value of $c = 0.1$ inches, is recommended for single-acting steam or diesel hammers. The value of c is equivalent to δ_0; however, it cannot be assumed a universal constant, since it represents the energy losses, especially those sensitive to energy absorption, motivated by the specific subsoil conditions and the ratio W_H/W_p. If the *Engineering News* formula is compared with the general energy formula 3-VIII.7, it may be noticed that a safety factor of 6 has been used.

Using the numerical values of the preceding example, one obtains

$$Q_a = \frac{2 \times 5000 \times 3}{0.083 + 0.1} = 163{,}900 \text{ lbs} = 74.3 \text{ ton}$$

Using the safety factor of 6 recommended in formula 7-VIII.7, we have $Q_{pu} = 446$ ton. Therefore, it may be seen that results obtained with Wellington's formula for $c = 0.1$ can be very misleading. Thus, although the very large safety factor proposed may sometimes give values on the safe side, the result is uneconomical since in the form given equation 3-VIII.7 calibrated from pile tests gives us a safety factor of 2 and a working load on the order of 100 ton, which is larger than the one allowed by means of the *Engineering News* formula.

The weight ratio of hammer to pile is an important consideration to obtain the maximum possible energy to assure a good penetration of the pile point in the firm stratum. An idea of the influence of the ratio W_H/W_p may be obtained, by means of the well known factor based on theoretical considerations during the hammer-pile impact, from Newton's principle of restitution, the loss is

$$(W_h \cdot h)\left[\frac{1 - n_e^2}{1 + \dfrac{W_H}{W_p}}\right] \qquad \text{8-VIII.7}$$

Here n_e is the restitution coefficient for perfectly elastic bodies $n_e = 1$. On the other hand, it may be seen that the larger the ratio W_H/W_p the smaller the factor. Therefore, expression 8-VIII.7 should be made as small as possible to reduce energy loss due to impact.

The reader is referred to other dynamic pile formulas such as those of Janbu, Ejtelwein, the Danish formula, and many others published in the literature. The foundation engineer, however, should recognize the limitations of pile driving formulas when used.

Jacking of piles is by far the best method when it proves economical. Nevertheless, a large dead weight is necessary to jack down the piles (Fig. 5-VIII.7). Jacking procedures are usually used when the building is under construction or in underpinning problems. It does not produce significant vibration, and adjacent property will be free from damage and noise. The author strongly believes that jacking techniques should be further developed in driving piles. Another advantage of jacking is that every pile driven to final bearing is a tested pile, and therefore field loading tests after driving may be eliminated. The control during driving is easier and more reliable than with hammer-driven or any other type of piles.

Cast-in-place piles, above and under the water table, require very careful control and supervision. In case of piles without casing, the foundation engineer should be sure no caving will take place during the process of filling up the boring. When

Fig. 5-VIII.7 A large dead weight to jack down the piles.

ramming the concrete at the bottom to obtain a larger solid base, there is always a danger of caving. This type of pile should be avoided if it is possible to use precast concrete piles or cased piles, except in those cases where the soil is firm and the foundation engineer can be sure that no caving of the excavation will take place. The caving phenomenon caused by soil back pressures should be prevented to avoid necking of the piles.

When conducting pile loading tests, it is always necessary to calibrate, either for point bearing or positive friction piles, the theoretical formulas for dynamic or static load capacity. Therefore, pile tests should always be performed before starting a job, and preferably before the foundation design is completed, to allow the foundation engineer the opportunity to make adjustments in the design for the sake of economy.

BIBLIOGRAPHY

ACI (1972). Suggested Design and Construction Procedures for Pier Foundations, American Concrete Institute, Committee 336, *Journal ACI*, August 1972.

ASCE (1946). Pile Foundations and Pile Structures, *ASCE Manual of Engineering Practice*, No. 27.

Barkan, D. D. (1957). Foundation Engineering and Drilling by the Vibration Method, *Proc. III ICOSOMEF,* Vol. II, p. 3, London.

Bjerrum, L., Johannessen, I. J. and Eide, O. (1969). Reduction of Negative Skin Friction on Steel Piles to Rock, *Proc. VII ICOSOMEF,* Vol. II, p. 27, Mexico City.

Brahma, S. P. and Brahma, C. S. (1966). Deep Foundations: Discussion, *Proc. VI ICOSOMEF,* Vol. III, p. 493, Montreal.

Broms, K. F., Amezs, A. W. and Rinck, J. (1969). The Negative Skin Friction Along the Shaft of a Foundation Pile, *Proc. VII ICOSOMEF,* Specialty Session. Universidad Nacional de Buenos Aires, Argentina.

Buisson, M., Ahu, J. and Habib, P. (1960). Le Frottement Négatif, *Annales de l'Institute Technique du Batiment et des Travaux Publics,* No. 145, Paris.

Buisson, M. (1962). *Cimentaciones sobre Pilotes,* México, Conference at the Department of Public Works.

Chellis, R. E. (1951). *Pile Foundations,* McGraw-Hill Book Company, Inc. New York.

Correa, J. J. (1961). The Application of Negative Friction Piles to Reduction of Settlement. *Proc. V ICOSOMEF,* Vol. II, p. 41, Paris.

Cummings, A. E., Kerkhoff, G. D. and Peck, R. B. (1948). Effects of Driving Piles into Soft Clay. *Proceedings ASCE,* Vol. 74, pp. 1553-1563.

De Beer, E. E. (1966). Berekening van de negatieve wrijving op palen, *Tijdschrift der openbaren werken van Belgie,* No. 6, p. 29.

De Beer, E. E. and Wallays, M. (1968). Quelques Problemes que Posent les Fondations sur Pieux dans les Zones Portuaires, *La Technique des Travaux,* Novembre-Decembre.

Eide, O., Hutchinson, J. N. and Landva, A. (1961). Short and Long-Term Loading of a Friction Pile in Clay. *Proc. V ICOSOMEF,* Vol. II, pp. 55-57, Paris.

Endo, M., Minou, B. and Kawasawi, T. (1969). Negative Skin Friction Acting on Steel Pipe Pile in Clay. *Proc. VII ICOSOMEF,* Vol. II, p. 85, Mexico City.

Fellenius, B. H. and Broms, B. B. (1969). Negative Skin Friction for Long Piles Driven in Clay. *Proc. VII ICOSOMEF,* Vol. II, p. 93, Mexico City.

Gibbs, H. J. and Holtz, W. G. (1957). Research on Determining the Density of Sands by Spoon Penetration Testing, *Proc. IV ICOSOMEF,* Vol. I, p. 35, London.

Golder, H. Q. (1957). A Note on Piles in Sensitive Clay, *Geotechnique,* Vol. VII.

Kerisel, J. (1961). Fondations Profondes en Milieux Sableux; Variation de la Force Portante Limite en Fonction de la Densité, de la Profondeur, du Diametre et de la Vitesse d'Enforcement. *Proc. V ICOSOMEF,* Vol. II, pp. 73-83, Paris.

Lambe, T. W. and Horn, H. M. (1965). The Influence on an Adjacent Building of Pile Driving for the M.I.T. Materials Center, *Proc. VI ICOSOMEF,* Vol. II, p. 281, Montreal.

Locher, H. G. (1965). Combined Cast-in-Place and Precast Piles for the Reduction of Negative Skin Friction Caused by Embankment Fill, *Proc. VI ICOSOMEF,* Vol. II, p. 290, Montreal.

Meyerhof, G. G. (1956). Penetration Tests and Bearing Capacity of Cohesionless Soils, *Journal of Soil Mechanics and Foundations Division,* Paper 866, SM1.

Mezenbach, E. (1961). The Determination of the Permissible Point-Load of Piles by Means of Static Penetration Tests, *Proc. V ICOSOMEF,* Vol. II, p. 99, Paris.

Reese, L. and Zeevaert, L. (1976). Pier Foundations, *Joint Meeting ADSC and SMMS*, Publ. SMMS, Mexico, D.F.

Sanglerat, G. (1965). Le Pénetromètre et la Reconnaissance des Sols. Dunod, Paris.

Seed, H. B., and Reese, L. C. (1957). The Action of Soft Clay Along Friction Piles, *Proceedings ASCE,* Vol. 122, p. 731.

Skempton, A. W., Yassin, A. G. and Gibson, R. E. (1953). Theorie de la Force Portante des Pieux dans le Sable, *Annales de L'Institute Technique du Batiment et des Travaux Publics,* No. 63–64, Mars-Avril.

Tomlinson, M. J. (1957). The Adhesion of Piles Driven in Clay Soils *Proc. IV ICOSOMEF,* Vol. II, pp. 66–71, London.

Van der Veen, C. (1953). The Bearing Capacity of a Pile. *Proc. III ICOSOMEF,* Vol. II, p. 84, Switzerland.

Weele, A. (1964). Negative Skin Friction on Pile Foundation in Holland, *Proceedings of the Symposium on Bearing Capacity of Piles*, Roorkee, India.

Zeevaert, L. (1949). Discussion on the Effect of Driving Piles in Soft Clay by Cummings, Kerkhoff and Peck, *Proceedings ASCE*, Vol. 74.

Zeevaert, L. (1957). Foundation Design and Behavior of Tower Latino Americana in Mexico City, *Geotechnique*, Vol. VII.

Zeevaert, L. (1957). Discussion on Negative Friction and Reduction of Point Bearing Capacity. *Proc. IV ICOSOMEF*, Vol. III, p. 188, London.

Zeevaert, L. (1959). Reduction of Point Bearing Capacity of Piles Because of Negative Friction, *Proceedings of the First Panamerican Conference on Soil Mechanics and Foundation Engineering*, Vol. III, pp. 1145–1152, Mexico City.

Zeevaert, L. (1962). Dynamic Design and Behavior of Friction Piles Compensated Foundations, *De Ingenieur*, No. 25, Bouw-en waterboukunde 13-B163, June.

Zeevaert, L. (1961). Piling and Piled Foundations, *Proc. V ICOSOMEF*, Division 3-B, Vol. II, pp. 836–842, Paris.

Zeevaert, L. (1964). General Considerations on Problems Related with Pile and Pier Foundations, *Conference on Deep Foundations*, Mexico City.

Zeevaert, L. (1980). *ISE-Interaccion Suelo-Estructura de Cimentaciones Superficiales y Profundas*, Publ. LIMUSA, México, D.F.

►IX◄ DESIGN OF COMPENSATED FRICTION PILE FOUNDATIONS

IX.1 GENERAL CONSIDERATIONS

Friction piles may be used to improve deep compressible soil deposits to support heavy raft foundations and reduce differential and total settlements; friction piles prove particularly efficient when the shear strength increases with depth and compressibility decreases. The mechanics of friction piles was treated in general terms in section VIII.5, on the basis of a mechanical model for different conditions in which friction piles may be used in foundations. The calculation of friction piles is simplified by means of the ultimate shear strength interaction between pile shaft and soil. The friction pile foundations are analyzed under static equilibrium conditions in conjunction with the foundation structure and load system.

The state of stress induced in the soil mass because of friction pile group action may be handled via the principle of superposition in conjunction with other stress change conditions taking place in the ground because of different types of loading applied. Friction piles produce a two-fold action in the soil mass: first, they restrict the deformation of the soil, reducing compressibility; and second, the piles transmit the stresses to deeper soil strata of lower compressibility. These two actions are translated into reduction of settlements even when the foundation is subjected to high loads. The bearing capacity of the foundation is also increased as the load is transferred to deeper and higher shear strength soils underlying the pile points, as described in section VIII.4.3.

From the use of long friction piles driven in deep compressible soil deposits, it may be concluded that the soil enclosed in the length of the piles suffers less compression due to the presence of the piles, in favor of reducing settlements of the foundation. Experience has verified the philosophy given in Chapter VIII that friction pile groups or fields to support buildings in compressible soil deposits should be distributed uniformly to reinforce the soil mass properly. The distance between pile centers is also important. Piles must not be driven very close together because soil disturbance would lead to reduction in the shear strength and increase in the compressibility (see section VIII.4.2). The disturbance may reach up to one pile diameter in well spaced piles. On the other hand, the driving of one pile should not disturb the driving of the others. High hydrostatic water excess pore pressures may cause the adjacent driven piles to emerge from the ground surface; however, this problem is overcome by means of an alternating driving program. In this procedure sufficient distance is allowed between piles during driving, and when pore pressures have substantially reduced in the soil mass, the intermediate piles are driven. The driving is performed from the center of the field toward the edges to allow time for the hydrostatic excess pore pressures to dissipate in the pile field. When sand drainage layers are found interbedded in the impervious soil mass, the problem of excess pore water pressures during driving becomes less critical. During driving, however, the volume of the piles introduced in the soil displaces instantly an approximately equal volume of soil, giving rise to an important driving heave (D-heave), particularly if the piles are driven close to each other. The average D-heave per unit length of pile may be expressed as follows:

$$\overline{\delta}_D = \alpha \frac{a_p}{\overline{a}} \qquad \text{1-IX.1}$$

where a_p is the volume per unit/length displaced by one pile and $\overline{a}$ is its nominal tributary area; the factor α is a function of the compressibility and permeability of the soil, and of the procedure and time taken to drive the piles.

In clay deposits without sand drainage strata and fast driving, the value of α may be as large as 0.8; for silty clay with sand strata this value may be found on the order of ¼, and much lower in case of loose sandy silt soils. The value of α, however, should be investigated in the field for each particular location. The minimum spacing recommended in clay deposits is n_0 = 3–4 diameters. Therefore, according to expression 1-IX.1 for average conditions in silty clay deposits, one can write approximately

$$\overline{\delta}_D = \frac{100\%}{4n_0^2} \qquad \text{2-IX.1}$$

Assume a pile field is driven in a deep silty clay deposit to a depth of 20 m. The piles are 50 cm in diameter and are spaced in a uniform grid at 2.5 m between centers (n_0 = 5). Using formula 2-IX.1, we obtain $\overline{\delta}_D$ = 1% and for the total length of the piles of 20 m, a heave may be expected on the order of δ_D = 0.01 × 2000 = 20 cm. This phenomenon is reduced considerably if a pre-excavation hole is performed for each pile. Assume that the area of the pre-excavated hole is ½ of the pile cross-sectional area. Then the heave may be reduced to about one half, or 10 cm. The edge piles will produce a heave at the perimeter of the pile field on the order of ¾ of that in the center, or about 7.5 cm. The D-heave, however, may cause damage to adjacent construction; therefore, it should be investigated before pile driving, in order to design for an allowable value (section I.3).

As pile driving proceeds, the excess pore water pressure dissipates, the effective stresses in the ground tend to return to their original values, and the D-heave decreases with the tendency of the soil to reoccupy the position it had before pile driving. Full recovery of the original position, however, cannot take place because the piles will oppose to the vertical downward movement. The result is that the piles work, after pile driving, under negative skin friction forces (*NF*) in their upper part, and under positive friction (*PF*) and point bearing resistance in their lower part.

The only way to gain experience in D-heave due to pile driving is by means of observations of this phenomenon in the field, installing before pile driving a sufficient number of reference points at the ground surface between the piles, at the central part and at the perimeter of the pile field. Thus the factor α may be investigated for certain specific subsoil conditions knowing the average spacing n_0 and the time program used for pile driving. To illustrate the problem, consider an investigation of this sort performed by the author (1945) for a point bearing pile foundation driven through soft silty clay in Mexico City to a depth of 32.5 m. The maximum D-heave observed was on the order of 35 cm at the center of the pile field, and on the order of 26 cm at the edges (Fig. 1-IX.1). The timber piles have an average diameter of 30 cm and are uniformly spaced at 1.49-m centers; hence $n_0 = 5$. Using formula 1-IX.1 we obtain $\alpha = 0.32$ for the center pile area and $\alpha = 0.24$ for the edge. Moreover, notice from the observations reported in Fig. 1-IX.1 that, after two months, in this particular case the ground surface returned to practically the same elevation it had before pile driving. From this observation it may be also concluded that due to the pore water excess pressures, the skin friction on the pile shaft immediately after pile driving was small, otherwise the ground surface would have retained a major part of the D-heave.

IX.2 FRICTION PILE RAFT FOUNDATIONS

When the upper part of the soil deposit exhibits high to very high compressibility and low shear strength, surface raft foundations are expected to have large settlements, often larger than allowed for such foundations. The problem may be improved supporting the raft foundation on friction piles. On the other hand, compressible soil deposits are usually subjected to consolidation, either because of continuous increase of load at the ground surface, or seepage forces produced by a drawdown of the piezometric water levels in deep seated water bearing strata (Chapter V). Therefore, ground surface subsidence should be always expected to take place on a long-term basis. This phenomenon will affect the friction pile design if it is not properly considered. Under ground surface subsidence conditions, the piles will be subjected to ultimate skin friction forces due to continuous relative movement between the stiff piles and the compressible soil. Two cases may be considered.

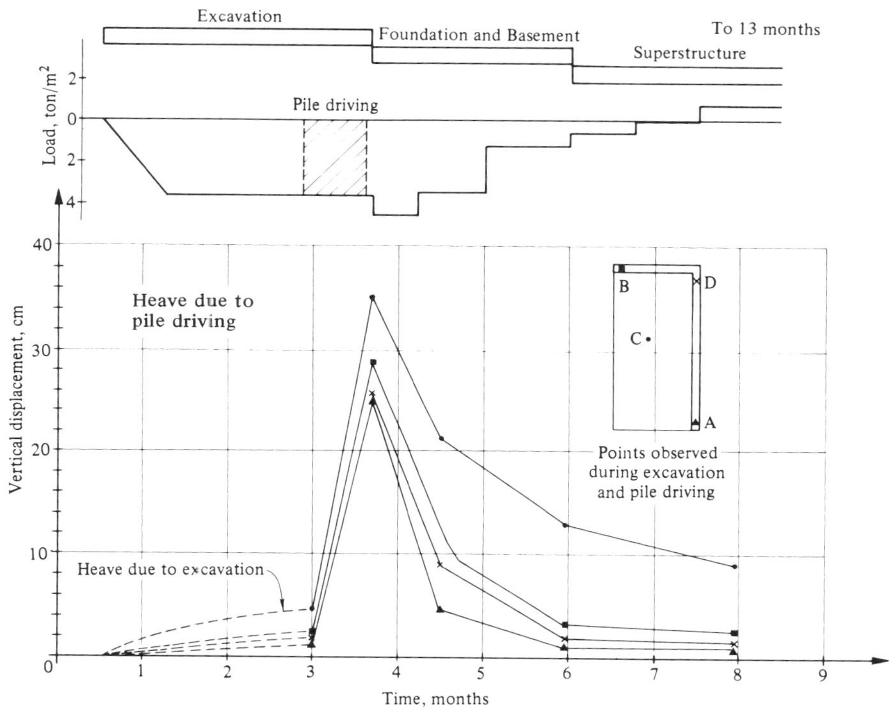

Fig. 1-IX.1 Heave observed during the construction of a pile foundation (1945).

Case 1. The piles are permitted to work under positive friction forces in their lower part, supporting the building load and the negative skin friction forces; hence:

$$Q_a + (NF)_{z_0} = (PF)_{z_0} \qquad \text{1-IX.2}$$

where Q_a is the allowable load of a pile in the group.

When such a design is contemplated, the problem can be solved as described in section VIII.5. The positive friction holding the load $Q_a + (NF)_{z_0}$ is calculated by the method described below. The final stresses induced in the soil mass within the pile field are given by curve 1-0-2-3′ in Fig. 1a-IX.2. From the foundation grade elevation to depth z_0, a relief of the vertical effective stresses will take place because of the *NF* phenomenon. The lower part, shaded stress area 2-3-4-4′-3′-2 represents an increase in the initial vertical effective stress σ_{0i}, causing settlement of the pile group according to

$$\delta_c = \sum_{z_2}^{d} m_{vi}(\sigma_i - \sigma_{0i})\Delta z_i + \sum_{d}^{f} m_{vi}(\sigma_i - \sigma_{0i})\Delta z_i \qquad \text{2-IX.2}$$

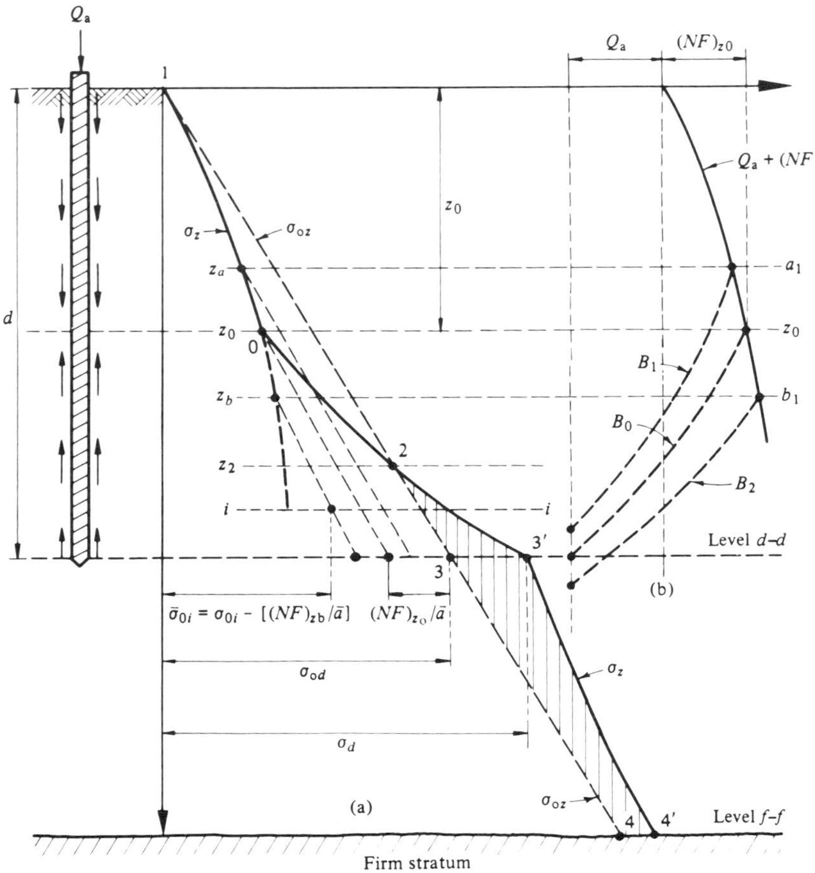

Fig. 1-IX.2 Piles supporting building load and negative skin friction.

The unit volume compressibility coefficient m_{vi} should be taken as discussed in section II.3. The part of the settlement δ_c caused by the increment of stresses $Q_a/\bar{a}_d$ underlying level d-d is expressed by the second term of formula 2-IX.2. It may be calculated by current methods, estimating the increment of stresses by the theory of elasticity (section III.1).

We estimate the stress changes in the soil mass within the piles by computing the negative and positive friction necessary for static equilibrium of the pile field subjected to load Q_a. The negative friction at the upper part of one pile in the group for $0 < z < z_0$ is calculated with algorithms given in Appendix C and section VIII.3 that read as follows:

$$\sigma_i = A_i\sigma_{0i} - B_i\sigma_{i-1} - C_i(NF)_{i-1}$$
$$(NF)_i = (\sigma_{0i} - \sigma_i)\bar{a}_i \qquad \text{3-IX.2}$$

and the load on the pile is

$$Q_i = (NF)_i + Q_a \qquad \text{4-IX.2}$$

The integration is performed from the ground surface with boundary conditions $NF_{i-1} = 0$ and $\sigma_{i-1} = 0$ at depth z_0; $(NF)_{z_0} = (\sigma_{0i} - \sigma_i)_{z_0} \cdot \overline{a}_{z_0}$.

The positive friction in the lower portion of the piles for $z_0 < z < d$ is calculated from the formulas

$$\begin{aligned} \sigma_i &= \overline{A}_i \overline{\sigma}_{0i} + \overline{B}_i \sigma_{i-1} + \overline{C}_i (PF)_{i-1} \\ (PF)_i &= (\sigma_i - \overline{\sigma}_{0i}) \overline{a}_i \end{aligned} \qquad \text{5-IX.2}$$

and the load on the pile

$$Q_i = [Q_a + (NF)_{z_0}] - (PF)_i \qquad \text{6-IX.2}$$

In this case $\overline{\sigma}_{0i}$ represents the initial stress obtained from computation of (NF) in the upper part of the piles to depth z_0, by means of formula 3-IX.2. The initial effective stresses to be used in 5-IX.2 (Fig. 1-IX.2) have the value:

$$\overline{\sigma}_{0i} = \sigma_{0i} - \frac{(NF)_{z_0}}{\overline{a}_0} \qquad \text{7-IX.2}$$

The integration will proceed from depth z_0 downward. At elevation z_0–z_0 we have $\overline{\sigma}_{0i-1} = \sigma_{i-1}$ and $(PF)_{i-1} = 0$, and the load on the pile $Q_{z_0} = Q_a + (NF)_{z_0}$ The value of $(NF)_{z_0}$ is unknown, therefore, the integration of (PF) is made for various probable values of $(NF)_{z_0}$ for an assumed depth z_0. After integration is performed, curves such as curve B for the pile load in the lower part of the piles are obtained (Fig. 1b-IX.2). From low and high values and by interpolation, the intersection where $(NF)_{z_0} + Q_a = (Q)_{z_0}$ may be obtained at depth z_0, and correspondingly we obtain the final effective stress distribution in the soil mass within the pile field. Below the points of the piles, level *d-d*, the stress increment on the overburden effective stresses is that due to $\sigma_d - \sigma_{0d}$ (Fig. 1a-IX.2).

For piles driven in clay, the positive friction is usually governed by the concept given in section VIII.4, based on the residual shear strength of the partially disturbed natural clay along the pile shaft: $s_{0z} = 0.3\, q_{uz}$. Because of ground surface subsidence, this type of foundation will emerge from the ground surface on a long-term basis since the soil will consolidate to a depth z_0 with respect to the rigid piles. If the rate of subsidence of the subsoil is found as an average of 0.2 cm/year per meter depth, and z_0 is 10 m, then the building will emerge from the ground surface at a rate of 2 cm/year. Furthermore, the piles in their lower part will penetrate the compressible soil deposit continuously. This is an important consideration that the foundation engineer should not overlook in this type of raft friction pile foundations design (section V.2).

Case II. This case is encountered when the piles are permitted to work under full positive friction (PF), carrying part of the total load of the building and permitting the ground surface to take a small unit load for stabilization purposes. A foundation of this type may prove economical since a smaller number of friction piles are required for equilibrium, since the piles are permitted to work under full skin friction load capacity. The total positive friction load $[PF]_d = Q_{fu}$ may be calculated with soil shear strength concepts and algorithms given in section VIII.4. Hence

$$\sigma_i = \overline{A}_i \sigma_{0i} + \overline{B}_i \sigma_{i-1} + \overline{C}_i (PF)_{i-1}$$
$$(PF)_i = (\sigma_i - \sigma_{0i}) \overline{a}_i \qquad 8\text{-IX.2}$$

and the load on the pile is

$$Q_i = Q_{fu} - (PF)_i \qquad 9\text{-IX.2}$$

The values of the coefficients given in formulas 3, 5 and 8-IX.2, respectively, may be found in Appendix C. They are

$$\overline{A}_i = \frac{\overline{a}_i}{\overline{a}_i - m_i \Delta z_i}, \quad A_i = \frac{\overline{a}_i}{\overline{a}_i + m_i \Delta z_i}$$

$$\overline{B}_i = \frac{m_i \Delta z_i}{\overline{a}_i - m_i \Delta z_i}, \quad B_i = \frac{m_i \Delta z_i}{\overline{a}_i + m_i \Delta z_i} \qquad 10\text{-IX.2}$$

$$\overline{C}_i = \frac{1}{\overline{a}_i - m_i \Delta z_i}, \quad C_i = \frac{1}{\overline{a}_i + m_i \Delta z_i}$$

The integration is started from the ground surface, where $(PF)_{i-1} = 0$ and $\sigma_{i-1} = 0$, and carried on to full depth level *d-d*. The minimum shear strength of the soil close to the pile shaft should be used, according to the philosophy established for this purpose in section VIII.4.2.

This type of raft friction pile foundation is used only in case of preconsolidated type soil deposits, where the increment of stress in the soil because of load transfer from pile to soil (curve σ_i) is always under the critical compressive stress σ_{bz} (section II.3). Otherwise, large compression of the soil deposit will take place in the soil mass between the piles and in the strata overlain by the pile field (Fig. 2-IX.2).

In case of soil deposits of the preconsolidated type, where $\sigma_{bi} - \sigma_{0i}$ is large before reaching the break of the compressibility curve at which the material becomes several times more compressible, it is possible to permit the ground surface to take a unit load q_a in addition to the load Q_{fu} carried by each pile. Therefore, such a raft friction pile foundation may be loaded with

$$q_a = \overline{q}_a + \frac{(PF)_d}{\overline{a}} \qquad 11\text{-IX.2}$$

in which $\overline{a} = \lambda\beta$ is the nominal tributary area at the foundation grade elevation, and

$$(PF)_d = Q_{fu}$$

To calculate the value of $(PF)_d$, it is necessary to consider, first, a load $\overline{q}_a$ applied at the ground surface (Fig. 3-IX.2). The stress distribution in the ground

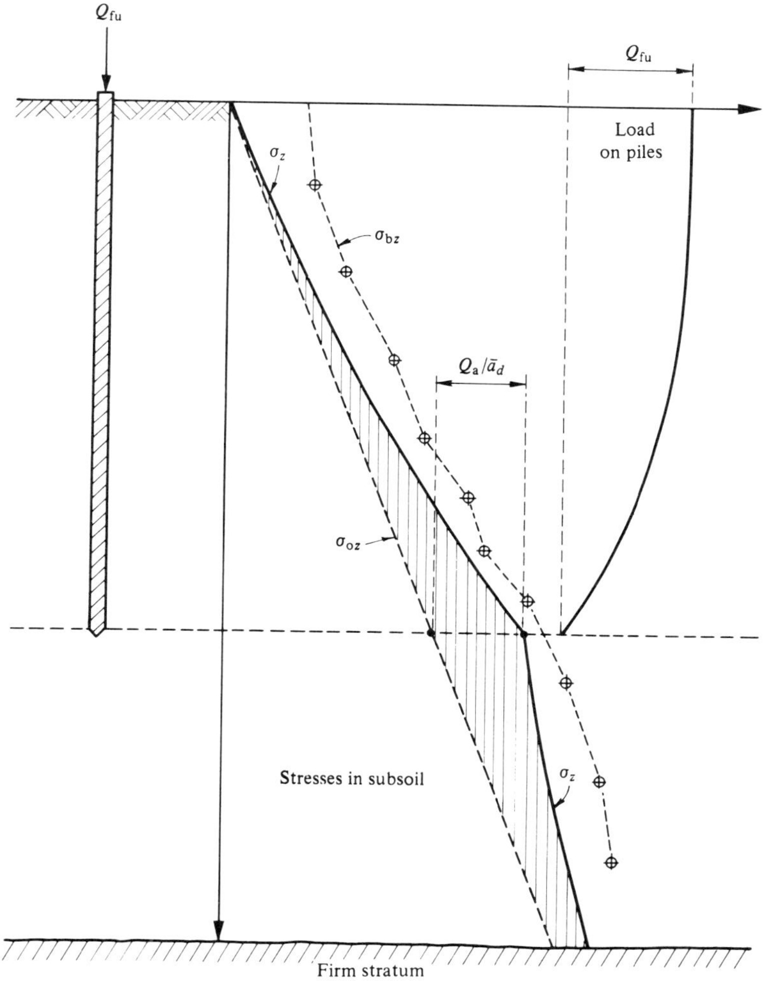

Fig. 2-IX.2 Piles under full positive friction.

is calculated by the theory of elasticity as described in section III.1, and curve $(\overline{\sigma}_{0i})_q$ is obtained as the initial stresses for integration in formula 8-IX.2. The boundary conditions in this case are $z = 0$, $(PF)_{i-1} = 0$ and $\sigma_{i-1} = \overline{q}_a$. The integration is followed to depth d, at which the total positive friction $(PF)_d$ is obtained.

The settlement of the foundation may be investigated integrating the total stress increment area 1-2-3-3″-2″-1′:

$$\delta_c = \sum_{0}^{f} m_{vi}(\sigma_i - \sigma_{0i})\Delta z_i \qquad \text{12-IX.2}$$

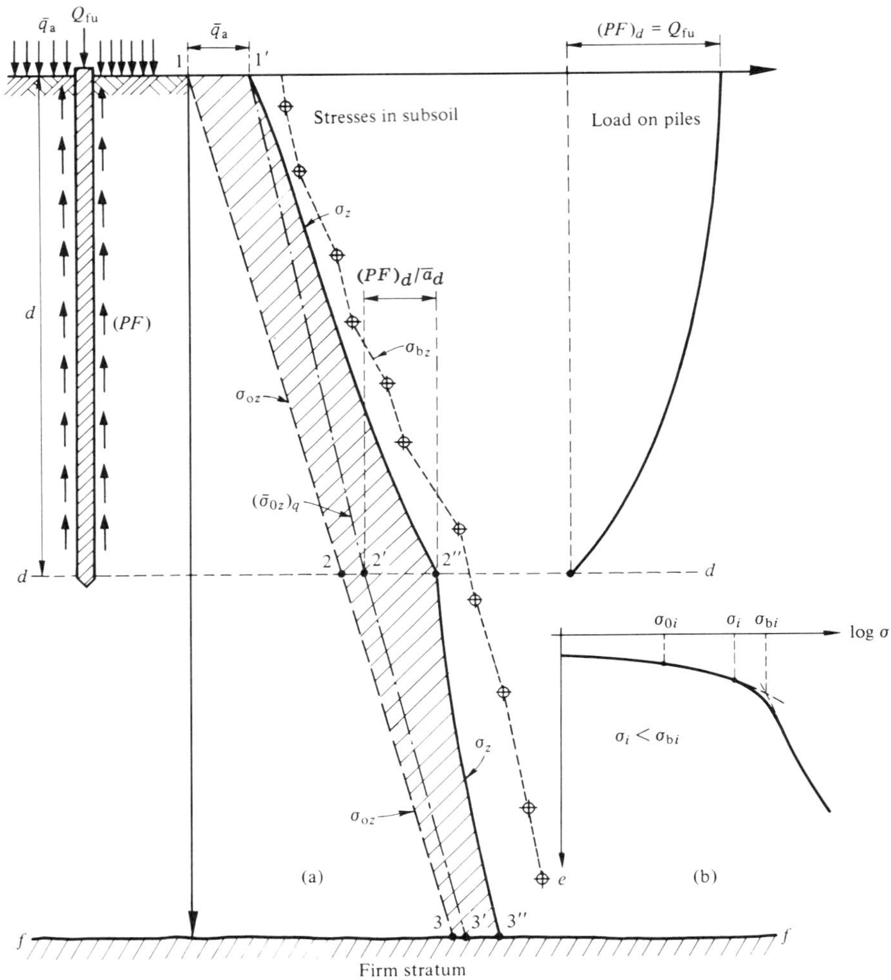

Fig. 3-IX.2 Friction pile foundation.

in which σ_{0i} is the initial overburden effective stress and σ_i is the final stress induced by $\overline{q}_a$ and $(PF)_d$ developed in the pile field, respectively (Fig. 3-IX.2).

It may be recognized that this type of pile foundation increases considerably the initial overburden vertical stresses, and large settlements may therefore be expected if the material is of very high compressibility extending to great depth. Moreover, in the design, one should observe that the final stresses σ_i induced in the ground should remain less than the critical stresses σ_b, in order to keep the compression of the soil layers in the flat portion of the compressibility curve in Fig. 3b-IX.2. Furthermore, in this type of foundation design, the friction piles are not very efficient, except when the compressibility of the soil deposit decreases strongly with depth. On the other hand, the advantage of this type of foundation is that it will follow the ground surface subsidence, after the nominal settlement δ_c expressed by 12-IX.2 has taken place. The piles under ultimate skin friction load will penetrate continuously the soil deposit; therefore, their tips should be located with amplitude away from any sand stratum that may interfere with their penetration. Otherwise, some piles may take a larger load, producing uneven reactions and tilting of the building. This type of foundation is not recommended, except in case of favorable compressibility properties of the subsoil strata and ideal stratigraphical conditions.

IX.3 COMPENSATED FOUNDATIONS WITH FRICTION PILES

3.1 General Considerations

Compensated foundations without piles have been treated in detail in Chapter VII. The general principles involved in the design of this type of foundation including friction piles are the same in general terms, except that more bearing capacity is given to the foundation by the presence of the piles. The E-heave due to excavation is also considerably reduced and, consequently, so is the subsequent settlement.

When the design of the building calls for deep basements, the E-heave at the bottom of the excavation may become very critical; therefore, friction piles when properly designed may be used to minimize the problem, permitting economy and reducing the possibility of damaging adjacent property, and also permitting a safer construction of the foundation structure. To fully benefit from the action of the friction piles, they should be driven from the ground surface before excavation is performed. Thus the skin friction action of the soil against the pile shaft causes a great part of the prestressed conditions of the subsoil to be retained, and this leads to a smaller stress relief due to excavation. The number of piles is an important factor in achieving the required design conditions for allowable heave and subsequent settlement (see section VIII.5). In a compensated friction pile foundation, the piles should be placed with a uniform distribution as in the case of raft friction pile foundations.

3.2 Heave Problem

The P-heave in designing pile foundations falls in the same category as that described in section VII.4, and may be treated in the same way. The S- and E-heave problem is considerably reduced owing to the presence of the piles; however, it may be important at the upper part of the soil deposit, when the bottom of the excavation is permitted to get flooded and clay minerals to adsorb water. The upper part of the soil softens with a corresponding reduction in its shear strength and increases in the compressibility properties, due to the swelling phenomenon induced by water adsorption in the clay minerals. Therefore, to reduce this undesirable phenomenon, mainly in cases of soils exhibiting swelling properties, the excavation is performed in two parts as described in section VII.7, in order to minimize the change in effective stresses. The first part of the excavation is performed in the upper soil deposits or fill of coarser materials without reaching, if possible, the underlying clay strata; the second part of the excavation is performed by substitution.

Assume (Fig. 1-IX.3) a compensated friction pile foundation is required with a deep excavation to be made in two stages. First, the piles are driven from the

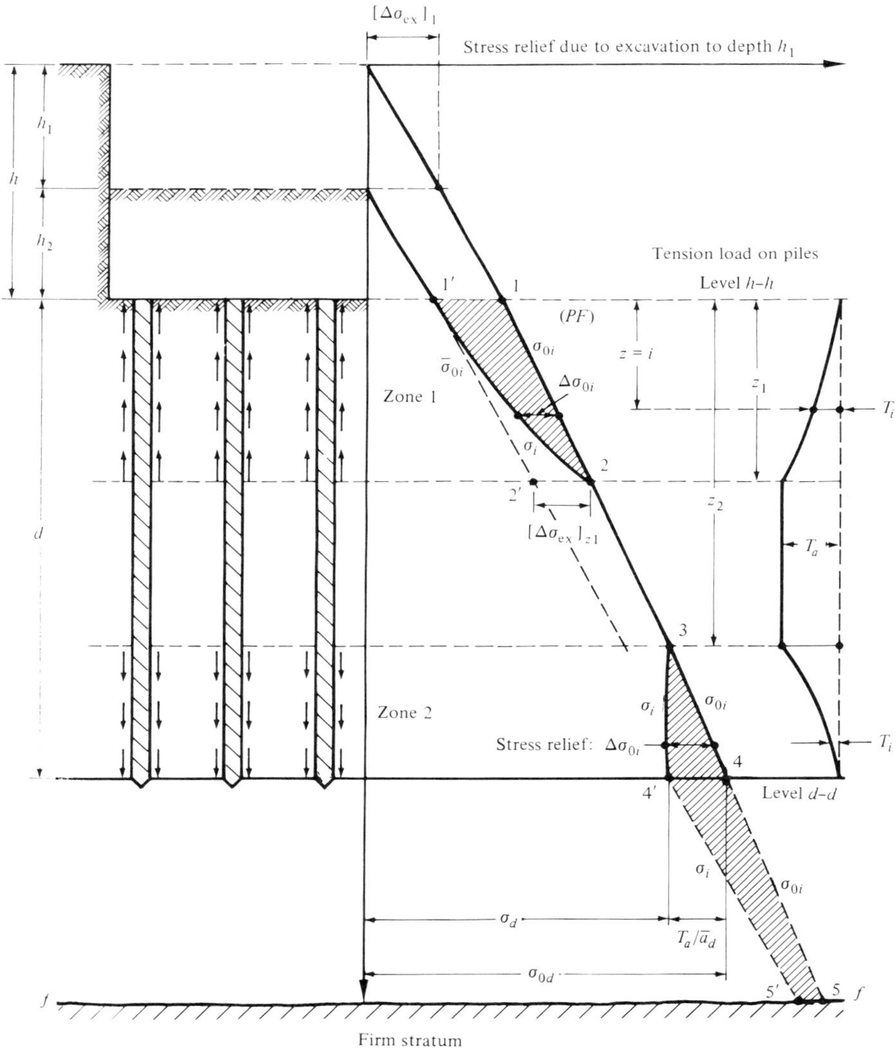

Fig. 1-IX.3 Excavation stress relief in the soil mass within a pile field.

ground surface to the selected depth, covering all the highly compressible soil strata. Since the excavation is made in two stages—first to depth h_1, and second by substitution to full depth h—the only stress relief will be that produced by full excavation to depth h_1; call it $[\Delta\sigma_{ex}]_1$. Therefore, a corresponding E-heave takes place, restricted, however, by the friction piles working under tension. As a first approximation, the piles are considered very stiff, if compared with the subsoil expansion properties. This assumption may be investigated later on, if it is found that the stiffness of the piles may have certain bearing in the E-heave, because of more stress relief permitted by the tension strain produced on the piles.

Under the expansion response forces due to excavation stress relief $[\Delta\sigma_{ex}]_1$, the upper part of the piles is pulled upward mobilizing the skin friction until static equilibrium is reached at depth z_1 (Fig. 1-IX.3). The residual stresses in the soil mass are calculated using the step by step procedure by means of

$$\begin{aligned} \sigma_i &= \overline{A}_i\overline{\sigma}_{0i} + \overline{B}_i\sigma_{i-1} + \overline{C}_i(PF)_{i-1} \\ (PF)_i &= (\sigma_i - \overline{\sigma}_{0i})\overline{a}_i \end{aligned} \qquad \text{1-IX.3}$$

The stress relief at any depth is $\Delta\sigma_{0i} = (\sigma_{0i} - \sigma_i)$. The integration is performed from the foundation grade elevation level $h - h$, where, for $z = 0$, the total friction of a pile in the field is $(PF)_{i-1} = 0$ and $\overline{\sigma}_{0i-1} = \sigma_{0i-1} - [\Delta\sigma_{ex}]_{i-1}$. The values of $\overline{\sigma}_{0i}$ are determined to depth z_1, by the theory of elasticity (Chapter III). The value of z_1 is determined when $\sigma_2 = \sigma_{02}$ at this level. The tension load on the pile has the value

$$T_a = (\sigma_{02} - \sigma_2)\overline{a}_2 \quad \text{or} \quad T_a = [\Delta\sigma_{ex}]_{z_1} \cdot \overline{a}_2 \qquad \text{2-IX.3}$$

At the head of the piles $T_{i-1} = 0$ and $z = z_1$, $T_i = T_a$ (Fig. 1-IX.3), at any other level $0 < z < z_1$, $T_i = (PF)_i$.

The tension force T_a has to be balanced at the lower part of the piles by the skin friction mobilized along the pile shaft from depth z_2 to d. This produces an effect of negative friction given by (Appendix C),

$$\begin{aligned} \sigma_{i-1} &= \overline{A}_{i-1}\sigma_{0i-1} + \overline{B}_{i-1}\sigma_i - \overline{C}_{i-1}(NF)_i \\ (NF)_{i-1} &= (\sigma_{0i-1} - \sigma_{i-1})\overline{a}_{i-1} \end{aligned} \qquad \text{3-IX.3}$$

The boundary conditions at level $d - d$ are $(NF)_d = T_a$ and $\sigma_d = \sigma_{0d} - T_a/\overline{a}_d$. The integration proceeds upward until reaching elevation $d - z_2$, where $(NF)_{i-1} = 0$ and $T_i = T_a$. The stress relief at any depth $z_2 < z < d$ in the subsoil because of this action is $\Delta\sigma_i = (\sigma_{0i} - \sigma_i)$ and the pile tension is $T_i = T_a - (NF)_i$. The stress relief area and pile load diagram are shown in Fig. 1-IX.3.

The E-heave due to stress relief in stratified subsoil conditions in an extensive pile field is estimated by means of expressions developed in Section VII.5:

$$\delta_e = \sum_0^{z_1} \rho_e\nu_c M_{ei}(\sigma_{0i} - \sigma_i)\Delta z_i + \sum_d^{d-z_2} \rho_e\nu_c M_{ei}(\sigma_{0i} - \sigma_i)\Delta z_i + \sum_d^{f} \rho_e\nu_c M_{ei}(\sigma_{0i} - \sigma_i)\Delta z_i + \delta_R \qquad \text{4-IX.3}$$

The value of ν_c is used only in highly stratified soil, where $\nu_c M_e = m_{ve}$ otherwise $\nu_c = 1$ may be used (Chapter II.3.5).

Usually, it is found that the first term in formula 4-IX.3 is considerably larger than the others. Notice that the E-heave is considerably reduced, as otherwise produced by a relief of stresses approximately equivalent to stress area 1-5-5′-1′, in case no piles were present.

The effect of the elongation of the piles due to elastic response in the intermediate zone z_2-z_1 should be investigated (Fig. 1-IX.3). Let us represent by δ_R the total elastic response of the soil, equal to the pile elongation in the same zone. The approximate expansion of the soil because of the additional stress relief $\Delta\sigma_{ep}$ due to the pile elongation is

$$\delta_R = \rho_e \nu_c M_{ea} \Delta\sigma_{ep}(z_2 - z_1) \qquad \text{5-IX.3}$$

Here M_{ea} represents the average elastic response strain modulus determined from full hysteresis loops in unconfined compression tests (section VII.5). The expansion factor ρ_e has been defined in section VII.5, and that of ν_c in section II.3.7. The pile elongation in the same zone is

$$\delta_R = \frac{T_a}{a_p E_p}(z_2 - z_1) \qquad \text{6-IX.3}$$

where a_p is the cross-sectional area of the pile and E_p is the tension modulus of elasticity. Equating expressions 5 and 6-IX.3, and solving for the stress relief in the soil, we obtain

$$\Delta\sigma_{ep} = \frac{T_a}{\rho_e \nu_c M_{ea} E_p \cdot a_p} \qquad \text{7-IX.3}$$

On the other hand, the change of tension in the pile is

$$\Delta T_a = \Delta\sigma_{ep} \cdot \overline{a} \qquad \text{8-IX.3}$$

Substituting, we obtain

$$\Delta T_a = \frac{T_a}{\rho_e \nu_c M_{ea} E_p} \cdot \frac{\overline{a}}{a_p} \qquad \text{9-IX.3}$$

and the soil expansion in zone $(z_2 - z_1)$

$$\delta_R' = \frac{T_a - \Delta T_a}{a_p E_p}(z_2 - z_1) \qquad \text{10-IX.3}$$

To give an idea of this correction, let us use values obtained in practice. From first calculation assuming the concrete piles to be infinitely stiff, the tension found is $T_a = 36{,}000$ kg in the intermediate zone $(z_2$-$z_1) = 1000$ cm. The cross-sectional area of the pile is 1500 cm^2, and the modulus of elasticity under tension $E_p = 150{,}000$ kg/cm^2. The expansion factor is on the order of $\rho_e = 0.2$. Under laterally confined conditions for a Poisson's ratio of $\nu = 0.35$, a value of $\nu_c = 0.6$ may be assigned (see section II.3.5 and Fig. 12-II.3). The linear response strain modulus of the soil in zone z_2-z_1, is found to be $M_{ea} = 0.02$ cm^2/kg. The nominal tributary area $\overline{a}$ of one pile in the field equals 40,000 cm^2. Therefore, applying expressions 9 and 10-IX.3, respectively, the following values are obtained: the change of tension in the pile ΔT_a = 2670 Kg, the relief of stress in the soil $\Delta \sigma_e$ = 0.067 kg/cm^2 and δ'_R = 0.15 cm. Thus, for this particular case, it is concluded that the correction in the pile tension and the elongation in zone $(z_2 - z_1)$ are negligible for practical purposes.

3.3 Load Reapplication

After the first stage of excavation is performed to depth h_1, as previously described, the new initial state of vertical stresses in the soil mass within the friction piles is given by curve 1'-2-3-4' in Fig. 2-IX.3. These stresses hereafter will be called $\overline{\sigma}_{0i}$. In the second stage of excavation, the construction of the foundation structure and superstructure takes place in trenches, by substitution of weights. Therefore, no more significant changes in stresses in the subsoil should take place. The effective stresses $[\Delta\sigma_{ex}]_2$ pertaining to the second stage of the excavation to depth h will be taken gradually by the weight of the rigid foundation structure and building. No further E-heave occurs at the foundation grade elevation when the substitution of weights is properly achieved, as described in section VII.7. Upon gradual application of stress $[\Delta\sigma_{ex}]_2$ by substitution of the foundation structure, the piles will start to take load by positive friction. The skin friction acting downward in the lower part of the piles, and producing the tension T_a, will be reversed up to a point where it will be cancelled, when the load on the piles reaches the value $[\Delta\sigma_{ex}]_1 \cdot \overline{a}$ (Fig. 1-IX.3). Hence, the stresses induced in the soil mass will turn back to the initial overburden vertical stresses represented by curve 3-4-5 in the lower part of the piles (Fig. 2-IX.3). Therefore, this procedure represents the design of a totally compensated friction pile foundation, in which the piles and the soil will support together a load equivalent to the original effective vertical stress encountered at elevation level h-h; hence

$$\sigma_{0h} = [\Delta\sigma_{ex}]_1 + [\Delta\sigma_{ex}]_2 \qquad \text{11-IX.3}$$

The value of $[\Delta\sigma_{ex}]_1$ is taken by the friction piles, and $[\Delta\sigma_{ex}]_2$ by the soil between the piles. If u_h represents the allowable uplift water pressure at the foundation grade elevation (depth h-h), then the total unit load that may be applied is

$$q_h = \sigma_{0h} + u_h \qquad \text{12-IX.3}$$

The settlement due to load reapplication will be only that pertaining to the recompression of stress area 3-4'-5'-5-4-3 (Fig. 2-IX.3) and may be calculated from the formula

$$\delta_s = \sum_{z_2}^{f} \rho_c m_{vi}(\sigma_{0i} - \overline{\sigma}_{0i})\Delta z_i \qquad \text{13-IX.3}$$

where ρ_c is the recompression factor described in section VII.5. It may be noticed that the recompression of stress area 1-1'-2 does not take place, since this stress

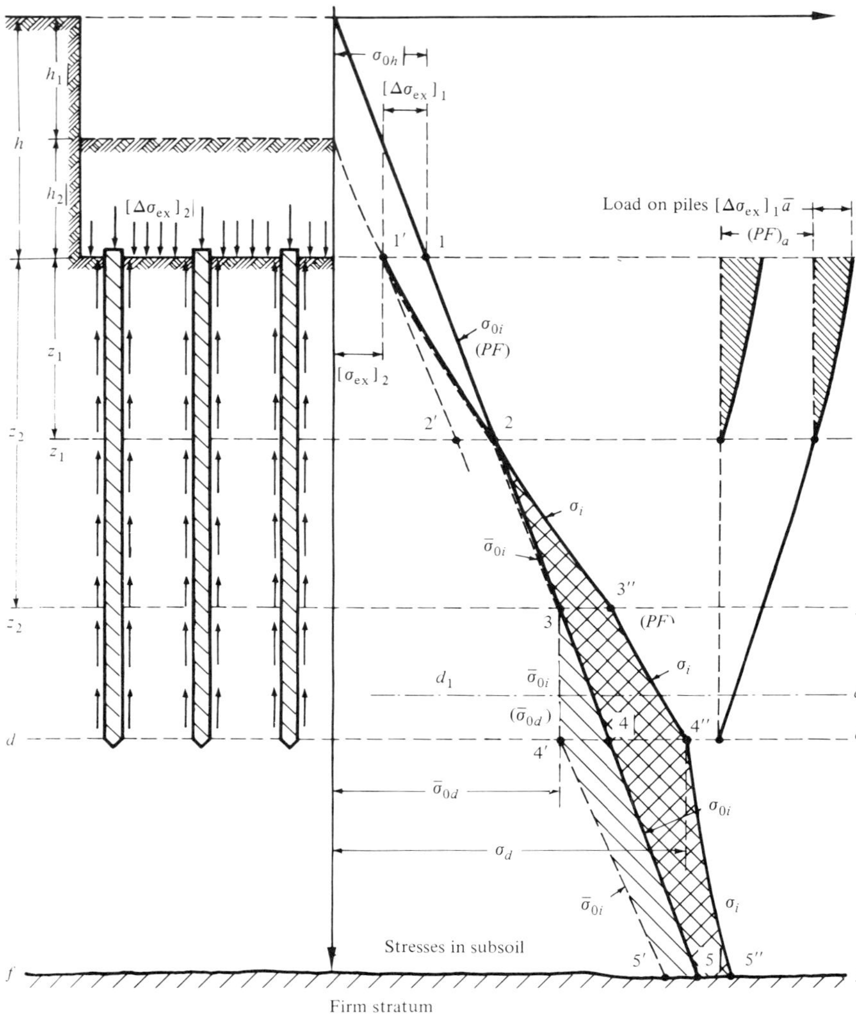

Fig. 2-IX.3 Stresses in soil mass for compensated friction pile foundation.

area is taken by the friction piles under the action of positive friction. Therefore, the settlement due to recompression of the soil strata of high compressibility at the upper part of the soil deposit to depth level z_1-z_1 is eliminated with this method. Since in nature the compressibility at the lower part of the soil deposits is usually smaller than in the upper soil layers, it may be recognized that a compensated friction pile foundation gives, for the same loading, considerably less settlement than a compensated foundation without piles.

More load on the piles may be assigned, however, if the allowable total settlement is fulfilled. This may be achieved, as stated before, when the lower part of the compressible soil deposit has considerably less compressibility than the upper part, and the critical stress of the soil is not passed. Compensated friction pile foundations are specially important in areas of ground surface subsidence. In such cases, it is convenient to design the foundation not emerging from the ground surface. Therefore, the piles are designed under full positive friction, which is fully mobilized because of the relative movement between the stiff piles and subsiding soil. For a foundation of this type we assume that the load on the piles is increased to such an extent that the positive skin friction is completely mobilized from elevation z_1 to the tip of the piles (level d-d, Fig. 2-IX.3). The positive friction is calculated step by step from the formula

$$\sigma_i = \overline{A}_i \overline{\sigma}_{0i} + \overline{B}_i \sigma_{i-1} + \overline{C}_i (PF)_{i-1}$$
$$(PF)_i = (\sigma_i - \overline{\sigma}_{0i})\overline{a}_i \qquad \text{14-IX.3}$$

Therefore, the ultimate load that the piles can support in addition to $\overline{a} \cdot \Delta\sigma_{\text{ex}}$ is obtained by integration from level $(1 - 1)$ downward to level $(d - d)$. The boundary conditions at $(z_1 - z_1)$ are $(PF)_{i-1} = (PF)_{z0} = 0$, $\sigma_{i-1} = (\Delta\sigma_{\text{ex}})_2$. The values of $\overline{\sigma}_{0i}$ are known and given by curve 2-3-4′-5′ (Fig. 2-IX.3). The integration is followed to a certain depth level $(d_1 - d_1)$, where the value of $(\sigma_i - \overline{\sigma}_{0d})$, is found, and the total positive friction $(PF)_{d_1} = (\sigma_1 - \overline{\sigma}_{0d_1})\overline{a}$. At this level if by 10-VIII.4 $0.3q_{ui} \leqslant K_\phi \sigma_1$, then $\Delta(PF) = 0.3\overline{\omega} \sum_{d_1}^{d} q_{ui}\Delta z_i$. Therefore, the total load that may be applied on the piles is

$$Q_{fu} = \overline{a}(\Delta\sigma_{\text{ex}})_1 + (PF)_{d_1} + 0.3\omega \sum_{d_1}^{d} q_{ui} \cdot \Delta z_i \qquad \text{15-IX.3}$$

and the total equivalent unit load the foundation can carry is

$$q_h = (\Delta\sigma_{\text{ex}})_2 + Q_{fu}/\overline{a} + u_h \qquad \text{16-IX.3}$$

The settlement that may be expected from this foundation design will be the recompression of stress area 3-4′-5′-5-4-3 and the compression of stress area 2-3-4-5-5″-4″-3″-2; therefore

$$\delta_s = \sum_{z_2}^{f} \rho_c m_{vi}(\sigma_{0i} - \overline{\sigma}_{0i})\Delta z_i + \sum_{z_1}^{f} m_{vi}(\sigma_i - \sigma_{0i})\Delta z_i \qquad \text{17-IX.3}$$

The value of m_v is used if the soil deposit is highly stratified or reinforced by piles; otherwise the value of M_z should be used (section II.3.5).

It may be noticed that the foundation design just described is an elasto-plastic design. The piles will work under plastic equilibrium, and the unit load applied to the soil at the foundation grade elevation represents the elastic response of the soil. Nevertheless, the critical stress σ_{bz} in the lower part of the deposit from level z_1-z_1 to f-f should be carefully observed to avoid large compressions of the soil strata, the same philosophy as described for the design of raft foundations. Therefore, compensated friction pile foundations may be used to advantage, especially in preconsolidated type soil deposits, where the compressibility decreases with depth, and in areas where ground surface subsidence is taking place.

In the case of sensitive normally consolidated soil deposits, the induced stresses should remain equal to or smaller than the overburden initial stresses σ_{0i}, that is to say, curve 2-3′-4″-5″ should as a maximum reach curve 1-2-3-4-5, representing the initial effective stresses.

3.4 Importance of Point Resistance

In the case of friction pile compensated foundations as described in previous paragraphs, resistance to penetration at the pile points should be expected since the foundation has to follow the ground surface subsidence. When the subsoil material underlying the pile points is a sensitive soil of medium to very high compressibility, the point resistance is negligible; however, it may be estimated under plastic flow conditions from

$$Q_{pu} = a_p \cdot \sigma_b \qquad \text{18-IX.2}$$

in which σ_b is the confined compressive critical stress of the soil at level d-d, and a_p is the cross-sectional area of the pile point.

In the case in which the subsoil material is of the semi-impervious type like silt and clayey silt with fine sand, the ultimate point resistance may be estimated by means of the following orthodox equation for bearing capacity:

$$Q_{pu} = a_p\{\alpha_1[cN_c + \sigma_d N_q]\}(D_r + 0.1) \qquad \text{19-IX.2}$$

The pile tips are designed to penetrate continuously in the subsoil forced by the ground surface subsidence and building weight. Therefore, the shear strength properties have to be selected carefully by means of consolidated-undrained tests performed in the triaxial chamber on undisturbed soil specimens. The test specimen is reconsolidated to the equivalent overburden state of effective stresses it had in the ground, and from there, tested under undrained conditions (section II.4). From this test, the shear parameters c_{cu} and ϕ_{cu} are determined. The value of D_r in 19-IX.3 for most medium compressible soils may be taken on the order of 0.3.

To give an idea of the order of magnitude of the point resistance in these cases, let us take information from professional practice in case of a clayey silt deposit underlying the pile tips. From the tests, the following values are obtained: $c_{cu} = 0.4$ kg/cm^2 and $\phi_{cu} = 20°$. From N-factors chart (Fig. 4-III.4 and section III.4.2b) $N_c = 23.5$ and $N_q = 8.3$. The confined effective stress at the depth of the pile tips including the effect of positive friction of the pile is $\sigma_d = 3.0$ kg/cm^2, and the pile cross-sectional area $a_p = 2000$ cm^2. Hence

$$Q_{pu} = (2000)\,1.2[0.4 \times 23.5 + 3 \times 8.3]\,(0.3 + 0.1) = 32\,900 \text{ kg}$$

The ultimate point bearing resistance thus obtained is added to the positive friction load Q_{fu}. Therefore, the total load the point-positive friction piles carry is

$$Q_{fp} = Q_{fu} + Q_{pu} \qquad \text{20-IX.3}$$

The stress diagram shown in Fig. 2-IX.3 will show the same, except that the vertical stresses between level *d-d* and *f-f* should be incremented, because of the point load of the piles at level *d-d* (see section VIII.5.3). The stress increment distribution owing to the point action of the piles may be determined from the theory of elasticity, and should be included in the second term of formula 17-IX.3, when calculating settlements.

IX.4 OVERTURNING MOMENTS

Earthquakes and high wind forces cause in the foundation structure overturning moments, increasing the reactions to a magnitude necessary for static balance. To estimate their effect in a friction pile compensated foundation, the same philosophy as discussed in section VII.8 may be used. The piles working under full positive friction load, however, are assumed not to admit load increments. Therefore, the increments induced by the overturning moments are taken by soil reaction against the foundation slab, as if the piles were not present. The rotation of the foundation structure and probable permanent tilt of the building originated by the overturning moment may be estimated from the safe side, by the methods already given for compensated foundations without piles. In the case of friction pile foundations, the contact soil stresses are usually small and remain under high elastic response properties; therefore they will behave more favorably under unbalanced forces than the compensated foundation without friction piles. Moreover, under transient loads produced by earthquakes, the piles will offer resistance, and therefore the approach recommended to investigate the foundation behavior due to the induced overturning moments is conservative.

IX.5 BENDING MOMENTS AND SHEARS

The calculation of the stability of the foundation structure with column loads, soil and pile reactions is performed similarly to that already described in Chapters IV, VI and VII. In case of friction piles, the load on the piles is considered not to change with the deflections of the foundation structure, since the friction piles are considered to work under plastic conditions. The part corresponding to the soil reaction against the rigid foundation structure may be investigated using the method of foundation moduli.

In the foundation structural design, however, it should be considered that the piles do not react to exactly the same plastic load. The limits should be fixed from the variation in the shear strength properties to be expected in the soil deposits to the depth the piles are embedded, and in the area they are driven. In practice, it is generally found a variation on the order of ±20% in uniform stratified subsoil, like compressible lacustrine and marine sediments to great depth. In general, the variations in the plastic load of the piles have little influence when the foundation is designed deep into the ground and with large stiffness, as is the case of compensated foundations, and when the stratigraphic conditions in the area covered by the piles may be assumed uniform. The uniformity of the subsoil should be investigated by means of penetration or cone test borings relating them with undisturbed sample borings made to determine the mechanical characteristics of the soil sediments. Important variations in subsoil conditions may produce uneven load capacities in the piles, inducing tilting of the building that should be investigated and evaluated taking into consideration the ratio of building height to foundation width.

BIBLIOGRAPHY

Pérez Carbajal, René (1964). *Análisis sobre el Comportamiento de un Edificio con dos Unidades de Diferente Peso, Cimentado sobre Arcilla de Alta Compresibilidad en la Ciudad de México*, Tesis profesional de Ingeniero Civil. Universidad Nacional Autónoma de México.

Zeevaert, L. (1957). Compensated Friction Pile Foundation to Reduce Settlement of Buildings on the Highly Compressible Volcanic Clay of Mexico City, *Proc. IV ICOSOMEF*, Vol. II, pp. 81–86, London.

Zeevaert, L. (1963). Foundation Problems Related to Ground Surface Subsidence in Mexico City, *Field Testing of Soils, ASTM Special Technical Publication*, No. 322, pp. 57–66.

▸X◂ DESIGN OF POINT BEARING PILES AND PIER FOUNDATIONS

X.1 GENERAL CONSIDERATIONS

The best foundation to support heavy and concentrated loads is a pile or pier foundation, for which a deep seated bearing stratum with mechanical characteristics of low compressibility and high shear strength is necessary. The acceptable depth of the stratum is limited, however, by the ability to drive the piles with reasonable verticality, to achieve the best possible distribution of load on the bearing stratum. The maximum inclination allowed is 2%. Therefore, if the minimum spacing of the piles at the bearing stratum is of not less than n_b diameters and the spacing at the elevation at which the piles are driven is n_0 diameters, then the critical length of the piles is approximately

$$h_0 - h_{ex} \leqslant 50\,(n_0 - n_b)\,d_0 \qquad \text{1-X.1}$$

where d_0 is the pile diameter, and h_{ex} is the depth that may be pre-excavated to pass without difficulty through the hard strata of the upper soil sediments, and at the same time, lead the piles vertically. Assume that the minimum allowable spacing at the bearing stratum is $n_b = 2.5$ diameters, and their spacing required at the surface $n_0 = 3.5$ diameters; if the diameter at the point is 0.35 m, the pile length from the driving level to the bearing stratum should not be longer than $l_e = 17.5$ m. The piles will be driven in pre-excavated holes to depth h_{ex} to overcome the driving resistance of the upper hard crust of the soil deposit. Hence, the convenient depth at which a hard stratum should be located to obtain a satisfactory job is $h_0 \leqslant l_e + h_{ex}$. Assuming the pre-excavation is performed to depth $h_{ex} = 6$ m, then $h_0 = 23.5$ m for this particular case. The values of l_e vs n_0 for different pile diameters are plotted in Fig. 1-X.1 to facilitate the investigation of the convenient depth for normal pile driving with reasonable verticality.

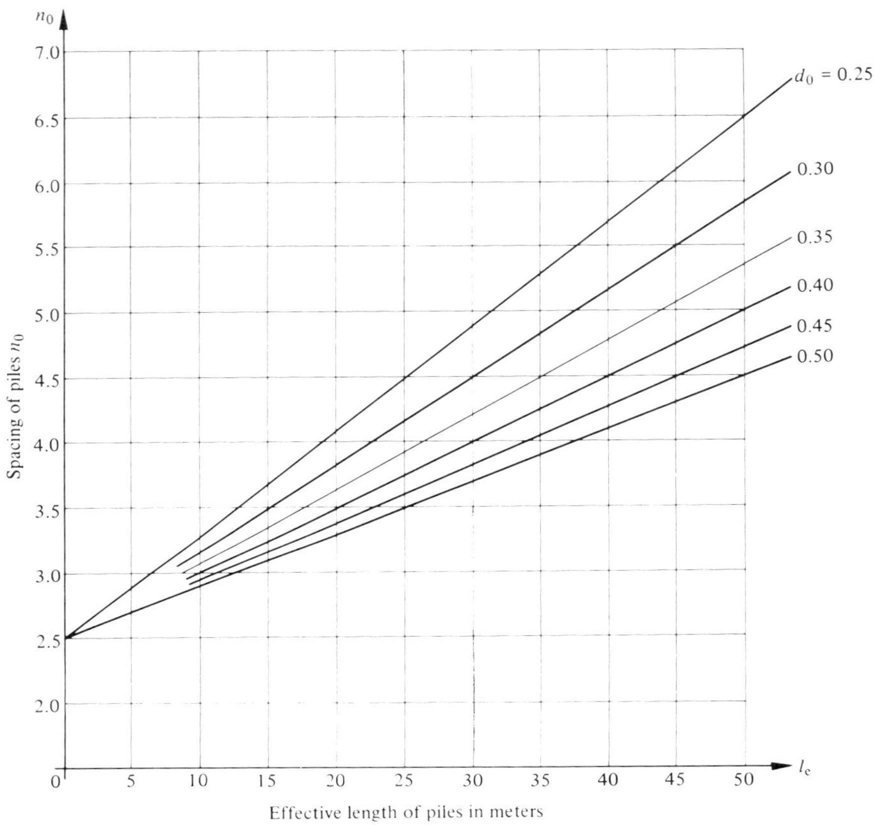

Fig. 1-X.1 Allowable pile spacing for point bearing piles.

Point bearing piles are used to carry the load through medium to very highly compressible layers that in the majority of the cases are subjected to ground surface subsidence (Chapter V). Furthermore, as already stated, during pile driving through these impervious soils, expansion takes place because of the volume displaced by the piles, and therefore immediately after the pile driving, negative skin friction takes place on the piles; this phenomenon cannot be overlooked. Moreover, piles are very slender elements driven to deep seated bearing ground; therefore, they are not able to take horizontal loads satisfactorily, even if they are strong and monolithic. The horizontal loads like wind and earthquakes when not properly considered in the foundation design, may produce important stresses in the pile heads, and damage or rupture may take place (Fig. 2-X.1). Shallow foundation structures on point bearing piles are particularly exposed to damage. This undesirable situation may be investigated in the foundation design as described in Chapter XII. The most convenient pile should be carefully selected from the practical and economical point of view, as already described in section VIII.7.

Fig. 2-X.1 Rupture of pile head due to high bending.

Piers are very desirable elements when firm ground of low and very low compressibility is encountered at depths not greater than about 12 m. The construction procedure of piers should be carefully considered, especially when the water table is close to the ground surface, since during excavation a strong upward water flow may take place, disturbing the natural mechanical conditions of the bearing stratum. On the other hand, in the case of piers, one is able to investigate directly the supporting soil material and take samples to verify the shear strength and compressibility under the base of each pier. Piers are very rigid elements subjected to drifting horizontal forces induced by strong earthquake ground motions, set in highly deformable soil sediments from the ground surface to the bearing stratum, causing overturning moments at their base and large shears and bending moments along their shaft. In case of ground surface subsidence or consolidation of the upper sediments, large negative skin friction is induced along the shafts, because of the relative movement between the upper soil deposits and the bearing stratum. The decision on whether to use piers or piles when a good bearing stratum is encountered is usually a matter of economy that, however, has to be investigated for every particular problem under consideration.

X.2 POINT BEARING PILE FOUNDATIONS

2.1 Typical Cases of Point Bearing Piles

The most common problems of point bearing pile foundations may be classified into three cases:

Case I: When point bearing piles or piers rest on a firm soil mass of low or very low compressibility extending to great depth.

Case II: When point bearing piles rest on a stratum of low to very low compressibility of limited thickness, underlain by an impervious or semi-impervious stratum of medium to high compressibility.

Case III: When the foundation load is placed at the ground surface using the point bearing piles to carry the surface load by skin friction along their shafts.

2.2 Case I

When the subsoil stratigraphy is such that the firm base of the upper soft sediments is encountered at an economical depth, the piles may be driven to the firm stratum with good point bearing capacity. They may be driven in groups under the column loads and with the minimum possible spacing, generally not smaller than 2.5 diameters. The convenient spacing at the head of the piles may be investigated by means of formula 1-X.1, expressed in the following form:

$$n_0 = \frac{h_0 - h_{ex}}{50 d_0} + 2.5 \qquad \text{1-X.2}$$

where $h_0 - h_{ex} = l_e$ represents the effective length of the piles. Hence, if the excavated depth is $h_{ex} = 3$ m, the firm bearing stratum is at 25 m depth, and the piles diameter is 0.4 m, then the minimum spacing at the foundation grade elevation is 3.6 diameters (Fig. 1-X.1). Hence, for maximum density the piles may be driven with a spacing of 1.44 m between centers. The point bearing capacity and the stability of the piles and pile groups should be determined as already discussed in sections III.3 and VIII.2, and the settlement of each group investigated according to procedures discussed for this purpose in preceding chapters. For point bearing capacity purposes, it is necessary to calculate the reduction in confining stress at the bearing stratum because of the negative skin friction, using the numerical step by step procedure with formulas given in section VIII.3:

$$\sigma_i = A_i \bar{\sigma}_{0i} - B_i \sigma_{i-1} - C_i (NF)_{i-1} \qquad \text{2-X.2}$$

and

$$(NF)_i = (\overline{\sigma}_{0i} - \sigma_i)\overline{a}_i$$

The integration is followed from the foundation grade elevation. The value of $\overline{\sigma}_{0i}$ is the effective vertical stress obtained after the relief of the vertical over-burden effective stresses because of excavation before driving the piles. At this elevation the boundary conditions are $(NF)_{i-1} = 0$ and $\sigma_{i-1} = 0$. The mechanical properties of the different layers have to be considered in the integration, since the subsoil may be highly stratified, containing layers ranging from clay to loose sand with variable index and shear strength properties along the pile shaft. At depth level $d - d$ (Fig. 1-X.2), we obtain the negative skin friction load $(NF)_d$ and a corresponding reduction in the confining stress $\overline{\sigma}_{0d} - \sigma_d = (NF)_d/\overline{a}_d$. Calling the safety factor G_s, the point of the piles should be able to take a load equal to

$$[Q_a + (NF)_d]\, G_s = Q_{pu} \qquad \text{3-X.2}$$

The quantity Q_{pu} is the ultimate point bearing capacity of the piles. Hence, the allowable load is given by

$$Q_a = \frac{1}{G_s} Q_{pu} - (NF)_d \qquad \text{4-X.2}$$

When the foundation structure is constructed deep into the ground, the permanent water uplift pressure u_h acting against the foundation slab should be included;

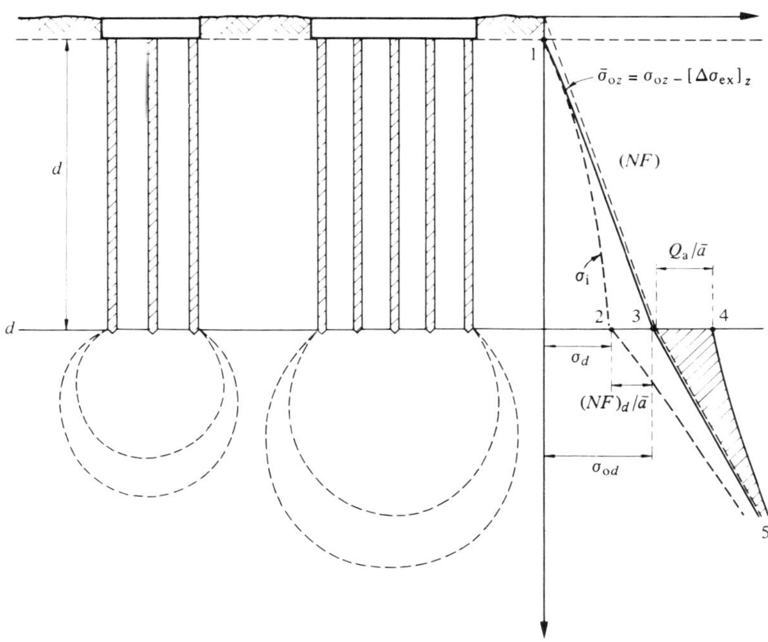

Fig. 1-X.2 Shallow foundation on point bearing piles.

hence the total allowable load per pile with tributary area $\bar{a}$ is

$$\bar{Q}_a = Q_a + u_h \bar{a} \tag{5-X.2}$$

2.3 Case II

The case of a hard stratum at a convenient depth, underlain by impervious or semi-impervious compressible series of soil sediments like silts, silty clays or clays, is idealized in Fig. 2-X.2, for a foundation requiring a deep excavation.

The hard stratum is able to take safely the point load of the piles. The ultimate point bearing capacity can be estimated by means of the index and mechanical

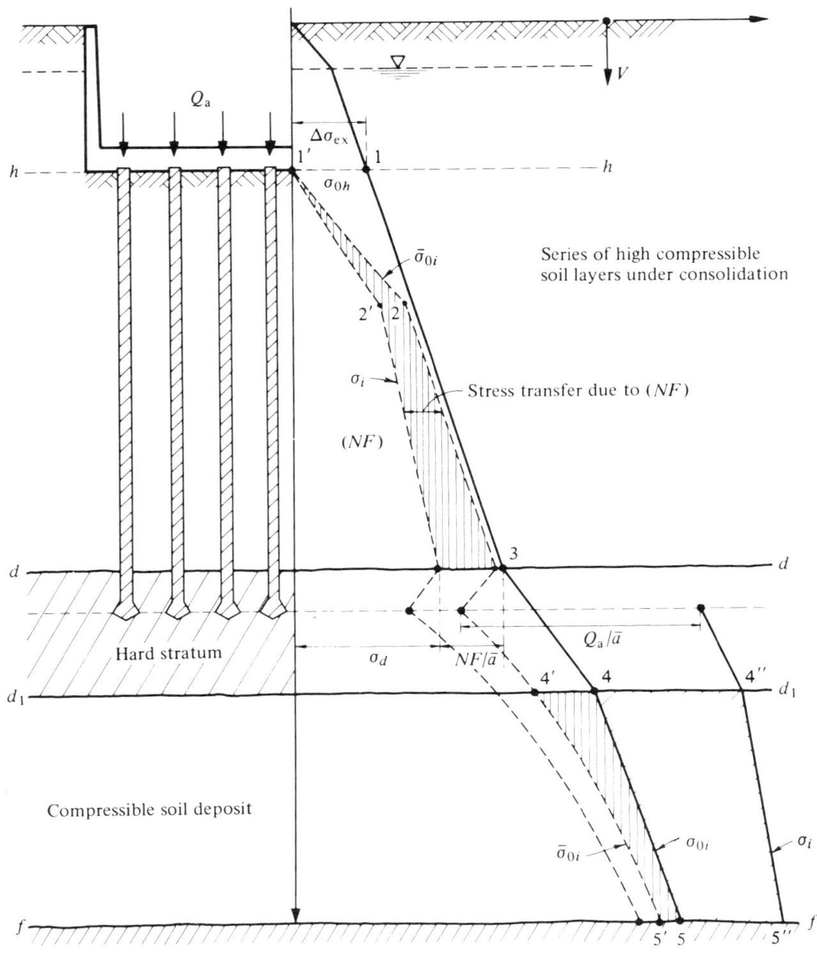

Fig. 2-X.2 Piles to a hard stratum underlain by compressible soil.

properties of the material and the confining stress on the bearing stratum, with methods described in sections III.4 and VIII.6 and 7. The series of soil sediments found overlying the hard stratum are assumed again, as in preceding cases, materials of medium to high or very high compressibility. These upper subsoil deposits may be considered under the phenomenon of ground surface subsidence, which, if not clearly present at the time the foundation is constructed, will take place eventually during the working life of the building. Therefore, in the design it is considered a continuous relative movement between piles and soil, and as in Case I, the piles will be subjected to negative friction, thus increasing the point load on them and reducing the confining stresses on the bearing stratum. From the point bearing capacity formula, it may be recognized that if the hard stratum is a cohesionless material, the confining stress and the relative density of the soil are decisive factors in the point resistance to penetration. Therefore, the two-fold action of negative friction on the piles is a very important factor in these cases.

Before calculating negative friction, it is necessary to determine the initial stresses in the ground on which the calculations should be based. This state of stresses in the soil mass depends on the procedure used for excavation and driving of the piles. It has been stated before that in order to reduce heave in deep excavations, and thereby minimize the danger of damage during construction of the foundation, it is convenient to drive the piles previously to the excavation (Chapter IX). Hence, the piles are driven through the upper soil crust in previously excavated borings to level $h - h$. This practice has a two-fold purpose: first, to avoid hard driving at the upper crust and reduce absorption of friction energy that may be used with advantage when driving the pile points in the hard layer; and second, the pre-excavated holes or borings serve as a guide to the piles, thus obtaining better verticality during driving according to allowances given by formula 1-X.2.

Consider the case of a deep excavation to be preceded by pile driving. After the piles are driven, the excavation may proceed in one or several stages, depending on the specific problem in hand and satisfying the vertical and horizontal allowable displacements set for the foundation problem (sections I.3 and VII.6). Assume that the excavation may be performed in one stage, because it was estimated from stability considerations that the S- and E-heaves are allowable due to the presence of the piles. The E-heave restricted by the presence of the piles is estimated from stress relief $\Delta\sigma_{ex}$ represented by area 1-2-1′, as already described in similar problems. The stress relief distribution in the subsoil because of excavation is shown by curve labeled $\overline{\sigma}_{0i}$ in Fig. 2-X.2. In this case the pulling force T_a on the piles due to the excavation relief is taken completely by the embedded point in the hard stratum. This phenomenon should be carefully investigated to be sure the piles will not pull loose from their driven position; an enlarged point at the pile bottom may be designed to help withstand the pulling force.

The vertical stresses induced within the pile field because of excavation stress relief will be used as the initial effective stresses to calculate the negative friction with the step by step procedure using the same algorithms given in 2-X.2.

The edge and corner piles have to be carefully investigated for *NF*, since on the outside the relief of stress due to excavation is not the same as that given by the curve labeled $\bar{\sigma}_{0i}$. Therefore, it is recommended for negative skin friction calculations to use in formula 2-X.2 for the initial stresses an intermediate value $(\frac{1}{2})(\sigma_{0i} + \bar{\sigma}_{0i})$ for edge piles, and $(\frac{1}{4})(3\sigma_{0i} + \bar{\sigma}_{0i})$ for corner piles. Hence, the negative skin friction on the perimeter piles will be larger than that obtained for interior piles. The settlement of the foundation will be that caused by the stress areas 4-4′-5′-5 and 4-4″-5″-5 (Fig. 2-X.2) added to the deformation of the hard stratum under the pile point loads:

$$\delta_s = \sum_{d_1}^{f} \rho_c m_{vi} (\sigma_{0i} - \bar{\sigma}_{0i}) \Delta z_i + \sum_{d_1}^{f} m_{vi} (\sigma_i - \sigma_{0i}) \Delta z_i + [\delta_p]_{d_1}^{d} \qquad \text{6-X.2}$$

The increment of stresses between level *d-d* and *f-f* is calculated by the theory of elasticity, as described in section III.1. The third term expressing the deformation in the hard stratum because of the high stresses induced by the pile points usually may be neglected; however, for semicompact cohesionless soils, it should be investigated by methods explained in section VIII.2.3.

Point bearing pile foundations, either Case I or Case II, have the liability that when piles pierce soil layers of high and very high compressibility under a state of consolidation, the foundation structure emerges from the ground surface as the surrounding soil settles owing to ground surface subsidence. The implications of this phenomenon have been extensively discussed in Chapter V (Fig. 17-V.3). As mentioned before, this phenomenon may be alleviated by a ground floor design which permits periodic lowering.

2.4 Case III

In the attempt to reduce the problem of buildings emerging from the ground surface because of the phenomenon of ground surface subsidence, a pile foundation system is used empirically in Mexico City (González Flores, 1950). From practical point of view, the method consists in permitting the head of the point bearing piles to pierce through the foundation slab. The load of the building is applied on the soil surface between the piles at the bottom of the excavation (Fig. 3-X.2). The piles take the load by skin friction. If q_a is called the unit load applied on the soil, each pile will take by skin friction a total load equal to $Q_{fu} = q_a \bar{a}$, where $\bar{a}$ is the nominal tributary area to the corresponding pile. The friction load on the pile is transmitted to the point and taken by the bearing stratum, where the piles are firmly supported, as in Cases I and II.

Since the goal is to make the building follow the ground surface subsidence, the piles should be allowed to work under fully mobilized skin friction. When the building is heavier, the load difference is applied at the head of the piles, by means of specially designed mechanical devices acting as jacks. The control devices may be also used to take differential vertical displacements occurring in each individual pile. The mechanical device, however, has to be maintained in good working condi-

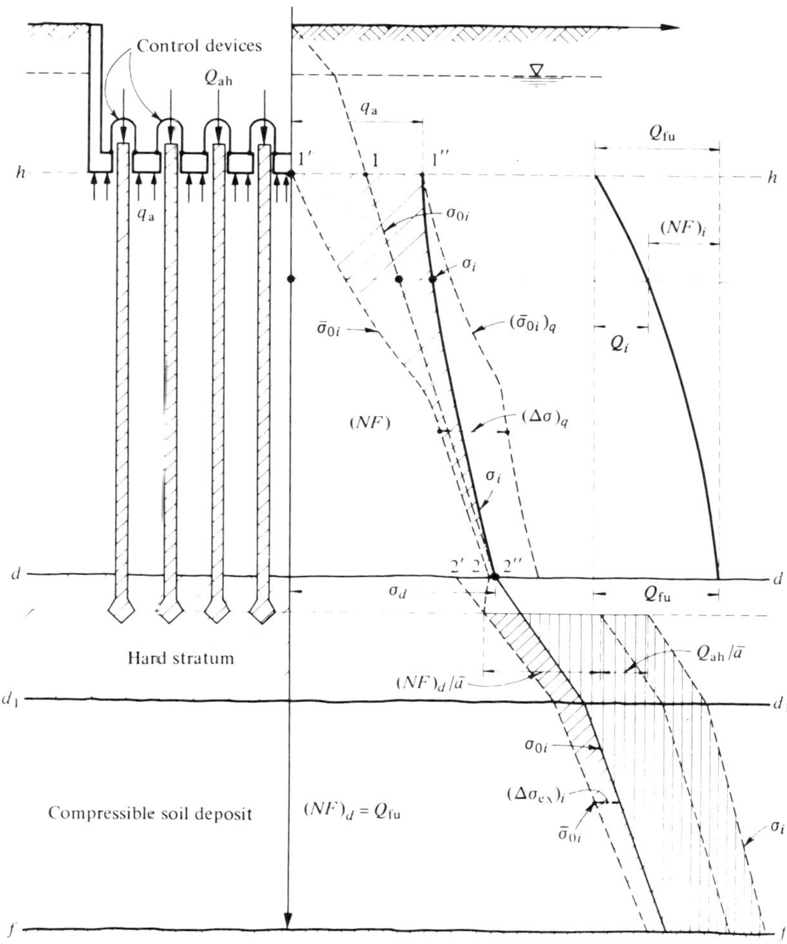

Fig. 3-X.2 Control device used to apply required load to piles.

tions throughout the life of the building, and will require periodic servicing as indicated by the observational behavior records of the foundation structure. Therefore, the total allowable load taken by the pile points is

$$Q_{pa} = Q_{fu} + Q_{ah} \tag{7-X.2}$$

where Q_{ah} is the additional load applied to the head of the piles by means of the control device.

The theoretical problem to be solved in this type of friction-controlled point bearing pile foundation is to find the value of q_a compatible with the total frictional force Q_{fu} that can be applied to the piles. The calculation may be performed using the step by step procedures for the solution of other skin friction problems described in preceding chapters.

The final value of Q_{fu} on a long-term basis is a function of the overburden effective stresses in the soil mass, and the stress induced by q_a applied on the soil at the foundation grade elevation and on the method used to perform the excavation. To illustrate this case, assume (Fig. 3-X.2) that piles have been driven to point bearing in a hard stratum at level *d-d* before excavation is made. They are provided with specially designed pile points to take the tension load caused by the stress relief due to the excavation. Moreover, the excavation will be performed in one stage, after a stability analysis has been made showing that no P-heave will occur, and the S- and E-heaves are under specified allowances. The piles are driven first and the excavation performed afterwards, the reduction in vertical effective stresses due to the presence of the piles is computed with the same procedure explained in section IX.3, shown by curve $\bar{\sigma}_{0i}$ in Fig. 3-X.2. When the unit load q_a is applied, the stresses in the soil within the piles are increased to $(\bar{\sigma}_{0i})_q$; however, the final vertical stresses in the soil mass within the piles after consolidation has taken place are governed by the presence of the pile field. The algorithms to calculate σ_i, the skin friction load F, and the pile load Q_i are the following:

$$\sigma_i = A_i(\bar{\sigma}_{0i-1})_q - B_i\sigma_{i-1} - C_i(NF)_{i-1} \qquad \text{8-X.2}$$

$$(NF)_i = [(\bar{\sigma}_{0i})_q - \sigma_i]\, a_i$$

The vertical stress increments due to q_a on $\bar{\sigma}_{0i}$ are calculated by the theory of elasticity (section III.2), and thus we obtain the $(\bar{\sigma}_{0i})_q$ values.

The load on the pile at any depth is

$$Q_i = Q_{fu} - (NF)_i \qquad \text{9-X.2}$$

but $Q_{fu} = q_a \cdot \bar{a}$ and, therefore

$$Q_i = q_a\bar{a} - (NF)_i \qquad \text{10-X.2}$$

Here $\bar{a}$ is the nominal tributary area for one pile loaded with q_a. Furthermore, the boundary conditions are taken at the foundation slab elevation, where $Q_{fu} = (NF)_{i-1} = q_a \cdot \bar{a}$ and $(\sigma_{0i-1})_q = q_a$. The integration follows downward step by step until at depth level *d-d*, corresponding to the top of the bearing stratum, the load $q_a\bar{a}$ should be fully taken by the pile skin friction. Hence at level *d-d* one should obtain $(NF)_i = 0$ and $\sigma_i = \sigma_d$ (point 2″, Fig. 3-X.2). The above condition is satisfied only if the value of q_a was properly assigned. Since the value of q_a is unknown, the calculation is performed by a trial-and-error procedure. A value of q_a is assumed, taking for instance as first approximation $q_a \geqslant \Delta\sigma_{ex}$ and performing the integration to depth level *d-d*, with formulas 8 and 9-X.2. If $(NF)_i$ vanishes before reaching depth level *d-d*, then q_a is increased to a value sufficient to pass level *d-d* with $(NF)_i > 0$. Thereafter, intermediate values of q_a are chosen until the proper value of q_a is found, such that $(NF)_i \doteq 0$ at level *d-d*. Thus the total skin friction for one pile in the field is $Q_{fu} = (NF)_d = q_a \cdot \bar{a}$. This procedure may be performed rapidly with a computer.

The shear strength along the pile shaft should be investigated by means of the equation

$$s_{0i} = \tfrac{1}{2} K_\phi(\sigma_i + \sigma_{i-1}) \qquad \text{11-X.2}$$

and compared with the value $c_{0i} = 0.3q_{ui}$ found in section VIII.4. For clay for which $c_{0i} < s_{0i}$ at depth z_0, the skin friction load from that depth to level *d-d* will be computed by means of the formula

$$\Delta Q_{fu} = \overline{\omega} \sum_{z_0}^{d} (0.3q_{ui})\Delta z_i \qquad \text{12-X.2}$$

When $c_{0i} = 0.3q_{ui}$ in the total length of the piles, then

$$q_a = \frac{\overline{\omega}}{\overline{a}} \sum_{0}^{d} c_{0i}\Delta z_i \qquad \text{13-X.2}$$

At any other depth z the effective stresses in the soil mass between the piles are

$$\sigma_z = \overline{\sigma}_{0z} + \frac{\overline{\omega}}{\overline{a}_z} \sum_{0}^{z} c_{0i}\Delta z_i \qquad \text{14-X.2}$$

The settlement taking place in this type of foundation may be estimated from the calculation of the average stress changes corresponding to the recompression stress areas up to the original overburden effective stresses (curve σ_{0i}) and to the final stress values σ_i (Fig. 3-X.2). Here again, the limiting conditions require the stress not to exceed the critical stress σ_{bz} of the highly sensitive soil strata. The settlement is given by

$$\delta_S = \sum_{d}^{h} \rho_c m_{vi}(\sigma_{0i} - \overline{\sigma}_{0i})\Delta z_i + \sum_{d}^{h} m_{vi}(\sigma_i - \sigma_{0i})\Delta z_i + \sum_{d_1}^{f} m_{vi}(\sigma_i - \sigma_{0i})\Delta z_i + [\delta_p]_{d_1}^{d} \qquad \text{15-X.2}$$

The value of $[\delta_p]_{d_1}^{d}$ has the same significance as that in formula 6-X.2, as it represents the deformation under the pile tips in the hard stratum; this, however, may be controlled at the pile head.

It may be observed from Fig. 3-X.2 that the compression in the upper part of the highly compressible soil deposits may be large, if the increment of stresses because of q_a exceeds the critical stress σ_{bz}. When this is expected, by means of the control devices load is transferred to the head of the piles. This action is translated, however, into a reduction of Q_{fu}, allowing the rest of the pile to work under negative friction. Eventually, if the entire Q_{fu} load is transferred to the head of the piles by the control devices, we have the foundation design for Cases I or II. Nevertheless, on a long-term basis, the upper part of the deposit may be permitted to consolidate gradually following the rate of ground surface subsidence by means of the mechanical control devices, until the full benefit of Q_{fu} is obtained and the foundation settles at the same rate as the ground surface. This type of foundation may be recommended when properly designed and serviced. Theoretically, however, the requirements of the point bearing piles necessary to support this type of foundation are the same as in Cases I and II, in order to facilitate the use of the control devices.

X.3 PIER FOUNDATIONS

When firm soil of very low compressibility, or consolidated rock is encountered at a shallow depth, the footings are lowered into open pit excavations and actually carved into the hard soil, transmitting the load by means of piers (Fig. 1-X.3). The supporting material should have such mechanical properties as to be able to withstand large bearing stresses with small vertical displacements. The most common use of these elements is found in piemont and volcanic detritus deposits, where each column support should be carefully investigated (section I.2). In such cases, the construction of the pier will, in general, not offer special problems such as those usually encountered in large bore holes or pit excavations; however, when the water table is above the base of the pier, important dewatering problems have to be overcome. This type of foundation has the advantage that the bearing soil or rock may be examined directly, therefore assuring that the mechanical properties of the soil for each element are adequate to support the load.

When the base of the pier is overlain by consolidating soil sediments, large negative skin friction takes place along the pier shaft, substantially increasing the load.

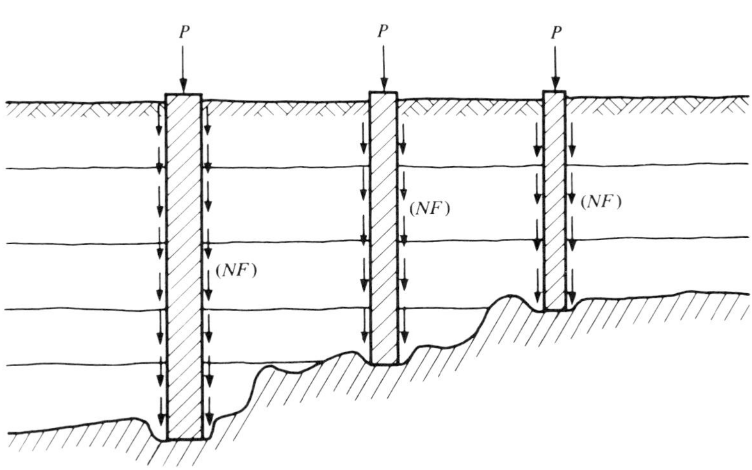

Fig. 1-X.3 Piers transmitting load to a hard stratum.

The shear parameters for calculating negative friction should be carefully selected. The pier, usually constructed of reinforced concrete, is poured against the soil, and therefore the potential surface of sliding for negative friction will be located through the natural ground. The values of c_d and ϕ_d will be required (section II.4). The negative skin friction $(NF)_d$ may be estimated by means of the approximate formula 13-VIII.3. The allowable load on the pier is

$$Q_a = \frac{1}{G_s} Q_{pu} - (NF)_d \qquad \text{1-X.3}$$

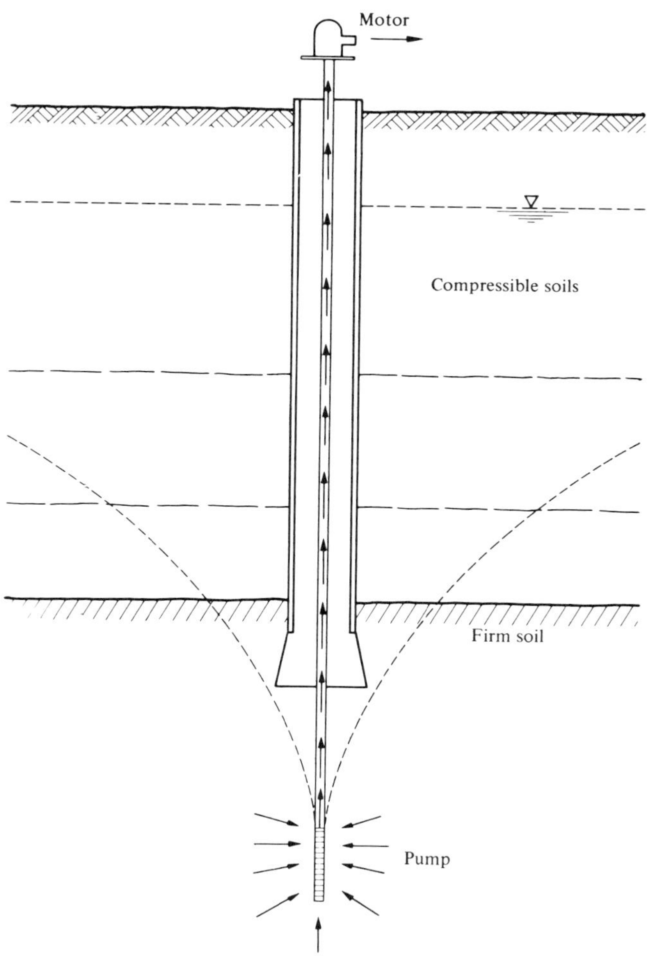

Fig. 2-X.3 Pumping to produce a downward water flow under piers.

in which G_s is the factor of safety for ultimate bearing capacity at the base of the pier. Common diameters of piers are from 0.90 m to several meters.

Piers often have to be excavated under water; hence the method of construction is important in calculating the negative skin friction and the point bearing capacity. The current construction methods for the excavation of piers are beyond the scope of this book. Nevertheless, whichever the method may be, the vertical water flow at the base of the excavation cannot be overlooked, since this phenomenon may produce upward seepage forces in the soil that will loosen or soften the materials affecting the bearing capacity and compressibility of the supporting soil under the base of the pier. This phenomenon may be avoided by driving pumping wells sufficiently deep from the base of the pier to produce a downward water flow (Fig. 2-X.3). One or several wells, depending on the pier diameter, may be required

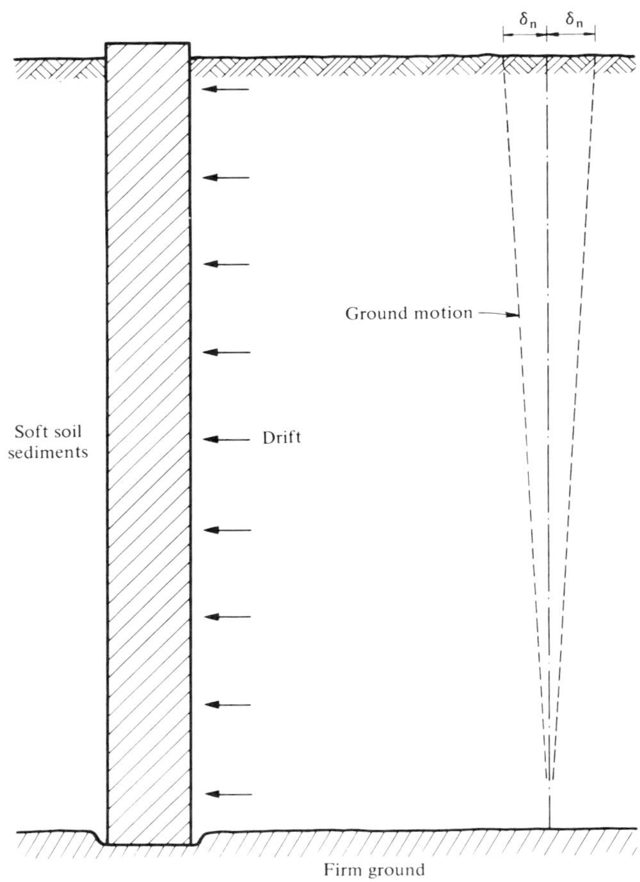

Fig. 3-X.3 Drifting forces on pier due to ground motion.

(section III.3.6). Pumping should be started prior to excavations. Complicated problems may be present in case the hard bearing deposit is overlain by fine sand, in which case the construction procedure calls for special methods and equipment. The use of wells or well points is always important to consider in excavations for piers.

During earthquakes, piers may be subjected to large overturning moments because of drifting forces exerted on them by the ground motion in the soft sediments overlying the base of the pier (Fig. 3-X.3). Pier elements are very rigid in contrast to slender piles and cannot easily follow the ground motion. This interesting problem of foundation seismic engineering in difficult subsoil conditions is discussed in Chapter XII.

X.4 OVERTURNING MOMENTS AND BASE SHEAR

4.1 Tilting of Foundation

Pile foundations are usually provided with heavy and stiff foundation structures, especially in Cases II and III, and even in Case I when the piles cover the foundation area in a uniformly spaced grid. The overturning moments in such cases may be transmitted by the foundation structure to the piles in a monolithic action, the assumptions being that the foundation slab, usually reinforced by heavy beams, remains plane during deformation and that the increments of load on the piles during the transient forces are proportional to the pile vertical displacements. Under this assumption, the load increment in a pile with coordinates x_i and y_i with respect to the principal axis for the corresponding layout of the piles is given by

$$\Delta Q_i = \pm \frac{O_{T_x}}{\Sigma x^2} x_i \pm \frac{O_{T_y}}{\Sigma y^2} y_i \qquad \text{1-X.4}$$

If the foundation structure may be considered flexible with respect to the stiffness of the piles, the load increments induced by the overturning moments in columns or walls bearing directly on the piles should be distributed in tributary piles of the corresponding support. The increment of ΔQ_i should be investigated for point bearing capacity, but generally if the static point bearing capacity of the piles has a nominal factor of safety, the transient load exerted in the foundation due to the overturning moments will cause elastic response of the piles and no permanent tilt problem of the building will be expected. (See Chapter XII in relation to foundation problems in seismic engineering.)

The base shear, however, is by far a more important problem in deep pile foundations. The study of this problem is treated in Chapter XII. It may be stated, however, that in case of shallow foundation structures, the horizontal forces exerted at the foundation grade elevation have to be taken properly by the pile heads, and therefore, their cross-section has to be specially designed for this purpose. In case of deep foundation structures, the passive earth pressure against the retaining walls usually is sufficient to withstand with an adequate degree of safety the horizontal forces. In any case, the forces induced by the base shear should be investigated to obtain a proper and safe response of the soil confining the foundation structure. If this is not the case, the specific problem from the point of view of structural design should be carefully considered by the foundation and structural engineers.

4.2 Tilting Control for Pile Foundations

An adequate factor of safety in the case of uneven settlement or probable tilting of the building can be achieved with jacking devices installed at the head of the perimeter piles, allowing reduction or increase of the load on these piles as necessary (Figs. 1 and 2-X.4). In such a design, the perimeter piles should be able safely to take loads at least twice as large as the interior piles. The rotation of the founda-

Fig. 1-X.4 Jacking device to adjust load on perimeter piles.

Fig. 2-X.4 Final arrangement at perimeter piles to control tilting of building.

tion is controlled, causing a counter-balancing moment, by increasing the load on the piles at one side and decreasing it on the other side of the foundation. The balancing loads are verified periodically at the time the tilting of the building is inspected. The tilting control piles are adjusted to the required values necessary to keep the building within the tilting allowances.

This procedure may be readily applied in all types of pile foundations, when because of nonuniformity and high compressibility properties of the subsoil strata a rotation of the foundation is expected under static loading. The use of tilting control piles are particularly useful in machine-bearing long strip foundations similar to those required for paper mills, where transverse rotation of the foundation structure is restricted to very small values. They are also useful in connection with compensated foundations or any other type of foundation structure subjected to eccentric loadings on compressible soil deposits. In extreme cases of overturning moments caused in slender buildings by earthquake or wind forces, the tilting control piles may be used to advantage in the economy and stability of the foundation subjected to these horizontal forces.

X.5 SHEARS AND BENDING MOMENTS

The determination of the shears and bending moments in a rigid foundation structure for tall buildings on pile foundations does not present a more complicated problem than the one already explained for other foundation types. As a matter of fact, the only expected complication is that concerning the statically indeterminate structural problem of the foundation structure itself. The reactions on the foundation slab are calculated with procedures described in Chapter VI. For this purpose, the foundation modulus or spring constant of the piles should be investigated. Let us consider a point bearing pile foundation. Let Q_a be the pile load under working conditions applied at the pile head. The modulus of deformation of the pile material at stress level Q_a is E_p and the pile length h. Further assume that Q_a stands for a line of n_p piles (Fig. 1-X.5). Therefore, the unit deformation of the piles will be

$$\bar{\delta}_\text{p} = \frac{h}{(n_\text{p} a_\text{p}) E_\text{p}} \qquad \text{1-X.5}$$

Call $\bar{\delta}_\text{sp}$ the unit vertical displacement at the point of the pile due to the compression of the bearing stratum under the pile point. This value is only a function of the stress-strain properties of the soil, the pile cross section and the important stresses induced in the bearing stratum take place in about three diameters of the pile point (section VIII.2.3). The pile points do not influence each other in this action when the bearing stratum compressibility reduces strongly with depth. If the strain modulus M_ep of the hard stratum in the neighborhood of the pile point is considered constant, then

$$\bar{\delta}_\text{sp} = \frac{1.9\, r_0}{(n_\text{p} a_\text{p})} \cdot M_\text{ep} \qquad \text{2-X.5}$$

Call the unknown reactions of the pile lines by X_a, X_i, X_j, X_b then the displacements at the pile heads, setting $\bar{\delta}_\text{p} + \bar{\delta}_\text{sp} = \bar{\delta}_\pi$ will be

$$\begin{bmatrix} \bar{\delta}_{\pi\text{a}} & \cdot & \cdot & \cdot \\ \cdot & \bar{\delta}_{\pi i} & \cdot & \cdot \\ \cdot & \cdot & \bar{\delta}_{\pi j} & \cdot \\ \cdot & \cdot & \cdot & \bar{\delta}_{\pi\text{b}} \end{bmatrix} \cdot \begin{vmatrix} X_\text{a} \\ X_i \\ X_j \\ X_\text{b} \end{vmatrix} = \begin{vmatrix} \delta_\text{a} \\ \delta_i \\ \delta_j \\ \delta_\text{b} \end{vmatrix}_\pi$$

If the foundation is rigid we may assume $\delta_\text{a} = \delta_i =$ constant, hence

$$[\bar{\delta}_{\pi i}]_D \cdot |X_i| = |\delta_\text{a}| \qquad \text{3-X.5}$$

and the values of the pile lines average reactions may be found solving the above equations for the unknowns X_i.

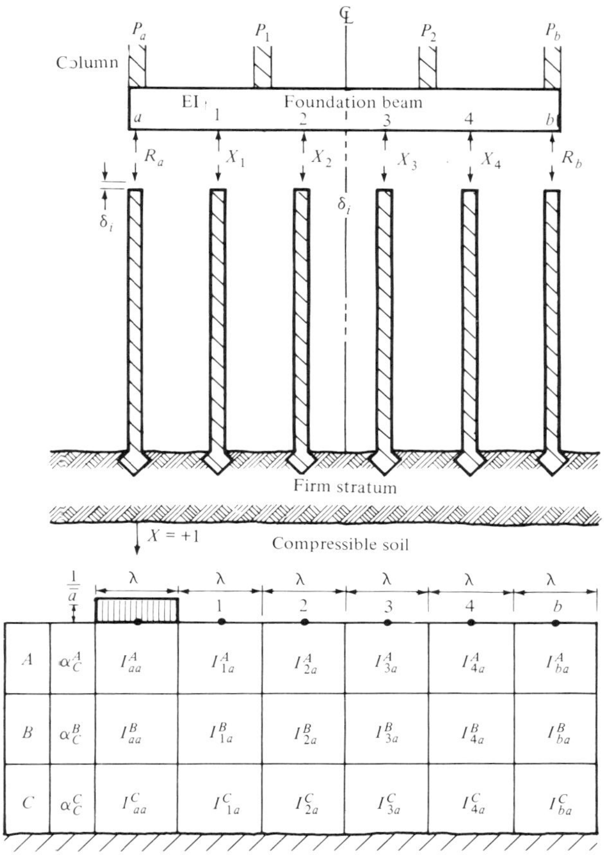

Fig. I-X.5 Pile foundation.

When a compressible soil sediment is underlying the hard stratum, then the displacement in this compressible soil deposit should be considered (Fig. 1-X.5). To take this in consideration because of the loading conditions imposed on the bearing stratum by the pile field, we calculate the influence values of the piles resting on the bearing stratum; therefore, δ_{ss} is a function of the mechanical and stratigraphical properties of the subsoil under the bearing stratum and on the load distribution given by the pile field. The method of calculating this effect has been discussed in Chapter VI using EMA which may be written in the form:

$$[\bar{\delta}_{ji}]^T \cdot |X_i| = |\delta_{ssi}| \qquad 4\text{-X.5}$$

The total vertical displacement at the head of the piles will be $|\bar{\delta}_{\pi i}| + |\bar{\delta}_{ssi}|$. Considering the foundation rigid as in the case of tall buildings, we may assume a constant vertical displacement of the foundation structure $\delta_a = \delta_\pi + \delta_{ss}$. Therefore, from equations 3-X.5 and 5-X.5 we can write the following expression for the solution of the foundation structure-pile interaction

$$\{[\bar{\delta}_{ji}]^T + [\bar{\delta}_{\pi i}]_D\} \; |X_i| = |\delta_a|. \qquad \text{5-X.5}$$

The system of simultaneous equations represented by matrix equation 5-X.5 may be solved for X_i. The procedure is to assume a reasonably close value of δ_a' and obtain X_i'. If W_e' is the weight of the foundation obtained for the solution when assuming a value δ_a', then

$$W_e' = X_a' + X_i' + X_j' + X_b'$$

However, the real weight is W_e, hence a correction on X_i' should be made multiplying these values by the ratio W_e'/W_e.

The load distribution in the pile lines is estimated calculating, in the same manner shown before, the pile reactions configuration which is adjusted to the total load at the line in question. With the loads on the foundation structure and reactions so determined, the shear forces and bending moments may be calculated. When the foundation structure flexibility should be introduced, the procedure explained in Chapter VI for these cases may be used. (See L. Zeevaert (1980), ISE).

BIBLIOGRAPHY

ACI (1972). Suggested Design and Construction Procedures for Pier Foundations, American Concrete Institute, Committee 336, *Journal ACI*, August.

ASCE (1946). *Pile Foundations and Pile Structures*, ASCE Manual of Engineering Practice, No. 27.

González Flores, M. (1959). Setting Vertical two Buildings, *First Panamerican Conference on Soil Mechanics and Foundation Engineering*, Vol. III, pp. 1215–1248.

Hagerty, D. J. and Peck, R. B. (1971). Heave and Lateral Movements due to Pile Driving, *Journal of SMFE, ASCE*, SM11.

White, Lazarus and Prentis, E. A. (1931). *Underpinning*, Columbia University Press.

Zeevaert, L. (1957). Foundation Design and Behavior of Tower Latino Americana in Mexico City, *Geotechnique*, Vol. VII, pp. 115–133, London.

Zeevaert, L. (1961). Foundation of Structures–(b) Piling and Piled Foundations. General Report. *Proc. VICOSOMEF*, Vol. II, pp. 836–843, Paris.

Zeevaert, L. (1980). ISE-Interaccion Suelo-Estructura de Cimentaciones Superficiales y Profundas, LIMUSA Publ. Co. México, D.F.

►XI◄ STABILITY OF DEEP EXCAVATIONS FOR FOUNDATIONS

XI.1 GENERAL CONSIDERATIONS

Heavy and important buildings are usually designed with one or several basements, therefore requiring deep excavations. In difficult subsoil conditions, the design of the excavation is as important as the foundation design itself. The design of the foundation cannot be separated from that of the excavation when analyzing the future behavior of a building, especially in cases of compensated foundations. We know that the behavior of the foundation depends highly on the method of excavation and on the degree to which the ground is disturbed during the excavation process, that is, on the magnitude of the change of effective stresses that occurs in the soil mass during excavation and construction of the foundation structure. Therefore, the excavation and construction of the foundation structure should be designed to minimize possible changes in effective stresses in the subsoil mass, within economical limits. To achieve this goal, the foundation engineer must obtain a good knowledge of the stratigraphy, the hydraulic conditions and mechanical properties of the soil strata. Therefore, from the very beginning, when planning subsoil investigations consideration should be given, in addition to the normal stratigraphical requirements, to an investigation of the hydraulic conditions by means of sufficient permanent piezometers installed at different depths to study, first, the prevailing hydraulic conditions for foundation and excavation design; and second, to use the same group of piezometers to control water level pressures during dewatering of the excavation. In addition to the compressibility and shear strength properties of the subsoil materials determined in the laboratory from the undisturbed sample borings, it is required to investigate the elastic response and swelling properties of the soils encountered at the proposed site and to a depth compatible with the size of the loaded area. With this information in hand, the foundation engineer will be able to estimate S- and E-heaves due to load relief (section VII.5). Moreover, it is necessary to investigate the foundation conditions and past behavior of the adjacent construction encountered close to the excavation in question, and an analysis should be performed of the probable movements that may occur during excavations and foundation structure construction of the proposed building (section I.3).

The excavation problem is, in general, specific for each foundation and site under consideration. Assume for further discussion and illustration a deep excavation represented by Fig. 1-XI.1. The foundation structure is required to be placed to depth level *d-d*. The excavation will be performed in two stages in order to minimize the E-heave. The stratigraphy of the subsoil is built up of series of highly compressible soil deposits interbedded with thin sand strata, extending to great depth. The shear strength increases with depth and the compressibility decreases. The water table is close to the ground surface, making it necessary to dewater the excavation. For this purpose and to retain the lateral earth pressures during excavation, an impervious sheet pile wall is constructed. The earth and water pressures on the wall will be taken by a supporting steel shoring system. The second stage of the excavation will be performed by substitution, excavating trenches for each concrete beam forming the foundation structure. The trenches will be braced under pressure to support the earth and water pressures exerted on the sheet-pile wall. The trenches will be excavated step by step, replacing the weight of the excavated

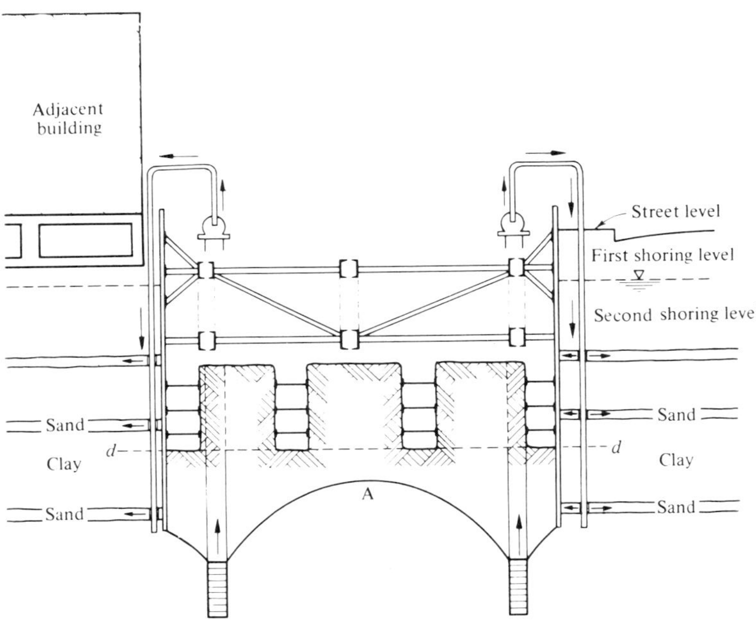

Fig. 1-XI.1 Excavation in two stages.

Fig. 2-XI.1 Detail of a pumping well used to dewater the excavation.

soil by the concrete beams. No change in vertical stresses in the subsoil will be allowed during the second stage of excavation, in order to minimize the E-heave.

The excavation is dewatered pumping from deep water wells, Figs. 2 and 3-XI.1. In each one of the wells a deep water pump is installed. The system of pumping wells is electrically controlled. The sequence of construction will be the following:

a) The sheet-pile wall is driven from the ground surface, and the pumping wells constructed to the required elevation.
b) The electric control system of the wells is installed. Outside of the sheet-pile wall, perimeter wells are driven to inject the water obtained from the interior pumping wells, used to dewater the excavation.
c) The water table is lowered by pumping from the wells, and the excavation for first shoring level is performed.

Fig. 3-XI.1 First shoring level installed; excavation of trenches begins.

d) The first shoring level is installed under pressure (Fig. 3-XI.1).
e) The water table is lowered further to be able to excavate to the second shoring level.
f) The excavation for the first stage is continued to depth D_1, and the second shoring level is installed under pressure (Fig. 4-XI.1).
g) The shoring levels previously installed are periodically inspected and maintained under pressure. Usually, hardwood wedges are used for this purpose.
h) The water table is lowered to the final position, curve A in Fig. 1-XI.1, obtaining the total weight of soil above water surface. This excess weight will compensate the effective weight of excavated soil performed in the first stage of excavation.
i) The trenches for the second stage are excavated, bracing them under pressure. The excavated soil weight is progressively replaced by the foundation structure concrete weight (Fig. 5-XI.1).
j) Once the foundation beams are constructed and sufficient weight is placed on them from the superstructure, the earth blocks between beam panels may be excavated step by step with an alternate program, and the construction of the foundation slab is carried on, until the total foundation structure is completed. Thereafter, the concrete box is formed by the foundation beams and

Fig. 4-XI.1 Second shcring level installed; pumping and excavation of trenches continues.

Fig. 5-XI.1 Second step in excavation in trenches.

reaction slabs are filled with water, used as ballast, thus not permitting change in effective stresses in the soil mass under the foundation area (Chapter VII).

k) The clay at the foundation grade elevation should be cut carefully in order not to alter the soil mechanical properties (Fig. 6-XI.1).

l) The retaining walls are constructed and the shoring is taken away.

m) The construction proceeds without interruption and the water table is permitted to rise an amount equal to the weight added by the building construction, until finally a point is reached at which the wells are cancelled, the concrete cells are dewatered, and water uplift pressure is permitted to act against the foundation slab. The water uplift pressure during the construction steps is controlled by valves and piezometric pipes installed in the pumping wells.

n) Eventually, the total load of the building is reached at the same time as the maximum water uplift pressure, obtaining thereafter final equilibrium.

Steps m and n may be changed according to the type of foundation design, and whether or not it is required at any time to maintain complete balance of the weight of the building under construction with the soil and water pressure reactions. During the construction process described above, piezometric and settlement observations are performed to control the behavior of the construction, and

Fig. 6-XI.1 The clay at the foundation grade elevation is carefully excavated.

therefore adjust differences that may deviate to certain degree from the design considerations.

The theoretical analysis of each one of the steps mentioned requires special attention to be able to design the sheet-pile wall, the shoring supports, the pumping levels (as excavation proceeds), and the stability of the excavation. The E-heave and the program of construction should be compatible with the calculations and design of the excavation. The different parts involved in the design are described in more detail in the following paragraphs.

XI.2 SHEET-PILE WALL

2.1 Lateral Support

In difficult subsoil conditions, when it is not possible to perform the excavation with stable slopes because of the location of adjacent buildings, or when the excavations must be made close to the sidewalk, it is necessary to use a sheet-pile wall for lateral support of the excavation banks. The sheet-pile wall should, first, have sufficient water tightness to avoid important percolation of water into the excavation; and second, it should be sufficiently rigid to support the earth and water pressures between horizontal shorings.

The sheet piles may be constructed of timber, concrete, steel or cast-in-place reinforced concrete. Every type of material used for a sheet-pile wall, however, has certain advantages and disadvantages, according to the characteristics of the problem under consideration. In selecting the sheet-pile wall material, one has to consider that during excavations this may deform strongly, as shown in Fig. 1-XI.3. In fact, upon excavation and placing of the shoring levels 1, 2, 3, 4 and 5 under pressure, the soil inside the excavation between sheet-piles will be compressed because of the change in effective horizontal stresses induced by the excavation, and hydraulic changes when pumping from the wells to dewater the excavation (see section VII.6). The lateral contraction, however, cannot be stopped completely with the rigidity of the sheet-pile. Nevertheless, it may be minimized placing the shoring levels under pressure immediately as the excavation has reached its corresponding level. Therefore, in most cases of deep excavations, the sheet-piles cannot be recovered because of the deformation set on them by the horizontal soil displacements. The process of pulling out the sheet-piles for recovery becomes very difficult; in most cases one should assume that this cannot be done and consider the possibility of keeping them as a permanent structure to support the water and earth pressures.

The rigidity given to the sheet-pile wall may help somewhat to achieve larger spacing between shoring levels, and facilitate the construction of the foundation structure. Very rigid sheet-piles are expensive and hard to drive; therefore, the problem involves the study of an economical design including the number of the necessary shoring levels in conjunction with the rigidity of the wall and soil engineering properties.

2.2 Timber

Timber sheet-piles of the Wakefield type (Figs. 1a and 2-XI.2) are constructed with two, three or four timber boards joined by bolts and properly spliced. They are bolted under axial compression to increase their rigidity. The effective moment of inertia, when well constructed, may be taken as one-half that given by the total cross-section formed by the timber boards. The timber sheet-pile has an advantage in that it forms an impervious wall. The timber swells when saturated with the ground water, forming a barrel-like water tight wall not permitting percolation into the excavation. The few places where water may seep through may be sealed, caulking the joints between sheet-pile elements. This type of sheet-pile is very

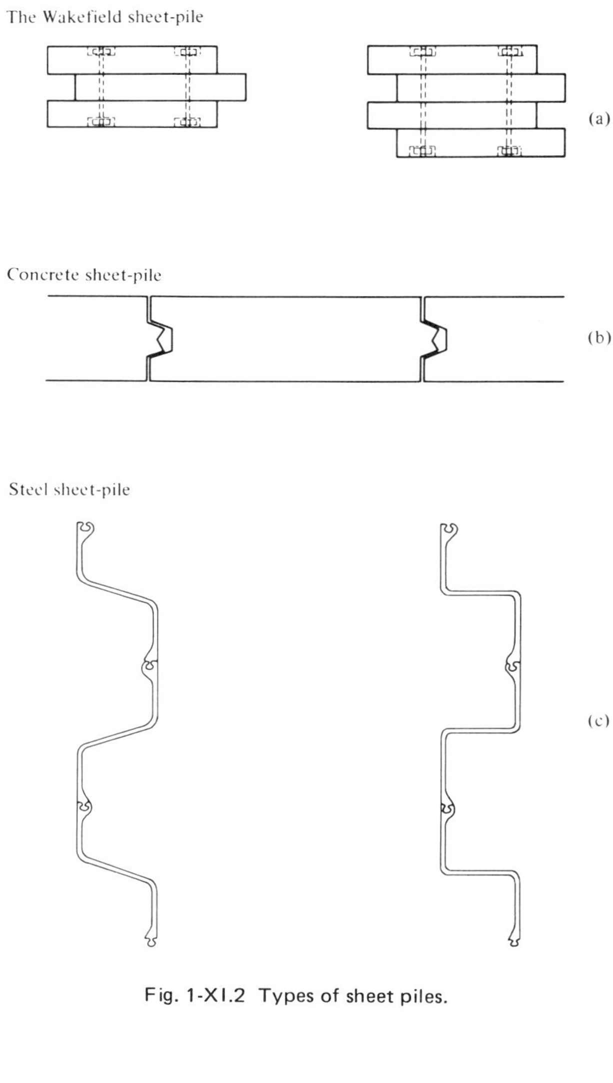

Fig. 1-XI.2 Types of sheet piles.

Fig. 2-XI.2 Construction of wakefield type timber sheet piles.

flexible, since usually, for economy, they have to be constructed with green timber. Therefore, more shoring levels may be required than with any other more rigid sheet-pile. Timber sheet-piles should be used only in soft soil deposits where driving is not hard; otherwise damage and distortion of the elements may be expected making driving difficult.

2.3 Concrete

Sheet-pile walls may be constructed using precast or prestressed concrete elements. In some instances, they may compete in price with timber sheet-piles. Concrete sheet-piles are considerably more rigid than timber piles (Fig. 1b-XI.2). Nevertheless, they cannot be made very thick because their weight makes them difficult to handle, especially when long elements are required for deep excavations. The vertical joints between sheet-pile elements have to be sealed, by injection of either cement-bentonite mixtures or any other economical material to seal them watertight. Concrete sheet-piles have the advantage that they may be used in the foundation design as a permanent retaining wall.

Concrete walls constructed by the process of displacing bentonite in pre-excavated trenches may be used with advantage. The concrete slurry walls may be reinforced as necessary. The large moment of inertia obtained favors a larger vertical spacing in the shoring system. Therefore, more economy and working space may be gained in supporting the sheet-pile wall than in any other wall system.

Prestressed concrete sheet-piles may be used; however, as in case of timber sheet-piles, they can only apply when soft driving is expected. When the subsoil contains sand and gravel layers, it is difficult to drive them; however, they may be jetted down when space is available and the adjacent construction is not damaged. Jetting may produce local subsidence of the ground surface. In case of sandy or gravelly clayey silty soils where driving is hard, the slurry walls are indicated. This procedure may be used provided the excavation construction is more economical than using the prestressed concrete or timber sheet-piles.

2.4 Steel

Commercial steel sheet-piles may be used to form the perimeter wall (Fig. 1c-XI.2). These elements, because of the small volume displacement they produce when driven in the soil mass, have a great advantage against any other type of sheet-piles. They may be easily driven in almost any kind of subsoil, from clay to fine gravelly soils. Their cost, however, may be prohibitive in some countries. The steel sheet-piles are not as impervious as those constructed of timber or concrete slurry walls. Nevertheless, they may be injected at the joints, as in the case of prestressed concrete piles, with substances like cement-bentonite mixtures, bitumen or any other inexpensive material that may fulfill the same purpose. The steel sheet-piles may be used also as permanent basement walls when the soil and water are not corrosive. The critical zone is located outside of the sheet-pile wall, at the water table; however, this zone may be properly protected. Within the foundation structure, proper maintenance is easily achieved.

In Mexico City's soft subsoil the author has made extensive use of Wakefield type timber sheet-piles with lengths up to 16 m, obtaining good results when the sheet-piles are fabricated with three or four timber boards and bolted together under pressure (Figs. 1a and 3-XI.2). Precast concrete sheet-piles may be used when their length is not greater than 8 m (26 ft). The concrete sheet-piles are considered as permanent structures in the design of the basement wall for the building under consideration. The method of the slurry walls and the use of steel sheet-piles is so far very expensive and not yet available in many countries. Therefore, the economics and characteristics of the problem in question are the principal objective to be considered in selecting the materials for sheet-pile walls. When possible, the wall should be designed with the largest rigidity, in order to use the minimum number of shoring levels and facilitate the construction of the foundation structure.

XI.3 PRESSURES ON THE SHEET-PILE WALL

The earth and water pressures acting on the sheet-pile wall may be estimated by current methods. The first shoring level installed under pressure, close to the ground surface, restricts the inward horizontal displacement at point A in

Fig. 3-XI.2 Driving of wakefield type timber sheet piles.

Fig. 1-XI.3. Nevertheless, as excavation proceeds, the sheet-pile wall will rotate about point A such that when installing the second shoring level the shear strength in the soil mass behind the wall is mobilized to the extent that the active earth pressure distribution in the lower part of the sheet-pile wall can be assumed approximately in accordance with that given by the Coulomb or Rankine theory. On the other hand, if the sheet-pile wall is of the rigid type and the shoring levels are carefully placed under pressure restricting the horizontal displacements, the earth pressure distribution against the sheet-pile wall assumes higher values at the upper shoring level A than the values obtained by means of Rankine's earth pressure theory. The real pressure distribution, however, depends on how much the sheet-pile wall is permitted to yield during the installation of the shoring levels. For safe design of the shoring system at the upper part of the sheet-pile wall, the earth

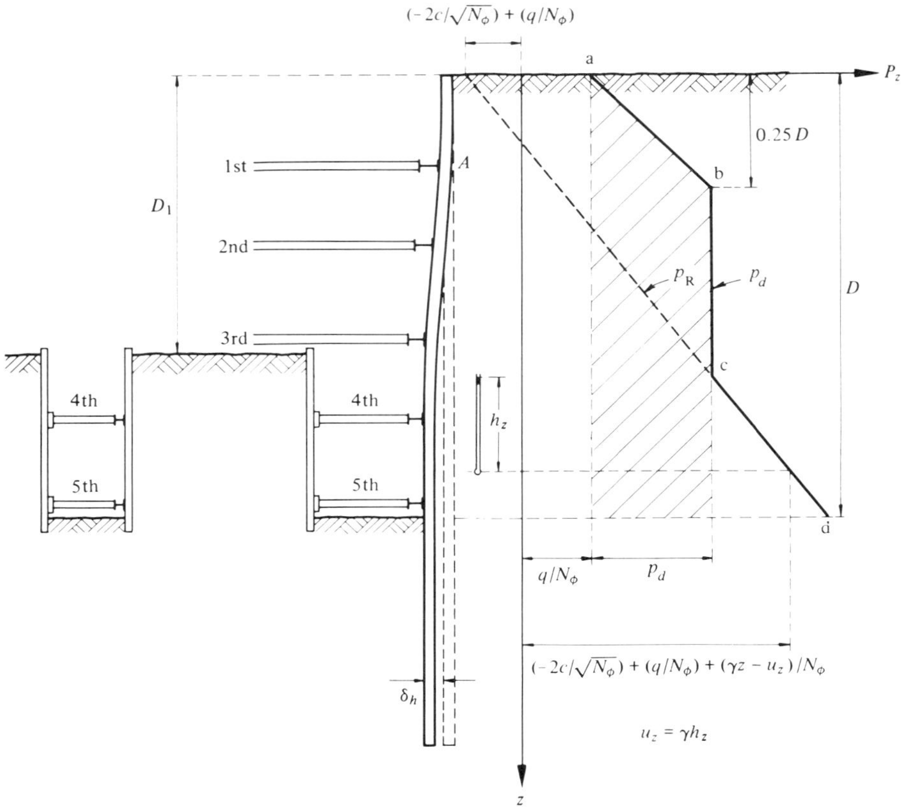

Fig. 1-XI.3 Earth pressures on sheet-pile wall.

pressure distribution may be assumed using the Terzaghi-Peck concept for a non-yielding wall.

The active earth pressure distribution when the sheet-pile wall rotates, which mobilizes the shear strength of the soil mass in the back of the wall, may be calculated by

$$p_z = -\frac{2c_z}{\sqrt{N_{\phi z}}} + \frac{1}{N_{\phi z}}\left(\sum_{1}^{z} \gamma_i \Delta z_i + q - u_z\right) \qquad \text{1-XI.3}$$

Here q is a surcharge applied at the ground surface. The value of γ_i is the average total unit weight for each layer of thickness Δz_i at depth z, and u_z is the hydraulic pressure at the same depth. The apparent angle of internal friction in clays is taken as $\phi = 0$, and in silts and sands, it is determined by consolidated-undrained tests (section II.4). The Rankine earth pressure distribution is shown labeled p_R in Fig. 1-XI.3. When the horizontal displacements are small in the lower shoring levels, because of a very rigid sheet-pile wall, the average earth pressure is calculated by means of

$$p_d = 1.14 \frac{E'_A}{D} \tag{2-XI.3}$$

in which the total active earth pressure E'_A without surcharge may be calculated from

$$E'_A = 1.1 \sum_0^D \left[-\frac{2c_z}{\sqrt{N_{\phi z}}} + \frac{1}{N_{\phi z}} \left(\sum_0^z \gamma_i \Delta z_i - u_z \right) \right] \Delta z \tag{3-XI.3}$$

Here c_z and $N_{\phi z}$ are average values for each stratum with thickness Δz. Assuming uniform unit weight and shear strength parameters from the ground surface to depth h, the earth pressure distribution against the sheet-pile wall for design purposes is shown qualitatively in Fig. 1-XI.3. The solid line abcd may be assumed to represent the largest possible active earth pressures that may take place on the sheet-pile wall during excavations and shoring installations, and may be used to calculate the brazing supporting system. The total pressure distribution against the sheet-pile wall, from the ground surface to depth D, is obtained by adding the water pressures u_z obtained from the flow net corresponding to the hydraulic conditions set by pumping from the excavation.

The critical bending of the sheet-pile wall due to the earth and water pressures should be investigated for each support level below depth D_1, corresponding to the first stage of excavation. For this purpose, the active earth pressure distribution is estimated from equation 1-XI.3. Assume that the stability conditions of the sheet-pile wall are required for the third shoring level, at the time of excavating the trench close to the sheet-pile wall to place the fourth shoring level (Fig. 2-XI.3). Also assume that the third level has been placed. The total active pressure E_a for section CE is computed, and the total hydrostatic pressure U in this section is obtained from hydrodynamic conditions, estimated from the flow net. Hence, the total pressure acting in section CE of the sheet-pile wall is $(E_a + U)$. This force is counteracted by a horizontal soil reaction in section DE at the inside of the sheet-pile wall below the bottom of the excavation. For cohesive soils, the unit soil reaction p_p may be assumed with a uniform distribution, and may be found from statics assuming the sheet-pile wall develops a plastic hinge at C, with a specific plastic moment M_{cp}. Therefore, taking moments about point C, the total horizontal soil reaction is

$$E'_p = \frac{E_a y_a + U y_u - M_{cp}}{y_p} \tag{4-XI.3}$$

and the unit soil reaction in section DE is

$$p_p = \frac{E'_p}{l - D} \tag{5-XI.3}$$

The safety factor, however, should be investigated to insure against failure or

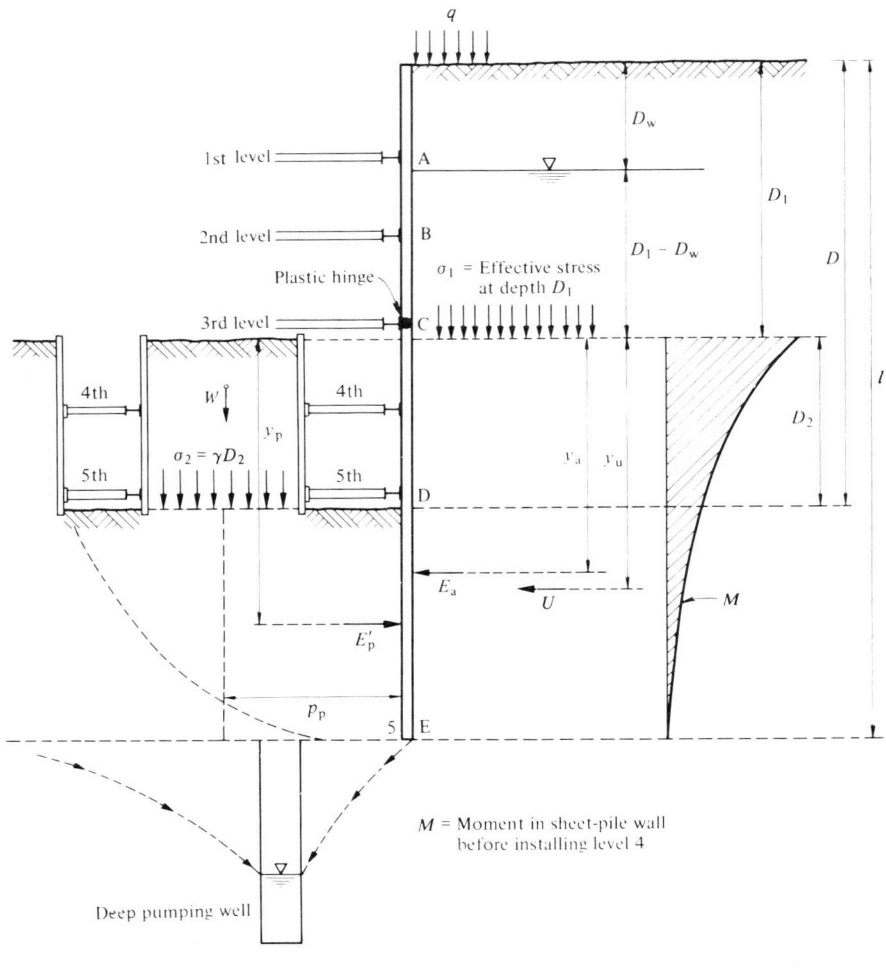

Fig. 2-XI.3 Stability of sheet-pile wall.

plastic flow toward the bottom of the excavation. For this purpose, the total passive earth pressure E_p should be estimated by current methods. The potential surface of sliding is assumed an arc of a circle with center at C, and the shear strength is evaluated for each stratum cut by this surface. In the case of clay, the shear strength may be taken as one-half of the unconfined compressive strength, in silts and clayey silts, the parameters c_{cu} and ϕ_{cu} should be determined for each one of the strata cut by the surface of sliding. The factor of safety is obtained from the formula

$$G_s = \frac{E_p}{E'_p} \qquad \text{6-XI.3}$$

The factor of safety thus determined should not be smaller than two, in order to avoid local plastic heave of the bottom of the excavation.

From expression 4-XI.3, it may be noticed that the larger the value of M_{cp} the less passive earth pressure is mobilized in the subsoil below the bottom of the excavation. By the same token, the horizontal sheet-pile displacements δ_h will be smaller; therefore, the lateral contraction is minimized.

XI.4 DEWATERING OF EXCAVATIONS

The construction of foundations below the water table requires special consideration. The dewatering system should be designed to keep the excavation under dry conditions 24 hours a day. For deep excavations like those necessary for compensated foundations, water extraction must be carefully controlled to comply with design conditions. For this purpose, piezometers are installed in sufficient number at various depths to detect the important piezometric water level changes, and be able to control by means of pumping from deep water wells the stability of the foundation excavation. Moreover, the water pumped out from the excavation is passed through sedimentation tanks before letting it flow into the city sewer system, or injecting it in the outside of the excavation. The sedimentation tanks are used to observe the amount of sediments carried out by the water pumped from the excavation and to avoid plugging the injection wells.

To achieve a water drawdown of the piezometric water elevation and stabilize the bottom of the excavation, the use of a system of deep wells is required. Their size can be from small diameter well points added if necessary by electro-osmosis, to large diameter wells, depending on the permeability of the subsoil material. In impermeable soils with interbedded thin sand strata, the use of wells up to 90 cm in diameter proves to be satisfactory. The pumping water wells should be provided with filters to prevent sand from being carried away. The use of wells may be illustrated assuming an excavation to be performed in two stages. Assume that the soil mass contains thin water bearing sand strata. The excavation is protected by a deep sheet-pile wall cutting the $(n - 1)$-st sand stratum. The wall is considered impervious. Assume the investigation of 8 wells as shown located in Figs. 1 and 2-XI.4. To reduce the uplift water pressure at the bottom of the excavation, it is necessary to reduce the water pressure in a sand stratum close to the sheet-pile wall base, and in the area under the excavation. In order to keep the change in effective stresses as small as possible outside the excavation area, the water pumped from the wells is injected into the sand strata outside the sheet-pile wall. This procedure has been used by the author (1949) to perform the 13 m deep excavation for the construction of the foundation structure of the Latino Americana Tower in Mexico City. With this method, the change in vertical effective stresses outside the excavation due to downward water flow is considerably reduced, thus effectively minimizing the subsidence of the ground surface and damage to adjacent buildings.

The procedure is illustrated by the drawing shown in Figs. 1 and 2-XI.4, where we assume that the unit weight of the soil is approximately constant with depth.

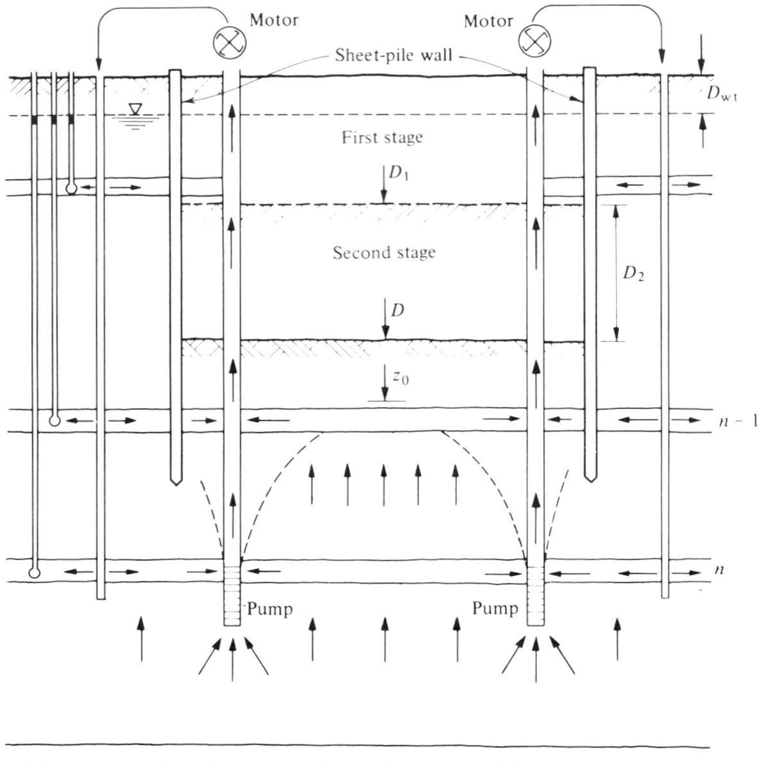

Fig. 1-XI.4 Dewatering of excavation.

The vertical effective stress corresponding to the elevation of the foundation slab is

$$\sigma_d = \gamma D_{wt} + \overline{\gamma}(D_1 - D_{wt}) + \overline{\gamma} D_2 \tag{1-XI.4}$$

in which γ is the average unit weight of the soil above the water table, and $\overline{\gamma}$ is the submerged unit weight.

After the first stage of the excavation is performed, the relief in vertical effective stress at depth D_1 is given by

$$[\Delta\sigma_{ex}]_1 = \gamma D_{wt} + \overline{\gamma}(D_1 - D_{wt}) \tag{2-XI.4}$$

In order to compensate for the relief in stress, the water pressure under the bottom of the excavation is reduced in the amount $(D_2 + z_0)\gamma_w = [\Delta\sigma_{ex}]_1 + (D_1 - D_{wt})\gamma_w$, where z_0 is the required depth of the water table under the bottom of the foundation grade elevation. Therefore, substituting into 1-XI.4, we obtain

$$\sigma_d \leqslant (D_2 + z_0)\gamma_w + \overline{\gamma} D_2 - (D_1 - D_{wt})\gamma_w \tag{3-XI.4}$$

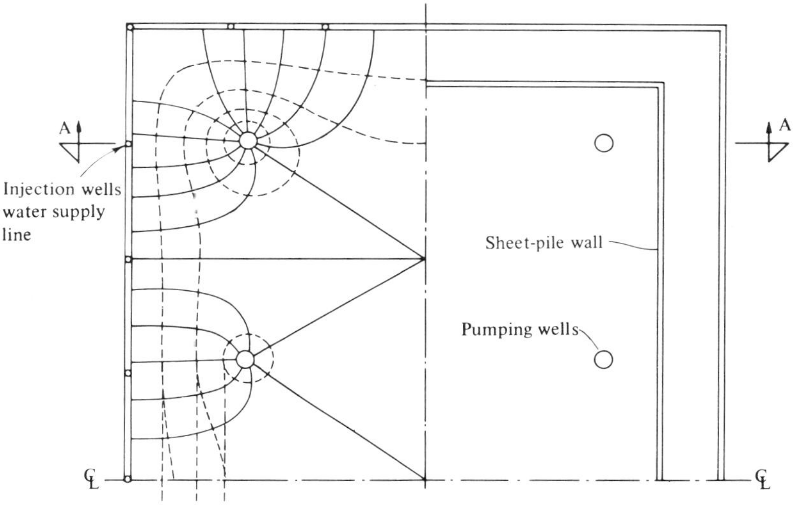

Water flow net for n-th sand stratum

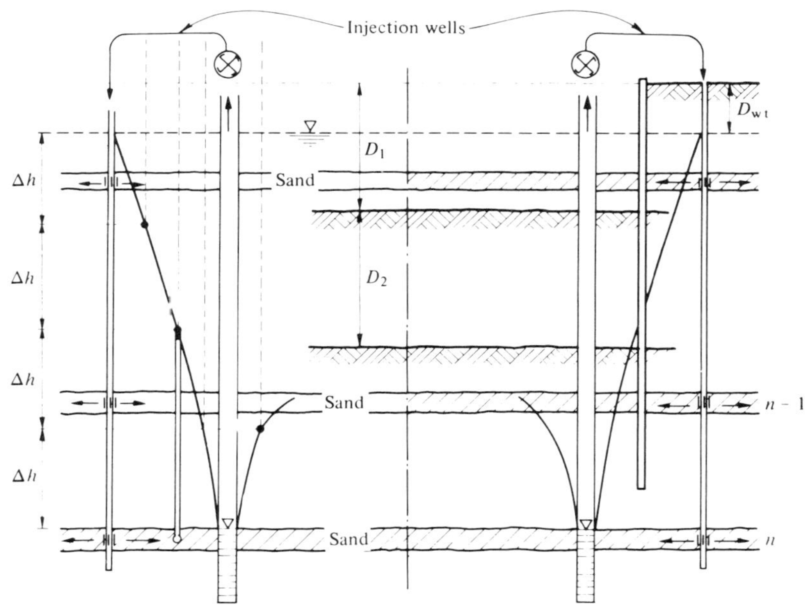

Cross section A–A

Fig. 2-XI.4 Water piezometric pressures in a sand stratum under bottom of excavation.

but considering that $\gamma = \overline{\gamma} + \gamma_w$, then

$$z_0 \geqslant \frac{1}{\gamma_w}(\sigma_d - \gamma D_2) + (D_1 - D_{wt}) \qquad \text{4-XI.4}$$

As an example, assume $D_1 = D_2 = 5.0$ m, $\gamma = 1.5$ ton/m^3, and $D_{wt} = 3$m. Hence, from 1-XI.4, $\sigma_d = 1.5 \cdot 3 + 0.5 \cdot 2 + 0.5 \cdot 5 = 8.0$ ton/m^2. Therefore, from 4-XI.4 we obtain $z_0 = 2.5$ m.

This calculation shows that if the water level is reduced to a piezometric water elevation of 2.5 m below the bottom of the excavation level *d-d*, then the effective pressures will remain the same or slightly larger than the initial effective stresses at the foundation grade elevation, and the E-heave because of stress relief $[\sigma_{ex}]_1$ will be considerably reduced.

The piezometric water elevations in the sand strata below the bottom of the excavation are investigated by means of flow nets with the method described in section III.3.6. The piezometric water levels outside the sheet-pile wall are held constant by injection of water under pressure into the sand strata up to the n-th stratum. Therefore, no sensible alteration of the effective vertical stresses takes place. The water pumped from the excavation is used for this purpose (Fig. 1-XI.4). The injection of water under pressure as required is controlled from observations in the piezometers installed outside of the sheet-pile wall. The boundary conditions to trace the flow net are established by a line of injection wells located at the perimeter of the excavation, giving a permanent water supply. From the flow net shown in Fig. 2-XI.4, obtained for eight wells representing horizontal flow conditions of the n-th stratum we get the following characteristics:

	N_f	N_p	R
Center wells	6	4	$\frac{3}{2}$
Corner wells	11	5	$\frac{11}{5}$

Hence, the total discharge is $Q = k_n \cdot S_0 \cdot t_n [4(\frac{11}{5}) + 4(\frac{3}{2})] = 14.8 k_n S_0 t_n$, where S_0 is the drawdown in the wells and k_n is the coefficient of permeability of the n-th sand stratum with thickness t_n (section III.3.6).

From the location of the equipotential lines obtained from the flow net, the piezometric water elevations are known for the n-th sand stratum, and the effective stresses can be calculated. It may be desirable to use a sufficient number of pumping wells to obtain a more uniform distribution in the reduction of the piezometric water elevations under the area of excavation. Deep water pumps have to be installed in each one of the wells, maintaining a uniform dynamic water elevation in them. All the pumps will be operated by an electrical automatic control system specially designed for this purpose, to maintain the water level in the wells at any desired elevation. Moreover, a system of alarms is provided to detect any defect that may take place during dewatering operations.

To illustrate the change in stresses in the subsoil mass at the center line of the excavated area because of pumping, assume the total pressures p_{0i}, and effective

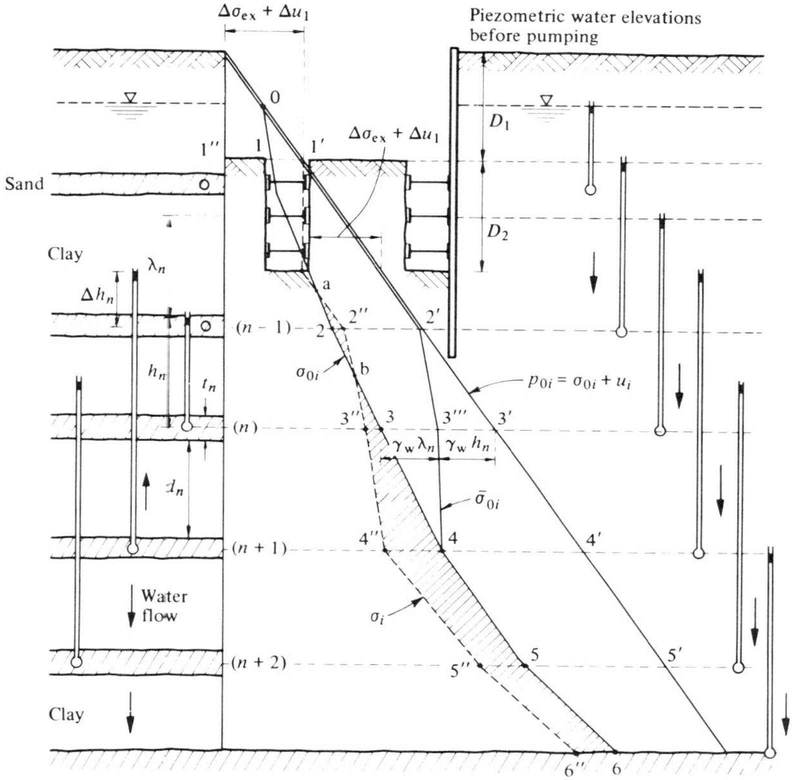

Fig. 3-XI.4 Change in effective stresses due to reduction of piezometric water levels.

overburden pressures σ_{0i} are known with depth (Fig. 3-XI.4). The value of u_i is determined from piezometric readings. Assume further that before excavation is performed, the piezometric water levels have been reduced to h_n in the n-th sand stratum, and to zero in the sand strata confined by the sheet-pile wall. In the sand strata $(n + 1)$, $(n + 2)$, ... the piezometric water levels remain unaltered. Under these conditions, the stresses in the ground will increase due to the piezometric water level reduction as shown by the curve labeled $\bar{\sigma}_{0i}$. At the n-th sand stratum, the increment in stress will be $(\gamma_w \lambda_n)$. It may be noticed that between sand strata n and $(n + 1)$, an upward water flow takes place with gradient $i = \Delta h_n/d_n$, where Δh_n is the piezometric water elevation drawdown, and d_n is the thickness of the n-th clay stratum. Therefore, the reduction in effective stresses because of the upward water flow has a linear variation from $(\gamma_w \lambda_n)$ at the n-th level to zero at the $(n + 1)$-st level, where no changes have taken place. This effect is compensated at the n-th stratum by the increased weight of the soil drained between the sheet-pile wall.

If this hydrodynamic condition prevails, the compressible soil deposit will consolidate within the stress area limited by the stress curves labeled $\overline{\sigma}_{0i}$ and σ_{0i}, namely: 0-1′-2′-3‴-4-3-2-1-0 in Fig. 3-XI.4. Now consider that excavation takes place to depth D_1; the soil mass experiences at this depth a total relief $[\Delta\sigma_{ex}]_1 + \Delta u_1$, and since the second stage of excavation is performed by substitution of weights of foundation and soil in trenches, the stress relief due to the first excavation stage is transmitted completely to depth D. At greater depth from the foundation grade elevation, the stress relief may be computed by means of the theory of elasticity, and deducted from the stress curve labeled $\overline{\sigma}_{0i}$. The final stresses producing vertical displacements in the soil mass owing to the hydraulic conditions imposed are at any depth $\Delta\sigma_{ex} + \Delta u_1 - \gamma_w \lambda$. The change in stresses producing vertical displacements at the foundation grade elevation will be those corresponding to the shaded stress areas in Fig. 3-XI.4. Hence

1. For the compression stress area: a-2″-b-2-a:

$$\delta_{c1} = \Sigma m_{vi} (\sigma_i - \sigma_{0i}) d_i$$

2. For the expansion stress area: b-3-4-5-6-6″-5″-4″-3″-b:

$$\delta_{e2} = - \Sigma (\rho_e \nu_c M_e)_i (\sigma_{0i} - \sigma_i) d_i$$

from which the total vertical displacement is

$$\delta_e = \delta_{c1} + \delta_{e2} \qquad \text{5-XI.4}$$

The value of 5-XI.4 should be kept as small as possible in order to minimize the heave when designing the dewatering of excavations by means of a deep pumping well system. For a totally compensated foundation, after the load has been applied and the water conditions are reestablished, the effective stresses will return to the initial value σ_{0i} (curve 1-2-3-4-5, Fig. 3-XI.4). Hence, the subsequent settlement is calculated from the same shaded areas by means of the following expressions (assuming totally confined conditions, $\nu_c < 1$)

1. For the expansion stress area: a-2″-b-2-a:

$$\delta_{e1} = - \Sigma (\rho_e \nu_c M_e) (\sigma_i - \sigma_{0i}) d_i$$

2. For the recompression stress area: b-3-4-5-6-6″-5″-4″-3″-b:

$$\delta_{c2} = \Sigma (\rho_c m_v)_i \cdot (\sigma_{0i} - \sigma_i) d_i$$

Adding these values we obtain the settlement

$$\delta_c = \delta_{e1} + \delta_{c2} \qquad \text{6-XI.4}$$

In practice, however, the calculated piezometric water elevations are used in the field as an approximation, and as excavation proceeds, leveling observations for heave and settlements must be made. From them, the dynamic level in the wells is controlled so as to minimize possible vertical displacements. The author's experience has been that, in this type of water control, such calculations performed usually must be corrected by 10 to 20%.

Because of the permeability characteristics of fine subsoil sediments extending to

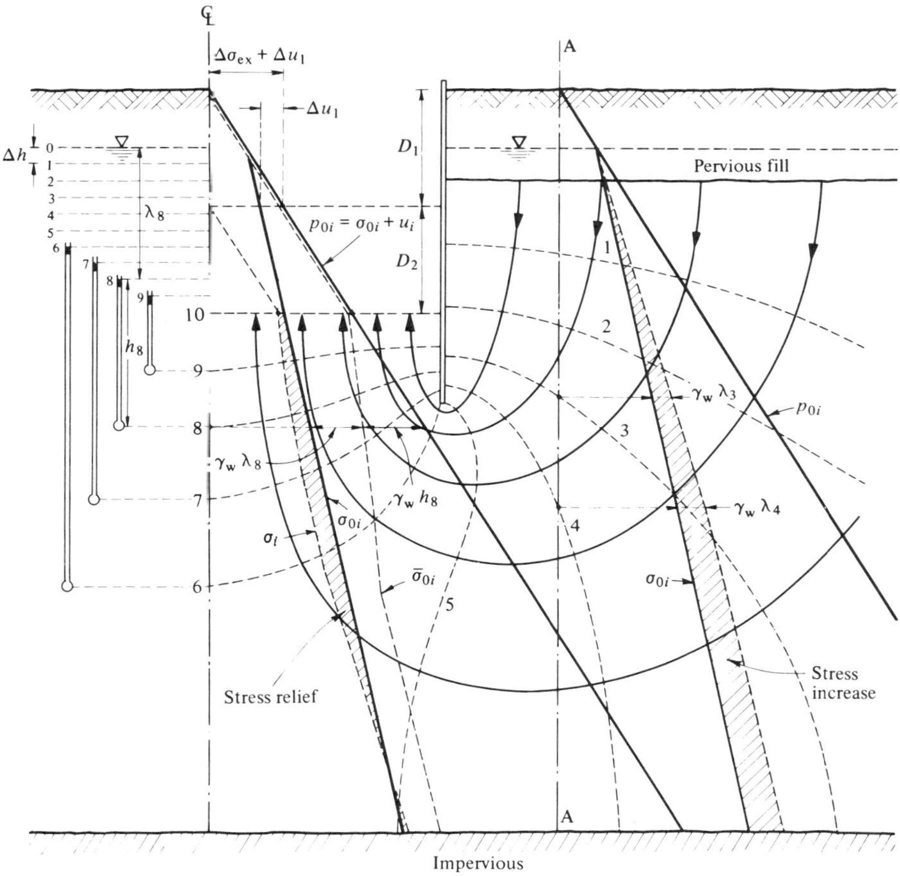

Fig. 4-XI.4 Case I—Change in effective stresses due to hydraulic conditions imposed by an impervious bottom stratum.

great depth not interbedded with sand strata, it is not possible to inject water under pressure into the soil mass close to the excavation to preserve the original overburden effective pressure. This problem is different than the one just treated. The stress changes in the soil mass are estimated by tracing the corresponding flow net according to the boundary conditions encountered (section III.3.2). Two cases may be present; the impervious soil mass is limited either by a stratum not permitting water percolation or by a pervious water bearing stratum (Figs. 4 and 5-XI.4).

To illustrate the problem, assume for simplicity that the soil mass is isotropic. The flow net is traced for Case I in Fig. 4-XI.4. From the equipotential lines obtained, the change in effective stress may be determined. These changes are shown in vertical sections, at the middle of the excavation and at section A-A close to the sheet-pile wall outside of the excavation. The flow net has been traced also for Case II when the boundary is a pervious stratum overlaid by impervious sediments.

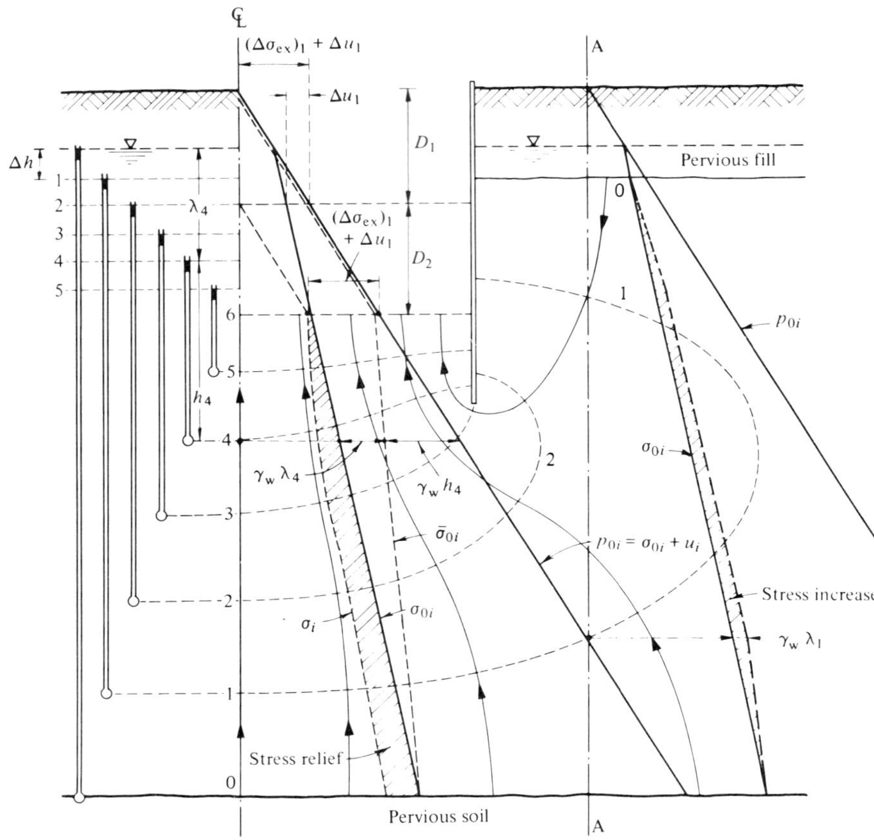

Fig. 5-XI.4 Case II—Change in effective stresses due to hydraulic conditions imposed by a pervious bottom stratum.

The effective stress changes are shown at the same sections mentioned before (Fig. 5-XI.4). The initial hydraulic conditions in both cases have been assumed hydrostatic—all piezometers before pumping have the same piezometric water elevation, corresponding to that of the water table. Pumping is performed in both cases from the bottom of the excavation at the depth corresponding to the second stage, since the second stage of excavation is performed in deep trenches. Under this condition, the saturated weight of the soil with thickness D_2 is acting fully at the foundation grade elevation. The stress relief in the subsoil is only that corresponding to the first stage of excavation. This is reduced from the stresses $\overline{\sigma}_{0i}$ corresponding to the stresses owing to the state of hydrodynamic conditions set by the flow net because of pumping. The study of the change of stress in the soil mass because of pumping and excavation may be performed, as required, at other vertical or horizontal sections, and used in the analysis of vertical displacements. The shaded areas shown in Figs. 4 and 5-XI.4 represent the stress changes. The vertical displacements are calculated, as previously described, from the change in stress and the stress-strain properties of the soil sediments.

From these examples, it may be noticed that Case II gives more relief of stress in the excavation area than Case I, shown in Fig. 4-XI.4. On the other hand, the vertical stress change outside of the sheet-pile wall for the same conditions given in Case II is smaller than for Case I. Hence, the subsidence of the ground surface will be larger for Case I than for Case II.

Under these subsoil conditions, the control of both subsidence in the ground surface outside of the excavation and heave at the bottom can be achieved by increasing the depth of the sheet-pile wall, as it may prove necessary to increase the path of percolation, thereby obtaining more equipotential lines. This case is illustrated by means of flow net shown in Fig. 6-XI.4, from which it may be noticed that the

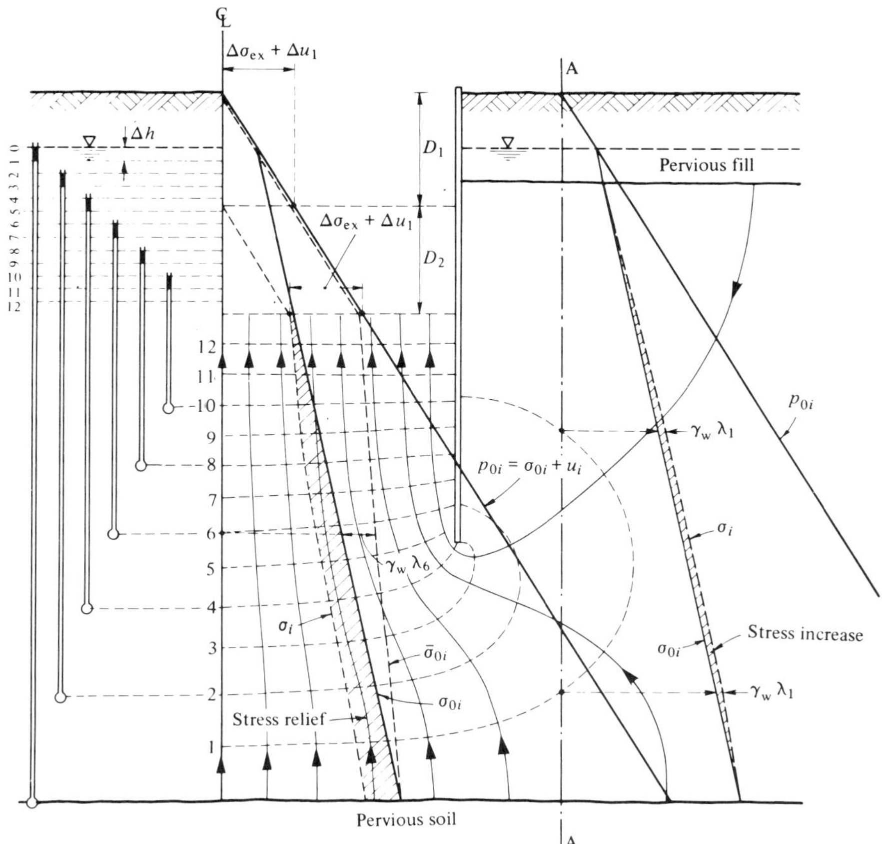

Fig. 6-XI.4 Change in effective stresses due to hydraulic conditions imposed by a deep sheet pile and a pervious bottom stratum.

stress increase in section A-A close to the sheet pile is considerably smaller. The stress change, however, under the center of the excavation area has practically remained the same as in Case II with a shorter sheet-pile wall. For soil-foundation structure interaction with the water table above the foundation grade elevation see L. Zeevaert (1980), ISE.

XI.5 STABILITY OF THE BOTTOM OF THE EXCAVATION

The stability of the bottom of the excavation may be estimated by current methods of stability of slopes or bearing capacity found in the literature. The most complicated case is that of stratified subsoil with cohesive sediments of different shear strength properties. The potential surface of sliding can be regarded as a circular arc (Fig. 1-XI.5) along which the shear strength is determined for each one of the strata cut by arc.

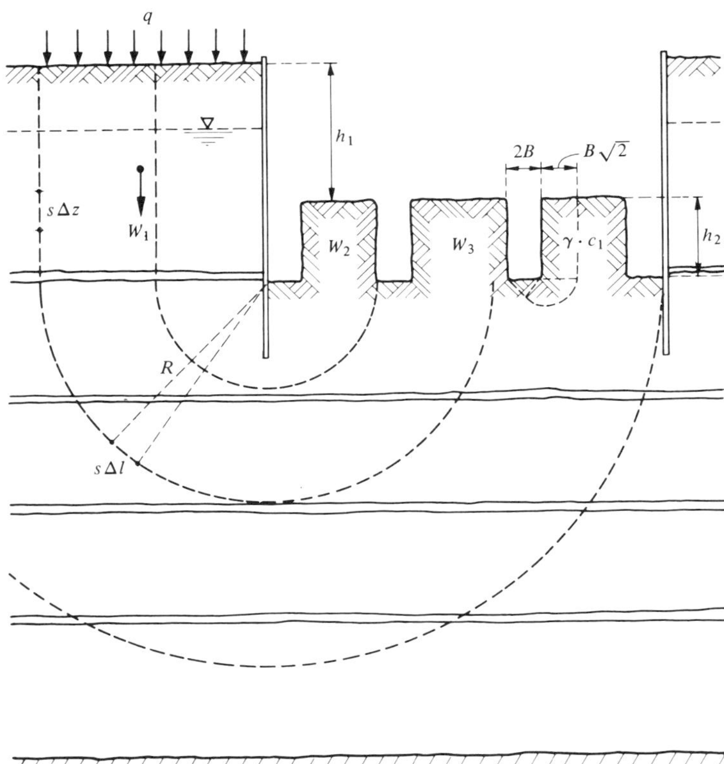

Fig. 1-XI.5 Stability of bottom with stratified soil.

For stability, it is necessary to investigate the potential surface of sliding tangent to the strata of higher strength. The largest surface of sliding will be that with a radius equal to the width of the excavation. The most unfavorable conditions of loading during the excavation process should be considered. When a second stage of excavation is used and this is performed in trenches reaching soft clay, the bottom stability of the trenches should be investigated. The bearing capacity method may be used for this purpose. The factor of safety is

$$G_s = \frac{5.7c}{h_2\left(\gamma - \frac{c_1}{\sqrt{2}B}\right)} \qquad \text{1-XI.5}$$

where c is the average cohesion of the clay to a depth B under the trench, and c_1 that above the bottom (Fig. 1-XI.5).

In case of fine low cohesive soils where the friction term in the shear strength equation is important for stability, the piezometric water elevations obtained from pumping the wells, as described in XI.4, should be considered in the analysis. The shear parameters should be those obtained from consolidated-undrained tests, according to the shear strength equation

$$s = c_{cu} + (\sigma - u) \tan \phi_{cu} \qquad \text{2-XI.5}$$

where u represents a decrease in the effective stresses σ because of upward flow of water toward the bottom of the excavation (see section XI.3). At points where the potential surface of sliding intersects a state of downward flow, no change in the effective stresses is considered.

The effective stresses along the potential surface of sliding should be carefully determined when performing the stability analysis. We obtain the effective stresses in the soil mass by deducting from the initial vertical effective stresses those due to excavation stress relief. The upward water flow toward the bottom of the excavation further reduces the vertical effective stresses, as already stated. This effect is estimated from the hydrodynamic conditions set in the soil mass due to the dewatering of the excavation, as already treated in preceding chapters. The downward water flow found outside of the sheet-pile wall has the long-term tendency to increase the effective stresses. During the excavation, however, this phenomenon is not considered to increase the vertical effective stresses in the soil mass, because of the low permeability characteristics of the cohesive soil sediments.

The value of the shear strength s is estimated for every point on the potential surface of sliding as previously described. The resisting moment is

$$M_r = R\,\{\Sigma s \Delta l + \Sigma s \Delta z\} \qquad \text{3-XI.5}$$

The driving moment M_D is calculated from the total weight of the soil mass and acting forces gravitating on the potential surface of sliding. Hence, the factor of safety

$$G_s = \frac{M_r}{M_D} \qquad \text{4-XI.5}$$

In clays and silty clays, the shear strength used for stability analysis is determined as one-half of the unconfined compressive strength of the soil profile. If such is the case, the changes in pore water pressures because of the hydrodynamic condition set during dewatering of the excavation have no bearing on the resisting moment.

In high plastic sensitive soils the factor of safety should not be smaller than 2, in order to avoid plastic flow toward the bottom of the excavation, thereby reducing the P-heave. Otherwise, the material under the foundation may be damaged and the compressibility increased; thus the subsequent settlement may be found considerably larger than expected. The plastic flow, if permitted, may produce excessive subsidence outside of the sheet-pile wall, and damage of the adjacent construction and public installations. The problem of plastic flow was discussed in sections VII.4 and II.4. The depth of the excavation is limited by the shear strength under the excavation area and the weight of the soil and adjacent construction.

BIBLIOGRAPHY

DiBiagio, E. and Bjerrum, L. (1957). Earth Pressure Measurements in a Trench Excavated in Stiff Marine Clay, *Proc. IV ICOSOMEF*, Vol. II, p. 196, London.

Kézdi, Arpád (1962). *Erddrucktheorien*, Springer-Verlag, Berlin.

Mansur, C. I. and Kaufman, R. I. (1962). Dewatering, in: *Foundation Engineering*, G. P. Leonards, editor. McGraw-Hill Book Company, Inc., New York.

Rendulic, Leo (1938). *Der Erddruck im Strassenbau und Brueckenbau*, Volk und Reich Verlag, Berlin.

Spangler, M. G. (1938). Lateral Pressures on Retaining Walls Caused by Superimposed Loads, *Proceedings of the Eighteenth Annual Meeting of the Highway Research Board*, Part II.

Terzaghi, K. (1936). Distribution of the Lateral Pressure of Sand on the Timbering of Cuts, *Proc. I ICOSOMEF*, Vol. I, pp. 211–215, Cambridge.

Terzaghi, K. (1941). General Wedge Theory of Earth Pressure, *Trans. ASCE*, Vol. 106, pp. 67–68.

Terzaghi, K. and Peck, R. B. (1967). *Soil Mechanics in Engineering Practice*, Second Edition, John Wiley & Sons, Inc.

Tschebotarioff, G. G. (1962). Retaining Structures, in: *Foundation Engineering*. McGraw-Hill Book Company, Inc., New York.

Veder, Ch. (1953). Procédé de Construction de Diaphragmes Impermeables a Grande Profondeur au Moyen de Boues Thixotropiques, *Proc. III ICOSOMEF*, Vol. II, p. 91, Switzerland.

White, L. and Prentis, E. A. (1950). *Cofferdams*, 2nd Editions, Columbia University Press, New York.

Zeevaert, L. (1957). Foundation Design and Behavior of Tower Latino Americana in Mexico City, *Geotechnique*, Vol. VII, pp. 115–133, London.

Zeevaert, L. (1980). *ISE-Interaccion Suelo-Estructura de Cimentaciones Superficiales y Profundas*, LIMUSA, Publ. Co., México, D.F.

Tower Latino Americana after thirty years of arduous work in the destructive earthquake area and difficult subsoil of Mexico City.

▸XII◂ INTRODUCTION TO EARTHQUAKE PROBLEMS IN BUILDING FOUNDATIONS

XII.1 GENERAL CONSIDERATIONS

In the course of history, man has learned to be afraid of earthquakes. Their effects have been estimated qualitatively with respect to damage, loss of life and panic, and today more accurately by means of seismic scales specifying the intensity of the ground motion. The seismic scale is intended to be understood by the layman, and is based mainly on the quantification of the damage produced in buildings and displacement of objects. In past literature and historical documents, earthquakes were classified as mild, strong, very strong and catastrophic. Not until early in the present century were the intensity scale and relative definitions revised in order to obtain greater precision in the determination of the earthquake intensity. The appreciation of damage in estimating intensity should be independent of the human factor of fear.

The source, or focus, of the earthquake is located by its epicenter, determined from its superficial location, and by its hypocenter, found from its depth. The most important earthquakes are originated in fault zones because of the sudden relative displacement of the upper part of the earth's crust. Earthquakes may also be originated by volcanic eruptions due to expulsion of explosive gases, by slides from collapse of large sub-surface cavities, or by other movements of the earth's crust. In the active volcanic areas where geological faults are encountered, earthquakes are more frequent since the internal pressures of the magma closer to the crust activate the relative displacement of the faults more often than in nonvolcanic areas. The mechanism for earthquake triggering at the fault zones is shown in Fig. 1-XII.1. First, a distortion takes place with the accumulation of potential energy, as indicated in the figure by the dashed lines. When the ultimate frictional resistance of the rock at the fault is reached, the potential energy stored is suddenly released into kinetic energy, giving rise to seismic motion. The rocks previously distorted have

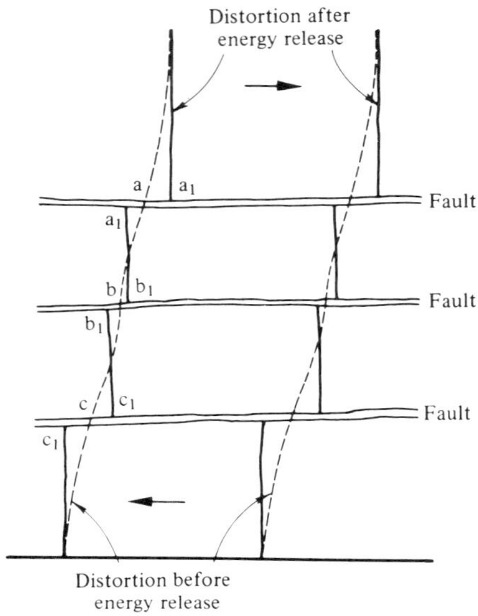

Fig. 1-XII.1 Shear distortion of earth crust.

the tendency to spring back to approximately the positions $(a_1 - a_1)$, $(b_1 - b_1)$, $(c_1 - c_1)$ . . . , and so momentarily experience a relaxation from high shear stresses.

The maximum motion takes place at the focus in the form of a strong vibration producing waves that travel in all directions. The kinetic energy of the waves is dissipated in the earth crust with distance from the source, and its magnitude is registered at various intensities at the locations through which the waves pass. The isoseismal lines, or lines of equal seismic intensity, give the seismologists the means to study and forecast the seismic configuration of certain regions with respect to the epicenters.

The structural geology of the seismic regions is important for study of the seismic epicentral zones, since in reality, an earthquake is not only the liberation of the energy at a certain specific point, but in a zone along one or several faulting lines of the earth crust, where the triggering action takes place. The liberation of the accumulated energy at a certain hypocenter may cause the release of more energy at another fault line in the same region or nearby. Therefore, when an earthquake takes place, one should expect a sequence of them of greater or less intensity until the energy at the zone in question has been totally or partially liberated, and static equilibrium is reached for some time. Necessarily, the earthquakes of larger magnitude (macroseisms) require longer periods to take place, since at the fault zone larger accumulation of potential energy is required before liberation. The geological characteristics of the fault zones are important in the absorption of energy because of the earth crust distortion. The fault zones encountered in a continuous movement do not accumulate a large quantity of potential energy. When for some mechanical reason a fault is prevented from continuing its normal relative displacement, a large amount of energy is stored until the shear strength at the fault is reached and the triggering action takes place, producing the vibrations at the earth crust. In areas of the earth where faulting zones are encountered, earthquakes of greater or less magnitude are likely. As mentioned before, in volcanic regions the faults and fractures may be activated more often because of the magmatic pressures.

The foundation engineer is interested in investigating the characteristics of the more active seismic zones and the possibility that strong motions may occur in areas that appear to be inactive. The intensity of the earthquake movements should be studied in the form of displacements and accelerations, useful to the seismic engineering profession. Therefore, it is important to record destructive earthquakes by means of accelerographs in conjunction with displacement meters installed in buildings. With this practice, the engineer may determine in an approximate quantitative way, the intensity of the earthquakes, based on which he can make a technical study of the behavior of the subsoil and buildings supported at the ground surface.

The seismic waves originated by strong earthquakes induce important dynamic forces in the structures of buildings, towers, bridges, piers, tunnels, and other engineering works, not to mention causing panic in the public. It is important to investigate the type of strong ground motion waves that produce destruction or damage. The detailed study of all types of waves that may be generated in a semi-infinite mass is beyond the scope of this book. Nevertheless, it is important for the foundation engineer to be aware of the effects of earthquake motions in the subsoil and foundations supported on the ground surface.

The seismologist studies the utilization of seismometers, either seismographs to measure displacements or accelerographs to measure accelerations produced by earthquakes. From the results is obtained important information concerning the ground surface accelerations, and from this the response spectra may be calculated based on one-degree-of-freedom structural systems. The results may be applied in seismic foundation and structural engineering. The discussion herein presented will be based on the professional experience of the author in Mexico City, since to understand these complex problems from an engineering point of view and the factors involved in seismic foundation engineering, the author believes it is well to use a case history for reference, in conjunction with the theories and reasoning needed to interpret and apply information obtained from the instruments. Therefore, the practical presentation of this chapter should be taken as an introduction to foundation engineering of strong ground motions in the subsoil, based on measurements of strong earthquakes recorded by means of accelerographs.

Important studies have been made in several countries, particularly in the United States of America and Japan, on the interpretation of field data in order to better comprehend this phenomenon that causes much panic, loss of life and destruction. Supported by private institutions and governments, engineers should investigate the importance of strong ground motions induced by earthquakes in order to foresee their effects in engineering works.

XII.2 EARTHQUAKE ENGINEERING CHARACTERISTICS

2 .1 Introduction to Seismic Waves

The disturbance originated at the focus induces in the soil mass a series of waves with accelerations, periods and velocities of different characteristics. The waves so produced are transmitted at large distances from the epicenter. At the point of observation the waves are screened according to their velocities as they arrive, some sooner than the others.

In order to study in a simple way the most important type of waves, let us consider the equations of motion for a differential element of the subsoil with unit mass ρ subjected to normal stresses σ_x, σ_y, σ_z, shear stresses τ_{xy}, τ_{xz}, τ_{yx}, angular distortions γ_{xy}, γ_{xz}, γ_{yz}, and displacements u, v, w, respectively (Fig. 1-XII.2). The dynamic equations from the theory of elasticity are

$$(\lambda + \mu)\frac{\partial \theta}{\partial x} + \mu \nabla^2 u = \rho \frac{\partial^2 u}{\partial t^2}$$

$$(\lambda + \mu)\frac{\partial \theta}{\partial y} + \mu \nabla^2 v = \rho \frac{\partial^2 v}{\partial t^2} \qquad \text{1-XII.2}$$

$$(\lambda + \mu)\frac{\partial \theta}{\partial z} + \mu \nabla^2 w = \rho \frac{\partial^2 w}{\partial t^2}$$

in which θ represents the volumetric expansion or compression produced by the in-

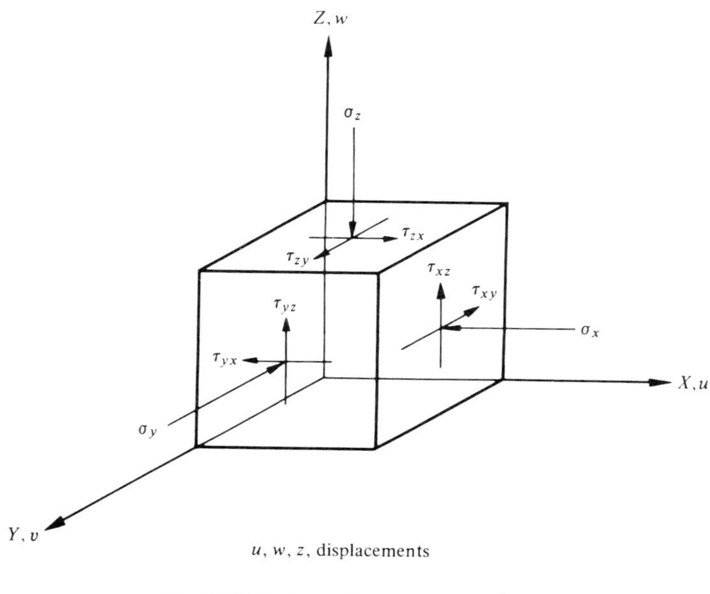

Fig. 1-XII.2 State of stresses at a point.

duced dynamic state of stresses, that is,

$$\theta = \epsilon_x + \epsilon_y + \epsilon_z = \frac{\partial u}{\partial x} + \frac{\partial v}{\partial y} + \frac{\partial w}{\partial z} \qquad \text{2-XII.2}$$

In equations 1-XII.2

$$\lambda = \frac{\nu E}{(1+\nu)(1-2\nu)} \qquad \text{3-XII.2}$$

in which E is the linear modulus of elasticity, under either tension or compression equal in the three directions for an isotropic material and defined by

$$\frac{\sigma}{\epsilon} = E \qquad \text{4-XII.2}$$

The elastic shear modulus or soil rigidity is defined as

$$\frac{\tau}{\gamma} = \mu \qquad \text{5-XII.2}$$

The relation between E and μ is

$$\mu = \frac{E}{2(1+\nu)} \qquad \text{6-XII.2}$$

in which ν is Poisson's ratio, considered constant in the elastic range and defined as

$$\epsilon_y, \epsilon_x = -\nu \frac{\sigma_z}{E} \qquad \text{7-XII.2}$$

Furthermore, in 1-XII.2 the symbol ∇^2 indicates the operation

$$\nabla^2 = \frac{\partial^2}{\partial x^2} + \frac{\partial^2}{\partial y^2} + \frac{\partial^2}{\partial z^2} \qquad \text{8-XII.2}$$

The displacements, strains, distortions and rotations are defined as follows:

strains:

$$\epsilon_x = \frac{\partial u}{\partial x}, \ \epsilon_y = \frac{\partial v}{\partial y}, \ \epsilon_z = \frac{\partial w}{\partial z} \qquad \text{9-XII.2}$$

distortions:

$$\gamma_{xy} = \frac{\partial u}{\partial y} + \frac{\partial v}{\partial x}$$

$$\gamma_{xz} = \frac{\partial u}{\partial z} + \frac{\partial w}{\partial x} \qquad \text{10-XII.2}$$

$$\gamma_{yz} = \frac{\partial v}{\partial z} + \frac{\partial w}{\partial y}$$

rotations:

$$\omega_x = \frac{1}{2}\left(\frac{\partial w}{\partial y} - \frac{\partial v}{\partial z}\right),$$
$$\omega_y = \frac{1}{2}\left(\frac{\partial u}{\partial z} - \frac{\partial w}{\partial x}\right), \qquad 11\text{-XII.2}$$
$$\omega_z = \frac{1}{2}\left(\frac{\partial v}{\partial x} - \frac{\partial u}{\partial y}\right)$$

In accordance with the above-mentioned definitions, one can study the equations of motion. In the first case we investigate the existence of an irrotational movement, for which one has to establish the condition that the values ω_x, ω_y, and ω_z are zero; therefore

$$\frac{\partial \theta}{\partial x} = \nabla^2 u, \quad \frac{\partial \theta}{\partial y} = \nabla^2 v, \quad \frac{\partial \theta}{\partial z} = \nabla^2 w \qquad 12\text{-XII.2}$$

and equations 1-XII.2 reduce to

$$(\lambda + 2\mu)\nabla^2 u = \rho \frac{\partial^2 u}{\partial t^2}$$
$$(\lambda + 2\mu)\nabla^2 v = \rho \frac{\partial^2 v}{\partial t^2} \qquad 13\text{-XII.2}$$
$$(\lambda + 2\mu)\nabla^2 w = \rho \frac{\partial^2 w}{\partial t^2}$$

The waves represented by the motion equations 13-XII.2 are known as irrotational waves that require the material to be compressible, since $\theta \neq 0$. Therefore, they are also known as waves of dilatation and compression.

The second case is present when the waves do not produce volume change; we have

$$\frac{\partial \theta}{\partial x} = \frac{\partial \theta}{\partial y} = \frac{\partial \theta}{\partial z} = 0 \qquad 14\text{-XII.2}$$

This implies exclusively an angular distortion and rotation of the material, in which case, one obtains from 1-XII.2 the following expressions:

$$\mu \nabla^2 u = \rho \frac{\partial^2 u}{\partial t^2}$$
$$\mu \nabla^2 v = \rho \frac{\partial^2 v}{\partial t^2} \qquad 15\text{-XII.2}$$
$$\mu \nabla^2 w = \rho \frac{\partial^2 w}{\partial t^2}$$

To study in general terms the wave equations 13-XII.2 and 15-XII.2, assume the following general form for one component

$$c \frac{\partial^2 \varphi}{\partial z^2} = \frac{\partial^2 \varphi}{\partial t^2} \qquad \text{16-XII.2}$$

The general solution is given by harmonic functions:

1) for time $f_1(\omega t + \theta)$

2) for displacement $f_2 \left(\frac{\pi z}{2L} + z_0\right)$

where ω is the circular frequency and L the wave length. If φ_0 is the amplitude displacement we can write a general solution of the form:

$$\varphi = \varphi_0 f_1(\omega t + \theta) f_2 \left(\frac{\pi z}{2L} + z_0\right) \qquad \text{17-XII.2}$$

Substituting in equation (16-XII.2) and performing algebraic operations we find

$$c \frac{(2\pi)^2}{L^2} = \omega^2$$

but $\omega = 2\pi/T$, where T is the period of the wave, hence the wave velocity is

$$\frac{L}{T} = \sqrt{c}$$

From expressions 13-XII.2 and 15-XII.2 we find:

1) velocity of the compressional waves: $v_d = \sqrt{\frac{\lambda + 2\mu}{\rho}}$ 18-XII.2

2) velocity of the shear waves: $v_s = \sqrt{\frac{\mu}{\rho}}$ 19-XII.2

Substituting the value of λ, (3-XII.2) in 18-XII.2, the velocity of the compressional waves is found to be

$$v_d = \sqrt{\frac{2(1 - \nu)}{1 - 2\nu}} \cdot v_s \qquad \text{20-XII.2}$$

Thus, in the generation zone two types of important body waves are produced with different velocities; they arrive at the place of observation at different times. The dilatation and compressional waves are known in seismology as P waves; they travel with considerably greater velocity than the distortion or shear waves named S. When these seismic body waves are refracted or reflected because of the subsoil stratigraphy, new P and S waves are generated again in the new medium. The difference in velocities between the P and S waves is used by seismologists to estimate the location of the epicenter, by means of recording their arrival at different seismological stations. The particles of the irrotational waves move in the direction of propagation requiring a change in volume of the soil. At a large distance from the epicenter they may be considered plain waves and they are a function of the coordinate in the direction of propagation. The motion of the shear waves particles is perpendicular to the direction of propagation, and the soil volume remains constant during the distortion it suffers due to this phenomenon.

Equations 13 and 15-XII.2 give an idea of the degree of complexity involved in studies related to seismic waves, especially when they are refracted and reflected at the ground surface, or soil interfaces. Solutions in specific isotropic cases for which the boundary conditions are known may be found mathematically in a relatively simple way. At the surface of the semi-infinite solid where there is no restriction on deformation, another type of wave may be generated similar to ocean waves; they are known as surface waves.

One can observe that the two systems of body waves generated in the interior of the mass by the seismic motion, at the focus, are independent from each other, since they travel at different velocities. It is interesting to note that the velocity of the irrotational waves is an important function of Poisson's ratio 20-XII.2. In rock, one may consider $\nu = 0.25$; in saturated soils, this value is close to 0.5, and therefore the velocity v_d is very large. The velocity of the shear waves is a function only of the rigidity of the soil and is independent of Poisson's ratio; hence, the practical precision obtained in problems where the shear wave velocity is present is greater than in those where it is necessary to determine Poisson's ratio for the soil mass.

If the subsoil is saturated, the displacements produced by the P waves are considerably less than those produced by the S waves. Therefore, in seismic engineering, the most important waves producing damage in unconsolidated subsoil sediments are the shear waves. When the waves approach soft subsoil deposits underlain by firm ground at great depth, the shear waves produce an oscillatory ground motion, and the P waves produce a trepidatory vertical motion. The phenomenon, however, produced either by the P or S waves, respectively, is studied in three components, that is to say, the waves are registered in the accelerograph during the earthquake in three directions, two horizontal and one vertical. In saturated nonconsolidated sediments one observes that the vertical component has no practical significance, when the epicenter is located at a great distance from the place of observation.

2.2 Magnitude

The magnitude of an earthquake at the focus may be assigned a value on the scale proposed by Richter. The magnitude represents the amount of potential energy liberated at the hypocenter, and is registered with a specific intensity at a certain distance. The seismological stations should be installed on firm rock, in order that the measurements be mutually consistent. The magnitude is measured by the amplitude of the waves recorded in conventional seismometers specially constructed for this purpose. The basic magnitude is defined by one wave of one micron amplitude at a distance of 100 km from the epicenter, and recorded with a seismometer having the following characteristics: damping period 0.8 sec, static magnification 2,800, damping of instrument 0.8. The magnitude is defined by

$$M_R = \log \frac{A}{A_0} \qquad \text{21-XII.2}$$

in which A is the earthquake amplitude at the same distance as the basic amplitude A_0 (Fig. 2-XII.2). Richter found that curves 1, 2, 3, etc., indicate the variation of the amplitudes for different earthquakes as a function of distance. It was observed that these curves are parallel between each other, and therefore the difference $\log A - \log A_0$ is independent of the distance. The conclusion was reached that the magnitude of an earthquake may be determined independently of its distance if the basic curve or conventional curve $\log A_0$ is known. The magnitude of the basic curve is zero according to definition given by 21-XII.2. From the logarithmic ratio, it may be seen that an earthquake with magnitude 3 will produce at the same distance amplitudes 1,000 times larger, and an earthquake of magnitude 6 will produce amplitudes one million times larger. In any event, an earthquake of magnitude 7 will have liberated, at the focus, 10 times more energy than one of magnitude 6.

The largest magnitude registered was on the order of 8.6 in South America, Asia and Japan; however in 1906, an earthquake was registered in Sanriku, Japan, with a magnitude of 8.9. The earthquakes of July 28, 1957, at the Pacific zone in front of

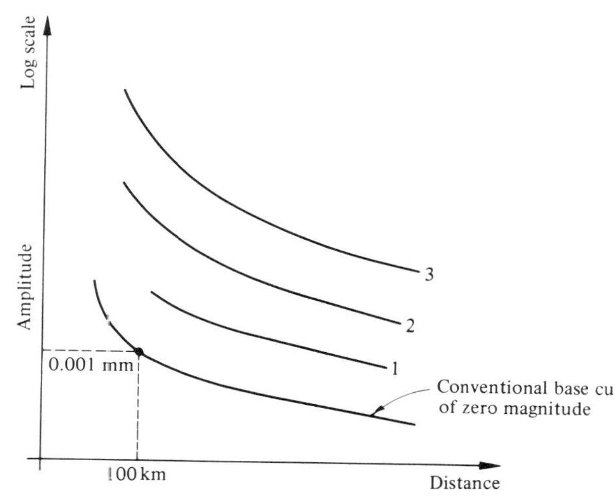

Fig. 2-XII.2 Measure of earthquake magnitude.

the Coast of the State of Guerrero in Mexico were reported between 7.5 and 8 magnitude, and those that occurred in May 1962 were estimated of magnitude 7 to 7.25 in the laboratory of seismological investigations at the California Institute of Technology.

2.3 Intensity

The intensity of an earthquake is the measure at the place of observation of the effect of the energy liberated at the focus. Therefore, it diminishes with the distance from the epicenter. The intensity is measured by the response obtained from measuring instruments like accelerographs or from the damage produced by the earthquake in objects and buildings at a specific location.

Intensity scales were introduced early in the last century. During subsequent years, other scales have been proposed to satisfy the deficiencies of the early scales. De Rossi and Forel proposed in 1883 an intensity scale used to study the earthquakes taking place in that epoch. In 1902, Mercalli adopted the Rossi-Forel scale to study Italian earthquakes and proposed a new one that bears his name. Mercalli's scale has been suffering changes and additions until finally what is known as the Modified Mercalli Scale was accepted in 1931, and is given condensed in Table 1-XII.2.

Cancani and Sieberg proposed an interesting scale that consists, mainly, in correlating the intensity scales with the probable maximum accelerations induced by the earthquake. Cancani in 1904 and Sieberg in 1912 tried to correlate the seismic scales of Mercalli with the maximum probable acceleration of the ground surface. Also Richter in later years, after careful studies of information from instrumentation in California by the U. S. Coast and Geodetic Survey, proposed an empirical formula to calculate the acceleration using the M. M. Scale. This expression is the following:

$$\log a = \tfrac{1}{3} I_{\mathrm{M}} - \tfrac{1}{2} \qquad \text{22-XII.2}$$

in which a is the ground acceleration in cm/sec^2 and I_{M} is the modified Mercalli estimated intensity.

Table 2-XII.2 shows a comparison of the M-Mercalli and Rossi-Forel scales, and correlations with those proposed for ground acceleration by Cancani-Sieberg and Richter.

Obviously, the intensity of the motion at the ground surface and at a certain depth in the soft subsoil deposit is a function of the mechanical, stratigraphical and hydraulic properties of the subsoil. Therefore, the intensity of an earthquake may vary considerably, depending on the engineering characteristics of the subsoil. Furthermore, the damage produced in buildings depends highly on their shape and structural design. It is concluded that in order to estimate quantitatively the destructive intensity of earthquakes on buildings, it is necessary to know the response spectrum in the area. The spectrum may be computed if the accelerogram of the earthquake is known. Moreover, from recent studies mentioned below, one can conclude that it is very important to determine the resonant periods of the ground.

TABLE 1-XII.2 Modified Mercalli Scale

I. Not felt except by a very few under especially favorable circumstances. (I Rossi-Forel Scale)

II. Felt only by a few persons at rest, especially on upper floors of buildings. Suspended objects may swing slightly. (I to II Rossi-Forel Scale)

III. Felt quite noticeably indoors, especially on upper floors of buildings, but many people do not recognize it as an earthquake. Standing motor cars may rock slightly. Vibration like passing truck. Duration estimated. (III Rossi-Forel Scale)

IV. During the day felt indoors by many, outdoors by few. At night some awakened. Dishes, windows, doors disturbed; walls make creaking sound. Sensation like heavy truck striking building. Standing motor cars rocked noticeably. (IV to V Rossi-Forel Scale)

V. Felt by nearly everyone; many awakened. Some dishes, windows, etc. broken; a few instances of cracked plaster; unstable objects overturned. Disturbances of trees, poles, and other tall objects sometimes noticed. Pendulum clocks may stop. (V to VI Rossi-Forel Scale)

VI. Felt by all; many are frightened and run outdoors. Some heavy furniture moved; a few instances of fallen plaster or damaged chimneys. Damage slight. (VI to VII Rossi-Forel Scale)

VII. Everybody runs outdoors. Damage negligible in buildings of good design and construction; slight to moderate in well-built ordinary structures; considerable in poorly built or badly designed structures; some chimneys broken. Noticed by persons driving motor cars. (VIII Rossi-Forel Scale)

VIII. Damage slight in specially designed structures; considerable in ordinary substantial buildings with partial collapse; great in poorly built structures. Panel walls thrown out of frame structures. Fall of chimneys, factory stacks, monuments, walls. Heavy furniture overturned. Sand and mud ejected in small amounts. Changes in well water. Persons driving motor cars disturbed. (VIII + to IX Rossi-Forel Scale)

IX. Damage considerable in specially designed structures; well designed frame structures thrown out of plumb; great in substantial buildings, with partial collapse. Buildings shifted off foundations. Ground cracked conspicuously. Underground pipes broken. (IX + Rossi-Forel Scale)

X. Some well-built wooden structures destroyed; most masonry and frame structures destroyed with foundations; ground badly cracked. Rails bent. Landslides considerable from river banks and steep slopes. Shifted sand and mud. Water splashed (slopped) over banks. (X Rossi-Forel Scale)

XI. Few, if any, (masonry) structures remain standing. Bridges destroyed. Broad fissures in ground. Underground pipe lines completely out of service. Earth slumps and land slips in soft ground. Rails bent greatly.

XII. Damage total. Waves seen on ground surfaces. Lines of sight and level distorted. Objects thrown upward into the air.

The results may be used in foundation and structural engineering to estimate the response forces induced in buildings and in their foundations.

In the present state of our knowledge, the use of the intensity scales like the M-Mercalli cannot be overlooked, since they are easily applied. Furthermore, there

TABLE 2-XII.2

Intensity		Acceleration in mm/sec^2		
M-M	Rossi-Forel	Cancani-Sieberg	Richter	Mexico, D.F.
I	I	5	7	
				5–10
II	I–II	10	13	
				10–25
III	III	25	30	
	IV-V			25–50
IV	IV	50	70	
				50–100
V	V–VI	100	150	100
VI	VI–VII	100–250	300	100–250
				250–500
VII	VIII	500	700	500
VIII	VIII–IX	500–1000	1500	
				500–2500
IX	IX	1000	3200	—
X	X	2500	6750	—
XI	X	5000	15000	—
XII	X	10000	32000	—

is not enough information on accelerographs located in different stratigraphical well defined subsoil conditions that would permit the elaboration of other seismic scales based on the response spectra for specific locations. Nevertheless, the M.M Scale should be used cautiously and with judgement, since it is mainly based on the appreciation of damage on buildings that, in general, have not been designed to resist earthquakes properly. In accordance with these scales, it may be considered that destructive earthquakes are those corresponding to grade VI or larger. If in a certain area the buildings have been designed to survive destructive earthquakes, evidently the destructive grade mentioned before should be increased and the M.M Scale revised. By the observation of the scales (Table 2-XII.2) it is concluded that the seismic design for grade IX would be the maximum possible for engineering purposes since grade X is considered catastrophic. The ground would not support construction in a satisfactory way, even if the construction could be designed for an earthquake of such magnitude. According to the Richter scale the acceleration would be on the order of 675 cm/sec^2, and for Cancani-Sieberg greater than 250 cm/sec^2. Evidently, there will be subsoils that will fail because of the shear forces induced in them before reaching these high accelerations, as will be shown further ahead in the case of the soft subsoil of Mexico City, in which the shear strength in unloaded ground may be reached with a ground acceleration on the order of 200 cm/sec^2, implying a maximum degree of IX according to the M-Mercalli's scale and ground accelerations.

It is concluded that in the Valley of Mexico the acceleration in the lacustrine zone is approximately two times larger than that registered in the firm ground surrounding the valley, on which the seismological station of Tacubaya, D. F., is located. Thus, at the station an earthquake may be recorded with intensity VI, while in the lacustrine zone the intensity may be on the order of VII.

2.4 Earthquake Recording

Two types of seismometers are considered to measure ground motions; those that contain a pendulum with a period considerably larger than the dominant period of the ground, and those where the period of the pendulum is considerably shorter than the dominant period. The first ones are called seismographs; with them the amplitudes of microseisms and teleseisms may be recorded to determine at great distances the magnitudes of the earthquakes. The second ones, the accelerographs, record the ground accelerations and they are very useful in seismic engineering. They usually can register macroseisms of grade V on the M-Mercalli scale and up to accelerations close to gravity and larger. In the present, one can obtain these instruments commercially from the U. S. A. and Japan.

Earthquakes recorded in Mexico City by means of the Japanese SMAC seismograph are shown in Figs. 3 and 4-XII.2. The vertical component not shown in the figure was found considerably smaller than the horizontal components. Apparently this may be due to the saturated conditions of the lacustrine silty clay deposits. From the record of an earthquake as shown, it is possible to obtain the maximum response accelerations of the ground and their dominant periods. Also, one can construct a spectrum of the number of waves against period, in order to determine the maximum dominant periods. The accelerations obtained from the accelerographs are applicable only to structures with periods considerably less than the dominant period of the ground. In fact, in soft ground where the wave periods are large, rigid and massive buildings move approximately in accordance with the ground motion. In case of flexible structures, however, the response of the building to the earthquake is a function of the absorbed energy during the ground motion and dissipation of this energy. The flexibility of the structure is measured by the period or periods of vibration of the building. The damping factor of the building is very important; it may depend highly on the secondary elements constructed into the buildings, like the type of floors, curtain and partition walls.

The accelerograms recorded for central part of Mexico City show a maximum ground acceleration on the order of 49 cm/sec^2 for a wave period of 2.5 sec, and 37 cm/sec^2 for a wave period of 2.1 sec at Alameda Park. For equals periods, respectively, about half this acceleration was recorded at a depth of 12 m in the rigid box foundation of the Latino Americana Tower. Table 3-XII.2 shows maximum ground accelerations and periods corresponding to those observed in May 1962 in the accelerograms.

2.5 Response Spectrum

The accelerograms represent a series of random impulses with variable periods and accelerations that are expressed mathematically by $a(\tau)$, where τ is the instant at

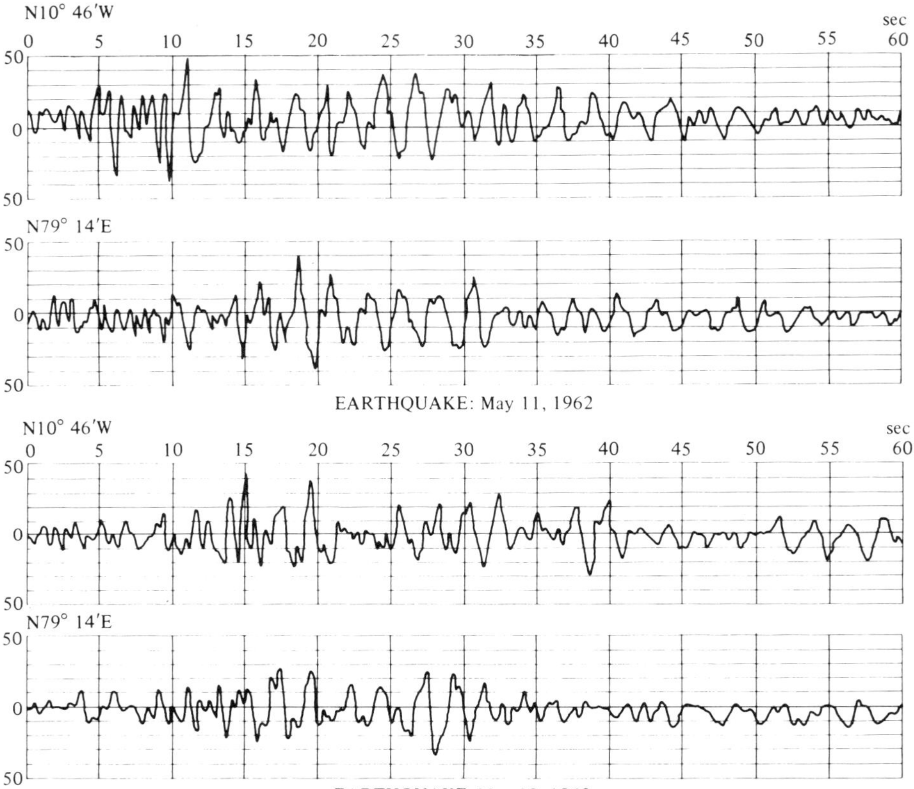

0, Seconds correspond to 33 seconds after seismograph started

Data from a photocopy of the original

Fig. 3-XII.2 Acceleration recorded at Alameda Park.

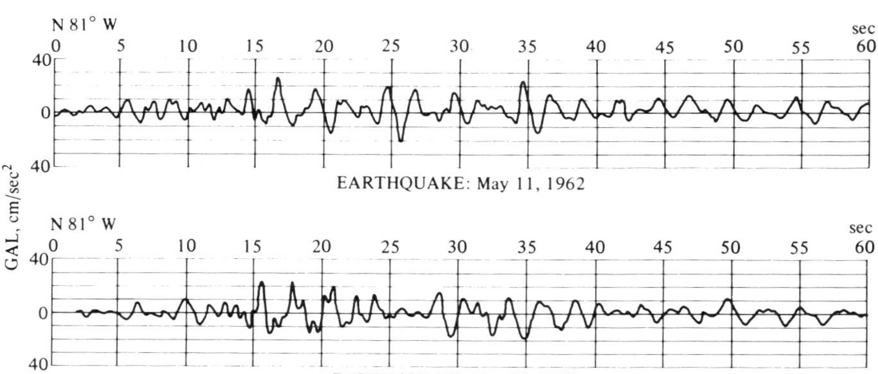

0, Seconds correspond to 33 seconds after seismograph started

Data from a photocopy of the original

Fig. 4-XII.2 Accelerations recorded at the basement of the Latino Americana Tower.

TABLE 3-XII.2

Date	AP-Site				Lat-Site			
	N 10° 46′ W		N 79° E		N 9° E		N 81° W	
1962	a_m	T	a	T	a_m	T	a	T
May 11	49	2.5	40	2.2	19	2.2	25	2.4
May 19	37	2.1	34	2.2	23	2.3	24	2.6

Note: a_m, maximum ground acceleration, is given in cm/sec^2, and T in sec.

which the acceleration is present from the start of the strong ground motion. The function mentioned above, however, cannot be represented by a simple mathematical formula. Furthermore, it varies according to the earthquake and the stratigraphical conditions and mechanical properties of the subsoil. This is shown by a comparison of the accelerograms recorded for the earthquakes of May 11 and 19, 1962, in central Mexico City (Figs. 3 and 4-XII.2). The maximum response for relative displacement of structures with one degree of freedom is given by

$$S_d = \frac{1}{\omega_d} \left[\int_0^t a(\tau) e^{-\zeta \omega_i (t-\tau)} \cdot \sin \omega_d (t - \tau) d\tau \right]_{max} \qquad \text{23-XII.2}$$

where:

t – instant of maximum S_d
ζ – damping coefficient of the structure
ω_i – free circular frequency of structure
ω_d – damped circular frequency, equal to $\omega_i \sqrt{1 - \zeta^2}$

Nevertheless, expression 23-XII.2 may show similar characteristics for locations with approximately the same stratigraphical and mechanical characteristics. The spectrum thus obtained will represent the maximum response for relative displacements of one-mass structures with damping. Therefore, the integration of 23-XII.2 is performed for each one-degree-of-freedom structure with period T, obtaining one point in the spectrum curve, that represents the maximum response for that specific period when the structure is subjected to the ground acceleration $a(\tau)$ of the earthquake. In fact, the horizontal maximum force induced in the mass of the structure with damping is:

$$V_{max} = M \omega_d^2 S_d \qquad \text{24-XII.2}$$

The value

$$\omega_d \cdot S_d = R_v \qquad \text{25-XII.2}$$

is known as the pseudovelocity response R_v for a structure with damped circular frequency ω_d, and the curve formed with the maximum response for periods T

representing various flexibilities of the one-degree-of-freedom structures is called the pseudovelocity spectrum.

Thus, one can write:

$$R_a = \omega_d \cdot R_v \qquad 26\text{-XII.2}$$

with acceleration units, and the curve thus formed for different values of T is called the pseudoacceleration spectrum, or equivalent acceleration to obtain maximum response of the mass M. Hence the shear force of the one-degree-of-freedom structure is

$$V_{max} = MR_a \qquad 27\text{-XII.2}$$

The relative displacement response spectra calculated from expression 23-XII.2 for one component of the earthquakes registered in Mexico City in 1962 at the Alameda Park and foundation of Latino Americana Tower, respectively, are shown in Figs. 5 to 8-XII.2. Their configuration is similar even though both earthquakes had different intensities and the accelerograms show different configurations. The pseudoacceleration spectra for the same components and earthquakes, respectively, are shown in Figs. 9 to 12-XII.2, from which we see that the response at the Alameda Park was twice as strong as at the foundation of the Tower, as shown also in the accelerograms. By means of the pseudoacceleration response spectra, an envelope spectrum may be constructed for general use in terms of R_a/a_m vs. T, and for critical damping $\zeta = 5\%$, 10% and 20%, respectively. The value of a_m represents the maximum average ground acceleration observed. The value R_a/a_m may be refered as the amplification factor, f_a. The spectra thus formed are shown in Fig. 13-XII.2. Other responses for intermediate values of critical damping may be interpolated. When the response spectra are known, the shear forces may be calculated for different modes of vibration of the structure. The seismic analysis may be performed using the structural method known as modal analysis, taking into account the coefficients of participation of each one of the modes when calculating the probable maximum forces to which the structure will be subjected.

When the amplification factor $f_a = R_a/a_m$ is known for a certain seismic region an envelope design spectrum may be constructed in terms of f_a vs. T_0/T_{s1}. Here T_0 represents the equivalent period of the structure and its foundation, and T_{s1} the dominant period of the subsoil from ground surface to firm ground. The author has had the opportunity to study the amplification factors against the critical damping for different locations, obtaining very similar values for this factor independent of the subsoil conditions. In Fig. 14-XII.2 the amplification factors for Mexico City and San Salvador have been plotted against the critical damping. It may be noticed that the amplification factors are very similar for the same critical damping, even though the subsoil at these locations is different from the geotechnical point of view. Mexico City (Chapter V) is founded on soft silty saturated lacustrine clay sediments, and San Salvador subsoil is formed of series of deposits of modified eolian volcanic glass interbedded with alluvial sediments.

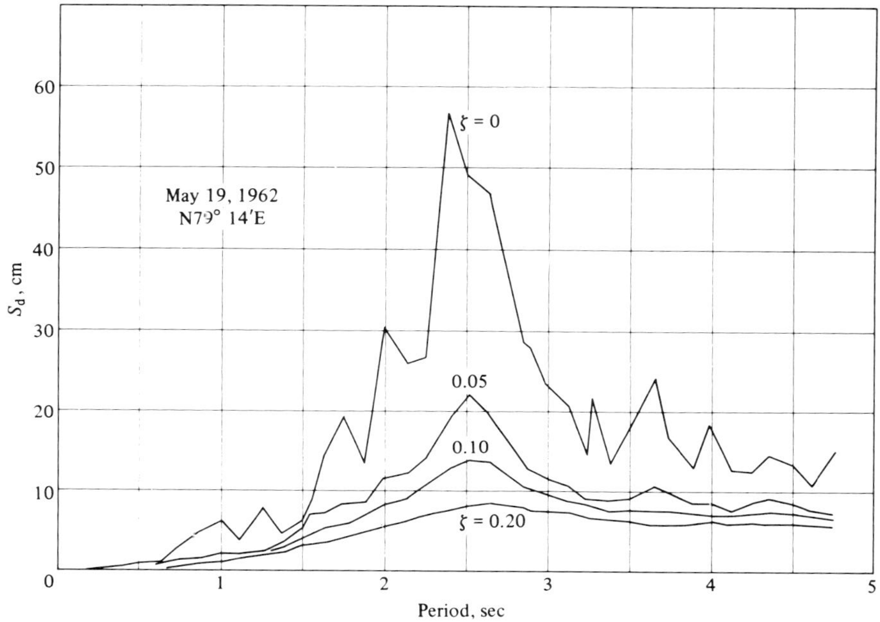

Fig. 5-XII.2 Relative displacement spectrum, Alameda Park (May 19, 1962).

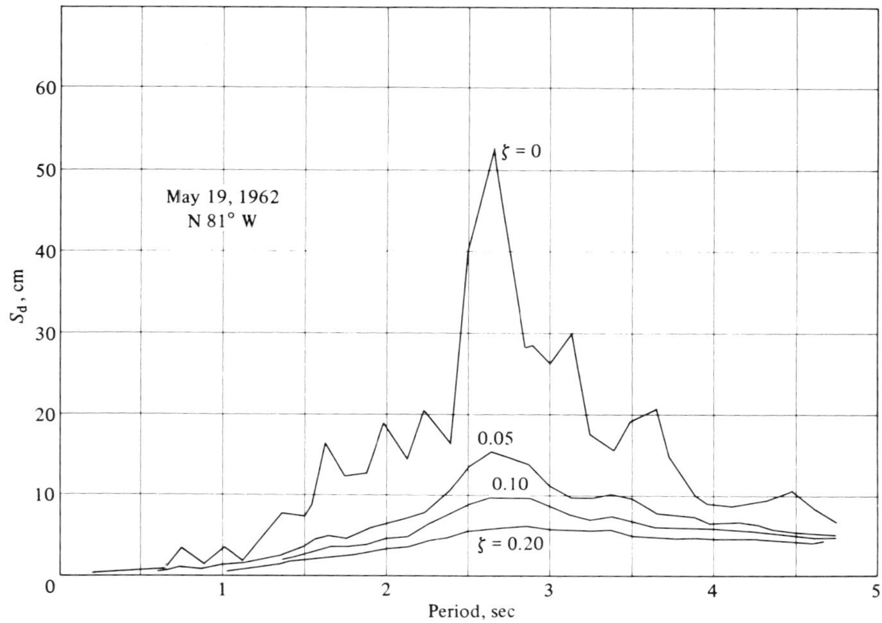

Fig. 6-XII.2 Relative displacement spectrum, Latino Americana Tower (May 19, 1962).

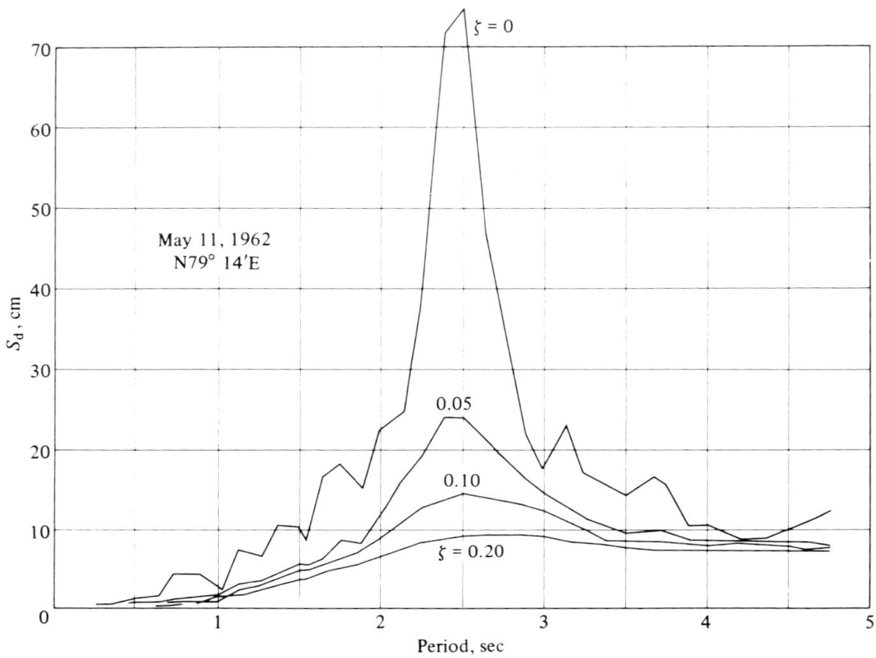

Fig. 7-XII.2 Relative displacement spectrum, Alameda Park (May 11, 1962).

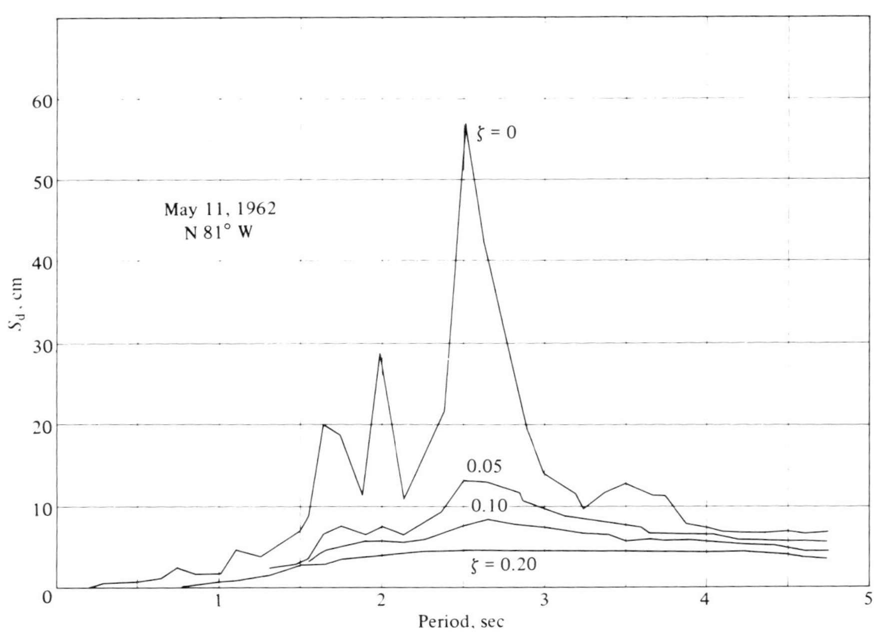

Fig. 8-XII.2 Relative displacement spectrum, Latino Americana Tower (May 11, 1962).

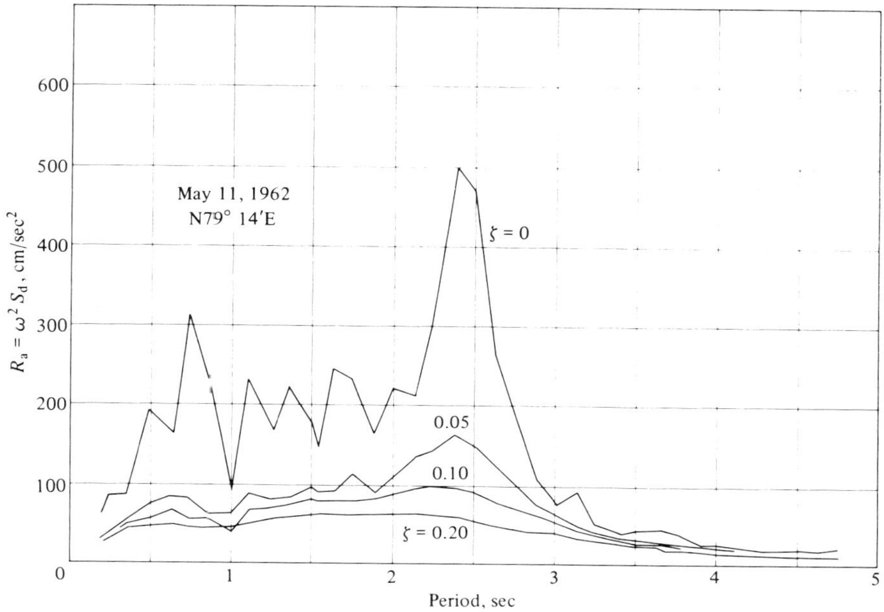

Fig. 9-XII.2 Pseudoacceleration spectrum, Alameda Park (May 11, 1962).

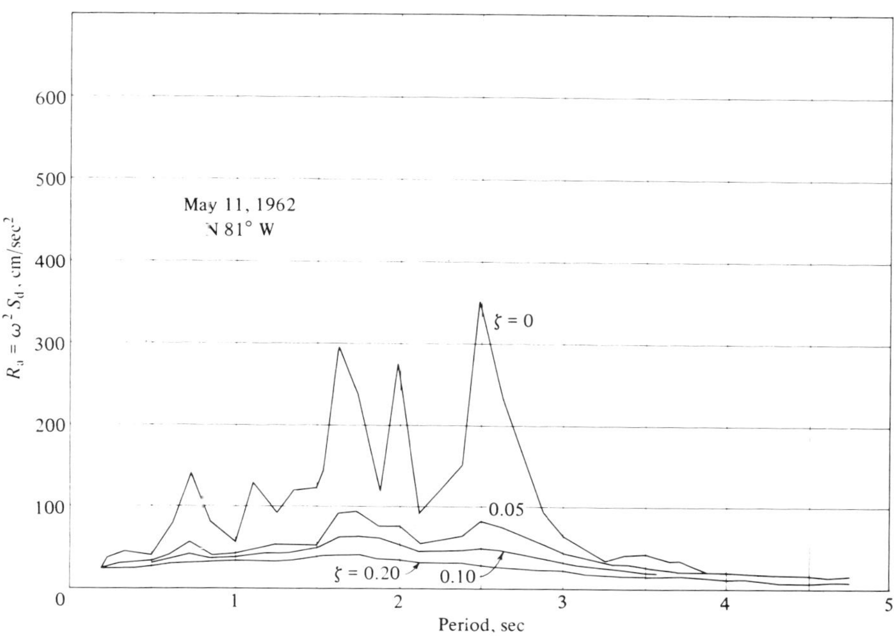

Fig. 10-XII.2 Fseudoacceleration spectrum, Latino Americana Tower (May 11, 1962).

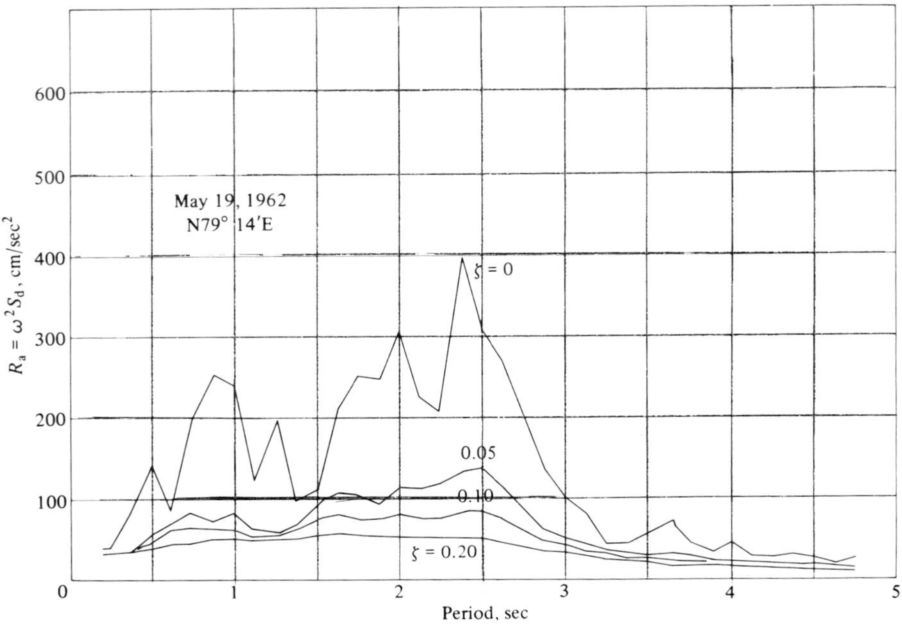

Fig. 11-XII.2 Pseudoacceleration spectrum, Alameda Park (May 19, 1962).

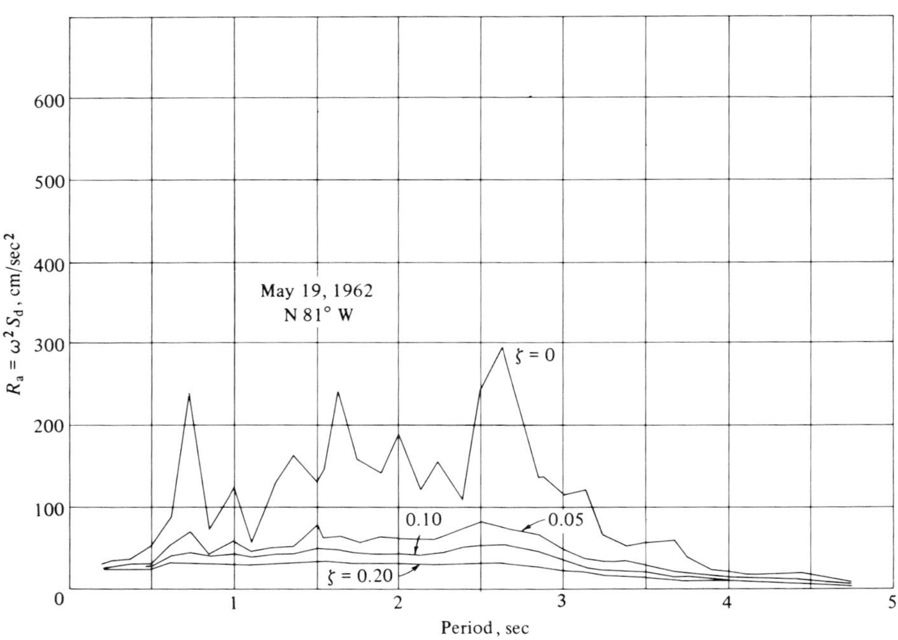

Fig. 12-XII.2 Pseudoacceleration spectrum, Latino Americana Tower (May 19, 1962).

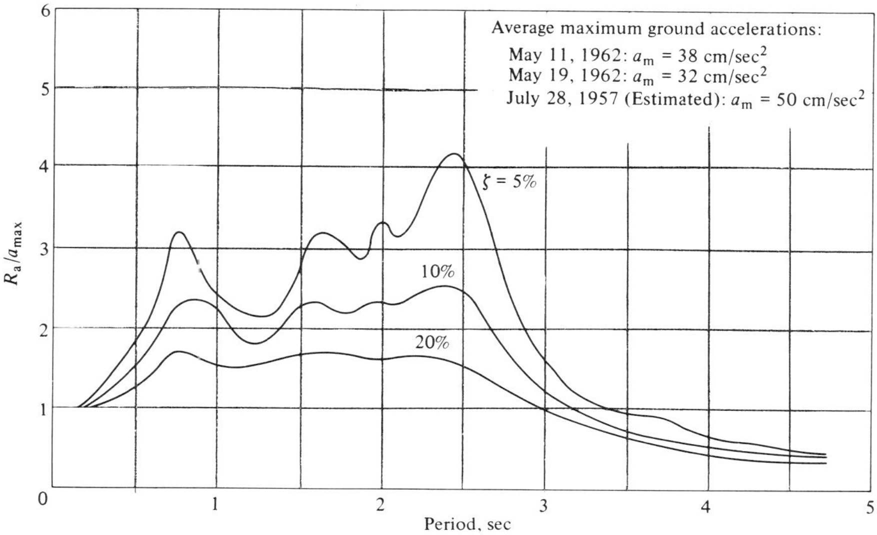

Fig. 13-XII.2 Envelope pseudoacceleration spectrum, Mexico City center.

These sediments are not saturated because the water table is deep seated from the ground surface.

From these two extreme subsoil conditions it may be concluded, upon more studies on the subject, that the amplification factors appear to be independent of the characteristics of the soil sediments overlying firm ground. From Fig. 14-XII.2 intermediate values may be selected to construct a design relative acceleration response spectrum as shown in Fig. 15-XII.2. Therefore, knowing the dominant period of the ground T_{s1} and the equivalent period of the structure and its founda-

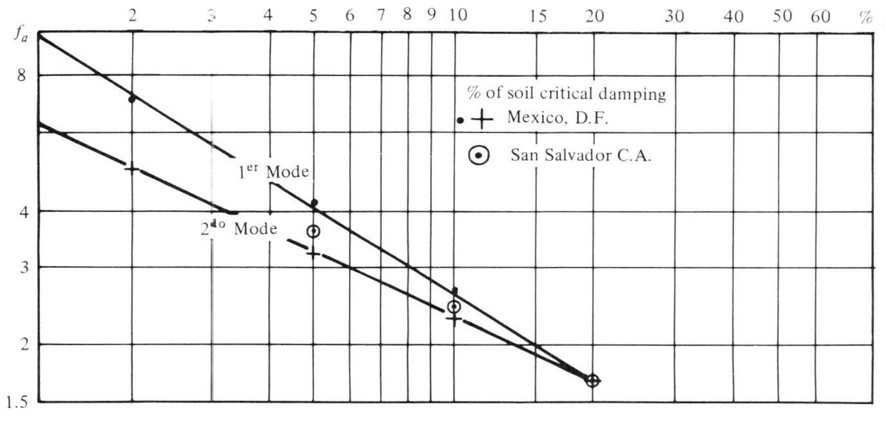

Fig. 14-XII.2 Amplification factors.

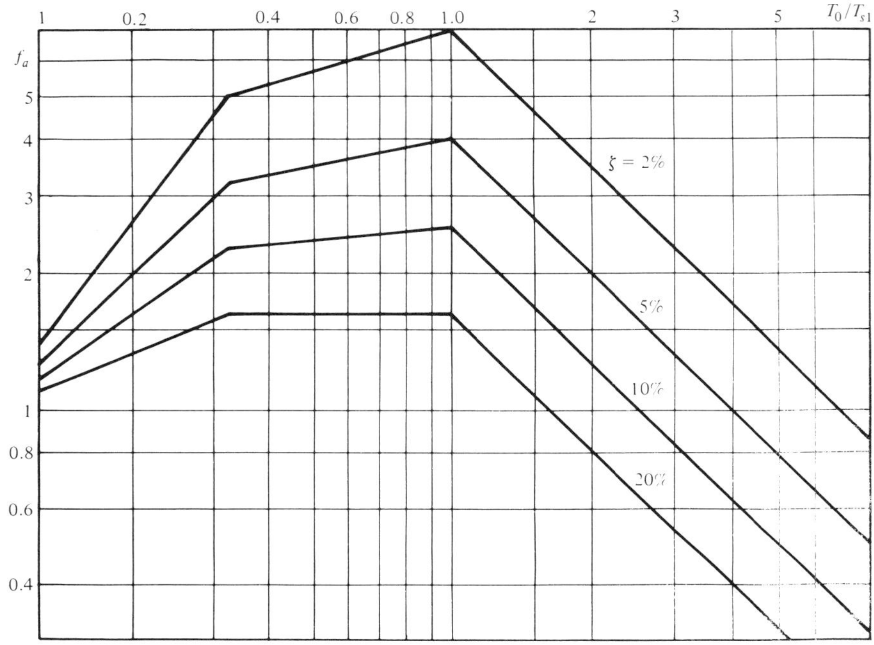

Fig. 15-XII.2 Seismic design spectrum.

tion T_0, the design spectrum may be entered with the ratio T_0/T_{s1} and the corresponding amplification factor f_a determined.

The seismic force at the center of mass M of the building and its foundation will be: $V_m = f_a \cdot a_m M$ in which a_m is the ground surface acceleration at the site in question.

XII.3 SUBSOIL BEHAVIOR

3.1 Basic Concepts

From observational records of the earthquakes of May 1962 recorded in Mexico City, and using his knowledge of the mechanical and stratigraphical properties, the engineer can now make general interpretations of the approximate behavior of the subsoil. In the case of the Valley of Mexico, these destructive earthquakes of May 1962 were reported as having a magnitude of 7 to $7\frac{1}{4}$, with epicenter 260 km distant and hypocenter depth on the order of 20 km, located geographically approximately 30 km southwest of Acapulco, Guerrero. The irrotational and shear waves produced by these earthquakes at the focus may be visualized reaching the Valley of Mexico as shown in Fig. 1-XII.3. The seismic waves are transmitted through the firm base of the basin of the Valley to the lacustrine saturated soft sediments. The longitudinal waves, however, are not transmitted with large amplitudes in the clay

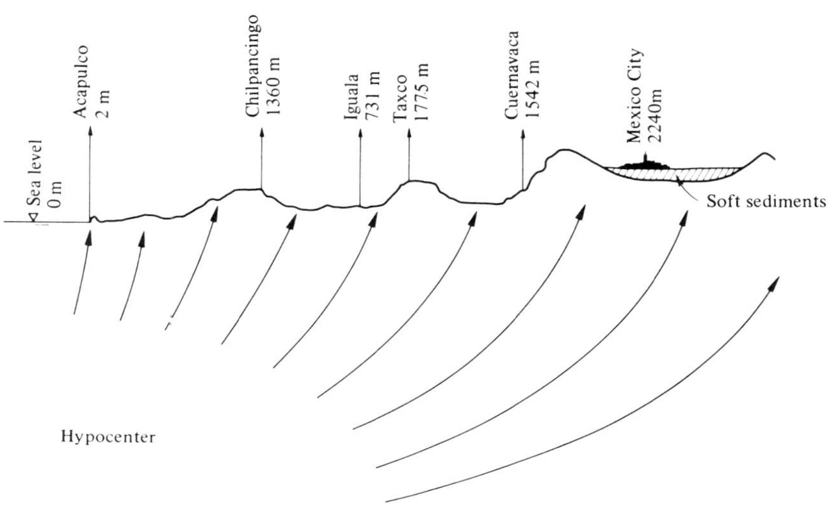

Fig. 1-XII.3 Seismic waves reaching the Valley of Mexico.

mass, since it is saturated and cannot change in volume instantaneously. Therefore, only the constant volume waves or shear waves produce large displacements in the sediments, because of distortion of the subsoil strata. At the surface, other waves may build up motivated by the reflection of the transverse waves; however, the surface waves in this particular case produce motions that are negligible compared with those produced by the shear waves, as is observed from the small interference they cause in the acceleration records.

The shear waves generated at the base of the saturated soft lacustrine sediments of high compressibility travel from this base to the surface. The equations of motion are

$$\mu \frac{\partial^2 u}{\partial z^2} = \rho \frac{\partial^2 u}{\partial t^2}$$

$$\mu \frac{\partial^2 v}{\partial z^2} = \rho \frac{\partial^2 v}{\partial t^2} \qquad \text{1-XII.3}$$

These equations represent the components of the motion produced by the shear waves traveling with velocity v_s in the vertical direction, and upward from the base of the firm deposits at the bottom of the basin to the ground surface. The shear waves produce an oscillatory movement recorded in the accelerographs by the horizontal components.

The boundary conditions may be established as follows:

$$\begin{aligned} &\text{a) for } t = 0, \quad u = v = 0 \\ &\text{b) for } z = 0, \quad u = v = 0 \\ &\text{c) for } z = D, \quad \frac{\partial u}{\partial z} = \frac{\partial v}{\partial z} = 0, \quad u = u_0, \quad v = v_0 \end{aligned} \qquad \text{2-XII.3}$$

The case may be studied for an isotropic elastic mass. The general solution of the differential equations 1-XII.3 may be written in the form

$$u(z,t) = f_1(z) \cdot f_2(t) \tag{3-XII.3}$$

For μ and ρ constant one obtains the following expression:

$$u(z,t) = [m_1 \cos cz + n_1 \sin cz] \times \left[m_2 \cos \left(c\sqrt{\frac{\mu}{\rho}} \right) t + n_2 \sin \left(c\sqrt{\frac{\mu}{\rho}} \right) t \right] \tag{4-XII.3}$$

From boundary conditions a and b, it is found that $m_1 = 0$ and $m_2 = 0$; thus

$$u(z,t) = n_1 n_2 \sin cz \cdot \sin \left(c\sqrt{\frac{\mu}{\rho}} \right) t \tag{5-XII.3}$$

Condition c expresses that the shear force equals zero for $z = D$, and therefore

$$cD = \frac{\pi}{2}(2n - 1); \; n = 1, 2, 3, \ldots \tag{6-XII.3}$$

from which the circular frequency for the n-th mode of vibration of the subsoil deposit of thickness D is obtained:

$$\omega_n = (2n - 1)\frac{\pi}{2D}\sqrt{\frac{\mu}{\rho}} \tag{7-XII.3}$$

and correspondingly the period is

$$T_n = \frac{4D}{(2n - 1)}\sqrt{\frac{\rho}{\mu}} \tag{8-XII.3}$$

Moreover, the value of $4D/(2n - 1)$ represents the wave length for the T_n period. For multiples of $T_n/4$, the maximum amplitude is obtained; therefore $n_1 n_2 = u_0$ and v_0, respectively.

The components u and v of the motion take the following configuration:

$$\sin (2n - 1)\frac{\pi}{2D} z \cdot \sin \left[2\pi \frac{(2n - 1)}{T_1} \cdot t \right] \tag{9-XII.3}$$

where T_1 is the first mode of vibration. From previous computations, it may be observed that if v_s is the velocity of propagation, the shear wave reaches its maximum amplitude for values of $T_1/4$ and $3T_1/4$, and consequently will travel in this time through the sediment in proportional distances corresponding to $D/(2n - 1)$.

The shear wave velocity is a dynamical characteristic property of the soil:

$$v_s = \sqrt{\frac{\mu}{\rho}} \tag{10-XII.3}$$

The periods at which magnification of stresses and displacements take place in the subsoil may be written in the following form for an isotropic soil mass:

$$T_n = \frac{4D}{(2n-1)\,v_s} \qquad \text{11-XII.3}$$

For $n = 1$, the first mode of vibration is obtained and for subsequent values $n = 2,3$, etc., the higher harmonics may be found in the proportion $\frac{1}{3}$, $\frac{1}{5}$, etc., of the first mode.

From equation 11-XII.3, we can draw the interesting conclusion that if T_a, D_a, and $(v_s)_a$ are the mechanical and geometrical properties for a certain location, and if T_b, D_b, and $(v_s)_b$ are those for another, then

$$\frac{T_a}{T_b} = \frac{D_a}{D_b} \cdot \frac{(v_s)_b}{(v_s)_a} \qquad \text{12-XII.3}$$

Considering that $\rho_a \doteq \rho_b$ one can write

$$\frac{T_a}{T_b} = \frac{D_a}{D_b} \cdot \frac{\sqrt{\mu_b}}{\sqrt{\mu_a}} \qquad \text{13-XII.3}$$

from which the following theorem may be stated: the resonant periods of vibration of the soil deposits are in direct proportion to their thicknesses and inversely proportional to the square root of their rigidities.

This conclusion is important because if one knows from observation the resonant periods and the engineering characteristics at one location, and one can estimate the stratigraphy and the average value of μ at other locations, it is possible to calculate approximately their corresponding periods from the above proportion (Fig. 2-XII.3). The earthquake response spectra, as described above, can be applied to the subsoil deposit considering it as a vibrator in itself to the waves transmitted through the soft lacustrine sediments from the firm base. The magnification of the phenomenon is given at wave periods corresponding to the peaks observed in the spectrum. From relative displacement and pseudoacceleration spectra for Alameda

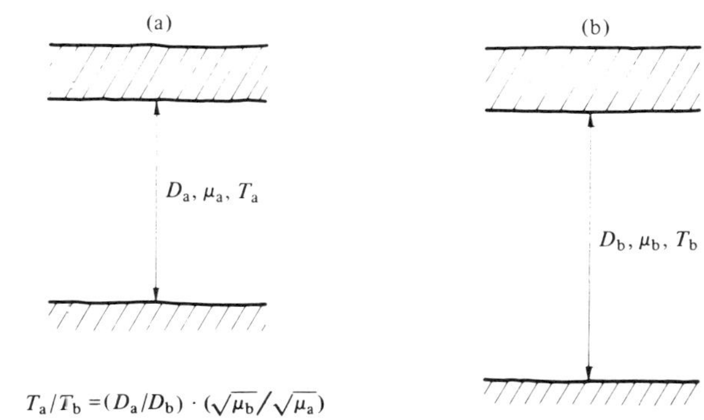

Fig. 2-XII.3 Relationship between resonant periods and rigidities and thicknesses of deposits.

Park in Mexico City, the largest peak is observed at periods $T = 2.35–2.55$ sec, and others may be detected corresponding to waves producing smaller magnification (section XII.2).

In practice, the engineering characteristics of the subsoil change with the stratigraphy, and consequently, with the different types of materials encountered; therefore, it is necessary to solve the differential equation of motion using a step by step procedure.

3.2 Resonant Periods of Vibration in Stratified Subsoil

The basic theoretical concepts of subsoil behavior stated before for an isotropic elastic mass may be applied to study a stratified subsoil. Hence, from the basic equation 11-XII.3, we can write

$$\Delta T_i = \frac{4d_i}{(v_S)_i} \qquad \text{14-XII.3}$$

where $(v_S)_i$ is the average shear wave velocity in a stratum of thickness d_i. The quantity ΔT_i represents the fundamental resonance period of one stratum experiencing wave reflection, that is to say, when a shear wave of length $4d_i$ passes through and is reflected in this stratum; higher harmonics may occur for wave lengths $4d_i/(2n - 1)$. The value of ΔT_i also represents the time required for a wave with shear velocity $(v_S)_i$ to travel the stratum with thickness d_i. Other resonant periods may be investigated adding the ΔT_i values for several strata:

$$T_n = 4 \sum_{1}^{i} \frac{d_i}{(v_S)_i} \qquad \text{15-XII.3}$$

The largest resonant periods are obtained by addition of all the soft soil strata confined between the firm base and the ground surface. Other modes of vibration may be important on occasion; they may be investigated by means of equation 15-XII.3. When the soil mass is stratified, the shear waves generated at the firm base traveling upward to the ground surface put into partial resonance a stratum or sum of strata having thicknesses compatible with the wave lengths traveling through and reflected within them. This phenomenon is likely to occur when several soft strata of approximately the same physical characteristics are limited at top and bottom by strata of higher rigidity. The response phenomenon is observed at the ground surface; it is investigated from the relative displacement or pseudoacceleration spectra computed from the ground acceleration obtained in the accelerograms recorded during the earthquake.

To investigate the response acceleration of the ground as a function of the wave period, that is, the ground acceleration spectrum, consider the ground as an oscillator with different degrees of freedom subjected to wave periods from shear waves with source at the firm base. Therefore, the peaks in the structural response spectra also correspond to the resonance or amplification produced by wave periods corresponding to wave lengths compatible with the thickness of the soft soil strata. To find the ground acceleration spectrum, the assumption is made that the amplification factor for a certain fraction of critical damping is the same for the ground as an oscillator as for structural one-degree-of-freedom oscillators. Hence, we define the amplification factor for the relative displacement spectrum, considering the ground as an oscillator, by

$$\frac{(x - y)_{gr}}{y_B} = \frac{(S_d)_{gr}}{y_B} = \psi_{gr}$$

in which y_B is the firm base maximum displacement. For the one-degree-of-freedom structural oscillators

$$\frac{(x - y)_{str}}{y_{gr}} = \frac{S_d}{y_{gr}} = \psi_{str}$$

where y_{gr} is the maximum ground displacement. Assuming $\psi_{gr} = \psi_{str}$, we find that the ground maximum displacement is

$$y_{gr} = \frac{(S_d)_{gr}}{\psi_{str}} \qquad \text{16-XII.3}$$

Multiplying 16-XII.3 by the square of the natural frequency of the particular wave under study, the ground acceleration spectrum may be represented by

$$G_a = \frac{(R_a)_{gr}}{\psi_{str}} \qquad \text{17-XII.3}$$

Therefore, dividing the values of S_d or R_a from the structural response spectra by ψ_{str} in proportion to the maximum wave peak value, we find approximately the maximum ground accelerations as a function of wave period. The critical damping coefficient, however, has to be decided for the soil in question.

The relative surface acceleration vs. wave period shown in Fig. 3-XII.3 for Mexico City may be interpreted as follows: The relative response ground surface acceleration G_a of the soil deposit, assuming 20% critical damping, is practically independent of the wave period in the range of 0.8 to 2.5 sec. For the same value of G_a in a subsoil with 5% critical damping, a wave period on the order of 0.8 sec will require a base acceleration of $a_B = 1.28\ G_a$. Furthermore, assuming the same firm base acceleration a_B, if the subsoil is assumed to have 20% critical damping, the same ground acceleration G_a is obtained for periods between 0.8 and 2.5 sec. However, if the soil has only 5% critical damping the ground response acceleration for wave periods of 0.8 and 2.5 sec has the ratio 0.78. The above discussion has the following practical meaning: If a wave of 2.5 sec period produces a maximum ground surface acceleration of 50 cm/sec^2 in a soil with 5% critical damping, then a wave with 0.8 sec period will produce only a response acceleration on the order of $0.78 \times 50 = 39$ cm/sec^2 at the ground surface. The same reasoning may be applied for other critical damping and wave periods. For the case of Mexico City we find

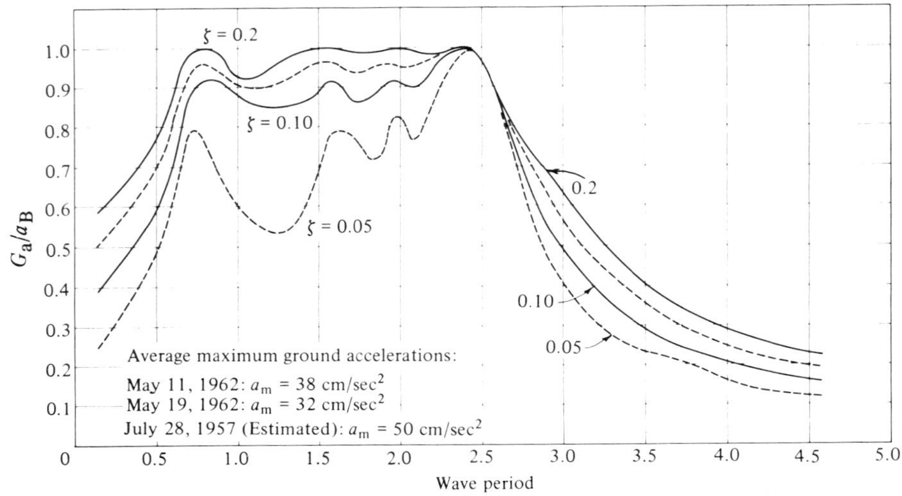

Fig. 3-XII.3 Ground acceleration spectrum, Mexico City center.

from accelerograms 3 and 4-XII.2 that the strongest wave train with about eight waves has nearly the same accelerations for periods ranging from 1.8 to 2.4 seconds. The ground acceleration spectrum shows this to occur for a fraction of critical damping between 10% and 20%, Fig. 3-XII.3. Therefore, it is justified to assign for Mexico City subsoil at the place investigated, an average critical damping coefficient on the order of 15%.

To illustrate the procedure to determine the probable periods in a stratified soft subsoil deposit, let us take the case history of Mexico City. The values of μ and v_s and stratigraphical characteristics of the subsoil are reported for central part of Mexico City in Table 1-XII.3. From these values, Table 2-XII.3 has been calculated from expression 15-XII.3. The periods investigated with the aid of the stratigraphical and dynamical properties of the materials are given in Table 3-XII.3.

Other combinations of strata may be investigated; however, from the stratigraphical point of view, those given in Table 3-XII.3 appear to be the most important. In fact, we note from the pseudoacceleration spectra (Fig. 13-XII.2) that the values given in Table 3-XII.3 agree fairly well with the peaks shown for the site investigated. The ground accelerations corresponding to the periods calculated may be obtained for critical damping of 15% from the ground acceleration spectrum (Fig. 3-XII.3) for a maximum average ground acceleration of 44–35 cm/sec^2. Notice that the accelerations obtained from the ground acceleration spectrum shown in Fig. 3-XII.3 also agree approximately with those of the accelerograms for the same periods in the strongest wave train (Fig. 3-XII.2). Moreover, the foundation engineer is interested in estimating the maximum horizontal relative displacements produced in the soil mass because of the strong ground motion.

Assume the average relative displacement of an element i of unit area and thickness d_i is $(\delta_i - \delta_{i+1})$ (Fig. 4-XII.3). The average unit shear is $(\frac{1}{2})(\tau_i + \tau_{i+1})$. There-

TABLE 1-XII.3 Engineering Physical Properties of Subsoil Materials

Depth	Classification	S_s	w_n, %	γ, ton/m³	μ, ton/m²
5.55	Fill	2.60	55	1.66	> 1000
9.15	Becerra	2.60	85	1.50	> 2000
15.80	Silty clay I	2.35	300	1.17	280
16.50	Clayey silty sand	2.60	42	1.77	700
19.80	Silty clay II	2.35	320	1.16	360
23.65	Silty clay III	2.35	200	1.24	470
27.20	Silty clay IV	2.35	273	1.18	470
29.10	Silty clay V_a	2.35	248	1.20	720
33.50	Silty clay V_b	2.35	300	1.17	720
38.20	Tarango sand I	2.60	42	1.77	> 4000
41.55	Tarango silty clay I_a	2.40	174	1.27	980
41.95	White volcanic glass	2.26	30	1.75	> 2000
45.25	Tarango silty clay I_b	2.40	174	1.27	1120
47.70	Tarango clayey sand I_c	2.40	168	1.28	2500
>47.70	Tarango sand II	2.65			>10000

Depth numbers indicate bottom of layer

S_s–specific gravity
w_n–natural water content
γ–unit weight
μ–rigidity of soil

TABLE 2-XII.3

Depth, m	d_i, m	γ, ton/m³	ρ, ton sec²/m⁴	μ, ton/m²	v_s m/sec	ΔT, sec
5.55	5.55	1.66	0.169	1000	76.87	0.289
9.15	3.60	1.50	0.153	2000	114.37	0.126
11.90	2.75	1.17	0.119	280	48.45	0.227
15.80	3.90	1.17	0.119	280	48.45	0.322
16.50	0.70	1.77	0.180	700	62.29	0.045
19.80	3.30	1.16	0.118	360	55.18	0.239
21.40	1.60	1.24	0.126	470	60.98	0.105
23.65	2.25	1.24	0.126	470	60.98	0.148
27.20	3.55	1.18	0.120	470	62.51	0.227
29.10	1.90	1.20	0.122	720	76.72	0.099
33.50	4.40	1.17	0.119	720	77.70	0.227
38.20	4.70	1.77	0.180	4000	148.89	0.126
41.55	3.35	1.27	0.129	980	87.00	0.154
41.95	0.40	1.75	0.178	2000	105.88	0.015
45.25	3.30	1.25	0.127	1120	93.75	0.141
47.70	2.45	1.28	0.130	2500	138.42	0.071

T, sec

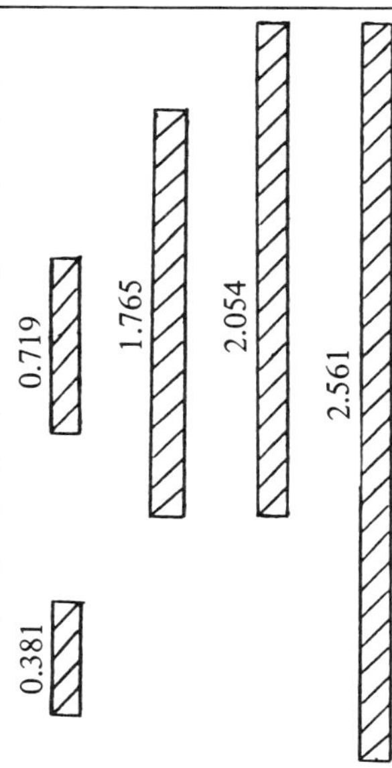

Indicates layers included in computation of T

TABLE 3-XII.3

Strata	Free Periods T	$\xi = 15\%$ G_a
1. From firm base at depth 47.70 m to ground surface	2.56	44–35
2. From hard layer at depth 33.50 m to ground surface	2.05	42–33
3. Between hard layer at depth 33.50 m and hard crust to 9.15 m	1.76	42–33
4. At the softest part of the deposit between depths 16.8 to 27.2 m	0.72	40–32
5. In silty clay deposit between hard layers depths 38.2 and 47.7 m	0.38	26–21
	sec	cm/sec^2

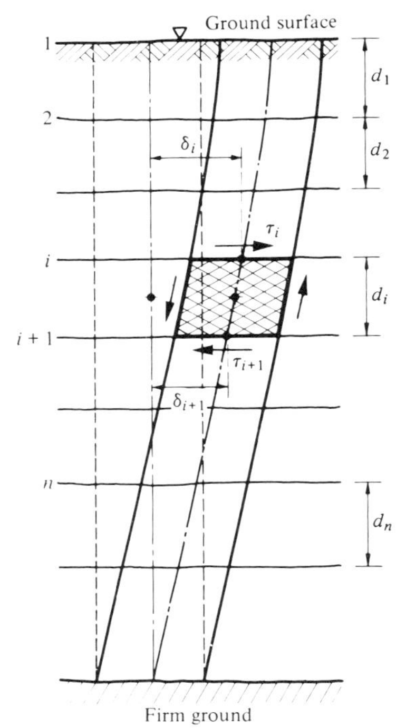

Fig. 4-XII.3 Horizontal relative displacement of an element.

fore, the angular shear distortion of the element is

$$\frac{\delta_i - \delta_{i+1}}{d_i} = \frac{\tau_i + \tau_{i+1}}{2\mu} \qquad \text{18-XII.3}$$

On the other hand, the maximum inertia force in the same element is

$$\tau_{i+1} - \tau_i = (\rho d_i)\omega_n^2 \cdot \tfrac{1}{2}(\delta_i + \delta_{i+1}) \qquad \text{19-XII.3}$$

in which ω_n is the circular frequency of the vibration producing the magnification effect in the stratified deposit of total thickness D_n. Combining the mechanical and dynamical conditions, respectively, we obtain the algorithms for the computation of the maximum horizontal displacements and the corresponding shear stresses in the soil mass for the ground motion induced by the shear waves:

$$\delta_{i+1} = A_i \delta_i - B_i \tau_i \qquad \text{20-XII.3}$$

$$\tau_{i+1} = C_i(\delta_i + \delta_{i+1}) + \tau_i \qquad \text{21-XII.3}$$

in which the coefficients have the following values:

$$A_i = \frac{1 - N_i}{1 + N_i}, \qquad B_i = \frac{1}{(1 + N_i)} \cdot \frac{d_i}{\mu_i}$$
$$C_i = \tfrac{1}{2}\rho d_i \cdot \omega_n^2, \quad N_i = \frac{\rho \cdot d_i^2 \cdot \omega_n^2}{4\mu_i} \qquad \text{22-XII.3}$$

From the value of N_i, it may be recognized that if d_i is made small $(1 + N_i) \doteq 1$, and for most practical purposes $A_i = 1$ and $B_i = d_i/\mu$. The computation of 20 and 21-XII.3 follows from the surface of the ground where boundary conditions are known: $\delta_i = \delta_{s0}$ and $\tau_i = 0$. The value $\delta_{s0} = G_{an}/\omega_n^2$, where G_{an} is calculated from the ground acceleration spectrum for wave period T_n. Hence, dividing the subsoil into small elements of thickness d_i, the integration follows step by step from the ground surface down, using equation 20-XII.3 to compute δ_{i+1}, and then computing from equation 21-XII.3 the value of τ_{i+1}. Thereafter, the values of δ_{i+2} and τ_{i+2} are calculated for the bottom of next layer d_{i+1}. The integration continues as far as the firm stratum, where the relative displacement should be zero, and the shear stress reaches its maximum value. If a value different from zero is found for the relative displacement at the interface of firm ground and soft soil deposit, then ω_n should be corrected to a value satisfying the boundary conditions. This method may be used also to compute the ω_n-values for the higher harmonics; it may be performed speedily with a computer.

To illustrate the step by step procedure of calculation with algorithms 20 and 21-XII.3, let us take the previous example (Table 2-XII.3), in which a value of the largest wave period producing resonance was found to be 2.56 sec; hence using as first approximation $\omega_1 = 2.42$ rad/sec and a maximum ground acceleration of 100 cm/sec^2, then $\delta_{s0} = 100/(2.42)^2 = 17.0$ cm. The numerical calculation may be

TABLE 4-XII.3 Example: Alameda Park. Relative Displacements and Shears Base at 47.70 m—First Mode

$\omega = 2.445\ \text{sec}^{-1}$ $\quad T = 2.57$ sec

Depth, m	d_i, m	ρ, ton sec^2/m^4	μ, ton/m^2	N_i, $\times 10^{-3}$	A_i	B_i, $\times 10^{-3}$	C_i	δ_i, m	τ_i, ton/m^2
0	0							0.167	0
5.5	5.5	0.169	1000	7.77	0.984	5.51	2.80	0.164	0.93
9.15	3.60	0.153	2000	1.48	0.997	1.80	1.65	0.162	1.47
11.90	2.75	0.119	280	4.80	0.990	9.77	0.98	0.146	1.77
15.80	3.90	0.119	280	9.66	0.981	13.80	1.39	0.119	2.14
16.50	0.70	0.18	700	0.19	1.000	1.0	0.38	0.117	2.23
19.80	3.30	0.118	360	5.33	0.989	9.12	1.16	0.095	2.47
21.40	1.60	0.126	470	1.03	0.998	3.40	0.60	0.087	2.58
23.65	2.25	0.126	470	2.03	0.996	4.78	0.85	0.074	2.72
27.20	3.55	0.120	470	4.81	0.990	7.52	1.27	0.053	2.88
29.10	1.90	0.122	720	0.91	0.998	2.64	0.69	0.045	2.95
33.50	4.40	0.119	720	4.78	0.990	6.08	1.57	0.027	3.06
38.20	4.70	0.180	4000	1.49	0.997	1.17	2.53	0.023	3.19
41.55	3.35	0.129	980	2.21	0.996	3.41	1.29	0.012	3.24
41.95	0.40	0.178	2000	0.02	1.000	0.20	0.21	0.012	3.24
45.25	3.30	0.127	1120	1.85	0.996	2.94	1.25	0.002	3.26
47.70	2.45	0.130	2500	0.47	0.999	0.98	0.95	0.000	3.26

$$\delta_{i+1} = A_i\delta_i - B_i\tau_i, \quad \tau_{i+1} = C_i(\delta_i + \delta_{i+1}) + \tau_i$$

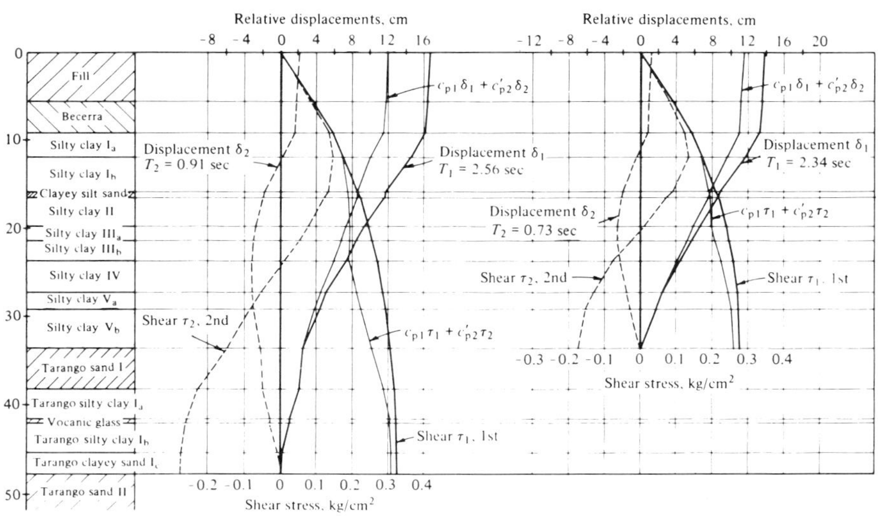

Fig. 5-XII.3 Relative displacements and shear stresses for 100 gal at the ground surface.

found in Table 4-XII.3 for the final results. The results obtained for the first and second modes of vibration calculated using the same procedure are plotted in Fig. 5-XII.3.

3.3 Contribution of Vibration Modes in the Ground Motion

To study the effect of the different vibration modes of the subsoil on ground motion, it is necessary to investigate the participation of each one of them, considering the maximum potential energy accumulated during vibration of one mode equal to the kinetic energy producing such vibration:

$$\text{P.E.} = \tfrac{1}{2}\, gA_n \Sigma \rho d_i \delta_{ni} \qquad \text{23-XII.3}$$

$$\text{K.E.} = \tfrac{1}{2}\, A_n^2 \omega_n^2 \Sigma \rho d_i \delta_{ni}^2$$

here ρ is the unit mass of the soil and δ_{ni} the vibration configuration for a specific mode, with participation $A_n \delta_n$. The factor A_n may be interpreted as a coefficient measuring the fraction of the potential energy necessary to produce the dynamic response of the mode considered; therefore

$$A_n = \frac{g}{\omega_n^2} \cdot \frac{\sum_0^D \rho_i d_i \delta_{ni}}{\sum_0^D \rho_i d_i \delta_{ni}^2} \qquad \text{24-XII.3}$$

The total shear force per unit area at the firm base is

$$V_{n\mathrm{B}} = A_n \sum_0^D \rho_i d_i \omega_n^2 \cdot \delta_{ni} \qquad \text{25-XII.3}$$

and substituting the value of A_n into 24-XII.3, we obtain

$$V_{n\mathrm{B}} = g \frac{\left(\sum_0^D \rho_i d_i \delta_{ni} \right)^2}{\sum_0^D \rho_i d_i \delta_{ni}^2} \qquad \text{26-XII.3}$$

On the other hand, if the total base shear force per unit surface may be defined as

$$V_{n\mathrm{B}} = (c_{\mathrm{p}})_n\, g \sum_0^D \rho_i d_i \qquad \text{27-XII.3}$$

then the participation of the n-th mode will be:

$$(c_p)_n = \frac{\left(\sum_0^D \rho_i d_i \delta_{ni}\right)^2}{\left(\sum_0^D \rho_i d_i \delta_{ni}^2\right)\left(\sum_0^D \rho_i d_i\right)} \qquad 28\text{-XII.3}$$

By means of this expression the coefficients c_p may be calculated. Their values for the first and second vibration modes in the case of Mexico City's subsoil at Alameda Parkare given in Table 5-XII.3.

The total response may be estimated by adding the individual responses for each mode of vibration T_n. Considering that $\Sigma c_{pn} \doteq 1$, we can write for the horizontal displacements at level i - i

$$\delta_i = \left[c_{p1}\,\delta_{s1} + c_{p2}\,\delta_{s2} + \ldots\ldots + \left(1 - \sum_1^{n-1} c_{pn-1}\right)\delta_{sn}\right]_i$$

and in the same manner for the shear stresses

$$\tau_i = \left[c_{p1}\,\tau_1 + c_{p2}\,\tau_2 + \ldots\ldots + \left(1 - \sum_1^{n-1} c_{pn-1}\right)\tau_n\right]_i \qquad 29\text{-XII.3}$$

The investigations show, however, that for practical purposes, only the first and second modes may be considered when calculating displacements and shear stresses in the soil mass; therefore, the value of c_{p2} may be taken as $(1 - c_{p1}) = c'_{p2}$. When this is the case, it only remains to calculate c_{p1}. The ground surface acceleration spectrum may be used to assign the maximum ground accelerations of the vibration modes considered. The higher responses are obtained for the full thickness of the deposit. For illustration purposes, the relative displacements and shear stresses for the stratigraphy found in central Mexico City are reported in Fig. 5-XII.3. They have been calculated on the basis of a ground surface acceleration of 100 cm/sec^2.

The determination of the shear stresses in the soil mass may be applied to investigate the stability of foundations. The shear stresses produced by the strong ground motion are added to the shears in the subsoil due to the static loads of the building (Chapters III to X) and compared with the shear strength of the soil under the foundation of the building. During strong ground motions, buildings founded in difficult subsoil conditions, in which the soil mass is heavily stressed to values close to the ultimate shear strength of the soil, have continued to settle after the ground

TABLE 5-XII.3

Base	T_1	c_{p1}	T_2	$(1 - c_{p1})$
Tarango Arena I	2.34	0.80	0.73	0.20
Tarango Arena II	2.56	0.67	0.91	0.33

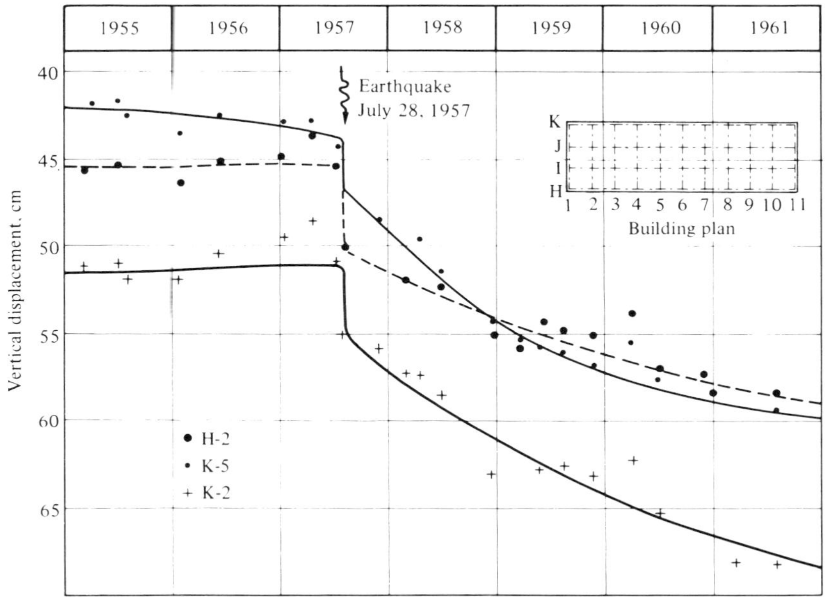

Fig. 6-XII.3 Effect of an earthquake on the settlement of a building.

motion has taken place. A case history is shown in Fig. 6-XII.3 for a heavy building placed on an undercompensated foundation, from which we see that the building, after being practically stabilized, started to settle strongly after the earthquake. The investigation of this case demonstrated that the static shear stresses added to the dynamic shears induced during the earthquake of July 1957 reached the ultimate shear strength of the silty clay deposit; therefore, partial damage took place in the soil skeleton structure, giving as a result an increase in the compressibility of the material that produced the consolidation phenomenon observed.

Finally, it is important to mention that the value of the shear modulus of elasticity should be determined, as accurately as possible, in good undisturbed soil samples. The tests should be performed under the confining stresses to which the soil is subjected in the ground at the depth the specimen is taken.

3.4 Problems Induced by Longitudinal Waves

The investigation of irrotational or longitudinal waves may be of importance to the foundation engineer in difficult subsoil conditions, such as in cohesionless fine saturated soil sediments. When the epicenter is close to the site under consideration and the hypocenter is deep, the vertical ground accelerations produced by these waves are important in the bracket of destructive earthquakes. The longitudinal waves, because of their characteristics of compression and dilatation, induce pore pressures in the saturated soil. The seismic pore water pressures reduce the shear strength of the soil, and consequently reduce the bearing capacity, producing a partial or total failure in the supporting subsoil.

In case of very strong ground motions, liquefaction may take place at the ground surface, when the existing effective stress is reached by the seismic pore water pressures according to shear strength equation $s = (p - u) \tan \phi$; $s = 0$ when $p = u$. In some instances when $u > p$, the phenomenon of boiling has been observed at the ground surface in loose fine sand sediments. In the case of road embankments, the stability of the supporting soil under the toe of the slopes may be considerably reduced up to the point of producing partial shear failure (Fig. 7-XII.3). The subsidence at the edge of the embankment develops longitudinal tension cracks damaging the road (Fig. 8-XII.3). In the case of foundations, the bearing capacity is reduced by the same phenomenon and subsidence of the footings takes place, producing large distortions in the structures, and collapse or damage beyond repair (Fig. 9-XII.3). The order of magnitude of the seismic pore water pressures in the soil mass may be analyzed using the equation of motion for longitudinal plane waves:

$$v_d^2 \frac{\partial^2 w}{\partial z^2} = \frac{\partial^2 w}{\partial t^2} \qquad \text{30-XII.3}$$

This differential equation may be solved for an isotropic soil mass in the same manner as the shear wave equation treated in section XII.3.1. Therefore, the vertical displacements during the ground motion are given by

$$w = w_0 \sin \frac{\pi}{2D} z \sin \frac{2\pi}{T_1} \cdot t \qquad \text{31-XII.3}$$

and the wave resonant period is

$$T_1 = \frac{4D}{v_d} \qquad \text{32-XII.3}$$

in which the longitudinal wave velocity v_d may assume two values:

1) for an infinitely long wave confined to zero lateral deformation:

$$v_d = \sqrt{\frac{2(1+\nu)}{\nu_c} \frac{\mu}{\rho}} \qquad \text{33-XII.3}$$

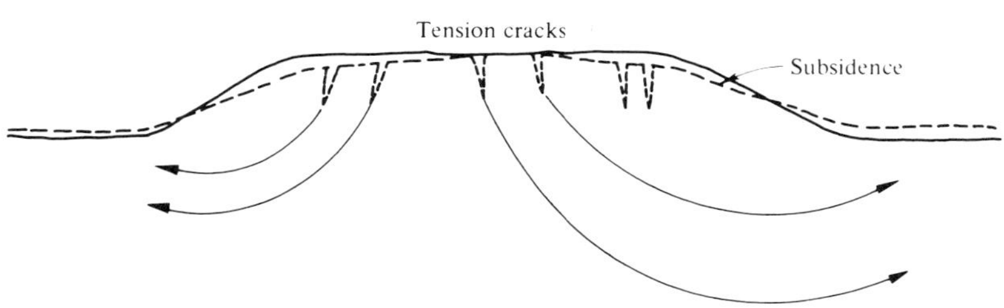

Fig. 7-XII.3 Shear failure in embankment due to earthquake.

Fig. 8-XII.3 Longitudinal tension cracks due to subsidence at edge of embankment.

where the confined modulus of elasticity is

$$E_c = \frac{2(1+\nu)}{\nu_c} \mu$$

2) for a plane wave not confined to zero lateral deformation:

$$v_d' = \sqrt{\frac{2(1+\nu)\mu}{\rho}} \quad \text{34-XII.3}$$

where the linear modulus of elasticity is

$$E = 2(1+\nu)\mu$$

In nature, the longitudinal waves cannot be considered infinite, and confined to zero lateral deformation. Therefore, the velocity v_d' in practice may have an intermediate value between v_d and v_d'. Now assume that it is required to find the seismic pore water pressures induced by the ground motion, and the source of the longitudinal waves is at the firm base at depth D from the ground surface. Therefore, differentiating equation 31-XII.3, the maximum pressure configuration is found as

$$p_z = E_c \frac{\partial w}{\partial z}$$

Fig. 9-XII.3 Tilted water tank due to reduction of the bearing capacity.

hence:

$$p_z = \left(v_d^2 \cdot \rho \, \frac{\pi \cdot w_0}{2D} \right) \cos \frac{\pi}{2D} \cdot z \qquad \text{35-XII.3}$$

Calling p_D the pressure amplitude at the firm base for $z = 0$, the stress configuration is given by

$$p_z = p_D \cos \frac{\pi}{2D} \, z \qquad \text{36-XII.3}$$

The value in the brackets of formula 35-XII.3 representing p_D may be calculated, considering that the maximum vertical surface displacement is generated by the maximum vertical response ground acceleration G_{av}; hence, from $\partial^2 w / \partial t^2$ in equation 31-XII.3 for $z = D$ and $t = T/4$, we obtain

$$w_0 = \frac{G_{av}}{(2\pi)^2} \cdot T_1^2 = \frac{G_{av}}{(2\pi)^2} \frac{16D^2}{v_d^2} \qquad \text{37-XII.3}$$

Moreover, substituting into 35-XII.3 the value of w_0, the pressure amplitude at the firm base is obtained as

$$p_D = \frac{2}{\pi} G_{av} \cdot D \cdot \rho \qquad \text{38-XII.3}$$

It is interesting to note from 38-XII.3 that the pressure amplitude is independent of the longitudinal wave velocity. When the material is considered saturated, the total pressures induced in the soil are equal to the seismic pore water pressures $p_z = u_z$. Therefore, the maximum seismic pore water pressure configuration is

$$u_z = \left(\frac{2}{\pi} G_{av} D \cdot \rho\right) \cos \frac{\pi}{2D} \cdot z \qquad \text{39-XII.3}$$

The intergranular pressures will be neutralized at the ground surface $z = D$ when the effective stress gradient represented by the submerged unit weight $\bar{\gamma}$ of the cohesionless sediment is

$$\left(\frac{\partial u_z}{\partial z}\right)_{z=D} \geqslant \bar{\gamma} \qquad \text{40-XII.3}$$

From formula 39-XII.3, this condition is obtained when the response ground acceleration is given by

$$G_{av} = \frac{\bar{\gamma}}{\rho} = \frac{\bar{\gamma}}{\bar{\gamma} + \gamma_w} g \qquad \text{41-XII.3}$$

where γ_w is unit weight of water.

In this case the upper part of the soil deposit enters into instantaneous liquefaction, exhibited by a loose cohesionless fine sediment with a saturated unit weight of 1.5 ton/m^3 when the vertical ground surface acceleration is on the order of $G_{av} = g/3$. Nevertheless, the supporting capacity of the soil may be reduced with much less ground acceleration. Assume a footing of width $2B$ placed at certain depth d from the ground surface. The water table is close to the ground surface, and according to the conventional formula the ultimate bearing capacity is

$$q_d = \bar{\gamma}[dN_q + 0.6\, BN_\gamma](D_r + 0.1) \qquad \text{42-XII.3}$$

in which the bearing capacity of a cohesionless soil is shown to be an important function of the submerged unit weight of the soil; if this is altered during the ground motion the capacity of the footing will be reduced. Assume the factor of safety used was of two; if $\bar{\gamma}$ is reduced by half, the footing will fail. From equation 40-XII.3 we can write the effective unit weight $\bar{\gamma}_e$ due to the reduction induced by the ground motion:

$$\bar{\gamma}_e = \bar{\gamma} - \Delta\bar{\gamma} = \bar{\gamma} - G_{av}\rho \sin \frac{\pi(D-d)}{2D} \qquad \text{43-XII.3}$$

from which the ground acceleration necessary to produce failure of the footing is given by

$$G_{av} = \frac{\bar{\gamma} - \bar{\gamma}_e}{\rho \cdot \sin \frac{\pi}{2D} (D - d)} \qquad \text{44-XII.3}$$

If in equation 42-XII.3 the submerged unit weight is reduced to $\bar{\gamma}_e = \bar{\gamma}/2$, at which the footing will suffer subsidence, then

$$G_{av} = \frac{\bar{\gamma} g}{2(\bar{\gamma} + \gamma_w) \sin \frac{\pi}{2D} (D - d)}$$

For the unit weight used before, and assuming a ratio of footing depth to deposit thickness of $d/D = 0.1$, the ground acceleration causing failure will be on the order of $g/6$.

This example demonstrates the importance of the seismic longitudinal waves in stability foundation problems, when these waves arrive from a firm base close to the ground surface. The problem is more critical, however, when the soil is stratified with very loose sediments at the upper part of the deposit. When the surface is loaded as in case of an embankment or a foundation loading the ground surface, the induced seismic pore pressures may reduce the stability to the point that plastic movements occur. The order of magnitude of the induced seismic pore water pressure in a stratified subsoil may be analyzed using the step by step procedure described for calculation of shear wave problems. Hence, the following algorithms may be used:

$$\begin{aligned} w_{i+1} &= \bar{A}_i w_i - \bar{B}_i p_i \\ p_{i+1} &= \bar{C}_i (w_i + w_{i+1}) + p_i \end{aligned} \qquad \text{45-XII.3}$$

where w are the relative vertical displacements, and the values of p represent the total pressures. The coefficients in 45-XII.3 have the values:

$$\bar{A}_i = \frac{1 - \bar{N}_i}{1 + \bar{N}_i}, \qquad \bar{B}_i = \frac{1}{1 + \bar{N}_i} \cdot \frac{d_i}{(v_d^2 \rho)}$$

$$\bar{C}_i = \frac{1}{2} \rho d_i \omega_n^2, \qquad \bar{N}_i = \frac{d_i^2 \cdot \omega_n^2}{4 v_d^2}$$

in longitudinal wave problems, usually $1 + \bar{N}_i \doteq 1$; hence $\bar{A}_i = 1$ and $\bar{B}_i = d_i/\rho v_d^2$.

The integration is performed from the base of the loaded area where the vertical ground response acceleration G_{av} may be assigned. Hence, at its base the initial pressure exerted by the loading mass m_0 per unit surface is $p_0 = m_0 G_{av}$; assuming a dominant period of the soil deposit including the loading mass, the initial vertical displacement may be calculated by means of

$$w_0 = \frac{G_{av}}{\omega_1^2} \qquad \text{46-XII.3}$$

The dominant period of the soft ground determined by the wave velocity method has the value

$$T_1 = 4 \sum_{1}^{n} \frac{d_i}{(v_d)_i} \qquad \text{47-XII.3}$$

This value may be used for the first cycle of calculations. The deposit is divided into layers of thickness d_i and the values of $\overline{A}_i, \overline{B}_i, \overline{C}_i$ and $\overline{N}_i$ are calculated for each one of them. The integration is performed step by step with algorithms 45-XII.3, to the firm base where the relative vertical displacement is zero and the vertical pressure reaches its maximum value. Nevertheless, the loaded surface imposes boundary conditions that change the circular frequency of the deposit. Therefore, the value of ω is improved in the calculations until the boundary condition at the base of the deposit is satisfied.

The total pressures thus obtained below the water table are equal to the seismic pore water pressures for a saturated sediment. They may be deducted from the initial effective stresses in the ground, and used to analyze the stability of the foundation along the potential surface of sliding under the loaded area.

3.5 Reduction of the Shear Strength and Bearing Capacity Because of Seismic Action

Fine cohesionless saturated semi-impervious sediments like fine sand and silty sand with mean diameter less than $D_{50} = 0.3$ mm, void ratio on the order of 0.8, and unit volume compressibility larger than $m_v = 0.015$ cm^2/kg, may be considered of "liquefaction" characteristics. An important reduction in the bearing capacity may be expected because of high pore water pressures induced by the seismic compressional and shear waves. The safe bearing capacity may be lost, causing the foundation to begin to sink into the ground, or reduced to such an extent that the building will tilt on one side of the foundation, because of the dual effect of the seismic overturning moment, and the reduction of the bearing capacity at the edges of the building foundation (Fig. 9 and 10-XII.3).

The problem may be approached from the practical engineering point of view assuming limiting conditions. Let us investigate the reduction of the shear strength s_l and the apparent angle of internal friction ϕ_l of the sediment because of an induced seismic pore water pressure u_{sis}. We call σ_{01} and σ_{03} the vertical and horizontal effective stresses at rest; $K_0 = \sigma_{03}/\sigma_{01}$. Furthermore, we assume in the range of the true angle of internal friction ϕ_d the following expression is valid:

$$K_0 = 1 - \sin \phi_d, \text{ (Jàky)} \qquad \text{48-XII.3}$$

The stress circle A represents the initial condition as shown in Fig. 11-XII.3. The in situ confining stress may be taken as follows: $\sigma_{0c} = \frac{1}{3}(1 + 2K_0)\sigma_{0i}$, and using 48-XII.3 we obtain

$$\sigma_{0c} = (1 - \tfrac{2}{3} \sin \phi_d)\sigma_{0i}$$

Fig. 10-XII.3 Tilting of building because of reduction in bearing capacity. *Courtesy of Lazarus White.*

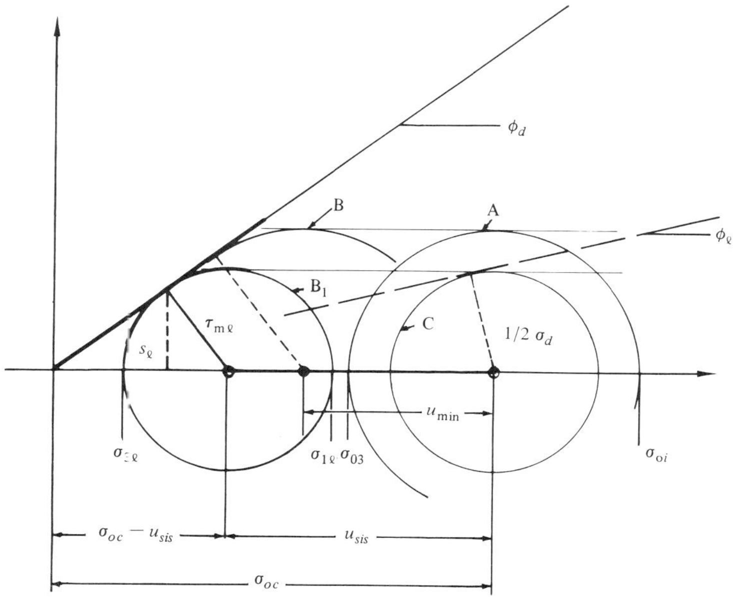

Fig. 11-XII.3 State of stress due to induced seismic pore water pressure.

The failure of the soil sediment may be obtained when the seismic pore water pressure u_{sis} induced by the strong ground motion is sufficiently large to make the effective stresses, circle B, meet the Coulomb-Mohr condition. Nevertheless, the pore water pressure should reach the values

$$u_{sis} \geqslant \tfrac{1}{2}(\sigma_{01} + \sigma_{03}) - \tfrac{1}{2}(\sigma_{01} - \sigma_{03})/\sin\phi_d$$

or

$$u_{min} = \tfrac{1}{2}(1 - \sin\phi_d)\sigma_{01} \qquad \text{49-XII.3}$$

The maximum seismic shear τ_l and shear strength s_l of the soil when the pore water pressure rises to $u_{sis} > u_{min}$, Fig. 11-XII.3, may be obtained from the Coulomb-Mohr condition circle B_1:

$$\tau_l = (\sigma_{0c} - u_{sis})\sin\phi_d$$

or

$$\frac{\tau_l}{\sigma_{0c}} = \left(1 - \frac{u_{sis}}{\sigma_{0c}}\right)\sin\phi_d \qquad \text{50-XII.3}$$

and the seismic shear strength

$$s_l = (\sigma_{0c} - u_{sis})\sin\phi_d\cos\phi_d$$

or

$$\frac{s_l}{\sigma_{0c}} = \left(1 - \frac{u_{sis}}{\sigma_{0c}}\right) \frac{1}{2} \sin \phi_d \qquad \text{51-XII.3}$$

From above expressions the conclusion is reached that the cohesionless sediment becomes fluid when $u_{sis} = \sigma_{0c}$. Any other value of $u_{sis} \leqslant K_0 \sigma_{01}/2$ or $u_{sis} \leqslant \frac{1}{2} \sigma_{03}$ will only reduce the shear strength of the soil.

From expression 51-XII.3 it may be recognized that to estimate the shear strength of the sediment during the seismic ground motion it is necessary to calculate the most probable pore water pressure ratio u_{sis}/σ_{0c} induced in the soil deposit due to the earthquake. This calculation may be performed from a practical point of view as explained in further paragraphs.

On many occasions it is necessary to estimate the safe bearing capacity of footings making use of the orthodox bearing capacity formula (Chapter III.4). For this purpose one is compelled to know the seismic apparent angle of internal friction ϕ_{sis} based on applied stresses. Hence from Fig. 11-XII.3, we obtain

$$\text{for circle B,} \quad \sin \phi_d = \frac{\sigma_d/2}{\sigma_{0c} - u_{sis}}$$

$$\text{for circle C,} \quad \sin \phi_l = \frac{\sigma_d/2}{\sigma_{0c}}$$

From above expressions we write, $\phi_l = \phi_{sis}$ and

$$\frac{\sin \phi_{sis}}{\sin \phi_d} = \left(1 - \frac{u_{sis}}{\sigma_{0c}}\right) \qquad \text{52-XII.3}$$

Here we noticed again that the apparent angle of internal friction vanishes when $u_{sis} = \sigma_{0c}$. The value σ_{0c}/u_{sis} may be considered a factor of safety for total liquefaction.

The accuracy of estimating the values of the seismic shear strength s_l and the apparent angle of internal friction ϕ_{sis} will depend on how precise the induced seismic pore water pressure u_{sis} may be estimated in the soil deposit during the strong ground motion. At present, no information is available concerning measurements of pore water pressures in the field induced by seismic waves that may be used to calibrate theoretical or semi-empirical concepts. We are compelled therefore, to perform a theoretical determination based on practical and simple working hypotheses.

Let us assume a uniform loose fine sand sediment supported on firm ground to a depth D. Furthermore, at the interface of soft soil and firm ground compressional and shear waves are developed traveling in the vertical direction to the ground surface. The waves traveling with wave velocity v have different lengths and periods and add to build up the maximum seismic pore water pressure. Considering the high damping of a soil mass of this type it is reasonable to consider only the first vibration modes of the soft soil deposit as important.

In order to establish a procedure to estimate the seismic pore water pressure, let us consider the soil deposit to be isotropic to full depth D. Under this condition we may study the effect of the compressional and shear waves separately. The equation of motion of the vertical component for these body waves may be written as follows:

$$v^2 \frac{\partial^2 \delta}{\partial z^2} = \frac{\partial^2 \delta}{\partial t^2} \qquad \text{53-XII.3}$$

in which v is the wave velocity, either for the shear wave v_s or for the compressional wave v_d, and δ correspondingly represents the horizontal or vertical displacement.

The solution of equation 53-XII.3 for the relative displacements is of the form

$$\delta_{nz} = \delta_{on} \cos (2n - 1) \frac{\pi}{2D} z \cdot \sin \left\{2\pi(2n - 1) \frac{t}{T_1}\right\} \qquad \text{54-XII.3}$$

The amplitude of the displacement for one mode of vibration is obtained when $t = T_1/4(2n - 1)$, where T_1 is the period of the fundamental mode for $n = 1$. The participation for each one of the modes may be found to be, for an isotropic soil mass: $c_{pn} = 8/(2n - 1)^2 \pi^2$ (Chapter XII.3.3). Let us consider a shear wave producing horizontal relative displacements u_{nz} in the n-th mode, thus,

$$u_{nz} = c_{pn}\bar{u}_{on} \cdot \cos (2n - 1) \frac{\pi}{2D} z \qquad \text{55-XII.3}$$

Here $\bar{u}_{on}$ is the amplitude of the motion at the ground surface defined by $\bar{u}_{on} = -a_m/\omega_n^2$. The value a_m is the maximum horizontal response ground surface acceleration observed during the earthquake, and ω_n is the circular frequency for that particular mode. Therefore, we write:

$$u_{nz} = -c_{pn} \frac{a_m}{\omega_n^2} \cdot \cos (2n - 1) \frac{\pi}{2D} \cdot z$$

The shear stress will be $\tau = \mu(\partial u/\partial z)$, in which μ is the soil rigidity. Hence,

$$\tau_{nz} = c_{pn} \frac{(2n - 1)\pi}{2D} \frac{a_m}{\omega_n^2} \mu \cdot \sin (2n - 1) \frac{\pi}{2D} z. \qquad \text{56-XII.3}$$

But, if μ is considered constant through depth $\mu = v_s^2 \rho$, and $v = 4D\omega_n/2\pi(2n - 1)$ (Chapter XII.3.1), and therefore,

$$\mu = \frac{4D^2 \omega_n^2}{\pi^2 (2n - 1)^2} \cdot \rho$$

and expression 56-XII.3 becomes

$$\tau_{nz} = (a_m \rho D) \frac{16}{(2n - 1)^3 \pi^3} \sin \frac{(2n - 1)\pi}{2D} z \qquad \text{57-XII.3}$$

The expression 57-XII.3 may be written in the following convenient form:

$$\tau_{nz} = \frac{a_m}{g} (\gamma_m \cdot z) \left\{ \frac{16}{(2n - 1)^3 \pi^3} \left(\frac{D}{z}\right) \sin \frac{(2n - 1)\pi}{2D} z \right\}$$

where γ_m represents the unit weight of the saturated soil mass, considered constant with depth. Therefore: $\gamma_m \cdot z = p_{oz}$, is the total overburden pressure. Here,

$$\psi_n \left(\frac{z}{D}\right) = \frac{16}{(2n - 1)^3 \pi^3} \left(\frac{D}{z}\right) \sin \frac{(2n - 1)\pi}{2D} z \qquad \text{58-XII.3}$$

and

$$\tau_{nz} = \left(\frac{a_m}{g}\right) p_{oi} \cdot \psi_n \left(\frac{z}{D}\right) \qquad \text{59-XII.3}$$

The maximum shear to be expected in the isotropic soil mass may be obtained from

$$\tau_{sis} = \left(\frac{a_m}{g}\right) p_{oi} \sum_1^n \psi_n \left(\frac{z}{D}\right) \qquad \text{60-XII.3}$$

Let

$$F_1 \left(\frac{z}{D}\right) = \sum_1^n \psi_n \left(\frac{z}{D}\right) \qquad \text{61-XII.3}$$

whence

$$\tau_{sis} = \left(\frac{a_m}{g}\right) p_{oi} \cdot F_1 \left(\frac{z}{D}\right) \qquad \text{62-XII.3}$$

From expression (57-XII.3) it may be seen that the third mode has little significance, therefore considering only the first two modes we write

$$F_1 \left(\frac{z}{D}\right) = \frac{2}{\pi} \frac{D}{z} \left\{ \frac{8}{\pi^2} \sin \frac{\pi}{2D} \cdot z + \left(1 - \frac{8}{\pi^2}\right) \frac{1}{3} \sin \frac{3\pi}{2D} \cdot z \right\}$$

or

$$F_1 \left(\frac{z}{D}\right) = 0.52 \frac{D}{z} \left\{ \sin \frac{\pi}{2D} z + 0.078 \sin \frac{3\pi}{2D} z \right\} \qquad \text{63-XII.3}$$

The value of function $F_1(z/D)$ may be found plotted in Fig. 12-XII.3 for the first two participating modes.

The solution given here for an isotropic subsoil deposit considering the shear wave is the same for the compressional wave, except that for the latter the value of the acceleration used should be the maximum vertical response ground surface acceleration. Calling p_{sis} the pressure induced by the compressional waves we establish the

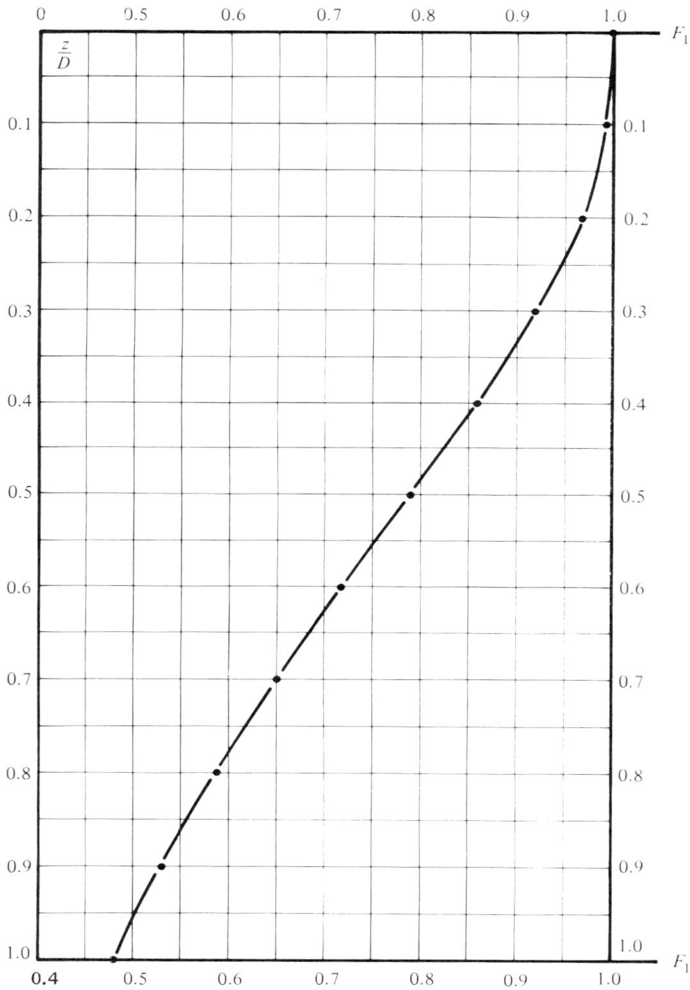

Fig. 12-XII.3 Seismic function $F_1\,(z/D)$.

following ratio:

$$\frac{\tau_{sis}}{p_{sis}} = \frac{G_{ah}}{G_{av}} \qquad \text{64-XII.3}$$

Here G_{ah} and G_{av} are the horizontal and vertical accelerations respectively.

In a loose fine sand of liquefaction characteristics during the short time the high accelerations take place, the saturated soil may be considered not to change in volume. Therefore, the applied stress may be assumed to be taken totally by the pore water pressure. In case of the compressional wave $p_{sis} = u_p$, and for the shear wave the induced diagonal compression equal to τ_{sis} will induce pore water pressures equal to u_τ, since the loose fine sand has the tendency to compact during the shearing action, as shown in Fig. 13-XII.3.

From above reasoning it is concluded that the total pore water pressure induced in the soil during the strongest action of the ground motion will be $u_{sis} = u_\tau + u_p$ and the pore water pressure ratio may be written as follows:

$$\frac{u_{sis}}{\sigma_{0c}} = \frac{G_{av} + G_{ah}}{g} \cdot \frac{p_{0i}}{\sigma_{0c}} \cdot F_1 \left(\frac{z}{D}\right) \qquad 65\text{-XII.3}$$

The maximum ground surface accelerations to be expected in the horizontal and vertical directions, respectively, should be considered with judgement.

In practice, however, a soil deposit is stratified, and the relative density and corresponding true angle of internal friction ϕ_d change with depth, as does the soil rigidity μ which is a function of the confining stress σ_{0c} and the angular distortion. Therefore, a more precise calculation of the seismic pore water pressure implies the knowledge of these two soil parameters. The calculation may be performed with algorithms given in Chapter XII (20 to 22-XII.3). Knowing the index properties of the soil profile and soil rigidity with depth, that is, $\mu = C_s \sigma_{0c}^n$, for the shear wave and $E_d = 2\mu(1 + \nu)$ for the compressional wave, we may calculate for the first two modes of vibration the seismic pore water pressure:

$$u_{sis} = c_{p1}(u_{sis})_1 + (1 - c_{p1})(u_{sis})_2 \qquad 66\text{-XII.3}$$

Equation 66-XII.3 gives the first approximation, since the confining stress σ_{0c} changes to $\bar{\sigma}_{0c} = \sigma_{0c} - u_{sis}$, and, therefore, the soil rigidity μ also changes. To comply with this situation the calculation should be repeated to find new values of u_{sis} at succesive stages until $\bar{\sigma}_{0c}$ remains without an important change. Further refinements on this subject may be obtained if μ is selected according to the shear distortion at which the soil may be expected to be stressed.

The phenomenon explained above will be illustrated by investigating the foundation stability for a building during an earthquake with horizontal maximum ground surface acceleration of 1.5 m/sec^2 and vertical maximum acceleration on the order of 0.75 m/sec^2. Therefore, the total acceleration will be $a_m = 1.5 + 0.75 = 2.25$ m/sec^2.

The building rests on a mat foundation 20 X 20 m, and its weight, including the mat, is 6.5 ton/m^2. The mat is placed at a depth of 0.5 m. The subsoil is a uniform loose silty sand with true angle of internal friction of 28° assumed constant with depth, therefore formulas previously developed apply. The water table is at a depth of 0.5 m. The bearing capacity under static conditions (section III.4.2) is:

a) Total failure $q_d = 1.2(\sigma_{0d} N_q + 0.6\bar{\gamma}_m B N_\gamma)(D_r + 0.1)$

b) Local failure $q_d = 1.2(\sigma_{0d} N_q)(D_r + 0.1)$ 67-XII.3

The bearing capacity factors for $\phi_d = 28°$ are $N_q = 18$, and $N_\gamma = 15$ respectively (Fig. 2-III.4, section III-4.2a). The potential surface of sliding is assumed located

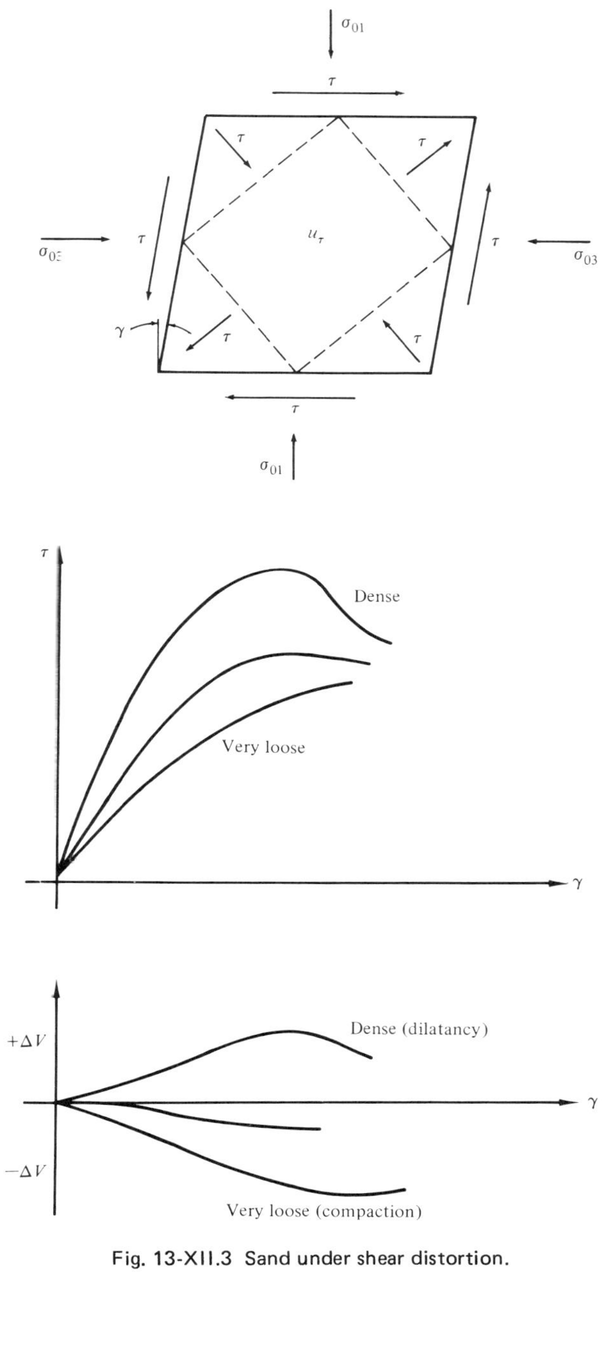

Fig. 13-XII.3 Sand under shear distortion.

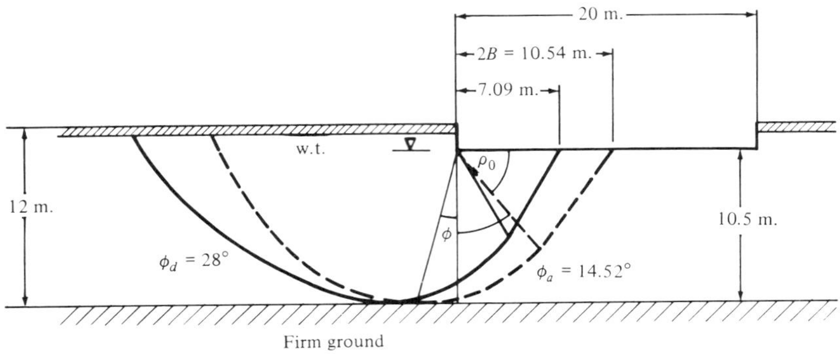

Fig. 14-XII.3 Potential surfaces of sliding for $\phi_d = 28°$ and $\phi_a = 14.52°$ respectively.

tangent to the firm ground at 12 m depth (Fig. 14-XII.3), from which the value of B in 67-XII.3 is found to be 3.54 m. Hence,

$$q_d = 1.2(0.90 \times 1.8 + 0.6 \times 0.80 \times 3.54 \times 15)(0.4 + 0.1)$$

From which we obtain:

a) Local failure 9.72 ton/m^2
b) Total failure at B 25.00 ton/m^2
c) Average ultimate bearing capacity 17.36 ton/m^2

The foundation stability may be considered satisfactory under static loading conditions, with a factor of safety on the order of 17.36/6.5 = 2.67.

To investigate the stability of the foundation under seismic conditions we consider the compressional and shear maximum acceleration waves acting at the same time interval, hence the total acceleration to consider is $a_m/g = 0.225$. Using algorithms 65-XII.3 and Fig. 12-XII.3 we calculate first the seismic pore water pressures u_{sis}, and u_{sis}/σ_{0c}. Using the values of u_{sis}/σ_{0c} due to the seismic action we calculate by means of 52-XII.3 the apparent angle of internal friction ϕ_{sis}, and investigate the foundation stability. The calculations have been tabulated in Table 6-XII.3, and are self explanatory.

The average value of the apparent angle of internal friction is found to be $\phi_{sis} = 14.52°$, for which we find $N_q = 4.0$ and $N_\gamma = 2.0$. The confining stress at 0.5 m depth is $\sigma_{0d} = 0.9$ ton/m^2. The potential surface of sliding for $\phi_{sis} = 14.52°$, tangent to firm ground at 12 m depth gives approximately $B = 5.27$ m (Fig. 14-XII.3). Hence,

$$q_d = 1.2(0.90 \times 4.0 + 0.6 \times 0.80 \times 5.27 \times 2.0)(0.4 + 0.1)$$

we obtain for

a) Local failure 2.16 ton/m^2
b) Total failure 5.20 ton/m^2
c) Average ultimate bearing capacity 3.60 ton/m^2

TABLE 6-XII.3

DEPTH	z/D	ton/m² p_{0i}	ton/m² σ_{0i}	ton/m² σ_{0c}	$F_1(z/D)$	ton/m² u_{sis}	u_{sis}/σ_{0c}	ϕ_{sis}.
ω.t 0	0	0	0	0	–	–	–	–
∇ 0.5	0.042	0.90	0.90	0.62	1.00	0.203	0.327	18.41
1.2	0.100	2.16	1.46	1.00	1.00	0.486	0.486	13.96
1.5	0.125	2.70	1.70	1.17	0.99	0.601	0.514	13.20
2.4	0.200	4.32	2.42	1.66	0.97	0.943	0.568	11.70
3.6	0.30	6.48	3.38	2.32	0.92	1.341	0.578	11.40
4.8	0.40	8.64	4.34	2.98	0.86	1.672	0.561	11.89
6.0	0.50	10.80	5.30	3.64	0.79	1.920	0.527	12.82
7.2	0.60	12.96	6.26	4.30	0.72	2.100	0.488	13.90
8.4	0.70	15.12	7.22	4.96	0.65	2.211	0.466	15.25
9.6	0.80	17.28	8.18	5.62	0.58	2.255	0.440	16.35
10.8	0.90	19.44	9.14	6.28	0.53	2.318	0.369	17.23
12.0	1.00	21.60	10.10	6.94	0.48	2.333	0.336	18.16

$\frac{a_m}{g} = 0.225, \quad \phi_d = 28°, \quad \gamma = 1.8 \text{ ton/m}^3.$ $\quad (\phi_{sis})_{av} = 14.52°$

$$\sigma_{0c} = \left(1 - \frac{2}{3}\sin\phi_d\right)\sigma_{0i}, \qquad \phi_{sis} = \sin^{-1}\left(1 - \frac{u_{sis}}{\sigma_{0c}}\right)\sin\phi_d$$

$$u_{sis} = \left(\frac{a_m}{g}\right) p_{0i} \cdot F_1\left(\frac{z}{D}\right)$$

Therefore, the factor of safety during seismic action is reduced to 3.60/6.5 = 0.57. From above analysis it is concluded that the building foundation will tilt and sink strongly on one side due to the dual effect of the seismic overturning moment and the high reduction of the bearing capacity of the fine silty sand under the foundation structure.

The procedure given in this section to estimate the reduction in bearing capacity due to strong ground motions is considered an approximate method from the practical foundation engineering point of view and is intended to give the foundation engineer a simple tool, that with judgement may be applied to investigate approximately a fine cohesionless soil of "liquefaction" characteristics. The horizontal and vertical response maximum accelerations at the ground surface for the strongest earthquake to occur at the site under study should be carefully assigned.

A discussion of different procedures proposed by other authors to investigate liquefaction of sands is out of the scope of this book. This is a complex problem and the reader is referred to published investigations on the subject of liquefaction by leading authorities in Canada (Linn, et al.), Japan (Ishihara, Yoshimi et al.), and the United States (Seed, Casagrande, Castro et al.).

XII.4 SHEAR MODULUS OF ELASTICITY

4.1 Basic Concepts

The elastic property of subsoil materials in the case of nonconsolidated sediments is important for the solution of dynamic problems in the subsoil. In previous articles, in discussion of the rheological properties of soils it was established that the soil has elasticity accompanied with plasticity and viscosity. Therefore, when the material is subjected to a state of stress, owing to the elastic elements of the soil we find static equilibrium and dynamic response. In the case of motion, the seismic shear waves are transmitted in the subsoil mass only if there is elastic response. The action of the visco-plastic elements, however, is to absorb the kinetic energy of the motion through the phenomenon of damping.

To illustrate the free vibration of an elastic material, consider a mass M coupled with an elastic spring of rigidity K (Fig. 1-XII.4). If we call δ the displacement at a certain time suffered by the mass M under vibratory harmonic motion, the elastic force in the spring will be $K\delta$. On the other hand, the inertia force during the movement will be $M\ddot{\delta}$, in which $\ddot{\delta}$ is the acceleration. Therefore, for dynamic equilibrium one can establish the following condition:

$$M\ddot{\delta} + K\delta = 0 \qquad \text{1-XII.4}$$

On the other hand, if the movement is simple harmonic, this may be represented by

$$\delta = \delta_0 \sin \omega_n \cdot t \qquad \text{2-XII.4}$$

here δ_0 is the amplitude of the movement, and ω_n the circular frequency with period T:

$$\omega_n = \frac{2\pi}{T} \qquad \text{3-XII.4}$$

Substituting into 1 and 2-XII.4, we obtain

$$-M\delta_0\,\omega_n^2 \sin \omega_n t + K\delta_0 \sin \omega_n t = 0$$

from which

$$\omega_n = \sqrt{\frac{K}{M}} \qquad \text{4-XII.4}$$

and the free period of vibration of the system is

$$T = 2\pi\sqrt{\frac{M}{K}} \qquad \text{5-XII.4}$$

Now assume that the system can absorb energy by means of a dash pot (Fig. 2-XII.4). The damping action may be represented by means of a Newtonian element:

$$\frac{\partial \delta}{\partial t} = \phi p$$

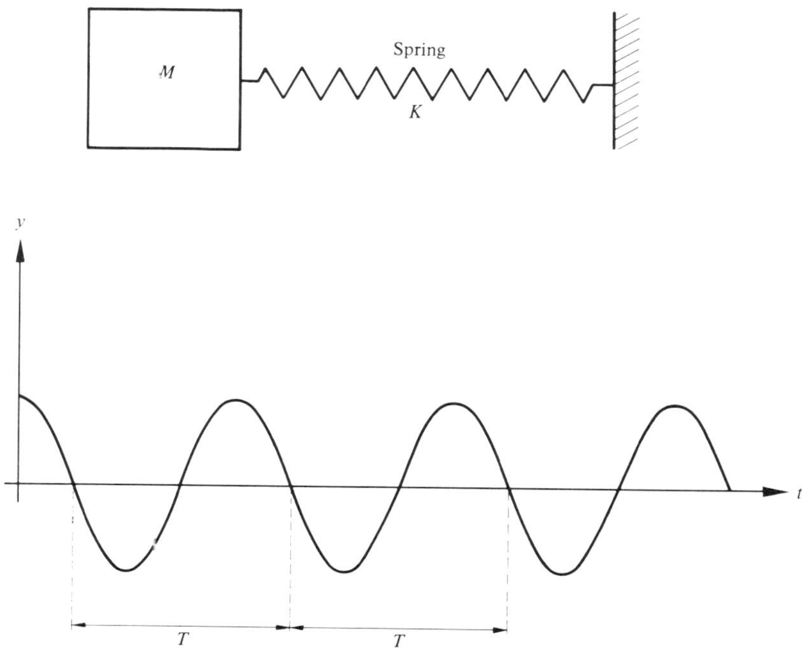

Fig. 1-XII.4 Free vibration of an elastic material.

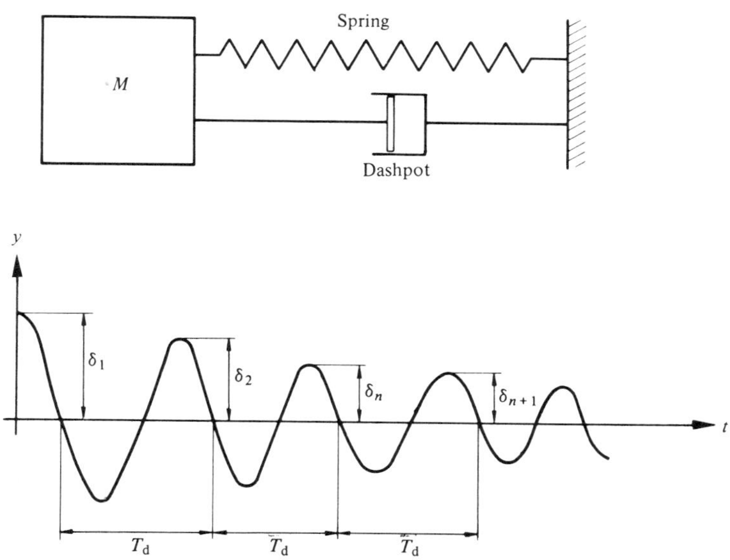

Fig. 2-XII.4 System absorbing energy.

in which the force applied on the viscous element is proportional to the velocity; therefore, we can also write $c \cdot \dot{\delta}$, in which c is a parameter property of the material representing the magnitude of the damping effect. Therefore, the dynamic equilibrium equation in this case is

$$M\ddot{\delta} + c\dot{\delta} + K\delta = 0 \qquad \text{6-XII.4}$$

The solution of this differential equation is a function of the value of c, according to whether this value is larger or smaller than the critical damping expressed by

$$c_c = 2\sqrt{KM} \quad \text{or} \quad c_c = 2M\omega_n \qquad \text{7-XII.4}$$

The ratio $c/c_c = \zeta$ represents a fraction of the critical damping.

If $\zeta < 1$, the equation of motion satisfying the differential equation is

$$\delta = e_N^{-\zeta\omega_n t}\,[A \sin \omega_d t + B \cos \omega_d t] \qquad \text{8-XII.4}$$

in which ω_d is the damped circular frequency and has the value

$$\omega_d = \omega_n \sqrt{1 - \zeta^2} \qquad \text{9-XII.4}$$

When $\zeta = 1$, there is no vibratory motion. If the mass is displaced from its static equilibrium, it will come back without oscillation to its original position.

For $\zeta < 1$, the vibration will be damped with maximum amplitudes according to Fig. 2-XII.4:

$$\delta = A e_N^{-\zeta\omega_n (nT_d)} \qquad \text{10-XII.4}$$

where T_d is the damped period of the vibration. The ratio between successive amplitudes is

$$\frac{\delta_n}{\delta_{n+1}} = \frac{e_N^{-\zeta\omega_n (nT_d)}}{e_N^{-\zeta\omega_n (n+1) T_d}} = e_N^{\zeta\omega_n T_d}$$

but $T_d/T_n = 1/\sqrt{1 - \zeta^2}$; therefore

$$\frac{\delta_n}{\delta_{n+1}} = e_N^{\frac{2\pi\zeta}{\sqrt{1-\zeta^2}}} \qquad \text{11-XII.4}$$

from which we obtain

$$\text{Log}\,\frac{\delta_n}{\delta_{n+1}} = \frac{2\pi\zeta}{\sqrt{1 - \zeta^2}} = \Delta \qquad \text{12-XII.4}$$

The value of Δ is known as the logarithmic decrement, and may be determined from successive amplitudes of the damped vibration (Fig. 2-XII.4). The coefficient of critical damping is

$$\zeta^2 = \frac{\Delta^2}{(2\pi)^2 + \Delta^2} \qquad \text{13-XII.4}$$

4.2 The Free Torsion Pendulum

The knowledge of the elastic properties of nonconsolidated subsoil materials is, at present, an important item of investigation in order to be able to solve seismic problems in the subsoil and earth masses. The determination of the shear modulus of elasticity is basic in the solution of these problems. The use of a simple instrument used by the author in his professional practice is described. The results obtained for different soils are reported at the end of this article, from which one can determine the phenomenological characteristics of the shear modulus of nonconsolidated sediments confined under a state of stress.

The determination of the linear modulus of elasticity E is difficult because volume changes in the specimen require precise determination of pore pressures. Moreover, the value of Poisson's ratio may influence importantly the results when calculating the shear wave velocity from the value of E. Poisson's ratio has the following approximate limiting values in soils: for saturated, nonorganic clay under transient loadings, $\nu = 0.5$; and for pervious coarse and nonsaturated soils, $\nu = 0.25$.

From the body waves study in XII.2, we can write:

1. From the shear wave velocity:

$$v_s^2 \rho = \mu$$

2. From the longitudinal wave velocity confined to zero lateral deformation:

$$v_d^2 \rho = \frac{2(1+\nu)}{\nu_c} \mu = E_c$$

3. From longitudinal wave velocity not confined to zero lateral deformation:

$$(v_d')^2 \rho = 2(1+\nu)\mu = E$$

where E_c is the confined modulus of elasticity equal to E/ν_c, and $E = 2(1+\nu)\mu$. The ratio of wave velocities is

$$\frac{v_d}{v_s} = \sqrt{\frac{2(1-\nu)}{(1-2\nu)}}, \quad \text{and} \quad \frac{v_d'}{v_s} = \sqrt{2(1+\nu)} \qquad \text{14-XII.4}$$

where it may be observed that theoretically for values of $\nu = 0.5$, in case of saturated soils confined to zero lateral deformation when no change in volume takes place, then $v_d/v_s \longrightarrow \infty$. On the other hand, when no volume change takes place, but lateral deformation is permitted, then $v_d'/v_s = 1.73$. For nonsaturated cohesionless materials where $\nu = 0.25$, one obtains that the irrotational wave velocity is on the order of 73%, higher than the shear wave velocity. Furthermore, investigations have shown that a value of $\nu = 0.42$–0.45 may be assigned as a limiting value in cohesive nonconsolidated soils, in which case $v_d/v_s = 2.7$–3.3. In organic silts or clays the material cannot be considered 100% saturated, owing to the presence of a small amount of gas. When the pore water is saturated with gas during vibration some of it is liberated, thus permitting a small change in volume in the saturated soil. The physical properties E/μ, v_d/v_s, v_d'/v_s and ν_c as functions of ν are plotted in Fig. 3-XII.4.

The determination of the value of the shear wave velocity is a function of the shear modulus of elasticity, and since its determination does not imply volume changes, it may be determined with precision. The free torsion pendulum test is the closest representation of the behavior of a soil sediment to this phenomenon. Moreover, during distortions, no excess pore water pressures take place, and thus complicated equipment is not required to determine the rigidity or shear modulus of elasticity. It is only necessary to measure the response of the elastic elements subjected to a free torsional vibration.

The determination may be achieved by making a slight modification to the conventional triaxial chamber (Fig. 4-XII.4). The chamber is used to subject the sam-

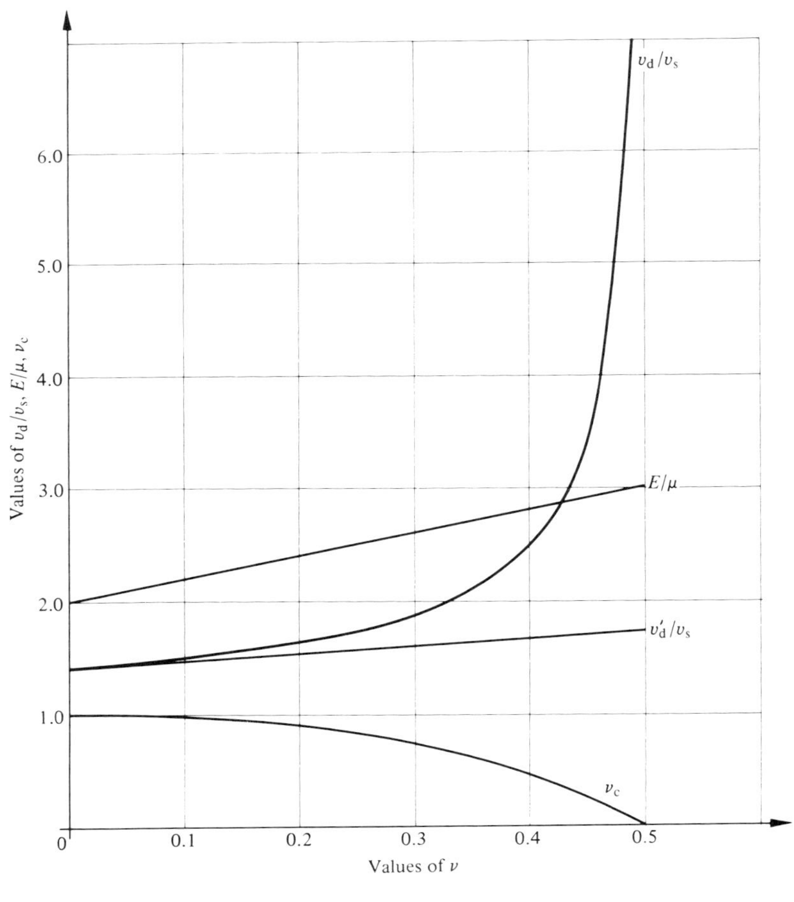

Fig. 3-XII.4 Plot of v_d/v_s, E/μ, and ν_c as functions of ν.

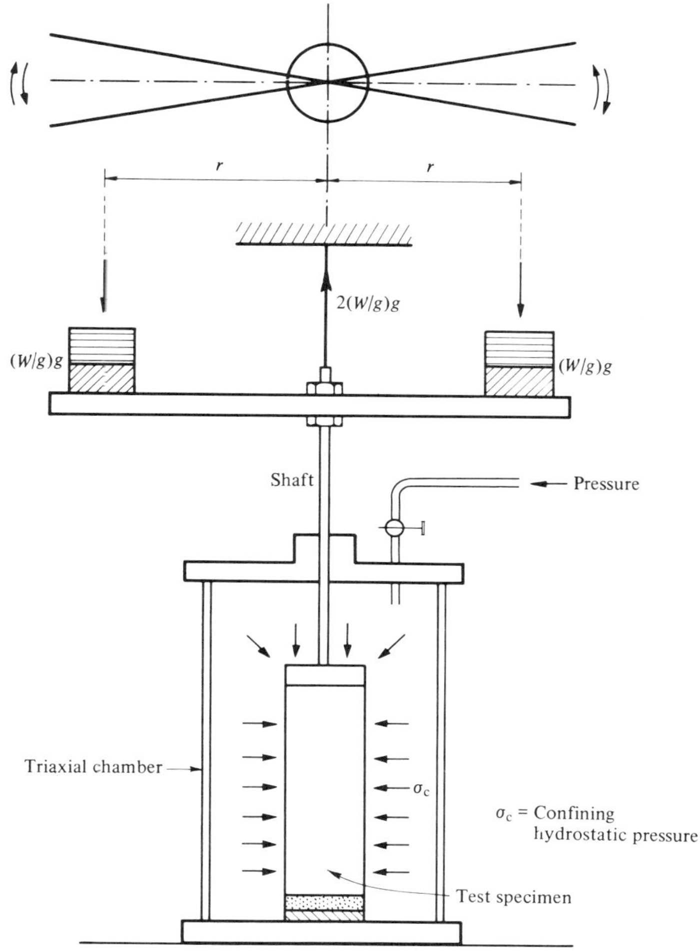

Fig. 4-XII.4 Free torsion pendulum test under confined conditions.

ple to different confining pressures. The sample is permitted to consolidate fully under a hydrostatic confining pressure at which this elastic property is required. A horizontal arm is attached as shown in Fig. 5-XII.4, on which two weights are placed, exactly symmetrical with the axis of the sample. The weight of the supporting arm and masses are balanced by holding them with a thin steel wire; therefore, no vertical load is applied to the sample. The arm is given a small impulse, permitting the system to vibrate freely. The vibration is produced by the response of the elastic elements in the soil, and is recorded in a special electric device. From the information obtained, the shear modulus of elasticity is calculated.

The theory of the instrument according to the sketch of the vibrating system

Fig. 5-XII.4 Torsion machine used to test the rigidity of the samples.

shown in Fig. 6-XII.4 is as follows: the spring constants of sample and instrument are called K_p and K_a, respectively. The torque given to the arm is

$$M_s = F \cdot D$$

Also we may write:

$$\begin{aligned} &\text{for the test specimen:} \quad & F \cdot D = K_p \cdot \theta_p \\ &\text{for the instrument:} \quad & F \cdot D = K_a \cdot \theta_a \end{aligned} \qquad 15\text{-XII.4}$$

The small impulse producing a rotation θ_s of the arm must satisfy the condition

$$\theta_s = \theta_p + \theta_a$$

and from 15-XII.4 we find

$$\theta_s = \frac{K_p + K_a}{K_p \cdot K_a} \cdot F \cdot D \qquad 16\text{-XII.4}$$

On the other hand, considering harmonic motion, the maximum circular acceleration of the arm is

$$\ddot{\theta}_s = \theta_s \cdot \omega_s^2 \qquad 17\text{-XII.4}$$

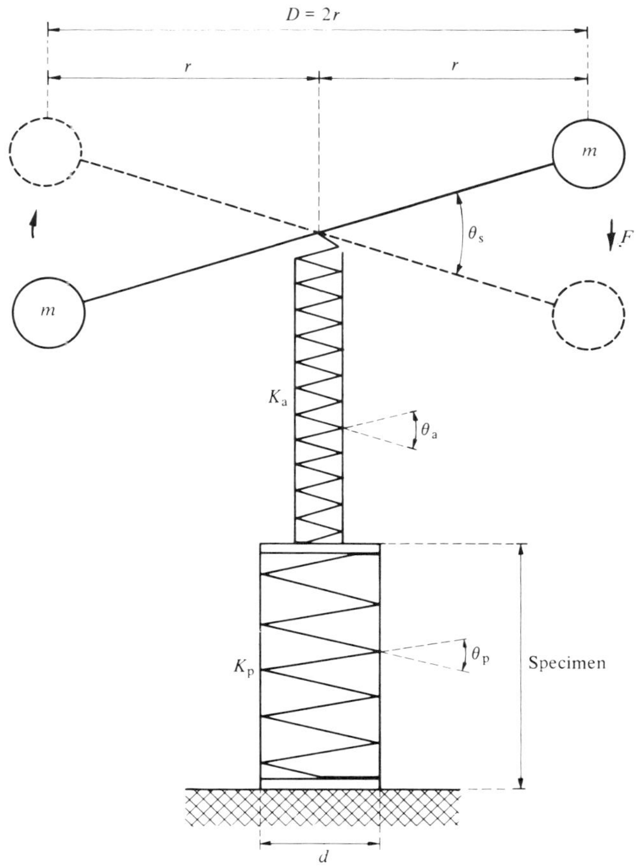

Fig. 6-XII.4 The torsion pendulum.

in which ω_s is the natural circular frequency of the soil-instrument system. Therefore, the dynamic maximum torque in the specimen may be expressed as follows:

$$M_s = \theta_s \omega_s^2 \, \Sigma m r^2 \tag{18-XII.4}$$

This may be written in the following form:

$$M_s = J_s \cdot \theta_s \cdot \omega_s^2 \tag{19-XII.4}$$

in which $J_s = \Sigma m r^2$ represents the polar moment of inertia of all the oscillating masses of the instrument and test specimen. From equations 16 and 19-XII.4 we obtain

$$\frac{K_p \cdot K_a}{K_p + K_a} \theta_s = J_s \theta_s \omega_s^2 \tag{20-XII.4}$$

and calling

$$K_s = \frac{K_p \cdot K_a}{K_p + K_a}$$

the following expression is obtained:

$$\omega_s^2 = \frac{K_s}{J_s} \qquad \text{21-XII.4}$$

The value of ω_s is measured in the test. The shear modulus of elasticity of the test specimen may be obtained from the theory of elasticity by

$$\mu = \frac{M_s \cdot h}{\theta_p I_p} \qquad \text{22-XII.4}$$

where $I_p = \pi d^4/32$ is the polar moment of inertia of the test specimen of height h, therefore, the spring constant of the sample is:

$$K_p = \frac{I_p}{h}\,\mu$$

Calling $C_p = I_p/h$, then $K_p = C_p\mu$. Here C_p is a function of the geometry of the test specimen.

From the calibration of the instrument, the value ω_a is known, and during the test ω_s is measured; therefore:

$$\begin{aligned} &\text{in the test specimen:} && K_p = C_p\mu \\ &\text{in the instrument:} && K_a = J_a\omega_a^2 \qquad \text{23-XII.4} \\ &\text{in the soil-instrument system:} && K_s = J_s\omega_s^2 \end{aligned}$$

According to formula 20-XII.4, we obtain

$$J_s \cdot \omega_s^2 = \frac{C_p\mu \cdot J_a\omega_a^2}{C_p\mu + J_a\omega_a^2}$$

Solving for μ:

$$\mu = \frac{\omega_a^2\,\omega_s^2}{\omega_a^2 J_a - \omega_s^2 J_s} \cdot \frac{J_s \cdot J_a}{C_p}$$

or

$$\mu = \frac{\omega_s^2}{1 - \dfrac{\omega_s^2}{\omega_a^2} \cdot \dfrac{J_s}{J_a}} \cdot \frac{J_s}{C_p} \qquad \text{24-XII.4}$$

Usually the value of J_a is made large enough so that $J_s \doteq J_a$. The mass polar moment of inertia of the sample is disregarded in comparison with that of the instrument system, hence:

$$\mu = \frac{\omega_s^2}{1 - \left(\dfrac{\omega_s}{\omega_a}\right)^2} G \qquad 25\text{-XII.4}$$

The value of

$$G = \frac{J_a}{C_p} = \frac{J_a}{I_p} h$$

is a constant characteristic of the geometry of the system. The value of ω_s may be increased or decreased as necessary (equation 21-XII.4) simply by changing the masses on the horizontal arm, aiming for more precision in equation 24-XII.4. The yoke of the instrument is constructed as rigidly as possible to obtain large values of ω_a.

From the test one obtains the damped free period of vibration T_{sd} (Fig. 7-XII.4). This value is related with the natural frequency by

$$\omega_{sd} = \omega_{sn} \sqrt{1 - \zeta_s^2} \qquad 26\text{-XII.4}$$

in which ζ_s is the fraction of critical damping. When ζ_s is close to unity, no vibration takes place; that is to say, after the impulse is given, the deformation taking place falls back to the rest position in a gradual and continuous motion without oscillation.

The value of ζ is given by formula 13-XII.4 and the logarithm decrement Δ by

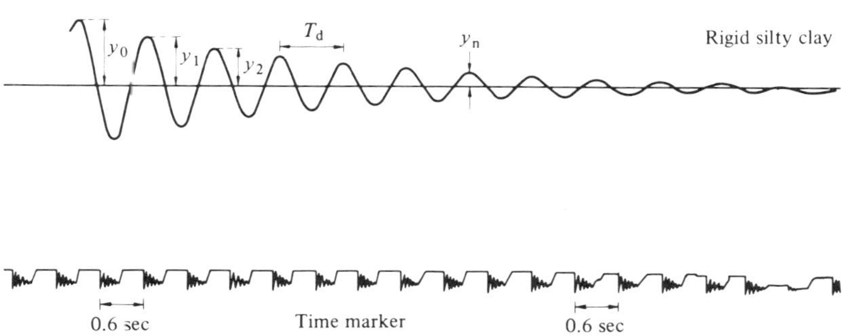

Fig. 7-XII.4 Results from tests with the free vibration torsion pendulum.

12-XII.4. Hence, the required natural frequencies are:

from calibration of the instrument:

$$\omega_a^2 = \frac{\omega_{ad}^2}{1 - \zeta_a^2} \qquad \text{27-XII.4}$$

from the test:

$$\omega_s^2 = \frac{\omega_{sd}^2}{1 - \zeta_s^2} \qquad \text{28-XII.4}$$

The values of ζ_s obtained from the test have no relationship with the real ζ_p for the soil, as this value is implicit in the frictional forces induced in the instrument during vibration.

The final formula to calculate μ is obtained by substituting into 25-XII.4 the natural frequencies for their corresponding values including damping (27 and 28-XII.4):

$$\mu = \frac{\omega_{sd}^2}{(1 - \zeta_s^2) - (1 - \zeta_a^2)\dfrac{\omega_{sd}^2}{\omega_{ad}^2}} \cdot G \qquad \text{29-XII.4}$$

When the instrument is designed in such a way as to make $(1 - \zeta_a^2)(\omega_{sd}^2/\omega_{ad}^2)$ negligible in the denominator of expression 28-XII.4, then for most practical purposes good accuracy is obtained with

$$\mu = \frac{\omega_{sd}^2}{1 - \zeta_s^2} G \qquad \text{30-XII.4}$$

The soil critical damping may be obtained approximately from the results of the tests. The coupled vibration in terms of the independent periods of vibration is as follows:

$$T_s^2 = T_p^2 + T_a^2$$
$$T_{sd}^2 \approx T_{pd}^2 + T_{ad}^2 \qquad \text{31-XII.4}$$

where the free period of vibration and the damped period are related by the critical damping:

$$T_s = T_{sd}(1 - \zeta_s^2)$$
$$T_p = T_{pd}(1 - \zeta_p^2) \qquad \text{32-XII.4}$$
$$T_a = T_{ad}(1 - \zeta_a)$$

Substituting 32-XII.4 into 31-XII.4 we obtain

$$T_{sd}^2(1 - \zeta_s^2)^2 = T_{pd}^2(1 - \zeta_p^2)^2 + T_{ad}^2(1 - \zeta_a^2)^2$$

Here the unknown is ζ_p^2, hence

$$(1 - \zeta_p^2)^2 = \frac{T_{sd}^2(1 - \zeta_s^2)^2 + T_{ad}^2(1 - \zeta_a^2)^2}{T_{pd}^2}$$

from which, after algebraic operations, we obtain

$$\zeta_p^2 = \frac{\zeta_s^2 - \zeta_a^2 T_{ad}^2/T_{sd}^2}{1 - T_{ad}^2/T_{sd}^2} \qquad \text{33-XII.4}$$

4.3 Results

The use of the instrument described is simple. The dynamic elastic properties of soils may be determined with enough accuracy for practical uses. The value of μ is found to increase with the confining pressure. Tests performed on dry 20-30 standard Ottawa sand by investigators at the University of Michigan and M. I. T. with different types of equipment are shown in Fig. 8-XII.4.

The tests performed by Richart and Hall at Michigan have shown that the shear wave velocity decreases with the increasing amplitude to which the sample was subjected in their torsion tests. The decrease was found to be as much as 15% when the double-amplitude increased from 1×10^{-5} to 2.5×10^{-3} radians. From these tests, it appears that in cohesionless soils the shear wave velocity decreases with the amplitude used in the torsion test.

Of the results of the free torsion pendulum tests illustrated in Fig. 8-XII.4, particularly noteworthy is a complete hysteresis loop, which shows clearly the effect on v_s when varying the confining pressure. The mean initial void ratio of the material studied was approximately 0.47. The results of these tests plot under the values found by Richart and Hall; it appears that the difference is due to the different amplitudes used in the tests, as the tests made using the free vibration torsion pendulum were carried out with double amplitudes on the order of 1×10^{-2} rad, considerably higher than those used by Richart and Hall.

From the hysteresis loop shown in Fig. 8-XII.4, it may be concluded also that in the case of cohesionless materials, the grain structure obtained when the test specimen is formed has an important influence on the test results. It appears that more

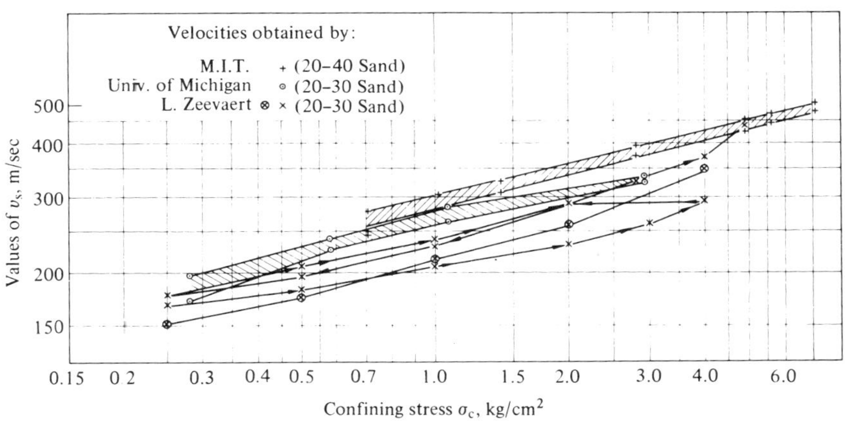

Fig. 8-XII.4 Results of dynamic torsion tests on dry Ottawa sand.

consistent results may be obtained after a stable structure is formed for a certain void ratio and range of confining pressures. This may be achieved if the sample is subjected to one or several loading and unloading cycles. From investigations performed to date in cohesionless soils, it appears necessary to determine the variation of the shear wave velocity caused by the following important factors:

a) amplitude used in the shear torsion tests;
b) method of applying the confining pressures at which the shear wave velocity is determined.

Numerous tests have been performed by the author and collaborators with the free vibration torsion pendulum in various materials. The value of the soil rigidity

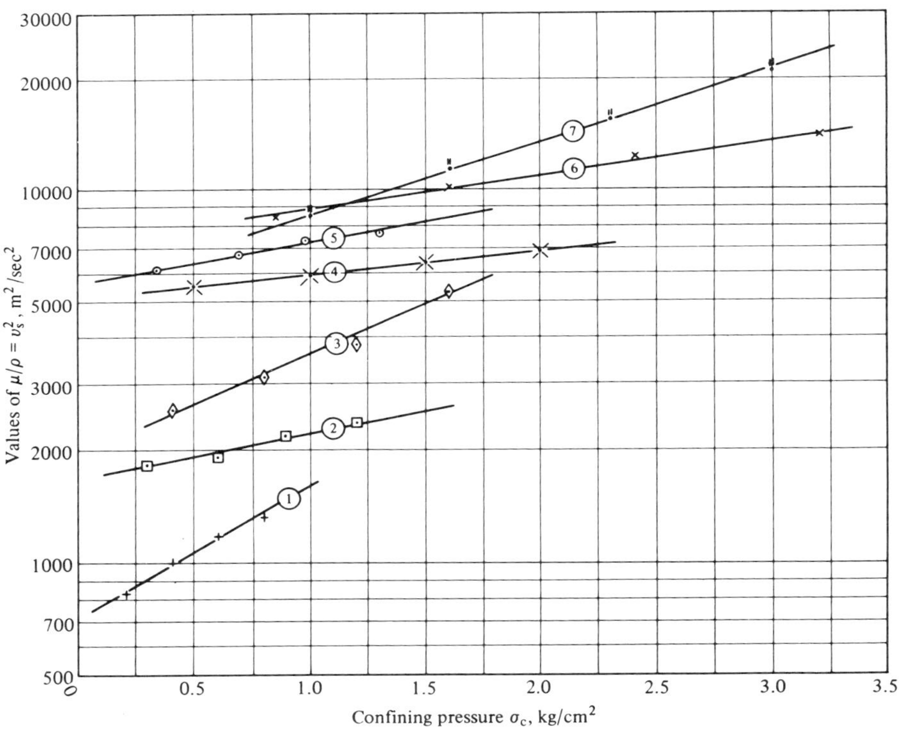

Fig. 9-XII.4 Shear modulus of elasticity for cohesive materials.

Curve	Material	Origin	q_u	w%
1	Clay, olive brown	Mexico City	0.70	366%
2	Clay, reddish brown	Mexico City	1.00	308%
3	Clayey silt, olive green	Mexico City	1.10	240%
4	Slightly silty clay	Mexico City	1.75	270%
5	Slightly silty clay	Mexico City	1.00	172%
6	Silty clay, olive green	Mexico City	2.60	132%
7	Silty clay, olive green	Mexico City	2.80	163%

μ is plotted against the volumetric confining effective stress. Undisturbed fine cohesive soils show a rectilinear relationship in semilog scales (Fig. 9-XII.4) and cohesionless soils in the bracket of fine to coarse sand show a rectilinear relationship in log-log scales (Fig. 10-XII.4). From this phenomenological behavior, the following laws may be stated:

for cohesive fine sediments $$\frac{\mu}{\rho} = \frac{\mu_0}{\rho} e_N^{n_c \sigma_c} \qquad \text{31-XII.4}$$

for sands $$\frac{\mu}{\rho} = C_s \sigma_c^{n_s} \qquad \text{32-XII.4}$$

Notice that in cohesive soils when the volumetric confining stress $\sigma_c = 0$, the soil rigidity corresponds to that determined under unconfined conditions. The value of

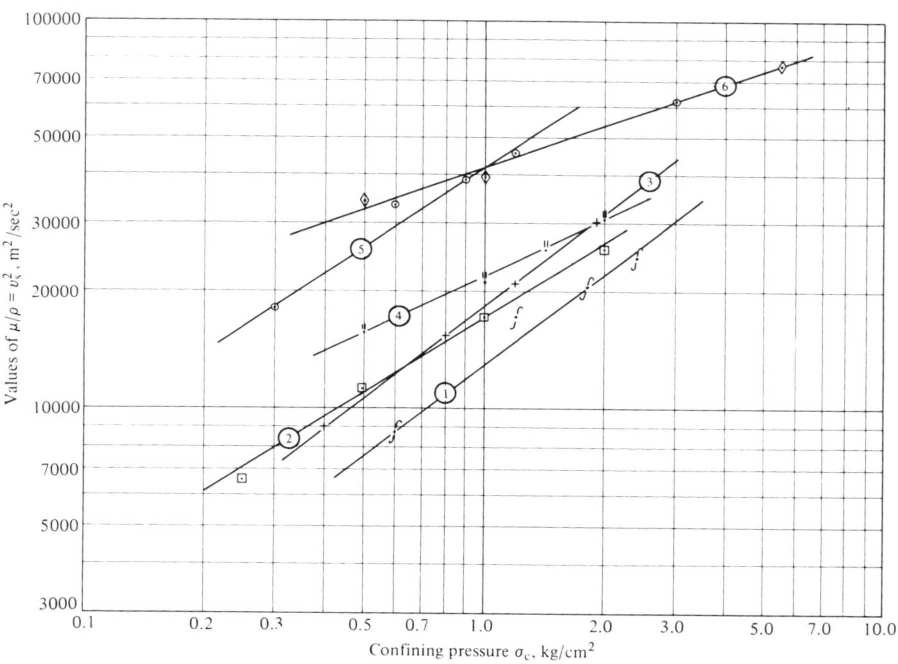

Fig. 10-XII.4 Shear modulus of elasticity for cohesionless materials.

Curve	Material	Origin
1	Fine silty sand, semicompact	Mexico, D.F.
2	Fine and medium sand, compact	Grijalva River, Mexico
3	Medium and fine sand, slightly silty	Acapulco, Mexico
4	Fine sand, slightly silty, compact	Mexico, D.F.
5	Fine and medium silty sand	San Salvador, C.A.
6	Fine and medium sand, compact	Ottawa, Canada

n_c does not change very much for the same soil stratum. Therefore, after making a characteristics test under the corresponding confined state of stress for each stratum in the soil profile, the value of n_c may be determined, and by means of more simple tests performed in unconfined and undisturbed soil specimens more values of μ_0 may be obtained. From these values, the average shear modulus of elasticity of each one of the strata is calculated for the corresponding state of confinement required.

XII.5 SEISMIC BEHAVIOR OF BUILDING FOUNDATIONS

5.1 Introduction

The contents of the preceding sections have practical application in estimating the response of foundations to strong ground motions produced by earthquakes, in cases in which the stratigraphical conditions of the subsoil are similar to those discussed for Mexico City (Fig. 1-XII.5). It was concluded that body waves hitting a deep seated interface of soft ground produce new irrotational and shear waves. It was also concluded that the compressional waves have a minor effect compared to that of shear waves. The waves induced at the ground surface may have some effect when the focus is close and shallow and the soft soil deposit is not very deep.

The ground motion, however, has to enter through the foundation structure. In buildings with raft foundations, compensated foundations and compensated friction pile foundations as described in previous chapters, the ground motion may be assumed transmitted to the building by shear forces developed at the interface between foundation slab and soil. Their magnitude is a function of the response of the building mass to the earthquake motion (Fig. 2-XII.5). On the other hand, in the case of pile foundations, two cases of transmission of the ground motion may

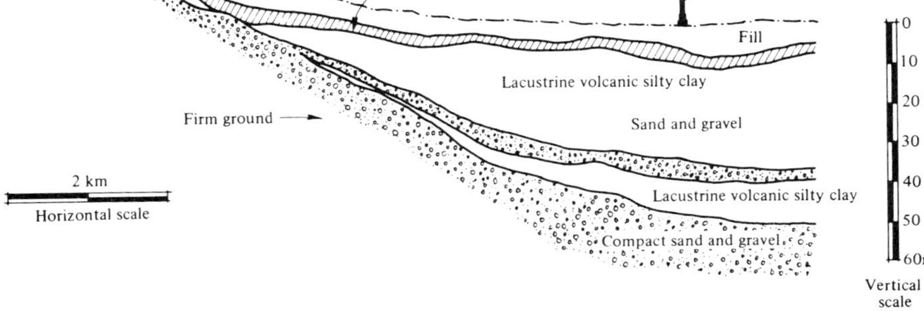

Fig. 1-XII.5 Basin of the Valley of Mexico–geologic profile E–W.

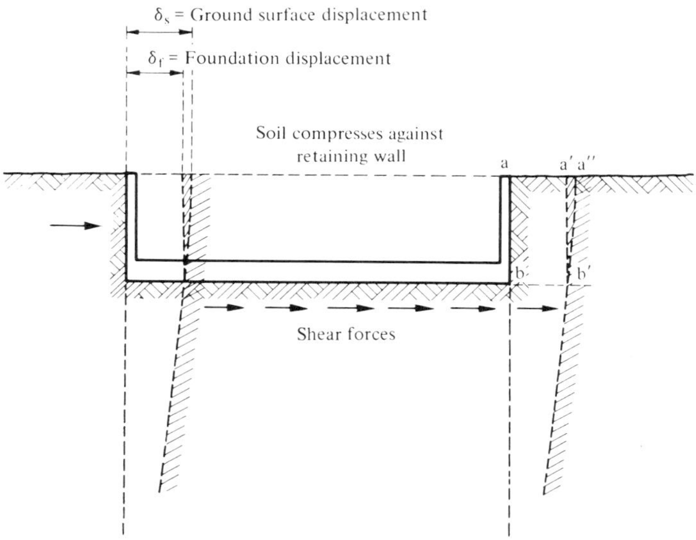

Fig. 2-XII.5 Shear forces developed between foundation slab and soil.

be present: (1) If the foundation structure is shallow, the piles or piers take the shear force (Fig. 3-XII.5); (2) When the foundation structure is deep seated in the soft soil mass, the ground motion is transmitted by passive earth pressure against the retaining wall (Fig. 4-XII.5).

The response forces induced in the building are at the same time a function of the foundation behavior. In heavy and slender buildings, rigid foundation structures are usually used in seismic areas. During ground motion, the building foundation is forced to displace horizontally and to rotate at its base. These effects are especially important to consider in an approximate evaluation of the response of the building to the strong ground motion. The quantification of this phenomenon, however, depends on the precision and uncertainties involved in investigating the response spectrum representing the earthquake ground motions. Nevertheless, in our present state of knowledge, the precision that may be obtained in most cases as a preventive measure proves sufficient from the practical engineering point of view. Therefore, it is the opinion of the author that the treatment of these problems should be as simple as possible, since complicated refinements are not yet justified. The foundation engineer should be satisfied to be able to estimate the order of magnitude of the forces involved, using them to design foundations in difficult subsoil conditions, and should report the results to the structural engineer for the design of the structural frame.

5.2 Foundation Response

The free periods of vibration of a building including its foundation structure must be known to estimate the response of the foundation and building subjected to the

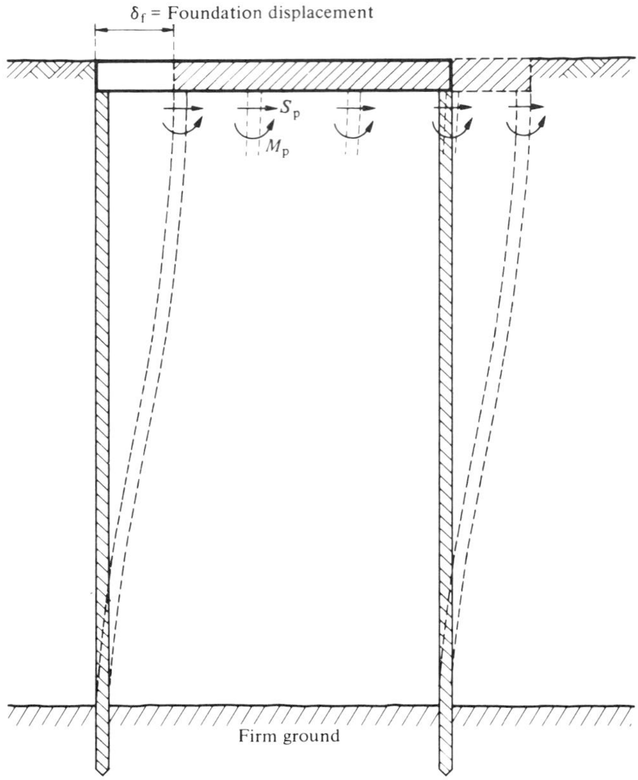

Fig. 3-XII.5 Shallow foundation structure, shear force taken by piles.

strong ground motion. Especially important is the rocking phenomenon of the foundation due to the overturning moments set by the horizontal inertia forces in the mass of the building by the earthquake. The rocking of the foundation increases the contact stresses in the soil at the foundation grade elevation. The stress increment has to be investigated to ensure that it does not exceed the elastic response behavior of the soil; otherwise large permanent tilting will result. The probable permanent tilt should be also estimated (section VII.8).

The problem will be discussed under the assumption that the overturning moment increases linearly the existing contact stresses at the foundation grade elevation, such that nearly elastic response is obtained. After the building has passed the earthquake, it will preserve its verticality or the permanent tilt will be negligible, because of minor plastic effects taking place in the soil that are unavoidable; furthermore, these should be expected to a certain degree.

Consider a tall building with any type of rigid foundation. The soil-foundation system has a rotational spring constant K_θ; hence by definition

$$O_T = K_\theta \cdot \theta \qquad \text{1-XII.5}$$

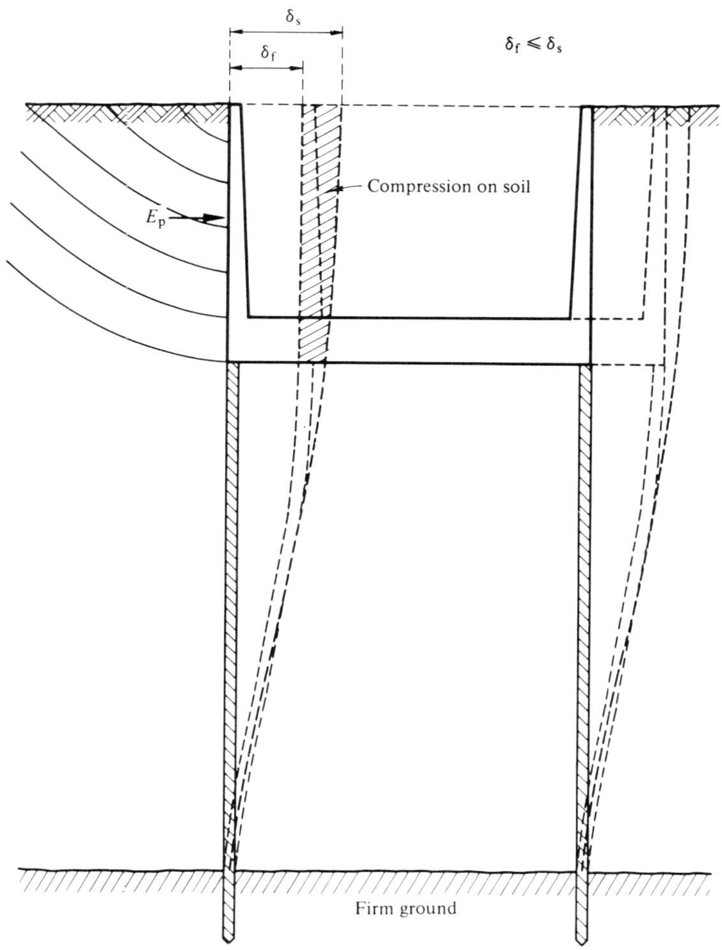

Fig. 4-XII.5 Shear force transmitted by passive earth pressure against retaining wall in a deep foundation structure.

in which O_T is the overturning moment induced by the inertia force on the mass of the building and applied at the center of mass located at height h_m from the foundation grade elevation. The value of θ is the amplitude angle of the rocking phenomenon (Fig. 5-XII.5). At the center of mass, two horizontal deflections may be recognized: δ_θ because of the rotation of the foundation, and δ_n caused by the elastic deflection of the structural frame of the building produced by the shear force:

$$V_B = K_n \delta_n \qquad \text{2-XII.5}$$

The quantity K_n represents the spring constant of the structural frame.

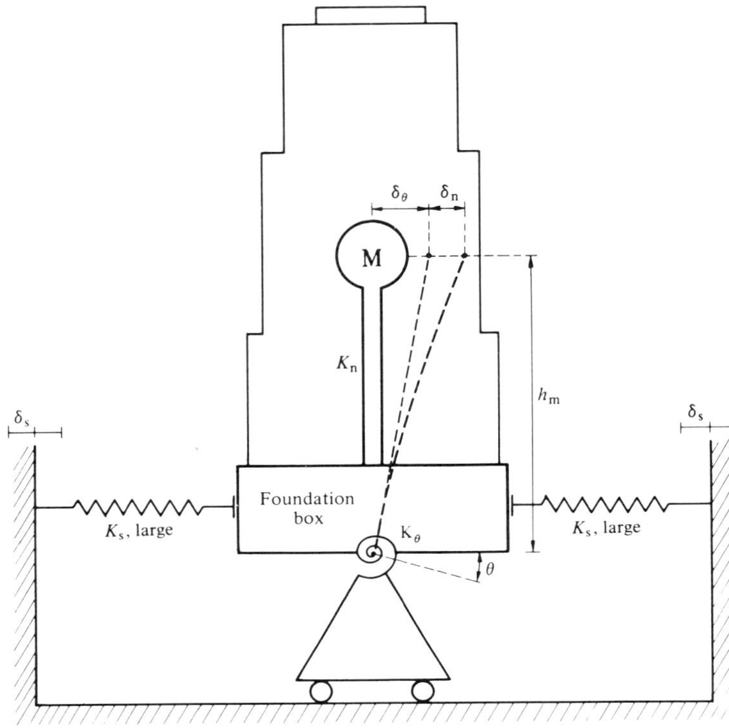

Fig. 5-XII.5 Rigid foundation soil interaction.

The compressional effect on the retaining wall may be neglected. Since in most cases in nature a rigid crust may be encountered at the upper part of the soft subsoil deposits, under these conditions the inertia force at the center of mass is

$$I_F = M \cdot \omega_0^2 (\delta_\theta + \delta_n) \qquad \text{3-XII.5}$$

in which ω_0 is the circular frequency of the system foundation-structural frame. At the foundation grade elevation

$$O_T = M \cdot \omega_0^2 (\delta_\theta + \delta_n) h_m$$

from which

$$\omega_0^2 = \frac{K_\theta}{M h_m^2} \left(\frac{\delta_\theta}{\delta_\theta + \delta_n} \right) \qquad \text{4-XII.5}$$

Furthermore, we find that the ratio of the deflections is given by

$$\frac{\delta_\theta}{\delta_n} = \frac{K_n}{K_\theta} \cdot h_m^2 \qquad \text{5-XII.5}$$

Substituting into 4-XII.5 we obtain

$$\frac{1}{\omega_0^2} = \frac{Mh_m^2}{K_\theta} + \frac{M}{K_n} \qquad 6\text{-XII.5}$$

The terms in the second member represent the inverse values of the independent circular frequencies of the system, namely:

a) the structural frame with no rocking effect:

$$\frac{1}{\omega_n^2} = \frac{M}{K_n} \qquad 7\text{-XII.5}$$

b) the rocking effect as if the structural frame were infinitely rigid:

$$\frac{1}{\omega_\theta^2} = \frac{M}{K_\theta} h_m^2 \qquad 8\text{-XII.5}$$

Therefore, substituting into 6-XII.5 we find

$$\frac{1}{\omega_0^2} = \frac{1}{\omega_\theta^2} + \frac{1}{\omega_n^2}$$

or in terms of the periods of vibration

$$T_0^2 = T_n^2 + T_\theta^2 \qquad 9\text{-XII.5}$$

from which it may be concluded that the rocking of the foundation increases the period of vibration of the structural frame in the proportion $\sqrt{1 + T_\theta^2/T_n^2}$.

This conclusion is important, since the response from the pseudoacceleration spectrum may be increased or decreased in contrast with that obtained with T_n depending on the location of the point with reference to the peak in the spectral response curve. Assuming a pseudoacceleration spectrum similar to that of Mexico City, if the period T_n is located to the left of the peak, the response acceleration increases when the rocking phenomenon is considered. On the other hand, the response acceleration decreases considerably because of the rocking phenomenon when the value of T_n is located to the right of the peak in the pseudoacceleration spectrum. The latter condition may be considered beneficial since the forces induced in the building structural frame will be smaller. For higher harmonics of the building equation 9-XII.5 may be also applied.

From the pseudoacceleration response spectrum, we obtain for T_n a value R_{an}. Therefore, the base shear has the value

$$V_{Bn} = M \cdot R_{an} \qquad 10\text{-XII.5}$$

and the overturning moment for that period is

$$O_{Tn} = MR_{an} \cdot h_m \qquad 11\text{-XII.5}$$

Calling c_{pn} the participation factor for the base shear induced by the response spectrum on each mode of period T_n, then the total base shear is

$$V_B = M\Sigma c_{pn} \cdot R_{an} \tag{12-XII.5}$$

Here $\Sigma c_{pn} = 1$. The total overturning moment is $O_T = V_B \cdot h_m$.

For practical foundation engineering purposes we assume that sufficient accuracy is obtained with the fundamental mode T_1 and the two subsequent harmonics corresponding to T_2 and T_3 of the building. The responses of the three modes considered do not take place at exactly the same time; however, we may use on the safe side

$$V_B = M\,[c_{p1}R_{a1} + c_{p2}R_{a2} + (1 - c_{p1} - c_{p2})R_{a3}] \tag{13-XII.5}$$

The c_{pn}-values may be calculated by means of the following formula:

$$c_{pn} = \frac{(\Sigma m_i \delta_{ni})^2}{(\Sigma m_i \delta_{ni}^2)M} \tag{14-XII.5}$$

in which m_i is the concentrated mass at the i-th floor elevation, δ_{ni} is the horizontal displacement at the same floor produced by the n-th mode of vibration, and M is the total mass of the building. The derivation of this expression is beyond the scope of this book. Nevertheless, it may be obtained by the same method used to derive formula 28-XII.3.

The seismic response of the foundation and building may be obtained if the value of T_0 is known, by entering the envelope design spectrum (Fig. 15-XII.2) to find the amplification factor f_a. The period or periods of vibration of the building are determined from modal analysis as mentioned above. However, to find the rocking period T_θ it is necessary to calculate the foundation modulus K_θ (8-XII.5). For this purpose, a calculation of the soil-foundation structure interaction is necessary. For illustration purposes, assume a rigid foundation for a tall building and consider only the rocking phenomenon. Referring to Chapter VI we can use the following matrix equation, EMA:

$$[\bar{\delta}_{ji}]^T \cdot |X_i| = |\delta_i| \tag{15-XII.5}$$

Notice that the values of $\bar{\delta}_{ji}$ should be calculated with $\alpha_d = d/2(1 + \nu)\mu$. If $\nu = 0.5$ in clay, then $\alpha_d = d/3\mu$ (Chapter VI).

In 15-XII.5 the values X_a, X_i, X_j, X_b represent the unknown reactions. However, since this is a symmetrical problem, $X_a = -X_b$ and $X_i = -X_j$, also $\delta_a = -\delta_b$ and $\delta_i = -\delta_j$. Furthermore, if we call θ the amplitude of the rocking angle of the rigid foundation (Fig. 6-XII.5) we may write:

$$\begin{aligned} \delta_a &= \theta x_a \\ \delta_i &= \theta x_i \\ \delta_j &= -\theta x_j \\ \delta_b &= -\theta x_b \end{aligned} \tag{16-XII.5}$$

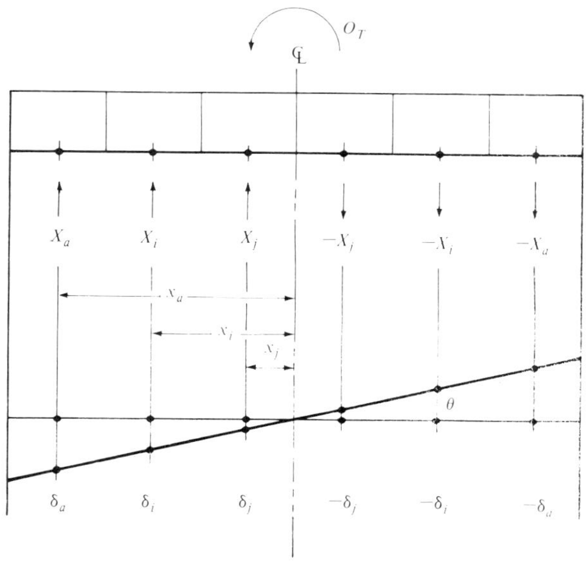

Fig. 6-XII.5 Rotation of rigid foundation.

in which x_a, x_i, x_j, x_b are the distances from the center of rotation to the unknown reactions X_a, X_i, X_j, X_b, respectively. Substituting the values of 16-XII.5 into 15-XII.5 we may write

$$[\bar{\delta}_{ji}]^T \cdot \left|\left(\frac{X_i}{\theta}\right)\right| = |x_i| \qquad \text{17-XII.5}$$

Expression 17-XII.5 may be solved as a system of simultaneous equations for the values (X_i/θ). The overturning moment is

$$O_T = \theta \sum (X_i/\theta) \cdot x_i$$

but the rocking foundation modulus is by definition $K_\theta = O_T/\theta$, therefore,

$$K_\theta = \sum_1^n (X_i/\theta) x_i \qquad \text{18-XII.5}$$

From above K_θ value the rocking period will be:

$$T_\theta = 2\pi h_m \sqrt{M/K_\theta}$$

From the value of $T_0^2 = T_n^2 + T_\theta^2$, the ratio T_0/T_{s1} may be calculated and entering the design acceleration spectrum (Fig. 15-XII.2) the amplification factor f_a is determined. The seismic force at the center of mass will be $V_m = f_a M a_m$, the seismic overturning moment

$$O_T = (f_a M a_m) h_m \qquad \text{19-XII.5}$$

and, finally, the amplitude of the rocking angle $\theta = O_T/K_\theta$, from which the seismic reactions are calculated, $X_i = (X_i/\theta) \cdot \theta$.

For illustration purposes, let us analyze in a simple way the case history of the Latino Americana Tower with a rigid box type foundation to 13 m depth supported on 361 point bearing piles. To calculate the response shear induced in the building by the earthquake of May 11, 1962, the average pseudoacceleration spectrum will be used for 5% and 10% critical damping obtained in terms of R_a/a_m (Fig. 13-XII.2). The quantity a_m represents the maximum average ground acceleration at Alameda Park registered in the accelerograms; a value of $a_m = 38$ cm/sec^2 will be used. The mass of the building is $M = 15.7$ ton sec^2/cm. Moreover, the foundation of the building is provided with a deep sheet-pile wall extending to a depth of 16 m from the ground surface. The sheet-pile wall working in its plane on the sides of the foundation provides high rigidity to horizontal soil displacements. Therefore, the ground acceleration used from the pseudoacceleration spectrum (Fig. 13-XII.2) should be corrected in proportion to the subsoil displacements at 16 m depth and ground surface shown calculated in Fig. 5-XII.3, from which it is found that the ground acceleration should be multiplied by 0.7 to obtain approximately the average ground acceleration at 16 m depth; hence $a_m = 26.6$ cm/sec^2. Moreover, the acceleration should be further reduced to make it applicable to the LAT site, since subsoil investigations showed this place to be on the order of one third stiffer than the Alameda Park site. Therefore, the ground acceleration at 16 m depth at the Tower site may be taken for seismic calculations with the value $26.6/\sqrt{1.33}$, or approximately 23 cm/sec^2, according to formula 13-XII.3. Note that from recorded accelerograms at the LAT site the average maximum acceleration registered is of the same order of magnitude as that calculated above (Fig. 4-XII.2).

Using average values from pseudo-acceleration spectra (Fig. 11-XII.2) calculation is given in Table 1-XII.5 for the response base shear, assuming no rocking effect or horizontal translation of the foundation. The building periods and participating factors were calculated from the mechanical and geometrical properties of the structural frame.

The base shear in this case is found to be approximately 518 ton for 5% critical

TABLE 1-XII.5

Mode	T_n sec	c_{pn}	R_{an}/a_m 5%	$c_{pn}R_{an}/a_m$ 5%
1st	3.66	0.61	0.95	0.58
2nd	1.54	0.22	3.0	0.66
3rd	0.98	0.17	2.4	0.41

$$\Sigma c_{pn} = 1.00 \qquad \frac{1}{a_m} \Sigma c_{pn} R_{an} = 1.65$$

for: $a_m = 20$ cm/sec^2 $\quad \zeta = 5\%$ $\quad V_B = 15.7\,(1.65)\,20 = 518$ ton damping. The center of mass of the building is located at $h_m = 67$ m from the foundation slab, therefore the overturning moment is 29 000 ton · m in the case of 5% critical damping.

Since the foundation is very rigid, the increment at the edge piles is

$$\Delta Q = \frac{O_T}{\Sigma x^2} \cdot x_e \qquad \text{15-XII.5}$$

For this building $\Sigma x^2 = 3.73 \times 10^4$ m², in the N-S direction, and $x_e = 16.7$ m. Therefore, for 5% damping the piles suffer a maximum increment of $\Delta Q = 15.5$ ton/pile.

In case of rotation of the pile foundation, the K_θ-value was estimated for the rigid foundation on point bearing piles of the Latin American Tower as $K_\theta = 1.6 \times 10^{10}$ ton · m/rad, hence from 8-XII.5:

$$T_\theta = 2\pi(6700)\sqrt{\frac{15.7}{1.60 \times 10^{10}}} = 1.31 \text{ sec}$$

Therefore, the new periods of vibration with account for the rocking effect of the building are given in Table 2-XII.5.

Notice from the T_0-values including rocking that the agreement of these periods with the resonance peaks shown in the pseudoacceleration spectrum obtained for the Tower given in Figs. 10 and 12-XII.2 are very close to the ones registered at about the same periods.

Now let us calculate the new base shear with the improved periods of vibration obtained in Table 2-XII.5. The average response spectrum is now used for 10% of critical damping; in this case it is better justified because of the rocking of the foundation structure. Assume that the same c_{pn} values apply. The calculation is given in Table 3-XII.5.

From the calculations in Table 3-XII.5, it may be seen that the rocking phenomenon of the Tower with 10% critical damping decreased the base shear to 107 ton, or 20%. Hence, the rocking effect in this case was beneficial because of the higher critical damping including the rocking phenomenon of the foundation. It may be concluded that 10% damping is a better figure to consider when the rocking effect is included. The shear force calculated from the deflections measured in the structural frame at the ground floor of the Tower was estimated at 350 ton for May 11, 1962. Moreover, when a foundation structure rests directly on the soft subsoil

TABLE 2-XII.5

	S-Frame		Rocking	Total
Mode	T_n	T_n^2	T_θ^2	T_0
1st	3.66	13.3	1.7	3.88
2nd	1.54	2.36	1.7	2.02
3rd	0.98	0.96	1.7	1.64

TABLE 3-XII.5

Mode	T_n	c_{pn}	$\frac{R_{an}}{a_m}$	$c_{pn} \frac{R_{an}}{a_m}$
1st	3.88	0.61	0.68	0.41
2nd	2.02	0.22	2.3	0.51
3rd and others	1.64	0.17	2.3	0.39

$\zeta = 10\%$ $\qquad \frac{1}{a_m} \Sigma c_{pn} R_{an} = 1.31$

for $a_m = 20$ cm/sec; $\quad V_B = 15.7\ (1.31)\ 20 = 411$ ton

sediments, a higher damping coefficient can be used that may vary between 10% and 20% in very soft sediments.

It was calculated in Table 2-XII.3, that the soil strata of the subsoil under the Tower gave periods on the order of 2.56, 2.05, 1.76, and 0.72 sec because of the wave reflections in intermediate strata. The LAT response spectrum showed resonant periods at about the same periods (Fig. 12-XII.2). From this observation, it may be concluded that calculations performed from the accelerograph recordings are in fair agreement with those calculated with subsoil engineering properties and shear wave theory discussed in XII.3. The calculations given above demonstrate the importance of the secondary resonance periods of the ground due to reflections in the different strata with the periods corresponding to the higher harmonics of the building. The possibility of occurrence of this phenomenon should not be overlooked when designing building foundations and structural frames under seismic forces.

5.3 Seismic Base Shear

As stated in previous paragraphs, the seismic response shear of a building is taken either by shear stresses at the foundation grade elevation in case of raft foundations, or by the passive earth pressure against the retaining walls in case of pile foundations with basements. Therefore, to estimate the factor of safety against horizontal sliding of the foundation, it is necessary to select properly the shear strength parameters.

Ground motions are applied to the soil mass for time periods on the order of only one second or less, and thus may be considered transient loadings. Under these conditions, the theoretical considerations discussed in section II.4 on shear strength properties may be used for judgment. In clayey soils, the concept of the shear strength obtained from unconfined compression tests performed in the laboratory under load application on the order of 0.3 to 1.0 sec is applicable. The shear strength, under these existing conditions, is slightly greater because of the viscous effects of rapid load application, than for a static determination.

In the case of clayey silty sand, silts and sandy silts cemented with clay minerals, the tests should be made in the triaxial chamber under consolidated undrained quick conditions. In practice, the subsoil is subjected to effective stresses because of the load of the building. When the strong motion takes place, the material has no time to drain; therefore, failure occurs quickly at constant volume. Hence, the consolidated undrained quick test applies in this case. The apparent shear strength parameters c_{cu} and ϕ_{cu} may be used to estimate the stability against the base shear forces obtained from the response of the building to the earthquake. The total passive forces so obtained should be given a factor of safety of not less than 3. The problem may be illustrated with a simple example. Consider as a first case a raft foundation on a silty clay where $q_u = 2.5$ kg/cm^2. The foundation area is 800 m^2, and the base shear estimated from the earthquake is on the order of 450 ton. From shear strength and area, the ultimate shear force at the interface between soil and foundation is $S_{ub} = 800 \cdot (1/2)(2.5) = 1000$ ton; hence, the factor of safety $G_s = 2.2$. Assume that this is inadequate when based on failure at the foundation slab. Assume further that the passive earth pressure on the upper part of the retaining wall should not be taken into consideration because rocking of the building makes its use uncertain. The only possible effective way to overcome this problem is to construct strong cut-off walls deep into the soil and long enough to take the passive earth pressure (Fig. 7-XII.5). In our example the depth to the foundation slab is z_1, and the depth to the toe of the cut-off walls z_2 (section II.4); hence the passive unit pressures in the case of impervious soil is

$$\text{at depth } z = z_1, \quad p_1 = q_{ud} + p_{v1}$$

$$\text{at depth } z = z_2, \quad p_2 = q_{ud} + p_{v2}$$

in which p_{v1} and p_{v2} are the total pressures at depths z_1 and z_2, respectively. Therefore

$$E_p = [q_{ud} + \tfrac{1}{2}(p_{v1} + p_{v2})]\,(z_2 - z_1)(2B) \qquad \text{16-XII.5}$$

and the ultimate shear strength force is

$$S_u = S_{ub} + E_p \qquad \text{17-XII.5}$$

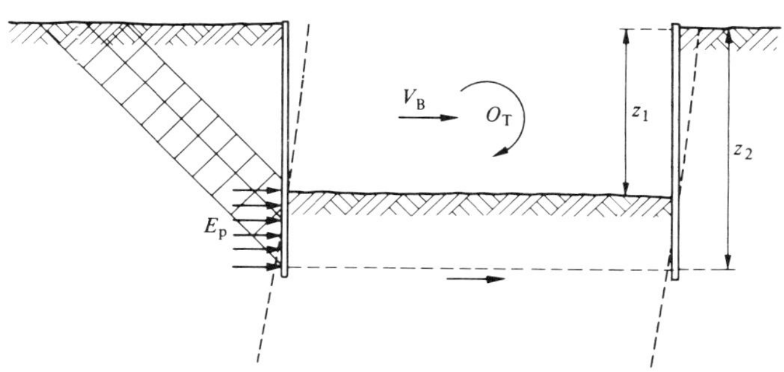

Fig. 7-XII.5 Cutoff walls to take passive earth pressure.

To continue with the example, assume the foundation depth is 5.0 m and its minimum width 25 m. The cut-off wall extends to 6.5 m depth. The water table is close to the ground surface. The soil has a saturated unit weight of 1.5 ton/m^3; hence p_{v1} = 7.5 ton/m^2 and p_{v2} = 9.75 ton/m^2. Therefore, according to 16-XII.5, $E_p = (2.5 + 8.68)(1.5)(25) = 420$ ton, which when added to the base shear resistance gives $G_s = (420 + 1000)/450 = 3.15$, which will now be adequate.

The second case is that of a building on point bearing piles, in which case the shear resistance at the foundation slab cannot be used, and the factor of safety is then computed on the assumption that the piles yield such that the full passive earth pressure on the retaining wall is applied. The unit passive pressures will be:

at the surface: $z = 0, \quad p_0 = q_{ud}$

at the toe of the cut-off wall: $z = z_2, \quad p_2 = q_{ud} + p_{v2}$

Hence, the total passive earth pressure is

$$E_p = [q_{ud} + \tfrac{1}{2} p_{v2}] \cdot z_2 \cdot (2B) \qquad \text{18-XII.5}$$

For our example

$$E_p = [2.5 + 4.87]\, 6.5 \times 25 = 1197 \text{ ton}$$

and $G_s = 2.6$. Nevertheless, if the factor of safety is not yet adequate, either the cut-off wall is extended farther, or the weight of the building is reduced, or both.

When the material is of the frictional type, usually raft foundations do not present a special problem; however, in the case of point bearing pile foundations, the problem is similar to the one already discussed, except that the apparent shear parameter c_{cu} and ϕ_{cu} obtained from consolidated quick tests at constant volume should be applied. The passive unit pressure

at the ground surface: $z = 0, \quad p_0 = 2c\sqrt{N_\phi}$

at the base of the foundation: $z = z_1, \quad p_1 = 2c\sqrt{N_\phi} + p_{v1} N_\phi$

Therefore, the total passive earth pressure is

$$E_p = [2c\sqrt{N_\phi} + \tfrac{1}{2} p_{v1} N_\phi]\, z_1 (2B) \qquad \text{19-XII.5}$$

For the problem under study, the consolidated-constant volume quick tests have the following average values for the parameters c_{cu} = 1.5 ton/m^2 and $\phi_{cu} = 16°$. Therefore $N_\phi = 1.76$ and the total passive force is

$$E_p = [3\sqrt{1.76} + \tfrac{1}{2}\, 7.5 \times 1.76] \times 5 \times 25 = 1320 \text{ ton}$$

Since the base shear induced by the ground motion is 450 ton, the factor of safety is on the order of three, and no cut-off wall is necessary. The design may continue on the basis that the ground motion will not cause a slip of the foundation with respect to the soil mass, as otherwise the building may be in danger of being damaged.

Moreover, the effect on the periods of vibration of the building because of deformation in the horizontal direction of the soil mass during transmission of the ground motions may be safely disregarded. If this phenomenon is present, it has the tendency to increase the periods of the foundation-structural system, thereby minimizing the response of the building to the ground motion.

XII.6 SEISMIC BEHAVIOR OF UNDERGROUND STRUCTURES

6.1 General Considerations

Underground structural elements constructed in soft subsoil deposits, such as piles and piers, and in general long and large pipe-like conduits, are subjected to distortions induced by strong ground motions. The slender and long flexible piles, pipes or shafts driven in the soil mass usually follow closely the soil displacements at the depth they are constructed. Nevertheless, stiff structural elements like large piers bearing on a hard stratum overlain by soft impermeable soils, offer resistance to the ground motion, and important drift forces may be induced in them.

Numerous reported cases of failure of load supporting elements have led to structural damage beyond repair. Therefore, it is the concern of the foundation engineer to be able to estimate the order of magnitude of the forces involved, and foresee the importance they may have in the design of the foundation and structural frame of the building, bridge or any other type of structure supported on such vertical elements piercing through soft soil sediments to great depth. Similarly, shear wave distortions may have important significance in the behavior of very long buildings and subsurface conduits, like sewer pipes, tunnels or similar structures.

An approximate analysis of the phenomenon involved will permit the foundation engineer to recognize the important factors in the design of these structural elements, and from there, take the necessary precautions in the design, making them work in the most efficient and economical manner. The stiffness of the underground structures is a very important factor in the induced moments and shears, and it is therefore highly desirable to design with the largest possible flexibility compatible with the load to be carried. Furthermore, the joints between different sections should be made strong to transmit the shear forces induced in them by the ground motion. Flexibility will better permit the structures to follow the subsoil displacements, and reduce to a minimum the seismic forces in them.

The maximum subsoil displacements during ground motion may be estimated knowing the stratigraphical and elastic response characteristics of the subsoil. The drift forces produced by the subsoil displacements against the piers are calculated establishing a model of structure-soil interaction.

6.2 Vertical Shafts, Piles and Piers

Consider a pile or pier driven to a point bearing on a strong stratum capable of supporting the vertical load transmitted by this structural member with an adequate factor of safety. Here a pile is considered to be an element up to 80 cm in diameter driven from the ground surface to the required depth, and a pier is considered to be an element cast in place with a diameter larger than 80 cm, bearing on firm ground.

The shear waves travelling through the soft soil deposit produce horizontal displacements. The maximum configuration or amplitude of the displacements may be computed knowing the shear modulus of elasticity of the ground and the response acceleration at the ground surface (section XII.3). The horizontal maximum displacements of the soil mass at any depth level $i - i$, we call δ_{si}. The pier is subjected at the top to a horizontal force P representing the base shear of the building induced by the seismic movement of the pier head. The inertia force due to the building load exerted at the head of the pier will cause horizontal deflections in the pier, opposite to the horizontal displacements δ_{si} of the soil. The above mentioned physical phenomenon will create soil reactions along the pier shaft and moments at the ends, top and bottom respectively. The displacement of pier and soil should be the same during the seismic movement. The problem, therefore, is to find the interacting soil reactions that put the pier under equilibrium. The mass forces due to the weight of the pier are small when compared with the soil mass involved in the motion, hence they will be neglected. The soil-pier interaction problem may be solved with the same philosophy as the soil-foundation structure interaction already treated for beam foundations in Chapter VI.

Let us consider the pier divided in any number of sections of length d_i. Calling $2r_0$ the diameter of the pier, each section will have a tributary area $\overline{a}_i = (2r_0)d_i$. At the center of each tributary area a soil reaction X_i is acting (Fig. 1-XII.6). The reactions at the ends of sections (a) and (b) and the end moments will be called R_a and R_b, and X_a and X_b, respectively. The soil reactions will induce, in the soil mass horizontal displacements we will call δ, and in the pile, structural displacements, we will call S and Δ. Following the general philosophy to establish the matrix interaction equations given in Chapter VI, we first analyze the horizontal displacements induced in the soil by the reactions R_a, X_i, X_j, R_b (Fig. 2-XII.6). For this purpose, we apply a unit reaction $X_i = +1$ at any section or a load per unit area of $1/\overline{a}_i$. Under these conditions we obtain in the soil mass horizontal displacements that we call $\overline{\delta}_{ji}$, that is, the horizontal displacement at point j because a load $X_i = +1$ is applied at point i. The seismic horizontal displacement matrix equation, we call HEMAS, and will read as follows:

$$[\overline{\delta}_{ji}] \cdot |X_i| = |\delta_i| \qquad \text{1-XII.6}$$

or

$$\begin{bmatrix} \overline{\delta}_{aa} & \overline{\delta}_{ai} & \overline{\delta}_{aj} & \overline{\delta}_{ab} \\ \overline{\delta}_{ia} & \overline{\delta}_{ii} & \overline{\delta}_{ij} & \overline{\delta}_{ib} \\ \overline{\delta}_{ja} & \overline{\delta}_{ji} & \overline{\delta}_{jj} & \overline{\delta}_{jb} \\ \overline{\delta}_{ba} & \overline{\delta}_{bi} & \overline{\delta}_{bj} & \overline{\delta}_{bb} \end{bmatrix} \cdot \begin{vmatrix} R_a \\ X_i \\ X_j \\ R_b \end{vmatrix} = \begin{vmatrix} \delta_a \\ \delta_i \\ \delta_j \\ \delta_b \end{vmatrix}$$

In matrix equation (1-XII.6), the columnar matrix $|\delta_i|$ represents the real horizontal displacements produced in the soil mass because of the application of horizontal reactions R_a, X_i, R_b.

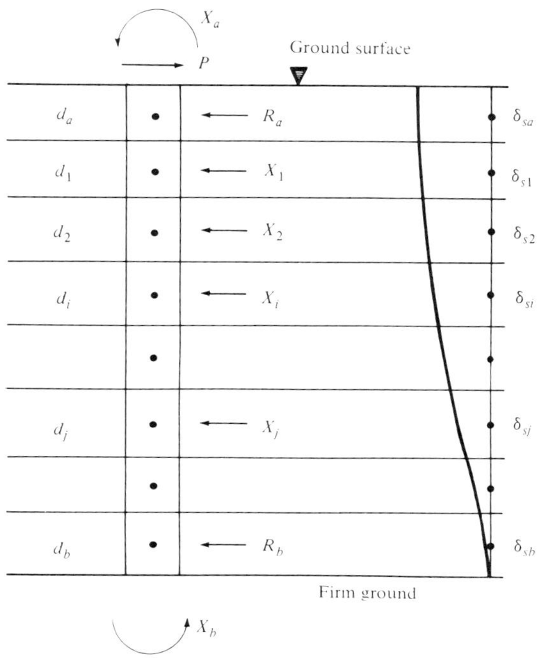

Fig. 1-XII.6 Pier soil reactions and seismic horizontal soil displacements.

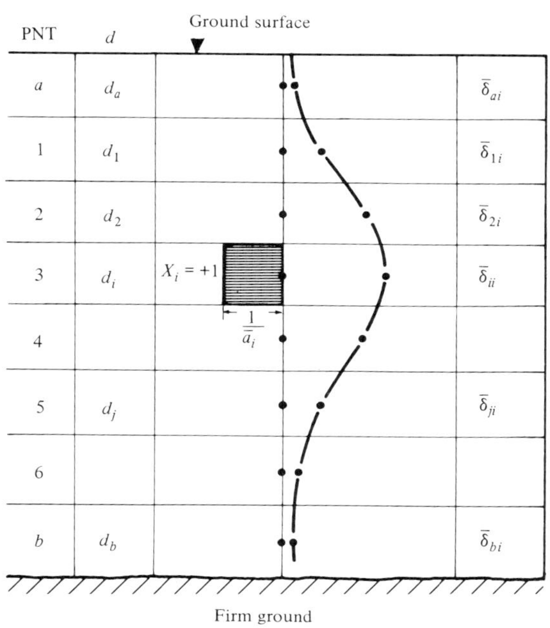

Fig. 2-XII.6 Horizontal displacements because of $X_i = +1$ applied in pier section d_i.

Now let us consider EMI from Chapter VI. We will call it here HEMISES or "seismic horizontal soil-structure interaction matrix equation"; this we may write in the following form:

$$\{[\bar{S}'_{ji}] + [\bar{S}''_{ji}] + [\bar{\delta}_{ji}]\} \cdot |X_i| = |\Delta^P_{io}| + \Delta^A_{io}| + |\Delta\delta_{si}| \qquad \text{2-XII.6}$$

The above expression is shown exploded in Fig. 3-XII.6, from which we may recognize two conditions:

a) Terms that are independent of the soil displacements produced by the soil-pier reactions X_i. This condition reads as follows:

$$[\bar{S}'_{ji}] \cdot |X_i| = |\Delta^P_{io}| + |\Delta\delta_{si}| \qquad \text{3-XII.6}$$

b) Terms that depend exclusively on the soil horizontal displacements produced by the soil-pier reactions and end moments. This fraction will read as follows:

$$\{[\bar{S}''_{ji}] + [\bar{\delta}_{ji}]\} \, |X_i| = |\Delta^A_{io}| \qquad \text{4-XII.6}$$

where $|X_i|$ represents the unknown values: end moment reactions X_a and X_b, and soil-pier horizontal reactions X_i, X_j.

The addition of 3-XII.6 and 4-XII.6 gives us the complete seismic matrix equation HEMISES Fig. 3-XII.6. The significance of the different matrices is explained below with the aid of Figs. 4 to 7-XII.6:

$[\bar{S}'_{ji}]$ symmetrical square matrix representing the flexibility coefficients of the pier under conditions $X_a = +1$, $X_i = +1$, and $X_b = +1$, respectively.

$$\overset{(1)}{\begin{bmatrix} \bar{\theta}'_{aa} & \bar{\theta}'_{ai} & \bar{\theta}'_{aj} & \bar{\theta}'_{ab} \\ \bar{S}'_{aa} & \bar{S}'_{ii} & \bar{S}'_{ij} & \bar{S}'_{ib} \\ \bar{S}'_{ja} & \bar{S}'_{ji} & \bar{S}'_{jj} & \bar{S}'_{jb} \\ \bar{\theta}'_{ba} & \bar{\theta}'_{bi} & \bar{\theta}'_{bj} & \bar{\theta}'_{bb} \end{bmatrix}} \cdot \begin{vmatrix} X_a \\ X_i \\ X_j \\ X_b \end{vmatrix} + \overset{(2)}{\begin{bmatrix} \bar{\psi}''_{aa} & \bar{\psi}''_{ai} & \bar{\psi}''_{aj} & \bar{\psi}''_{ab} \\ \bar{S}''_{ia} & \bar{S}''_{ii} & \bar{S}''_{ij} & \bar{S}''_{ib} \\ \bar{S}''_{ja} & \bar{S}''_{ji} & \bar{S}''_{jj} & \bar{S}''_{jb} \\ \bar{\psi}''_{ba} & \bar{\psi}''_{bi} & \bar{\psi}''_{bj} & \bar{\psi}''_{bb} \end{bmatrix}} \cdot \begin{vmatrix} X_a \\ X_i \\ X_j \\ X_b \end{vmatrix}$$

$$+ \overset{(3)}{\begin{bmatrix} \frac{1}{K_{\theta a}} & \cdots & & \\ \vdots & \bar{\delta}_{ii} & \bar{\delta}_{ij} & \vdots \\ & \bar{\delta}_{ji} & \bar{\delta}_{jj} & \\ & & \cdots & \frac{1}{K_{\theta B}} \end{bmatrix}} \cdot \begin{vmatrix} X_a \\ X_i \\ X_j \\ X_b \end{vmatrix} = \begin{vmatrix} \theta^P_{ao} \\ \Delta^P_{io} \\ \Delta^P_{jo} \\ \theta^P_{bo} \end{vmatrix} + \begin{vmatrix} \psi^A_0 \\ \Delta^A_{io} \\ \Delta^A_{jo} \\ \psi^A_0 \end{vmatrix} + \begin{vmatrix} -\frac{\delta_{sa}}{h} \\ \delta_{si} - \frac{\delta_{sa}}{h} \cdot z_i \\ \delta_{sj} - \frac{\delta_{sa}}{h} \cdot z_j \\ -\frac{\delta_{sa}}{h} \end{vmatrix}$$

Fig. 3-XII.6 HEMISES matrix equation.

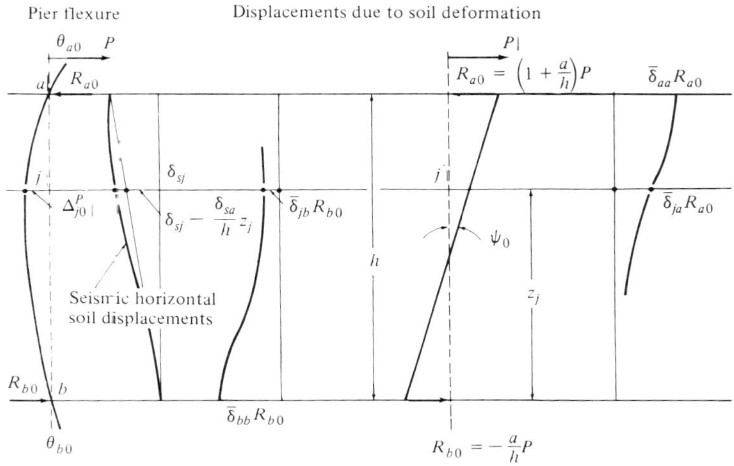

Fig. 4-XII.6 Conditions $X_a = 0$, $X_j = 0$ and $X_b = 0$.

$|\Delta^P_{i0}|$ columnar matrix representing the horizontal displacement of the pier due to force P applied at the head under condition $X_a = 0$, $X_i = 0$ and $X_b = 0$.

$|\Delta\delta_{si}|$ columnar matrix representing the rotations and differential horizontal displacements due to the seismic soil displacements in the length of the pier.

$[\bar{S}''_{ji}]$ square matrix representing the horizontal displacements produced in the soil mass by the end reactions applied at end sections (a) and (b) respectively, under conditions: $X_a = +1$, $X_i = +1$ and $X_b = +1$.

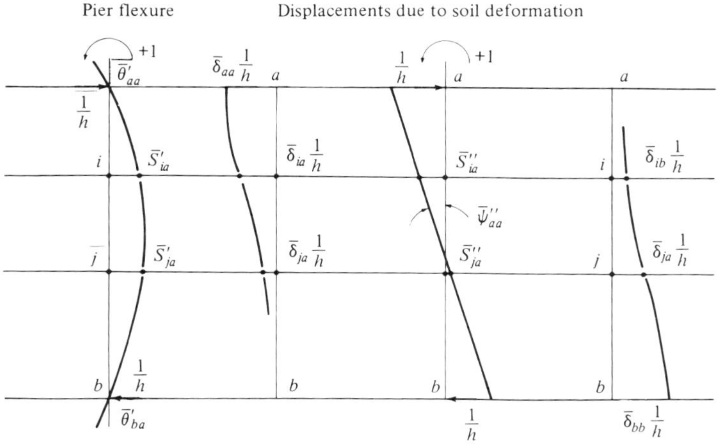

Fig. 5-XII.6 CONDITION $X_a = +1$.

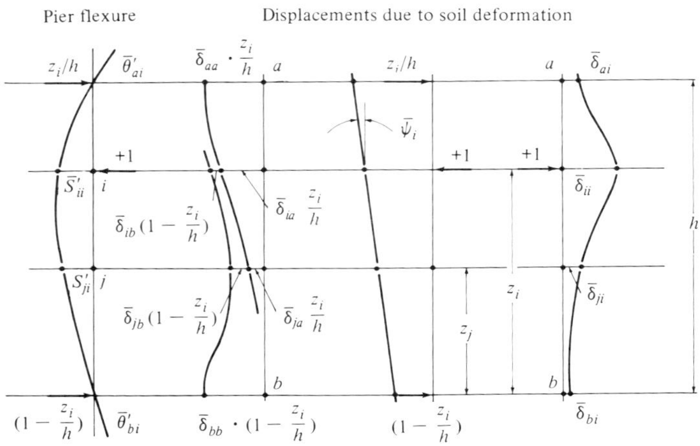

Fig. 6-XII.6 CONDITION $X_i = +1$.

$[\bar{\delta}_{ji}]$ square matrix representing the unit horizontal soil displacements produced by soil-pier horizontal reactions under condition $X_i = +1$.

$|\Delta_{i0}^A|$ columnar matrix representing the soil horizontal displacements due to reactions R_{a0} and R_{b0} at sections (a) and (b) due to conditions $X_a = 0$, $X_i = 0, X_b = 0$.

The matrices in expression 3-XII.6 are readily known. The matrices in expression 4-XII.6, however, have to be investigated for their values, except for $[\bar{\delta}_{ji}]$ which is

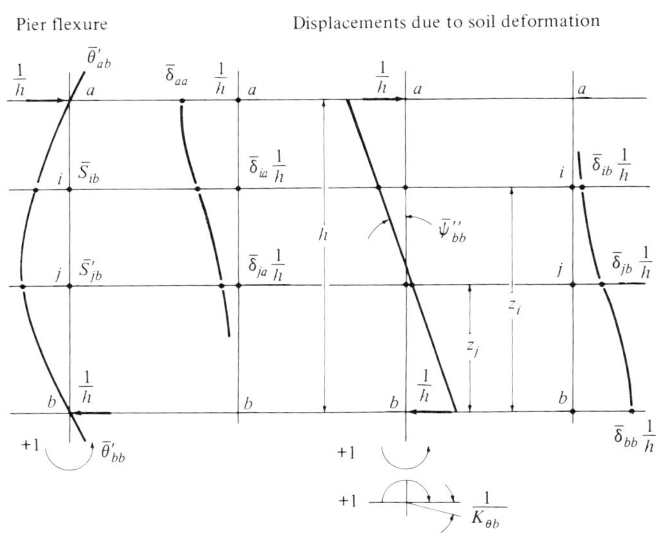

Fig. 7-XII.6 CONDITION $X_b = +1$

known from HEMAS (1-XII.6). To find their values as a function of the unit soil displacements and X_i values we proceed by introducing HEMAS into the unknown part of HEMISES (4-XII.6). Let us calculate from HEMAS (1-XII.6) the values of δ_a and δ_b, considering that for sections (a) and (b) $\delta_{ab} \doteq 0$ and $\delta_{ba} \doteq 0$; hence,

$$\delta_a = \bar{\delta}_{aa} R_a + \sum \bar{\delta}_{ai} X_i$$
$$\delta_b = \bar{\delta}_{bb} R_b + \sum \bar{\delta}_{bi} X_i \qquad \text{5-XII.6}$$

The equilibrium condition (Fig. 7-XII.6) requires that

$$R_a = \left(1 + \frac{a}{h}\right) P - \sum \frac{z_i}{h} X_i - \frac{X_a + X_b}{h}$$

and

$$R_b = -\frac{a}{h} P - \sum \left(1 - \frac{z_i}{h}\right) X_i + \frac{X_a + X_b}{h}$$

But in the condition $X = 0$ (Fig. 4-XII.6) the end reactions have the values $R_{ao} = (1 + a/h)P$ and $R_{bo} = -(a/h)P$. Therefore, we may write

$$R_a = R_{ao} - \sum \frac{z_i}{h} X_i - \frac{X_a + X_b}{h}$$
$$R_b = R_{bo} - \sum \left(1 - \frac{z_i}{h}\right) X_i + \frac{X_a + X_b}{h} \qquad \text{6-XII.6}$$

Substituting the reactions R_a and R_b in 5-XII.6 we obtain

$$\delta_a = \bar{\delta}_{aa} R_{ao} - \sum \bar{\delta}_{aa} \frac{z_i}{h} X_i - \bar{\delta}_{aa} \frac{X_a + X_b}{h} + \sum \bar{\delta}_{ai} X_i$$
$$\delta_b = \bar{\delta}_{bb} R_{bo} - \sum \bar{\delta}_{bb} \left(1 - \frac{z_i}{h}\right) X_i + \bar{\delta}_{bb} \frac{X_a + X_b}{h} + \sum \bar{\delta}_{bi} X_i \qquad \text{7-XII.6}$$

Considering the rotation of the pier as a rigid body we deduce the following condition for point j:

$$\delta_j = \delta_a \frac{z_j}{h} + \delta_b \left(1 - \frac{z_j}{h}\right) \qquad \text{8-XII.6}$$

And, substituting the values of δ_a and δ_b from 7-XII.6 and arranging terms, we obtain

$$\delta_j = \bar{\delta}_{aa} \frac{z_i}{h} R_{ao} + \bar{\delta}_{bb} \left(1 - \frac{z_j}{h}\right) R_{bo} - \sum \left\{ \bar{\delta}_{aa} \frac{z_j z_i}{h^2} + \bar{\delta}_{bb} \left(1 - \frac{z_j}{h}\right)\left(1 - \frac{z_i}{h}\right) \right.$$
$$\left. - \bar{\delta}_{ai} \frac{z_j}{h} - \bar{\delta}_{bi} \left(1 - \frac{z_j}{h}\right) \right\} X_i - \left\{ \bar{\delta}_{aa} \frac{z_j}{h} - \bar{\delta}_{bb} \left(1 - \frac{z_j}{h}\right) \right\} \frac{X_a + X_b}{h} \qquad \text{9-XII.6}$$

On the other hand, we can obtain from HEMAS (1-XII.6) the value of δ_j, hence

$$\delta_j = \bar{\delta}_{ja} R_a + \sum \bar{\delta}_{ji} X_i + \bar{\delta}_{jb} R_b$$

Here we substitute the values of R_a and R_b from 6-XII.6 and arranging terms we obtain

$$\delta_j = \bar{\delta}_{ja} R_{ao} + \bar{\delta}_{jb} R_{bo} - \sum \left\{ \bar{\delta}_{ja} \frac{z_i}{h} + \bar{\delta}_{jb} \left(1 - \frac{z_i}{h}\right) + \bar{\delta}_{ji} \right\} X_i$$

$$- (\bar{\delta}_{ja} - \bar{\delta}_{jb}) \frac{X_a + X_b}{h} \qquad \text{10-XII.6}$$

In order to establish the compatibility of horizontal displacements we equalize expressions 9-XII.6 and 10-XII.6, and after algebraic operations we obtain

$$\frac{1}{h} \left\{ \bar{\delta}_{aa} \frac{z_j}{h} - \bar{\delta}_{bb} \left(1 - \frac{z_j}{h}\right) - (\bar{\delta}_{ja} - \bar{\delta}_{jb}) \right\} X_a + \sum \left\{ \bar{\delta}_{aa} \frac{z_j z_i}{h^2} \right.$$

$$\left. + \bar{\delta}_{bb} \left(1 - \frac{z_j}{h}\right) \left(1 - \frac{z_i}{h}\right) - \bar{\delta}_{ai} \frac{z_j}{h} - \bar{\delta}_{bi} \left(1 - \frac{z_j}{h}\right) - \bar{\delta}_{ja} \frac{z_i}{h} - \bar{\delta}_{jb} \left(1 - \frac{z_i}{h}\right) \right\} X_i$$

$$+ \sum \bar{\delta}_{ji} X_i + \frac{1}{h} \left\{ \bar{\delta}_{aa} \frac{z_j}{h} - \bar{\delta}_{bb} \left(1 - \frac{z_j}{h}\right) - (\bar{\delta}_{ja} - \bar{\delta}_{jb}) \right\} X_b$$

$$= \left(\bar{\delta}_{aa} \frac{z_j}{h} - \bar{\delta}_{ja} \right) R_{ao} + \left\{ \bar{\delta}_{bb} \left(1 - \frac{z_j}{h}\right) - \bar{\delta}_{jb} \right\} R_{bo} \qquad \text{11-XII.6}$$

From expression 11-XII.6 it may be recognized that the values of the coefficients of the unknowns X_a, X_i and X_b and the independent terms in matrix equation 4-XII.6, line j, when the pier is acting as a rigid body are:

1) Coefficients of X_a and X_b for $X_a = +1$ and $X_b = +1$;

$$\bar{S}''_{ja} = \bar{S}''_{jb} = \frac{1}{h} \left\{ \bar{\delta}_{aa} \frac{z_j}{h} - \bar{\delta}_{bb} \left(1 - \frac{z_j}{h}\right) - (\bar{\delta}_{ja} - \bar{\delta}_{jb}) \right\} \qquad \text{12-XII.6}$$

2) Coefficients of X_i for $X_i = +1$;

$$\bar{S}''_{ji} + \bar{\delta}_{ji} = \left\{ \bar{\delta}_{aa} \frac{z_j z_i}{h^2} + \bar{\delta}_{bb} \left(1 - \frac{z_j}{h}\right) \left(1 - \frac{z_i}{h}\right) - \bar{\delta}_{ai} \frac{z_j}{h} \right\}$$

$$- \bar{\delta}_{bi} \left(1 - \frac{z_j}{h}\right) - \bar{\delta}_{ja} \frac{z_i}{h} - \bar{\delta}_{jb} \left(1 - \frac{z_i}{h}\right) + \bar{\delta}_{ji} \qquad \text{13-XII.6}$$

3) Independent terms for $X = 0$;

$$\Delta^A_{jo} = \left(\bar{\delta}_{aa} \frac{z_j}{h} - \bar{\delta}_{ja} \right) R_{ao} - \left\{ \bar{\delta}_{bb} \left(1 - \frac{z_j}{h}\right) - \bar{\delta}_{jb} \right\} R_{bo} \qquad \text{14-XII.6}$$

From above expressions we write for the j line corresponding to matrix equation 4-XII.6

$$\bar{S}'_{ja} X_a + (\bar{S}''_{ji} + \bar{\delta}_{ji}) X_i + (\bar{S}''_{jj} + \bar{\delta}_{jj}) X_j + \bar{S}''_{jb} X_b = \Delta^A_{jo} \qquad \text{15-XII.6}$$

and, for the j line of HEMISES according to Fig. 2-XII.6

$$(\bar{S}'_{ja} + \bar{S}''_{ja}) X_a + \sum (\bar{S}'_{ji} + \bar{S}''_{ji} + \bar{\delta}_{ji}) X_i + (\bar{S}'_{jb} + \bar{S}''_{jb}) X_b = \Delta^P_{io} + \Delta^A_{io} + \left(\delta_{sj} - \frac{\delta_{sa}}{h} z_j\right) \qquad \text{16-XII.6}$$

The end rotation coefficients of lines (a) and (b) in matrix equation HEMISES induced by end moments X_a and X_b, respectively, may be obtained algebraically using a similar procedure as applied to obtain expression 15-XII.6, or we can determine these values physically from Figs. 4, 5, 6 and 7-XII.6. Furthermore, we will call $K_{\theta a}$ the rigidity per unit moment of the foundation structure joining the pier. This value may be obtained by conventional methods in structural engineering. The foundation rotation modulus at the base of the pier of diameter $2r_0$ will be called; $K_{\theta b}$.

From the Theory of Elasticity (Fröhlich 1953), the rotation of a circular rigid plate subjected to a moment M, and supported on an homogeneous and isotropic semi-infinite mass with modulus of elasticity E, is given by:

$$\theta = (1 - \nu^2) \frac{3}{4} \frac{M}{ER_0^3}$$

from which the foundation rotation modulus is

$$K_{\theta b} = \frac{4ER_0^3}{3(1 - \nu^2)}$$

Using dynamic mechanical properties of the subsoil $E = 2(1 + \nu)\mu$ and for a semi-impervious soil we consider $\nu = 0.5$, hence we can write

$$K_{\theta b} = 4R_0^3 \mu \qquad \text{17-XII.6}$$

Finally, we obtain the algorithms to calculate by columns the coefficients and independent terms of HEMISES for a pile or pier:

$$\begin{bmatrix} \bar{\theta}_{aa} & \bar{\theta}_{ai} & \bar{\theta}_{aj} & \bar{\theta}_{ab} \\ \bar{S}_{ia} & \bar{S}_{ii} & \bar{S}_{ij} & \bar{S}_{ib} \\ \bar{S}_{ja} & \bar{S}_{ji} & \bar{S}_{jj} & \bar{S}_{jb} \\ \bar{\theta}_{ba} & \bar{\theta}_{bi} & \bar{\theta}_{bj} & \bar{\theta}_{bb} \end{bmatrix} \cdot \begin{vmatrix} X_a \\ X_i \\ X_j \\ X_b \end{vmatrix} = \begin{vmatrix} \theta_{ao} - \dfrac{\delta_{sa}}{h} \\ \Delta_{io} + \left(\delta_{si} - \dfrac{\delta_{sa}}{h} z_i\right) \\ \Delta_{jo} + \left(\delta_{sj} - \dfrac{\delta_{sa}}{h} z_j\right) \\ \theta_{bo} - \dfrac{\delta_{sa}}{h} \end{vmatrix} \qquad \text{18-XII.6}$$

1) First column X_a coefficients, moment at the head of the pier

$$EI\overline{\theta}_{aa} = \left(a + \frac{h}{3}\right) + \frac{EI}{h^2}(\overline{\delta}_{aa} + \overline{\delta}_{bb}) + \frac{EI}{K_{\theta a}}$$

$$EI\overline{S}_{ia} = -\frac{h^2}{6}\left(\frac{z_i}{h} - \frac{z_i^3}{h^3}\right) + \frac{EI}{h}\left\{\overline{\delta}_{aa}\frac{z_i}{h} - \overline{\delta}_{bb}\left(1 - \frac{z_i}{h}\right) - (\overline{\delta}_{ia} - \overline{\delta}_{ib})\right\} \quad \text{19-XII.6}$$

$$EI\overline{\theta}_{ba} = -\frac{h}{6} + \frac{EI}{h^2}(\overline{\delta}_{aa} + \overline{\delta}_{bb})$$

2) Column i, coefficient of X_i, soil-pier reactions on the shaft of the pier:

$$EI\overline{\theta}_{ai} = -\frac{h^2}{6}\left(\frac{z_i}{h} - \frac{z_i^3}{h^3}\right) + \frac{EI}{h}\left\{\overline{\delta}_{aa}\frac{z_i}{h} - \overline{\delta}_{bb}\left(1 - \frac{z_i}{h}\right) - (\overline{\delta}_{ai} - \overline{\delta}_{bi})\right\}$$

for $z_j > z_i$

$$EI\overline{S}_{ji} = \frac{h^3}{6}\left(1 - \frac{z_j}{h}\right)\frac{z_i}{h}\left\{1 - \left(\frac{z_i}{h}\right)^2 - \left(1 - \frac{z_j}{h}\right)^2\right\} + EI\left\{\overline{\delta}_{aa}\frac{z_j z_i}{h^2}\right.$$
$$+ \overline{\delta}_{bb}\left(1 - \frac{z_j}{h}\right)\left(1 - \frac{z_i}{h}\right) - \overline{\delta}_{ai}\frac{z_j}{h} - \overline{\delta}_{bi}\left(1 - \frac{z_j}{h}\right) - \overline{\delta}_{ja}\frac{z_i}{h}$$
$$\left. - \overline{\delta}_{jb}\left(1 - \frac{z_i}{h}\right) + \overline{\delta}_{ji}\right\} \quad \text{20-XII.6}$$

and for $z_i > z_j$

$$EI\overline{S}_{ji} = \frac{h^3}{6}\frac{z_j}{h}\left(1 - \frac{z_i}{h}\right)\left\{1 - \left(1 - \frac{z_i}{h}\right)^2 - \left(\frac{z_j}{h}\right)^2\right\} + EI\left\{\overline{\delta}_{aa}\frac{z_j z_i}{h^2}\right.$$
$$+ \overline{\delta}_{bb}\left(1 - \frac{z_j}{h}\right)\left(1 - \frac{z_i}{h}\right) - \overline{\delta}_{ai}\frac{z_j}{h} - \overline{\delta}_{bi}\left(1 - \frac{z_j}{h}\right)$$
$$\left. - \overline{\delta}_{ja}\frac{z_i}{h} - \overline{\delta}_{jb}\left(1 - \frac{z_i}{h}\right) + \overline{\delta}_{ji}\right\} \quad \text{21-XII.6}$$

3) Last column, X_b coefficients, moment at the base of the pier

$$EI\overline{\theta}_{ab} = -\frac{h}{6} + \frac{EI}{h^2}(\overline{\delta}_{aa} + \overline{\delta}_{bb})$$

$$EI\overline{S}_{ib} = \frac{h^2}{6}\frac{z_i}{h}\left(1 - \frac{z_i}{h}\right)\left(2 - \frac{z_i}{h}\right) + \frac{EI}{h}\left\{\overline{\delta}_{aa}\frac{z_i}{h} - \overline{\delta}_{bb}\left(1 - \frac{z_i}{h}\right)\right.$$
$$\left. - (\overline{\delta}_{ia} - \overline{\delta}_{ib})\right\} = EI\overline{\theta}_{bi} \quad \text{22-XII.6}$$

$$EI\overline{\theta}_{bb} = \frac{h}{3} + \frac{EI}{h^2}(\overline{\delta}_{aa} + \overline{\delta}_{bb}) + \frac{EI}{K_{\theta b}}$$

4) Column of independent terms:

$$EI\theta_{ao} = \left\{\frac{a^2}{2} + \frac{ah}{3} + \frac{EI}{h}\left\{\bar{\delta}_{aa}\left(1 + \frac{a}{h}\right) + \bar{\delta}_{bb}\frac{a}{h}\right\}\right\}P - \frac{EI}{h}\delta_{sa}$$

$$EI\Delta_{io} = \left\{-\frac{ah^2}{6}\left(\frac{z_i}{h} - \frac{z_i^3}{h^3}\right)\right\}P + \left\{\left(\bar{\delta}_{aa}\frac{z_i}{h} - \bar{\delta}_{ia}\right)\left(1 + \frac{a}{h}\right) - \left\{\bar{\delta}_{bb}\left(1 - \frac{z_i}{h}\right) - \bar{\delta}_{ib}\right\}\frac{a}{h}\right\}(EI)P + \left(\delta_{si} - \frac{\delta_{sa}}{h}z_i\right)(EI) \qquad \text{23-XII.6}$$

$$EI\theta_{bo} = \left\{-\frac{ah}{6} + \frac{EI}{h}\left\{\bar{\delta}_{aa}\left(1 + \frac{a}{h}\right) + \bar{\delta}_{bb}\frac{a}{h}\right\}\right\}P - \frac{(EI)\delta_{sa}}{h}$$

The calculation of the coefficients and independent terms in HEMISES (18-XII.6) with algorithms given in 19 to 23-XII.6, is greatly facilitated using format given in Fig. 8-XII.6.

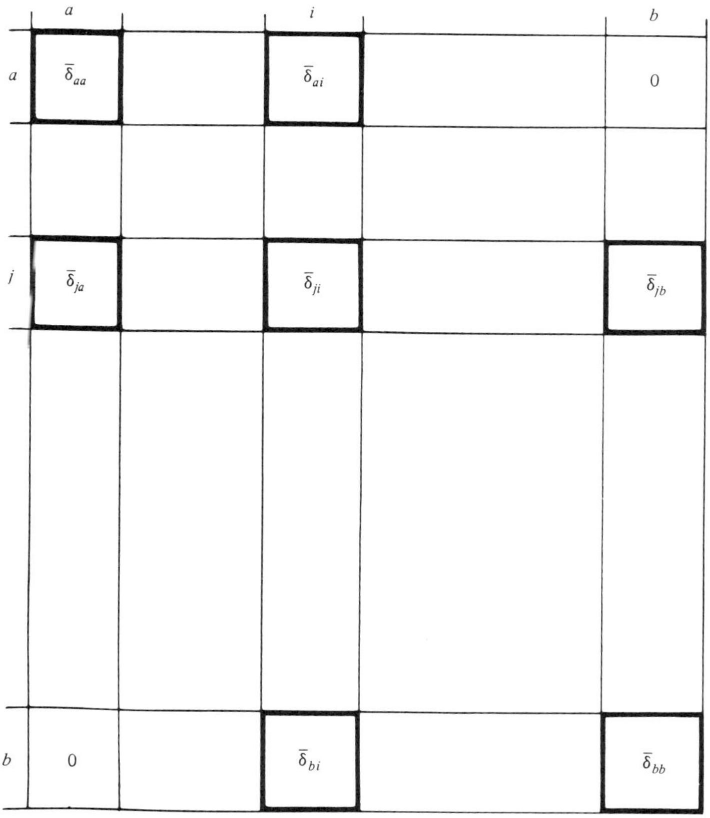

Fig. 8-XII.6 Format to form HEMISES coefficients.

The values $\overline{\delta}_{ji}$ in HEMAS are a function of the horizontal stress distribution in the soil mass and of the dynamic stress-strain properties of the soil, which is determined for each stratum pierced by the pile or pier. The horizontal stress distribution, however, in the case of stratified soil mass, may only be calculated approximately. The author proposes a practical method to calculate the $\overline{\delta}_{ji}$ values on the safe side as described in the following paragraphs (L. Zeevaert, 1980, ISE).

Assume there is available an expression to calculate the influence values $\overline{\delta}_{ji}$ with reasonable accuracy for uniformly loaded vertical rectangular areas in the interior of the semi-infinite mass, representing the pier sections with tributary area $\overline{a}_i = (2r_0)\,d_i$ (Fig. 9-XII.6). The tributary areas will be loaded with a unit or equivalent uniform horizontal load per unit area $1/\overline{a}_i$ applied in section i. Hence, we obtain influence stress values I_{ji} at the center line passing through each tributary area and at distances $n\Delta x$ away from the pile shaft and for each soil layer defining a soil stratum with dynamic strain modulus M_{ej}. The horizontal displacement at point j due to unit load $1/\overline{a}_i$ applied on section i is computed by

$$\overline{\delta}_{ji} = M_{ej} \sum_{1}^{n} I_{ji}^{n} \Delta x \qquad \text{24-XII.6}$$

where $\sum I_{ji}\Delta x$ represents the stress area under section j due to unit load applied on section i. In the integration we consider Δx = constant and take integration steps

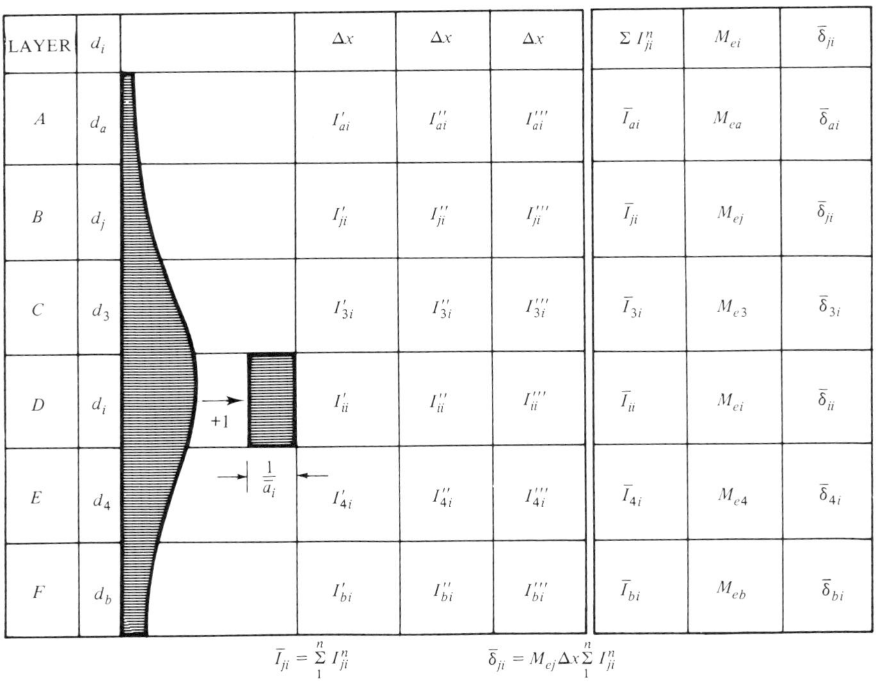

Fig. 9-XII.6 Influence displacements for unit horizontal stress applied in pier section d_i.

up to a value of I_{ji}^{n+1} of 5% of $\Sigma_1^n I_{ji}^n$. Hence we may write

$$\bar{\delta}_{ji} = M_{ej} \Delta x \sum_1^n I_{ji}^n \qquad \text{25-XII.6}$$

For the seismic problem we use

$$M_{ej} = \frac{1}{3\mu_j}$$

where μ_j is the soil dynamic rigidity for stratum j. Finally we can write

$$\bar{\delta}_{ji} = \frac{\Delta x}{3\mu_j} \sum_1^n I_{ji}^n \qquad \text{26-XII.6}$$

With the unit horizontal displacements $\bar{\delta}_{ji}$ for each pier section loaded with $1/\bar{a}_i$ at section i (Fig. 9-XII.6) we form the HEMAS square matrix for all the pile sections and strata considered as shown below:

$$\begin{array}{c} \text{Layer} \\ \text{a} \\ i \\ 3 \\ j \\ 5 \\ \text{b} \end{array} \quad \begin{array}{c} \\ \begin{bmatrix} \bar{\delta}_{aa} & \bar{\delta}_{ai} & \bar{\delta}_{a3} & \bar{\delta}_{aj} & \bar{\delta}_{ab} \\ \bar{\delta}_{ia} & \bar{\delta}_{ii} & \bar{\delta}_{i3} & \bar{\delta}_{ij} & \bar{\delta}_{ib} \\ \bar{\delta}_{3a} & \bar{\delta}_{3i} & \bar{\delta}_{33} & \bar{\delta}_{3j} & \bar{\delta}_{3b} \\ \bar{\delta}_{ja} & \bar{\delta}_{ji} & \bar{\delta}_{j3} & \bar{\delta}_{jj} & \bar{\delta}_{jb} \\ \bar{\delta}_{5a} & \bar{\delta}_{5i} & \bar{\delta}_{53} & \bar{\delta}_{5j} & \bar{\delta}_{5b} \\ \bar{\delta}_{ba} & \bar{\delta}_{bi} & \bar{\delta}_{b3} & \bar{\delta}_{bj} & \bar{\delta}_{bb} \end{bmatrix} \end{array} = [\bar{\delta}_{ji}]$$

In case of piles and piers we find that the influence horizontal displacements in the HEMAS matrix are very small for sections away from section i where the unit load $1/\bar{a}_i$ is applied, therefore we obtain, in general, a diagonal band matrix neglecting the terms of small order. Usually, sufficient accuracy is obtained by considering a three band matrix for $[\bar{\delta}_{ji}]$. When integrating $\bar{I}_{ji} = \Sigma_1^n I_{ji}^n$ it is recommended to use $n \geqslant 5$, that is, five Δx sections away from the pile shaft, also we may consider $\Delta x = 3r_0$ where $2r_0$ is the diameter of the pile. Therefore, the distances from the pile shaft to obtain the I_{ji} values are $1.5\, r_0$, $4.5\, r_0$, $7.5\, r_0$, $10.5\, r_0$ and $13.5\, r_0$.

The author has obtained from the theory of elasticity and the images method (ISE, L. Zeevaert, 1980), the following approximate expression for the influence coefficients according to Fig. 10-XII.6.

$$I_{ji} = \frac{3}{2\pi}\left[\left(\sin\alpha_0 - \frac{\sin^3\alpha_0}{3}\right)\left\{(\psi_1 - \psi_2) + \frac{1}{2}(\sin 2\psi_1 - \sin 2\psi_2)\right\}\right.$$
$$\left. + \left(\sin\alpha_0' - \frac{\sin^3\alpha_0'}{3}\right)\left\{(\psi_1' - \psi_2') + \frac{1}{2}(\sin 2\psi_1' - \sin 2\psi_2')\right\}\right] \qquad \text{27-XII.6}$$

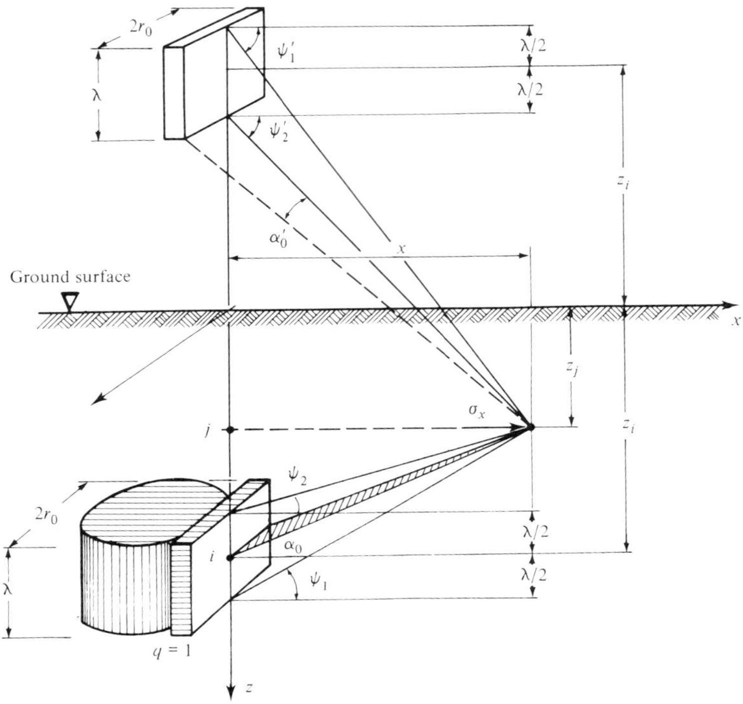

Fig. 10-XII.6 Horizontal stress at a point due to a uniformly loaded vertical rectangular area.

here the angular arguments are:

$$\alpha_0 = \tan^{-1}\frac{r_0}{\sqrt{(z_i - z_j)^2 + x^2}}, \quad \psi_1 = \tan^{-1}\frac{(z_i - z_j) + \lambda/2}{x}, \quad \psi_2 = \tan^{-1}\frac{(z_i - z_j) - \lambda/2}{x}$$

$$\alpha_0' = \tan^{-1}\frac{r_0}{\sqrt{(z_i + z_j)^2 + x^2}}, \quad \psi_1' = \tan^{-1}\frac{(z_i + z_j) + \lambda/2}{x}, \quad \psi_2' = \tan^{-1}\frac{(z_i + z_j) - \lambda/2}{x}$$

To illustrate the calculation procedure let us solve a problem. Suppose that a foundation on piers is required for a building to be constructed in a seismic area. The piers will be constructed one under each column and will reach firm ground at a depth of 18.0 m. The piers have a constant diameter $2r_0$ = 1.2 m. The top of the pier will be constructed monolithically with the foundation beams and ground floor slab. The rotational rigidity of the foundation structure has been estimated to be $K_{\theta a} = 8.454 \times 10^4$ ton · m/rad. The bottom of the piers will penetrate the firm ground 2 m and the base will be bell shaped with a diameter of 2.4 m. The geometry of the problem and soil-dynamics properties of the soil are given in Fig. 11-XII.6. The ground surface has been assigned a surface response acceleration of 1 m/sec^2. The seismic base shear at the pier head is estimated in

Fig. 11-XII.6 Rupture of pile due to high bending moments.

57.4 ton. The dynamic modulus of elasticity of the concrete of the piers is estimated to be $E = 2.60 \times 10^5$ ton/m^2, and the value of $EI = 2.646 \times 10^4$ ton · m^2. The calculations given here were performed with a desk computer. The reader may verify the results of the different steps of calculation using the algorithms given in this section.

The first step is to calculate the seismic soil horizontal displacements δ_{si} due to the ground surface acceleration of 1.0 m/sec^2. Using soil information given in Fig. 12-XII.6 we compute the period of the soil deposit using the shear wave velocity method (section XII-3.2), as given in Table 1-XII.6. Furthermore, we compute the maximum horizontal soil displacements to be expected in the subsoil using the method explained in Chapter XII, section 3.2 with algorithms 20 to 22-XII.3. The ground surface displacement is calculated by $\delta_{s0} = a_m/\omega^2$, in which ω is the domi-

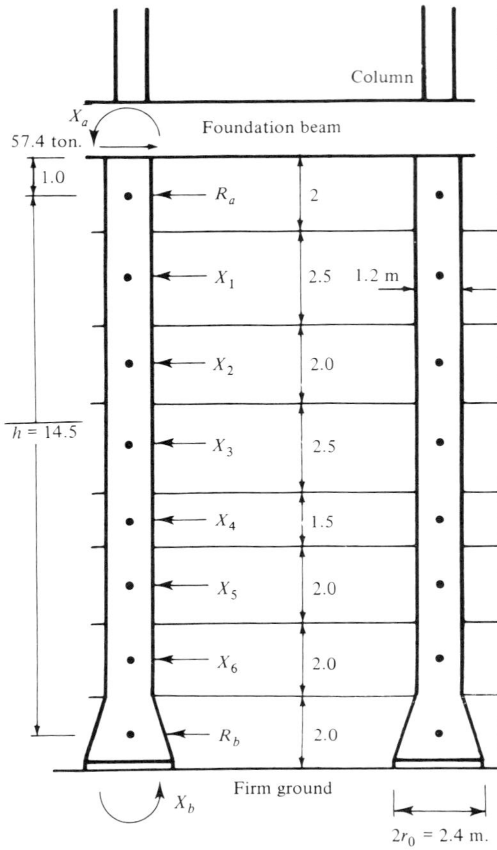

DEPTH

DEPTH	z_i	γ	q_u	μ
0	18.0	ton/m^3	ton/m^2	ton/m^2
1.5				
3.5	14.50	1.65	18.0	1500
6.0	12.25	1.32	12.0	380
8.0	10.0	1.20	6.0	280
10.5	7.75	1.40	7.5	650
12.0	5.75	1.20	20	2500
14.0	4.0	1.40	90	850
16.0	2.0	1.45	10.0	900
18.0	0	1.60	25	2000
		1.75	40	6000

Fig. 12-XII.6 Soil information for pier foundation.

nant natural frequency of the deposit to firm ground; also $\delta_{so} = a_m T^2/(2\pi)^2$. From Table 1-XII.6 the first approximation for the dominant period of the ground is $T = 0.97$ sec, hence $\omega = 6.464$ rad/sec. The horizontal displacements at the center of the strata considered may be found in Table 2-XII.6.

The second step is to calculate the HEMAS matrix for the values in Table 3-XII.6. Using the HEMAS matrix of $\overline{\delta}_{ji}$ values (format Fig. 8-XII.6 and algorithms 19 to 23-XII.6), HEMISES is calculated as given in Table 4-XII.6. Upon solving for the unknowns the following values are found in Table 5-XII.6.

With the values given above and seismic base shear at the head of the pile of 57.4 ton, the reactions R_a and R_b are calculated. The final results are given in Table 6-XII.6.

Knowing the soil reactions on the pier and end moments, the shear forces and bending moments in the pier are calculated (Table 7-XII.6) and plotted (Fig. 14-XII.6).

TABLE 1-XII.6

Depth	d_i	γ	μ	v_s	$4\Sigma\Delta T$
1.5	1.5	1.65	1500	94.44	0.064
3.5	2.0	1.65	1500	94.44	0.148
6.0	2.5	1.32	380	53.14	0.336
8.0	2.0	1.20	280	47.84	0.504
10.5	2.5	1.40	650	67.49	0.652
12.0	1.5	1.20	2500	142.96	0.694
14.0	2.0	1.40	850	77.176	0.797
16.0	2.0	1.45	900	78.032	0.900
18.0	2.0	1.60	2000	110.736	0.972
m	m	ton/m^3	ton/m^2	m/sec	sec

TABLE 2-XII.6

$\omega = 6.464$ rad/sec

Depth	d_i	γ	μ	δ_{si}	τ_{si}	$(\delta_{si})_{av}$
0	–	–	–	–	0	–
1.5	1.5	1.65	1500	2.381	0.252	2.387
3.5	2.0	1.65	1500	2.325	0.582	2.383
6.0	2.5	1.32	380	1.845	0.875	2.085
8.0	2.0	1.20	280	1.165	1.029	1.505
10.5	2.5	1.40	650	0.742	1.171	0.954
12.0	1.5	1.20	2500	0.670	1.226	0.706
14.0	2.0	1.40	850	0.374	1.288	0.522
16.0	2.0	1.45	900	0.085	1.316	0.230
18.0	2.0	1.60	2000	0.005	1.319	0.043
m	m	ton/m^3	ton/m^2	$\times 10^{-2}$ m	ton/m^2	$\times 10^{-2}$ m

TABLE 3-XII.6

$2r_0 = 1.20$ m, $n = 5$, $\Delta x = 3r_0 = 1.80$ m

HEMAS MATRIX

PNT	d_i	$\bar{a}_i$	M_{ej}	DEPTH	a	1	2	3	4	5	6	b
a	2.0	2.40	2.222	1.00	1.712	0.292	0.085	0.033	0.015	0.008	0.005	0.003
1	2.5	3.00	8.772	3.25	1.105	5.174	0.877	0.260	0.106	0.054	0.027	0.014
2	2.0	2.40	11.905	5.50	0.454	1.247	7.818	1.204	0.366	0.167	0.075	0.036
3	2.5	3.00	5.128	7.75	0.075	0.152	0.494	2.970	0.568	0.203	0.079	0.035
4	1.5	1.80	1.333	9.75	0.009	0.016	0.042	0.165	0.996	0.194	0.053	0.020
5	2.0	2.40	3.922	11.50	0.015	0.024	0.055	0.158	0.533	2.568	0.454	0.136
6	2.0	2.40	3.704	13.50	0.008	0.011	0.023	0.058	0.144	0.429	2.425	0.429
b	2.0	2.40	1.667	15.50	0.002	0.003	0.005	0.011	0.025	0.058	0.193	1.091
	m	m^2	$\times 10^{-4}$ m^2/ton	m	$\times 10^{-4}$ m^2/ton							

TABLE 4-XII.6

$a = 1.0$ m, $h = 14.5$ m, $EI = 2.646 \cdot 10^4$ ton · m², $K_{\theta a} = 8.454 \times 10^4$ ton · m/rad, $K_{\theta b} = 4.147 \cdot 10^4$ ton · m/rad, $P = 57.4$ ton

Ordinate			12.25	10.0	7.75	5.75	4.0	2.0		
PNT	δ_{si}	X_a	X_1	X_2	X_3	X_4	X_5	X_6	X_b	IT
a	2.383	6.182	-8.295	-12.534	-13.310	-11.702	-8.984	-4.835	-2.381	282.605
1	2.085	-7.716	31.329	30.027	30.324	26.433	20.223	10.776	5.338	-413.809
2	1.505	-12.595	31.719	68.658	56.489	48.758	37.333	20.126	9.905	-649.218
3	0.954	-13.312	31.149	55.195	72.493	61.657	48.019	26.118	12.844	-721.096
4	0.706	-11.703	27.112	48.460	60.869	62.519	49.334	27.390	13.453	-631.892
5	0.522	-8.964	20.562	37.482	48.012	50.248	48.897	25.429	12.036	-478.637
6	0.230	-4.792	11.038	20.061	26.040	27.475	25.403	21.597	7.708	-265.228
b	0.043	-2.381	5.486	9.966	12.846	13.454	12.016	7.665	5.507	-162.245
	m × 10⁻²									

The soil reactions are analyzed to find out if the soil will not fail under the reactions encountered. The limiting stress allowed on the soil to obtain elastic response with factor of safety of two is approximately (Nadai, 1931): $q_m = \pi s_m/2$. Under quick undrained conditions $s_m = q_u/2$, in which q_u is the consistency of the cohesive soil; therefore

$$q_{max} \geqslant \pi q_u/4 \quad \text{or} \quad R_a \leqslant \frac{\pi}{2}(d_a r_o) q_u$$

From Fig. 12-XII.6 we find at section (a) $q_u = 18$ ton/m², hence $R_a = 33.93$ ton against 60.95 ton in the same section. The design of the pier has, therefore, to be corrected (L. Zeevaert, 1980, ISE.). We assume $R_a = 33.93$ ton and repeat the calculation with same method employed before. The new support (a) will be at sec-

TABLE 5-XII.6

X_a	X_1	X_2	X_3	X_4	X_5	X_6	X_b
39.348	-1.721	0.415	-2.183	0.513	2.315	1.450	-14.716
ton · m	ton	ton	ton	ton	ton	ton	ton · m

TABLE 6-XII.6

X_a	R_a	X_1	X_2	X_3	X_4	X_5	X_6	R_b	M_b
39.348	60.952	-1.721	0.415	-2.183	0.513	2.315	1.450	4.341	-14.716
ton · m	ton	ton	ton	ton	ton	ton	ton	ton	ton · m
	2.0	2.5	2.0	2.5	1.5	2.0	2.0	2.0	$d_j m$
	30.746	-0.688	0.208	-0.873	0.342	1.158	0.725	2.171	Q_l, ton/m

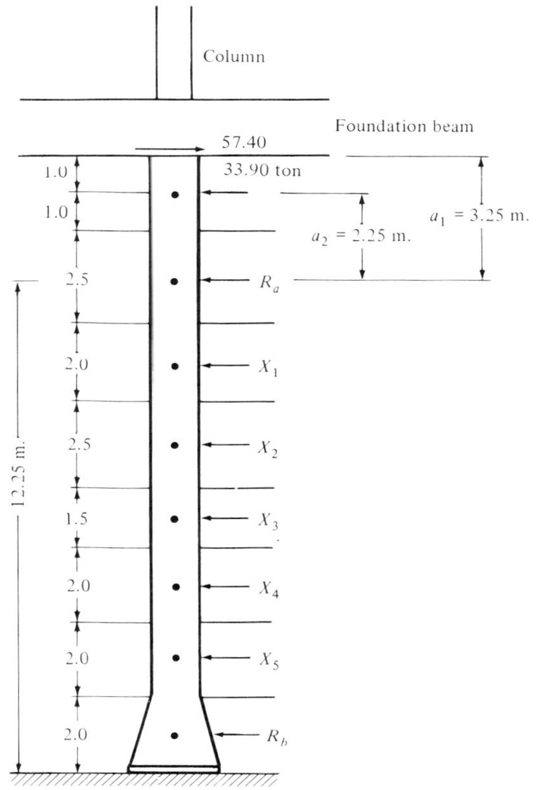

z_i m	δ_{si} m
12.25	2.085
10.0	1.505
7.75	0.954
5.75	0.706
4.0	0.522
2.0	0.230
0	0.043

Fig. 13-XII.6 New conditions with plastic reaction at top.

tion 1, and X_1 now becomes R_a. Furthermore, we will now have five unknown horizontal reactions and unknown end moments X_a and X_b, respectively. The new forces and geometry are considered as shown in Fig. 13-XII.6. The displacements produced by the constant reaction R_1 = 33.93 ton must be deducted from the δ_{si} values for soil seismic displacements. This is obtained multiplying the first column of the HEMAS matrix Table 3-XII.6 by 33.930 ton.

The values obtained are given in Table 8-XII.6 where the new HEMAS matrix is shown. Using values of the above mentioned table and algorithms 19 to 23-XII.6, the coefficients and the independent terms may be calculated for HEMISES as given in Table 9-XII.6 from which the system of simultaneous equations may be solved. The result is given in Table 10-XII.6.

The bending moments for this condition are calculated with soil reactions and end bending moments obtained Table 11-XII.6 and may be found plotted in Fig. 14-XII.6. The importance of the ultimate response reaction at the upper end section (a) is in itself evident.

TABLE 7-XII.6

X_a = 39.348 ton · m

P = –57.4 ton

PNT	X	z_i	d_i	Q_i	V	M
					–57.4	39.348
a	R_a = • 60.952	14.50	2.0	30.4760		
					3.552	–14.500
1	• –1.721	12.25	2.5	–0.6884		
					1.831	–7.771
2	• 0.415	10.0	2.0	0.2075		
					2.246	–3.694
3	• –2.183	7.75	2.5	–0.8732		
					0.063	–0.808
4	• 0.513	5.75	1.5	0.3420		
					0.576	–0.329
5	• 2.315	4.0	2.0	1.1575		
					2.895	3.138
6	• 1.450	2.0	2.0	0.7250		
					4.341	10.370
b	R_b •← –4.341	0	2.0	–2.1705		
					0.00	14.711
	–14.716					
	ton · m	m	m	ton/m	ton	ton · m

TABLE 8-XII.6

δ_{si}	33.93 × δ_{ai}	δ'_{si}	PNT	z_i	a	1	2	3	4	5	b
2.085	0.375	1.710	a		5.174	0.877	0.260	0.106	0.054	0.027	0
1.505	0.154	1.351	1	10.0	1.247	7.818	1.204	0.366	0.167	0.075	0.036
0.954	0.025	0.929	2	7.75	0.152	0.494	2.970	0.568	0.203	0.079	0.035
0.706	0	0.706	3	5.75	0.016	0.042	0.165	0.996	0.194	0.053	0.020
0.522	0	0.522	4	4.0	0.024	0.055	0.158	0.533	2.568	0.454	0.136
0.230	0	0.230	5	2.0	0	0.023	0.058	0.144	0.429	2.425	0.429
0.043	0	0.043	6	0	0	0	0	0.025	0.058	0.193	1.091
m $\times 10^{-2}$	m $\times 10^{-2}$	m $\times 10^{-2}$		m				$\times 10^{-4}$			

TABLE 9-XII.6

h = 12.25 m, EI = 2.646 × 10^4 ton · m^2, $K_{\theta a}$ = 8.454 × 10^4, $K_{\theta b}$ = 4.147 × 10^4 ton · m/rad, a_1 = 3.25 m, a_2 = 2.25 m, P = 57.4 R_a = 33.93 ton

PNT	z_i	δ'_{si}	X_a	X_2	X_3	X_4	X_5	X_6	X_b	$(IT)_1$ +	$(IT)_2$ =	IT
a	12.25	1.710	7.757	-6.132	-8.925	-8.771	-7.089	-3.953	-1.931	1112.711	-480.906	631.805
1	10.0	1.351	-6.204	39.077	27.918	24.358	18.958	10.329	5.046	-717.100	194.484	-522.616
2	7.75	0.929	-8.889	26.910	46.100	39.750	31.611	17.540	8.543	-1220.467	358.364	-862.103
3	5.75	0.706	-8.753	24.208	38.756	44.260	35.923	20.477	9.935	1291.954	426.024	-865.930
4	4.0	0.522	-7.092	19.072	31.436	36.661	38.784	20.389	9.434	-1211.481	380.252	-831.229
5	2.0	0.230	-3.897	10.374	17.331	20.370	20.086	18.849	6.353	-611.817	208.619	-403.198
b	0	0.043	-1.931	5.118	8.512	9.916	9.411	6.297	4.832	-333.052	72.577	-260.475
	m	10^{-2} m										

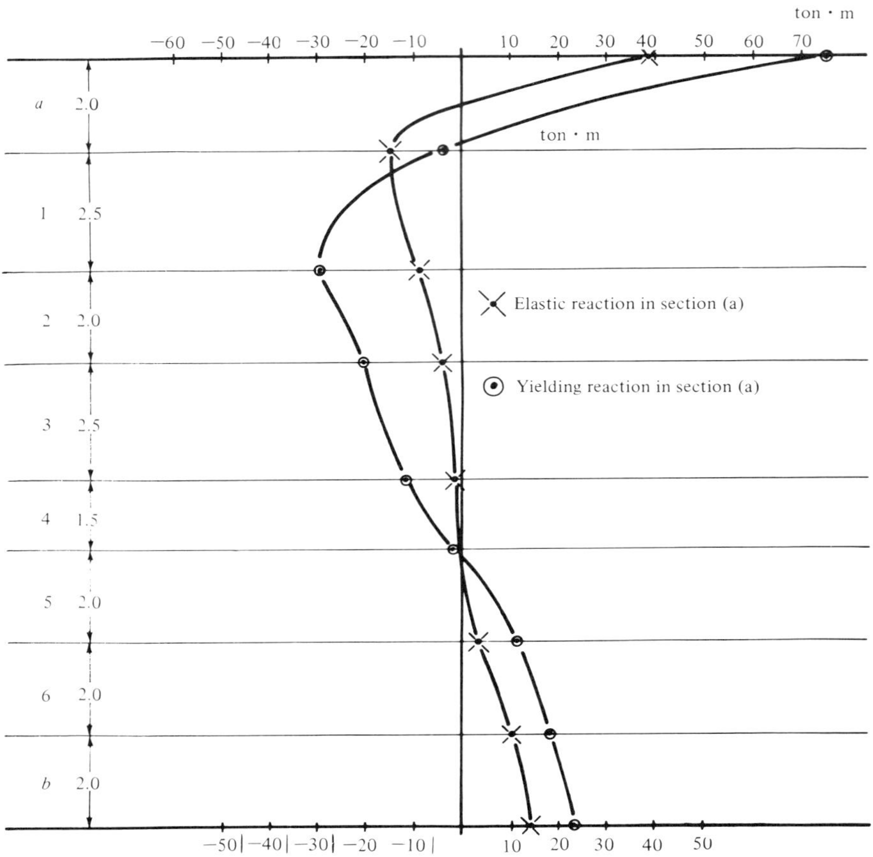

Fig. 14-XII.6 Bending moments of pier.

TABLE 10-XII.6

X_a	R_a	R_1	X_2	X_3	X_4	X_5	X_6	R_b	X_b
ton · m									ton · m
76.396	33.930	26.622	2.473	-2.636	9.355	-13.778	8.325	-6.896	-24.526
d	2.0	2.5	2.0	2.5	1.5	2.0	2.0	2.0	m
Q_l	16.965	10.651	1.237	-1.054	6.237	-6.890	4.163	3.448	ton/m

6.3 Underground Pipes and Tunnels

Large diameter sewer pipes and tunnels are underground installations that if constructed in soft subsoil deposits, may be subjected to important bending moments and shears during strong ground motions. To study from a practical engineering point of view the behavior of these structures, consider a soft subsoil deposit of great depth excited during a strong earthquake. Shear waves travel from the firm ground supporting stratum to the ground surface, producing shear displacements in the subsoil mass. The waves propagate with shear velocity v_s. The important wave periods producing large magnification of the soil displacements are those with wave lengths compatible with the total thickness of the soft soil deposit. The maximum ground acceleration may be obtained from the ground response acceleration spectrum for the corresponding periods to be analyzed (Fig. 3-XII.3).

Therefore, if T_n is a period of the soft soil deposit, the wave length is $L_n = v_s T_n$. The soil particles on a longitudinal line along the underground installations are displaced relative to each other and at a certain instant exhibit a definite configuration approximating a sinusoidal wave shape (Fig. 15-XII.6). The horizontal displacement equation representing the average configuration may be assumed to be

$$(\delta_{sz})_x = \delta_{nz} \cdot \sin \frac{2\pi}{L_n} \cdot x \qquad \text{28-XII.6}$$

in which δ_{nz} is the maximum amplitude displacement of the soil at depth z corresponding to resonant wave period T_n and wavelength L_n.

In case of an isotropic soft soil deposit, the first mode or fundamental period of vibration is determined by $T_n = 4D/v_s$ for the total thickness D of the deposit. From maximum ground acceleration for period T_n, the amplitude of the relative

TABLE 11-XII.6

z_i		↶ X_a = 76.496 ton · m	d_i	R	V	M
15.50		P = −57.4 ton →			−57.400	76.426
14.50	a	•← 33.930 ton	2	16.965		
					−23.470	−4.374
12.25	1	•← 26.622	2.5	10.649		
					−3.153	−29.771
10.0	2	•← 2.473	2.0	1.237		
					5.627	−20.992
7.75	3	•← 2.636	2.5	−1.054		
					−2.992	−10.219
5.75	4	•← 9.355	1.5	6.237		
					12.347	−1.285
4.0	5	•→ −13.778	2.0	−6.890		
					−1.433	12.199
2.0	6	•← 8.325	2.0	4.163		
					6.893	17.659
0	b	•→ −6.891	2.0	−3.446		
					0.000	24.553
m		↺ X_b = 24.526 ton · m	m	ton	ton	ton · m

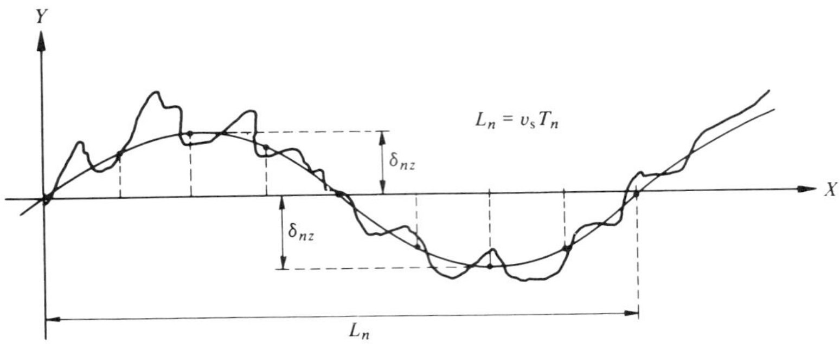

Fig. 15-XII.6 Relative displacements of a longtiudinal line along underground installations.

displacements at any depth level is calculated from

$$\delta_{nz} = G_a \cdot \frac{T_n^2}{(2\pi)^2} \cos \frac{\pi}{2D} z \qquad \text{29-XII.6}$$

The value of z represents the depth from the ground surface. The maximum relative displacements of the soil mass along the installation may be estimated by means of equation 28-XII.6. For long installations, one can safely assume that the structure is forced to follow the soil maximum horizontal displacement configuration. Therefore, from the imposed displacements the shear, bending moments and load to which the structure may be subjected are approximately determined by means of the following equations:

$$EI \frac{d^2\delta}{dx^2} = \text{Moment}$$

$$EI \frac{d^3\delta}{dx^3} = \text{Shear} \qquad \text{30-XII.6}$$

$$EI \frac{d^4\delta}{dx^4} = \text{Load}$$

From the deflection configuration of the conduit at depth z, assumed equal to that of 28-XII.6, the maximum bending occurring at $x = (1/4)L_n$ and $(3/4)L_n$ is

$$M_m = -EI \cdot \delta_{nz} \left(\frac{2\pi}{L_n}\right)^2 \qquad \text{31-XII.6}$$

The maximum shear will occur at $x = 0, L_n/2$ and L_n:

$$S_n = EI \cdot \delta_{nz} \left(\frac{2\pi}{L_n}\right)^3 \qquad \text{32-XII.6}$$

and the load distribution is

$$Q_1 = -EI \cdot \delta_{nz} \left(\frac{2\pi}{L_n}\right)^4 \sin \frac{2\pi}{L_n} \cdot x$$

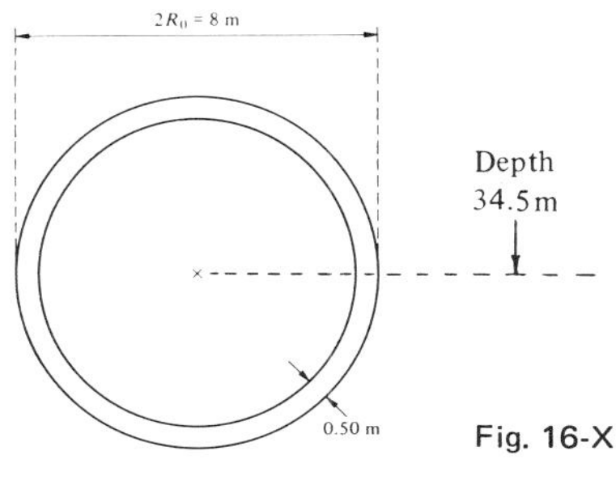

Fig. 16-XII.6 Cross-sectional view of large conduit.

From the above values, the approximate behavior of the installation is investigated. The following illustrates a practical application. Assume (Fig. 16-XII.6) that a long and large conduit installation of circular cross-section with 8.0 m diameter and concrete wall thickness of 0.50 m is to be constructed in a soft subsoil deposit at 34.5 m depth. The soil deposit is considered isotropic and has a thickness of 45 m. The cross-sectional area of the pipe is $A = 11.75$ m^2, and the moment of inertia $I = 83.3$ m^4. The concrete used is estimated to have a modulus of elasticity of $E = 2 \times 10^6$ ton/m^2. Therefore, $EI = 1.66 \times 10^8$ ton · m^2, the subsoil investigated has the following two important modes of vibration: $T_1 = 2.4$ sec and $T_2 = 0.8$ sec. From the ground acceleration spectrum, for these periods we find a corresponding maximum ground acceleration on the order of 100 cm/sec^2 and 75 cm/sec^2, respectively. The displacements and wavelengths are shown in Table 12-XII.6 for shear wave velocity of $v_s = 75$ m/sec and $\mu = 800$ ton/m^2.

With the information of Table 12-XII.6 concerning maximum displacements at 34.5 m depth and wavelengths, the maximum shears and bending moments in the structure may be calculated (Table 13-XII.6).

From the results obtained for this simple example, it may be concluded that the first vibration mode induces high bending stresses in the concrete pipe, and therefore damage may take place during an earthquake of the assumed magnitude if the pipe is not properly reinforced. The second mode introduces the highest shears, that may also damage the pipe. To minimize the high bending moments and shear forces and achieve a more economical design, the pipe is provided with joints

TABLE 12-XII.6

Conduit depth 34.5 m, deposit depth 45 m; $v_s = 75$ m/sec

T_n	ω_n^2	G_a	δ_{no}	$\cos \frac{\pi}{2} \frac{z}{D}$	$(\delta_n)_{35}$
2.4	6.85	100	0.146	0.356	0.0520
0.8	61.7	75	0.0122	0.356	0.00434
sec	1/sec^2	cm/sec^2	m		m

Surface amplitude $\delta_{no} = G_a \cdot \dfrac{T_n^2}{(2\pi)^2}$

TABLE 13-XII.6

T_n	$(\delta_n)_{35}$	L_n	$EI(\delta_n)_{35}$	$\frac{2\pi}{L_n}$	$\left(\frac{2\pi}{L_n}\right)^2$	$\left(\frac{2\pi}{L_n}\right)^3$	Moment M	Moment $\pm f_c$	Shear S	Shear τ_{ave}
2.4	0.0520	180	863	3.49	12.2	42.5	10510	50.5	367	3.12
0.8	0.00434	60	72	10.47	109.6	1148	7890	37.9	826	7.03
sec	m	m	ton · m³ × 10⁴	m⁻¹ × 10⁻²	m⁻² × 10⁻⁴	m⁻³ × 10⁻⁶	ton · m	kg/cm²	ton	kg/cm²

$EI = 1.66 \times 10^8$ ton · m², $A = 11.75$ m², $I = 83.3$ m⁴, $S = 20.8$ m³

placed at every $\lambda = 15$ m (Fig. 17-XII.6) and designed with a shear force equivalent to the one induced in the ground by the seismic motion, namely $S_{35} = \tau_{35} \cdot A_g$, where τ_{35} is the average shear stress in the soil at depth 10 m, and A_g the gross area of the pipe. The value of τ is given by

$$\tau = \left(\frac{\partial \delta}{\partial x}\right) \mu$$

and from 28-XII.6, the maximum shear force at depth 34.5 m is

$$\tau_{35} = (\delta_{n1})_{35} \frac{2\pi}{L_n} \mu$$

Substituting numerical values for the case of $T = 2.4$ sec, we obtain $\tau_{35} = 1.45$ ton/m². The gross area of the pipe is 50 m², and therefore the minimum shear force at the joint is $S_m = 73$ ton.

Long buildings subjected to the shear wave ground motions in the longitudinal direction may be studied in the same manner, as just described for underground

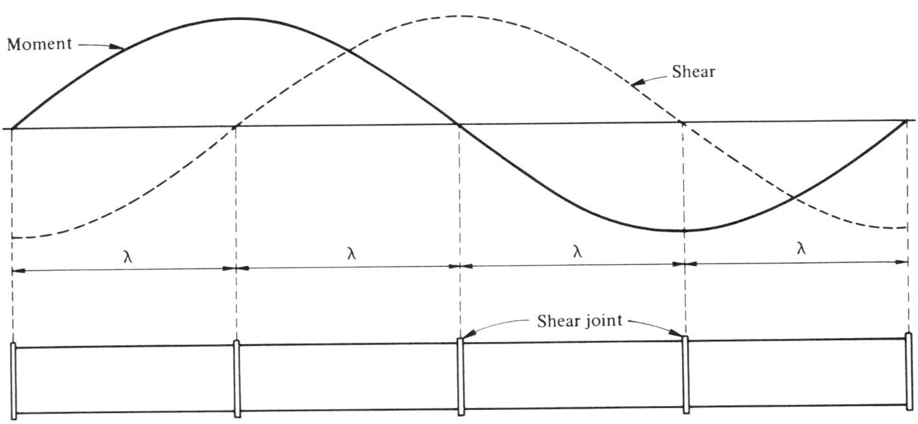

Fig. 17-XII.6 Underground pipe with joints.

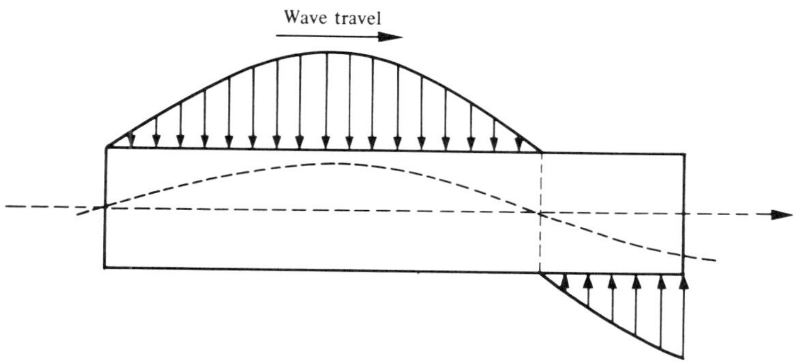

Fig. 18-XII.6 Torsion induced in long building due to shear waves traveling along the length of the building.

conduits. Moreover, since they are of finite length, they are induced to vibrate under torsion because of the horizontal unbalanced forces that the shear waves develop against the foundation structure (Fig. 18-XII.6). The effects produced by the torsional vibration in the structure and foundation cannot be overlooked.

BIBLIOGRAPHY

Barkan, P. D. (1960). *Dynamics of Bases and Foundations*, Translated from Russian by G. P. Tschebotarioff, McGraw-Hill Book Company, Inc., New York.

Biot, M. A. (1943). Analytical and Experimental Methods in Engineering Seismology, *ASCE*, Vol. 108, pp. 365–384.

Clough, R. W. (1960). Dynamic Effects of Earthquakes, *Journal of Structural Division*, ST4.

Freeman, J. R. (1932). *Earthquake Damage and Earthquake Insurance*, McGraw-Hill Book Company, Inc., New York.

Hall, J. R. and Richart, F. E. (1963). Effect of Vibration Amplitude on Wave Velocities in Granular Materials, *Proceedings II Panamerican Conference on Soil Mechanics and Foundation Engineering*, Vol. 1, p. 145.

Housner, W. G. (1954). Geotechnical Problems of Destructive Earthquakes, *Geotechnique*, Vol. IV, pp. 153–162.

Housner, W. G. (1961). Vibration of Structures Induced by Seismic Waves, *Shock and Vibration Handbook*, Harris and Creede, Vol. 3, p. 50–51, McGraw-Hill Book Company, Inc., New York.

Hudson, D. C. (1962). Some Problems in the Application of Spectrum Techniques to Strong Motion Earthquake Analysis, *Bulletin Seismological Society of America*, Vol. 52, No. 2, p. 417.

Jennings, P. C. (1962). *Velocity Spectra of the Mexican Earthquakes of 11 May and 19 May 1962*, California Institute of Technology.

Marsal, R. J. (1961). Behavior of a Sandy Uniform Soil During the Jaltipan Earthquake, Mexico, *Proc. V ICOSOMEF*, Vol. 1, p. 229, Paris.

Neumann, F. (1960). A Lateral Force Formula Based on Seismological Concepts, *The Trend in Engineering*, Vol. 12, No. 3, July.

Richart, F. E., Hall, J. R. and Woods, R. D. (1970). *Vibration of Soils and Foundation*, Prentice Hall.

Richter, Ch. F. (1958). *Elementary Seismology*, W. H. Freeman and Company.

Seed, H. B. and Idrizz, M. I. (1972). Simplified Procedure for Evaluating Soil Liquefaction Potential, *Journal SMCE*, SM9.

Terzaghi, K. (1943). *Theoretical Soil Mechanics*, John Wiley and Sons, Inc.

Timoshenko, S. and Goodier, J. N. (1934). *Theory of Elasticity*, McGraw-Hill Book Company, Inc., New York.

Zeevaert, L. (1943). Ecuaciones de Condición en las Estructuras Hiperestáticas, *Revista Irrigación en México*, Septiembre.

Zeevaert, L. (1962). *Mediciones y Cálculos Sísmicos Durante los Temblores Registrados en la Ciudad de México en Mayo de 1962*, Publicado por la Sociedad Mexicana de Ingeniería Sísmica.

Zeevaert, L. (1962). The Development of Shear Displacement Meters and Accelerometers to Measure Earthquake Forces in Buildings, *Proceedings 31st Annual Convention, Structural Engineers Association of California*, Coronado, October 4–6.

Zeevaert, L. (1963). *The Effect of Earthquakes in Soft Subsoil Conditions*, SEAOC Convention, Yosemite, October 2–5.

Zeevaert, L. (1964). Strong Ground Motions Recorded During Earthquakes of May 11 and 19, 1962 in Mexico City, *Seismological Society of America*, Vol. 54, No. 1, pp. 209–231, February.

Zeevaert, L. (1964). Características de los Temblores en Ingeniería Sísmica, *Conference de Ingeniería Sísmica patrocinada por la Sociedad Mexicana de Ingeniería Sísmica*.

Zeevaert, L. (1966). Soil Properties, Discussion. *Proc. VI ICOSOMEF*, Vol. III, p. 370, Montreal, Canada.

Zeevaert, L. (1967). Free Vibration Torsion Tests to Determine the Shear Modulus of Elasticity of Soils, *Proceedings III Panamerican Conference on Soil Mechanics and Foundation Engineering*, Vol. I, p. 111, Caracas, Venezuela.

Zeevaert, L. (1975). Earthquake Effects in Compensated Foundations. *Short Course on Analysis and Design of Foundations for Tall Buildings*, August 2, 1975. Lehigh University, Bethlehem, Pennsylvania.

Zeevaert, L. (1975). Foundations Problems in Earthquake Regions. *Short Course on Analysis and Design of Foundation for Tall Buildings*, August 8, 1975. Lehigh University, Bethlehem, Pennsylvania.

Zeevaert, L. (1977). Pile Foundation Design Problems in Earthquake Areas. *Pile Tips Seminar*, APF Associated, Pile & Fitting Corp. 1977, 262 Rutherford Blvd., Clifton, N.J. 07014.

Zeevaert, L. (1977). Seismic Rocking Problems of Rigid Compensated Foundations. *Earthquake Resistant Reinforced Building Construction Workshop*, July 11-15, 1977, Berkeley, California.

Zeevaert, L. (1978). Algunos conceptos Sísmicos de San Salvador, El Salvador, C.A. *Central American Conference on Earthquake Engineering, Enero 1978.* Publicado por la Universidad Centroamericana José Simeón Cañas, San Salvador, El Salvador, y Lehigh University, Bethlehem, Pennsylvania, U.S.A.

Zeevaert, L. (1980). *ISE-Interacción Suelo-Estructura de Cimentaciones Superficiales y Profundas Sujetas a Cargas Estáticas y Sismicas*, Editorial LIMUSA, México, D.F.

Zeevaert, L. (1981). Theory and Practice of the Free Vibration Torsion Pendulum División Estudios de Postgrado, Facultad de Ingenieria, U.N.A.M.

FROM *PROCEEDINGS ON EARTHQUAKE BLAST EFFECTS OF STRUCTURES*, LOS ANGELES, CALIFORNIA, JUNE 1962:

Housner, G. W. Spectrum Intensities of Strong-Motion Earthquakes, pp. 20–36.

Neumann, F. Some Generalized Concepts of Earthquake Motion, pp. 8–19.

Rinne, J. Building Code Provisions for Aseismic Design, pp. 291–308.

FROM *PROCEEDINGS OF FIRST WORLD CONFERENCE ON EARTHQUAKE ENGINEERING*, BERKELEY, CALIFORNIA, JUNE 1956:

Hudson, D. E. The Response Spectrum Technique, p. 4.1.

Kanai, K., Takahasi, R. and Kawasumi, H. Seismic Characteristics of Ground, p. 31.1.

Roberts and Cloud, W. U.S.C. and G. S. Instruments, p. 2.1.

Takahasi, R. SMAC Strong Motion Seismometer, p. 3.1.

Zeevaert, L. and Newmark, N. M. Aseismic Design of Latino Americana Tower in Mexico City, p. 35.1.

FROM *PROCEEDINGS OF SECOND WORLD CONFERENCE ON EARTHQUAKE ENGINEERING*, TOKYO AND KYOTO, JAPAN, JULY 11–18, 1960:

Binder, R. W. and Wheeler, W. T. Building Code Provisions for Aseismic Design. Vol. III, pp. 1843–1876.

Cloud, W. K. and Carder, D. S. Ground Motions Generated by Underground Nuclear Explosions. Vol. III, pp. 1609–1632.

Figueroa, A. J. Some Considerations About the Effect of Mexican Earthquakes. Vol. III, pp. 1553–1561.

Kanai, K. An Empirical Formula for the Spectrum of Strong Earthquake Motion. Vol. III, pp. 1541–1552.

Murayama, S. and Shibata, T. On the Dynamic Properties of Clay. Vol. I, pp. 297–310.

Neumann, F. A Broad Formula for Estimating Earthquake Forces on Oscillators, Vol. II, pp. 849–862.

Rosenblueth, E. The Earthquake of July 28, 1957 in Mexico City. Vol. I, pp. 359–379.

Zeevaert, L. Base Shear in Tall Buildings During Earthquake of July 28, 1957 in Mexico City. Vol. II, pp. 983–996.

APPENDIX A: LIST OF SYMBOLS

In general, the symbols are defined in the book. The list given below is intended to serve only as a guide for the reader.

LATIN LETTERS

a	Coefficient, parameter; linear dimension; tributary area
$\overline{a}$	Nominal tributary area in pile group = $\lambda \cdot \beta$
a_m	Maximum acceleration
a_p	Cross-sectional area of pile
a_v	Coefficient of compressibility
A	Area; amplitude
b	Coefficient, parameter; time
B	Linear dimension, footing halfwidth
c	Coefficient, parameter; cohesion
c_v	Coefficient of consolidation in Terzaghi's theory
c_{vp}	Apparent coefficient of consolidation for intergranular plasto-viscosity
C	Coefficient, parameter, factor
C_r	Relative consistency of clay
d	Thickness of stratum; dimension; excavation depth; pile diameter
D	Thickness of strata
D_a	Equivalent or average cross section of soil strata
D_e, D_{10}	Effective diameter
D_r	Relative density of cohesionless soil
e	Void ratio
E	Modulus of elasticity
E_a	Active earth pressure
E_p	Passive earth pressure

f_ϕ, f_{vp}	Settlement functions
$F(T_v)$	Primary consolidation in Terzaghi's function
g	Gravity acceleration equal to 9.81 m/sec^2
G_a	Ground acceleration
G_s	Factor of safety
h	Piezometric head; height; depth
h_{ex}	Excavated depth
h_m	Height of center of mass
H	Thickness; height
i	Hydraulic gradient
i_c	Critical gradient
$i_{\alpha\beta}$	Influence settlement coefficient at point with coordinates α, β
I	Influence coefficient with subscript Bq for Boussinesq, Wd for Westergaard, Fh for Fröhlich; moment of inertia
I_F	Inertia force
I_M	Modified Mercalli earthquake intensity
I_p	Centroidal moment of inertia of test specimen
I_w	Plasticity index
J	Mass polar moment of inertia
k	Coefficient of permeability; unit foundation modulus
$\overline{k}_s$	Average unit foundation modulus for one foot square plate
K	Constant parameter; foundation modulus for tributary area; permeability; spring constant parameter
K_l	Foundation modulus per unit length
K_v	Unit bulk modulus
K_0	Ratio of effective stresses at rest
K_θ	Rotational linear foundation modulus
l	Linear dimension; length of pile
l_e	Effective length of pile
l_s	Penetration length of pile in sand stratum
L_w	Liquid limit
m	Mass of an element; parameter, factor
m_t	Coefficient of unit intergranular viscous volume compressibility
m_v	Coefficient of unit volume compressibility
M, M_t	Moment; torsional moment in foundation structure; building mass
M_d	Driving moment
M_e	Elastic expansion or response strain modulus
M_{ep}, M_{vp}	Linear strain modulus with subscripts ep (elasto-plastic) and vp (visco-plastic)
M_p	Plastic moment; moment at pile head
M_r	Resisting moment
M_R	Richter earthquake magnitude
M_0	Moment at pier or pile head
n	Porosity of soil; number defining pile spacing; number of terms
N_c	Influence value for one curvilinear rectangle in stress net, for confining stress at a point
N_f	Number of flow channels
N_p	Number of equipotential drops

N_χ	Influence value for a curvilinear rectangle in stress nets for vertical stress at a point
N_c, N_q, N_γ	Bearing capacity factors
NF	Negative skin friction in piles
N_ϕ	Average flow number depending on angle of internal friction
O_T	Overturning moment
p	Unit load; unit total pressure; unit earth pressure
p_p	Unit horizontal passive earth pressure
P	Load
PF	Positive skin friction in piles
P_w	Plastic limit
q	Water flow per unit volume; unit load; contact stress in foundation
q_{ad}	Allowable unit load bearing capacity
q_d	Ultimate unit load bearing capacity
q_{ep}	Limiting stress at edge of footing
q_m	Deviatory stress
q_u	Unconfined compressive strength; natural soil consistency
Q	Total discharge; concentrated load; total load
Q_a	Allowable load of pile
Q_{dy}	Dynamic load of pile
Q_{fu}	Ultimate load in friction pile
Q_i	Pile load at any depth; $z = i$
Q_{pu}	Ultimate load in point bearing piles
r	Radius, ratio of quantities
R	Radius, reaction force
R_a	Pseudoacceleration response spectrum
R_c	Characteristic value of flow net
R_q	Loading pile-soil ratio
R_s	Sichardt's empirical distance in water wells
R_v	Pseudovelocity response spectrum
s	Unit shear strength
s_s	Specific gravity
S	Shear force; section modulus
S_0	Dynamic water depression in water wells
S_d	Relative displacement response spectrum
S_p	Shear force at pile head
t	Time; sand stratum thickness
T	Torsional moment; tension force; period of vibration
T_d	Damped period of vibration
T_v	Time factor in primary theory of consolidation
u	Hydraulic pressure; pore water pressure, horizontal displacement in x axis of coordinates
u_d	Uplift water pressure at depth d
$U\%$	Average degree of primary consolidation in Terzaghi's Theory
v	Velocity; flow velocity
v_d	Compressional or irrotational wave velocity
v_s	Shear wave velocity
v	Horizontal displacement in y axis of coordinates

V	Volume; shear force
V_B	Base shear due to earthquake
V, V_p	Shear force at pile or pier head
w	Natural water content; vertical displacement in z axis of coordinates
W	Weight of building
x, y, z	Coordinates, abscissas
z_t	Linear transformation of coordinate z
Z	Settlement function used in plasto-viscous phenomenon

GREEK LETTERS

α	Coefficient, parameter, factor; ratio; angle
α_1, α_2	Footing bearing capacity shape factors
β	Coefficient, parameter; linear dimension; angle
γ	Unit weight; rate of angular shear distortion
δ	Vertical displacement; settlement; penetration of pile, parameter with length units
δ_e	Elastic heave; E-heave in excavation
δ_s	Swelling heave, S-heave in excavation
δ_c	Settlement of foundation
$\underline{\delta_D}$	Vertical heave due to soil displacement when driving piles, D-heave
δ_D	D-heave per unit length
δ_R	Elongation of pile
δ_{pp}	Vertical displacement under pile tip
δ_{ps}	Vertical displacement in subsoil under pile field or group
Δ	Stands for: increment, decrement, expressing change in function or quantity, logarithmic decrement
$\Delta\epsilon_{ss}$	Volumetric change in strain in SS pores of soil
$\Delta\epsilon_{vp}$	Volumetric change in strain in VP pores of soil
$\Delta\delta_{ij}$	Increment of vertical displacements between horizontal planes at depths $z = i$, and $z = j$
$\Delta\sigma_{ex}$	Effective stress relief in excavation
$\Delta\theta_e$	Elastic rotation change in foundation
$\Delta\theta_{ep}$	Change in elasto-plastic rotation
$\epsilon_e, \epsilon_{ep}, \epsilon_{vp}$	Strain, with subscripts e (elastic), ep (elasto-plastic), vp (visco-plastic)
ζ	Fraction or critical damping
η	Water viscosity; depth factor; parameter
θ	Rotation; volumetric expansion; or compression
κ	Coefficient; ratio of two quantities
λ	Drop in piezometric water level; ratio of depth to width of foundation; dimension; scale factor
μ	Constant; soil rigidity or shear modulus of elasticity
ν	Poisson's ratio
π	3.1416 geometric ratio of circumference to diameter of a circle
ρ	Unit mass; density
ρ_c	Stress relief compression factor
ρ_e	Stress relief expansion factor

σ	Stress, intergranular stress
σ_a	Average effective stress
σ_b	Critical compressive effective stress
σ_c	Volumetric confining effective stress
σ_i	Final vertical effective stress at depth $z = i$
σ_{0i}	Initial effective overburden vertical effective stress
$\sum_1^n$	Summation of terms from 1 to n
τ	Shear stress; time parameter
ϕ	Angle of internal friction, used to define a function; flow line; fluidity in soil
ϕ_a	Apparent linear fluidity in saturated soil
Φ	Flow function
χ	Stress distribution factor
Ψ	Equipotential function
ω	Rotation, circular frequency
$\overline{\omega}$	Pile perimeter
ω_d	Damped circular frequency
$\omega_x, \omega_y, \omega_z$	Rotations around coordinate axes x, y and z

APPENDIX B: INFLUENCE STRESS NET AND CHARTS

The influence stress nets and charts given here can be used by the foundation engineer in evaluating vertical stresses in the ground for cases in which settlements should be estimated. The method of construction and use of the stress nets is given in Chapter III.

The influence stress net given in Fig. 1-B is used to estimate the volumetric confining stress at a point in the semi-infinite isotropic elastic solid. The influence stress net in Fig. 5-B gives values to estimate the vertical stress at a point for Boussinesq solution of the semi-infinite elastic solid, and the stress nets in Figs. 2, 4, 5, 6 and 7-B may be used in the semi-infinite mass when the strain modulus assumes variations with depth according to the Fröhlich concept.

The corresponding stress net may be selected for vertical stress determination at a point according to the experience of the foundation engineer (see sections III.1 and 2). The influence stress net in Fig. 3-B gives the influence values for vertical stresses for the Westergaard solution on the semi-infinite solid strongly reinforced by a series of horizontal nondeformable strata.

The influence charts given in Figs. 8 and 9-B are based on Westergaard's solution of the vertical stress at a point in the semi-infinite elastic solid under the assumptions developed in section VIII.3 (Fig. 9-VIII.3). They are used to investigate the tributary effective areas with depth in single piles or pile groups, to be used in skin friction pile problems (section VIII.3.2). Their use is explained in section VIII.3.3.

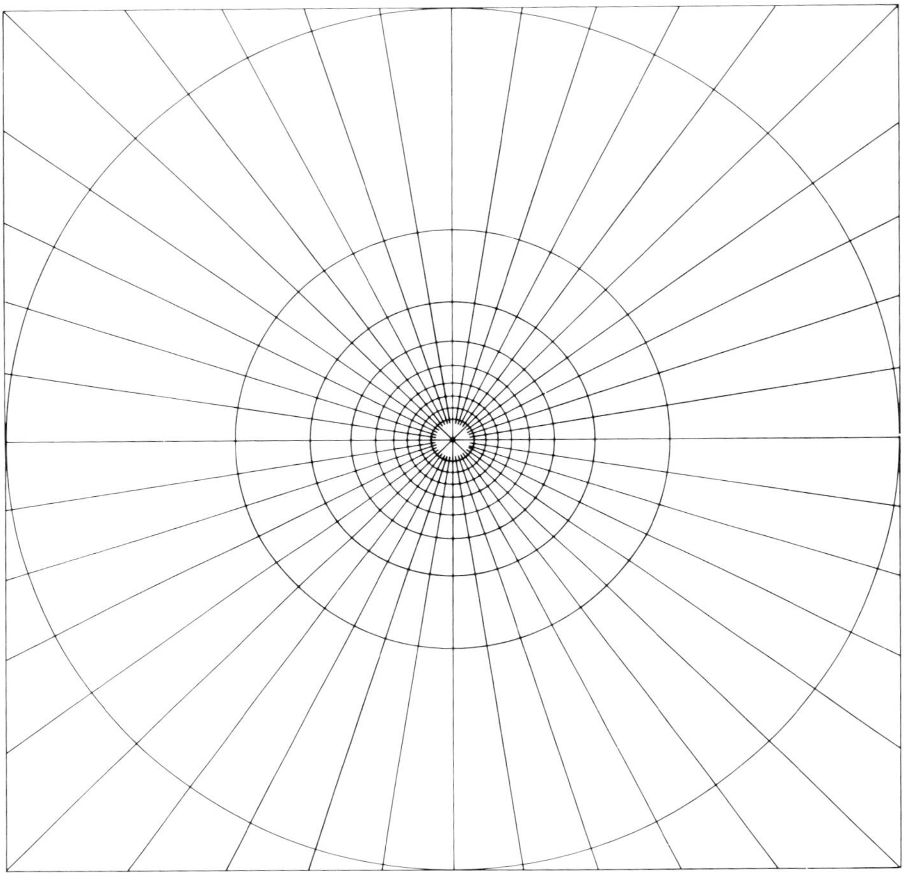

$N_\chi = 0.0025$

Depth scale

Fig. 1-B Influence stress net $\chi = 1$ for increments of volumetric confining stresses in a semi-infinite solid, $\nu = 0.5$. For $\nu \neq 0.5$, use factor $2(1 + \nu)/3$.

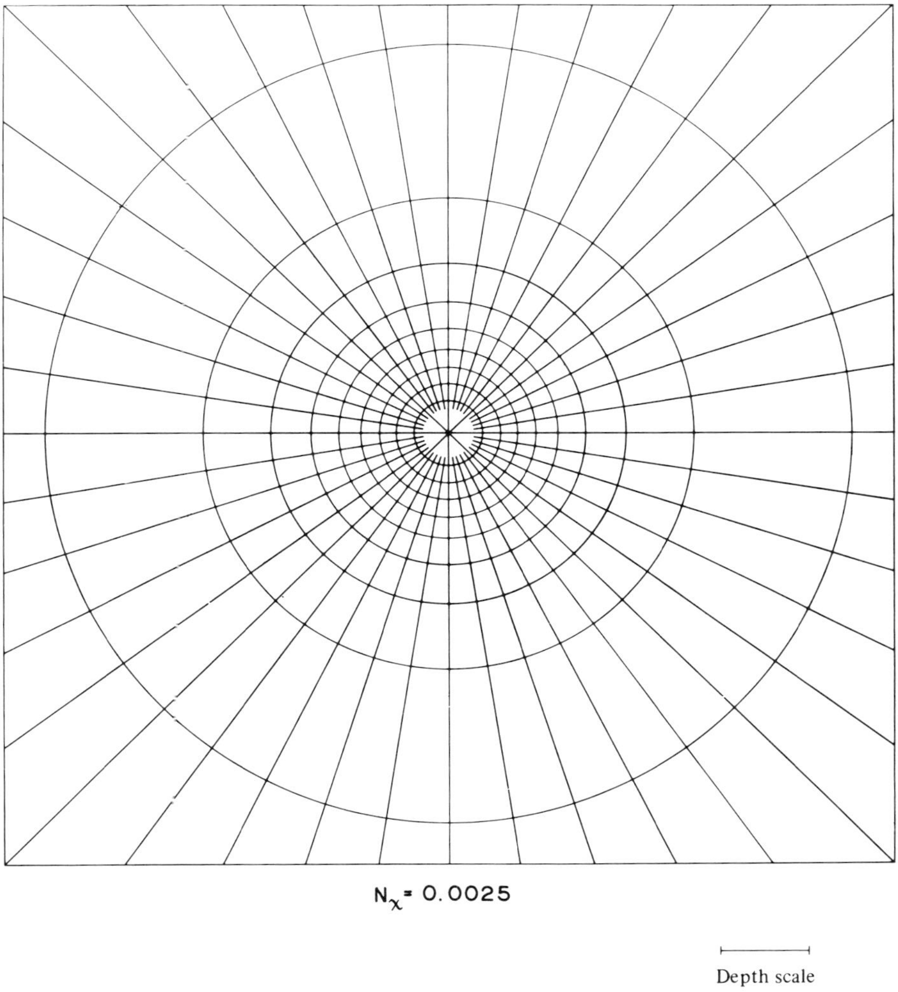

Fig. 2-B Influence stress net $\chi = 1.5$ for vertical stresses approximately equal to Westergaard solution for a semi-infinite layered soil mass.

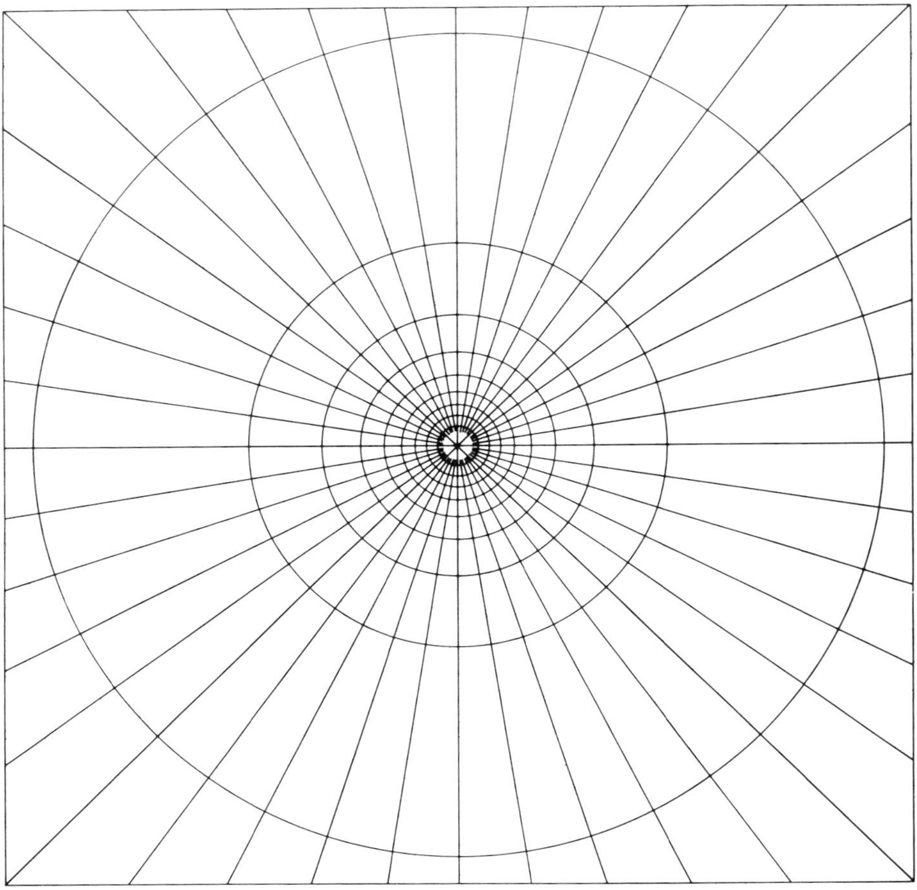

$N_{\chi} = 0.0025$

Depth scale

Fig. 3-B Influence stress net for vertical stresses. Westergaard solution for a semi-infinite layered soil mass.

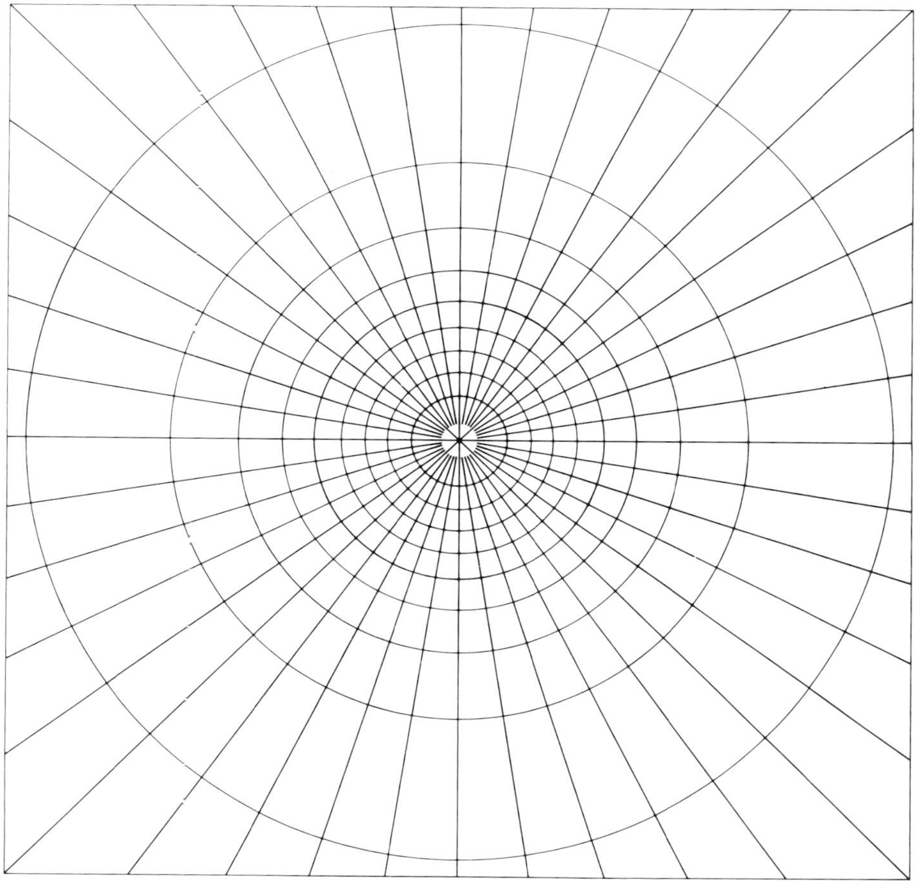

$N_\chi = 0.0025$

Depth scale

Fig. 4-B Influence stress net $\chi = 2$ for vertical stresses in a semi-infinite solid. Intermediate between an isotropic solid and a high layered soil mass.

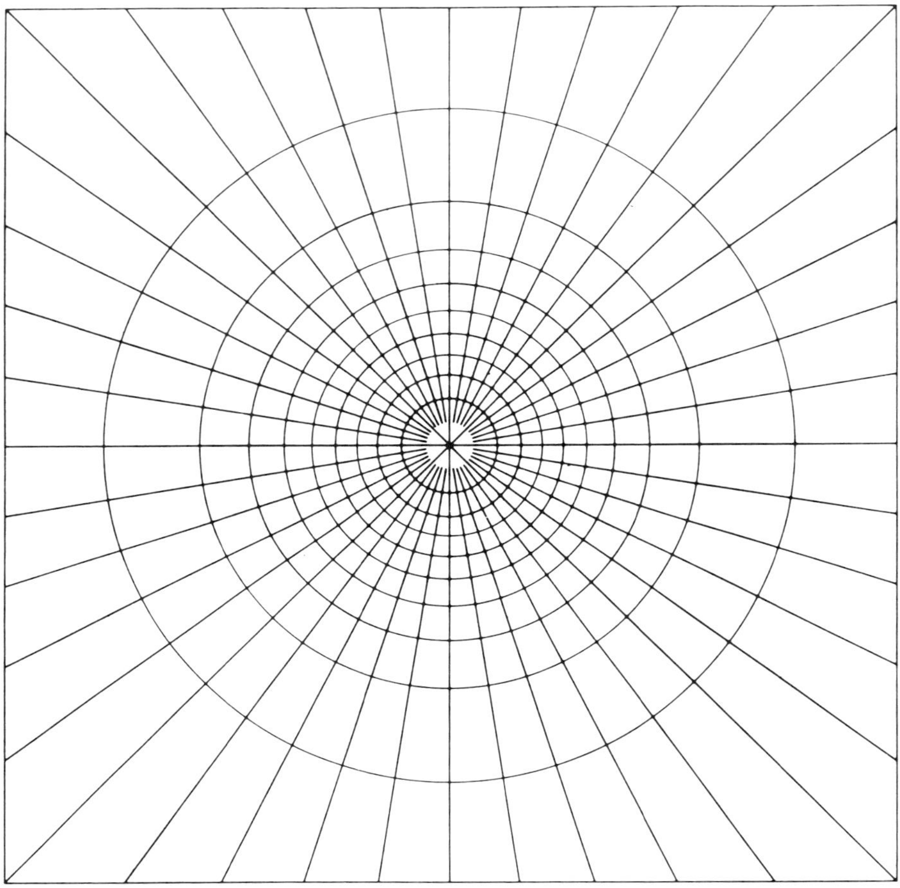

$N_{\chi} = 0.0025$

Depth scale

Fig. 5-B Influence stress net $\chi = 3$ equal to Boussinesq stress solution in isotropic semi-infinite mass.

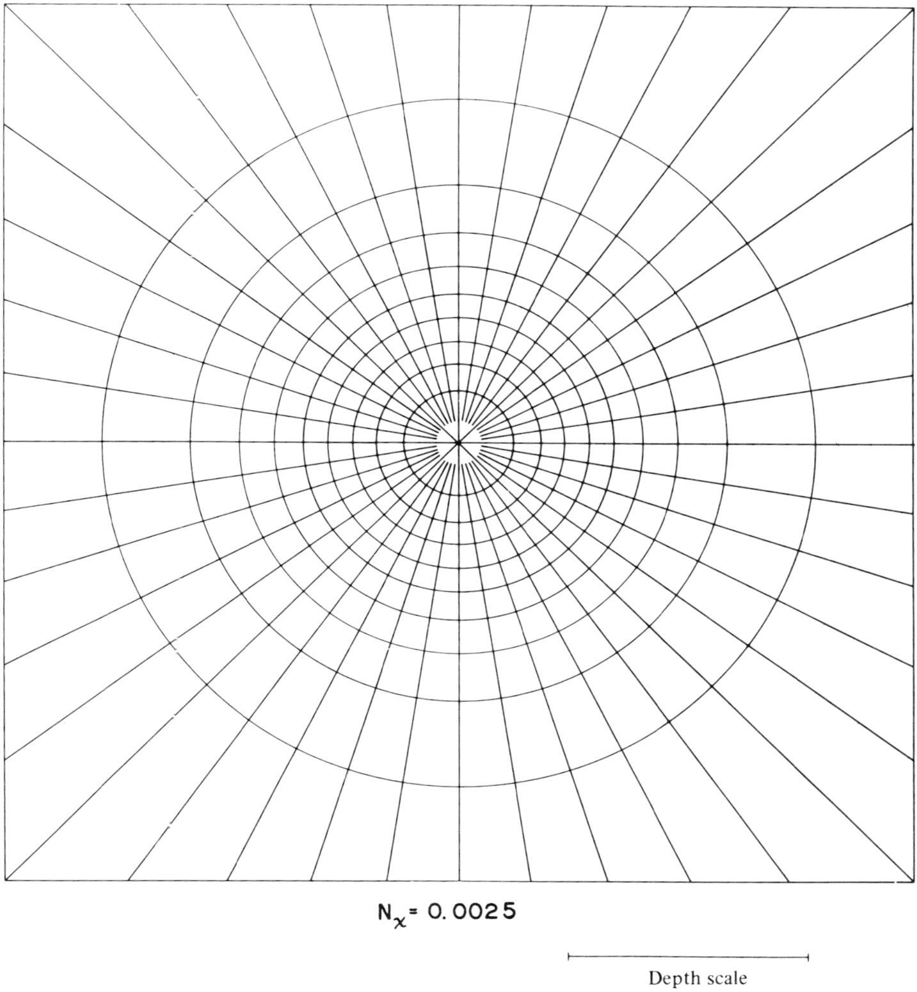

Fig. 6-B Influence stress net $\chi = 4$, equivalent to Frölich solution for semi-infinite mass and strain modulus decreasing with depth.

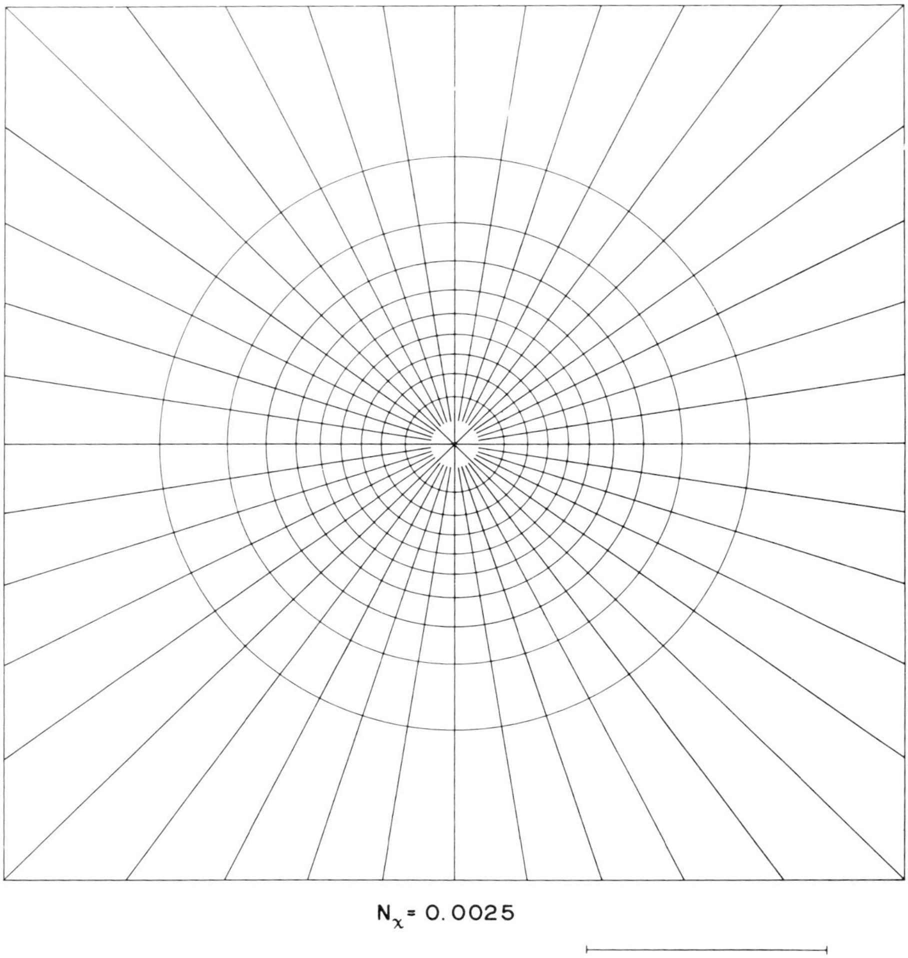

Fig. 7-B Influence stress net $\chi = 5$, for Frölich solution when strain modulus decreases very strongly with depth.

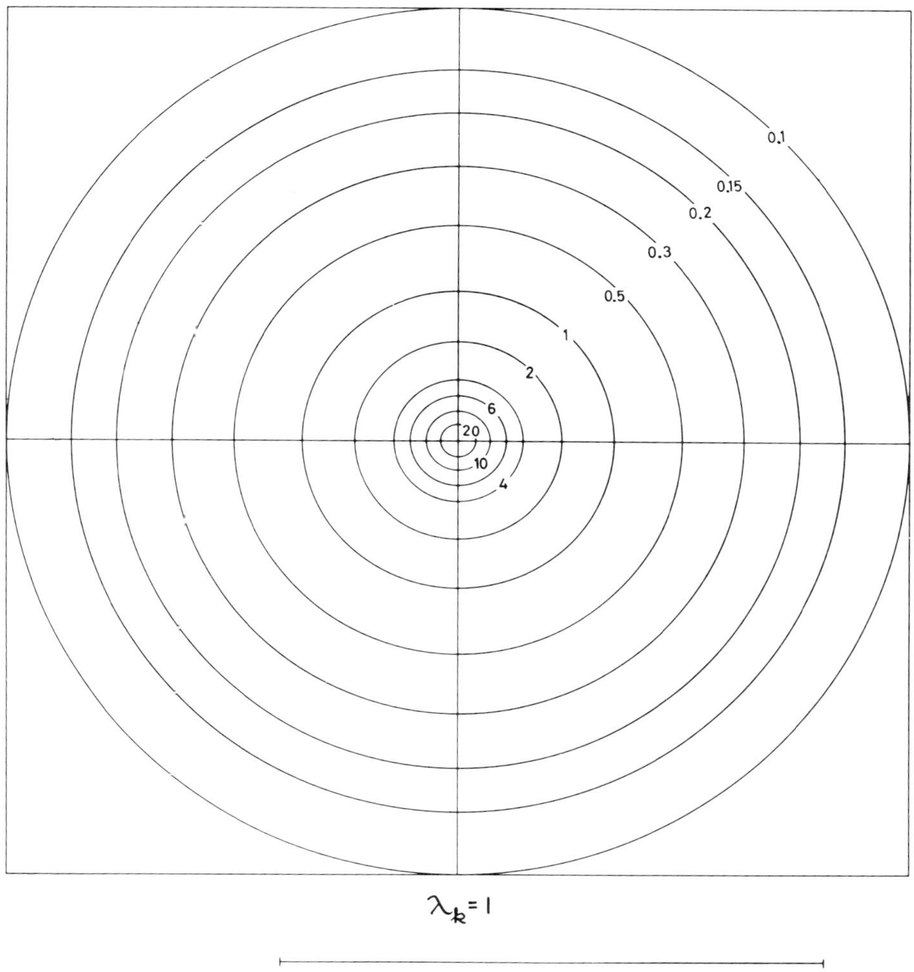

Fig. 8-B Friction pile influence chart to calculate equivalent tributary area in pile groups for case I: $s = kz$.

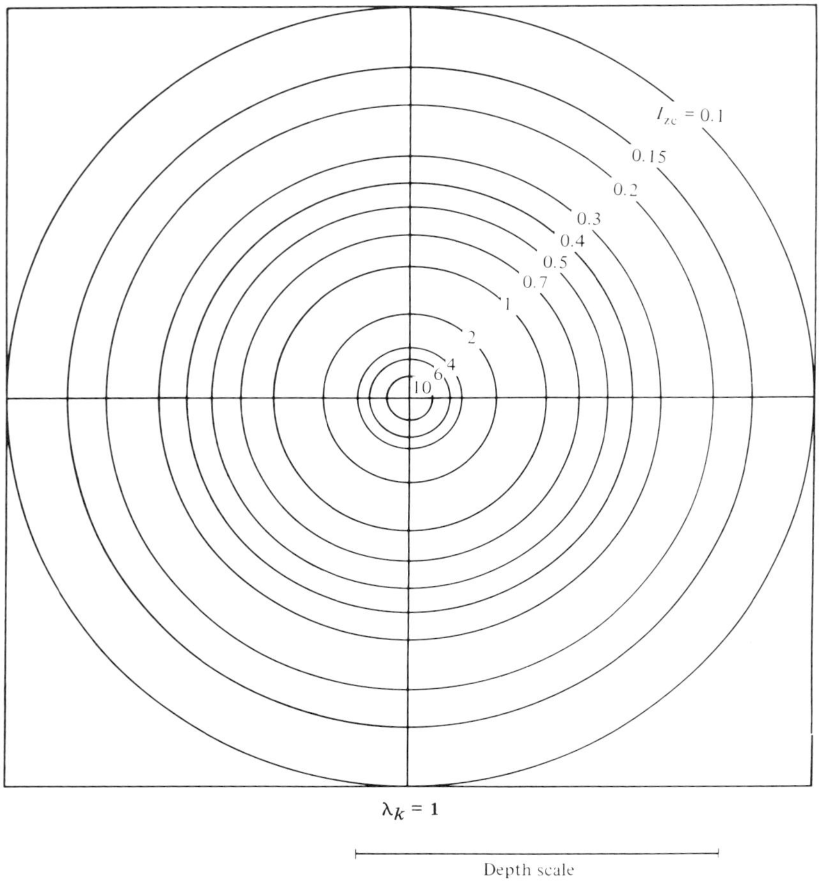

Fig. 9-B Friction pile influence chart to calculate equivalent tributary area in pile groups for case II: s_0 = constant.

APPENDIX C: INTEGRATION FORMULAS FOR SKIN FRICTION PROBLEMS IN PILE FIELDS

The algorithms for the step by step integration of skin friction problems treated in Chapters VIII, IX and X may be obtained for negative friction (*NF*), and for positive friction (*PF*). The assumed boundary conditions for which the integration is started should be known. When establishing the formulas, it is assumed that the *NF* phenomenon decreases the overburden effective stresses and that the *PF* phenomenon increases them. The general conditions for either case are established with the aid of Fig. 1-C, calling F the total friction at the considered elevation:

at level $(i - 1)$

$$F_{i-1} = (\sigma_{i-1} - \sigma_{0i-1})\bar{a}_{i-1} \qquad \text{1-C}$$

at level i

$$F_i = (\sigma_i - \sigma_{0i})\bar{a}_i \qquad \text{2-C}$$

For plastic equilibrium at the pile shaft in section Δz_i we obtain

$$F_i - F_{i-1} = \tfrac{1}{2}(\sigma_i + \sigma_{i-1})K_{\phi i}2\pi\bar{r}_0 \cdot \Delta z_i$$

and calling $m_i = \pi\bar{r}_0 K_{\phi i}$, we can write

$$F_i - F_{i-1} = m_i(\sigma_i + \sigma_{i-1})\Delta z_i \qquad \text{3-C}$$

With formulas 1, 2, and 3-C, the following cases may be considered.

1C. NEGATIVE FRICTION ($\sigma_i < \sigma_{0i}$)

Case I. The integration starts from the head of the pile where boundary conditions are known, namely: $(NF)_{i-1}$, σ_{0i-1} and σ_{i-1}. The unknowns are $(NF)_i$ and σ_i. Substituting equation 2-C into 3-C we obtain

$$(\sigma_{0i} - \sigma_i)\bar{a}_i - (NF)_{i-1} = m_i(\sigma_i + \sigma_{i-1})\Delta z_i$$

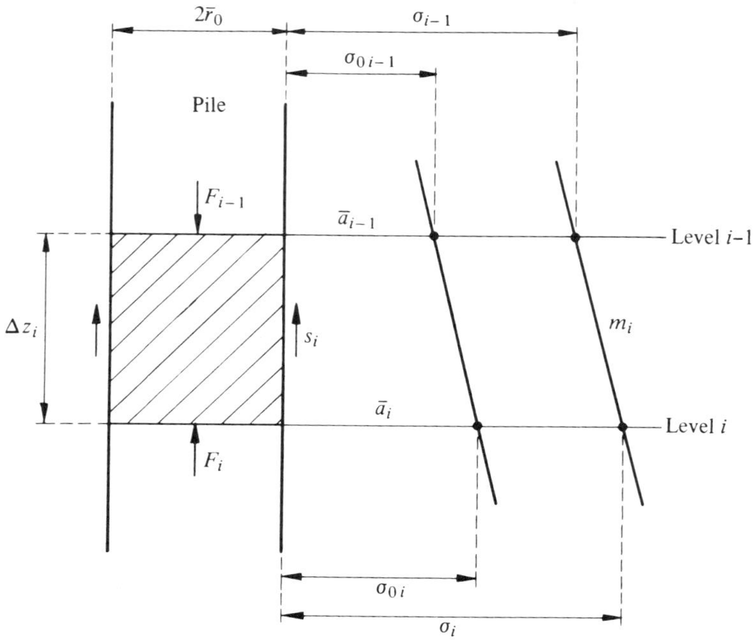

Fig. 1-C Effective stresses on pile shaft for use in algorithm formulas.

and solving for σ_i we get

$$\sigma_i = \frac{\bar{a}_i}{\bar{a}_i + m_i \Delta z_i} \sigma_{0i} - \frac{m_i \Delta z_i}{\bar{a}_i + m_i \Delta z_i} \sigma_{i-1} - \frac{1}{\bar{a}_i + m_i \Delta z_i} (NF)_{i-1}$$

If we define

$$A_i = \frac{\bar{a}_i}{\bar{a}_i + m_i \Delta z_i}, \quad B_i = \frac{m_i \Delta z_i}{\bar{a}_i + m_i \Delta z_i}, \quad C_i = \frac{1}{\bar{a}_i + m_i \Delta z_i} \qquad \text{4-C}$$

then

$$\sigma_i = A_i \sigma_{0i} - B_i \sigma_{i-1} - C_i (NF)_{i-1}$$

and 5-C

$$(NF)_i = (\sigma_{0i} - \sigma_i) \bar{a}_i$$

Case II. The integration starts from the tip of the piles, where boundary conditions $(NF)_i$, σ_{0i} and σ_i are known. The unknown values are $(NF)_{i-1}$ and σ_{i-1}. Substituting 1-C into 3-C we obtain

$$(NF)_i - (\sigma_{0i} - \sigma_{i-1}) \bar{a}_{i-1} = m_i (\sigma_i + \sigma_{i-1}) \Delta z_i$$

Solving for σ_{i-1}

$$\sigma_{i-1} = \frac{\bar{a}_{i-1}}{\bar{a}_{i-1} - m_i \Delta z_i} \sigma_{0i-1} + \frac{m_i \Delta z_i}{\bar{a}_{i-1} - m_i \Delta z_i} \sigma_i - \frac{1}{\bar{a}_{i-1} - m_i \Delta z_i} (NF)_i$$

and defining

$$\overline{A}_{i-1} = \frac{\overline{a}_{i-1}}{\overline{a}_{i-1} - m_i \Delta z_i}, \quad \overline{B}_{i-1} = \frac{m_i \Delta z_i}{\overline{a}_{i-1} - m_i \Delta z_i}, \quad \overline{C}_{i-1} = \frac{1}{\overline{a}_{i-1} - m_i \Delta z_i} \qquad \text{6-C}$$

we can write

$$\sigma_{i-1} = \overline{A}_{i-1}\,\sigma_{0i-1} + \overline{B}_{i-1}\,\sigma_i - \overline{C}_{i-1}\,(NF)_i$$

and 7-C

$$(NF)_{i-1} = (\sigma_{0i-1} - \sigma_{i-1})\,\overline{a}_{i-1}$$

2C. POSITIVE FRICTION ($\sigma_i > \sigma_{0i}$)

Case III. The integration starts from the pile head where the boundary conditions $(PF)_{i-1}$, σ_{0i-1} and σ_{i-1} are known. The unknowns are $(PF)_i$ and σ_i. Substituting 2-C into 3-C we obtain

$$(\sigma_i - \sigma_{0i})\overline{a}_i - (PF)_{i-1} = m_i(\sigma_i + \sigma_{i-1})\Delta z_i$$

and solving for σ_i we get

$$\sigma_i = \frac{\overline{a}_i}{\overline{a}_i - m_i \Delta z_i}\,\sigma_{0i} + \frac{m_i \Delta z_i}{\overline{a}_i - m_i \Delta z_i}\,\sigma_{i-1} + \frac{1}{\overline{a}_i - m_i \Delta z_i}\,(PF)_{i-1}$$

Defining

$$\overline{A}_i = \frac{\overline{a}_i}{\overline{a}_i - m_i \Delta z_i}, \quad \overline{B}_i = \frac{m_i \Delta z_i}{\overline{a}_i - m_i \Delta z_i}, \quad \overline{C}_i = \frac{1}{\overline{a}_i - m_i \Delta z_i} \qquad \text{8-C}$$

we can write

$$\sigma_i = \overline{A}_i\sigma_{0i} + \overline{B}_i\sigma_{i-1} + \overline{C}_i(PF)_{i-1}$$

and 9-C

$$(PF)_i = (\sigma_i - \sigma_{0i})\overline{a}_i$$

Case IV. The integration starts from the tip of the piles field where the known boundary conditions are $(PF)_i$, σ_{0i} and σ_i. The unknowns within the pile field are $(PF)_{i-1}$ and σ_{i-1}. Therefore substituting expression 1-C into 3-C we find

$$(PF)_i - (\sigma_{i-1} - \sigma_{0i-1})\,\overline{a}_{i-1} = m_i \Delta z_i\,(\sigma_i + \sigma_{i-1})$$

and solving for the unknown values of σ_{i-1}, we find

$$\sigma_{i-1} = \frac{\overline{a}_{i-1}}{\overline{a}_{i-1} + m_i \Delta z_i}\,\sigma_{0i-1} - \frac{m_i \Delta z_i}{\overline{a}_{i-1} + m_i \Delta z_i}\,\sigma_i + \frac{(PF)_i}{\overline{a}_{i-1} + m_i \Delta z_i}$$

Calling

$$A_{i-1} = \frac{\overline{a}_{i-1}}{\overline{a}_{i-1} + m_i \Delta z_i}, \quad B_{i-1} = \frac{m_i \Delta z_i}{\overline{a}_{i-1} + m_i \Delta z_i}, \quad C_{i-1} = \frac{1}{\overline{a}_{i-1} + m_i \Delta z_i} \qquad \text{10-C}$$

by substitution we finally obtain

$$\sigma_{i-1} = A_{i-1}\,\sigma_{0i-1} - B_{i-1}\,\sigma_i + C_{i-1}\,(PF)_i$$

and 11-C

$$(PF)_{i-1} = (\sigma_{i-1} - \sigma_{0i-1})\,\overline{a}_{i-1}$$

The value of $\overline{a}_i$ varies with depth, and is estimated by means of influence values (Fig. 11-VIII.3) and the influence charts given in Appendix B. When the piles are closely spaced, at a certain depth the effective tributary area reaches the nominal tributary area $\overline{a} = \lambda \cdot \beta$, where λ and β are the pile spacing. Therefore, at greater depth the equivalent tributary area $\overline{a}_i = \overline{a}$ and is assumed constant in the algorithms for the step by step integration.

The parameters used in coefficients A, B and C are:

$\overline{a}_i$ = effective tributary area at depth level $z = i$
$m_i = \pi \overline{r}_0 K_{\phi i}$, in cm
$\overline{r}_0$ = effective radius: $1.05\,r_0$ for NF, and $1.1\,r_0$ for PF
$K_{\phi i} = \tan \phi_r / \overline{N}_\phi$
ϕ_p = angle of friction between soil and coated pile
$\overline{N}_\phi = \dfrac{1 + \sin^2 \phi_r}{1 - \sin^2 \phi_r}$
ϕ_r = angle of internal friction of remolded soil close to the pile shaft.

APPENDIX D: CONVERSION TABLES FROM METRIC (CGS) TO THE ENGLISH SYSTEM

All units given in this book are those of the metric (CGS) system. The tables of units given below are meant to aid the reader unfamiliar with the metric system and to save time in converting the numerical examples, tables and figures into the English (FPS) system.

CONVERSION TABLE FROM THE METRIC TO THE ENGLISH SYSTEM

	LINEAR	Inches	Feet	Yards
mm	Millimeter 1000 Micron (μ)	0.03937		
cm	Centimeter = 10 millimeters	0.3937	0.0328	
m	Meter = 100 centimeters	39.37	3.281	1.094
km	Kilometer = 1000 meters		3280.8	1093.6
	AREA	**(Inches)^2**	**(Feet)^2**	**(Yards)^2**
mm^2	Square millimeter	0.00155		
cm^2	Square centimeter = 100 mm^2	0.155	0.00107	
m^2	Square meter = 10 000 cm^2	1550	10.764	1.196
	VOLUME	**(Inches)^3**	**(Feet)^3**	**(Yards)^3**
mm^3	Cubic millimeter			
cm^3	Cubic centimeter = 10^3 mm^3	0.0610		
m^3	Cubic meter = 10^6 cm^3	61 023	35.31	1.308
	1 Liter = 10^6 mm^3 = 1000 cm^3	61.023	0.03531	
	1 US Gallon = 3.785 liters			

	WEIGHT	Pounds	Short tons
g	Gram	0.002205	
kg	Kilogram = 1000 g	2.205	
ton	Metric ton = 1000 kg	2 204.63	1.102
	Note: In the English system, 1 ton = 2000 lb.		

UNIT WEIGHT	lb/in^3	lb/ft^3	ton/ft^3
kg/cm^3	36.13	62 428	31.214
ton/m^3		62.428	0.03121

UNIT STRESS OR UNIT PRESSURE	lb/in^2	lb/ft^2	ton/ft^2
g/cm^2 = 0.001 kg/cm^2 = 0.01 ton/m^2	0.01422	2.0482	
kg/cm^2 = 10 ton/m^2	14.223	2048.2	1.024
ton/m^2 = 0.1 kg/cm^2	1.422	204.8	0.1024

VELOCITY	in/sec	ft/sec	ft/min
cm/sec	0.3937	0.0328	1.968
m/sec = 100 cm/sec	39.37	3.281	196.8

ACCELERATION	in/sec^2	ft/sec^2
m/sec^2 = 100 cm/sec^2	39.37	3.281
1 cm/sec^2 = 1 gal		

MASS	$\frac{lb \cdot sec^2}{in^3}$	$\frac{lb \cdot sec^2}{ft^3}$	$\frac{ton \cdot sec^2}{ft^3}$
$\frac{kg \cdot sec^2}{cm^3} = 1000 \frac{ton \cdot sec^2}{m^3}$	36.13	62 428	31.214

RATES	ft^3/sec	ft^3/hr	US gal/hr	US gal/day
cm^3/sec				
m^3/sec = 1000 liter /sec	35.31			
m^3/hr		35.31	264.2	6 350
m^3/day				264.2

MOMENT	lb · in	ton · ft
kg · cm	0.8679	
ton · m = 1000 kg · m = 10^5 kg · cm	86 796	3.615

MOMENT OF INERTIA	in^4	ft^4
$cm^4 = 10^{-3}\ m^4$	0.024	1.158×10^{-6}

MASS POLAR MCMENT OF INERTIA	lb · in · sec²	ton · ft · sec²
kg · cm · sec²	0.8679	
ton · m · sec²		3.615

Water viscosity: $\eta = 10.3 \times 10^{-6}$ g · sec/cm² at 20° C
g · sec/cm² = 980.7 poises = 980 700 millipoises
g · sec/cm² = 930.7 dyne · sec/cm² = 0.0142 lb · sec/in²
cm²/sec = 0.155 in²/sec
cm²/kg = 0.0703 in²/lb = 0.9765 ft²/ton
In Hazen's formula $k = CD_{10}^2$, the value of C has the units 1/cm · sec = 2.54/in · sec = 30.48/ft · sec
In Sichardt's formula the value of C_s has the units $(\text{sec/cm})^{1/2} = (2.54\ \text{sec/in})^{1/2}$
In formula 43-II.3, $M_z = M_0 e_N^{-n\sigma_c}$, the value of n has the units cm²/kg and M_z, M_0 · 1 cm²/kg = 0.0703 in²/lb = 0.9765 ft²/ton
The fluidity parameter ϕ_a in formula 64-II.3 has the units of cm²/kg · sec = 0.0703 in²/lb · sec
In Chapter IV, Table 1-IV.5:

k, the unit foundation modulus, has the units
kg/cm³ = 1000 ton/m³ = 36.127 lb/in³ = 6.243 × 10⁴ lb/ft³ = 31.214 ton/ft³

K_1, the linear foundation modulus, has the units
kg/cm² = 10 ton/m² = 14.223 lb/in² = 2048.2 lb/ft² = 1.024 ton/ft²

K, the foundation modulus for a certain specific tributary area, has the units
kg/cm = 0.1 ton/m = 5.60 lb/in = 67.197 lb/ft = 0.0336 ton/ft

C_r = relative consistency = $(L_w - w_n)/I_w$, in clay; w_n = natural water content
D_r = relative density = $(e_{max} - e_n)/(e_{max} - e_{min})$, in cohesionless soil;
e_{max} = maximum void ratio
e_{min} = minimum void ratio
e_n = natural void ratio
$N_\phi = (1 + \sin\phi)/(1 - \sin\phi)$, where ϕ is the angle of internal friction
$\overline{N}_\phi = (1 + \sin^2\phi)/(1 - \sin^2\phi)$
$K_\phi = \tan\phi_r/\overline{N}_\phi$, where ϕ_p is the angle of friction between pile and coating
log a = common logartithm
Log a = e_N natural logarithm = 2.7182

APPENDIX E: NUMERICAL EXAMPLES TO CHAPTERS IV, VII, AND VIII

EXAMPLE A.IV CALCULATION OF A SEMIFLEXIBLE FOUNDATION

Assume the subgrade reactions and bending moments in the longitudinal direction are required for a semiflexible foundation. The foundation structure is placed at a depth of 2 m and covers an area of $24 \times 14.4 = 345.6$ m^2. The foundation structure is formed of a grid of beams assumed to act monolithically with the foundation and ground floor slabs. The cross-sectional rigidity of the foundation structure in the transverse direction has a value of $EI = 6 \times 10^6$ ton · m^2. The column loads acting on the foundation structure, and the subsoil compressibility are given in Fig. 1-A.IV.

By means of the method described in Chapter VI, the foundation area is divided into small tributary areas as shown in Fig. 2-A.IV. Considering a uniform reaction distribution of $q_a = 2000/345.6 = 5.787$ ton/m^2, the reaction on the tributary areas $\bar{a} = 34.560$ m^2 will be equal to $34.560 \times 5.787 = 200$ ton in each one of these areas.

To calculate the stress influence values I_{ji} due to a unit load applied on tributary band area $\bar{a}$, we consider a stratified soil with $\chi = 2$ and use either stress net Fig. 4B, Appendix B, or formula 30-III.2. The I_{ji} values obtained are shown in TABLE 1-A.IV where also the values of the strata unit compression factors α_c are given. Therefore, the unit vertical displacements for unit load $q = +1$ at band (a) are obtained from the following expression:

$$[I_{ji}^N]^T \cdot |\alpha_c^N| = |\bar{\delta}_{ji}|$$

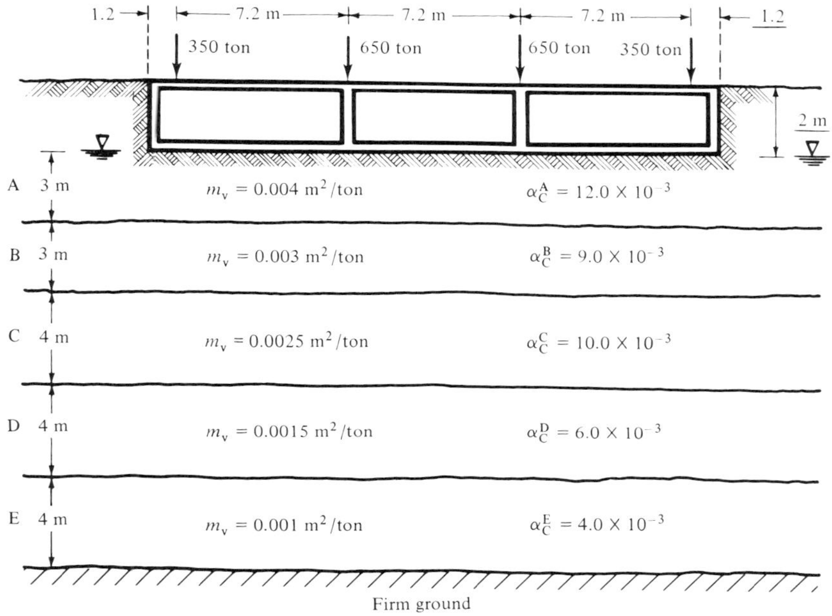

Fig. 1-A.IV Cross section of foundation.

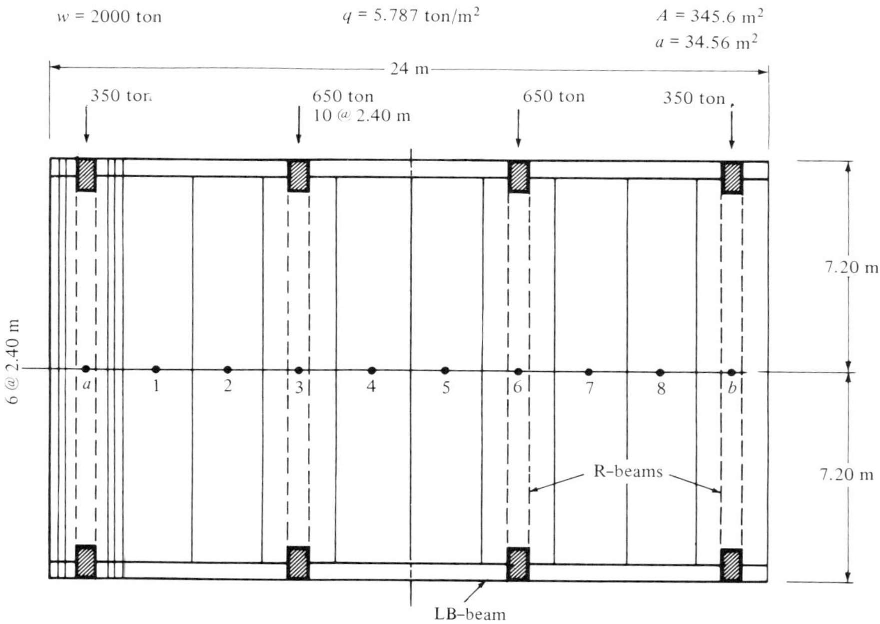

Fig. 2-A.IV Foundation plan.

TABLE 1-A.IV

$\chi = 2, B = 7.20$ m

	x	+1	2.4	4.8	7.20	9.60	12.00	14.40	16.80	19.20	21.60	$\times 10^{-3}$ m^3/ton
Stratum	Ave. Depth	1	2	3	4	5	6	7	8	9	10	α_c
A	1.5	0.622	0.146	0.021	0.006	0.002	0.001	0	0	0	0	12.0
B	4.5	0.240	0.167	0.074	0.031	0.014	0.007	0.004	0.002	0.001	0.001	9.0
C	8.0	0.116	0.101	0.069	0.042	0.024	0.014	0.009	0.005	0.004	0.002	10.0
D	12.0	0.062	0.058	0.047	0.035	0.025	0.017	0.012	0.008	0.006	0.004	6.0
E	16.0	0.038	0.036	0.032	0.027	0.021	0.016	0.012	0.009	0.007	0.005	4.0

The values of $\overline{\delta}_{ji}$ are given in Table 2-A.IV. With the values of $\overline{\delta}_{ji}$ induced by all the tributary areas a, 1, 2, · · · b, the EMA settlement matrix equation is formed, as given in Table 3-A.IV.

Because of the symmetry of the problem, EMA is reduced to Table 4-A.IV. If all the reactions are equal; $X_a = X_i = 5.787 \times 34.56 = 200$ ton, then the settlements under the center of each one of the tributary areas are

$$
\begin{aligned}
\delta_a &= \delta_b = 12.0 \times 10^{-2} \text{ m} \\
\delta_1 &= \delta_8 = 14.7 \times 10^{-2} \text{ m} \\
\delta_2 &= \delta_7 = 15.8 \times 10^{-2} \text{ m} \\
\delta_3 &= \delta_6 = 16.4 \times 10^{-2} \text{ m} \\
\delta_4 &= \delta_5 = 16.6 \times 10^{-2} \text{ m} \\
\delta_{av} &= 15.10 \times 10^{-2} \text{ m}
\end{aligned}
$$

These settlements will take place only if the foundation is 100% flexible and all the reactions are equal. However, we will assume a foundation structure with a stiffness $EI = 6.0 \times 10^6$ ton · m^2 in the longitudinal direction given by the section of the

TABLE 2-A.IV

$$
[I_{ji}]^T \times \begin{vmatrix} 12.0 \\ 9.0 \\ 10.0 \\ 6.0 \\ 4.0 \end{vmatrix} = \overline{\delta}_{ji} \begin{vmatrix} 11.308 \\ 4.757 \\ 2.018 \\ 1.089 \\ 0.624 \\ 0.381 \\ 0.246 \\ 0.152 \\ 0.113 \\ - \end{vmatrix} \times \frac{1}{\overline{a}} = \overline{\delta}_{ji} \begin{vmatrix} 32.720 \\ 13.764 \\ 5.839 \\ 3.151 \\ 1.806 \\ 1.102 \\ 0.712 \\ 0.440 \\ 0.327 \\ - \end{vmatrix}
$$

m^3/ton 10^{-3}	m^3/ton 10^{-3}	$1/m^2$	m/ton $\times 10^{-5}$

TABLE 3-A.IV

	a	1	2	3	4	5	6	7	8	b
a	32.720	13.764	5.839	3.151	1.806	1.102	0.712	0.440	0.327	—
1	13.764	32.720	13.764	5.839	3.151	1.806	1.102	0.712	0.440	0.327
2	5.839	13.764	32.720	13.764	5.839	3.151	1.806	1.102	0.712	0.440
3	3.151	5.839	13.764	32.720	13.764	5.839	3.151	1.806	1.102	0.712
4	1.806	3.151	5.839	13.764	32.720	13.764	5.839	3.151	1.806	1.102
5	1.102	1.806	3.151	5.839	13.764	32.720	13.764	5.839	3.151	1.806
6	0.712	1.102	1.806	3.151	5.839	13.764	32.720	13.764	5.839	3.151
7	0.440	0.712	1.102	1.806	3.151	5.839	13.764	32.720	13.764	5.839
8	0.327	0.440	0.712	1.102	1.806	3.151	5.839	13.764	32.720	13.764
b	—	0.327	0.440	0.712	1.102	1.806	3.151	5.839	13.764	32.720

$\cdot\ |X_i| = |\delta_i|$

$\times 10^{-5}$

box type foundation structure. The foundation structure will be considered as a beam in the longitudinal direction.

For CONDITION $X_i = 0$, the beam is considered supported at (a) and (b). The deflections of the beam at any point i will be called Δ_{io}^{P} and those due to the yielding of the supports Δ_{io}^{A}, hence according to Fig. 3-A.IV, the coordinates will be: for the position of the load, P_n; α_n; β_n; for the position of the reactions, X_i; ξ_i, ψ_i.

We call:

$$\alpha_n = \frac{a_n}{l}, \quad \beta_n = \frac{b_n}{l}, \quad \xi_i = \frac{x_i}{l}, \quad \text{and} \quad \psi = \frac{y_i}{l}$$

CONDITION $X_i = 0$

Deflections to the right of the load P_n for $\xi_i \leqslant \alpha_n$ are

$$EI\Delta_{io}^{PR} = \frac{l^3}{6}\sum_{1}^{n} P_n\beta_n\xi_i(1 - \beta_n^2 - \xi_i^2) \qquad \text{1-A.IV}$$

Deflections to the left of the load P_n for $\psi_i \leqslant \beta_n$ are

$$EI\Delta_{io}^{PL} = \frac{l^3}{6}\sum_{1}^{n} P_n\alpha_n\psi_i(1 - \alpha_n^2 - \psi_i^2) \qquad \text{2-A.IV}$$

and

$$\Delta_{io}^{P} = \Delta_{io}^{PL} + \Delta_{io}^{PR}$$

TABLE 4-A.IV

$$\begin{bmatrix} 32.720 & 14.091 & 6.279 & 3.863 & 2.908 \\ 14.091 & 33.160 & 14.476 & 14.476 & 4.957 \\ 6.279 & 14.476 & 33.822 & 15.570 & 8.990 \\ 3.863 & 6.941 & 15.570 & 35.871 & 19.603 \\ 2.908 & 4.957 & 8.990 & 19.603 & 46.484 \end{bmatrix} \cdot \begin{vmatrix} X_a \\ X_1 \\ X_2 \\ X_3 \\ X_4 \end{vmatrix} = \begin{vmatrix} \delta_a \\ \delta_1 \\ \delta_2 \\ \delta_3 \\ \delta_4 \end{vmatrix}$$

$\times 10^{-5}$

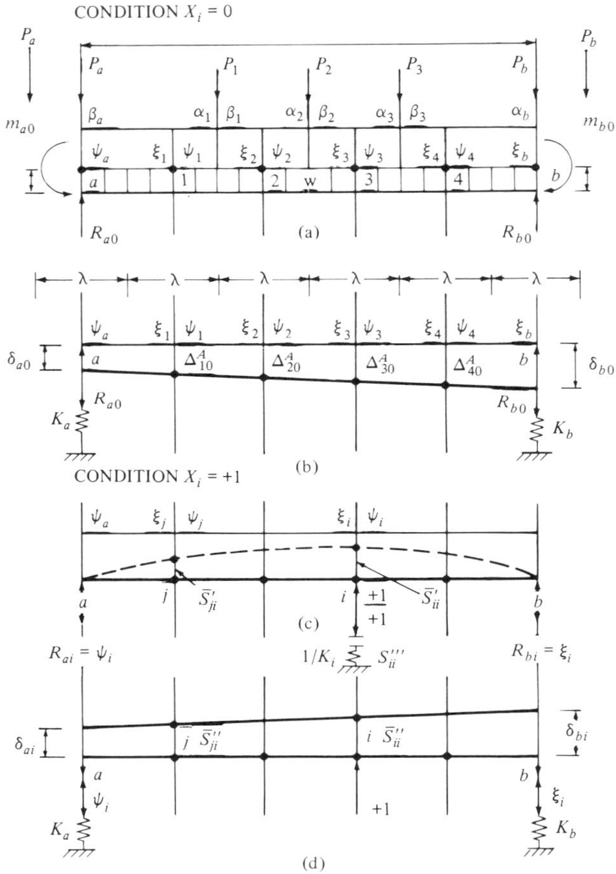

Fig. 3-A.IV Displacements of foundation beam.

CONDITION $X_i = +1$

for $\xi_j \leqslant \xi_i$

$$(EI)\bar{S}'_{ji} = \frac{l^3}{6}\,\xi_j\psi_i(1 - \psi_i^2 - \xi_i^2) \qquad \text{3-A.IV}$$

for $\psi_j \leqslant \psi_i$

$$(EI)\bar{S}'_{ji} = \frac{l^3}{6}\,\psi_j\xi_i(1 - \xi_i^2 - \psi_j^2) \qquad \text{4-A.IV}$$

Using algorithms 14, 15 and 16-VI.3 for

CONDITION $X_i = 0$

$$(EI)\Delta^A_{i0} = \{(\bar{\delta}_{\text{aa}}\psi_j - \bar{\delta}_{j\text{a}})R_{\text{a0}} + (\bar{\delta}_{\text{bb}}\xi_i - \bar{\delta}_{j\text{b}})R_{\text{b0}}\}\,(EI) \qquad \text{5-A.IV}$$

TABLE 5-A.IV

$EI = 6 \times 10^6$ ton · m²

Loads: 350 ton (at a), 650 (at 3), 650 ton (at 6), 350 (at b)

R_{a0} = 1000 ton; R_{b0} = 1000

	a	1	2	3	4	5	6	7	8	b	
$\xi_i 0$		0.111	0.222	0.333	0.444	0.556	0.667	0.778	0.889	1.0	$\times 10^4$
$EI(\Delta_{i0}^{PL} + \Delta_{i0}^{PR})$		7.816	14.749	19.914	22.577	22.577	19.914	14.749	7.8160		$\times 10^4$
$EI \cdot \Delta_{i0}^{A}$		111.774	158.646	173.142	178.872	178.872	173.142	158.646	111.774		
$EI \cdot \Delta_{i0}$		119.590	173.395	193.056	201.449	201.449	193.056	173.395	119.590		$\times 10^4$

CONDITION $X_i = +1$

$$(EI)(\bar{S}''_{ji} + \bar{S}'''_{ji}) = \{\bar{\delta}_{aa}\psi_j\psi_i + \bar{\delta}_{bb}\xi_j\xi_i - \bar{\delta}_{ai}\psi_j - \bar{\delta}_{bi}\xi_j - \bar{\delta}_{ja}\psi_i - \bar{\delta}_{jb}\xi_i + \bar{\delta}_{ab}\psi_j\xi_i + \bar{\delta}_{ba}\psi_i\xi_j + \bar{\delta}_{ji}\}(EI) \quad \text{6-A.IV}$$

Using the above algorithms we can calculate, with the aid of EMA unit displacement matrix, the coefficients of the unknown reactions and the independent terms forming matrix soil-structure interaction equation EMISE (Tables 5 and 6-A.IV, respectively), from which EMISE itself is obtained as given in Table 7-A.IV. Due to symmetry, this may be reduced as given in Table 8-A.IV. The system of simultaneous equations is solved for the unknowns R_a, X_i, R_b given in Table 9-A.IV. The average settlements in the tributary band areas may be obtained from the values of R_a, X_i and R_b using EMA given in Table 4-A.IV; their values may be found in Table 9-A.IV. The bending moments obtained with these soil reactions and loads on the foundation beam structure are shown in Fig. 4-A.IV. For comparison the bending moments of the foundation are also shown when the soil reactions are considered uniform. The importance of the results may be readily recognized.

The subgrade reactions in the transverse direction may be obtained approximately

TABLE 6-A.IV

POINT	1	2	3	4	5	6	7	8
$\xi_i =$	0.111	0.222	0.333	0.444	0.556	0.667	0.778	0.889
$EI\bar{S}'_{j1}$	32.711	55.466	67.498	70.337	65.440	54.434	38.839	20.185
$EI(\bar{S}''_{j1} + \bar{\delta}_{j1})$	2066.248	1270.807	856.826	684.643	556.758	445.594	338.753	195.632
$EI\bar{S}_{j1}$	2098.959	1326.273	924.324	754.980	622.198	500.028	377.592	215.817
$EI\bar{S}'_{j2}$	55.466	100.209	125.803	133.011	124.760	104.279	74.619	38.839
$EI(\bar{S}''_{j2} + \bar{\delta}_{j2})$	1270.807	2691.397	1590.687	1087.573	863.807	698.193	547.643	338.753
$EI\bar{S}_{j2}$	1326.273	2791.606	1716.490	1220.584	988.567	802.472	622.262	377.592
$EI\bar{S}'_{j3}$	67.498	125.803	165.722	180.359	171.841	144.945	103.864	54.434
$EI(\bar{S}''_{j3} + \bar{\delta}_{j3})$	856.826	1590.687	2773.646	1625.781	1107.339	878.254	698.193	445.594
$EI\bar{S}_{j3}$	924.324	1716.490	2939.368	1806.140	1279.180	1023.204	802.027	500.028
$EI\bar{S}'_{j4}$	70.337	133.01	180.359	204.718	200.564	171.841	124.760	65.440
$EI(\bar{S}''_{j4} + \bar{\delta}_{j4})$	684.643	1087.573	1625.781	2777.902	1625.378	1107.339	863.807	556.758
$EI\bar{S}_{j4}$	754.980	1220.583	1806.140	2982.620	1825.942	1279.180	988.567	622.198
$EI\bar{S}'_{j5}$	65.440	124.760	171.841	200.564	204.718	180.539	133.011	70.337
$EI(\bar{S}''_{j5} + \bar{\delta}_{j5})$	556.758	863.807	1107.339	1625.378	2777.902	1625.785	1087.573	684.643
$EI\bar{S}_{j5}$	622.198	988.567	1279.180	1825.942	2982.620	1806.230	1220.584	754.980
$EI\bar{S}'_{j6}$	54.434	103.864	144.945	171.841	180.359	165.722	125.803	67.498
$EI(\bar{S}''_{j6} + \bar{\delta}_{j6})$	445.594	698.193	878.254	1107.339	1625.781	2773.646	1590.687	856.826
$EI\bar{S}_{j6}$	500.028	802.057	1023.199	1279.180	1806.140	2939.368	1716.490	924.324
$EI\bar{S}'_{j7}$	38.839	74.619	104.619	124.760	133.011	125.803	100.209	55.466
$EI(\bar{S}''_{j7} + \bar{\delta}_{j7})$	338.753	547.643	698.193	863.807	1087.573	1590.687	2691.397	1270.807
$EI\bar{S}_{j7}$	377.592	622.262	802.812	988.567	1220.584	1716.490	2791.606	1326.273
$EI\bar{S}'_{j8}$	20.185	38.839	54.434	65.440	70.337	67.337	55.466	32.711
$EI(\bar{S}''_{j8} + \bar{\delta}_{j8})$	195.632	338.753	445.594	556.758	684.643	856.826	1270.807	2066.248
$EI\bar{S}_{j8}$	215.817	377.592	500.028	622.198	754.980	924.163	1326.273	2098.959

TABLE 7-A.IV

EMISE	1	2	3	4	5	6	7	8			×10⁴
1	2098.959	1326.273	924.324	754.980	622.198	500.028	377.592	215.817	X_1		119.590
2	1326.273	2791.606	1716.490	1220.584	988.567	802.472	622.262	377.592	X_2		173.395
3	924.324	1716.490	2939.368	1806.140	1279.180	1023.204	802.472	500.028	X_3		193.056
4	754.980	1220.583	1806.140	2982.620	1825.942	1279.180	988.567	622.198	· X_4	=	201.449
5	622.198	988.567	1279.180	1825.942	2982.620	1806.140	1220.584	754.980	X_5		201.449
6	500.028	802.472	1023.204	1279.180	1806.140	2939.368	1716.490	924.324	X_6		193.056
7	377.592	622.262	802.812	988.567	1220.584	1716.490	2791.606	1326.273	X_7		173.395
8	215.817	377.592	500.028	622.198	754.980	924.163	1326.273	2098.959	X_8		119.590

$X_1 = X_8, X_2 = X_7, X_3 = X_6, X_4 = X_5$

TABLE 8-A.IV

$$\begin{bmatrix} 2314.776 & 1703.865 & 1424.352 & 1377.178 \\ 1703.865 & 3413.868 & 2518.962 & 2209.151 \\ 1424.352 & 2118.962 & 3962.572 & 3085.320 \\ 1377.178 & 2209.151 & 3085.320 & 4808.562 \end{bmatrix} \cdot \begin{vmatrix} X_1 \\ X_2 \\ X_3 \\ X_4 \end{vmatrix} = \begin{vmatrix} 119.590 \\ 173.395 \\ 193.056 \\ 201.449 \end{vmatrix} \times 10^4$$

TABLE 9-A.IV

R_a	X_1	X_2	X_3	X_4	℄
297.945	172.957	179.475	174.887	174.736	ton
8.621	5.005	5.193	5.060	5.056	Ave. load/m^2
124.102	72.065	74.781	72.870	72.807	Load/m
14.50	14.61	14.74	14.84	14.89	Ave. settlement in cm.

by computing the reactions configuration. The influence values under the conditions assumed above are given in Fig. 5-A.IV. Dividing the values of $\overline{\delta}_{ji}$ by the tributary band areas in the transverse direction 24 X 2.4 = 57.6 m^2 we obtain the influence values of EMA in Table 10-A.IV. EMISE is calculated as explained for calculations in the longitudinal direction using algorithms given before for the beam as a rigid body, that is to say, Δ_{i0}^A for CONDITION $X_i = 0$, and $(\overline{S}_{ji}'' + \overline{\delta}_{ji})$ for CONDITION $X_i = +1$.

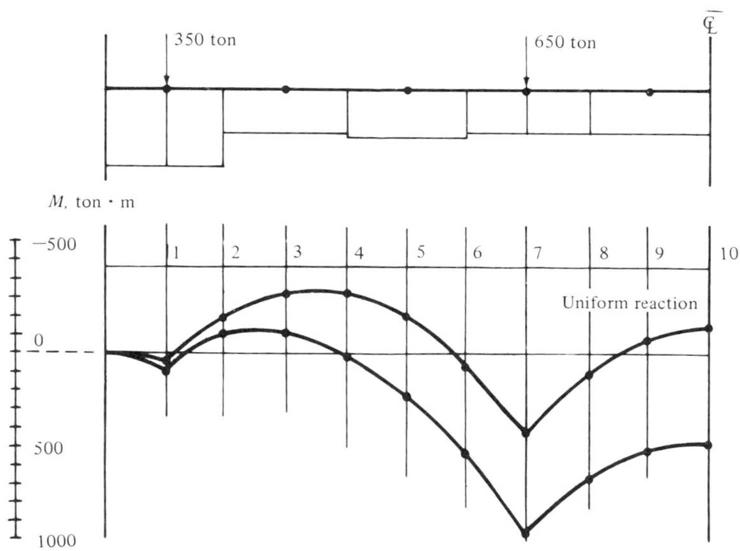

Fig. 4-A.IV Bending moments.

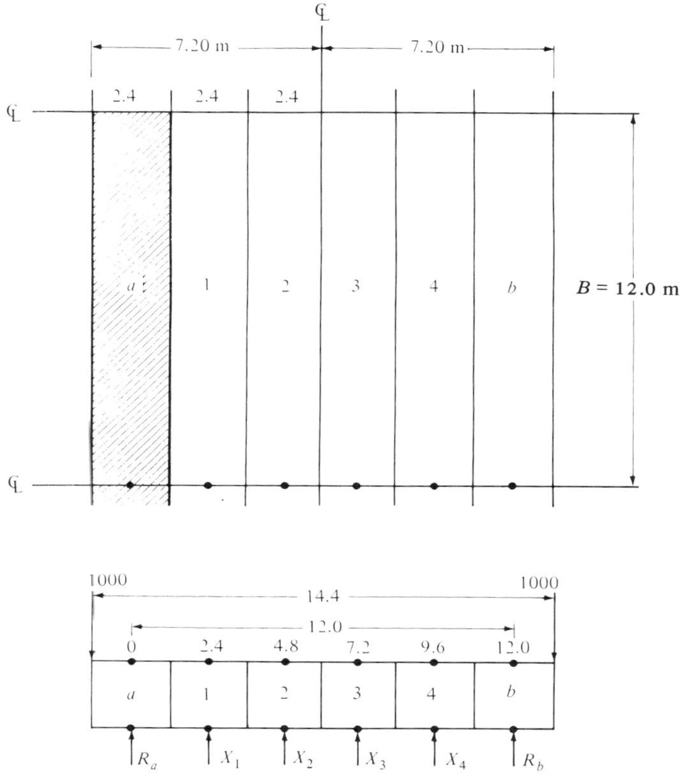

STRM	DEPTH	(hatched load) = 1.0						α_C^N	$\bar{\delta}_{j1}$
A	1.5	0.624	0.148	0.023	0.007	0.003	0.001	12.0	11.823
B	4.5	0.253	0.179	0.083	0.037	0.018	0.009	9.0	5.237
C	8.0	0.136	0.119	0.084	0.053	0.032	0.019	10.0	2.425
D	12.0	0.081	0.076	0.063	0.048	0.035	0.025	6.0	1.391
E	16.0	0.053	0.051	0.046	0.039	0.031	0.024	4.0	0.852
								$\times 10^{-3}$ m^3/ton	0.529 $\times 10^{-2}$ m

Fig. 5-A.IV Influence values in the transverse direction.

TABLE 10-A.IV

	a	1	2	3	4	b
a	20.526	9.092	4.210	2.415	1.479	0.918
1	9.092	20.526	4.092	4.210	2.415	1.479
2	4.210	9.092	20.526	9.092	4.210	2.415
3	2.415	4.210	9.092	20.526	9.092	4.210
4	1.479	2.415	4.210	9.092	20.526	9.092
b	0.918	1.479	2.415	4.210	9.092	20.526

$\times 10^{-2}$ $= [\bar{\delta}_{ji}]^T$

Therefore, for CONDITION $X_i = 0$

Deflection of Beam:

1) $(EI)\,\Delta_{io}^{P} = -\dfrac{l^2}{2}\, m_{ao}\, \psi_i \xi_i = -8.640\ \psi_i \xi_i \times 10^4$ m

2) Soil Displacements $(EI)\Delta_{io}^{A}$, same as 5-A.IV. The value of $EI = 24 \times 4.617 \cdot 10^5 = 10.0 \times 10^6$ ton · m², $R_{ao} = R_{bo} = 1000$ ton, and $m_{ao} = m_{bo} = 1200$ ton · m. The calculations of the independent terms are given in Table 11-A.IV.

TABLE 11-A.IV

POINT	1	2	3	4	
ξ_i	0.20	0.40	0.60	0.80	
$EI\Delta_{io}^{P}$	−1.382	−2.074	−2.074	−1.382	$\times 10^4$
$EI\Delta_{io}^{A}$	99.550	139.010	139.010	99.550	$\times 10^4$
$EI\Delta_{io}$	98.268	136.936	136.936	98.268	$\times 10^4$
ξ_i	0.20	0.40	0.60	0.80	
$EI\bar{S}'_{j1}$	14.746	20.736	18.432	10.598	
$EI(\bar{S}''_{ji} + \bar{\delta}_{j1})$	1934.488	1068.876	594.324	298.012	
$EI\bar{S}_{ji}$	1949.234	1089.612	612.756	308.610	
$EI\bar{S}'_{j2}$	20.736	33.178	31.334	18.432	
$EI(\bar{S}''_{j2} + \bar{\delta}_{j2})$	1068.876	2421.552	1267.848	594.324	
$EI\bar{S}_{j2}$	1089.612	2454.730	1299.182	612.756	
$EI\bar{S}'_{j3}$	18.432	31.334	33.178	20.736	
$EI(\bar{S}''_{j3} + \bar{\delta}_{j3})$	594.324	1267.848	2421.552	1068.876	
$EI\bar{S}_{j4}$	612.756	1299.182	2454.730	1089.876	
$EI\bar{S}'_{j4}$	10.598	18.432	20.736	14.746	
$EI(\bar{S}''_{j4} + \bar{\delta}_{j4})$	298.012	594.324	1068.876	1934.488	
$EI\bar{S}_{j4}$	308.610	612.756	1089.612	1949.234	

CONDITION $X_i = +1$

1) Deflection of Beam

$$\xi_j \leqslant \xi_i \quad \text{same as (3-A.IV)}$$

$$\psi_j \leqslant \psi_i \quad \text{same as (4-A.IV)}$$

2) Soil Displacements

$$(EI)(\bar{S}''_{ji} + \bar{\delta}_{ji}) \text{ same as (6-A.IV)}$$

With algorithms mentioned above and with the help of the EMA matrix for $\bar{\delta}_{ji}$ values, the EMISE coefficients may be calculated as given in Table 11-A.IV, from which the unknown values of X_1 and X_2 may be obtained, also the values of reactions $R_a = R_b$, hence

X_1	X_2	IT
2257.844	1702.368	98.268×10^4
1702.368	3753.912	136.936×10^4

and

	R_a	X_1	X_2	X_3	X_4	R_b
Reactions	502.185	243.400	254.415	254.415	243.400	502.185
	Average reaction 333.331 ton					
factors	1.507	0.730	0.763	0.763	0.730	1.507

Using the factors of configuration thus computed and the average band reactions obtained for the longitudinal direction, the subgrade reactions are obtained, Table 12-A.IV.

TABLE 12-A.IV

LONGITUDINAL DIRECTION	CONFIGURATION FACTORS			
q_{mi}	1.507	0.730	0.763	PNT
8.621	12.992	6.293	6.578	a
5.005	7.543	3.654	3.819	1
5.193	7.826	3.791	3.962	2
5.063	7.625	3.694	3.861	3
5.056	7.619	3.691	3.858	4
PNT	a	1	2	

TRANSVERSE DIRECTION

EXAMPLE B.IV RIGID BOX TYPE FOUNDATION

We are required to determine the subgrade reactions at the foundation grade elevation of a rigid compensated box type foundation for a tall building.

The relative vertical displacements within the foundation structure may be considered negligible when compared with the soil vertical displacements. The foundation area of the building is of 20 m × 20 m = 400 m². The building weight is of 4000 ton including basements and foundation structure; therefore the average soil reaction is 10 ton/m². The compressible soil deposit is 20 m thick. The average unit stratum moduli M_z for certain times after construction, and for the strata considered are given in Fig. 1-B.IV, where also the calculation of the unit stress influence values has been recorded.

From the information given above, the unit vertical displacements induced by each one of the tributary areas loaded with $q_i = +1$ may be calculated by means of the following expression:

$$|\bar{\delta}_{ji}| = [I_{ji}]^T \times |\alpha_i^N| \qquad \text{1-B.IV}$$

With the $\bar{\delta}_{ji}$ values obtained with 1-B.IV, the EMA matrix equation given in Table 1-A.IV is formed. Hence,

$$[\bar{\delta}_{ji}]^T \cdot |q_i| = |\delta_i| \qquad \text{2-B.IV}$$

By symmetry we obtain

$$\begin{matrix} 1 & 2 & 3 \end{matrix}$$
$$\begin{bmatrix} 2.744 & 1.799 & 1.236 \\ 1.799 & 3.041 & 2.350 \\ 1.236 & 2.350 & 4.145 \end{bmatrix} \times 10^{-3} \cdot \begin{vmatrix} q_1 \\ q_2 \\ q_3 \end{vmatrix} = \begin{vmatrix} \delta_1 \\ \delta_2 \\ \delta_3 \end{vmatrix}$$

Assuming $q_i = \text{const} = 10$ ton/m² we obtain the settlements at the center of the tributary areas when the foundation structure is considered 100% flexible. These values in centimeters are:

$$\delta_1 = 5.799 \text{ cm}, \quad \delta_2 = 7.190 \text{ cm}, \quad \delta_3 = 7.731 \text{ cm},$$

from which the average settlement $\delta_a = 6.90$ cm is obtained. Using this value in $|\delta_i|$ of EMA we solve for the unknowns q_i and obtain:

$$q_1' = 17.501 \text{ ton/m}^2, \quad q_2' = 6.239 \text{ ton/m}^2, \quad q_3' = 7.891 \text{ ton/m}^2$$

The values given above produce a total reaction of $W_e' = \bar{a} \, \Sigma \, q_i' = 4217.028$ ton. However, the total weight of the building is $W_e = 4000$ ton. Therefore, we conclude that the δ_a used to calculate q_i' was not the exact value and we have to correct the q_i' and δ_a in the proportion 4000/4217.028 = 0.949 to obtain the final values of the average reactions in the tributary band areas considered. Hence,

$$q_1 = 16.60 \text{ ton/m}^2, \quad q_2 = 5.918 \text{ ton/m}^2, \quad q_3 = 7.485 \text{ ton/m}^2$$

and the uniform settlement is $\delta_a = 6.545$ cm.

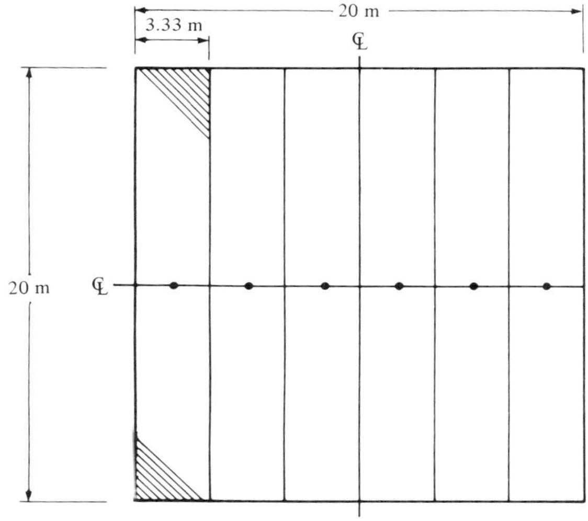

$\chi = 2.0$
$B = 10$ m
depth = 20 m
$W_e = 4000$ ton
$A = 400$ m^2
$q_a = 10$ ton/m^2

$q_i = +1$

PNT	z_i	d_i	1 (x = 0)	2 (x = 3.33)	3 (x = 6.67)	4 (x = 10.0)	5 (x = 13.33)	6 (x = 16.67)	M_{zi}	α_C^N	$\bar{\delta}_{j1}$	
1	0.5	1.0	0.958	0.018	0.001	–	–	–	3.00	1.50	2.597	1
2	2.0	2.0	0.638	0.141	0.020	0.005	0.002	0.001	3.54	7.08	1.548	2
3	3.90	1.8	0.385	0.189	0.052	0.017	0.007	0.003	6.45	11.61	0.802	3
4	6.10	2.6	0.246	0.169	0.073	0.030	0.014	0.007	10.00	26.00	0.434	4
5	9.00	3.2	0.155	0.126	0.076	0.040	0.021	0.011	9.54	30.53	0.251	5
6	11.80	2.4	0.106	0.094	0.066	0.041	0.025	0.015	7.50	18.00	0.147	6
7	14.80	3.6	0.075	0.069	0.054	0.038	0.026	0.017	6.30	22.68		
8	18.30	3.4	0.053	0.050	0.042	0.033	0.024	0.017	4.50	15.30		
	m	m	ton/m^2						m^2/ton 10^{-4}	m^3/ton 10^{-4}	m 10^{-3}	

Fig. 1-B.IV Rigid foundation.

TABLE 1-B.IV

$$
\begin{matrix} 1 & 2 & 3 & 4 & 5 & 6 \end{matrix}
$$

$$
\begin{bmatrix}
2.597 & 1.548 & 0.802 & 0.434 & 0.251 & 0.147 \\
1.548 & 2.597 & 1.548 & 0.802 & 0.434 & 0.251 \\
0.802 & 1.548 & 2.597 & 1.548 & 0.802 & 0.434 \\
0.434 & 0.802 & 1.548 & 2.597 & 1.548 & 0.802 \\
0.251 & 0.434 & 0.802 & 1.548 & 2.597 & 1.548 \\
0.147 & 0.251 & 0.434 & 0.802 & 1.548 & 2.597
\end{bmatrix} \times 10^{-3} \cdot \begin{vmatrix} q_1 \\ q_2 \\ q_3 \\ q_4 \\ q_5 \\ q_6 \end{vmatrix} = \begin{vmatrix} \delta_1 \\ \delta_2 \\ \delta_3 \\ \delta_4 \\ \delta_5 \\ \delta_6 \end{vmatrix}
$$

In order to obtain the distribution of reactions in the transverse direction we calculate the configuration factors for the transverse bands, and since this is a symmetrical case we use the configuration already calculated. Hence,

$$
f_1 = \frac{16.60}{10} = 1.66, \quad f_2 = \frac{5.918}{10} = 0.592, \quad f_3 = \frac{7.485}{10} = 0.748
$$

The final calculation of the subgrade reactions for the rigid foundation is given in Table 2-B.IV. The soil reactions are given in ton/m^2, and act uniformly in the square areas 3.33 m × 3.33 m. Only one quadrant of the foundation reactions is shown.

TABLE 2-B.IV

PNT	Average Reaction	CONFIGURATION FACTOR 1 1.660	2 0.592	3 0.748
1	16.60	27.56	9.83	12.42
2	5.92	9.83	3.50	4.43
3	7.48	12.42	4.43	5.60

EXAMPLE A.VII PONTOON STRIP FOUNDATION

A pontoon foundation is formed with several rigid structural foundation sections joined together to form a long strip foundation. It is used to control flexibility of the foundation and reduce large bending moments, thus yielding economy in the foundation structural design. An example for the calculation of a foundation of this type, according to the theory described in section VII.9 may be performed in two steps: in the first step we compute the strip foundation considering it infinitely stiff in comparison with the subsoil compressibility; second, the foundation is divided into three rigid sections, for which purpose two joints are introduced, assuming one of Type III and another of Type IV, thus forming a pontoon foundation. The solution is found by means of corrections introduced because of pontoon action to the reactions, shears and bending moments obtained in the first step of calculation in which the foundation structure was considered as a rigid body. Finally the pontoon solution is investigated for allowable differential settlements.

First Step

(a) Foundation area: 12 m wide, 60 m long, area 720 m^2.
(b) The compressible soil is considered isotropic, with thickness 20 m and coefficient of unit volume compressibility $m_v = 0.01$ cm^2/kg corresponding to the stress level induced in the soil mass by the foundation loads.
(c) The loads are distributed as shown in Fig. 1-A.VII. The average load per unit area is 10 ton/m^2.

The stress influence values I_{ji} at the center and under each tributary band area are determined at different elevations in the compressible soil deposits with stress net $\chi = 3$ given in Appendix B. A unit load is applied at a certain time to one tributary band, and using the compressibility properties of the subsoil materials, we calculate the influence settlement coefficients $\bar{\delta}_{ji}$ (Chapter VI) at points 1, 2, 3, . . . 7 at the center of each tributary band (Fig. 2-A.VII). For point 1 we obtain

$$|\bar{\delta}_{ji}| = [I_{ji}]^T \cdot |\alpha^N|$$

In the same manner we compute the values for a unit load applied to other bands *i*. With values $|\bar{\delta}_{ji}|$ so obtained for all bands, the square matrix $[\bar{\delta}_{ji}]^T$ in EMA may be formed as shown in Table 1-A.VII. In this particular case, owing to the different size of the tributary band areas, the above mentioned square matrix is not symmetrical.

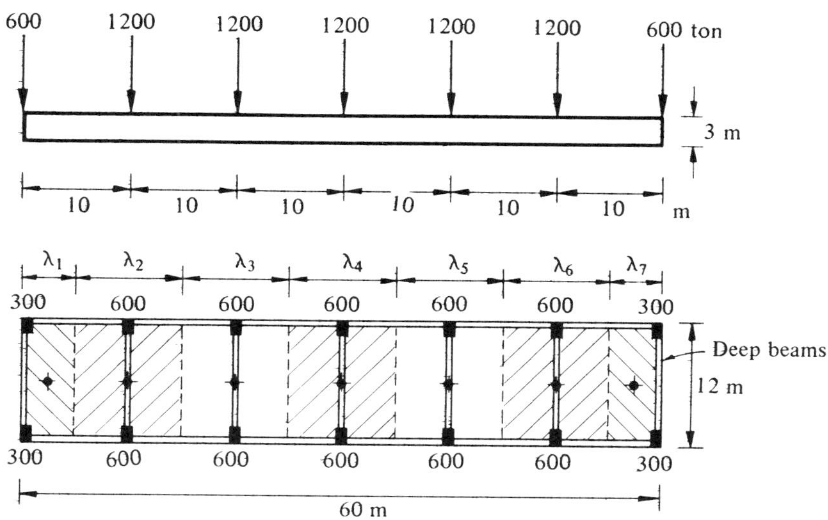

Fig. 1-A.VII Loads and tributary bands.

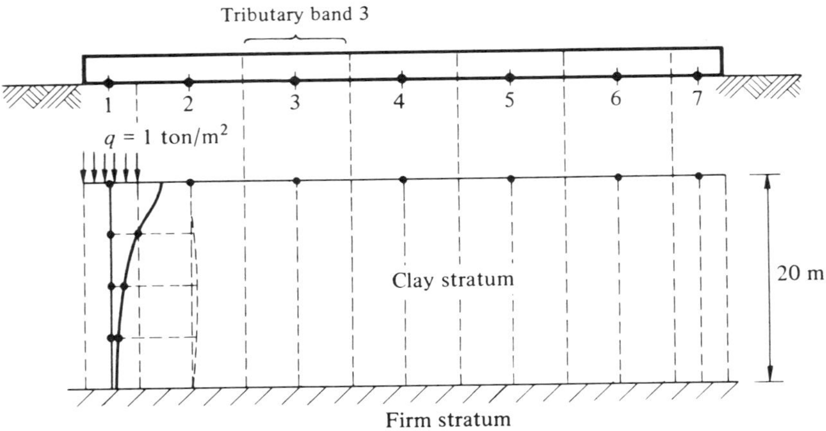

Fig. 2-A.VII Stresses under tributary areas.

When we know the loads per unit surface q_i we can compute by EMA (Table 1-A.VII) the vertical displacements δ_i for points 1 to 7. Since this is a symmetrical problem, that is to say, $q_1 = q_7$, $q_2 = q_6$ and $q_3 = q_5$, we may reduce the EMA settlement matrix equation as given in Table 2-A.VII.

In the case of a uniform unit reaction of 10 ton/m^2, the settlement at the center of each one of the tributary areas is

$$\delta_1 = \delta_7 = 10.0 \text{ cm}$$

$$\delta_2 = \delta_6 = 12.4 \text{ cm}$$

$$\delta_3 = \delta_5 = 12.93 \text{ cm}$$

$$\delta_4 = 13.02 \text{ cm},$$

when the foundation structure is considered rigid $\delta_1 = \delta_2 = \delta_i = \delta'_a$ and constant. The average value of the vertical displacement is

$$\delta'_a = 2(5.0 \times 10 + 10 \times 12.4 + 10 \times 12.93 + 5 \times 13.02)/60$$

$$\delta'_a = 12.28 \text{ cm}$$

TABLE 1-A.VII Influence Matrix $[\bar{\delta}_{ji}]^T$ for Settlement Calculations

$$\begin{bmatrix} 0.6858 & 0.2678 & 0.0368 & 0.0077 & 0.0021 & 0.0009 & 0.0002 \\ 0.1130 & 0.9445 & 0.1508 & 0.0241 & 0.0055 & 0.0015 & 0.0005 \\ 0.0165 & 0.1508 & 0.9445 & 0.1508 & 0.0241 & 0.0055 & 0.0010 \\ 0.0037 & 0.0241 & 0.1508 & 0.9445 & 0.1508 & 0.0241 & 0.0037 \\ 0.0010 & 0.0055 & 0.0241 & 0.1508 & 0.9445 & 0.1508 & 0.0165 \\ 0.0005 & 0.0015 & 0.0055 & 0.0241 & 0.1508 & 0.9445 & 0.1130 \\ 0.0002 & 0.0009 & 0.0021 & 0.0077 & 0.0368 & 0.2678 & 0.6858 \end{bmatrix}$$

For details of operation see procedure Fig. 2-IV.6.

TABLE 2-A.VII

$$\begin{bmatrix} 0.6860 & 0.2687 & 0.0389 & 0.0077 \\ 0.1135 & 0.9460 & 0.1563 & 0.0241 \\ 0.0175 & 0.1563 & 0.9686 & 0.1508 \\ 0.0074 & 0.0482 & 0.3016 & 0.9445 \end{bmatrix} \cdot \begin{vmatrix} q_1 \\ q_2 \\ q_3 \\ q_4 \end{vmatrix} = \begin{vmatrix} \delta_1 \\ \delta_2 \\ \delta_3 \\ \delta_4 \end{vmatrix}$$

Using the average value $\delta_a' = 12.28$ cm for a rigid foundation we calculate the soil unit reactions from EMA (Table 2-A.VII), from which we obtain:

TABLE 3-A.VII

Unit Reactions	ton/m²	X_i	Corrected unit reaction
$q_1' = q_7'$	13.58	1629.6	13.37
$q_2' = q_6'$	9.55	2292.0	9.40
$q_3' = q_5'$	9.43	2263.2	9.28
q_4'	9.40	1128.0	9.26

With q_i' values obtained for assumed average vertical displacement the reactions add to a total load of 7312.8 ton, hence a correction of 7200/7312.8 = 0.98 should be made to q_i' values to obtain the final values reported in Table 3-A.VII.

It may be recognized from the foundation assumed as a rigid body that high bridging action takes place with high bending moments on the order of 3200 ton · m at the center of the foundation. The bending moments, however, may be reduced for economy introducing joints at points 3 and 5. The joints at these points will be called (a) and (b) respectively, Fig. 3-A.VII.

Second Step: Pontoon Action

A joint of Type III not taking bending moments will be used at (a), and at (b) one of Type IV taking a plastic moment $M_p = 800$ ton · m. Therefore, the reduced moment at joint (a) will be 2670 ton · m and at joint (b) 2670 – 800 = 1870 ton · m. The differential settlement shall not exceed 1:500 in pontoon sections 1 and 2, rotating as rigid bodies, and 1:1000 in the third section, Fig. 3-A.VII.

From section VII.9 the general joint equation reads

$$a_{n-1} \Delta S_{i-1} + (b_{n-1} + b_n) \Delta S_i + a_n \Delta S_{i+1} = c_{n-1} + c_n \qquad \text{5-VII.9}$$

and we obtain for joint (a):

$$(b_1 + b_2) \Delta S_a + a_2 \Delta S_b = c_1 + c_2$$

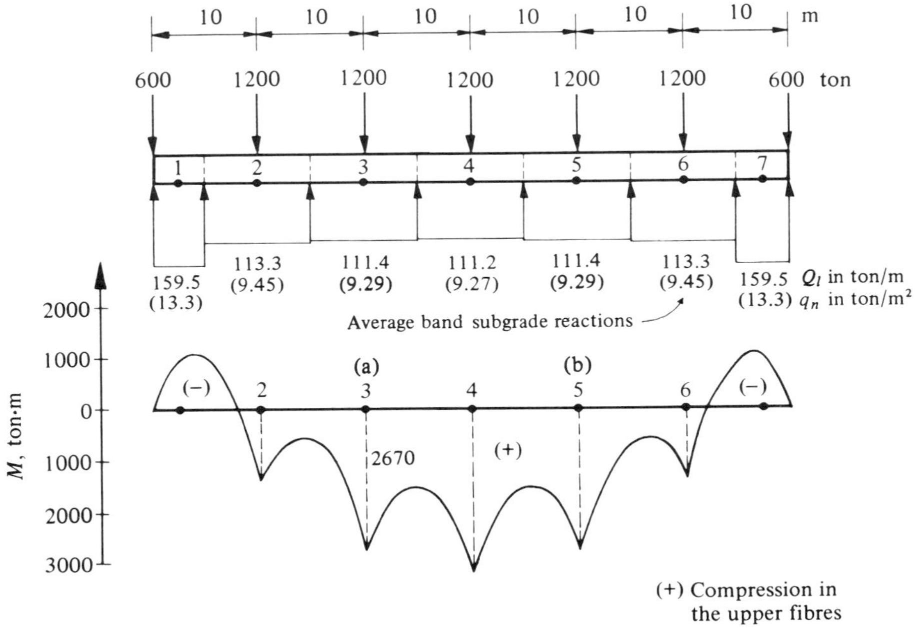

Fig. 3-A.VII Loads and bending moments for rigid strip foundation.

and for joint (b):

$$a_2\,\Delta S_a + (b_2 + b_3)\,\Delta S_b = c_2 + c_3$$

The coefficients calculation is reported in Table 4-A.VII.

The equations to compute the change in reactions at the joint are

$$0.0333\ \ \Delta S_a + 0.00833\,\Delta S_b = -2.338$$

$$0.00833\,\Delta S_a + 0.0333\ \ \Delta S_b = \ \ 3.338$$

from which $\Delta S_a = -101.6$ ton and $\Delta S_b = 125.6$ ton. Using equations 1-VII.9 (section VII.9) for soil reactions, the changes in subgrade reactions at the edges of the pontoon sections are calculated and given in Table 5-A.VII.

Assuming a 12-m-wide strip foundation, the changes in total subgrade reactions for tributary band areas given in ton/m may be found (Fig. 4-A.VII). Accordingly, the moment corrections and final bending moments for the pontoon strip structure foundation are plotted in Fig. 5-A.VII, where it may be seen that the bending moments have been reduced considerably, allowing greater economy in the foundation structural design. With final band average unit reactions (Fig. 4-A.VII) the total and differential settlements may be investigated on every tributary band using the influence settlement matrix $[\bar{\delta}_{ji}]^T$ given in Table 1-A.VII. The results are plotted in Fig. 6-A.VII. When the foundation is considered rigid in the transverse direction, the settlements thus obtained may be reduced to 85%. The rotations of the pontoon sections are found to be below the specified allowable values.

TABLE 4-A.VII Coefficients for Pontoon Equations

Data \ Pontoon	1	(a) 2	(b) 3	Units
$A_n = 2B \times 2l_n$	240	240	240	m^2
$S_n = \frac{1}{3} A_n \cdot l_n$	800	800	800	m^3
$a_n = 2/A_n$	0.00833	0.00833	0.00833	$1/m^2$
$b_n = 4/A_n$	0.0167	0.0167	0.0167	$1/m^2$
ΔM_i	−2670	2670 −1870	1870	ton-m
$\Delta M_i + \Delta M_{i+1}$	−2670	800	1870	ton-m
$c_n = \dfrac{\Delta M_i + \Delta M_{i+1}}{S_n}$	−3.338	1.00	2.338	ton/m^2

TABLE 5-A.VII Pontoon Corrections

Data \ Pontoon	1	2	3	Units
ΔS	0 −101.6↓	↑−101.6 125.6↑	↓125.6 0	ton
$(\Delta S_i - \Delta S_{i+1})/A_n$	0.423	−0.946	0.523	ton/m^2
$\mp(\Delta M_i + \Delta M_{i+1})/S_n$	±3.34	∓1.00	∓2.34	ton/m^2
$\pm(\Delta S_i + \Delta S_{i+1}) \cdot l_n/S_n$	∓1.27	±0.30	±1.57	ton/m^2
Δq_i	2.49 −1.647	−1.646 −0.246	−0.247 1.29	ton/m^2

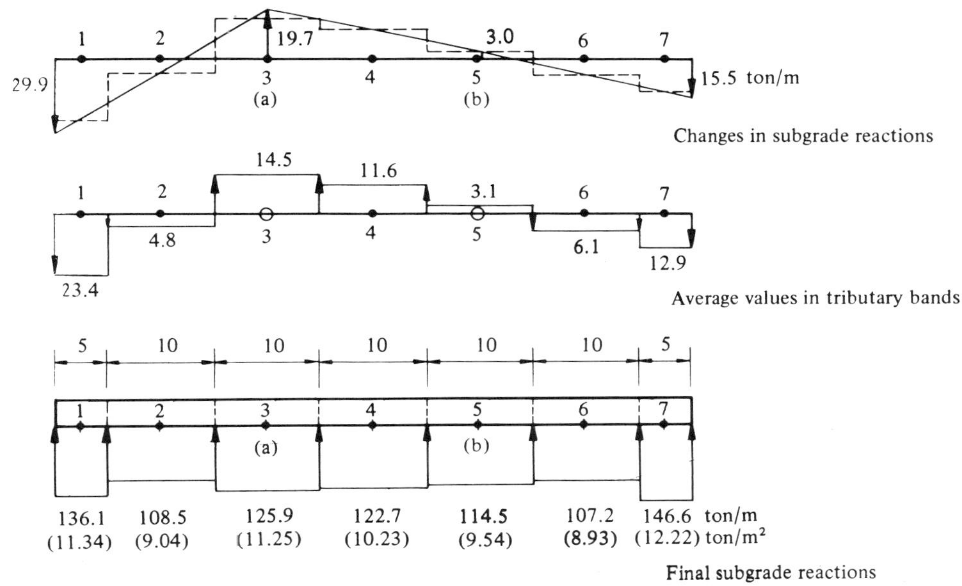

Fig. 4-A.VII Changes in subgrade reaction due to joints introduced in rigid strip foundation.

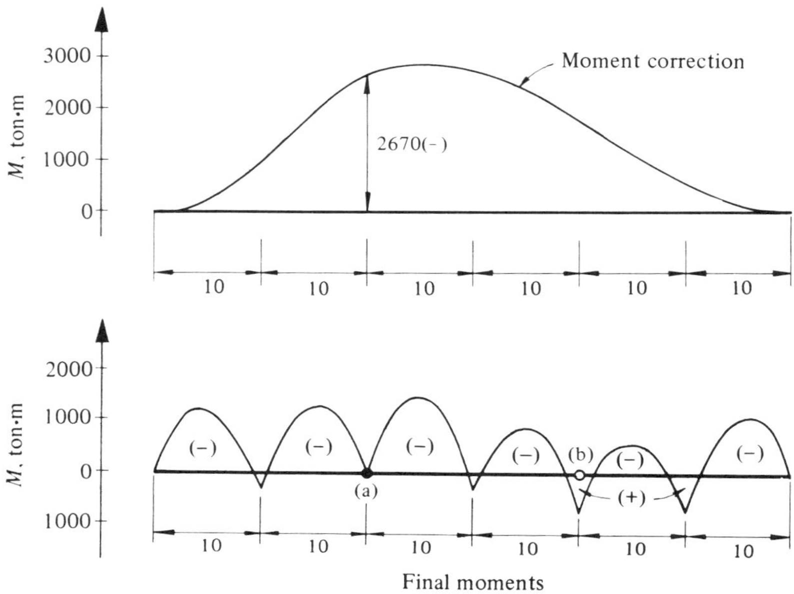

Fig. 5-A.VII Moment correction and final moments.

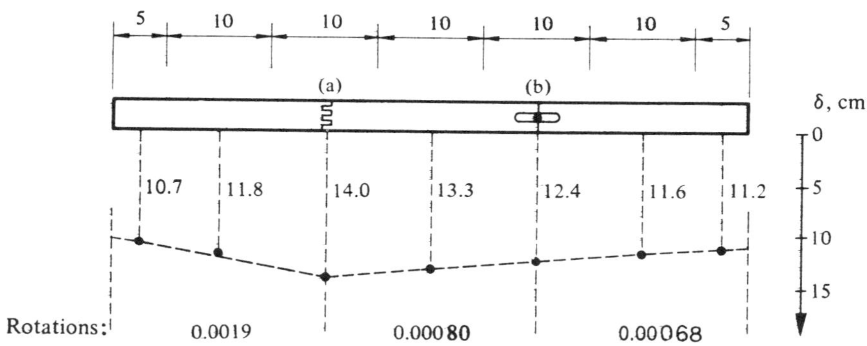

Fig. 6-A.VII Deflections and differential settlements due to pontoon action.

TABLE 6-A.VII

x, m	$x/6$	$\sqrt{1-(x^2/36)}$	q_x, ton/m^2
1	0.167	0.986	6.5
2	0.333	0.943	6.8
3	0.50	0.866	7.4
4	0.667	0.745	8.6
5	0.833	0.553	11.6
5.7	0.95	0.315	20.0
6.0	–	–	20.0

TABLE 7-A.VII Final Shears and Bending Moments in Tributary Band 2

x	x/B	V_1	M_1	ΔV_x	ΔM_x	V_{ax}	M_{ax}	V_x	M_x
m		ton	ton · m	ton	ton · m	ton	ton · m	ton	ton · m
0	0	0	0.195	0	−316.3	0	1621.8	0	1305.5
1	0.167	0.046	0.187	−24.9	−303.3	90.1	1576.8	65.2	1273.5
2	0.333	0.090	0.163	−48.7	−264.4	180.2	1441.6	131.5	1177.2
3	0.50	0.128	0.124	−69.2	−201.1	270.3	1216.4	201.1	1015.3
4	0.667	0.150	0.076	−81.1	−123.3	360.4	901.0	279.3	777.7
5	0.833	0.132	0.032	−71.4	− 51.9	450.5	495.6	379.1	443.7
6	1.0	0	0	0	0	540.6	0	540.6	0

$\Delta V_x = q_a B V_1\left(a, \frac{x}{B}\right)$ (7-IV.4) $\qquad q_a B = 90.1 \times 6 = 540.6$ ton

$\Delta M_x = \frac{1}{2} q_a B^2 M_1\left(a, \frac{x}{B}\right)$ (11-IV.4) $\qquad \frac{1}{2} q_a B^2 = \frac{1}{2} \times 90.1 \times 36 = 1621.8$ ton·m

The shears and bending moments in the transverse direction representing the tributary band areas may be estimated by means of the procedure explained in sections IV.3 and 4, considering each pontoon structure infinitely stiff. An example will be given for one of the transverse tributary bands. Let us consider band 2, Fig. 4-A.VII, with average subgrade reaction of 9.04 ton/m^2. The plastic yield at the edge is assumed to be $q_{ep} = 20$ ton/m^2. Therefore $q_a/q_{ep} = 9.04/20 = 0.45$. From Figs. 4 and 5-IV.3, we find $a = 0.945$ and $C_s = 0.71$; from formula 19-IV.3, Table 5-A.VII, we have

$$q_x = \frac{0.71(9.04)}{\sqrt{1 - x^2/36}}, \quad \text{for } x \leqslant 5.7 \text{ m}$$

The moments and shears may be obtained by correcting the solution based on uniform subgrade reaction, via the method explained in section IV.4, and with argument $a = 0.945$ and the aid of Figs. 2 and 3-IV.4. The values obtained are given in Table 7-A.VII, and are plotted in Fig. 7-A.VII. The other transverse foundation band areas may be investigated in the same way.

EXAMPLE B.VII CASE HISTORY OF HEAVE FOR DEEP OVERCOMPENSATED FOUNDATION

We present here an example of E-heave calculation and compare it with reported vertical displacements observed during and after the construction period for the foundation of a large building in Mexico City. The foundation structure with plan dimensions of 55.5 × 74.1 m is placed at a depth of 10 m into the soft lacustrine sediments of very high compressibility of Mexico City. The building is supported on a rigid box type foundation, and is designed with two basements and five stories above the ground surface. The total weight of the building including the foundation structure is of 10.2 ton/m^2. At the foundation grade elevation to 10 m depth the

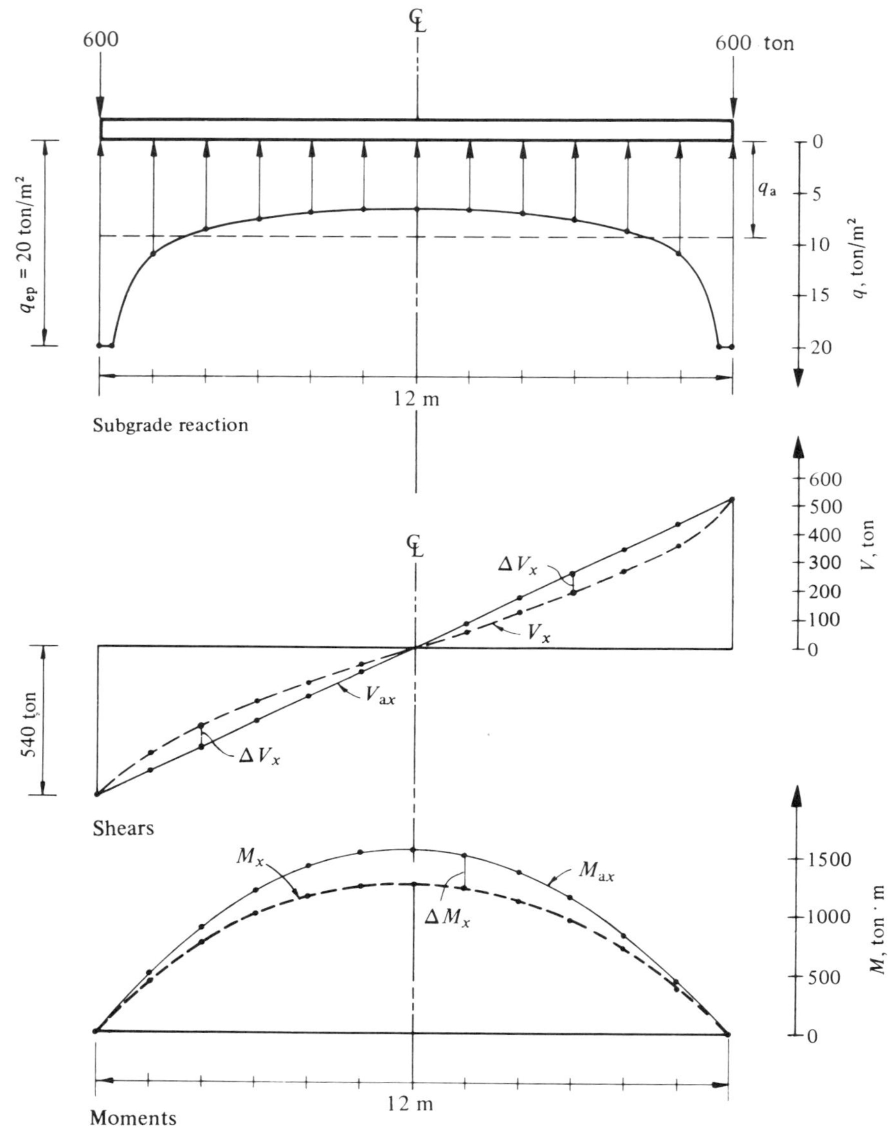

Fig. 7-A.VII Transverse bands; subgrade reaction; shear and bending moments.

overburden effective stress including the weight of old buildings is 9.5 ton/m^2, with water piezometric pressure of 8.2 ton/m^2. Therefore, the total available pressure is 17.7 ton/m^2. Hence, the foundation is designed overcompensated with a permanent stress relief of 7.5 ton/m^2, from the original overburden effective stresses.

The required excavation to a depth of 10 m proceeds in two stages via the method described in Chapter VII. The first stage of excavation to 4.2 m depth produced an effective stress relief of 5.8 ton/m^2, and the second stage due to excavation in trenches and concrete beam construction produced an increment of stress equal to 0.73 ton/m^2. Therefore, the total stress relief during foundation beam construction was on the order of 5.07 ton/m^2, at a depth of 10 m. The recorded values of the vertical displacements and the piezometric water level observa-

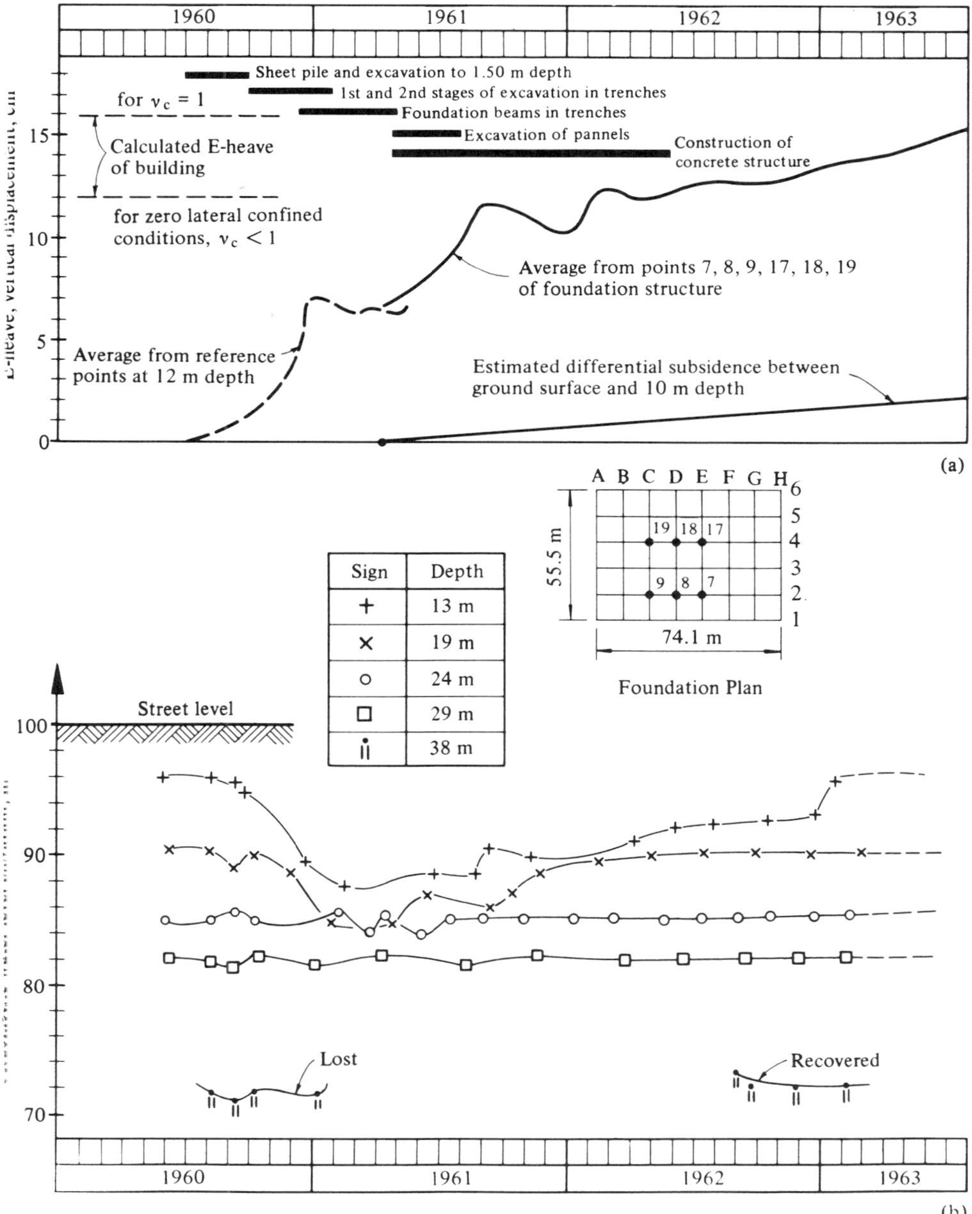

Fig. 1-B.VII E-heave and piezometric observations in central part of excavated area.

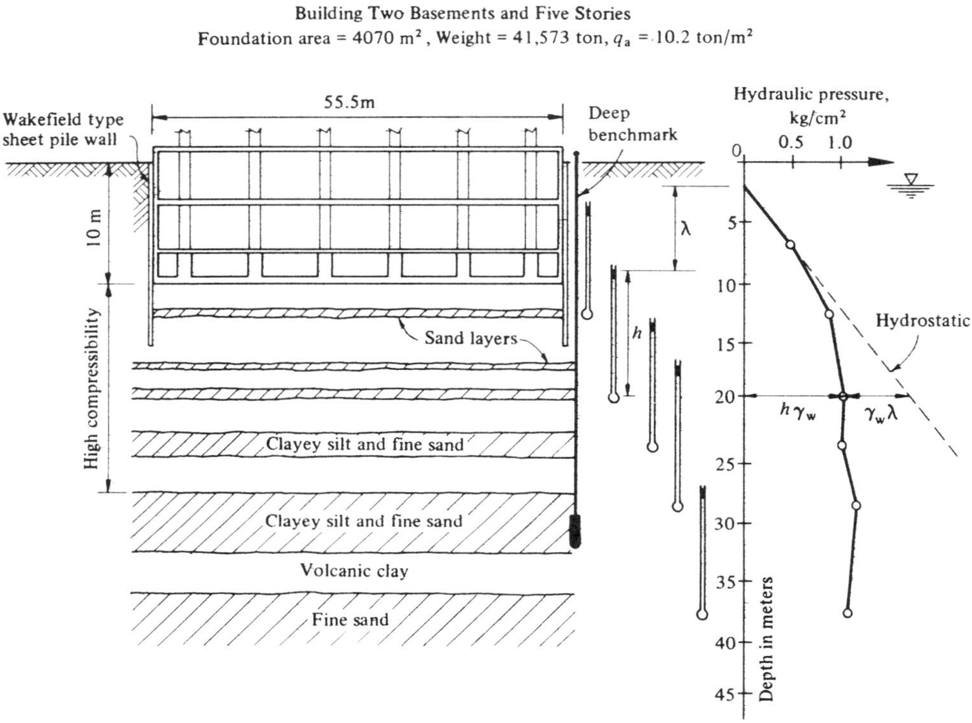

Fig. 2-B.VII Stratigraphical and initial hydraulic conditions.

tions during excavations are reported in Fig. 1-B.VII. The E-heave observed is the average for the central part of the excavation, its value reaching 14.5 cm about one year after the total weight of the completed building is applied. At the time the foundation was constructed the general ground surface subsidence observed in that area was on the order of 8 cm/year, measured with reference to a deep benchmark installed at a depth of 32 m into a compact silty fine sand stratum.

The initial hydraulic conditions observed at the site are shown in Fig. 2-B.VII. In Fig. 3-B.VII the detailed stratigraphic conditions and the vertical effective stresses for the initial and final conditions are reported. The E-heave is calculated from the elastic response of the soil mass due to the total stress relief. For this purpose, the elastic response strain modulus was investigated in the laboratory for each one of the soil strata encountered, from hysteresis loops obtained in unconfined compression test specimens. The formula used for heave calculation considering zero horizontal displacements is

$$\delta_e = \sum_{1}^{n} [\nu_c \rho_e M_e \cdot \Delta\sigma_e \cdot d]_i$$

The terms are defined in Chapter VII.

To reduce the E-heave during excavations, the effective stresses in the soil mass

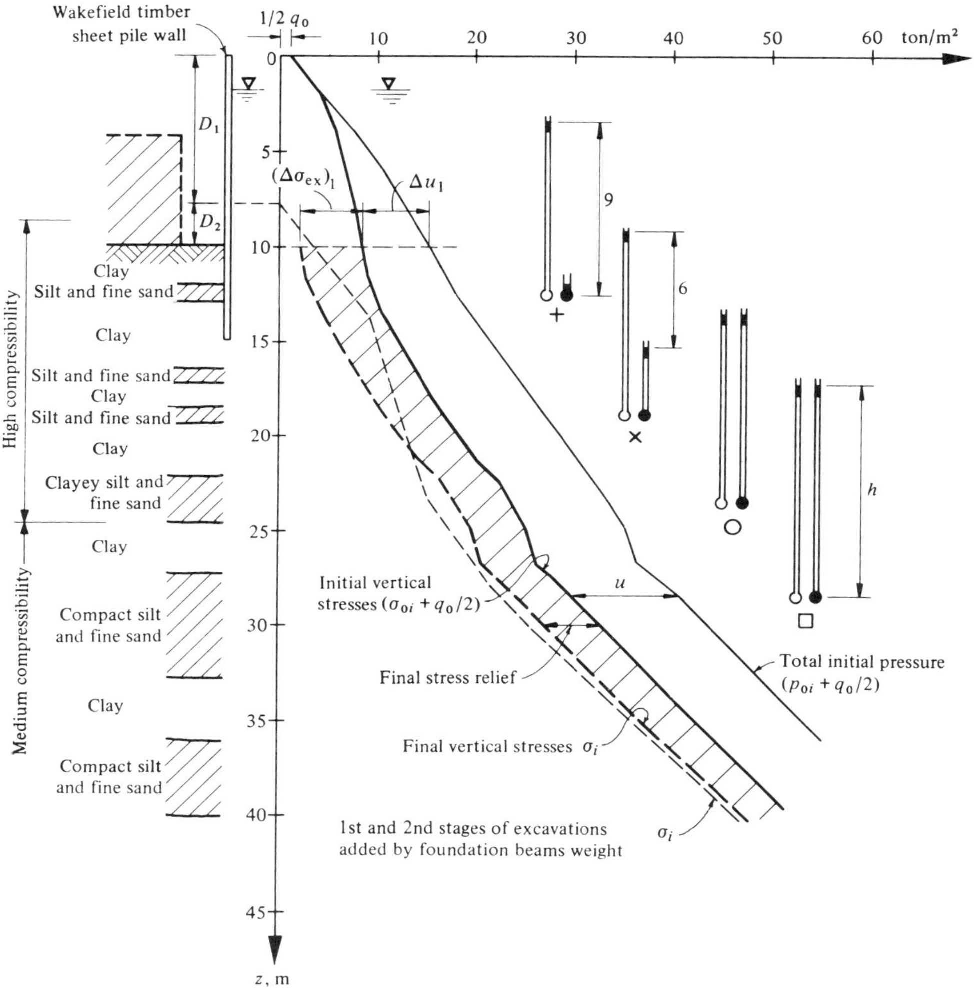

Fig. 3-B.VII Conditions of initial and final effective stresses.

were increased under the foundation grade elevation at 10 m depth. To achieve this, the piezometric water elevations were lowered using nine deep water pumping wells, 1.2 m in diameter, excavated to a depth of 20 m. The water level reduction obtained in the soil mass under the foundation grade elevation during excavation is reported in Fig. 1-B.VII(b). This was done gradually until reaching the final excavation design conditions.

For the E-heave calculation it was assumed, however, that before installing reference points at 12 m depth to record the vertical displacements during excavation, a heave of the site had already taken place during demolition of the old buildings, corresponding to estimated stress relief on the order of one-half the weight removed: $\frac{1}{2}q_0 = 1$ ton/m^2. Hence, the basic initial overburden effective stresses for heave calculation were taken as $\sigma_{0i} + \frac{1}{2}q_0$, shown in Figs. 2 and 3-B.VII.

During excavations the measurement of the vertical displacements of six reference points in the central part of the foundation gives an average value of 7.5 cm with respect to the elevation at which the reference points were installed at 12 m depth. The E-heave remained practically constant during the subsequent five months.

The foundation structure was constructed in trenches corresponding to the second stage of excavation by substitution of the soil weight by concrete. After March 1961, the E-heave vertical displacement measurements were taken from the elevation of the foundation structure at 10 m depth and added to the previous E-heave observations (Fig. 1-B.VII).

During excavation of the soil panels between foundation beams, and as construction of the building proceeded, the water levels were permitted to rise gradually until final equilibrium with the building weight was reached at the end of 1962 (Fig. 1-B.VII(b)) and the final stress relief in the soil mass under the foundation grade elevation was obtained (curve σ_i, Fig. 3-B.VII).

An example of the E-heave calculation for the final effective stress conditions is given in Table 1-B.VII.

The calculations affected by ν_c for horizontally confined conditions give a final E-heave of 12 cm. The actual conditions of the soil beneath the foundation structure is between zero lateral deformation and no restriction to lateral deformation. Therefore, the E-heave was also calculated for the latter condition, yielding 16.0 cm for $\nu_c = 1$.

TABLE 1-B.VII E-Heave Calculations Due to Permanent Stress Relief in Soil Mass

Strata		M_e	d_i	$\sigma_{0i} + \frac{1}{2} q_0$	σ_i	$\Delta\sigma_{ex}$	ν_c	ρ_e	δ_e
m	m	cm²/kg	cm	kg/cm²	kg/cm²	kg/cm²			cm
10.0	11.5	0.0455	150	0.85	0.22	0.63	0.74	0.861	2.738
11.5	12.0	0.0328	50	0.89	0.27	0.62	0.74	0.835	0.628
12.0	12.9	0.0244	90	0.95	0.33	0.62	0.74	0.808	0.814
12.9	13.3	0.0385	40	0.99	0.38	0.61	0.74	0.785	0.546
13.3	14.1	0.0290	80	1.05	0.44	0.61	0.74	0.762	0.798
14.1	15.0	0.0313	90	1.14	0.53	0.61	0.74	0.731	0.930
15.0	15.8	0.0250	80	1.23	0.62	0.61	0.74	0.704	0.635
15.8	16.5	0.0200	70	1.30	0.69	0.61	0.74	0.685	0.433
16.5	17.2	0.0122	70	1.40	0.79	0.61	0.74	0.660	0.254
17.2	18.1	0.0264	90	1.50	0.89	0.61	0.74	0.638	0.683
18.1	18.6	0.0286	50	1.57	0.96	0.61	0.82	0.623	0.446
18.6	19.4	0.0122	80	1.67	1.06	0.61	0.74	0.604	0.266
19.4	21.3	0.0143	190	1.81	1.21	0.60	0.82	0.576	0.768
21.3	22.4	0.0098	110	2.0	1.41	0.59	0.82	0.543	0.283
22.4	24.8	0.0075	240	2.23	1.64	0.59	0.74	0.514	0.403
24.8	26.7	0.0091	190	2.50	1.93	0.57	0.82	0.477	0.385
26.7	27.4	0.0091	70	2.62	2.05	0.57	0.74	0.466	0.125
27.4	33.0	0.0050	560	2.80	2.25	0.55	0.74	0.443	0.504
33.0	36.0	0.0083	300	3.85	3.30	0.55	0.74	0.378	0.382
Σ									12.021
For $\nu_c = 1$; $\delta_e = 16.00$ cm									

The observed E-heave about one year after the total weight of the building was placed (Fig. 1-B.VIIa) is between the two calculated E-heave values.

However, in later years a residual heave has been observed up to 20% more than the calculated E-heave, which undoubtedly is due to swelling of the clay because of water adsorption due to the decompression of the silty clay (section VII.5), and also because of a small difference of velocities of the general ground surface subsidence between the surface of the ground and the bottom of the foundation structure at 10 m depth after completion of the building (Fig. 1-B.VII(a)).

EXAMPLE VIII CALCULATION OF FRICTION PILES

The use of the formulas given in Chapter VIII for skin friction pile problems is illustrated by the following examples.

Negative Skin Friction: Example 1

A pile field with 64 uniformly spaced piles is under investigation for point penetration conditions with a load of 180 ton per pile. The soil mass is under consolidation. The effective stresses induced in the soil mass are required to calculate settlements, and the load on the piles is also required to design the pile element. The piles are spaced on centers 3 X 3 m and have a diameter of 0.40 m. The piles are driven into a highly stratified subsoil in a series of silty clay and clayey silt fine sand compressible deposits. The pile points reach a deep, uniform, medium density silty fine sand stratum at a depth of 33 m. The stratigraphy, geometry, mechanical and hydraulic properties of the subsoil are given in Fig. 1-VIII. The piezometric water elevations decrease continuously.

For illustration, the study of an interior pile in the group is presented. The investigation of the equivalent tributary area at different depths is given in Table I-VIII. (See section VIII.3.3.)

From depth 13 m, the nominal tributary area will apply, equal to 8.87 m^2. The final column in the table shows the areas to be used for calculation of negative skin friction, according to algorithms

$$\sigma_i = A_i \sigma_{0i} - B_i \sigma_{i-1} - C_i (NF)_{i-1} \qquad \text{1-VIII}$$

$$(NF)_i = (\sigma_{0i} - \sigma_i)\bar{a}_i \qquad \text{2-VIII}$$

For the selected tributary areas the values of A_i, B_i and C_i shown in Table 2-VIII may be calculated, using the expressions

$$A_i = \frac{\bar{a}_i}{\bar{a}_i + m_i \Delta z_i}; \qquad B_i = \frac{m_i \Delta z_i}{\bar{a}_i + m_i \Delta z_i}; \qquad C_i = \frac{1}{\bar{a}_i + m_i \Delta z_i}$$

where $m_i = \pi \bar{r}_0 K_{\phi_r}$ and $K_{\phi_r} = \tan \phi_r / \bar{N}_{\phi_r}$. The quantity ϕ_r is the angle of internal friction of the remolded soil close to the pile shaft.

The integration is performed step by step. At the pile head $\sigma_{i-1} = 0$ and $(NF)_{i-1} = 0$. At the first increment $\sigma_i = A_i \sigma_{0i}$, this value is substituted into 2-VIII to obtain $(NF)_i$; from σ_i and $(NF)_i$ the next value

$$\sigma_{i+1} = A_{i+1} \sigma_{0i+1} - B_{i+1} \sigma_i - C_{i+1} (NF)_i$$

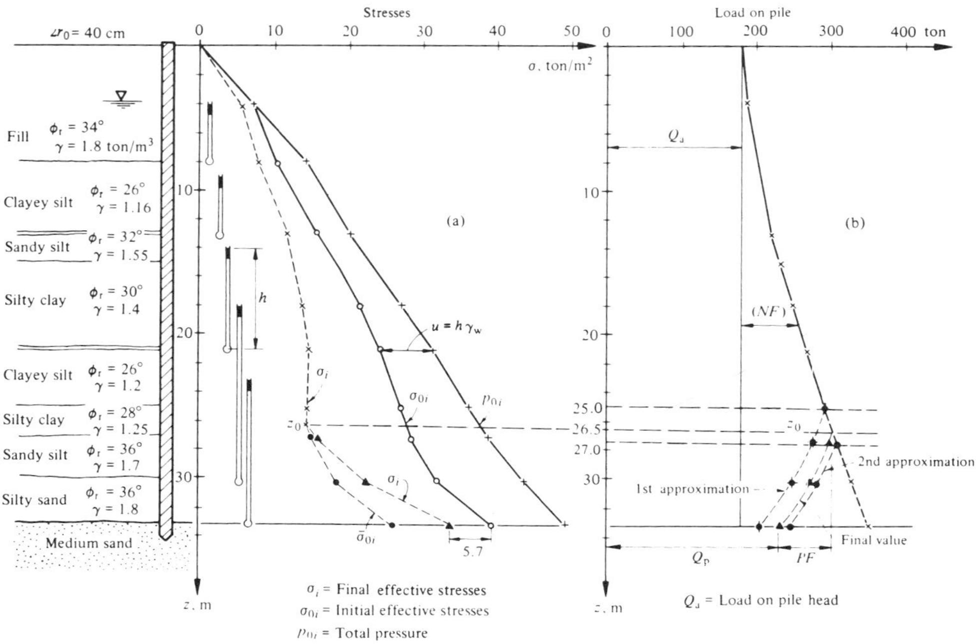

Fig. 1-VIII Pile subjected to *NF*, *PF*, and point bearing.

is calculated and then $(NF)_{i+1}$, from expression 2-VIII. The procedure is repeated for the total length of the pile. The calculation is self explanatory as shown in Table 2-VIII.

At the tip of the pile, the negative friction load is $(NF)_{33} = 172.6$ ton and the reduction of the overburden confining pressure is 19.45 ton/m^2. The point penetration into the medium density fine sand stratum is investigated by means of

TABLE 1-VIII Tributary Areas

z, m	z/r_0	I_{z_0}	$\sum_1^n I_{zr}$	$I_{z_0} + \sum_1^n I_{zr}$	a_{ez}, m^2	$\overline{a}_{ez}$, m^2	$\overline{a}$, m^2
4	20	10.8	0.75	11.55	4.352	4.226	4.226
8	40	24.3	4.86	29.16	6.895	6.769	6.769
13	65	42.0	14.27	56.27	9.435	9.309	8.874

$\overline{a}_{ez} = (a'_{ez}) - a_p$ $\qquad \overline{a}_0 = 9.00 - 0.126 = 8.874$ m^2

Data: $2r_0 = 40$ cm; $r_0 = 0.2$ m; $\qquad \overline{\omega} = 1.05\pi \times 0.40 = 1.319$ m

$$a_p = \pi(0.2)^2 = 0.126\text{m}^2 \qquad a'_{ez} = \frac{\pi z^2}{I_{z0} + \sum_1^n I_{zr}}$$

TABLE 2-VIII (*NF*) Step by Step Calculation

z, m	$\bar{a}_i$, m^2	K_ϕ	m_i m	Δz, m	σ_{0i}, ton/m²	A_i	B_i	C_i m^{-2}	σ_i, ton/m²	NF, ton
0	BOUNDARY CONDITIONS				0				0	0
4	4.226	0.353	0.233	4	7.2	0.819	0.181	0.194	5.9	5.5
8	6.769	0.353	0.233	4	10.4	0.879	0.121	0.130	7.71	18.2
13	8.874	0.33	0.218	5	16.2	0.89	0.109	0.100	11.76	39.4
15	8.874	0.35	0.231	2	18.55	0.95	0.049	0.107	12.83	50.8
18	8.874	0.346	0.228	3	21.62	0.928	0.072	0.105	13.84	69.0
21	8.874	0.346	0.228	3	24.7	0.928	0.072	0.105	14.72	88.6
25	8.874	0.33	0.218	4	27.28	0.911	0.089	0.103	14.44	114.0
27	8.874	0.34	0.224	2	28.67	0.952	0.048	0.107	14.37	126.9
30	8.874	0.353	0.233	3	32.10	0.927	0.073	0.104	15.45	147.7
33	8.874	0.353	0.233	3	39.5	0.927	0.073	0.104	20.05	172.6

the bearing capacity formula

$$Q_{pu} = 1.2\,[(\sigma_{0i} - \Delta\sigma_d)N_q\,]\,(D_r + 0.1)\,a_p$$

using the numerical values $\phi_d = 36^\circ$, bearing capacity factor $N_q = 75$ from Fig. 4-III.4, and relative density $D_r = 0.5$ we obtain

$$Q_{pu} = 1.2\,[20.05 \times 75] \times 0.6 \times 0.126 = 137 \text{ ton}$$

From this result, it is concluded that due to the high negative friction the point is forced to penetrate continuously into the sand stratum at a rate higher than the ground surface subsidence. Hence, positive friction has to develop at the lower part of the pile to help resist the negative friction and pile load of 180 ton. Therefore, the following limiting equilibrium conditions should be investigated:

$$Q_{pu} + (PF)_{z_0} = (NF)_{z_0} + Q_a \qquad \text{3-VIII}$$

In this equation, the unknowns Q_{pu}, and $(PF)_{z_0}$ may be found by a procedure of trial and error. The value of Q_{pu} is a function of the final confining pressure (section IX.2). (See Case VII of Chapter VIII.) A first approximation is investigated for a depth $z_0 = 25$ m, where the total $NF = 114.0$ ton is found; therefore, the load on the pile at this depth is $Q_a + NF = 294$ ton. From this depth, positive friction calculations are performed with the algorithms

$$\sigma_i = \bar{A}_i\bar{\sigma}_{0i} + \bar{B}_i\sigma_{i-1} + \bar{C}_i(PF)_{i-1}$$

$$(PF)_i = (\sigma_i - \bar{\sigma}_{0i})\,\bar{a}_i$$

starting integration at a depth of 25 m, where $\bar{\sigma}_{0i} = \sigma_{0i} = 14.44$ ton/m² and $(PF)_{i-1} = 0$. The tributary areas from this depth remain equal to $\bar{a} = \lambda\beta - a_p = 8.874$ m². The values of $\bar{\sigma}_{0i}$ at further depths, to be used in the calculation in the case of a pile field where $\bar{a}_{ez}$ remain constant, are taken as $(\bar{\sigma}_{0i})_z = (\sigma_{0i})_z - (\sigma_{0i} - \sigma_i)_{z_0}$ (Chapter IX, Fig. 1-IX.2). The calculation is shown in Table 3-VIII; in this case a load of $(PF)_{25} = 90.6$ ton is calculated. With the result obtained, the

TABLE 3-VIII *(PF)* in Lower Section of Pile

z, m	$\bar{a}_i$, m^2	m_i m	Δz m	σ_{0i} ton/m^2	$\bar{\sigma}_{0i}$ ton/m^2	$\overline{A}_i$	$\overline{B}_i$	$\overline{C}_i$ m^{-2}	σ_i, ton/m^2	PF ton
25	8.874			27.28	14.44				14.44	0
27	8.874	0.235	2	28.67	15.83	1.06	0.056	0.119	17.52	15.0
30	8.874	0.244	3	32.1	19.26	1.09	0.09	0.123	24.41	45.7
33	8.874	0.244	3	39.5	26.66	1.09	0.09	0.123	36.87	90.6

$\bar{\omega} = 1.1 \times 2\pi \times 0.2 = 1.38$ m; $m_i = \frac{1}{2}\bar{\omega}K_\phi = \frac{1}{2} \times 1.382 K_\phi = 0.691 K_\phi$

penetration point load is estimated by means of

$$Q_{pu} = 1.2\,[36.87 \times 75]\;0.6 \times 0.126 = 250 \text{ ton}$$

Hence

$$Q_{pu} + (PF)_{25} = 250 + 90.6 = 340.6 \text{ ton}$$

Therefore

$$Q_a + (NF)_{25} < 340.6 \text{ ton}$$

from which it may be seen that the assumed value of z_0 has been underestimated.

A larger value of z_0 is now selected to obtain $Q_a + (NF)_{z0}$, and calculations are performed in a similar manner as before. From Table 2-VIII, for $z_0 = 27$ m, $(NF) =$ 126.9 ton. The (PF) is now calculated similarly as shown in Table 3-VIII, and plotted in Fig. 1-VIII, showing that for this value of z_0 the equilibrium condition is overestimated since $(PF)_{27} = 64.6$ ton, and $Q_{pu} = 220$ ton, therefore

$$Q_{pu} + (PF)_{27} = 284.6 \text{ ton} < Q_a + (NF)_{27} = 306.9 \text{ ton}$$

By interpolation we obtain the final values $z_0 = 26.5$ m, $(NF)_{26.5} = 124$ ton; hence $Q_a + (NF)_{26.5} = 304$ ton. From calculations similar to those given in Table 3-VIII a value of $(PF)_{26.5} = 71.4$ ton is obtained, with corresponding confining stress at the pile points of $\sigma_d = 33.8$ ton/m^2. Using the bearing capacity formula the corresponding load at the pile point is calculated: $Q_{pu} = 229.8$ ton. Therefore $Q_{pu} + (PF)_{26.5} = 301.2$ ton, satisfying approximately the plastic equilibrium equation 3-VIII for the acting and resisting loads on the pile, respectively.

The final effective stresses within the pile field are shown in Fig. 1-VIII, where σ_i represents the final vertical effective stress due to positive friction, in the upper section of the pile for the final condition considered. From the pile point elevation the stresses may be investigated to greater depth, in the soil within pile field with $\sigma_d - \sigma_{0d} = -5.7$ ton/m^2 and with point loads on the piles of $Q_{pu} = 229.8$ ton.

Example 2

The accuracy of the method given may be investigated from actual tests. Information on field investigations has been selected from the literature reported in ICOSOMEF 1969, Vol. II, p. 85, by Endo, Minou, Kawasaki and Shibata in the paper "Negative Skin Friction Acting on Steel Pipe Pile in Clay." The investigation reports the skin friction obtained from instrumented steel pipe piles for a period of more than two years in a thick alluvial deposit in Tokyo, where ground surface subsidence on the order of 8.0 cm/year was observed due to reduction in the piezometric water levels. Pile labeled cE43 was taken for comparison of measurements with calculations. The pile geometry and soil mechanical and hydraulical properties are found in Fig. 2-VIII, where the results of the loads on the pile measured because of negative friction are also shown. The calculation was performed on ultimate skin friction conditions, considering an angle of internal friction of $\phi_r = 32°$ for the clayey sandy silts and silty sands. The calculations with algorithms for negative friction gave the results plotted in Fig. 2-VIII, which agree fairly well with the authors' calculations. The positive friction, however, does not show good agreement. This was calculated with the report that the tip of the pile takes a point load of 150 ton and the steel pile is stressed only to 830 kg/cm^2. With this observation, a theoretical value for the negative friction is computed of 415 ton at a depth of 36 m. This load on the pile produces very high stresses that may have induced yielding, as reported. The reported value of the negative friction load at 31 m depth induced a 1700 kg/cm^2 stress in the pipe. If the pile had not yielded, it appears that the negative friction could have reached a maximum value on the order of 415 ton with a corresponding stress in the steel of 2400 kg/cm^2 (Fig. 2-VIII) at an approximate depth of 36 m.

The order of magnitude of the negative friction load to a depth of 31 m as measured in the test may be calculated with the approximate formula 13-VIII.3 given in Chapter VIII:

$$(NF)_d = \frac{\overline{\omega} K_\phi}{1 + \dfrac{\overline{\omega} K_\phi}{3\overline{a}} d} \int_0^d \sigma_{0z}\, dz$$

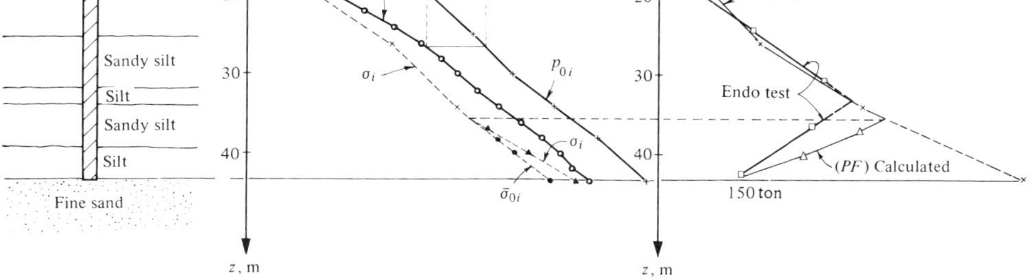

Fig. 2-VIII Stratigraphy, effective stresses and *NF* from pile test. (After Endo, Minou, and Kawasaki, ICOSOMEF 1969.)

in which $\bar{\omega} = 2.01$ m, $K_\phi = 0.353$, and $\bar{a} = \pi(12r_0)^2 = 42.0$ m^2. The stress area represented by the integral to 31 m may be obtained from Fig. 2-VIII: $A_\sigma = 502$ ton/m; hence

$$(NF)_{31} = \frac{2.01 \times 0.353}{1 + \dfrac{2.01 \times 0.353}{3 \times 42} \times 31} \times 502 = \frac{0.709}{1.174}\, 502 = 303 \text{ ton}$$

Positive Friction: Example 3

The results of positive friction pile tests are reported in Fig. 3-VIII. The steel piles were pushed into a highly sensitive and compressible soil mass to a depth of 31.3 m in Mexico City. The total area of the site was excavated to a depth of 1.5 m to clean the ground surface from old footings. Therefore, the overburden stress was reduced owing to the excavation before the piles were driven. To perform the tests the soil around the pile was removed to a depth of 4.7 m, giving an effective pile length of 26.6 m. The piles were allowed to rest for a period of 5 weeks before the test was started. The soil stratigraphy and the soil engineering characteristics are shown in Fig. 4-VIII, as well as the position for which the piles were tested at the site.

Data

$$r_0 = 0.209 \text{ m};\quad \bar{\omega} = 1.1 \times 0.418\pi = 1.44 \text{ m}$$

$$a_p = \pi r_0^2 = 0.138 \text{ m}^2$$

$$\phi_r = 26^\circ;\quad \text{and}\quad K_\phi = \frac{\tan \phi_r}{\bar{N}_{\phi_r}} = 0.33$$

$$m = \tfrac{1}{2}\, \bar{\omega} K_\phi = \tfrac{1}{2}\,(1.44 \times 0.33) = 0.238$$

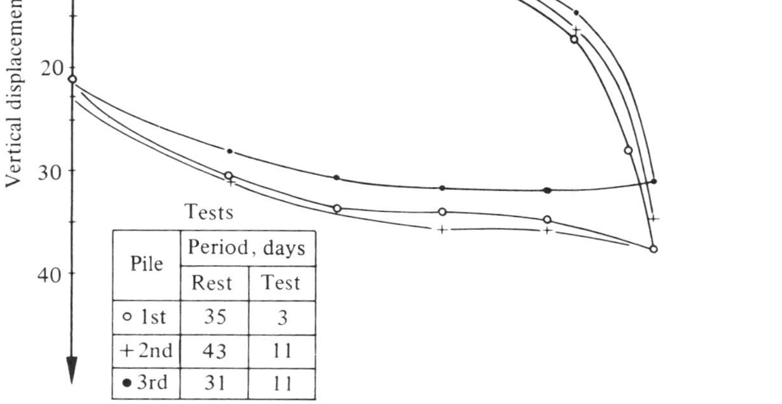

Fig. 3-VIII Loading tests on positive friction piles.

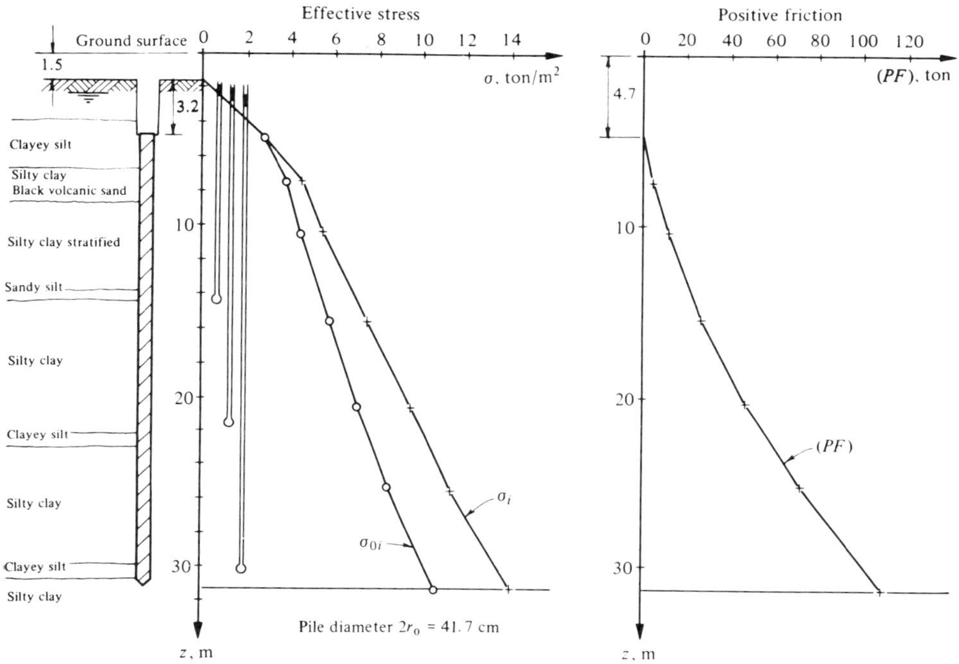

Fig. 4-VIII Subsoil stresses and load in positive friction pile.

From the information given, the approximate friction load on the pile may be computed from equations 16 and 17-VIII.2 in Chapter VIII:

$$\Delta\sigma_{0d} = \frac{\overline{m}}{1 - \overline{m}}(\sigma_{0i} + \sigma_{0d})$$

and

$$(PF)_d = \overline{a}(\sigma_d - \sigma_{0d}) = \overline{a}\Delta\sigma_{0d}$$

in which

$$\overline{m} = \frac{\overline{\omega}K_\phi}{2\overline{a}}\, l_s; \quad \overline{a} = \pi(12r_0)^2 = 19.76 \text{ m}^2$$

$$\overline{m} = \frac{1.44 \times 0.33}{2 \times 19.76}\, 26.6 = 0.32; \sigma_{0i} = 2.8 \text{ ton/m}^2; \sigma_{0d} = 10.70 \text{ ton/m}^2$$

$$\Delta\sigma_{0d} = \frac{0.32}{0.68}\, 13.50 = 6.353 \text{ ton/m}^2; \quad (PF)_d = 19.76 \times 6.353 = 125.5 \text{ ton}$$

With formula 13-VIII.3 a better approximation may be obtained:

$$(PF)_d = \frac{\overline{\omega}K_\phi}{1 - \dfrac{\overline{\omega}K_\phi}{3\overline{a}}\, d} \int_{h_e}^{d} \sigma_{0z}\, dz$$

in which the stress area from Fig. 4-VIII is:

$$A_\sigma = \int_{h_e}^{d} \sigma_{0z}\, dz = 184.7 \text{ ton/m}$$

Hence

$$(PF)_d = \frac{1.44 \times 0.33}{1 - \dfrac{1.44 \times 0.33}{3 \times 19.76}\, 26.6} \times 184.7 = 111.6 \text{ ton}$$

The point load may be estimated investigating the soil properties under the pile tip located on a highly sensitive compressible silty clay represented by the compressibility curve of Fig. 5-VIII, from which a critical compressive strength $\sigma_b = 15$ ton/m^2 may be found; hence $Q_{pu} = 0.138 \times 15 = 2.07$ ton. Therefore, from

$$\begin{aligned} &\text{1st approximate formula:} \quad Q_u = 125.5 + 2.07 = 127.6 \text{ ton} \\ &\text{2nd approximate formula:} \quad Q_u = 111.6 + 2.07 = 113.7 \text{ ton} \end{aligned}$$

It is interesting to investigate, however, the values that may be obtained applying algorithms for positive friction, integrating from the head of the pile:

$$\sigma_i = \bar{A}_i \sigma_{0i} + \bar{B}_i \sigma_{i-1} + \bar{C}_i (PF)_{i-1}$$

$$(PF)_i = (\sigma_i - \sigma_{0i})\bar{a}_i$$

where

$$\bar{A}_i = \frac{\bar{a}_i}{\bar{a}_i - m_i \Delta z_i}; \quad \bar{B}_i = \frac{m_i \Delta z_i}{\bar{a}_i - m_i \Delta z_i}; \quad \bar{C}_i = \frac{1}{\bar{a}_i - m_i \Delta z_i}$$

The equivalent tributary areas at various depths are calculated for this case by means of

$$\bar{a}_{ez} = \frac{(2h_e + z)\pi z^2}{zI_{zk} + h_e I_{zc}\sqrt{2}}$$

Here $h_e = 3.2$ m is the depth of the pile head from the ground surface, that is, from the bottom of the excavation performed in the total area of the site. The values of I_{zk} and I_{zc} are obtained from Fig. 11-VIII.3 in Chapter VIII. The step by step calculation is shown in Table 4-VIII, from which a value of $PF = 118.9$ ton is obtained; the total computed load including the point resistance is 121.0 ton. The total load measured in the tests was on the order of 110 ton.

It is important also to investigate the order of magnitude of the time the soil takes to consolidate against the pile shaft. According to procedure given in Chapter VIII, Table 1-VIII.3, the time factor

$$T_v = \frac{(c_{vr})_a}{H^2}\, t$$

where $H = 0.4 r_0$, but since the pile is drained only on one side $H = 0.4 \times 21 = 8.4$ cm.

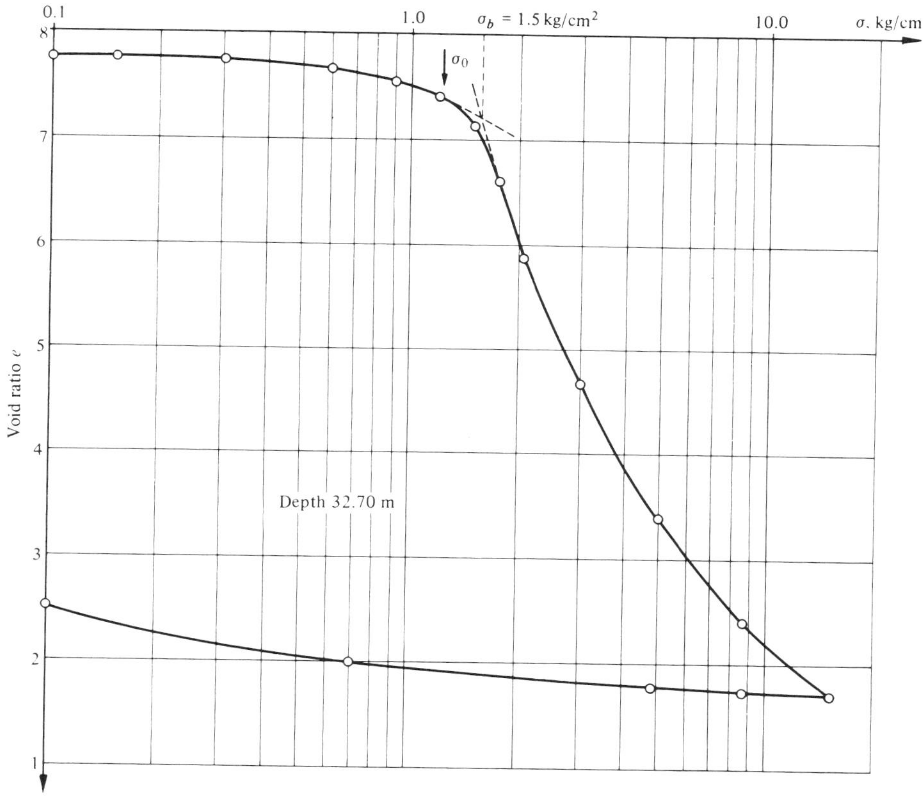

Fig. 5-VIII Compressibility curve for silty clay at the pile point; depth 32.7 m.

TABLE 4-VIII (*PF*) Step by Step Calculation

Depth	z	z/r	I_{zk}	I_{zc}	$\bar{a}_{ez}$	Δz	$\sigma_{0\,i}$	$\bar{A}_i$	$\bar{B}_i$	$\bar{C}_i$	σ_i	PF
4.7	0						2.8				2.8	0
7.3	2.6	12.47	6	8	3.69	2.6	3.8	1.203	0.201	0.326	5.13	4.92
10.3	5.6	26.86	15.3	19.0	6.89	3.0	4.4	1.116	0.116	0.162	6.30	13.10
15.3	10.6	50.84	32.0	36.0	11.95	5.0	5.8	1.111	0.111	0.093	8.36	30.61
20.5	15.6	74.82	46.0	53.0	17.57	5.0	7.1	1.072	0.073	0.061	10.09	52.51
25.3	20.6	98.80	66.0	72.0	21.36	5.0	8.5	1.059	0.059	0.050	12.22	79.50
31.3	26.6	127.58	85.0	95.0	27.26	6.0	10.7	1.055	0.055	0.039	15.06	118.88
	m				m²	m	ton/m²			m⁻²	ton/m²	ton

The effective stresses and *PF* are plotted in Fig. 4-VIII.
The initial effective stresses are calculated from excavation surface at 1.5 m depth.

TABLE 5-VIII Values of ϕ_r' and K_ϕ' From Average Degree of Consolidation

t_d, days	T_v	$U\%$	$F(T_v)\tan\phi_r$	ϕ_r'	K_ϕ'
5	0.471	0.74	0.362	20°	0.29
10	0.943	0.92	0.45	24°	0.32
30	2.828	0.995	0.487	25.9°	0.33

The smallest value of c_v for the undisturbed condition of the soil along the pile shaft is $c_v = 0.001$ cm²/sec, and in the remolded state $c_{vr} = 0.00004$ cm²/sec. Therefore, the average value is

$$(c_{vr})_a = \frac{2}{\frac{1}{4}\cdot 10^5 + 10^3} = 7.7 \times 10^{-5}\ \text{cm}^2/\text{sec}$$

The time factor is

$$T_v = \frac{7.7 \times 10^{-5}}{(8.4)^2} = 0.0943\ t_d$$

where t_d is given in days. With the values of T_v the average percent of consolidation may be found from Fig. 34-II.3, and Table 5-VIII may be formed.

From the preceding calculation, it may be seen that the soil, 4–6 weeks after the piles were driven, had sufficient time to consolidate against the pile shaft. The duration of the test was also sufficiently slow to permit the full effective stresses to act against the pile shaft. The above statement may be further observed from the results of the pile tests, since pile tests of 11 days duration gave the same test results as pile test of 3 days duration (Fig. 3-VIII).

The result of other pile investigations for positive friction are given in Table 6-VIII.

TABLE 6-VIII (*PF*) Pile Tests

Pile	Calculated Loads Q_{fu}	Q_{pu}	$(Q_u)_{pf}$	Test Load $(Q_u)_{pf}$
No 1	118.9	2.1	121.0	110
No 2	105.4	12 .1	117.5	120
No 3	70.3	2.5	72.8	75
No 4	96.6	13.6	110.2	118
No 5	129.4	3.3	132.7	136
No 6	182.0	13.8	195.8	200

Points of piles 2, 4 and 6 were located in clayey silt, and those of piles 1, 3 and 5 in highly sensitive silty clay. Loads in tons.

INDEX